MODERN PRESTRESSED CONCRETE

DESIGN PRINCIPLES AND CONSTRUCTION METHODS

MODERN PRESTRESSED CONCRETE

DESIGN PRINCIPLES AND CONSTRUCTION METHODS

Third Edition

James R. Libby

James R. Libby and Associates
San Diego, California

CBS PUBLISHERS & DISTRIBUTORS PVT. LTD.

New Delhi • Bangalore • Pune • Cochin • Chennai (India)

ISBN : 978-81-239-1580-7

First Indian Edition : 1986
Reprint : 2007

This edition has been published in India by arrangement with
Van Nostrand Reinhold Company, USA

Published by Satish Kumar Jain and produced by V.K. Jain for
CBS Publishers & Distributors Pvt. Ltd.,
CBS Plaza, 4819/XI Prahlad Street, 24 Ansari Road, Daryaganj,
New Delhi - 110002, India. • Website: www.cbspd.com
e-mail: delhi@cbspd.com, cbspubs@vsnl.com, cbspubs@airtelmail.in
Ph.: 23289259, 23266861, 23266867 • Fax: 011-23243014

Branches:

- ***Bengaluru:*** Seema House, 2975, 17th Cross, K.R. Road,
  Bansankari 2nd Stage, Bangalore - 560070 Ph.: 26771678/79
  Fax: 080-26771680 • e-mail: bangalore@cbspd.com
- ***Pune:*** Bhuruk Prestige, Sr. No. 52/12/2+1+3/2,
  Narhe, Haveli (Near Katraj-Dehu Road by Pass), Pune-411051
  Ph.: +91-20-32404169 • Fax: 020-24464059
  e-mail: pune@cbspd.com
- ***Kochi:*** 36/14, Kalluvilakam, Lissie Hospital Road,
  Cochin - 682018, Kerala • e-mail: cochin@cbspd.com
  Ph.: 0484-4059061-65 • Fax: 0484-4059065
- ***Chennai:*** 20, West Park Road, Shenoy Nagar, Chennai - 600030
  e-mail: chennai@cbspd.com Ph.: 044-26260666-26202620
  Fax: 044-45530020

Printed at :
Nikunj Print Process, Delhi

Preface

This book has been written with a dual purpose. It is intended for use as a reference book by practicing engineers as well as a textbook by students of prestressed concrete. It is the fourth generation of books on this subject written by the author. It is believed to accurately present the current state of the art of prestressed concrete design and construction methods in a style that is easily understood.

The first Chapter is an introduction to the fundamental principles of prestressing as applied to concrete. Chapters 2 and 3 are devoted to important considerations of the principal materials utilized in prestressed concrete. The information contained in Chapters 2 and 3 should prove to be valuable reference material for the student and the practicing engineer.

The theoretical considerations, both analytical and empirical, currently accepted as being the correct approach to the design and analysis of prestressed concrete members, are covered in detail in Chapters 4 through 9.

Chapter 10 is devoted to cracking and other defects the author has observed in prestressed concrete members during thirty five years of practice as an engineer. It is believed others can benefit from the information contained in this Chapter.

The last five Chapters contain information that is more general and more of a practical nature. Both the student and the practioner should find useful data in the discussion of connections in Chapter 13 as well as in other portions of these Chapters.

This book has been written with frequent references to the American Concrete Institute publication "Building Code Requirements for Reinforced Concrete" (ACI 318-83). Less frequently reference is made to the design standards contained in a publication of the American Association of State Highway and Transportation Officials entitled "Standard Specification for Highway Bridges". Each of these references are important standards in the use of prestressed concrete.

Two engineers, Wendy Thornton and Donald R. Libby, have the author's sincere thanks for their help in reviewing and checking the contents of this book.

JAMES R. LIBBY

San Diego, California

Contents

3 | CONCRETE FOR PRESTRESSING

4 | BASIC PRINCIPLES FOR FLEXURAL DESIGN

5 | CRACKING LOAD, ULTIMATE MOMENT, SHEAR, AND BOND

8 | CONTINUITY IN PRESTRESSED CONCRETE FLEXURAL MEMBERS

9 | DIRECT STRESS MEMBERS, TEMPERATURE AND FATIGUE

10 | CRACKING AND OTHER DEFECTS—THEIR CAUSE AND REMEDY

11 | ROOF AND FLOOR FRAMING SYSTEMS

12 | BRIDGE CONSTRUCTION

13 | CONNECTIONS FOR PRECAST MEMBERS

14 | PRE-TENSIONING EQUIPMENT AND PROCEDURES

15 | POST-TENSIONING SYSTEMS AND PROCEDURES

16 | ERECTION OF PRECAST MEMBERS

1 Prestressing Methods

1-1 Introduction

Prestressing can be defined as the application of a predetermined force or moment to a structural member in such a manner that the combined internal stresses in the member, resulting from this force or moment and from any anticipated condition of external loading, will be confined within specific limits. Prestressing concrete, which is the subject of this book, is the result of applying this principle to concrete structural members, with a view toward eliminating or materially reducing the tensile stresses in the concrete.

The prestressing principle is believed to have been well understood since about 1910, although patent applications relating to types of construction which involved the principle of prestressing date back to 1888 (Ref. 1).* The early attempts at prestressing were abortive, however, owing to the poor quality of materials that were available in the early days as well as to a lack of understanding of the action of creep in concrete. Eugene Freyssinet, the eminent French engineer, is generally regarded as the first to discover the nature of creep in concrete and to realize the necessity of using high-quality concrete and high-tensile

*Numbers in the text refer to the references at the end of each chapter.

strength steel to ensure that adequate prestress is retained. Freyssinet applied prestressing in structural applications during the early 1930's. The history and evolution of prestressing are controversial and not well documented, and for this reason, they are not discussed further in this book. The reader who is interested may find additional historical details in the references (Refs. 2 and 3).

Many experiments have been conducted to demonstrate that prestressed concrete has properties that differ from those of reinforced concrete. Diving boards and "fishing poles" have been made of prestressed concrete to demonstrate the ability of this material to withstand large deflections without cracking. The more significant fact, however, is that prestressed concrete has proved to be economical in buildings, bridges, and other structures (under conditions of span and loading) that would not be practical or economical in reinforced concrete.

Prestressed concrete was first used in the U.S. (except in tanks) in the late 1940's. At that time, most U.S. engineers were completely unfamiliar with this mode of construction. Design principles of prestressed concrete were not taught in the universities, and the occasional structure that was constructed with this new material received wide publicity.

The amount of construction utilizing prestressed concrete has become tremendous and certainly will increase in the future. The contemporary structural engineer must be well informed on all facets of prestressing concrete.

1-2 General Design Principles

Prestressing, in its simplest form, can be illustrated by considering a simple, prismatic flexural member (rectangular in cross section) prestressed by a concentric force, as shown in Fig. 1-1. The distribution of the stresses at midspan are as indicated in Fig. 1-2. It is readily seen that if the flexural tensile stresses in the bottom fiber, due to the dead and live loads, are to be eliminated, the uniform compressive stress due to prestressing must be equal in magnitude to the sum of these tensile stresses.

There is a time-dependent reduction in the prestressing force, due to the creep and shrinkage of the concrete and the relaxation of the prestressing steel (*see* Chapters 2 and 3). If no tensile stresses are to be permitted in the concrete, it

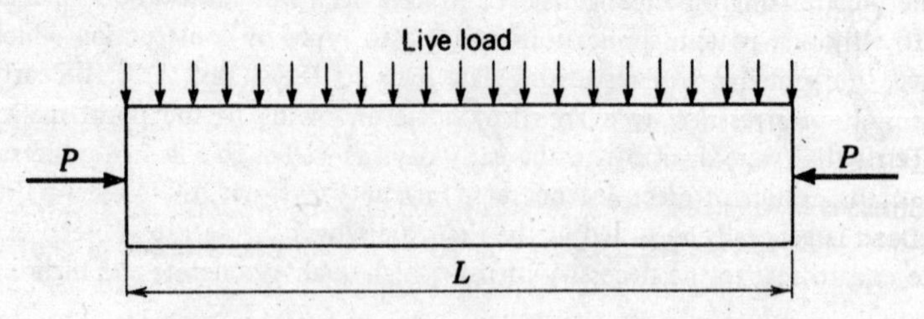

Fig. 1-1 Simple rectangular beam prestressed concentrically.

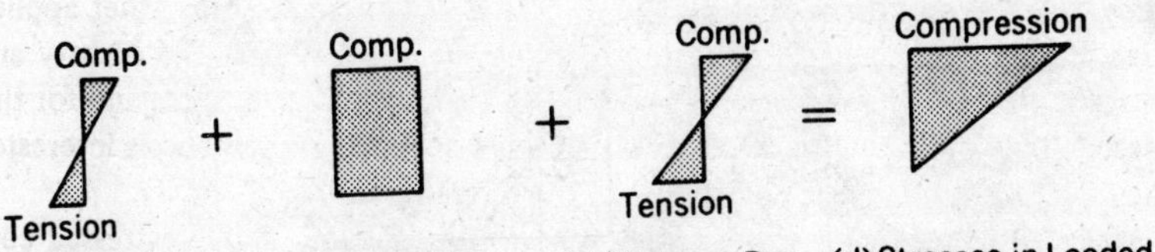

Comp. Tension

(a) Stresses Due to Dead Load

Comp.

(b) Stresses Due to Prestressing

Comp. Tension

(c) Stresses Due to Live Load

Compression

(d) Stresses in Loaded Prestressed Member

Fig. 1-2 Distribution of stresses at *midspan* of a simple beam concentrically prestressed.

is necessary to provide an initial prestressing force that is larger than would be required to compensate for the flexural stresses resulting from the external loads alone. These losses, which are discussed in detail in Sec. 6-2, generally result in a reduction of the initial prestressing force by 10 to 30%. Therefore, if the stress distributions shown in Fig. 1-2 are desired after the loss of stress has taken place (under the effects of the final prestressing force), the distribution of stresses under the initial prestressing force would have to be as shown in Fig. 1-3.

Prestressing, with the concentric force just illustrated has the disadvantage that the top fiber is required to withstand the compressive stress due to prestressing in addition to the compressive stresses resulting from the design loads. Furthermore, because prestressing must be provided to compress the top fibers, as well as the bottom fibers, if sufficient prestressing is to be supplied to eliminate all of the flexural tensile stresses, the *average* stress due to the prestressing force (P/A) must be equal to the maximum flexural tensile stress resulting from the design loads.

If this same rectangular member were prestressed by a force applied by a point one-third of the depth of the beam from the bottom of the beam, the distribution of the stresses due to prestressing would be as shown in Fig. 1-4. In this case, as in the previous example, the final stress in the bottom fiber due to prestressing should be equal in magnitude to the sum of the tensile stresses resulting from the design loads. By inspection of the two stress diagrams for prestressing (Figs. 1-2b and 1-4), it is evident that the average stress in the beam, prestressed

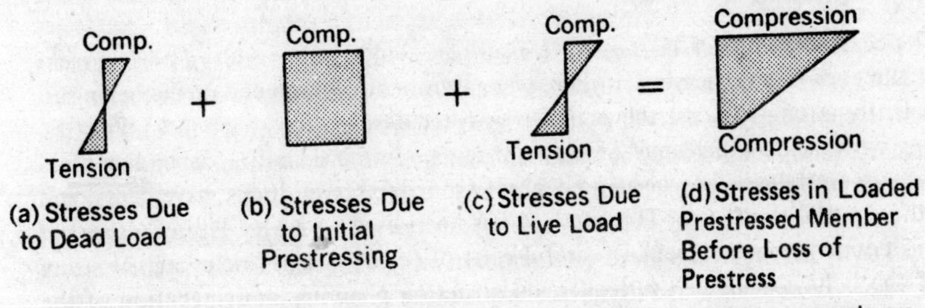

Fig. 1-3 Distribution of stresses at *midspan* of a simple beam under initial concentric prestressing force.

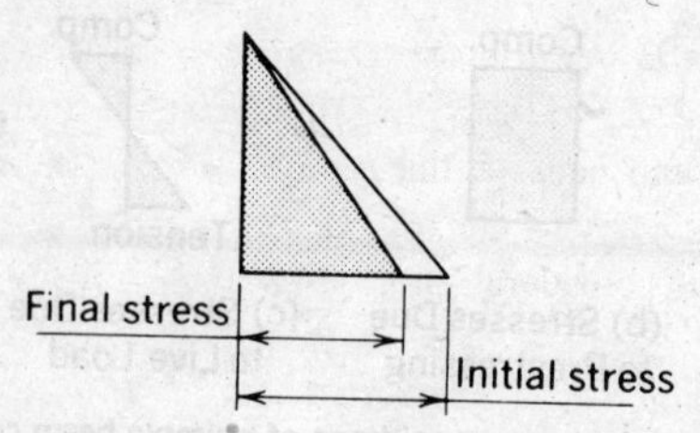

Fig. 1-4 Distribution of stresses due to prestressing force applied at lower third point of rectangular cross section.

with the force at the third point, is only one-half of that required for the beam with concentric prestress. Therefore, the total prestressing force required to develop the desired prestressing of the second example will be only one-half of the amount required in the first example. In addition, the top fiber is not required to carry any compressive stress due to prestressing when the force is applied at the third point.

The economy that results from applying the prestressing force eccentrically is obvious. Further economy can be achieved when small tensile stresses are permissible in the top fibers—these tensile stresses may be due to prestressing alone or to the combined effects of prestressing and any external loads that may be acting at the time of prestressing. This is because the required bottom-fiber prestress can be attained with a smaller prestressing force, which is applied at a greater eccentricity under such conditions. This principle is treated in greater detail in subsequent chapters.

In many contemporary applications of prestressing, the flexural tensile stresses due to the applied loads are not completely nullified by the prestressing; nominal flexural tensile stresses are knowingly permitted under service load conditions. Economy of construction is the motivation for this practice also. The use of flexural tensile stresses under service load conditions is considered in detail in this book.

1-3 Prestressing with Jacks

The prestressing force in the above examples could be the result of placing jacks at the ends of the member, if there were abutments at each end of the beam sufficiently strong to resist the prestressing force developed by such jacks. Prestressing with jacks, which may or may not remain in the structure, depending upon the circumstances, has been used abroad on dams, dry docks, pavements, and other special structures. This method has been used to a very limited degree in this country, since extremely careful control of the design (including the study of the behavior under overloads), construction planning, and execution of the construction is required if the results obtained are to be satisfactory. Furthermore, the loss of prestress resulting from this method is much larger than when other methods are used (*see* Sec. 3-12), unless frequent adjustments of the jacks are made. This is because the concrete is subjected to constant strain in this method rather than to nearly constant stress as is the case in other methods. For these reasons, and because the type of structures to which this method of prestressing can be applied are very limited and beyond the scope of usual generalities, subsequent consideration of this method is not given in this book (*see* Ref. 5).

1-4 Pre-tensioning

Another method of creating the necessary prestressing force is referred to as pre-tensioning. Pre-tensioning is accomplished by stressing steel wires or strands, called tendons, to a predetermined amount, and then, while the stress is maintained in the tendons, placing concrete around the tendons. After the concrete has hardened, the tendons are released and the concrete, which has become bonded to the tendons, is prestressed as a result of the tendons attempting to regain the length they had before they were stressed. In pre-tensioning, the tendons are usually stressed by the use of hydraulic jacks. The stress is maintained during the placing and curing of the concrete by anchoring the ends of the tendons to abutments that may be as much as 500 ft or more apart. The abutments and appurtenances used in this procedure are referred to as a pre-tensioning bed or bench. In some instances, rather than using pre-tensioning benches, the steel molds or forms that are used to form the concrete members are designed in such a manner that the tendons can be safely anchored to the mold after they have been stressed. The results obtained with each of these methods is identical, and the factors involved in determining which method should be used are of concern to the fabricator of prestressed concrete, but do not usually affect the designer.

The tendons used in pre-tensioned construction must be relatively small in diameter, since the bond stress between the concrete and the tendon is relied upon to transfer the stress from the tendon to the concrete. It should be recognized that the ratio of bond area to cross-sectional area for a circular wire or bar is

$$\frac{\text{Bond Area}}{\text{Cross-Section Area}} = \frac{4L}{d} \tag{1-1}$$

in which d is the diameter and L is the length. For a unit length, it will be seen from Eq. 1-1 that the ratio, which is also the ratio of the bond area available to the force the tendon can withstand, decreases as the diameter increases. Hence, when solid wires are used in pre-tensioning, a large number of small-diameter wires are normally used. Strands, composed of several wires twisted around a straight center wire, are also widely used in pre-tensioning.

Pre-tensioning is a major method used in the manufacture of prestressed concrete in this country. The basic principles and some of the methods currently used domestically were imported from Europe, but much has been done here to develop and adapt the procedures to the North American market. One of the developments in this country has been the use of pre-tensioned tendons that do not pass straight through the concrete member, but which are deflected or draped into a trajectory that approximates a curve. This procedure was first used on light roof slabs, but it is now used on large structural members. The use of deflected, pre-tensioned tendons is considered a practical method for the fabrication of precast bridge girders. In other countries, this method is used to a much lesser extent.

Although many of the devices used in pre-tensioned construction are patented, the basic principle is in the public domain. A detailed discussion of the construction procedures and equipment used in pre-tensioned construction is given in Chapter 14.

1-5 Post-tensioning with Tendons

When a member is fabricated in such a manner that the tendons are stressed and each end is anchored to the concrete section *after* the concrete has been cast and has attained sufficient strength to safely withstand the prestressing force, the member is said to be post-tensioned. In this country, when using post-tensioning, a common method used in preventing the tendon from bonding to the concrete during placing and curing of the concrete is to encase the tendon in a mortar-tight, metal tube (or flexible metal hose) before placing it in the forms. The metal hose or tube is referred to as the sheath or duct and remains in the structure. After the tendon has been stressed, the void between the tendon and the sheath is filled with grout. In this manner, the tendon becomes bonded to the concrete section and corrosion of the steel is prevented.

In other cases relatively stiff metal tubes are cast in the concrete without having the tendons inserted in them. Later, when the concrete has cured and is ready to be prestressed, the tendons are inserted in the tubes and post-tensioned.

Rather than use a metal tube or hose, a rubber hose (which may be inflated with air or water or may be stiffened during the placing of the concrete by putting a metal rod in it) has been used to form a hole through the concrete section. After the concrete is sufficiently set, the rubber tube is removed by pulling from one end. The tendon is inserted in the duct that was formed by the tube. The construction is completed by stressing the tendon, anchoring each end, and grouting it in place.

Another method of preventing post-tensioning tendons from becoming bonded to the concrete is to coat the tendons with grease or a bituminous material, after which, the tendon is wrapped in waterproof paper or plastic. Tendons of this type are not pressure grouted after stressing. This type of post-tensioning is usually referred to as unbonded construction.

Post-tensioning offers a means of prestressing on the job site. This procedure may be necessary or desirable in some instances. Very large building or bridge girders that cannot be transported from a precasting plant to the job site (due to their weight, size, or the distance between plant and job site) can be made by post-tensioning on the job site. Post-tensioning is used in precast as well as in cast-in-place construction. In addition, fabricators of pre-tensioned concrete will frequently post-tension the members for small projects on which the number of units to be produced does not warrant the expenditures required to set up pre-tensioning facilities. There are other advantages inherent in post-tensioned construction—these will be discussed later.

In post-tensioning, it is necessary to use some type of device to attach or anchor the ends of the tendons to the concrete section. These devices are usually referred to as end anchorages. The end anchorages, together with the special jacking and grouting equipment used in accomplishing the post-tensioning by one of the several available methods, are generally referred to as post-tensioning systems. Many of the systems used in the U.S. were invented and developed in Europe or were modeled after such a system. The various sytems are or were patented; this somewhat deterred the early use of the method. Post-tensioning systems and their use are discussed in detail in Chapter 15.

1-6 Pre-tensioning vs Post-tensioning

It is generally considered impractical to use post-tensioning on very short members, because the elongation of a short tendon (during the stressing) is small and would require very precise measurement by the workmen. In addition, many of the post-tensioning systems do not function well with very short tendons. A number of short members can be made in series on a pre-tensioning bench without difficulty and without the necessity of precise measurement of the elongation of the tendons during stressing, since relatively long tendon lengths result from making a number of short members in series.

It has been pointed out that very large members may be more economical when cast-in-place and post-tensioned or when precast and post-tensioned near the job site, rather than attempting to transport and handle large pre-tensioned structural elements.

Post-tensioning allows the tendons to be placed, with little difficulty, through the structural elements on smooth curves of any desired path. Pre-tensioned tendons can be employed on other than straight paths, but not without expensive plant facilities and somewhat complicated construction procedures.

The cost of post-tensioned tendons, measured in either cost per pound of prestressing steel or in cost per pound of effective prestressing force, is generally significantly greater than the cost of pre-tensioned tendons. This is due to the larger amount of labor required in placing, stressing, and grouting (where applicable) post-tensioned tendons and to the cost of the special anchorage devices and stressing equipment. A post-tensioned member may require less total prestressing force than an equally strong pre-tensioned member. For this reason, care must be excercised when comparing the relative cost of these modes of prestressing.

The basic shape of an efficient pre-tensioned flexural member may be different from the most economical shape that can be found for a post-tensioned design. This is particularly true of moderate- and long-span members and somewhat complicates any generalization about which method is best under such conditions.

Post-tensioning is generally regarded as a method of making prestressed concrete at the job site, yet post-tensioned beams are often made in precasting plants and transported to the job site. Pre-tensioning is often thought of as a method of manufacturing that is limited to permanent precasting plants. Yet on very large projects where pre-tensioned elements are to be utilized, it is not uncommon for the general contractor to set up a temporary pre-tensioning plant at or near the job site. Each method of making prestressed concrete has particular theoretical and practical advantages and disadvantages, which will become more apparent after the principles are well understood. The final determination of the mode of prestressing that should be used on any particular project can only be made after careful consideration of the structural requirements and the economic factors that prevail for the particular project.

1-7 Linear vs Circular Prestressing

The subject of prestressed concrete is frequently divided into linear prestressing, which includes the prestressing of elongated structures or elements such as beams, bridges, slabs, piles, etc., and circular prestressing, which includes pipe, tanks, pressure vessels, and domes. There are no generally recognized criteria for the design and construction of circularly prestressed structures. The theory of such construction is relatively simple and is adequately covered in the literature (*see* Refs. 6 through 14). This book has been confined to the structural design and analysis of linear prestressed structures and the methods of prestressing used in this type of construction.

1-8 Application of Prestressed Concrete

Prestressed concrete, when properly designed and fabricated, can be virtually crack-free under normal service loads as well as under moderate overload. This is believed to be an advantage in a structure that is exposed to an especially corrosive atmosphere. Prestressed concrete efficiently utilizes high-strength concretes and steels and is economical even with long spans. Reinforced concrete flexural members cannot be designed to be crack-free, cannot efficiently utilize high-strength materials, and are not economical on long spans.

A number of other statements can be made in favor of prestressed concrete, but there are bona fide objections to the use of this material under specific conditions. An attempt is made to point out these criticisms in subsequent chapters. Among the more significant points to be kept in mind about this material are that, in many structural applications, prestressed concrete is more economical in first cost than other types of construction and, in many cases, if the reduced maintenance costs that are inherent with concrete construction are taken into account, prestressed concrete offers the most economical solution for the structure. This fact has been well confirmed by the very rapid increase in the use of prestressed concrete that has taken place in the United States since 1953. It is well known that the advantage of real economy outweighs all intangible advantages that may be claimed, except for very special conditions. Precautions that must be observed in designing and constructing prestressed concrete structures differ from those required for reinforced concrete structures. These precautions are discussed in Chapter 10.

Illustrations of prestressed structures and structural elements are given in Chapters 11 and 12, where the various types of building and bridge construction are described and compared.

REFERENCES

1. "Proceedings of the Conference Held at the Institution," *The Institution of Civil Engineers*, 5-10 (Feb. 1949).
2. Abeles, P. W., *Principles and Practice of Prestressed Concrete*, pp. 18-20, Crosby Lockwood & Son, Ltd., London, 1949.
3. Dobell, C., "Patents and Code Relating to Prestressed Concrete," *Journal, American Concrete Institute*, **46**, 713-724 (May 1950).
4. Libby, James, R., "The Elastic Design of Simple Prestressed Concrete Beams," *Proc. Western Conf. on Prestressed Concrete*, 71 (1952).
5. Guyon, Y., *Prestressed Concrete*, pp. 4, 49, John Wiley & Sons, Inc., New York, 1953.
6. Dobell, C., "Prestressed Concrete Tanks," *Proc. First U.S. Conference on Prestressed Concrete*, 9 (1951).
7. Hendrickson, J. G., "Prestressed Concrete Pipe," *Proc. First U.S. Conference on Prestressed Concrete*, 21 (1951).
8. Kennison, H. F., "Prestressed Concrete Pipe–Discussion," *Proc. First U.S. Conference on Prestressed Concrete*, 25 (1951).
9. "Proceedings of the Conference Held at the Institution," *The Institution of Civil Engineers*, 52-56 (Feb. 1949).
10. Timoshenko, S., *Theory of Plates and Shells*, McGraw-Hill Book Company, Inc., New York, 1940.
11. "Circular Concrete Tanks Without Prestressing," Bulletin, Portland Cement Association.
12. "Design of Circular Domes," Bulletin, Portland Cement Association.
13. Crom, J. M., "Design of Prestressed Tanks," *Proc. A.S.C.E.*, **37** (Oct. 1950).
14. ACI Committee 344. "Design and Construction of Circular Prestressed Concrete Structures," *Journal, American Concrete Institute*, **67**, No. 9, 657-672 (Sept. 1970).

2 Steel for Prestressing

2-1 Introduction

It was stated in Sec. 1-2 that the loss in prestress which results from the effects of steel relaxation and the shrinkage and creep of the concrete is generally from 10 to 30% of the initial prestress. The computation of the losses of prestress due to the various causes is discussed in detail in Sec. 6-2, but it is important for the designer of prestressed concrete to be aware that the greater portion of the loss of prestress is normally attributed to be the result of the shrinkage and creep of the concrete. It should also be recognized that this fact accounts for the necessity of using high-strength steel, with a relatively high initial stress, in the construction of prestressed concrete.

The shrinkage and creep of concrete produce inelastic volume or length changes. Because the tendons are anchored to the concrete, either by bond or by end anchorages, the length changes in the concrete result in a length change in the tendons. Furthermore, since the steel used for prestressing is fundamentally an elastic material at the stress levels employed in normal designs, the reduction of the stress in the tendons, which results from the length changes in the concrete, is equal to the product of the elastic modulus of the steel and the unit length change in the concrete.

It is essential the loss of prestress be a relatively small portion of the total prestress, in order to attain an economical and practical design. The elastic modulus of steel is a physical property which, for all practical purposes, cannot be altered or adjusted by manufacturing processes. In a similar manner, the inelastic volume changes of concrete of any particular quality are physical properties that cannot be eliminated using practical construction procedures. Therefore, the product of these two factors is normally beyond the control of the designer.

It can be shown that the normal loss of prestress is generally of the order of 15,000 to 50,000 psi. It is apparent that if the loss of prestress is to be a small portion of the initial prestress, the initial stress in the steel must be very high and of the order of 100,000 to 200,000 psi. If a steel having a yield point of 40,000 psi were used to prestress concrete and if this steel were stressed initially to 30,000 psi, the entire prestress could be lost, as was the case indeed in the early attempts at prestressing with low-strength steel and concrete of poor quality.

It should be pointed out that research has been conducted into the use of other materials, such as fiber glass and aluminum alloys, for prestressing concrete. Some of these materials have elastic moduli that are of the order of one-third of that of steel. If such materials could be safely and economically used, the loss of prestress would be reduced to approximately one-third of the loss obtained with steel tendons. Hence, the loss of prestress could possibly be ignored in normal design practice if these materials were employed as tendons. There are, however, many problems to be studied and overcome before these materials can be used safely and with economy. The use of tendons having a low elastic modulus and plastic deformations different from those of steel would result in members having post-cracking deflection and ultimate strength characteristics different from those obtained when steel tendons are used. These problems will be apparent from the subsequent discussion of the desirable physical properties of the steel that is used in prestressing.

There are several basic forms of high-strength steel that are used currently in domestic prestressed work. In general, these can be divided into three groups: uncoated stress-relieved wires, uncoated stress-relieved strand, and uncoated high-strength steel bars. Each of these types of steel is described briefly in the following sections. For a more detailed description of the method of manufacture, chemical composition, and physical properties of these materials, the reader should consult the applicable ASTM Specifications and the references listed at the end of this chapter.

Other types of wire, such as straightened "as-drawn" wire and oil-tempered wire, are used for prestressing in other parts of the world. Experience in Europe has shown that oil-tempered wire may, under certain circumstances, be more susceptable to stress corrosion (*see* Sec. 2-8) than the types of steels commonly employed in North America. In addition, "as-drawn" wire generally exhibits greater relaxation than stress-relieved wire and strand of the types employed domestically. Hence, caution should be exercised by the engineer who uses materials that do not conform to usual ASTM Standards; some adjustment in the design may be necessary. These materials are not considered here since they are not normally used in the U.S. The interested reader can find descriptions of these materials in the references listed at the end of this chapter.

2-2 Stress-Relieved Wire

Cold-drawn stress relieved wire, which is commonly used in post-tensioned construction and rarely used in pre-tensioned members, is manufactured to conform to the "Standard Specifications for Uncoated Stress-relieved Wire for Prestressed Concrete" (ASTM Designation A 421). These specifications provide that the wire shall be made in two types (BA and WA), depending upon whether it is to be used with button- or wedge-type anchorages (*see* Chapter 15). Other major requirements in these specifications include the minimum ultimate tensile strength, the minimum yield strength, the minimum elongation at rupture, as well as diameter tolerances. The principal strength requirements of ASTM A 421 are summarized in Table 2-1. A supplement to these specifications covers low-relaxation wire (*see* Sec. 2-7).

Typical stress-strain curves for uncoated, stress-relieved wires are shown in Fig. 2-1. It should be noted that the stress-strain curves for the two wire diameters shown are similar in shape and that the ultimate tensile strength is higher for a wire of smaller diameter. It should be pointed out that ASTM A 421 requires a minimum elongation of 4.0% when measured in a gage length of 10 in., which means the steel is quite ductile and has a "plastic range" of considerable magnitude. (The "plastic range" is not shown in Fig. 2.1.) During recent years, the use of solid wire has diminished greatly while the use of strand (Sec. 2-3) has increased substantially. It is expected that this trend will continue, due to economic considerations, in spite of the fact that stress-relieved wire performs very well in prestressed concrete.

TABLE 2-1 Properties of Stress-Relieved Wire for Prestressed Concrete Required by ASTM A 421.

	Min. Tensile Strength (psi)		Min. Stress at 1% Extension (psi)*	
Nominal Diameter (in)	Type BA	Type WA	Type BA	Type WA
0.192		250,000		212,500
0.196	240,000	250,000	204,000	212,500
0.250	240,000	240,000	204,000	204,000
0.276	235,000	235,000	199,750	199,750

*Measured according to procedures specified in ASTM A 421.

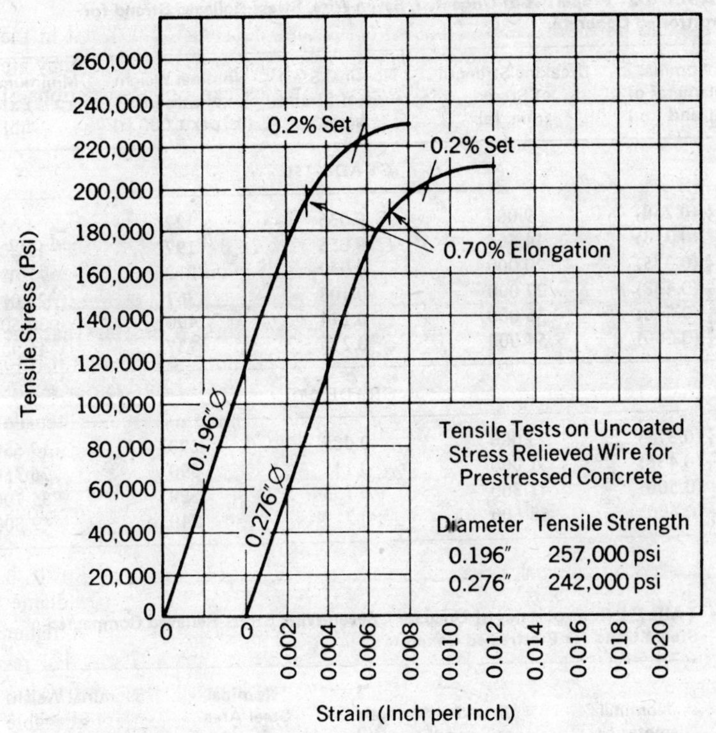

Fig. 2-1 Typical stress-strain curves for wires in elastic range. (*Source: C. F. & I. Steel Corp., Trenton, N.J.*)

2-3 Stress-Relieved Strand

Stress-relieved strand is made to conform to the requirements of "Standard Specifications for Uncoated Seven-Wire Stress-Relieved Strand for Prestressed Concrete" (ASTM Designation A 416). Basic strength, area, and weight requirements for seven-wire strands, as provided by ASTM Designation A 416, are given in Table 2-2 for the two grades of strands covered therein.

Seven-wire strands are made by twisting 6 wires, on a pitch of between 12- and 16-wire diameters, around a slightly larger, straight central wire. The strands are stress-relieved after being stranded. Typical stress-strain curves for seven-wire strands commonly used in pre-tensioning and in multistrand post-tensioning tendons are shown in Figs. 2-2 and 2-3. It is interesting to note that seven-wire strands are available in two grades. Grade 250 strand has a nominal ultimate tensile strength of 250,000 psi. Grade 270 strand has slightly larger wires than Grade 250 strand and has a nominal ultimate tensile strength of 270,000 psi.

TABLE 2-2 Properties of Uncoated, Seven-Wire, Stress-Relieved Strand for Prestressed Concrete.

Nominal Diameter of Strand (in.)	Breaking Strength of Strand (min. lb)	Nominal Steel Area of Strand (sq in.)	Nominal Weight of Strands (lb. per 1 000 ft)	Minimum Load at 1% Extension (lb)
		GRADE 250		
$\frac{1}{4}$ (0.250)	9 000	0.036	122	7 650
$\frac{5}{16}$ (0.313)	14 500	0.058	197	12 300
$\frac{3}{8}$ (0.375)	20 000	0.080	272	17 000
$\frac{7}{16}$ (0.438)	27 000	0.108	367	23 000
$\frac{1}{2}$ (0.500)	36 000	0.144	490	30 600
$\frac{3}{5}$ (0.600)	54 000	0.216	737	45 900
		GRADE 270		
$\frac{3}{8}$ (0.375)	23 000	0.085	290	19 550
$\frac{7}{16}$ (0.438)	31 000	0.115	390	26 350
$\frac{1}{2}$ (0.500)	41 300	0.153	520	35 100
$\frac{3}{5}$ (0.600)	58 600	0.217	740	49 800

TABLE 2-3 Properties of Uncoated, Seven-Wire, Stress-Relieved Compacted Steel Strand for Prestressed Concrete

Nominal Diameter (in.)	Breaking Strength of Strand (min. lb)	Nominal Steel Area (in.^{2})	Nominal Weight of Strand (per 1000 ft lb)
$\frac{1}{2}$	47,000	0.174	600
0.6	67,440	0.256	873
0.7	85,430	0.346	1176

A supplement to ASTM Designation 416 covers low-relaxation strand (*see* Sec. 2-7). The elastic and plastic characteristics of seven-wire strand are quite similar to those of stress-relieved wire, as can be seen in comparing Figs. 2-1 and 2-2.

Another type of uncoated, stress-relieved, seven-wire strand for prestressing concrete is covered by ASTM Designation A 779. This material is "compacted" by having been drawn through a die after having been stranded and hence having a cross-sectional shape, as shown in Fig. 2-4. This type of strand has a greater steel cross-sectional area, for any nominal diameter, as compared to "uncompacted" seven-wire strand. The result is a larger ultimate tensile strength for the compacted strand of a particular diameter.

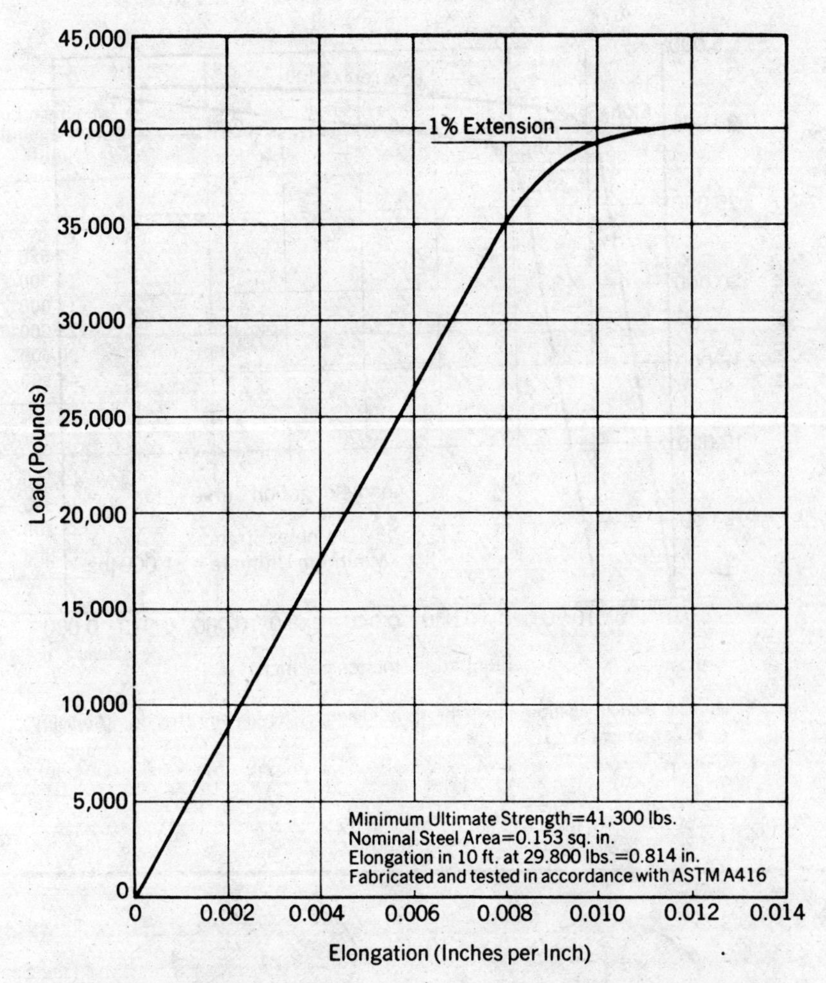

Fig. 2-2 Typical load-elongation curve for $\frac{1}{2}$-in. diameter, seven-wire strands. (*Courtesy C. F. & I. Steel Corp.*)

Basic strength, area, and weight properties for compacted strand are given in Table 2-3.

2-4 High-Tensile Strength Bars

Both smooth (type I) and deformed (type II) high-tensile, alloy steel bars are available in nominal diameters from $\frac{3}{4}$ to $1\frac{3}{8}$ in. Bars of other sizes are available

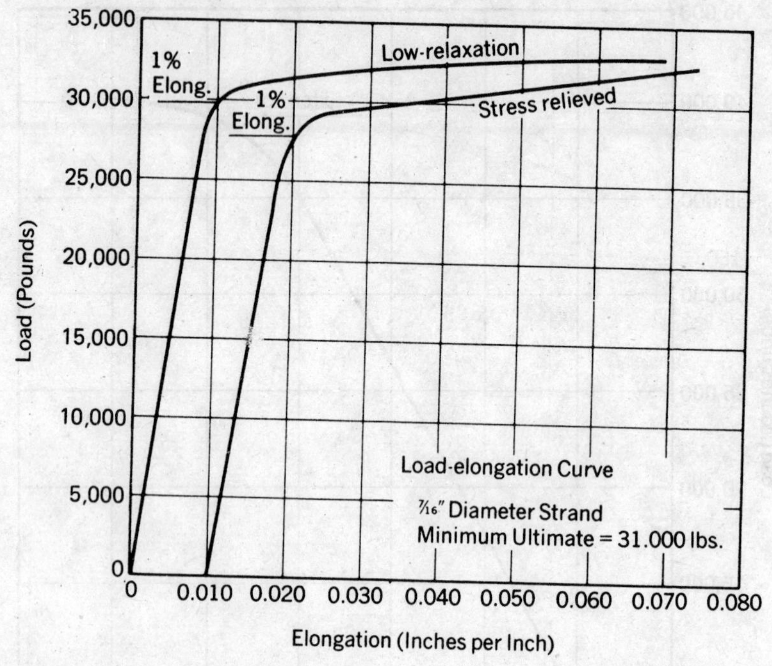

Fig. 2-3 Typical load-elongation curve for $\frac{7}{16}$-in. diameter, seven-wire strands. (*Courtesy C. F. & I. Steel Corp.*)

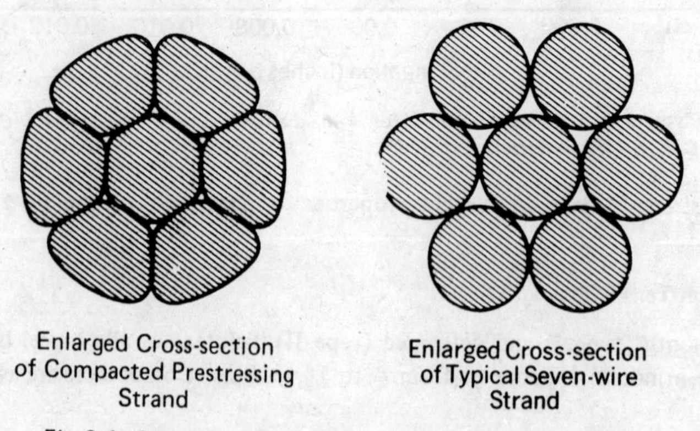

Fig. 2-4 Comparison of compacted to typical seven-wire strand.

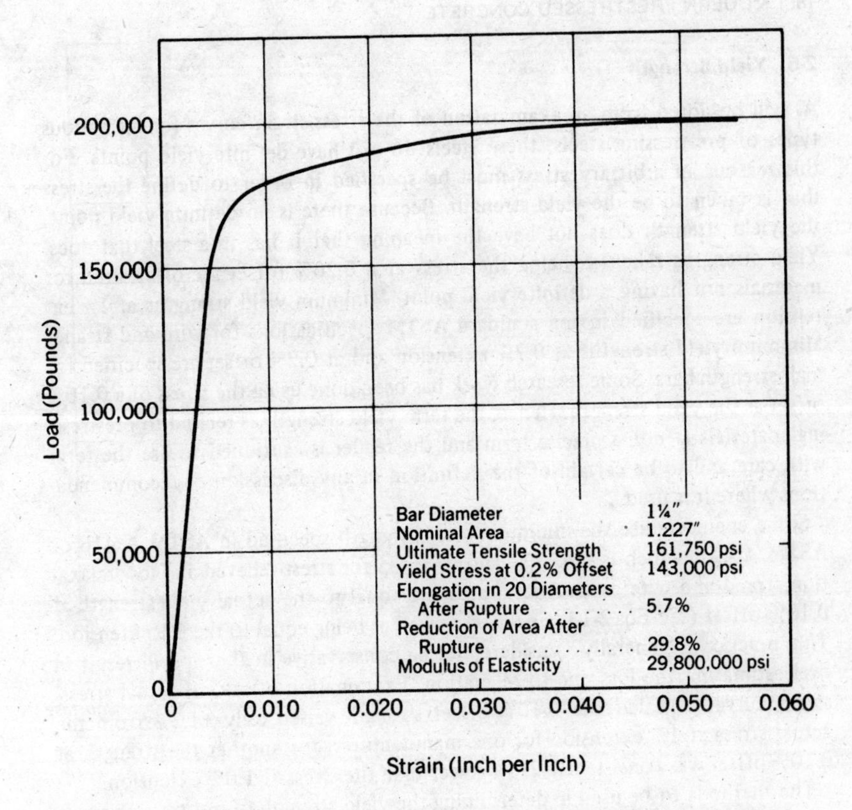

Fig. 2-5 Typical load-strain curve for a prestressing bar. (*Courtesy of Stressteel Corp.*)

by special arrangement with some bar manufacturers. The bars are made from an alloy steel and conform to "Standard Specification for Uncoated High-Strength Steel Bar for Prestressing Concrete," ASTM Designation A 722.

The bars are generally cold-stretched in order to raise the yield point and to render the bars more elastic at stress levels below the yield point. After cold-stretching, they are frequently stress-relieved in order to improve the ductility and stress-strain characteristics. Principal minimum requirements from ASTM A 722 include a minimum tensile strength of 150,000 psi, minimum yield strengths for smooth and deformed bars of 85% and 80% of the minimum ultimate tensile strength respectively, and a minimum elongation after rupture (in a gage length equal to 20 bar diameters) of 4.0%. Bars are sometimes produced with properties exceeding the minimum requirements of ASTM A 722.

A typical load-strain curve for a high-tensile strength bar is given in Fig. 2-5.

2-5 Yield Strength

As will be noted from an examination of the stress-strain curves for the various types of prestressing steels, these steels do not have definite yield points. For this reason, an arbitrary stress must be specified in order to define the stress that is taken to be the yield strength. Because there is no definite yield point, the yield strength does not have the meaning that it has in a steel that does. Yield strengths taken as being the stress at a 0.20% offset are often used for materials not having a definite yield point. Minimum yield strengths at 1% extension are specified in the standard ASTM specifications for wire and strand. Minimum yield strengths at 0.7% extension and at 0.2% offset are specified for high-strength bars. Some research work has been done using the stress of a 0.10% offset as the yield strength. Hence, the term yield strength as related to prestressing materials is not a precise term and the reader is cautioned to use the term with care and to be certain of the definition in any discussion or recommendations where it is used.

Some engineers use the minimum yield strength specified in ASTM A 416 or ASTM A 421, which is 0.85 f_{pu}^{*} and 0.90 f_{pu} for stress-relieved and low-relaxation strand and wire, respectively, as being equal to the actual yield strength at 0.10% offset (see Eq. 2-1 in Sec. 2-7) as well as being equal to the 1% extension. This practice is generally considered to be conservative in that it will result in over-estimating the loss due to relaxation. Examination of several actual stress-strain curves revealed the 0.10% offset strength varied only +1.3% from the actual stress at 1% extension for one manufacturer; for another the strength at 0.10% offset was consistently 2–3% lower than the stress at 1.0% extension.

The methods to be used in determining the yield strength of prestressing steels is given in "Standard Specifications and Definitions for Mechanical Testing of Steel Products" (ASTM Designation A 370). Yield strength requirements for the various types of steels used in prestressed concrete are specified in the ASTM specifications applicable to each type of steel.

2-6 Plasticity

Plasticity at very high stress levels is as essential in prestressing steel as it is in reinforcing steel. The object is to ensure that ultimate bending moments will be reached only after large and very apparent plastic deformations have taken place. The use of brittle steel could result in a sudden failure similar to that which is characteristic of an over-reinforced concrete member. In order to avoid this possibility, the normal practice is to specify that the prestressing steel will have a minimum elongation at rupture of 3.5–4.0%, depending upon the type of steel used and the method that is used to measure the elongation at rupture (*see* ASTM A 416, ASTM A 421, ASTM 722, and ASTM 779).

*f_{pu} is defined as the minimum guaranteed ultimate tensile strength of the prestressing steel.

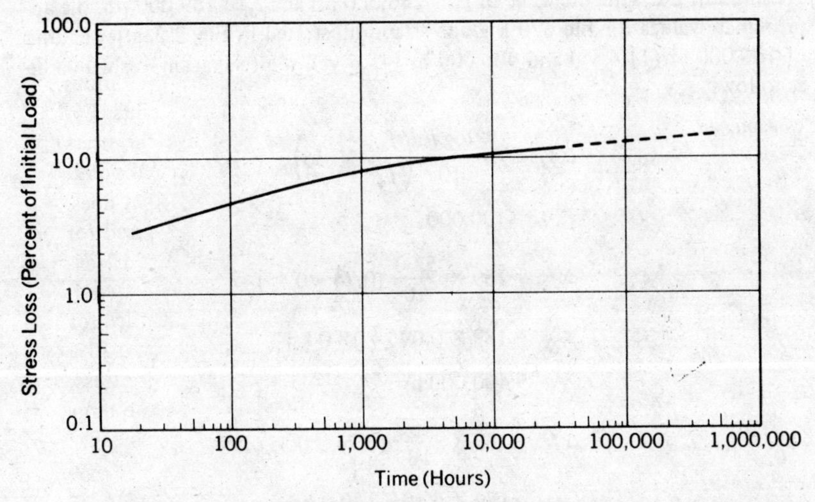

Fig. 2-6 Stress loss vs time for a stress-relieved, prestressed-concrete wire initially loaded at 70% of the guaranteed ultimate tensile strength and held at constant length at 85.0° F. *(Courtesy C. F. & I. Steel Corp.)*

2-7 Relaxation and Creep

Relaxation is defined as the loss of stress in a material that is placed under stress and held at a constant strain, whereas creep is defined as the change in strain for a member held under constant stress. Although tendons are not subjected to constant strain or to constant stress, it is generally agreed the condition more closely approximates a condition of constant strain, and hence, relaxation studies are made to evaluate the loss of prestress that can be attributed to the inelastic behavior of the steel. A typical relaxation curve for stress-relieved prestressing wire is shown in Fig. 2-6.

For stress-relieved wire and strand, the loss of stress due to relaxation at normal temperatures can be estimated with sufficient accuracy for design purposes using the following relationship:

$$\Delta f_{sr} = f_{si} \frac{\log t'}{10}\left(\frac{f_{si}}{f'_y} - 0.55\right) \qquad (2\text{-}1)$$

where Δf_{sr} = the relaxation loss at time t' hours after prestressing, f_{si} = the initial stress and f'_y = 0.10% offset stress for the steel under consideration (Ref. 1). The logarithm of the time t' is to the base 10. This relationship is only applicable when the ratio of f_{si}/f'_y is equal to or greater than 0.55.*

*Note that f'_y is used here rather than f_{py} which is the notation used for the specified yield strength of prestressing steel in ACI 318-83 (Ref. 12). This is due to the difference in definitions of the two yield strengths.

Using Eq. 2-1 with the value of f'_y = 256,000 psi and f_{si} = 189,000 psi, the approximate values for the 270 k grade strand illustrated in Fig. 2-2, after t' time of 100,000 hr (11.4 yr) and 400,000 hr (45.6 yr), the relaxation loss would be as follows

$$\Delta f_{sr} = f_{si} \frac{\log t'}{10} \left(\frac{f_{si}}{f'_y} - 0.55 \right)$$

$$t' = 100{,}000 \text{ hr}$$

$$\Delta f_{sr} = 189 \times \frac{5.00}{10} (0.74 - 0.55)$$

$$= 189 \times 0.095 = 18.0 \text{ ksi}$$

$$t' = 400{,}000 \text{ hr}$$

$$\Delta f_{sr} = 189 \times \frac{5.60}{10} (0.74 - 0.55)$$

$$= 189 \times 0.106 = 20.0 \text{ ksi}$$

Assuming Eq. 2-1 accurately relates the relaxation of prestressing steel, the efficiency of the steel at various levels of initial stress after 50 years of service can be studied through the use of Fig. 2-7. From this plot it will be seen that the increase in effective stress f_{se} is nearly equal to the increase in initial stress f_{si} up to the point where $f_{si} = 0.60 f'_y$. Above this stress, the efficiency is progressively reduced. For an initial stress ratio of 0.90, virtually no gain in effective stress is realized for increases in initial stress.

In the fabrication of pre-tensioned concrete members, the tendons are stressed to an initial level, held at that elongation for a period of time, and then released. At the time the prestressing force is transferred to the concrete (tendons released), the stress in the tendons is less than the initial stress. This is due to the relaxation that has taken place in the interval between stressing and releasing, as well as to the elastic shortening of the concrete that takes place upon transferring the prestressing force to the concrete. The effect of relaxation of the steel can be estimated at time t'_n, assuming the tendon was stressed at time zero and released at time t'_r, using the following relationship

$$\frac{f_{st}}{f'_{si}} = 1 - \left(\frac{f'_{si}}{f'_y} - 0.55 \right) \left(\frac{\log t'_n - \log t'_r}{10} \right) \tag{2-2}$$

in which f'_{si} is the effective stress in the steel after release of the tendons and f_{st} is the stress in the steel at time t'_n. In Eq. 2-2, the stresses are in ksi and time is in hours. In using Eq. 2-2, one must compute the relaxation loss occurring in the steel from the time of initial stressing until the time of release, using Eq. 2-1.

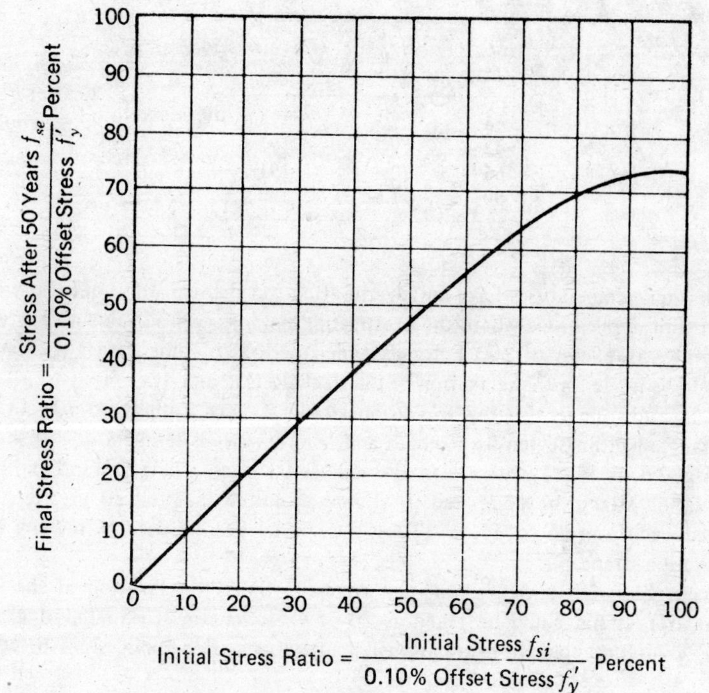

Fig. 2-7 Final stress ratio vs initial stress ratio.

This relaxation loss should be added to the loss due to elastic shortening of the concrete and the sum subtracted from the initial steel stress to determine the value of f'_{si} for use in Eq. 2-2.

ILLUSTRATIVE PROBLEM 2-1 During the production of pre-tensioned members for a large project, the normal production cycle provided for an 18 hour period between the time the tendons were stressed and released. Occasionally, due to weekends, holidays, or other events beyond the control of the contractor, the time interval between stressing and releasing the tendons was as great as 90 hrs. Determine the stress in the stress-relieved prestressing steel at transfer for time intervals between stressing and release of 18, 42, 66, and 90 hr if f_{pu} = 270 ksi, f'_y = 240 ksi and the tendons are always stressed to 0.75 f_{pu} initially.

SOLUTION: From Eq. 2-1

$$\Delta f_{sr} = 202.5 \frac{\log t}{10}\left(\frac{202.5}{240} - 0.55\right) = 5.948 \log t$$

t (hr)	Δf_{sr} (ksi)	f_s at transfer (ksi)
18	7.5	195.0
42	9.7	192.8
66	10.8	191.7
90	11.6	190.9

Elevated temperatures have an adverse effect on the relaxation of prestressing steel. For applications where the prestressing tendons will be subjected to temperatures in excess of 100°F for extended periods of time, larger allowances should be made for the relaxation of the steel (Refs. 7 and 8).

Relaxation curves for low-relaxation strand stressed initially to 70% of the guaranteed ultimate tensile strength and held at various temperatures is shown in Fig. 2-8. In Fig. 2-9, the relaxation curves for stress-relieved strand and low-relaxation strand are compared. It should be noted that the 50 yr stress loss is reduced from about 15 to 3% by the special process used in making low-relaxation strand.

From Figs. 2-8 and 2-9, one can conclude that the relaxation of the low-relaxation strand could be taken as 20–25% of that for stress-relieved strands (Eq. 2-1) for applications at normal temperatures. For applications in which

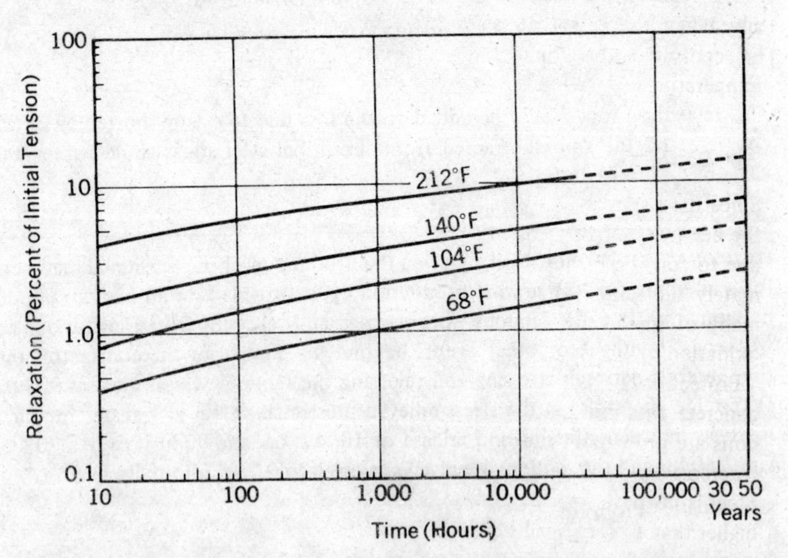

Fig. 2-8 Relaxation vs time curve for low-relaxation strand stressed initially to 70% of the guaranteed ultimate tensile strength and held at various temperatures. (*Courtesy of C. F. & I. Steel Corp.*)

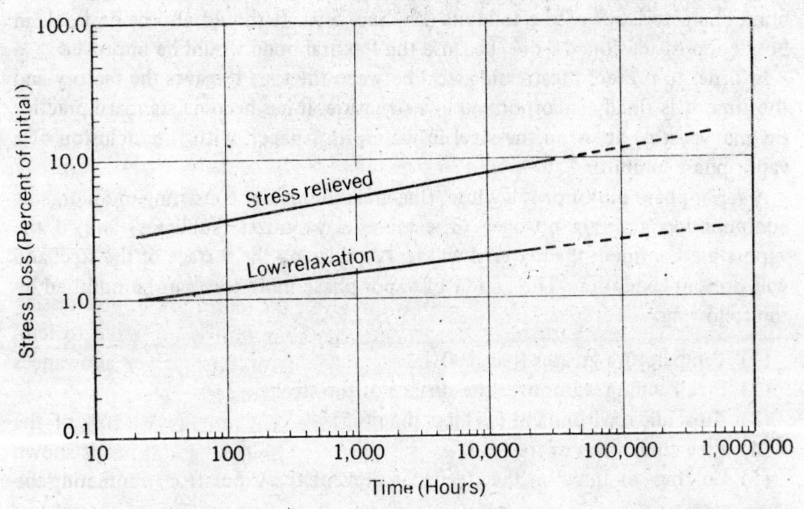

Fig. 2-9 Relaxation vs time curves for stress-relieved and low-relaxation seven-wire strands held at constant length at 85°F. Initial stress was 70% of the guaranteed ultimate tensile strength.

the tendons may be exposed to temperatures above 100°F, only the low-relaxation-type strand should be used.

The relaxation of high-tensile strength bars at normal temperatures is generally taken to be 3%. It is recommended that the manufacturers of the bars be consulted concerning the relaxation of their specific product at other temperatures.

2-8 Corrosion

Since the strength of a prestressed-concrete flexural member is dependent upon the adequacy of the tendons, it is essential that the tendons do not deteriorate due to corrosion. Prestressing steels are subject to normal oxidation in approximately the same degree as structural-grade steels. Because the tendons are normally of small proportions, it is essential they be protected against significant oxidation.

Protection against corrosion is effected in pre-tensioned construction by the concrete that surrounds the tendons. In post-tensioned construction, the tendons are protected by grout that is injected around the tendons after stressing or by grease or bituminous materials placed on the tendons, as was explained earlier. Steel is not attacked by oxygen when the pH of the surrounding environment is higher than 8, as is usually provided by concrete made with Portland cement.

It must be emphasized that research has shown that a light, hard oxide on the tendons is desirable in pre-tensioned members, since such an oxide improves the bond characteristics of the tendons (*see* Sec. 5-9). It should also be desirable in bonded, post-tensioned work, because the flexural bond would be improved.

In order to protect prestressing steel between the time it leaves the factory and the time it is finally incorporated in a structure, it has become standard practice (in many areas) to wrap the steel in waterproof paper, with the inclusion of a vapor-phase inhibitor.*

A vapor-phase inhibitor is a white, fine-grained powder consisting of an organic compound containing nitrogen. The material vaporizes (sublimes) and, if the vapors are confined, the material will recrystallize on the surface of the steel and will prevent oxidation. The action of vapor-phase inhibitors can be nullified by the following:

(1) Temperature greater than 160°F.
(2) Free-running water over the surface of the steel.
(3) An acidic environment (pH less than 6.5).
(4) Free circulation of fresh air.
(5) Coatings or films on the steel that prevent the vapor from contacting the steel.
(6) The powder not being in the immediate vicinity of the steel (further than 12 in.).
(7) An environment containing a high concentration of chlorides.

The material will work in either air or water, providing the above conditions do not exist to nullify the action.

Post-tensioned construction is occasionally protected against the effects of corrosion by using galvanized tendons. This procedure is not used frequently because galvanized tendons are not as strong as bright tendons of the same size, due to the fact that some of the diameter of a galvanized tendon is composed of low-strength zinc. As a result, galvanized tendons are materially more expensive than bright tendons of equal strength. (For equal diameters, galvanized, seven-wire strands are approximately 15% lower in strength and 10% more in cost.) Furthermore, the various types of anchorage devices used in post-tensioning with the parallel-wire systems either cannot anchor galvanized wire, due to the low coefficient of friction, or cannot be used without damaging the zinc coating. For these reasons, the use of galvanized wire is generally considered to be impractical with parallel-wire systems. The use of galvanized, large-diameter strand is feasible under some conditions. Galvanized, seven-wire strand can be used in some of the more modern anchorage devices, but they rarely are due to the cost.

Galvanized wire and strand is not used in pretensioned concrete because the concrete adequately protects the steel against corrosion.

A type of corrosion referred to as "pitting corrosion" is the cause of some deterioration (and even failures) of prestressed concrete structures. Calcium

*A convenient way to specify the protective packaging of prestressing materials is to use the provisions of "Standard Recommended Practices for Packaging Marking and Loading Methods for Steel Products for Domestic Shipment" (ASTM Designation A700).

chloride or sodium chloride in the concrete or grout is generally considered to be the cause of this type of corrosion. It is for this reason that chlorides must never be permitted, even in very small amounts, in the concrete or grout used in prestressed-concrete construction. (*See* Ref. 4.)

Prestressing steels, particularly wires and strands, are very susceptible to a type of deterioration that is called "stress corrosion." This type of corrosion has occurred relatively infrequently. Stress corrosion is characterized by a breakdown of the cementitious portion of the steel, resulting in fine cracks in the steel. These fine cracks render the steel nearly as brittle as glass. Since little is known about this type of corrosion, there is no way to be certain that it will not occur during construction of a prestressed member. It is known that nitrates (not to be confused with the rust-inhibiting nitrites), chlorides, sulfides, and some other agents can result in stress corrosion under certain conditions. It is also known that the steel is more susceptible to this type of corrosion when highly stressed; this accounts for the name "stress corrosion."

Another cause of delayed failure, which can occur in high-strength steels, is called "hydrogen embrittlement." This phenomenon, which apparently results when steel is exposed to hydrogen ions but not to hydrogen molecules, is characterized by a decrease in ductility and tensile strength. Hydrogen embrittlement may be promoted by electroplating steel with cadmium or zinc, as well as from corrosion and electrical currents. Confining the prestressing steel in an environment having a pH greater than 8 is thought to be the best protection against the absorption of hydrogen.

It is interesting to note that aluminum powder, which causes the release of hydrogen gas (molecular hydrogen), has been used for many years as an expansive additive for the grouting of post-tensioned tendons. This practice has apparently not been harmful, because failures have not been reported in structures so constructed. Additives which obtain expansion by the release of nitrogen gas are also used.

2-9 Application of Steel Types

The same basic steel can be used in pre-tensioning and post-tensioning, but in the former it is necessary that the individual tendons are not so large that they cannot be adequately bonded to the concrete, since this bond is relied upon to transfer the prestressing force from the steel to the concrete. In post-tensioning, as has been explained, end anchorages are used to transfer the prestressing force to the concrete—the grouting (when used) is relied upon to protect the steel against corrosion and to develop flexural bond stress, i.e., bond stresses resulting from changes in the externally applied loads. Bond stresses are discussed in detail in Secs. 5-9 and 5-10. It should be mentioned here that, while in Europe it is customary to use wires up to 0.276 in. in diameter as pre-tensioning tendons, the usual practice in this country has been to use the uncoated, seven-wire strands described in Sec. 2-3. Little or no use of high-tensile alloy bars has been made in pre-tensioning in this country, although favorable results have been obtained experimentally in Europe with bars up to $\frac{5}{8}$ in. in diameter. (*See* Ref. 5.)

2-10 Idealized Tendon Material

One may wish to consider the properties an ideal material for prestressing concrete would have. Some characteristics are desirable from one standpoint and not from another. For example, high tensile strength, coupled with a low elastic modulus, permits a high strain under initial stress, which minimizes the losses of stress due to the inelastic properties of the concrete. On the other hand, a high tensile strength results in a small area of tendon being required and this, coupled with a low modulus of elasticity, could result in very high deflections upon the application of an overload that would cause cracking. In actuality, the steels we currently have available generally result in designs that are efficiently balanced in "serviceability" and "strength" characteristics. Perhaps steels with somewhat higher strengths could be used efficiently. Steels without any relaxation loss would obviously be advantageous.

It is also possible that materials of very high strength and low elastic modulus will eventually be used in combination with non-prestressed mild reinforcing in order to achieve efficient and economical construction.

The major desirable physical characteristics of the material to be used for prestressing tendons can be summarized as follows:

(1) High strength that allows high prestressing stresses.
(2) Elastic up to high stress levels.
(3) Plastic at very high stress levels.
(4) Low elastic modulus at time of stressing in order to minimize the loss of prestress.
(5) High elastic modulus after bonding in order to contribute to stiffness of the member.
(6) Low creep or relaxation losses at the stress levels normally employed in prestressing and at elevated temperatures.
(7) Resistant to corrosion.
(8) Small diameter or relatively large surface area of the individual tendons to achieve good bond characteristics.
(9) Absence of dirt and lubricants on the surface.
(10) Straightness to facilitate handling and placing.

From the above description one will see there is no known material that has all of these desirable qualities. The high-strength steels currently used possess most of the desirable characteristics.

2-11 Allowable Steel Stresses

The two most significant design criteria for prestressed concrete in the United States are the "Standard Specifications for Highway Bridges," which is published by the American Association of State Highway and Transportation Officials (AASHTO), (Ref. 11), and "Building Code Requirements for Reinforced Concrete (ACI 318)," published by the American Concrete Institute, (Ref. 12).

The stresses permitted in the prestressing steel in the AASHTO Specification are as follows:

(1) Temporary stress before losses due to creep and shrinkage 0.70 f_{pu} (Overstressing to 0.80 f_{pu} for short periods of time may be permitted provided the stress, after transfer to concrete in pre-tensioning or seating of anchorage in post-tensioning, does not exceed 0.70 f_{pu}).
(2) Stress at design load (after losses) 0.80 f_{py} (f_{py} = yield strength).

The steel stresses permitted by ACI 318 are as follows:

(a) Due to tendon jacking force 0.94 f_{py} but not greater than 0.85 f_{pu}, or maximum value recommended by manufacturer of prestressing tendons or anchorages.
(b) Pretensioning tendons immediately after prestress transfer 0.82 f_{py} but not greater than 0.74 f_{pu}.
(c) Post-tensioning tendons, at anchorages and couplers, immediately after tendon anchorage 0.70f_{pu}.

From this it will be seen that stresses as high as 0.80 f_{pu} are permitted by both criteria during jacking. This stress is of a temporary nature. In addition, both criteria limit the initial stress to 0.70 f_{pu}. Initial stress is defined as the stress in pretensioning tendons immediately after transfer (release). In post-tensioned tendons, initial stress is defined as the stress in the tendons immediately after anchoring.

The AASHTO Specification also restricts the effective stress (initial stress minus the loss of prestress) to 0.60 f_{pu}. This requirement was in ACI 318-63, but was not included in ACI 318-71. There is no justification for restricting the effective stress to 0.60 f_{pu}. It is the author's opinion that this requirement will eventually be eliminated from the AASHTO Specification.

REFERENCES

1. Magura, Sozen and Siess, "A Study of Stress Relaxation in Prestressing Reinforcement" Civil Engineering Studies, Structural Research Series 237, University of Illinois, Urbana, Illinois (Sept. 1962).
2. Cahill, T., "The Development of Stabilized Wire and Strand," *Wire and Wire Products* (Oct. 1964).
3. Bannister, J. L., "Steel Reinforcement and Tendons for Structural Concrete," *Concrete*, **2**, No. 8 (Aug. 1968).
4. Szilard, Rudolph, "Corrosion and Corrosion Protection of Tendons in Prestressed Concrete Bridges," *Journal of the American Concrete Institute, Proceedings*, **66**, No. 1 (Jan. 1969).

5. Base, G. P., "An Investigation of Transmission Length in Pretensioned Concrete," Research Report No. 5, Cement and Concrete Association, London, 1958.
6. Bannister, J. L. Private Communication.
7. de Strycker R., "The Influence of Temperature and Variations of Stress on the Creep of Prestressing Steels," *Revue de Metallurgie*, Paris, **56**, No. 1, 49–54 (Jan. 1959).
8. Papsdorf and Schwier, "Creep and Relaxation of Steel Wire, Particularly at Slightly Elevated Temperatures," *Stahl und Eisen*, **78**, No. 14, 937-947 (July 10, 1958).
9. Monfore, G. E. and Verbeck, G. J., "Corrosion of Prestressed Concrete Wire in Concrete," *Journal of the American Concrete Institute*, **32**, No. 5, 491–515 (Nov. 1960).
10. Podolny, W. Jr. and Melville, T., "Understanding the Relaxation in Prestressing," *Journal of the Prestressed Concrete Institute*, **14**, No. 4, 43–54 (Aug. 1969).
11. American Association of State Highway and Transportation Officials, *Standard Specifications for Highway Bridges*, 12th ed., Washington, D.C., 1977.
12. ACI Committee 318, *Building Code Requirements for Reinforced Concrete* (ACI 318-83), American Concrete Institute, Detroit, 1971.

CHAPTER 2–PROBLEMS

1. A prestressed member requires an effective prestressing force of 980 kips. Determine the area of steel required for stress-relieved strand, low-relaxation strand, and high-tensile bars, if the strain change in the concrete surrounding the prestressing steel is 700×10^{-6} in./in. owing to concrete elastic shortening, shrinkage, and creep. Assume a 25 year useful life of the structure, normal temperatures during service, and that the elastic modulus of steel is 29,000 ksi. Use $f_y' = 256$ ksi and $f_{pu} = 270$ ksi for strand. Assume the tendons are stressed to $0.70\, f_{pu}$ initially.

SOLUTION:

From Eq. 2-1, $t' = 25 \times 365 \times 24 = 219{,}000$ hr

$$\text{taking } f_{si} = 0.70 \times 270 = 189 \text{ ksi}$$

Stress-relieved strand $\Delta f_{sr} = (189)\, \dfrac{\log 219000}{10}\left(\dfrac{189}{256} - 0.55\right) = 19.0$ ksi

Low-relaxation strand $\Delta f_{sr} = 0.25 \times 19.0 = 4.8$ ksi

High-tensile bars $\Delta f_{sr} = 0.70 \times 150 \times 0.03 = 3.2$ ksi

Concrete strain loss $= 700 \times 10^{-6} \times 29 \times 10^3 = 20.3$ ksi

SUMMARY:

Steel Type	Initial Stress (ksi)	Relaxation Loss (ksi)	Concrete Strain Loss (ksi)	Total Loss (ksi)	Area Steel Req.'d (sq in.)
Stress-relieved strand	189	19.0	20.3	39.3	6.55
Low-relaxation strand	189	4.8	20.3	25.1	5.98
High-tensile bars	105	3.2	20.3	23.5	12.02

2. For a pre-tensioned stress-relieved strand that is released 24 hours after having been stressed to 175 ksi, determine the stress remaining in the tendon after 50 years, if f_{pu} = 250 ksi and f'_y = 225 ksi. Assume the elastic shortening is 170 × 10^{-6} in./in. and the deferred deformation of the concrete (creep and shrinkage) is 825 × 10^{-6} in./in. Use an elastic modulus for the steel of 30,000 ksi.

SOLUTION:

At 24 hr

$$\Delta f_{sr} = 175 \frac{\log 24}{10}\left(\frac{175}{225} - 0.55\right) = \quad 5.5 \text{ ksi}$$

$$\text{Elastic shortening loss} = 30{,}000 \times 170 \times 10^{-6} = \frac{5.1 \text{ ksi}}{10.6 \text{ ksi} \cong 11.0 \text{ ksi}}$$

f'_{si} = 175 - 11 = 164 ksi

$$\frac{f_{st}}{164} = 1 - \left(\frac{164}{225} - 0.55\right)\left(\frac{\log 438000 - \log 24}{10}\right) = 0.924$$

Deferred strain loss = 30,000 × 825 × 10^{-6} = 25 ksi

Therefore, f_{se} = 164 × 0.924 - 25 = 126 ksi

3 Concrete for Prestressing

3-1 Introduction

It is presumed that the reader is familiar with the basic physical properties of portland-cement concrete, which is the principal constituent of prestressed concrete. It is important that a proper concrete be employed in prestressed construction—only the factors that are particularly important in this type of construction are considered here. General data pertaining to the factors affecting the physical properties of concrete can be found in the references (*see* Refs. 1, 2, and 3).

Although concretes having 28-day cylinder compressive strengths of 5000 to 6000 psi are relatively easily obtained in most localities today, such concrete cannot be employed effectively in reinforced-concrete flexural members. This is not true in the case of prestressed concrete where the use of 5000-psi concrete is common and efficient, due to the reduction in the dead weight and cost of the members (which is derived from the use of stronger concretes). Furthermore, as has been explained, volume changes of concrete affect, to a very significant degree, the amount of prestressing that is lost. Since the high-strength concretes generally undergo substantially smaller volume changes than the lower-strength concretes, their use is desirable, if not necessary, in many prestressing applications.

Volume changes in concrete are affected by many variables, but in practice, the control of these variables is generally limited to specifications which govern the amount of water used in the concrete mixture, the types and proportions of the aggregates, the type and amount of cement in the mixture, the use of admixtures, and the method and duration of the curing. The water content of the concrete mixture is kept as low as possible, since by so doing, the shrinkage of the concrete is reduced, the strength is increased, and the creep is reduced. All of these are desirable effects. Care must be taken to ensure that there is sufficient water in the mixture to avoid honeycomb and permit the concrete to be properly placed and compacted.

3-2 Cement Type

Although concrete containing modern, high early-strength (type III) portland cement yields shrinkages that are slightly greater than those obtained with normal portland cement (type I), there is evidence that the combined loss of prestress, due to all changes in concrete volume, is less when high early-strength cement is used (Ref. 4). The very high shrinkages that were associated with type III cements of 40 to 50 years ago have been substantially reduced as a result of improved cement-manufacturing techniques (Ref. 1). In addition, in precast-prestressed concrete products, the length changes due to shrinkage and the high heat of hydration generated with type III cement can take place virtually without restraint, and, as a result, the cracking associated with these phenomena in cast-in-place construction does not occur.

In applications where type III cement is not used, due to the higher cost or to the unimportance of early strength, type I portland cement is recommended. In structures that are to be subjected to exposure to sea water or moderately reactive aggregates, the use of type II (modified) portland cement is considered a good practice.

Cement used in prestressed-concrete work should conform to ASTM C 150. There is some evidence that high alumina cement should not be used in prestressed concrete, since some of these cements contain significant quantities of sulfides that can undergo changes which may lead to hydrogen embrittlement of the prestressing steel (Ref. 5).

3-3 Admixtures

Admixtures, which make the concrete mixture more plastic, retard the initial set, accelerate the final set, and reduce the amount of water required in the mixture, are often used in prestressed work. Admixtures frequently facilitate placing and handling of the concrete, as well as obtaining the desired high early strengths.

Admixtures that entrain air are also used in prestressed concrete to provide resistance to deterioration due to freezing and thawing.

Admixtures used in prestressed concrete construction generally are specified to be either Type A or D, as defined in ASTM Designation C 494 "Standard Specification for Chemical Admixtures for Concrete." These can be further defined as either lignosulfonates, organic acids, or polymers.

In recent years admixture Types F and G (ASTM C 494), which are water-reducing, high range and water-reducing, high range and retarding admixtures respectively, have been introduced and are expected to be widely used in the future.

The addition of an air-entraining agent is considered to increase slightly the shrinkage and creep of concrete. Because these admixtures also permit a reduction in the mixing water, without adversely affecting the workability, it is generally accepted that the shrinkage and creep characteristics of air-entrained and non-air-entrained concrete *of equal workability* are equal.

The use of admixtures is considered a good practice where job conditions can be improved thereby; however, care must be exercised to ensure that admixtures containing calcium chloride or other chlorides are not used, since the chloride ion may result in pitting or stress corrosion of the prestressing tendons. A number of the well known admixtures commonly used in modern concrete practice contain chloride ions and should not be used in prestressed concrete.

3-4 Slump

European literature emphasizes the desirability of using low-slump concrete in the manufacture of prestressed products. Experience in this country would indicate that the use of no-slump concrete should generally be confined to products that can be made on a vibrating table or vibrating pallets and, perhaps, to shallow members that are of such cross-sectional shape that all areas of the member are readily accessible to internal vibrators. For average prestressed members that are too large to be produced on a vibrating table or that have large bottom flanges that cannot be readily vibrated with internal vibrators, it has been found that good results are obtained when the slump of the plastic concrete is between 2 and 4 in. There are established plants in the U.S. that manufacture precast, prestressed members with no slump concrete, but the majority of plants do not.

3-5 Curing

The best method of curing concrete is by keeping the concrete surfaces thoroughly wet, as long as possible. This applies to prestressed concrete just as it does to other modes of concrete construction. Because of the necessity of obtaining high early strength in prestressed concrete, in order to permit rapid re-use of the manufacturing facilities, steam curing at atmospheric pressure is often employed to accelerate the hardening of the concrete. Steam curing is discussed in detail in Sec. 3-13.

Hot water has also been successfully used in the manufacture of precast members. In some instances, hot water is circulated through pipes that are close to the concrete elements being cured (while the exposed surfaces are kept wet) and, hence, the system heats concrete in much the same way that hot water is used to heat dwellings or buildings. Hot water is sometimes used in lieu of cold water in much the same way that cold water is used in curing—matts or burlap are placed on the exposed concrete surfaces, thus keeping them wet with hot water (150° F±). Each of these methods can be satisfactory and each offers certain advantages.

Heat, applied by any means, can be used to accelerate the curing of concrete, but it is extremely important that the concrete be kept wet during the curing period.

3-6 Concrete Aggregates

The aggregates used in the manufacture of normal concrete members are usually satisfactory for use in prestressed concrete. However, because of the higher strengths required for prestressed concrete, difficulty has been experienced, in some localities, in finding suitable natural aggregates for prestressed construction. Where a choice of aggregates is available, the selection should be made after considering the ease of obtaining the necessary strength, as well as the magnitude of the elastic and inelastic volume changes that might be expected with the different types available. Lightweight aggregates of the expanded shale or clay type have been used with good results in this country. Care must be exercised in employing lightweight aggregates in prestressed concrete in order to assure that a reasonable estimate of the volume changes is taken into account when estimating the loss of prestress.

Normal concrete aggregates should conform to the requirements of "Standard Specifications for Concrete Aggregates," ASTM Designation C 33. Aggregates for lightweight prestressed concrete should conform to "Standard Specifications for Lightweight Aggregates for Structural Concrete," ASTM Designation C 330.

3-7 Strength of Concrete

A principal reason concrete with a minimum 28-day cylinder compressive strength of the order of 4000 to 6000 psi is used in prestressed-concrete work is that concrete of this quality exhibits lower volume changes than concretes of lower quality. Another reason for using higher strength concretes is that efficient use can generally be made of such concretes in the flexural design of prestressed concrete (this is not the case in reinforced concrete design).

In some areas it is difficult to consistently produce concrete of high quality with local materials. The designer of prestressed-concrete structures should carefully investigate this problem on each project undertaken. It is generally possible to prepare reasonably economical designs with concretes of moderately high strength, and it is better to anticipate this problem and provide for it in the design stage rather than struggle with what may be an almost impossible situation during construction.

Whenever possible, and always on major jobs, the concrete mixes used should be trial batched and laboratory tested before use on the job. The mixes employed in the work should have laboratory strengths 10 to 20% higher than that required by the job specification.

A general equation for predicting the compressive strength of concrete at any age has been proposed (Refs. 6 & 16). This relationship is

$$f'_{ct} = \frac{t}{a + \beta t} \cdot f'_c \tag{3-1}$$

in which f'_{ct} is the compressive strength at age t in days, a and β are constants, and f'_c is the compressive strength at the age of 28 days. The average values of a and β have been found to be as shown in Table 3-1. The variation in compressive strengths using Eq. 3-1 and the values of the constants a and β given in Table 3-1 are shown in Fig. 3-1.

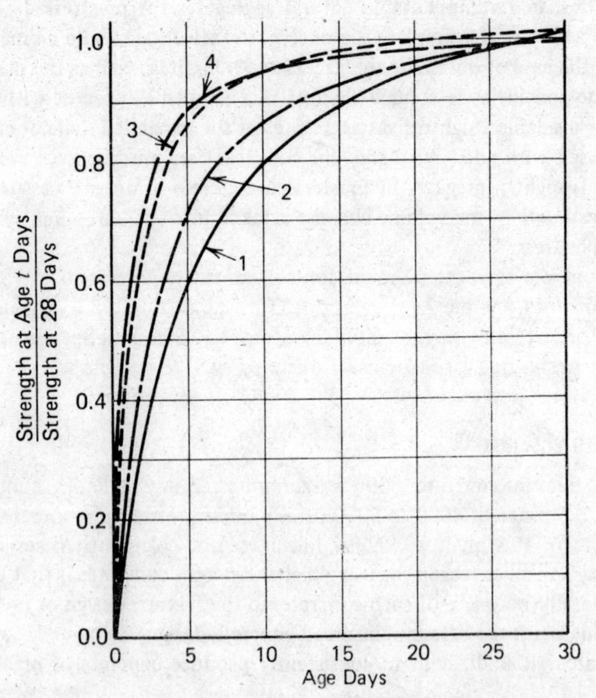

Fig. 3-1 Strength-time curves for concrete as predicted by Eq. 3-1 for the concretes listed in Table 3-1.

TABLE 3-1

Concrete Curing	Cement Type	a	β	Curve in Fig. 3-1
Moist cured	I	4.00	0.85	1
Moist cured	III	2.30	0.92	2
Steam cured	I	1.00	0.95	3
Steam cured	III	0.70	0.98	4

On important projects, the strength-time curve should be developed for the particular concrete being used. By so doing, the work can be accurately planned in advance and low strength concrete that occurs can be detected at an early age.

The 28-day compressive strength of concrete as a function of the water-cement ratio for air-entrained and non-air-entrained concrete is illustrated in Fig. 3-2. In Fig. 3-3, the 28-day compressive strength of concrete is shown as a function of the voids-cement ratio. These curves are useful in estimating the quantities of cement and water that must be used to achieve a desired concrete strength.

The compressive strength that will eventually be achieved by moist-cured con-

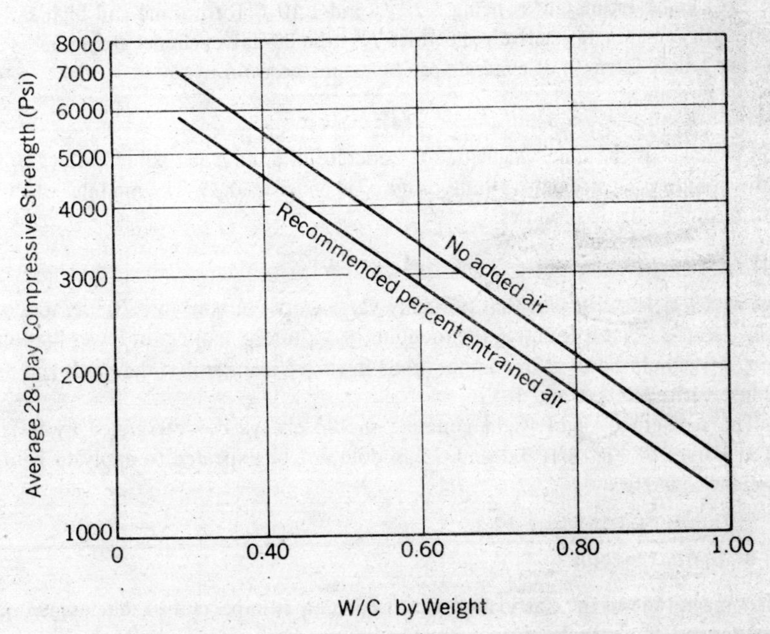

Fig. 3-2 Strength in relation to water-cement ratio for air-entrained and non-air-entrained concrete. Strength decreases with an increase in water-cement ratio; or with the water-cement ratio held constant, use of air entrainment decreases the strength by about 20%. (*See* Ref. 2.)

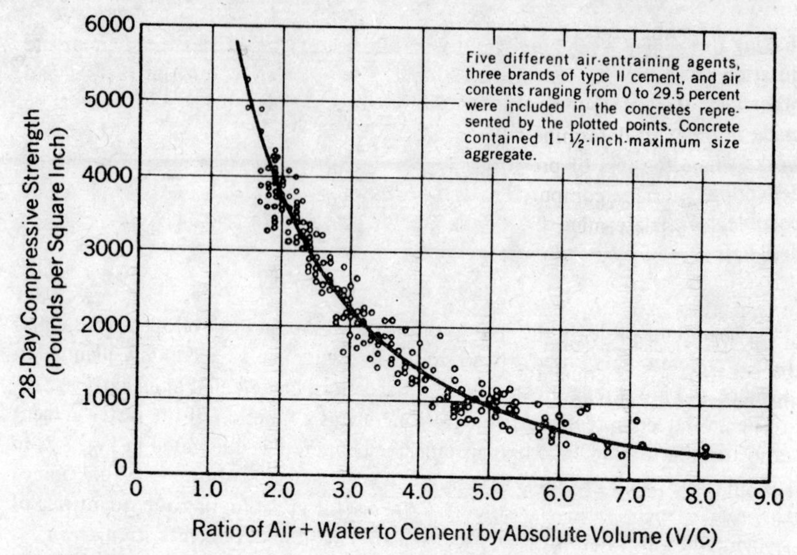

Fig. 3-3 Compressive strength of concrete in relation to voids-cement ratio. (*See* Ref. 2.)

crete can be estimated as being 1.20 f_c' and 1.10 f_c' for normal and high early strength cements, respectively, in which f_c' is the 28-day cylinder strength.

The tensile strength of concrete can be estimated from

$$f_t = g_t \sqrt{w(f_c')_t} \tag{3-2}$$

in which w is the unit weight of the concrete in pounds per cubic foot, $(f_c')_t$ is the compressive strength at time t and g_t is equal to 0.33. The modulus of rupture can be taken as follows:

$$f_r = g_r \,[w(f_c')_t]^{1/2} \tag{3-3}$$

in which g_r is a constant that normally varies between 0.60 to 0.70. Because of the variation that is found in the modulus of rupture, the upper or lower limit of Eq. 3-3 should be used in computations in such a manner that the result will be conservative (Refs. 6 and 16).

The strength of lightweight concrete should always be determined by tests. The curves of Fig. 3-1, 3-2 and 3-3 should not be expected to apply to lightweight concrete.

3-8 Elastic Modulus

The magnitude of the elastic modulus of concrete is important to the designer of prestressed concrete, because it must be known when computing the deflections and the losses of prestress. Unfortunately, the elastic modulus of concrete is a function of many variables, including the type and amount of ingredients used in making the concrete (cement, aggregates and water), as well as the manner and duration of curing the concrete, age at the time of loading, rate of loading and other factors (Ref. 1). When possible, it is recommended that measurements be made in order to determine the magnitude of the elastic modulus on important works where the loss of prestress may be very critical or for structures on which deflections must be computed with the highest possible precision. When it is not possible to predetermine the elastic modulus for the concrete to be used in a design, it may be assumed to be

$$E_{ct} = g_{ct} \, [w^3 (f_c')_t]^{1/2} \tag{3-4}$$

in psi, for values of w (unit weight of the concrete) between 90 and 155 lb/cu ft. the constant g_{ct} is equal to 33. For normal weight concrete, the relationship may be taken as

$$E_{ct} = 57{,}000 \, (f_c')_t^{1/2} \tag{3-5}$$

It should be noted that $(f_c')_t$ in this relationship is the compressive strength of the concrete, as determined by tests of standard 6 X 12 in. cylinders made in accordance with ASTM Designation C 192 and tested in accordance with ASTM C 39 at the age of t days.

It should be pointed out that Eq. 3-4 is intended to give a result that approximates the value that would be obtained if the concrete were tested in accordance with ASTM Designation C 469. Because the value of the elastic modulus thus obtained is intended to be the chord modulus at a stress of 40% of the ultimate strength, a higher value for the elastic modulus (as much as 10%) could be anticipated in applications where the concrete is stressed to lower levels.

In prestressing concrete, the prestressing force is often transferred to the concrete at a relatively early age (one to fourteen days, depending upon the materials and method of curing used). Hence, at the time of stressing, the concrete frequently has a strength that is somewhat less than the minimum specified at the age of 28 days. Equation 3-4 gives a means of approximating the modulus of elasticity of the concrete at a given age by relating it to the cylinder strength, which varies with age.

Equation 3-4 can be altered to better represent known properties of a particular concrete by substituting an empirically determined number for the integer 33 contained therein. This is particularly of interest when writing general-purpose computer programs in which the elastic modulus of concrete is a variable.

ILLUSTRATIVE PROBLEM 3-1 A post-tensioned beam is to be stressed when the concrete strength is 4000 psi. The specifications also provide that the minimum cylinder compressive strength at the age of 28 days shall be 5000 psi. Compute the elastic modulus that should be used in computing instantaneous deflections and stress losses at the time of stressing and at the age of 28 days if (1) the concrete is normal concrete and (2) the concrete is lightweight—weighing 100 lb/cu ft.

(1) $$E_{ct} = 57{,}000\,(f'_c)_t^{1/2}$$

At time of stressing:

$$E_{ct} = 57{,}000 \times \sqrt{4000} = 3{,}605{,}000 \text{ psi}$$

At 28 days:

$$E_c = 57{,}000 \times \sqrt{5000} = 4{,}030{,}000 \text{ psi}$$

(2) $$E_{ct} = 100^{1.5} \times 33 \times (f'_c)_t^{1/2} = 33{,}000\,(f'_c)_t^{1/2}$$

At time of stressing:

$$E_{ct} = 33{,}000 \times \sqrt{4000} = 2{,}090{,}000 \text{ psi}$$

At 28 days:

$$E_c = 33{,}000 \times \sqrt{5000} = 2{,}330{,}000 \text{ psi}$$

Creep of concrete, which is discussed in detail in Sec. 3-11, is defined as the increase in strain that occurs when a concrete member or specimen is subjected to constant stress. Because the elastic modulus of concrete is equal to stress divided by strain, an increase in strain has the effect of decreasing the modulus of elasticity. The elastic modulus that has been corrected to take into account the effect of creep at some particular time is referred to as the "effective modulus" or "reduced modulus." This can be expressed mathematically as follows:

$$\text{Elastic modulus} = E_c = \frac{\text{Stress}}{\text{Elastic Strain}} \tag{3-6}$$

$$\text{Effective modulus} = E'_c = \frac{\text{Stress}}{\text{Elastic Strain} + \text{Creep Strain}} \tag{3-7}$$

For the ultimate effective modulus, Eq. 3-7 can be written

$$E_{cu} = \frac{E_c}{1 + v_t} \tag{3-8}$$

where v_t is the creep ratio equal to the ratio of the ultimate creep strain to the elastic strain, or

$$v_t = \frac{\text{Ultimate Creep Strain}}{\text{Elastic Strain}} \tag{3-9}$$

v_t is a function of many variables, but principally of the relative humidity, concrete quality and age of concrete when loaded. Methods of estimating v_t are given in Sec. 3-11.

The reduced modulus is frequently used in computing deflections of reinforced and prestressed concrete, as well as the losses of prestress in prestressed-concrete members. These are discussed in greater detail in Secs. 6-2 and 6-3.

3-9 Shrinkage

The shrinkage of concrete is an important factor to the designer of prestressed concrete for several reasons. As has been stated previously, the shrinkage of the concrete contributes to the loss of prestress. The magnitude of the shrinkage must also be known with reasonable accuracy when the deflection of prestressed members are being computed with the more sophisticated methods. The deflection of composite prestressed-concrete members cannot be computed without knowing the shrinkage characteristics of each of the concretes involved. In addition, the magnitude of the concrete shrinkage must be estimated in order to evaluate secondary stresses (due to volume changes) that may result.

The effects of concrete shrinkage in prestressed-concrete structures are considerably different than those in reinforced-concrete structures. In reinforced concrete, the shrinkage strains are resisted by compressive stresses in the reinforcing steel, whereas in prestressed concrete, the prestressing steel is always in tension and causes compressive strains that add to the shrinkage strains in the concrete. In addition, reinforced-concrete structural members are normally cracked with many closely spaced minute cracks that tend to relieve the effect of shrinkage stresses. The designer of prestressed-concrete structures must give particular attention to the effects of shrinkage, creep, and temperature variations. If these movements are restrained, forces of very high magnitude can result with the very real possibility of serious structural and non-structural damage. This subject is discussed in greater detail in Chapter 10.

The shrinkage of concrete is known to be related to the loss of moisture. It has been demonstrated that concrete will expand if subjected to 100% relative humidity or if submerged in water. The amount of shrinkage obtained in dry storage can also vary widely. Shrinkage of concrete is known to be a function of the following:

A. Composition of the cement.
B. Physical properties of the aggregate.
C. Maximum aggregate size.
D. Quantity of water.
E. Method and duration of the curing.
F. Temperature and humidity of atmosphere during service.
G. Volume to surface ratio of the concrete.
H. Admixtures.

A considerable amount of data is available in the literature relative to the effect of each of these variables. The discussion that follows is of a general nature but is considered sufficiently accurate for most design purposes. The designer of prestressed concrete should recognize that he can often control shrinkage to some degree through careful consideration of the materials and methods specified for each project. The effect of each of the above variables can be summarized as follows.

A. Cement. High early strength portland cement (type III) would normally be expected to have a shrinkage that is 10% higher than normal portland cement (type I) or modified portland cement (type II) (Ref. 1). In addition, a cement which may exhibit a large amount of shrinkage may have a total shrinkage that is 100% greater than that of a cement, which due to its chemical composition, exhibits a small amount of shrinkage. This is an extreme range, however, and it may be beneficial to investigate the cements available in any locality, in order to determine if any of the commonly used cements have exceptionally high or low shrinkage characteristics. As has been previously stated, there is some evidence that the use of a high early strength cement of good quality may result in a concrete that will exhibit somewhat lower total volume changes in prestressed construction than that which would be obtained with normal cement (type I) (Refs. 1, 4 and 11).

B. Aggregates. The physical properties of the larger aggregate particles have considerable influence on the shrinkage of concrete. This is due to the fact that the concrete aggregate reinforces the cement paste and resists its contraction. Aggregates with higher elastic moduli are stiffer and hence restrict the contraction of the paste to a greater degree. Aggregates that have a low volume change in themselves, due to drying, generally result in concrete with lower shrinkage. Concretes containing aggregates of quartz, limestone, dolomite, granite or feldspar are generally low in shrinkage, whereas those containing sandstone, slate, trap rock or basalt may be relatively high in shrinkage. Therefore, if aggregates of the latter type, or gravels containing a large portion of such minerals, are used, an allowance should be made for a relatively high shrinkage value. Concretes made with soft, porous sandstone may shrink 50% more than concretes made with hard dense aggregates (Refs. 1 and 8).

Lightweight concrete aggregates manufactured by expanding clay or shale have been used to a significant extent in prestressed-concrete structures. High-quality expanded shale or clay aggregates that are not crushed after burning, and hence are coated and less absorptive than crushed material, have been reported as having drying shrinkage characteristics that are approximately of the same magnitude and rate as were found with normal aggregates (Ref. 12). Other research has indicated that lightweight aggregates may have shrinkages as much as 50% greater than normal aggregates at the age of 300 days (Ref. 13). When the use of lightweight aggregates is contemplated, the designer should investigate the shrinkage characteristics of the actual concrete mix proposed.

C. Aggregate Size. Aggregate size also has a marked effect in the magnitude of concrete shrinkage. This is due to two reasons. The first of these is that the larger particles offer more restraint to the shrinkage of the mortar. For this reason, increasing the aggregate size from $\frac{3}{4}$ to $1\frac{1}{2}$ in. would be expected to reduce the shrinkage as much as 20%. The second factor is that in changing maximum aggregate size from $\frac{3}{4}$ to $1\frac{1}{2}$ in., about 10% less water is required to achieve the same consistency, and this reduction in water will account for a reduction in shrinkage of as much as 20%. The result is that a change in aggregate size from $\frac{3}{4}$ to $1\frac{1}{2}$ in. may result in a reduction in the shrinkage by about 40% (Ref. 9).

D. Water. The amount of water in a concrete mix is the most important single factor affecting shrinkage of concrete. The shrinkage of concrete made with a particular aggregate has been found to vary directly with the water content of concrete, as is illustrated in Fig. 3-4. Therefore, the amount of water used in prestressed-concrete construction should be kept to the minimum amount required for the consistency necessary for proper placing and compaction. The water required to obtain the necessary plasticity in a concrete mix is a function, among other things, of the amount of mortar (cement + sand) in the mix. For this reason, it is desirable to keep the quantity of mortar as low as practicable.

E. Curing. There is little if any concrete shrinkage during curing, if the concrete is kept sufficiently moist to prevent the loss of moisture. Some investigators report that ultimate shrinkage is unaffected by an increase in the duration of curing time (Ref. 9). There is evidence that curing concrete at an elevated temperature (atmospheric pressure steam curing) will result in a reduction in shrinkage by as much as 30% (Refs. 10 and 17). The acceleration in curing that is obtained from steam curing apparently leads to a more complete hydration of the cement. Hence, less free water remains available for evaporation and the shrinkage is reduced. Atmospheric pressure steam curing has resulted in reductions of shrinking between 10 and 30% for type I cement and 25 to 40% for type III cement, when compared to specimens that were moist cured for 6 days (Ref. 11).

F. Temperature and Humidity. Temperature and humidity each affect the rate and magnitude of concrete shrinkage. Higher temperatures during service would be expected to result in somewhat greater shrinkage, because the higher temperatures would result in greater moisture loss. Variations in temperature and humidity results in higher shrinkage (and creep) than is obtained under constant conditions. Therefore, estimates of shrinkage made on laboratory tests may be low (Ref. 22).

The relative humidity during service has a marked effect on shrinkage, with lower humidities resulting in greater shrinkages. Relationships for estimating concrete shrinkage as a function of relative humidity have been proposed by Branson and Christiason (Ref. 6) as correction factors to the shrinkage strain at

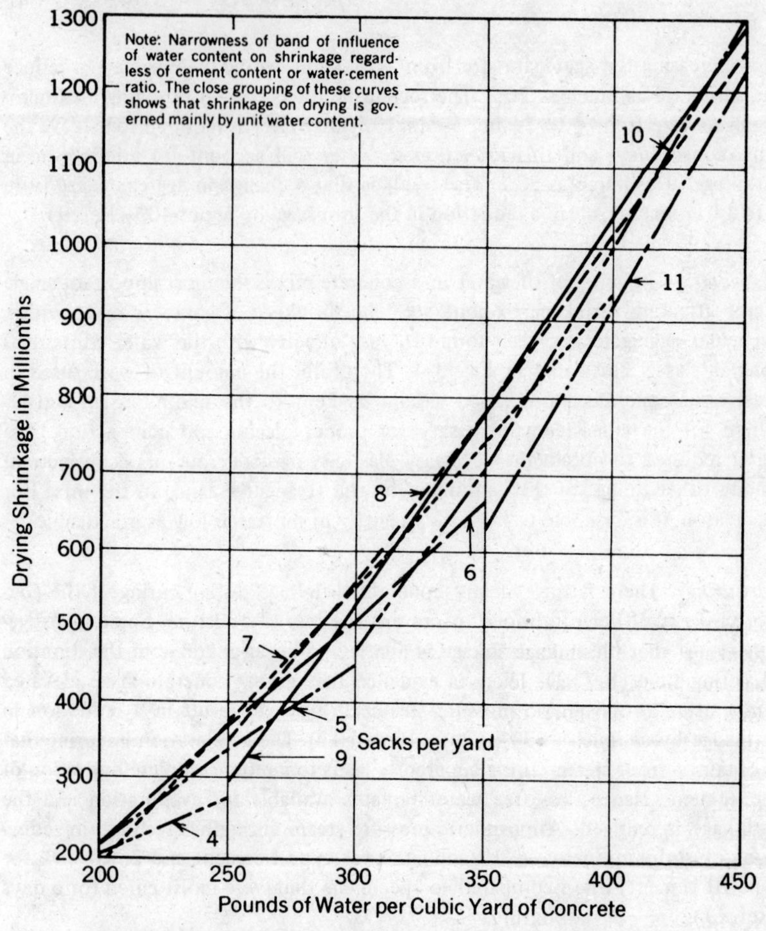

Fig. 3-4 The interrelation of shrinkage and water content (Ref. 2).

40% relative humidity. For relative humidities between 80 and 100% the relationship is:

$$\frac{\text{Shrinkage at relative humidity } H\ (\%)}{\text{Shrinkage at 40\% relative humidity}} = 3.00 - 0.030\ \mathrm{H} \qquad (3\text{-}10)$$

and between 40 and 80% the relationship is:

$$\frac{\text{Shrinkage at relative humidity H } (\%)}{\text{Shrinkage at 40\% relative humidity}} = 1.40 - 0.010\ \mathrm{H} \qquad (3\text{-}11)$$

These relationships are shown plotted in Fig. 3-5. No recommendations are given for humidities lower than 40%.

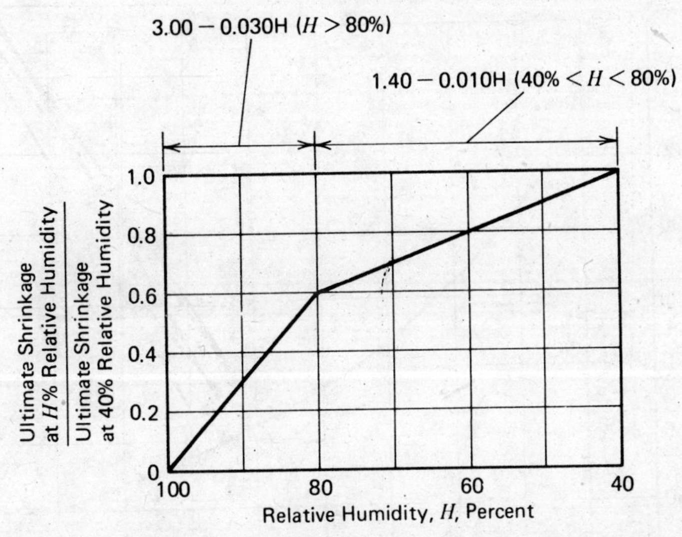

Fig. 3-5 Variation in concrete shrinkage as a function of relative humidity.

G. Size of Member. The size of a member affects the magnitude and rate of shrinkage. Since shrinkage is caused by evaporation of moisture from the surface, members that have low volume-to-surface ratios will be expected to shrink more, as well as more rapidly, than members having high volume-to-surface ratios. The curve of Fig. 3-6 shows the relationship between the shrinkage size coefficient and the volume-to-surface ratio based on a specimen having a volume-to-surface ratio of 1.5 in., which is the value for a cylindrical specimen having a diameter of 6 in. This curve is useful in estimating the effect of member size when test data for specimens having a diameter of 6 in. are available.

H. Admixtures. Admixtures may increase, decrease or have practically no effect on the amount of concrete shrinkage. The more commonly used admixtures in prestressed work are of the water-reducing and water-reducing and retarding type (classified in ASTM C 494 as types A and D, respectively). Admixtures of these types can be further classified according to their general chemical composition and, as such, are categorized as lignosulfonates, organic acids or polymers. Test data are available that would indicate that the lignosulfonates tend to increase shrinkage (from 5 to 50%) when compared with the control concrete (a concrete without admixture but having the same slump). In the same tests, organic acid types of admixtures showed shrinkages from 89 to 117% of the control concrete, while the polymer type admixtures revealed shrinkages of from 98 to 112% of the control concrete. Calcium chloride, which should not be used as an admixture in prestressed concrete (*see* Sec. 2-8), increases concrete shrinkage about one third (Ref. 15).

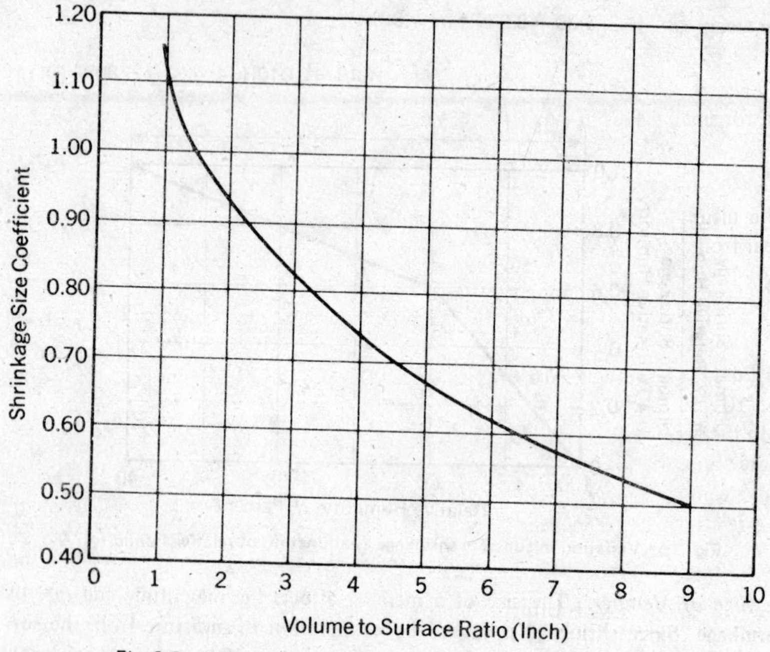

Fig. 3-6 Shrinkage size coefficient vs volume-to-surface ratio.

3-10 Estimating Shrinkage

The best method of estimating the amount of shrinkage of concrete that should be used in any structural design is through the use of shrinkage tests. Established precasting plants and firms engaged in supplying ready-mixed concrete, cement, or aggregates should have shrinkage test results available for typical concrete mixtures obtainable in the localities they serve. In the event such data are not available, the designer must either make tests or use his judgment in estimating the unrestrained shrinkage of concrete for his particular conditions. A conservative estimate is recommended if tests are not made.

Long term shrinkage can be estimated from short-term tests using relationships developed by Brooks and Neville (Ref. 23). The relationships for predicting the shrinkage strain of one year (ϵ_{s365}) based upon the shrinkage measured at the age of 28 days (ϵ_{s28}) are:

For moist cured concrete

$$\epsilon_{s365} = 347 + 1.08\,\epsilon_{s28} \tag{3-12}$$

or, for small values of ϵ_{s28} (less than 100×10^{-6})

$$\epsilon_{s365} = 52\,\epsilon_{s28}^{\;0.45} \tag{3-13}$$

for steam cured concrete

$$\epsilon_{s365} = 243 + 1.51\,\epsilon_{s28} \tag{3-14}$$

or, for small values of ϵ_{s28} (less than 100×10^{-6})

$$\epsilon_{s365} = 27\,\epsilon_{s28}^{0.57} \tag{3-15}$$

The ultimate shrinkage (ϵ_{su}) can then be estimated from the shrinkage at one year from:

$$\epsilon_{su} = \frac{\epsilon_{s365}(1.06\,\epsilon_{s365} - 192)}{1.085\,\epsilon_{s365} - 265} \tag{3-16}$$

In Eqs. 3-12 through 3-16 the shrinkages are expressed in millionths inches per in. (10^{-6}).

In the absence of experimental data, Branson and Christiason suggest using ultimate shrinkage strains of 800 and 730 millionths inches per in. for moist-cured and steam-cured concrete at relative humidity of 40%, respectively (Ref. 6).

A relationship for the shrinkage at time t (days) after drying commences is:

$$\frac{\text{Shrinkage Strain at Time } t}{\text{Ultimate Shrinkage Strain}} = \frac{t^{\alpha}}{f + t^{\alpha}} \tag{3-17}$$

in which α and f are parameters that can be determined experimentally for any particular concrete. When it is not possible to determine the values of α and f experimentally, the value of α can be assumed to be 1 and the value of f can be assumed to be 35 and 55 for moist-cured concrete at any age after 7 days and for steam-cured concrete at any age after 1 to 3 days, respectively. The shrinkage of moist-cured concrete from the age of one day, in terms of the seven day shrinkage can be determined from Fig. 3-7.

ILLUSTRATIVE PROBLEM 3-2 Estimate the one-year and ultimate shrinkages of a moist-cured concrete if the shrinkage at 28 days is 300 millionths inches per in. and the concrete is stored at a constant humidity.

SOLUTION: From Eq. 3-12

$$\epsilon_{s365} = 347 + 1.08 \times 300 = 671 \text{ millionths inches per in.}$$

and from Eq. 3-16

$$\epsilon_{su} = \frac{671\,(1.06 \times 671 - 192)}{1.085 \times 671 - 265} = 752 \text{ millionths inches per in.}$$

Note that using Eq. 3-13 rather than Eq. 3-12, the one-year shrinkage is:

$$\epsilon_{s365} = 52 \times 300^{0.45} = 677 \text{ millionths inches per in.}$$

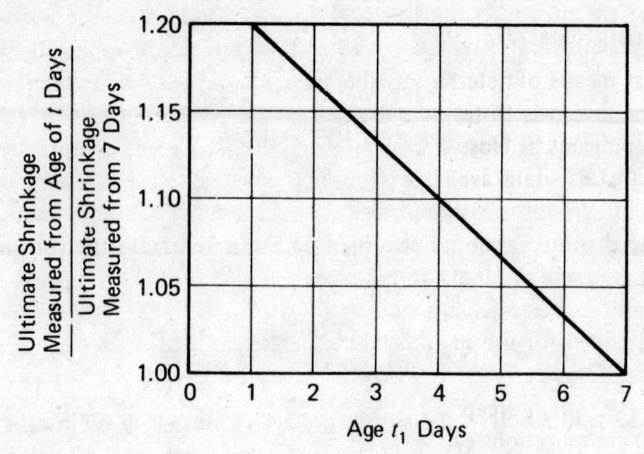

Fig. 3-7 Ultimate shrinkage factors for shrinkage strain from ages less than 7 days, moist-cured concrete.

ILLUSTRATIVE PROBLEM 3-3 Using the recommendations of Art. 3-10, determine the shrinkage strain of a steam-cured concrete between the ages of 28 and 600 days. The concrete is to be stored at an ambient humidity of 60%.

SOLUTION:

ϵ_{su} at 40% rel. humidity = 730 millionths inches per in.

ϵ_{su} at 60% rel. humidity = 0.80 × 730 = 584 millionths inches per in.

$$\epsilon_{s28} = \left(\frac{28}{55 + 28}\right) 584 = 197 \text{ millionths inches per in.}$$

$$\epsilon_{s600} = \left(\frac{600}{55 + 600}\right) 584 = 535 \text{ millionths inches per in.}$$

$$\Delta\epsilon_s = 338 \text{ millionths inches per in.}$$

3-11 Creep of Concrete

The creep of concrete is defined as the strain change which takes place in concrete that is subjected to constant stress and, as is the case with shrinkage, is associated with the loss of moisture from the concrete. Unlike shrinkage, creep is affected by the stress level in the concrete as well as by the "maturity" of the concrete. Maturity is defined as the age, or degree of hydration of the concrete, at the time of loading. Relaxation of concrete is defined as the loss of stress in concrete that is subjected to a constant strain. (*See* Sec. 3-12.) Prestressed concrete is not subjected to constant stress or strain, but rather to changing conditions as a result of variations in external loads, relaxation of the prestressing tendons, shrinkage, and creep of the concrete. However, for purposes of computing loss of prestress, the loss due to the plasticity of the concrete is more accurately estimated using creep, rather than relaxation curves for the concrete.

The best means of determining the creep characteristics of concrete to be used in any one design is to test the concrete under conditions that approximate the service conditions as closely as possible (Ref. 24). Progressive prestressing plants should have test data available on concrete of the type normally used in their products for the information and guidance of engineers contemplating their use. Because special tests are not always possible and because many producers of concrete and concrete products do not have test data on the concrete they normally produce, the engineer needs a method to estimate creep. The discussion that follows is intended for use in approximating the creep at various ages during varied conditions of service.

The rate at which creep takes place for the purposes of losses of prestress and deflection computations can be assumed to be

$$\frac{\text{Creep Strain at Time } t}{\text{Ultimate Creep Strain}} = \frac{t^{\psi}}{d + t^{\psi}} \tag{3-18}$$

in which t is time in days, reckoned from the time that stress is applied, and ψ and d are parameters that can be determined experimentally for each particular concrete. Values of 0.60 and 10 are recommended for ψ and d respectively (Refs. 6 and 16), for cases where experimental data are not available.

Creep is thought to be the effect of two factors. The first of these is basic creep and is independent of moisture movement. The second is drying creep, which results from the concrete losing moisture to its environment. Only the drying creep is affected by member size and shape and it is believed drying creep has no additional effect after about 3 months. The relationship between the creep size coefficient and the volume-to-surface ratio is shown in Fig. 3-8, from which it will be seen that the creep size coefficient is equal to 1 for members with a volume-to-surface ratio of 10. This is an indication that drying creep has a negligible effect on more massive members.

Creep strain is generally expressed in terms of the initial elastic strain through the use of the "creep ratio." The creep ratio is defined in Eq. 3-9 as the ultimate creep strain divided by the elastic strain. Recommended curves for use in estimating the creep ratio for normal water-cured concrete are given in Figs. 3-9 and 3-10 as a function of the average humidity conditions to which the concrete will be exposed in service. The limits for the creep ratio for concrete of prestressing quality can be expressed mathematically as follows:

$$\text{Upper Limit of Creep Ratio} = 1.75 + 2.25\,\frac{100 - H}{65} \tag{3-19}$$

$$\text{Lower Limit of Creep Ratio} = 0.75 + 0.75\,\frac{100 - H}{50} \tag{3-20}$$

in which H is the mean humidity to which the concrete is exposed in service.

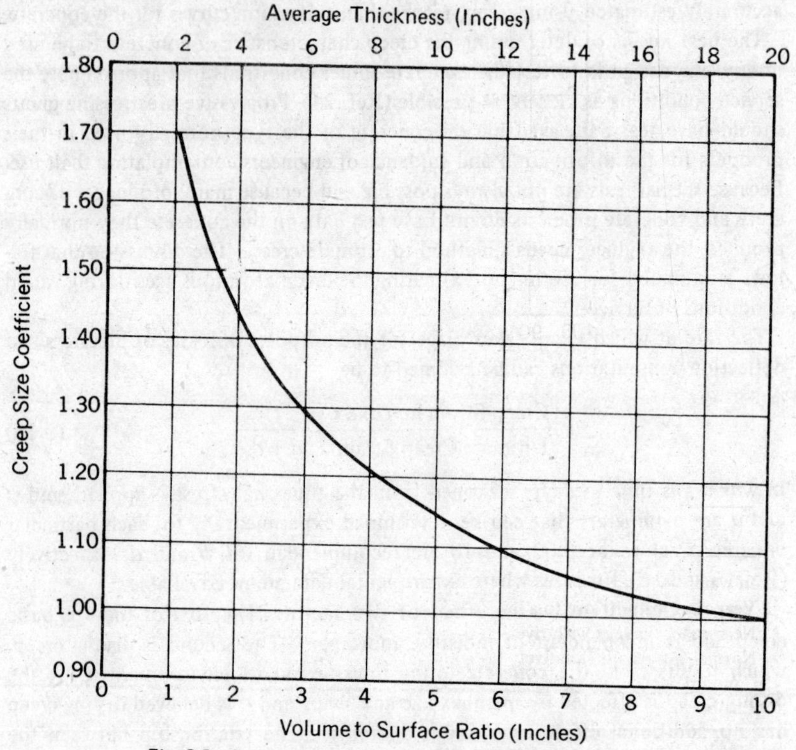

Fig. 3-8 Creep size coefficient vs volume-to-surface ratio.

For concrete of ordinary quality, the relationships for the upper and lower limits of the ultimate creep ratio are as follows:

$$\text{Upper Limit } v_u = 2.00 + 2.00\,\frac{100 - H}{50} \tag{3-21}$$

$$\text{Lower Limit } v_u = 1.00 + 1.00\,\frac{100 - H}{50} \tag{3-22}$$

For concrete having a slump of 4 in. or less, an average thickness of 6 in. or less (minimum thickness of 6 in. or less equals a volume to surface ratio of 3 in.), and an ambient relative humidity H of 40% or more, the following relationship may be used for computing the creep ratio C_u (Ref. 6):

$$v_u = 2.35\,(1.27 - 0.0067\,H) \tag{3-23}$$

Equation 3-23, which is plotted in Fig. 3-9, represents values for loading ages of 7 days and 1–3 days for moist-cured and steam-cured concretes, respectively.

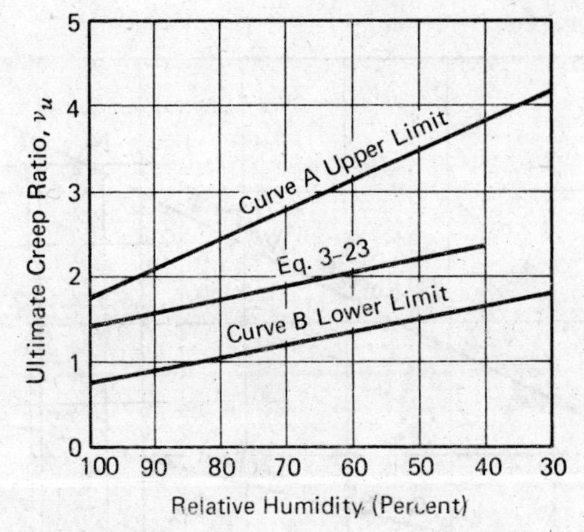

Fig. 3-9 Creep ratio ν_u vs relative humidity for water-cured concrete of prestressing quality (normal aggregates).

TABLE 3-2. Data for Estimating Humidity.

Service Conditions	Approx. Humidity (H)
In water	100%
Very near water (close to a large body of water)	90%
Near water (valleys with rivers in dense forests)	70%
Normal conditions (relatively dry climate, high mountains)	50%
Enclosed buildings (heated in winter)	40%

Humidity during service can be approximated from the data in Table 3-2. Local climatological data for most cities of the United States are available.* These data include averages and extremes for temperature and relative humidity throughout the year.

The amount of creep strain is also a function of the relative compressive strength of the concrete at the time the concrete is subjected to stress. This factor can be approximated in terms of the 28-day compressive strength, as well as in terms of the age at loading. These approximate relationships are shown in Figs. 3-11(a) and 3-11(b), respectively. Expressed mathematically, Fig. 3-11(a) is as follows:

$$M_c = 1.80 - 1.28\left(\frac{f'_{ci}}{f'_c} - 0.375\right) \tag{3-24}$$

*U.S. Department of Commerce, Weather Bureau, "Local Climatological Data." Available from the Superintendent of Documents, Government Printing Office, Washington, D.C. 20402.

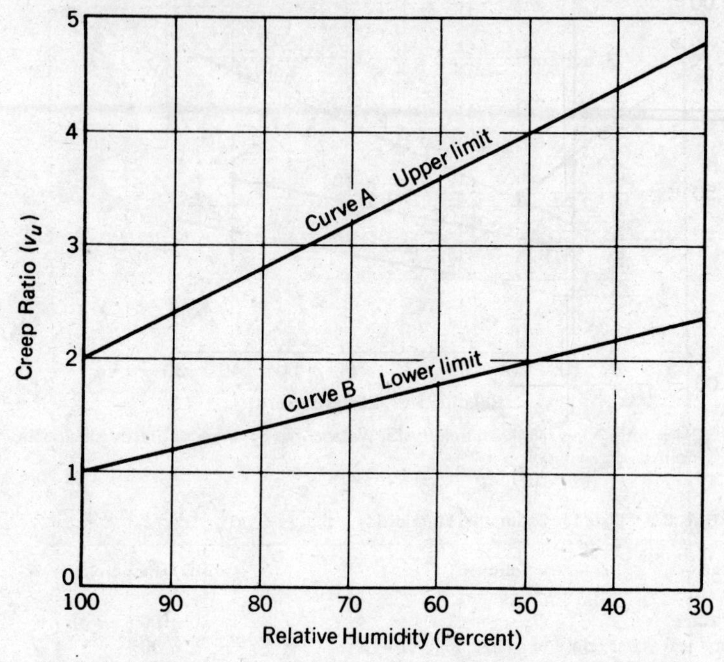

Fig. 3-10 Creep ratio v_u vs relative humidity for concrete of ordinary quality.

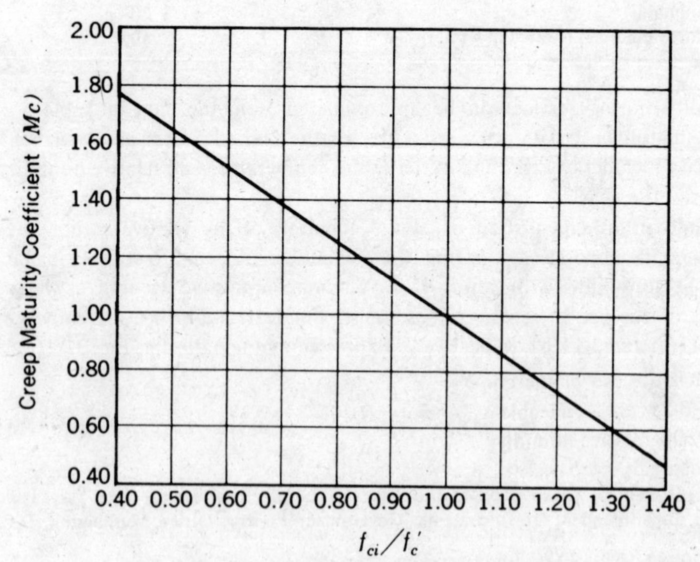

Fig. 3-11(a) Creep maturity coefficient vs f'_{ci}/f'_c.

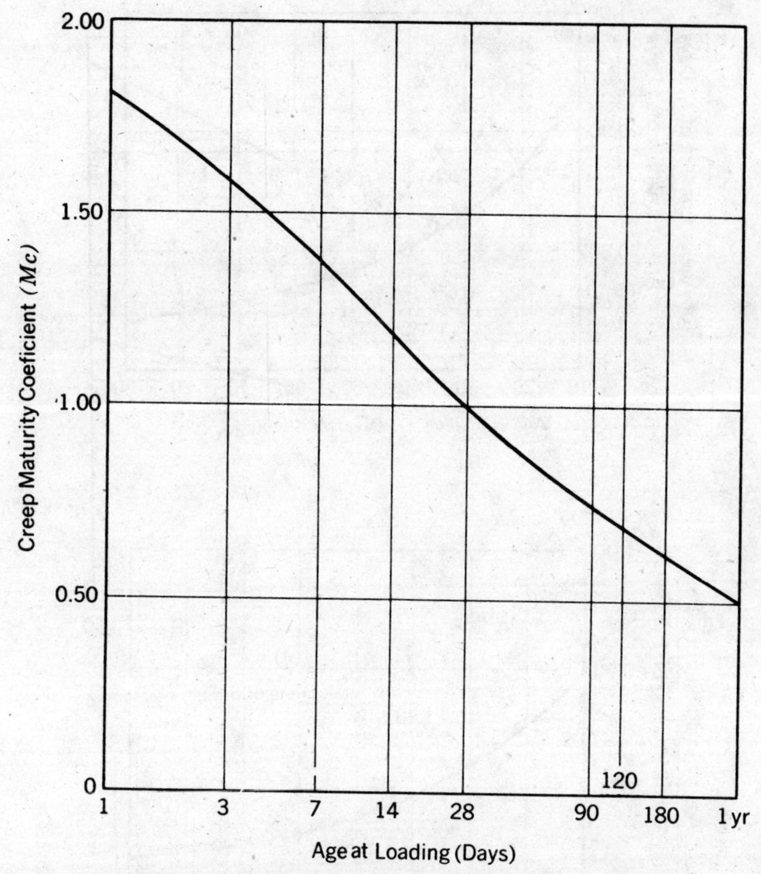

Fig. 3-11(b) Creep maturity coefficient vs age at loading, in days.

in which M_c is the maturity coefficient, f'_{ci} is the concrete strength at the time of stressing, and f'_c is the cylinder strength at the age of 28 days. The relationship of Fig. 3-11(b) can be expressed as:

$$M_c = 1.80 - 0.238 (\log_e t) \quad (3\text{-}25)$$

in which t is the age of the concrete in days. It should be recognized that Eqs. 3-24 and 3-25 are applicable to concrete made with type I cement cured at 70°F and 50% relative humidity.

Recommended creep maturity coefficients for moist-cured and steam-cured concretes are given in Figs. 3-12(a) and 3-12(b), respectively (Ref. 6). The curves are expressed as the ratio of the creep ratio at time t to that at 7 days and 1-3 days for moist-cured and steam-cured concrete, respectively.

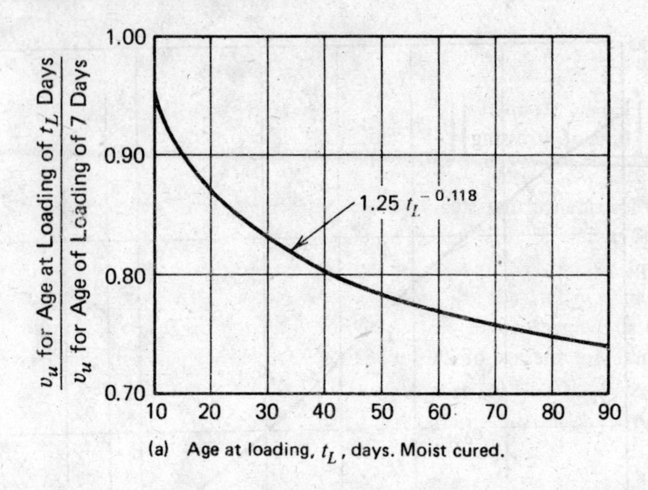

(a) Age at loading, t_L, days. Moist cured.

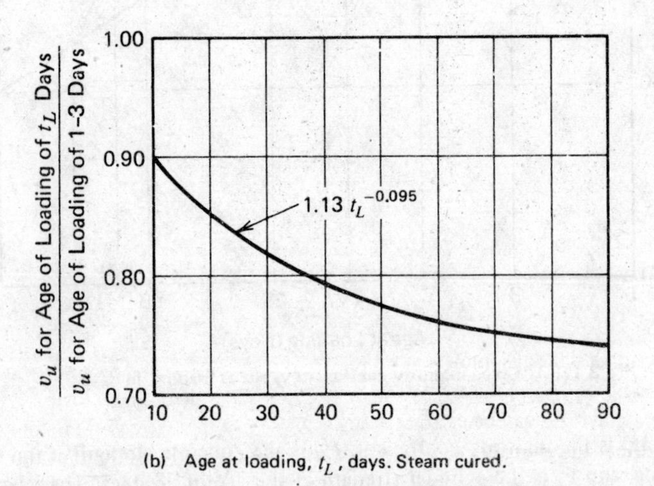

(b) Age at loading, t_L, days. Steam cured.

Fig. 3-12 Variation in creep ratio for various ages at loading.

It must be pointed out that wide variations in the amount of creep that is experienced can result from the type of aggregate employed, the design mix (lower quality concrete would be expected to creep more than the higher quality concrete normally used in prestressed work), and the quality or chemical composition of the cement. The prudent designer will consider these factors in his work.

The total ultimate creep strain can be computed using the above factors as follows:

Ultimate Creep Strain =

$$\begin{bmatrix}\text{Elastic Strain at}\\ \text{time of Stressing}\end{bmatrix}\begin{bmatrix}\text{Creep Ratio}\end{bmatrix}\begin{bmatrix}\text{Creep Maturity}\\ \text{Coefficient}\end{bmatrix}\begin{bmatrix}\text{Creep Size}\\ \text{Coefficient}\end{bmatrix} \quad (3\text{-}26)$$

There is evidence that steam curing may reduce creep as much as 50% (Refs. 10 and 17). The factors that affect shrinkage of concrete, as described in Sec. 3-9, except as described herein, are considered to affect creep in about the same degree, and the designer should consider them in estimating creep strain.

Creep at the age of one year can be estimated from the amount obtained experimentally at the age of 28 days (Ref. 23). The relationship for specific creep at the age of one year ($\epsilon'_{c\,365}$), in terms of the 28 day value ($\epsilon'_{c\,28}$) is as follows:

$$\epsilon'_{c\,365} = 0.127 + 1.70\,\epsilon'_{c\,28} \quad (3\text{-}27)$$

in which the strains are expressed in millionths inches per inch per psi. Ultimate specific creep (ϵ'_{cu}) can be estimated from known or estimated values of creep at 1 year (ϵ_{c365}) by one of the following:

$$\epsilon'_{cu} = \frac{1.15\,t\,\epsilon'_{c\,365}}{0.396 + t} \quad (3\text{-}28)$$

$$\epsilon'_{cu} = \frac{1.45\,t^{0.6}\epsilon'_{c\,365}}{0.107 + t^{0.6}} \quad (3\text{-}29)$$

ILLUSTRATIVE PROBLEM 3-4 Estimate the creep strain after 14, 28, and 365 days of loading for a concrete subjected to a constant stress of 1000 psi under the following conditions: $f'_c = 5000$ psi, $E_c = 57000\,(f'_c)_t^{1/2}, (f'_c)_t = \dfrac{t}{3.08 + 0.89t}(f'_c)$, age at loading = 9 days, moist curing, humidity to be constant at 65%, volume to surface ratio of 3 in., constants c and d for Eq. 3-17 are 0.60 and 10, respectively, and the creep ratio can be estimated from Eq. 3-23. Use the curve in Fig. 3-12(a) for the effect of maturity.

SOLUTION: At 9 days, $(f'_c)_t = \dfrac{(9)(5000)}{3.08 + (0.89)(9)} = 4060$ psi

$$E_{ct} = 57000\sqrt{4060} = 3.63 \times 10^6 \text{ psi}$$

$$\epsilon_i = \frac{1000}{3.63 \times 10^6} = 275 \times 10^{-6} \text{ in./in.}$$

Creep ratio: $v_u = 2.35\,[1.27 - 0.0067 \times 65] = 1.96$

Ultimate creep strain $\epsilon_u = 1.96 \times 275 \times 10^{-6} = 539 \times 10^{-6}$

Correction factor for age of loading = $(1.25)\,(9)^{-0.118} = 0.965$

Adjusted creep strain = $0.965 \times 539 \times 10^{-6}$

$= 520 \times 10^{-6}$ in./in.

Creep strain at time $t = \dfrac{t^{0.6}}{10 + t^{0.6}} (520 \times 10^{-6})$

t (Days)	Creep Strain (in./in.)
14	171×10^{-6}
28	221×10^{-6}
365	403×10^{-6}

ILLUSTRATIVE PROBLEM 3-5 Solve I.P. 3-4 using Eq. 3-26, assuming the curves of Figs. 3-8, 3-9 (curve B) and 3-11(a) apply.

Creep size coefficients = 1.30

Creep ratio $= 0.75 + 0.75 \dfrac{(100 - 65)}{50} = 1.28$

Creep maturity coefficient $= 1.80 - 1.28 \left(\dfrac{4060}{5000} - 0.375\right) = 1.24$

$\epsilon_{cu} = [275 \times 10^{-6}] [1.28] [1.24] [1.30] = 567 \times 10^{-6}$ in./in.

t (Days)	Creep Strain (in./in.)
14	186×10^{-6}
28	241×10^{-6}
365	440×10^{-6}

ILLUSTRATIVE PROBLEM 3-6 A concrete having an elastic modulus of 4×10^6 psi is subjected to a constant stress of 1000 psi at a constant temperature and humidity. The creep strain measured after 28 days of loading was found to be 235×10^{-6} in./in. Using Eqs. 3-27 and 3-29, estimate the ultimate creep strain and that at the age of 1 year.

SOLUTION:

Specific creep at 28 days, $\epsilon'_{c28} = \dfrac{235}{1000} = 0.235$ millionths inches per inch per psi

$\epsilon'_{c\,365} = 0.127 + 1.70 \times 0.235 = 0.527$ millionths inches per inch per psi

Assuming ultimate specific creep is obtained at 1500 days.

$$\epsilon'_{cu} = \frac{1.45\,(1500)^{0.6}\,0.527}{0.107 + (1500)^{0.6}} = 0.763 \text{ millionths inches per inch per psi}$$

Therefore, the creep strains at one year and at 1500 days (ultimate) are estimated to be as follows:

$$\epsilon_{c\,365} = 1000\,\epsilon'_{c\,365} = 527 \times 10^{-6} \text{ in./in.}$$

$$\epsilon_{cu} = 1000\,\epsilon'_{c\,1500} = 763 \times 10^{-6} \text{ in./in.}$$

3-12 Relaxation of Concrete

When concrete is subjected to a constant strain, as is the case when a concrete member is stressed by jacking and anchoring (shimming) against rigid abutments, a loss of stress takes place with the passage of time. The stress relaxation of concrete is affected by the factors that affect the creep of concrete, in about the same manner.

The rate of stress loss can be estimated from the curve shown in Fig. 3-13 (Ref. 27). Because the loss of stress due to relaxation can be very large, in comparison with the final stress, the design of structures that will be affected by this phenomenon must be done with great care. Provision should normally be made for checking the stress retained in the structure, from time to time, and adjusting it when necessary. The relaxation properties of the concrete to be used in important structures should be determined for the conditions of service by experimental means when stressing by constant strain is contemplated.

3-13 Low-Pressure (Atmospheric Pressure) Steam Curing

In the manufacture of structural concrete products, it is often desirable or necessary to accelerate the early hydration of the cement in the products in order that a rapid re-use can be made of the manufacturing facilities. In the case of precast reinforced concrete, it may be necessary to obtain a concrete strength of 1000 to 2000 psi at the age of 24 hr or less so that the products can be safely stripped, moved to storage for further curing, and the forms can be re-used. Concrete strengths from 3000 to 4500 psi are required in the manufacture of pre-tensioned concrete before the pre-tensioning tendons can be released, the products removed from the prestressing bench, and the bench re-used. Since forms for structural concrete and pre-tensioning benches represent rather large capital investments that are tied up while the concrete gains these required strengths, it is apparent that a reasonable expenditure for accelerating the hydra-

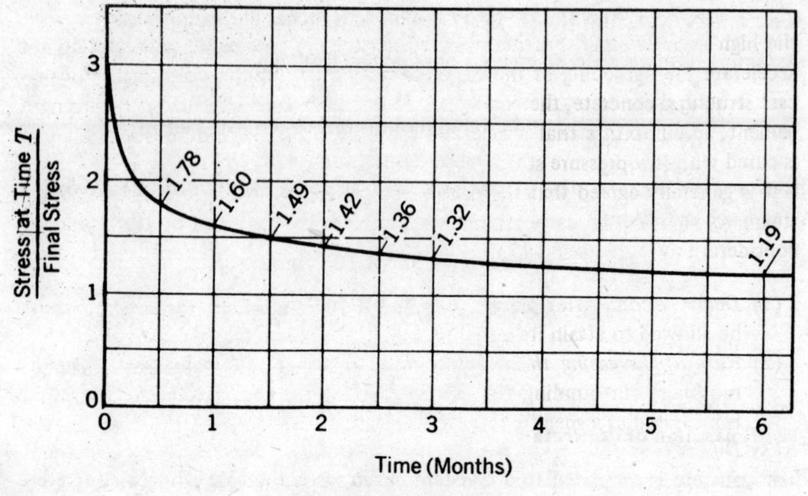

Fig. 3-13 Stress loss curve.

tion of the cement can be justified if the time required for the concrete to attain the adequate strength can be sufficiently reduced.

Low-pressure or atmospheric-pressure steam curing, which is referred to simply as steam curing in this book, is often employed to accelerate the curing of concrete. Well executed steam curing can result in more than 60% of the standard cured 28-day compressive strength being attained in 24 hr. This method consists of confining the concrete products in hot, nearly saturated air at atmospheric pressure by isolating the products from the normal atmosphere in an enclosure into which steam is injected.

Another process that employs steam at elevated pressure and is referred to as high-pressure steam curing (or as autoclaving) is more effective than low-pressure steam curing and is being used rather extensively in concrete-block-manufacturing plants. High-pressure steam curing requires that the concrete products be placed in a steel pressure vessel in which the pressure can be increased above atmospheric. Because of this, this method is considered impracticable for large, structural concrete produ particularly those that are made on pre-tensioning benches.

Hot water and hot oil are used in some applications, in which case, the hot fluid is pumped through longitudinal cavities in the forms or through pipes in or on the casting beds, thus heating the concrete products. This method can give results similar to those obtained by steaming, if the products are kept moist during heating.

Chemical admixtures, designed to accelerate the hydration are available; however, many of these contain calcium chloride, which cannot be safely used in prestressed concrete, due to the danger of the chloride ion causing corrosion of the high-strength steel. Furthermore, admixtures in themselves generally do not accelerate the hardening of the concrete sufficiently. In the manufacture of precast structural concrete, the concrete is frequently made with high early-strength cement, an admixture that accelerates the set is used, and the resulting concrete is cured with low-pressure steam, as described above.

It is generally agreed that the optimum curing cycle to be used in employing steam to increase the early strength of concrete is influenced by the following considerations (Refs. 1 and 17):

(1) *Delay period:* After placing and vibrating the concrete, the concrete must be allowed to attain its initial set before the steam is applied.
(2) *Rate of increasing the concrete temperature:* The temperature of the atmosphere surrounding the concrete and hence the concrete temperature is increased at a specific rate to a maximum temperature.
(3) *Duration of maximum temperature:* The maximum temperature is generally maintained for a specific period of time.
(4) *Rate of cooling:* The temperatures of the concrete and the atmosphere surrounding it are reduced slowly.

The normal American procedure is to employ a delay period of 2 to 6 hr, depending upon the type of cement being used. The longer delay periods are used with the slower setting cements and when higher maximum temperatures are used. Specifications on steam curing usually require that the temperature be increased at a rate less than 1°F per min. to the maximum temperature, which is not greater than 165°F. At this rate the maximum temperature will be reached 1 or 2 hr after steaming is commenced, depending upon the ambient temperature. Most steam-curing facilities do not permit the temperature to be increased at a precise rate, and, as a result, the temperature is usually increased in a few small increments over a period of time. The maximum temperature is maintained between 140°F and 165°F for a period that varies from 10 to 20 hr. The products are then allowed to cool more or less slowly, depending upon the practice at the particular plant.

The procedure described above, which is the general practice in the manufacture of structural concrete products as well as concrete block and pipe, is based upon research that is reviewed in a very brief discussion of steam curing in the *Concrete Manual* published by the Bureau of Reclamation. In this discussion it is stated, "A delay of 2 to 6 hr prior to steam curing will result in higher strength at 24 hr than would be obtained if steam curing were commenced immediately after filling of the forms as was the case in the tests from which the data plotted in Fig. 173 were derived." The Fig. 173 referred to is reproduced here as Fig. 3-14, in which it is seen that these tests would lead the casual reader to the conclusion that temperatures of from 100 to 165°F will not adversely affect the 28-day strengths, whereas temperatures of 185 and 195°F seem to give 3-day strengths that are very adversely affected by the increase of the tem-

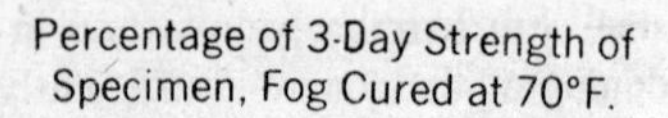

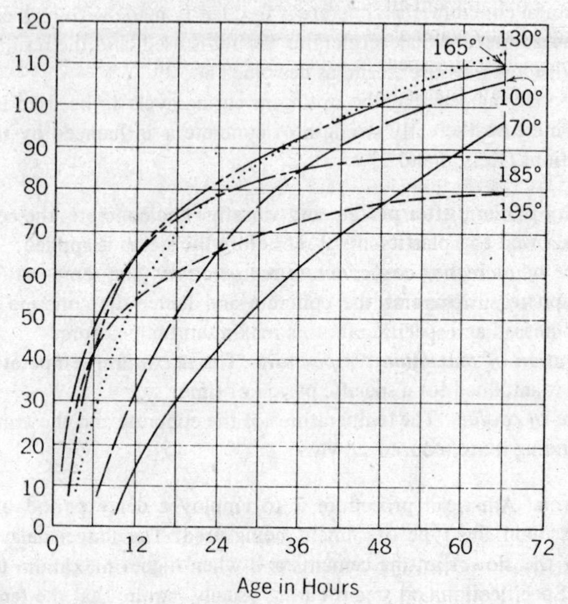

Fig. 3-14 Effect of early steam curing at various temperatures. Steam curing was started immediately after specimens were cast. Compressive strength at 3 days of specimen fog cured at 70°F was 2000 psi. The mix was composed of 1.37 bbl of type II cement per cubic yard and the water-cement ratio was 0.55 (Source: *Concrete Manual*, U.S. Bureau of Reclamation.)

perature above 165°F. The large adverse effect on the 3-day strength illustrated in the figure is the result of the steam being applied without the desired delay period. More recent research has revealed that the higher temperatures can be used to advantage if the delay period is adequate, without any adverse effects upon the 28-day cylinder strengths (Ref. 18).

It is recommended that tests be performed as a means of determining the optimum curing cycle that should be used under specific conditions. By employing the trial and error method, one can determine the delay period, maximum temperature, and time required at maximum temperature that will yield optimum results.

3-14 Cold Weather Concrete

Frequently, in fabricating precast, prestressed members for state or federal projects during the winter months, manufacturers are required to conform to standard specifications that were originally written for job-site winter concrete. Such specifications frequently provide that concrete cannot be placed when the temperature of the ambient air reaches a particular minimum value and that the aggregates and mixing water shall be heated in such a manner that the plastic concrete does not go below a specific value. Although these specifications may be necessary for concrete that is to be placed and allowed to cure without having the surrounding air artificially heated, such specifications frequently have a deterimental effect on the concrete used in plant-produced products that are to be steam cured. There is no question that the aggregates used in precasting plants should be kept sufficiently warm to prevent ice or frost from being in the plastic concrete and that the plastic concrete should not be allowed to freeze. However, it is known that higher concrete strengths are obtained for concretes mixed and placed at lower temperatures than for concretes that are mixed and placed at higher temperatures. This is attributed to the fact that cool plastic concrete mixes require less water for workability than do warmer mixes. The use of very-hot mixing water can have a very serious detrimental effect on the strength of concrete, and in particular, on the early strength.

3-15 Allowable Concrete Flexural Stresses

The two most significant design criteria for prestressed concrete in the United States are the "Standard Specifications for Highway Bridges," which is published by The American Association of State Highway and Transportation Officials (AASHTO) and "Building Code Requirements for Reinforced Concrete" (ACI 318) published by the American Concrete Institute.

The concrete stresses permitted in the AASHTO Specification are (Ref. 25):

1.6.6(B) (1) Temporary stresses before losses due to creep and shrinkage:

Compression

Pretensioned members $0.60f'_{ci}$

Post-tensioned members $0.55f'_{ci}$

Tension

Precompressed tensile zone. No temporary allowable stresses are specified. See Article 1.6.6(B) (2) for allowable stresses after losses.

Other Areas

In tension areas with no bonded reinforcement

200 psi or $3\sqrt{f'_{ci}}$

Where the calculated tensile stress exceeds this value, bonded reinforcement shall be provided to resist the total tension force in the concrete computed on the assumption of an uncracked section. The maximum tensile stress shall not exceed $7.5\sqrt{f'_{ci}}$

(2) Stress at service load after losses have occurred:

Compression $0.40f'_c$

Tension in the precompressed tensile zone

(a) For members with bonded reinforcement $6\sqrt{f'_c}$

For severe corrosive exposure conditions, such as coastal areas $3\sqrt{f'_c}$

(b) For members without bonded reinforcement 0

Tension in other areas is limited by the allowable temporary stresses specified in Article 1.6.6(B) (1).

(3) Cracking Stress*

Modulus rupture from tests or if not available:

For normal weight concrete $7.5\sqrt{f'_c}$

For sand-lightweight concrete $6.3\sqrt{f'_c}$

For all other lightweight concrete $5.5\sqrt{f'_c}$

(4) Anchorage bearing stress:

Post-tensioned anchorage at service load 3000 psi

(but not to exceed $0.9\,f'_{ci}$)

The concrete stresses permitted in ACI 318 (Ref. 26) are:

18.4 Permissible stresses in concrete—Flexural members

18.41—Stresses in concrete immediately after prestress transfer (before time-dependent prestress losses) shall not exceed the following:

(a) Extreme fiber stress in compression . . . . $0.60\,f'_{ci}$

(b) Extreme fiber stress in tension except as permitted in (c) . . . . $3\sqrt{f'_{ci}}$

(c) Extreme fiber stress in tension at ends of simply supported members . $6\sqrt{f'_{ci}}$

Where computed tensile stresses exceed these values, bonded auxiliary reinforcement (non-prestressed or prestressed) shall be provided in the tensile zone to resist the total tensile force in concrete computed with the assumption of an uncracked section.

18.4.2—Stresses in concrete at service loads (after allowance for all prestress losses) shall not exceed the following:

(a) Extreme fiber stress in compression . . . . $0.45\,f'_c$

(b) Extreme fiber stress in tension in precompressed tensile zone . . $6\sqrt{f'_c}$

(c) Extreme fiber stress in tension in precompressed tensile zone of members (except two-way slab systems) where analysis based on transformed cracked sections and on bilinear moment-deflection relationships show that immediate and long-time deflections comply with requirements of Section 9.5.4, and where cover requirements comply with Section 7.7.3.2 . . . . $12\sqrt{f'_c}$

***The total amount of prestressed and non-prestressed reinforcement shall be adequate to develop an ultimate load in flexure at the critical section at least 1.2 times the cracking load calculated on the basis of the modulus of rupture.**

18.4.3–Permissible stresses in concrete of Section 18.4.1 and 18.4.2 may be exceeded if shown by test or analysis that performance will not be impaired.

In the above, f'_c is defined as the specified compressive strength of the concrete in psi and f'_{ci} is the compressive strength of the concrete at the time of initial prestress.

Comparison of these allowable stresses will show that the requirements of AASHTO are more conservative. This is reasonable because bridge structures are subjected to more severe conditions of service (i.e., fatigue, weather, temporary overloads, etc.).

REFERENCES

1. Troxell, G. D., Davis, H. E. and Kelley, J. W., "Composition and Properties of Concrete." McGraw-Hill Book Company, Inc., New York, 1956.
2. "Concrete Manual." U.S. Bureau of Reclamation, 1952.
3. Guyon, op. cit. pp. 53–66.
4. Glanville, W. H. "Studies in Reinforced Concrete III–The Creep or Flow Concrete Under Load." Tech. Paper 12, Department of Scientific and Industrial Research, Building Research (England).
5. Szilard, Rudolph. "Corrosion and Corrosion Protection of Tendons in Prestressed Concrete Bridges." *Journal of the American Concrete Institute, Proceedings*, **66,** No. 1 (Jan. 1969).
6. Branson, D. E. and Christiason, M. L., "Time Dependent Concrete Properties Related to Design–Strength and Elastic Properties, Creep and Shrinkage," Special Publication 27, p. 257–277, American Concrete Institute, Detroit, 1971.
7. Scordelis, Branson and Sozen. "Deflection of Prestressed Concrete Members." *Journal of the ACI, Proceedings*, **60,** No. 12 (Dec. 1963).
8. Troxell, Raphael and Davis. "Long-Time Creep and Shrinkage Tests of Plain and Reinforced Concrete." *A.S.T.M. Proceedings*, **59** (1958).
9. Carlson, Roy W. "Drying Shrinkage of Concrete as Affected by Many Factors." *A.S.T.M. Proceedings*, **38,** Part (1938).
10. Klieger, Paul. "Some Aspects of Durability and Volume Change of Concrete for Prestressing." Portland Cement Association, Research Department Bulletin 118 (Nov. 1960).
11. Hanson, J. A. "Prestress Loss As Affected by Type of Curing." *Journal of the Prestressed Concrete Institute*, 9, No. 2, 69–93 (Apr. 1964).
12. "Lightweight Aggregate Concrete." Housing and Home Finance Agency, Washington, D.C. (Aug. 1949).
13. Furr, H. L. and Sinno, R. "Creep in Prestressed Lightweight Concrete." Research Report No. 69-2, Texas Transportation Institute, Texas A. & M. University, College Station, Texas.
14. Leonhardt, Fritz. "Prestressed Concrete Design and Construction." Wilhelm Ernst and Sohn, Berlin, 1964.
15. Ivey, D. L. and Hirsch, T. J. "Effects of Chemical Admixtures in Concrete and Mortar." Research Report 70-3, Texas Transportation Institute, Texas A. & M. University, College Station, Texas.

16. ACI Committee 209, "Prediction of Creep, Shrinkage, and Temperature Effects in Concrete Structures," ACI Special Publication SP-76, American Concrete Institute, Detroit, 1982.

17. "Recommended Practice for Atmospheric Pressure Steam Curing of Concrete." Reported by ACI Committee 517, *Journal of the American Concrete Institute*, **66,** No. 8 (Aug. 1969).

18. Saul, A. G. A. "Principles Underlying the Steam Curing of Concrete at Atmospheric Pressure." *Magazine of Concrete Research*, **2,** 127 (Mar. 1951).

19. Plowman, J. M. "Maturity and the Strength of Concrete." *Magazine of Concrete Research*, **8,** No. 22, 13 (Mar. 1956).

20. Schorer, H. "Prestressed Concrete, Design Principals and Reinforcing Units." *ACI Journal*, **39,** No. 4, 493-528 (July 1943).

21. Ross, A. D. "Creep of Concrete Under Variable Stress." *ACI Journal Proceedings*, **29,** No. 9, 739-758 (Mar. 1958).

22. Fintel, Mark and Khan, Fazlur R. "Effects of Column Creep and Shrinkage in Tall Structures–Prediction of Inelastic Column Shortening." *Journal of the American Concrete Institute*, **66,** No. 12, 957-67 (Dec. 1969).

23. Brooks, J. J. and Neville, A. M. "Estimating long-term creep and shrinkage from short-term tests." Magazine of Concrete Research, **27,** No. 90, 3-12. (Mar. 1975).

24. Neville, A. M. and Liszka, W. Z. "Accelerated Determination of Creep of Lightweight Aggregate Concrete." *Civil Engineering and Public Works Review*. **68,** No. 803, 515-519 (June 1973).

25. American Association of State Highway and Transportation Officials, *Standard Specifications for Highway Bridges*, 12th ed., Washington, D.C., 1977.

26. ACI Committee 318, *Building Code Requirements for Reinforced Concrete* (ACI 318-83). American Concrete Institute, Detroit, 1983.

27. Guyon, Y. "Prestressed Concrete," p. 63, John Wiley & Sons, Inc., New York, 1953.

CHAPTER 3–PROBLEMS

1. Prepare a plot showing the variations in concrete tensile strength and modulus of rupture as predicted by Eqs. 3-2 and 3-3. Use the product of concrete strength and unit weight as the ordinate with values of f_c' from 2000 psi to 10,000 psi and unit weights from 100 to 160 pcf. Use values of f_t' and f_r as the abscissa.

SOLUTION:

wf_c'	$\sqrt{wf_c'}$	$\frac{1}{3}\sqrt{wf_c'}$	$0.60\sqrt{wf_c'}$	$0.70\sqrt{wf_c'}$
200,000	447	149	268	313
600,000	774	258	464	542
1,000,000	1000	333	600	700
1,400,000	1183	394	710	828
1,800,000	1341	447	805	938

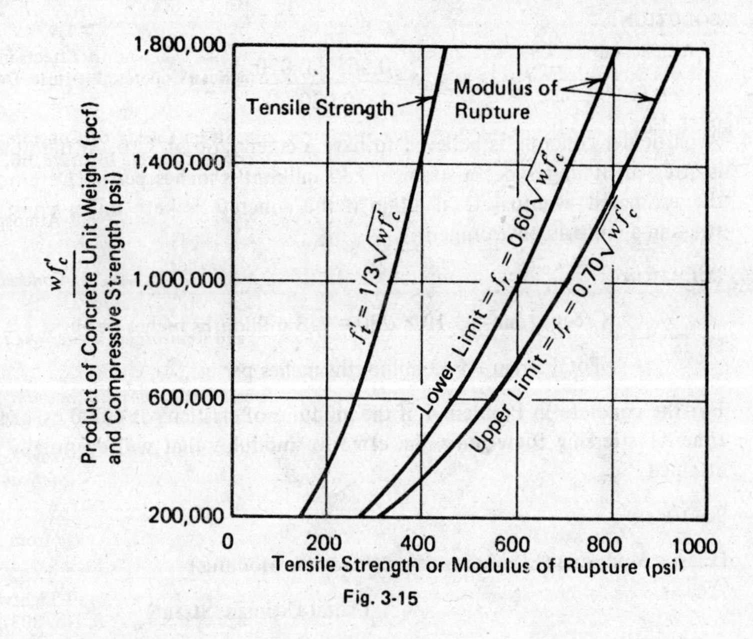

Fig. 3-15

2. Prepare a plot of $(f_c')_t$ vs. E_{ct} using Eq. 3-4 for values of the variable w (unit weight of concrete) of 100, 120, 140, and 160 pcf. Use $(f_c')_t$ as the ordinate with values from 2000 to 10,000 psi.

SOLUTION:

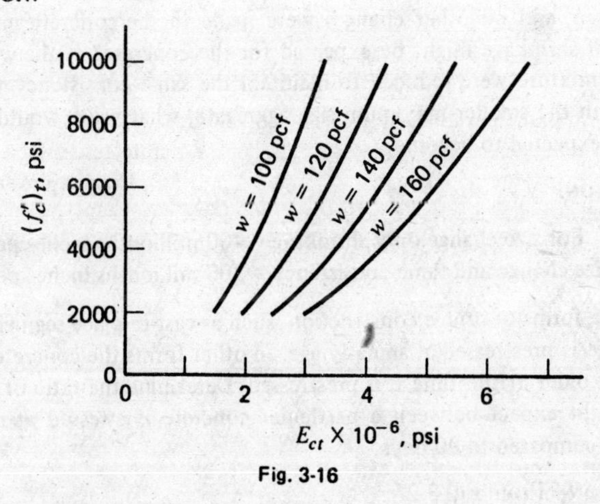

Fig. 3-16

3. **A piece of concrete 10 ft long is stressed in compression to 500 psi. The initial (elastic) shortening is 0.020 in. After 2 years the total shortening is 0.065 in. Determine the creep ratio at the age of two years.**

SOLUTION:

$$\text{Creep Ratio} = \frac{0.065 - 0.020}{0.020} = 2.25$$

4. A particular concrete is believed to have a creep ratio of 3.10. At the time it is stressed, it undergoes a strain of 230 millionths inches per in. Determine the estimated eventual total strain if the concrete is kept under constant stress in a uniform environment.

SOLUTION:

Creep strain = 3.10 × 230 = 713 millionths inches per in.

Total strain = 943 millionths inches per in.

5. For the concrete in Problem 4, if the modulus of elasticity is 4,000 ksi at the time of stressing, determine the effective modulus that will eventually be attained.

SOLUTION:

(Elastic Modulus) (Elastic Strain) = (Effective Modulus) · (Total Ultimate Strain)

$$E_c' = \frac{4{,}000 \times 230 \times 10^{-6}}{943 \times 10^{-6}} = 976 \text{ ksi}$$

6. A concrete made with a maximum sized aggregate of $\frac{3}{4}$ in. has a shrinkage of 500 millionths inches per inch. If the maximum aggregate size were increased to 1.5 in, and no other changes were made in the concrete mixture, what value of shrinkage might be expected for the concrete? If the water content of the mixture were reduced to maintain the same consistency the concrete had with the smaller maximum size aggregate, what value would the shrinkage be expected to become?

SOLUTION:

For size change only shrinkage = 400 millionths inches per inch
For size change and same consistency = 300 millionths inches per inch

7. In some forms of bridge construction, such as cast-in-place segmental bridges, concrete is prestressed at an early age. In other forms the concrete may be 90 days or older at the time it is prestressed. Determine the ratio of creep strain one might expect between a particular concrete if stressed at the age of 7 days as compared to 90 days.

SOLUTION: From Eq. 3-25,

$$\text{Ratio} = \frac{1.33}{0.73} = 1.82$$

8. Concrete pavements for airfields have been prestressed by activating hydraulic jacks installed in joints at the ends of the pavements, end abutments being provided to hold the jacks firmly in place. If such a pavement were stressed to 300 psi by the jacks which were then permanently locked in their extended position, assuming no friction between the pavement and the subgrade (not a correct assumption!), determine the stress one would expect to remain in the pavement after one, three, six and an infinite number of months.

Time Month	Stress Remaining (psi)
1	160
3	132
6	119
∞	100

(From Fig. 3-13)

4 Basic Principles for Flexural Design

4-1 Introduction

The basic principles and mathematical relationships used in the design and analysis of prestressed-concrete flexural members are not unique to this type of construction. Virtually all of the fundamental relationships are based upon the normal assumptions of elastic design, which form the basis of the study of the strength of materials. Although the form in which the relationships appear in a discussion of prestressed concrete may be somewhat modified to facilitate their application, the student of engineering should have little difficulty in understanding these modified relationships.

Two major forms of design problems are encountered by the engineer engaged in the design of prestressed concrete flexural members. These are frequently referred to as the *review* of a member and as the *design* of a member.

The review of a member consists of the determination of the concrete flexural stresses and deflections under service conditions of loading and prestressing, in order to confirm their compliance with the design criteria. In addition, the strength of the member in bending, shear, and bond must be determined to equal or exceed the minimum strength requirements of the design criteria. It should be apparent that in order to review a member as described here, the dimensions of the concrete section, the properties of the materials, the amount and eccentricity of the prestressing steel, the amount of nonprestressed reinforcing, as well as the amount of web reinforcing, must be known.

The design of a member consists of selecting and proportioning a concrete section in which the stresses in the concrete do not exceed the permissible values under any combination of service loads and prestressing. Design also includes the determination of the amount and eccentricity of the prestressing force required for the specific section. The design of a member must include a study of the flexural strength which the section can develop under design load, and the determination of the amount of non-prestressed flexural reinforcing that may be required. Additionally, a study of the shear stresses must be made and the amount of web reinforcing required for adequate shear strength under design loads must be determined. Consideration of tendon development lengths, both for flexural strength and, in the case of pretensioned tendons, for transfer length, is included in the design of a member. It must be emphasized that the design of a flexural member is normally a trial and error procedure. The designer must assume a concrete section and compute the prestressing force and eccentricity that are required to confine the concrete stresses within the allowable limits under all conditions of service loads. After compliance with service loading criteria has been confirmed, a complete strength analysis must be made. (Some designers make the strength design before the service load design.) In the design of a member, several adjustments of the trial section are normally required before a satisfactory solution is found.

This chapter is devoted to the consideration of the fundamental principles pertaining to the determination of the concrete stresses due to prestressing, the determination of the prestressing force and eccentricity required for a specific distribution of stresses due to prestressing, the consideration of the pressure line in simple flexural members that are loaded in the elastic range, and other topics related to flexural analysis and design. The problems given in this chapter are confined to the "review" type. The procedures used in preparing preliminary designs by the trial and error procedure are treated in Sec. 7-8.

The elastic analysis and design of prestressed flexural members can be done rapidly and accurately only after the fundamental theorems and axioms have been thoroughly mastered. Many of the operations discussed in this chapter can be done more rapidly by the use of the simple expedients treated in Chapter 7. These "classical" methods should be well understood, however, before attempting the use of the expedients. The design and analysis of continuous prestressed members, which are treated in Chapter 8, also require complete familiarity with the principles presented in this chapter.

4-2 Mathematical Relationships for Prestressing Stresses

The stresses due to prestressing alone are generally combined stresses due to a direct load eccentrically applied. Therefore, these stresses are computed using the following well-known relationship for combined stresses

$$f = \frac{P}{A} \pm \frac{My}{I} \tag{4-1}$$

in which f is the fiber stress at the distance y from the centroidal axis, P is the axial force, A is the area of the cross section, M is the moment acting on the section, and I is the moment of inertia of the cross section. In this book compressive stresses are taken to be positive.

Because the moment due to the prestressing is equal to the prestressing force multiplied by the eccentricity of this force (i.e., $M = Pe$), and because the square of the radius of gyration is equal to the moment of inertia divided by the area of the cross section ($r^2 = I/A$), the above relationship can be rewritten

$$f = \frac{P}{A}\left(1 \pm \frac{ey}{r^2}\right) \tag{4-2}$$

Adopting the sign convention that vertical dimensions (i.e., y and e) are positive when above the centroid of the section and using y_t and y_b to denote the distances from the centroidal axis to the top and bottom fibers, respectively, the top and bottom fiber stresses for a prestressing force applied eccentrically are expressed by

$$f_t = \frac{P}{A}\left(1 + \frac{ey_t}{r^2}\right) \tag{4-3}$$

$$f_b = \frac{P}{A}\left(1 + \frac{ey_b}{r^2}\right) \tag{4-4}$$

where f_t and f_b are the stresses in the top and bottom fibers due to the prestressing alone, respectively. A positive stress is compressive in the above relationships.

These relationships are the same for the stresses resulting from the initial and the final prestressing forces (*see* Sec. 1-2). In computing these stresses, one would of course use the initial prestressing force when computing the initial stresses and the final force when computing the final stresses. Frequently, the designer assumes a ratio between the final and the initial prestressing forces for design purposes, because the reduction of the prestressing force cannot be accurately estimated until the design is nearly complete (*see* Sec. 6-2). Therefore, if the designer bases his computation on the final prestressing force and has assumed that the total loss will be 15% of the initial force, for example, the stresses resulting from the initial prestressing force can be determined by dividing the final stresses by 0.85.

The experienced designer generally prefers to design with the final prestressing force assumed to be from 75 to 90% of the initial force. A comprehensive study of the losses of stress cannot be made until the basic design is finalized. If, when this study is made, it is found the loss will be greater than assumed, the initial prestressing force can be increased so that the final force will be satisfactory. In addition, strength requirements of design criteria frequently control the amount of flexural reinforcement that is required; serviceability requirements may or may not control the amount of prestressing required. The advantage of this procedure will be apparent after consideration of the data presented in Chapter 7.

ILLUSTRATIVE PROBLEM 4-1 Compute the stresses due to prestressing alone in a beam with a rectangular cross section 10 in. wide and 12 in. high that is prestressed by a final force of 120 k at an eccentricity of +2.5 in. State whether the stresses are compressive or tensile. Compute the stresses due to the initial prestressing force, if the ratio between the final force and the initial force is 0.85.

SOLUTION:

$$A = 120 \text{ sq in.}, \quad I = \frac{10 \times 12^3}{12} = 1440 \text{ in.}^4, \quad r^2 = \frac{1440}{120} = 12 \text{ sq in.}$$

Final stresses:

$$f_t = \frac{120{,}000}{120}\left(1 + \frac{2.5 \times 6}{12}\right) = +2250 \text{ psi (compression)}$$

$$f_b = \frac{120{,}000}{120}\left(1 + \frac{2.5 \times -6}{12}\right) = -250 \text{ psi (tension)}$$

Initial stresses:

$$f_t = \frac{+2250}{0.85} = +2650 \text{ psi (compression)}$$

$$f_b = \frac{-250}{0.85} = -294 \text{ psi (tension)}$$

ILLUSTRATIVE PROBLEM 4-2 Compute the prestressing force and eccentricity that would be necessary in the beam of Problem 4-1, in order to obtain a bottom-fiber compression of 2400 psi and a top-fiber tension of 350 psi, by equating the relationships for stresses due to prestressing in the top and bottom fibers.

$$f_t = \frac{P}{A}\left(1 + \frac{ey_t}{r^2}\right) = -350 = \frac{P}{120}\left(1 + \frac{e \times 6}{12}\right)$$

$$f_b = \frac{P}{A}\left(1 + \frac{ey_b}{r^2}\right) = +2400 = \frac{P}{120}\left(1 + \frac{e \times -6}{12}\right)$$

$$2400 - \frac{2400\, e \times -6}{12} = -350 + \frac{350\, e \times 6}{12}$$

$$1025\, e = -2750$$

$$e = -2.68 \text{ in.}$$

From Eq. 4-4,

$$P = \frac{120 \times 2400}{1 + \frac{-2.68 \times -6}{12}} = 123{,}000 \text{ lb}$$

The familiar principle of superposition is used to determine the combined effect of the prestressing and the other loads that may be acting simultaneously on a prestressed beam. Although it is possible to write a single equation that will accurately define the stress at any particular point in a beam, for "hand" calculations it is normally less confusing if the effect of each load (or prestressing) is computed separately and the net effect is determined by algebraically adding the effects of the several loads.

ILLUSTRATIVE PROBLEM 4-3 Compute the net initial and final concrete stresses in the extreme top and bottom fibers at the midspan of a beam that is 10 in. wide, 12 in. deep and on a span of 25 ft. The beam is to support an intermittent, uniformly distributed live load of 0.45 k/ft and is to be prestressed with a final force of 120 k positioned with an eccentricity of −2.50 in. The ratio between the final and initial prestressing forces is assumed to be 0.85.

SOLUTION: The initial and final stresses due to prestressing in the top and bottom fibers are opposite to those in I. P. 4-1 because eccentricity of the prestressing force is negative; these stresses are as shown below in the tabulation of combined stresses.

The section modulus of the section is equal to I/y or 240 in.^{3} for the top fiber and −240 in.^{3} for the bottom fiber.

Stresses due to the dead load of the beam alone:

$$w_{DL} = \frac{120}{144} \times 0.150 = 0.125 \text{ k/ft}, \quad M_{DL} = 0.125 \times \frac{(25)^2}{8} = 9.77 \text{ k-ft}$$

$$f_t = + \frac{9.77 \times 12{,}000}{240} = +488 \text{ psi (compression)}$$

$$f_b = \frac{9.77 \times 12{,}000}{-240} = -488 \text{ psi (tension)}$$

Stresses due to live load alone:

$$w_{LL} = 0.45 \text{ k/ft}, \quad M_{LL} = 0.45 \times \frac{(25)^2}{8} = 35.2 \text{ k-ft}$$

$$f_t = \frac{+35.2 \times 12{,}000}{240} = +1760 \text{ psi (compression)}$$

$$f_b = \frac{35.2 \times 12{,}000}{-240} = -1760 \text{ psi (tension)}$$

Combined stresses:

	Top Fiber	*Bottom Fiber*
Initial prestress	- 294 psi	+2650 psi
Beam dead load	+ 488 psi	- 488 psi
Initial prestress plus dead load	+ 194 psi	+2162 psi
Live load	+1760 psi	-1760 psi
Initial prestress plus total load	+1954 psi	+ 402 psi
Final prestress	- 250 psi	+2250 psi
Dead load of beam	+ 488 psi	- 488 psi
Final prestress plus dead load	+ 238 psi	+1762 psi
Live load	+1760 psi	-1760 psi
Final prestress plus total load	+1998 psi	+ 2 psi

4-3 Pressure Line in a Beam with a Straight Tendon

At any section of a beam, the combined effect of the prestressing force and the externally applied load will result in a distribution of concrete stresses that can be resolved into a single force. *The locus of the points of application of this force in any beam or structure is called* the pressure line.

This can be illustrated by considering a rectangular beam prestressed by an eccentric, straight tendon, as is shown in Fig. 4-1. Such a beam would have a distribution of stresses due to prestressing alone at every cross section, as is shown in Fig. 4-2(a). It is readily seen that the force resulting from the distribution of internal prestressing stresses (C) is equal in magnitude to the prestressing force. In addition, it is applied at the same point as the prestressing force at every section, because the prestressing force and the eccentricity are both constant throughout the length of the beam.

If a uniform load, of such magnitude that results in the bottom-fiber prestress being nullified at midspan, is applied to the beam, the resulting stress distribution would be as indicated in Fig. 4-2(b) and the pressure line at this point would then be applied at a point $d/6$ above the centroidal axis of the beam. At

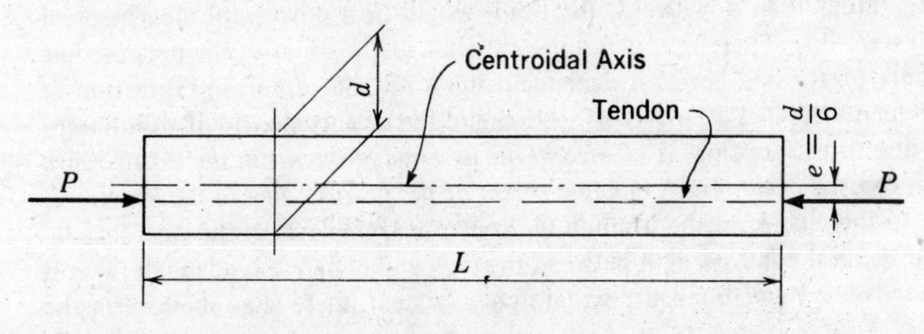

Fig. 4-1 Simple rectangular beam prestressed by an eccentric straight tendon.

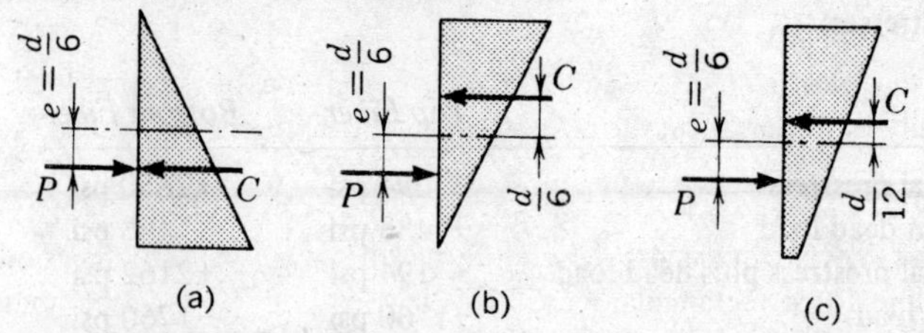

Fig. 4-2 Stress distributions and pressure-line locations for a simple rectangular beam prestressed with a straight eccentric tendon (a) due to prestressing alone, (b) at midspan under full design load, and (c) at quarter point under full design load.

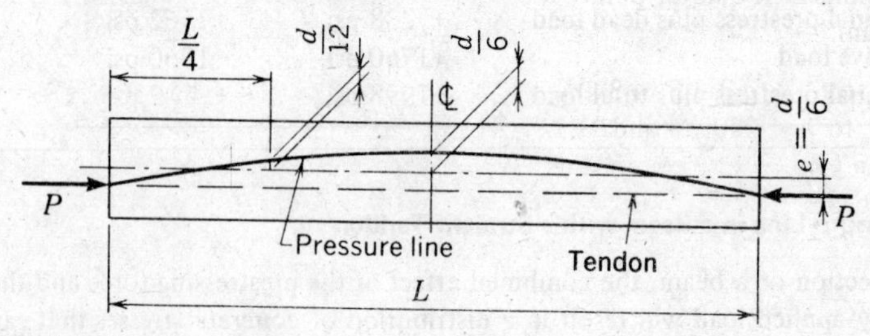

Fig. 4-3 Location of pressure line in a simple beam of rectangular cross section, prestressed by a force at $e = d/6$ and under uniform load resulting in zero bottom-fiber stress at the midspan.

the quarter point of this beam, under the same loading conditions, the stresses due to the external load are only 75% as much as those at midspan. The stress distribution resulting from the combination of prestressing and the flexural stresses due to the external load would be as shown in Fig. 4-2(c). At this point the pressure line is located at a distance of $d/12$ above the centroidal axis. At the support, because there are no flexural stresses resulting from the external load, the pressure line remains at the level of the steel. Plotting the location of the pressure line for this loading reveals that it is a parabola with its vertex at the center of the beam, as shown in Fig. 4-3.

In a similar manner, it can be shown that a larger uniform load would result in the pressure line being moved up even higher, and for a uniform load applied upward rather than downward, the result would be a downward movement of the pressure line. Therefore, it is apparent that the location of the pressure line in simple prestressed beams is dependent upon the magnitude and direction of the moments applied at any cross section and the magnitude and distribution of stress due to prestressing: *A change in the external moments in the elastic range of a prestressed beam results in a movement of the pressure line in the beam.*

Due to the change in the strain in the concrete at the level of the steel (assuming the flexural bond strength between the steel and concrete is adequate, as it is in pre-tensioned and bonded post-tensioned beams), there is an increase in the stress in the prestressing steel as a result of applying an external load. This is occasionally of importance; the effect is normally disregarded (*see* Sec. 4-11).

ILLUSTRATIVE PROBLEM 4-4 Compute and draw to scale the location of the pressure line for a rectangular beam 10 in. wide and 12 in. deep that is prestressed with a force of 120 k at a constant eccentricity of -2.5 in. and that is supporting a 15 k concentrated force at midspan of a span of 10 ft. Use an exaggerated vertical scale in a sketch, and dimension the location of the pressure line at the midspan, quarter point, and end of the beam. Neglect the dead weight of the beam.

SOLUTION: From I.P. 4-3 the stresses due to final prestressing of 120 k are known to be -250 psi and +2250 psi and the section moduli are known to be +240 in.^{3} and -240 in.^{3} for the top and bottom fibers, respectively. At the end of the beam, moment = zero. Therefore, the pressure line is at $e = -2.50$ in.

At the midspan:

$$M = \frac{PL}{4} = \frac{15 \text{ k} \times 10 \text{ ft}}{4} = 37.5 \text{ k ft}$$

$$f = \frac{37.5 \times 12{,}000}{\pm 240} = \pm 1875 \text{ psi}$$

Stress distribution at midspan:

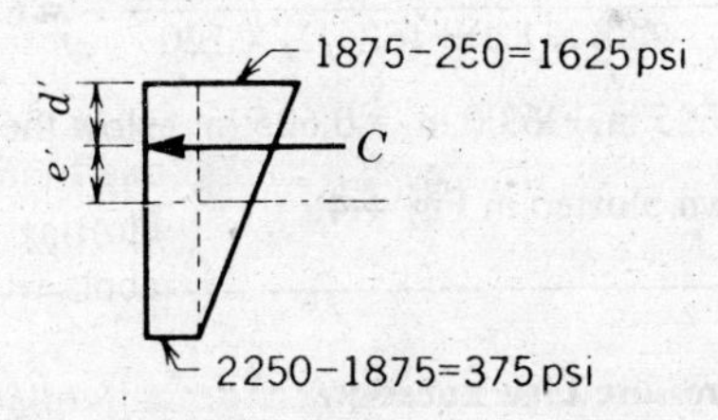

The distance from the top fiber to the resultant force in the section is:

$$d' = \frac{6 \times 375 \times 120 + (1250/2) \times 120 \times 4}{375 \times 120 + (1250/2) \times 120} = 4.75 \text{ in.}$$

The eccentricity of the resultant force is:

$$e' = 6.00 \text{ in.} - 4.75 \text{ in.} = 1.25 \text{ in.}$$

At the quarter point, the moment due to the external load is only one-half that at the midspan. Therefore, the flexural stresses due to the applied load are only one-half of those at midspan, or ±938 psi.

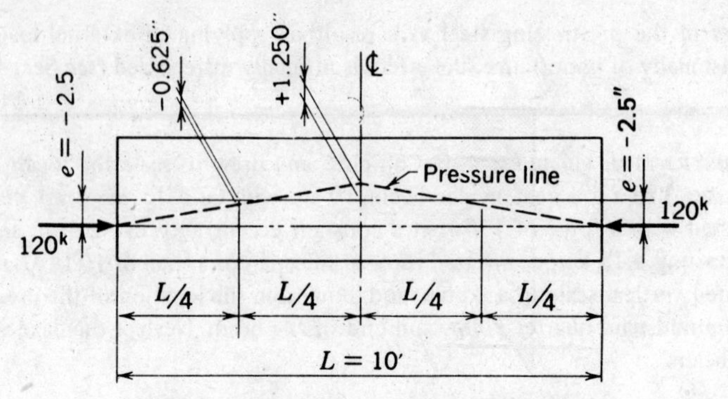

Fig. 4-4 Location of pressure line, Problem 4-4.

Stress distribution at quarter point:

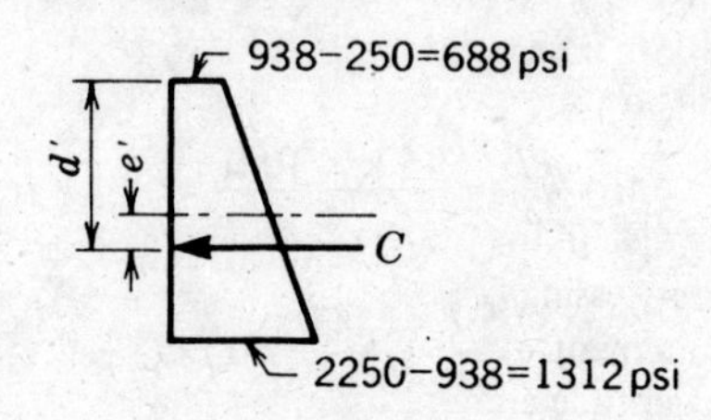

$$d' = \frac{6 \times 688 \times 120 + (624/2) \times 120 \times 8}{688 \times 120 + (624/2) \times 120} = 6.625 \text{ in.}$$

$$e' = 6.625 \text{ in.} - 6.00 \text{ in.} = 0.625 \text{ in. below the centroid.}$$

The results are shown plotted in Fig. 4-4.

4-4 Variation in Pressure-Line Location

If tensile stresses are not permitted in the bottom fibers of a simple prestressed concrete beam, when it is subjected to service loads, the distribution of stresses will be as shown in Fig. 4-5. Also shown in Fig. 4-5 is the cross section of the beam. The force C is the resultant of the stresses in the concrete (pressure line) and it obviously must be equal in magnitude and opposite in direction to the prestressing force P; the horizontal forces acting on the cross section must be in equilibrium. In addition, from Eq. 4-4 developed in Sec. 4-2, we can write the relationship for the stress in the bottom fibers as follows:

$$f_b = \frac{C}{A}\left(1 + \frac{e' y_b}{r^2}\right) = 0$$

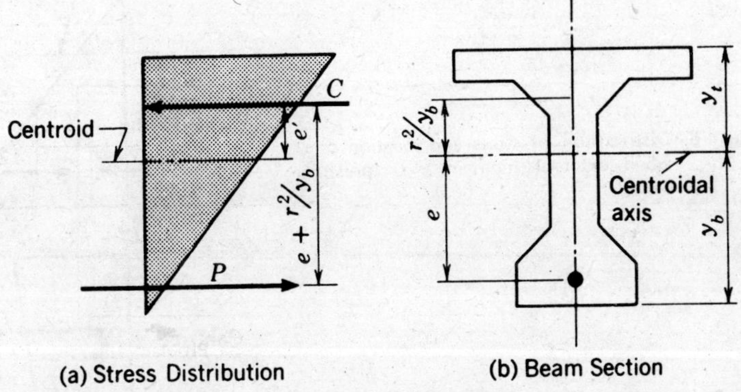

Fig. 4-5 Relationship between prestressing force, pressure line, and section properties of a beam having zero stress in bottom fiber under design load.

from which we obtain

$$e' = -r^2/y_b$$

The eccentricity e' of the resultant C should not be confused with the eccentricity of the prestressing force.

Another requirement of equilibrium is that the internal and external moments are equal in magnitude and opposite in direction at every section. It follows that the total external moment the beam is resisting at this section is *numerically* equal to

$$M_T = M_{DL} + M_{LL} = C(e + r^2/y_b) = P(e + r^2/y_b) \tag{4-5}$$

in which e is the eccentricity of the prestressing force. In Eq. 4-5, both terms e and r^2/y_b are positive because they are vectors measured upward from the point of application of the prestressing force P.

The above example further illustrates that prestressed beams, functioning in the elastic range, resist the moment due to externally applied loads by the movement of the resultant of the stresses in the concrete, rather than by an increase in the prestressing stress, as was brought out in Sec. 4-3. From Eq. 4-5, it is apparent that if $M_T = 0$, the product of C multiplied by the quantity $(e + r^2/y_b)$ must also be equal to zero and the concrete stresses would be distributed as shown in Fig. 4-6. If the external moment (M_T) were some value less than that which nullifies the precompression of the bottom fibers, the force C would be applied above the location of the prestressing steel at a distance (d) equal to

$$d = \frac{M_T}{P} \tag{4-6}$$

This condition is illustrated in Fig. 4-7. It should be noted that positive moments cause positive, and hence upward in direction, values of d.

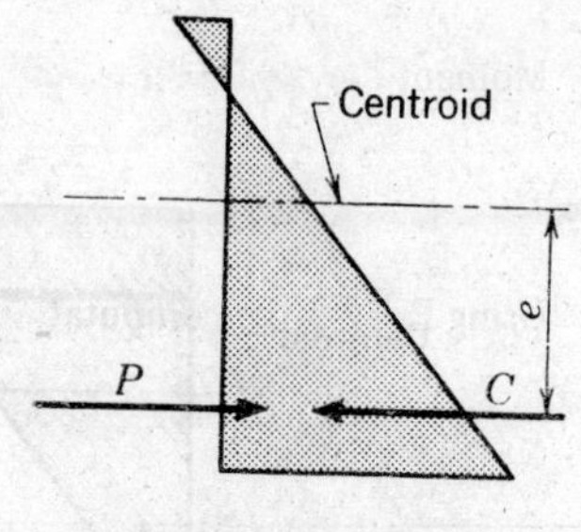

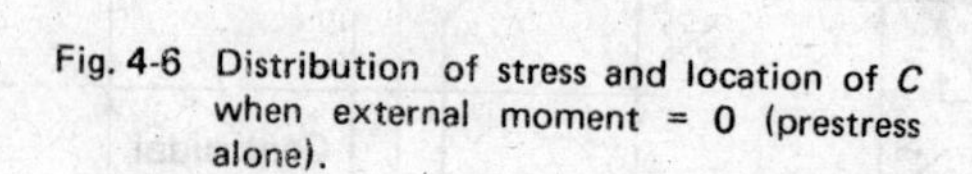

Fig. 4-6 Distribution of stress and location of C when external moment = 0 (prestress alone).

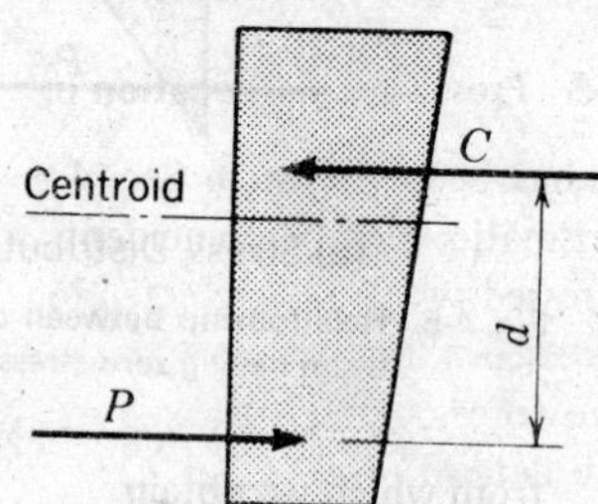

Fig. 4-7 Distribution of stress and location of resultant C when external moment is of nominal magnitude.

The relationship given by Eq. 4-5 is extremely useful in the preliminary design of beams as well as in checking the final design. Because the value of $(e + r^2/y_b)$ is normally of the order of 65% of the depth of the beam section (it varies between the approximate limits of 33 to 80% for different cross sections) for a given superimposed moment, the designer can assume a dead weight for the beam and estimate the prestressing force required for different depths of construction. The use of this relationship is illustrated in Prob. 4-5 and in Sec. 7-8, in the discussion of preliminary design.

ILLUSTRATIVE PROBLEM 4-5 Compute the maximum concentrated load that can be applied at the midspan of a beam that is 10 in. wide, 12 in. deep, prestressed with 120 k at an eccentricity of −2.5 in., and is to be used on a span of 10.0 ft center to center of bearings without tensile stresses resulting in the bottom fibers.

SOLUTION: (Using the basic relationships for flexural design) From I.P. 4-1 and I.P. 4-3, the section properties and the stresses due to prestressing are known. Hence,

$$\text{Maximum allowable moment} = f_b \times S_b = \frac{-2250 \times -240}{12000}$$

$$= 45.0 \text{ k-ft}$$

$$\text{Moment due to dead load} = \frac{wl^2}{8} = \frac{120}{144} \times 0.15 \times \frac{(10)^2}{8} = 1.56 \text{ k-ft}$$

$$\text{Moment due to concentrated load} = \frac{PL}{4} = 45.0 \text{ k-ft} - 1.6 \text{ k-ft} = 43.4 \text{ k-ft}$$

$$P = \frac{43.4 \times 4}{10} = 17.4 \text{ k}$$

Using Eq. 4-5 the computation of the moment at which $f_b = 0$ becomes

$$M_T = 120 \text{ k} \frac{2.5 \text{ in.} + 2.0 \text{ in.}}{12} = 45.0 \text{ k-ft}$$

4-5 Pressure-Line Location in a Beam with a Curved Tendon

It has been shown in Sec. 4-3 that the pressure line for prestressing alone in a prismatic beam is coincident with the prestressing force when the beam is prestressed with a straight tendon. This can also be demonstrated for a beam prestressed with a tendon that changes slope, as shown in Fig. 4-8. By inspection, the forces acting on the concrete at the point where the tendon changes slope are determined to be as indicated in Fig. 4-9. The forces acting on the concrete are shown at their respective points of application in the free-body diagram of Fig. 4-10. In order to determine where the pressure line is acting at the center of the beam, the conditions of statics at point A are investigated. The sum of the vertical forces is equal to zero because $P \sin \alpha$ is acting downward at the end of the beam and upward at the quarter point. The sum of the horizontal forces indicates the force R must be equal to P, because

$$\Sigma H = P \cos \alpha + (P - P \cos \alpha) - R = 0 \therefore R = P$$

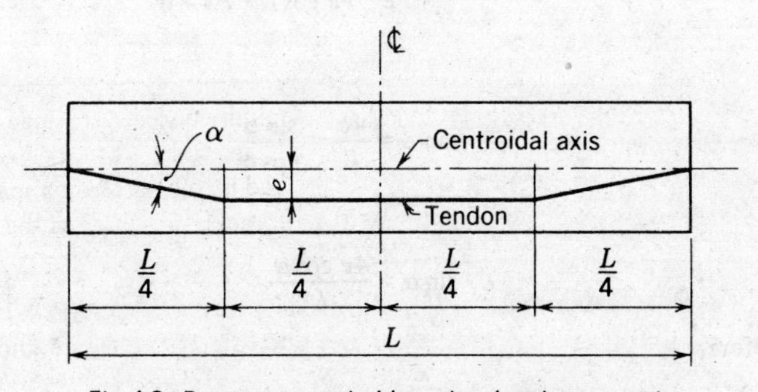

Fig. 4-8 Beam prestressed with tendon that slopes at ends.

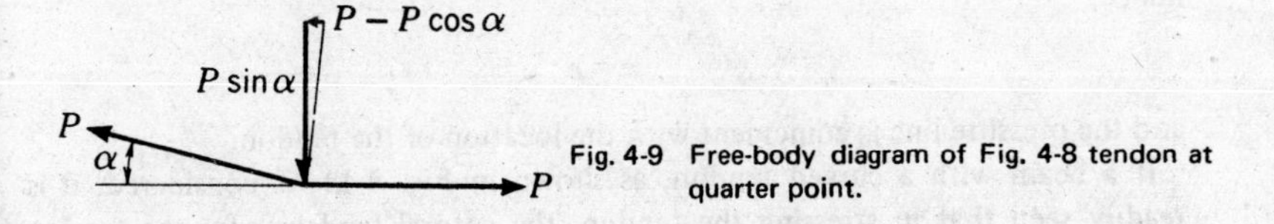

Fig. 4-9 Free-body diagram of Fig. 4-8 tendon at quarter point.

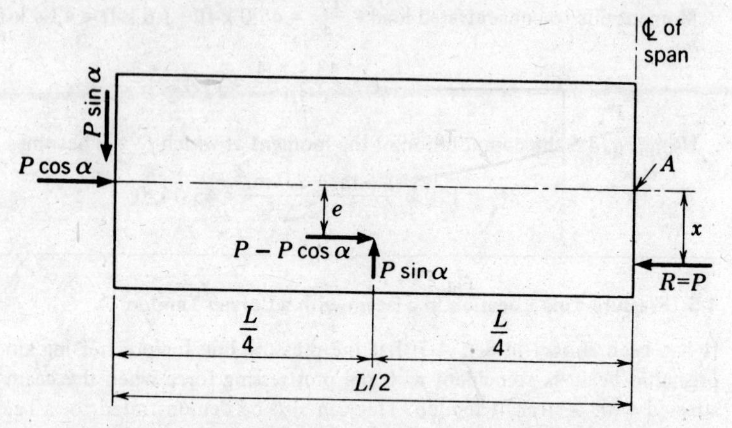

Fig. 4-10 Free-body diagram for half of the beam shown in Fig. 4-8.

To determine the distance from the centroidal axis to the point of application of the force R (and hence the location of the pressure line at the center of the beam), moments are taken about point A as follows:

$$\Sigma M_A = (P \sin \alpha)\frac{L}{2} + (P - P \cos \alpha)e - (P \sin \alpha)\frac{L}{4} - Px = 0$$

and

$$\frac{PL \sin \alpha}{4} + Pe - Pe \cos \alpha - Px = 0$$

but

$$\tan \alpha = \frac{4e}{L} = \frac{\sin \alpha}{\cos \alpha}$$

and

$$\sin \alpha = \frac{4e \cos \alpha}{L}$$

therefore,

$$Pe \cos \alpha + Pe - Pe \cos \alpha - Px = 0$$

hence,

$$x = e$$

and the pressure line is coincident with the location of the tendon.

If a beam with a curved tendon, as shown in Fig. 4-11, is considered, it is readily seen that in stressing the tendon, the natural tendency for the tendon

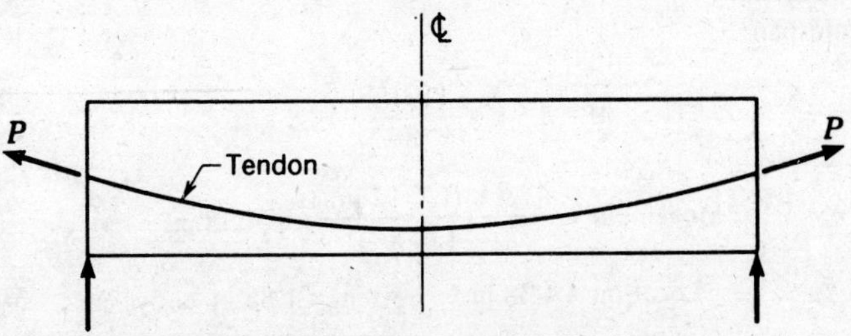

Fig. 4-11 Simple beam with curved tendon.

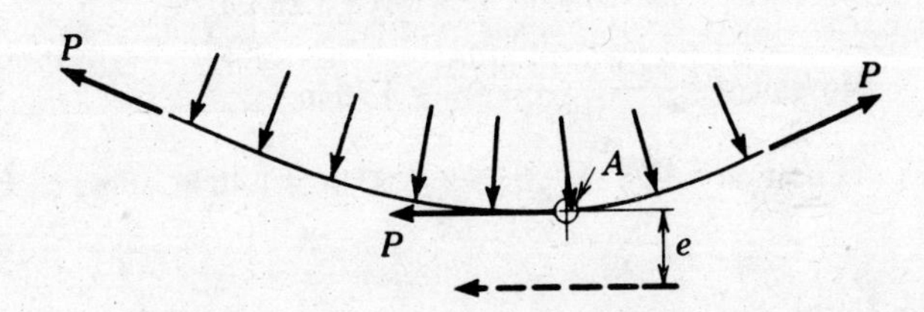

Fig. 4-12 Free-body diagram of portion of curved prestressing tendon.

to straighten out is resisted by the concrete. If a short segment of the tendon is studied as a free body, as shown in Fig. 4-12, forces must be present normal to the tendon (neglecting friction) in order to prevent this straightening. If friction is neglected, the force acting throughout the tendon is uniform, and since the tendon is flexible, it cannot support any bending moments. Therefore, at every point such as point A, the force in the tendon is equal to P and is located at the location of the tendon. If the force were not coincident with the tendon at A, but were located at some distance from A (as shown by the dashed vector), the tendon would have to withstand the moment Pe caused by this eccentricity.

From this analysis then, it can be concluded that *the pressure line for prestressing alone in a simple beam, prestressed with a curved tendon, is coincident with the path of the tendon*, since the forces in the concrete must be equal and opposite to those in the steel in order to maintain equilibrium. Furthermore, it can be shown that the pressure line moves when an external load is applied to a beam with a curved tendon, just as it does in a beam with a straight tendon.

ILLUSTRATIVE PROBLEM 4-6 Compute and plot to scale the location of the pressure line for the 10×12 in. rectangular beam, if prestressed with a force of 120 k, which is on a parabolic curve and which has an eccentricity of -2.5 in. and zero, at the midspan, and end, respectively, if the beam is spanning 10 ft and subjected to a uniformly distributed load of 3.5 k/ft. Neglect dead load of the beam.

SOLUTION:

At midspan:

$$M = 3.5 \text{ k} \times \frac{(10)^2}{8} = 43.8 \text{ k-ft}$$

$$\text{Movement} = \frac{43.8 \text{ k-ft} \times 12 \text{ in./ft}}{120 \text{ k}} = 4.38 \text{ in.}$$

$$\text{Location} = 4.38 \text{ in.} - 2.50 \text{ in.} = 1.88 \text{ in. above c.g.s.}$$

At quarter point:

$$M = 0.75 \times 43.8 \text{ k-ft} = 32.8 \text{ k-ft*}$$

$$\text{Movement} = \frac{32.8 \times 12 \text{ in./ft}}{120 \text{ k}} = 3.28 \text{ in.}$$

$$\text{Location} = 3.28 \text{ in.} - 0.75 \times 2.50 \text{ in.} = 1.40 \text{ in. above c.g.s.}$$

At end:

(*See* Fig. 4-13.)

ILLUSTRATIVE PROBLEM 4-7 Calculate the maximum, uniformly distributed load that can be applied to the beam of Problem 4-6, if the span is 10 ft and the bottom-fiber stress is zero at midspan.

SOLUTION: Under the loaded condition, the pressure line will be r^2/y_b above the centroid (the upper limit of the Kern zone).

$$r^2/y_b = 2.0 \text{ in.}, \quad e = -2.5 \text{ in.}$$

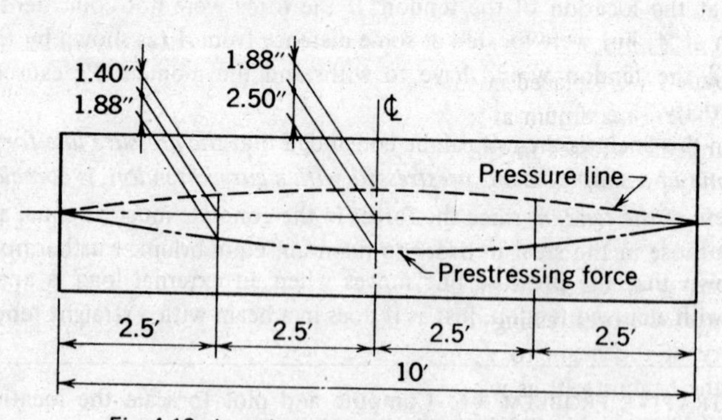

Fig. 4-13 Location of pressure line for Problem 4-6.

*Note that the ordinate of a second degree parabola at half-way between midspan and the end of the span is 0.75 times to ordinate at the midspan.

$$M_T = 120 \text{ k } \frac{4.5 \text{ in.}}{12} = 45.0 \text{ k-ft}$$

$$w_{max} = \frac{45 \times 8}{(10)^2} = 3.60 \text{ k/ft}$$

4-6 Advantages of Curved or Draped Tendons

When a beam, such as is shown in Fig. 4-14, is prestressed by a straight tendon, it deflects upward or cambers. It is apparent that the dead weight of the beam itself is acting at the time of the prestressing, since, as the beam cambers, the soffit of the beam is no longer in contact with the soffit form, except at the extremities of the beam. From this consideration, it can be concluded that the actual stresses existing in the concrete at any point in the beam, at the time of prestressing, is equal to the algebraic sum of the stresses caused by the prestressing and the dead weight of the beam itself.

The variation in the stresses, along the length of the beam in the extreme top and bottom fibers, for a beam prestressed with straight tendons is also illustrated in Fig. 4-14. If it is assumed that, for the concrete in the beam under consideration, the maximum permissible bottom-fiber compressive stress is +2000 psi and the maximum permissible top-fiber tensile stress is -200 psi—the beam as illustrated is prestressed as highly as possible. Assuming no tensile stresses are to be allowed in the bottom fiber under the total load, it will be seen that 1500 psi or 75% of the total prestressing stresses in the concrete at the midspan of the beam are "reserved" for the superimposed loads and 25% are used in carrying the dead weight of the beam alone. Furthermore, the maximum stresses that limit the capacity of the beam occur at the end of the beam, where there are no dead-load flexural stresses, rather than near the center, where the total load flexural stresses are maximum.

If the tendon were placed in the member on a parabolic curve such that the eccentricity were maximum at midspan of the beam and minimum at the ends of the beam, the stresses in the top and bottom fibers would vary along the length of the beam, as illustrated in Fig. 4-15. It will be seen, from an examination of these stress distributions, that the maximum stresses resulting from prestressing in both the top and bottom fibers occur at midspan of the beam. Furthermore, it is apparent that, by careful selection of the magnitude and eccentricity of the prestressing force, it is possible to eliminate the reduction in the capacity of the beam to withstand a superimposed load due to the dead weight of the beam itself, as was the case in the previous example. This can be explained in terms of the pressure line as follows: The prestressing force can be applied lower at midspan of the beam than at the ends, without exceeding the permissible stresses, since the dead-load moment of the beam is acting in a direc-

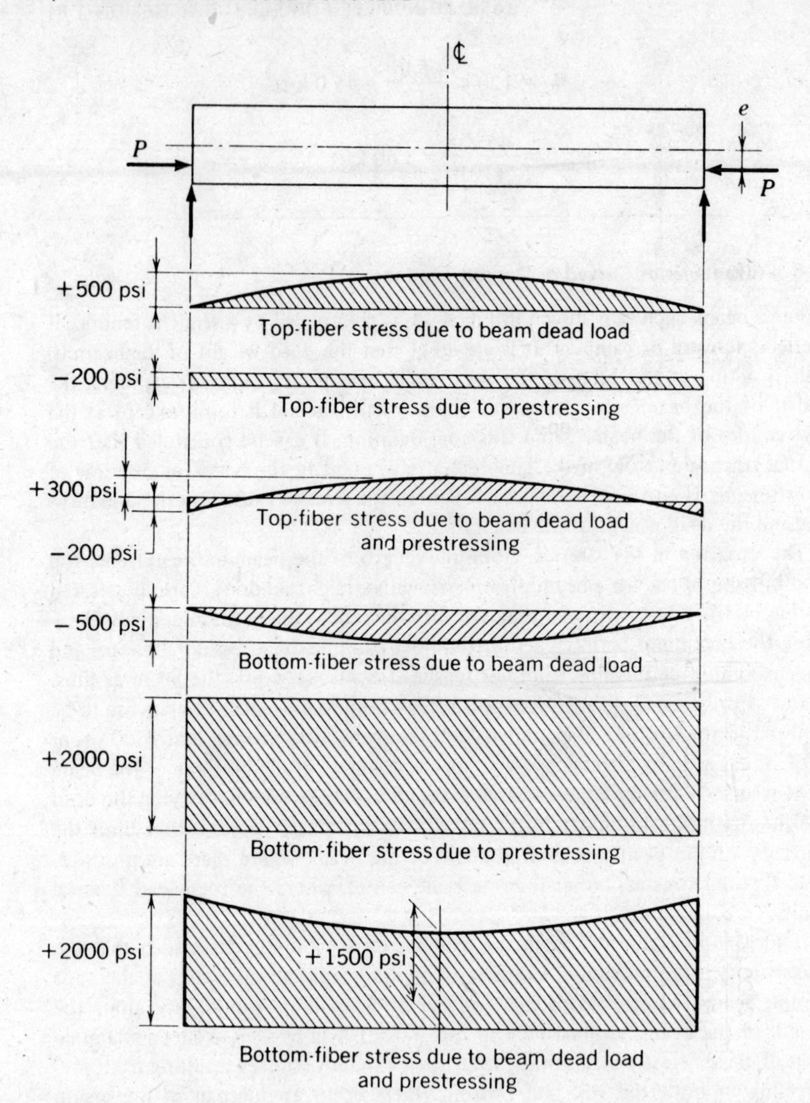

Fig. 4-14 Stress distribution of top and bottom fibers of simple prismatic beam prestressed with a straight tendon.

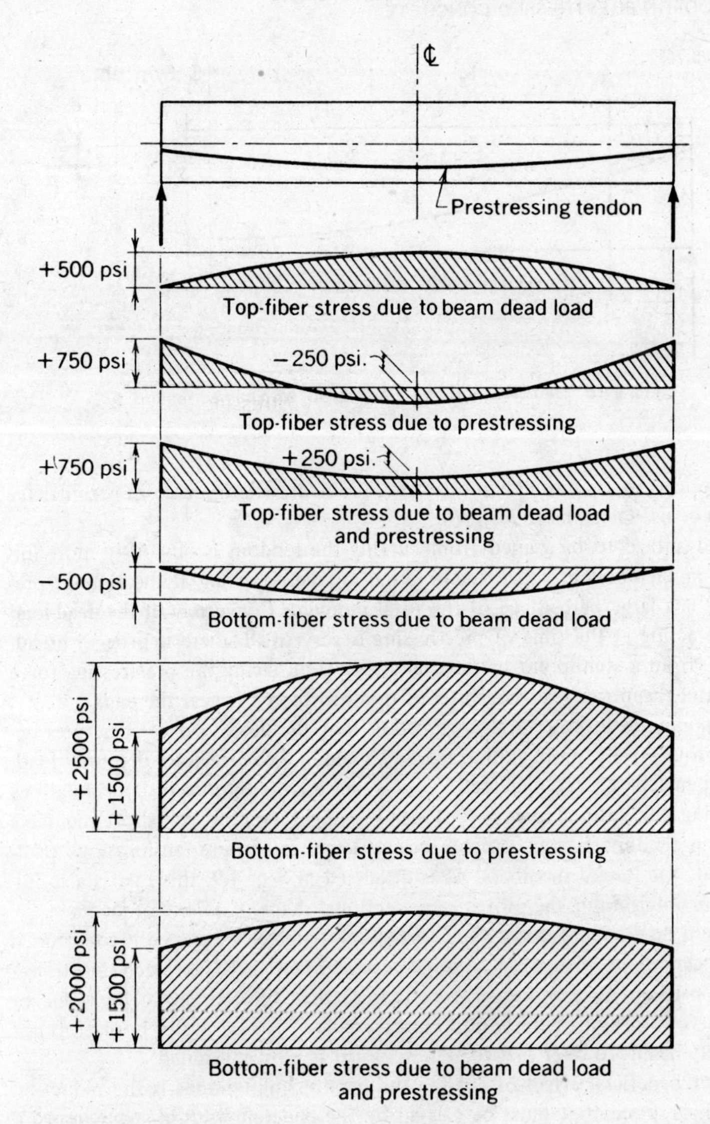

Fig. 4-15 Stress distribution in the top and bottom fibers of a simple prismatic beam prestressed with a curved tendon.

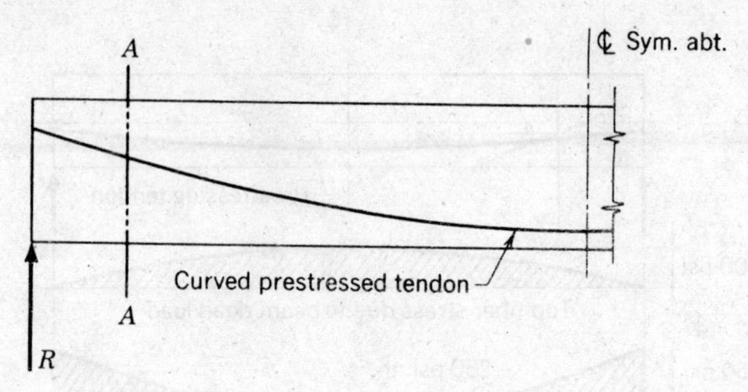

Fig. 4-16 Half-elevation of a simple beam with curved tendon.

tion opposite to that of the prestressing moment. The increase in eccentricity that can be used is equal to M_{DL}/P.

The advantage to be gained from curving the tendons is obviously more important in members in which the external moment existing at the time of prestressing is a large percentage of the total moment. Conversely, if the dead-load moment acting at the time of prestressing is very small, there is little or no advantage (from a standpoint of flexural stresses) in having the prestressing force at a greater eccentricity at the center of the span than it is near the ends.

It is axiomatic in structural engineering that the dead load of structures becomes progressively more important and greater, in respect to the total load, as the span lengths are increased. This is one of the important considerations influencing the normal practice of using straight tendons for short members and using tendons having variable eccentricity, either pre-tensioned or post-tensioned, for longer members. As is discussed in Sec. 4-9, this fact is also important in determining the proper cross-sectional shape of a flexural member.

It should be recognized (*see* Sec. 14-6) that deflected or draped pre-tensioned tendons cannot be placed on smooth curves. They are often placed on trajectories consisting of a series of straight lines that approximate a parabolic or other curve form. When the term "curved tendon" is used in this book, it is not necessarily meant to infer that the tendon must be post-tensioned.

Another beneficial effect of curving the prestressing tendons is the reduction of the shear force that must be carried by the concrete section (*see* Sec. 5-7). This can be illustrated by considering a beam having a curved prestressing tendon that is sloped at an angle of α to the horizontal at the point under consideration, as is illustrated in Fig. 4-16. Inspection of a free-body diagram for this condition, as illustrated in Fig. 4-17, will reveal that the prestressing force P can be resolved into two components: $P \sin \alpha$, which acts vertically upward, and $P \cos \alpha$, which acts horizontally. If the total shear force at the end of the free body

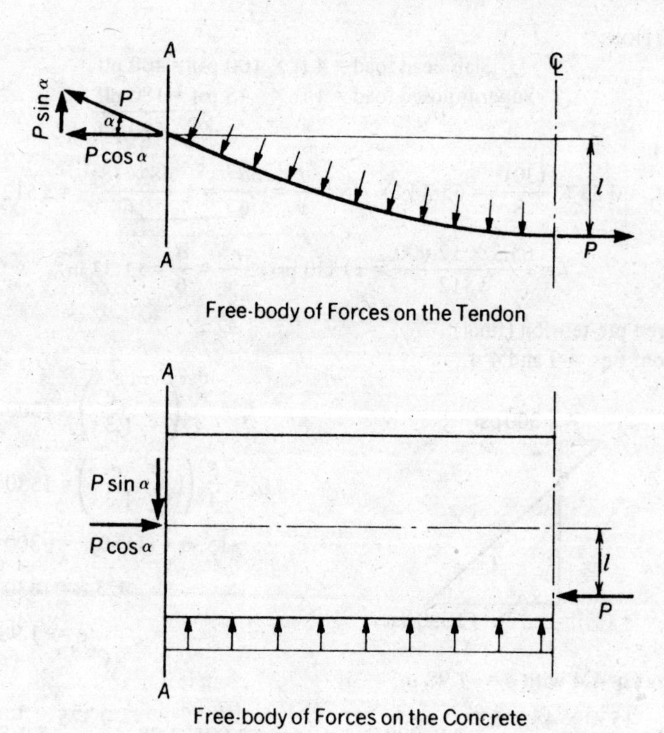

Fig. 4-17 Free-body diagrams for curved tendon and concrete section.

due to external loads is V, the concrete must resist the amount $V - P \sin \alpha$, since the tendon is exerting an upward force equal to $P \sin \alpha$ between the center of the span and the end. If the tendon were not curved, the entire shear force V would have to be carried by the concrete section alone.

ILLUSTRATIVE PROBLEM 4-8 Determine the prestressing force and eccentricity required to prestress a slab, 4 ft wide, 8 in. deep, that is to be used on a span of 30 ft and is to be simply supported. Maximum final compression in the bottom fibers is 2000 psi and maximum allowable, final, top-fiber tensile stress is -300 psi. The slab is solid sand and gravel concrete, and the superimposed load is 45 psf. Minimum cover for the tendons is $1\frac{1}{2}$ in. Assume the tendons have a $\frac{3}{8}$ in. nominal diameter and have a final prestressing force of 11 k each. If the tendons are straight and in one row, how many are required and how thick is the concrete cover? If the tendons can be placed on a parabolic path in such a manner that the cover at midspan is $1\frac{1}{2}$ in., how many tendons are required?

SOLUTION:

$$\text{Slab dead load} = 4 \text{ ft} \times 100 \text{ psf} = 400 \text{ plf}$$
$$\text{Superimposed load} = 4 \text{ ft} \times \ 45 \text{ psf} = \underline{180 \text{ plf}}$$
$$580 \text{ plf}$$

$$M_t = 0.58 \times \frac{(30)^2}{8} = 65.3 \text{ k-ft} \quad S = \frac{I}{y} = \frac{bd^2}{6} = \pm \frac{48 \times (8)^2}{6} = \pm 512 \text{ in.}^3$$

$$f = \frac{65.3 \times 12{,}000}{\pm 512} = \pm 1530 \text{ psi} \quad \frac{r^2}{y} = \frac{d}{6} = \pm 1.33 \text{ in.}$$

Desired pre-tension (final):
From Eqs. 4-3 and 4-4,

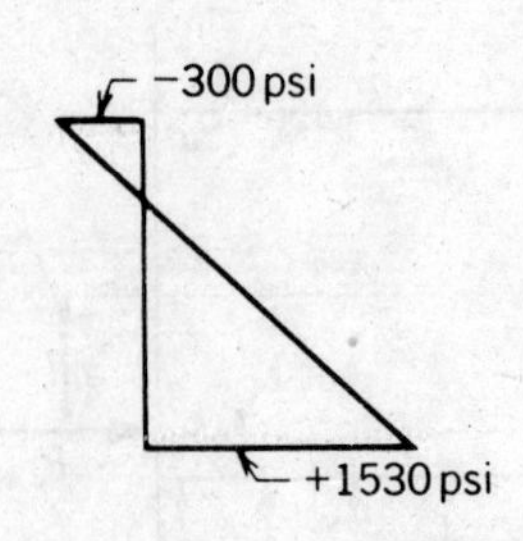

$$f_t = \frac{P}{A}\left(1 + \frac{e}{1.33}\right) = -300 \text{ psi}$$

$$f_b = \frac{P}{A}\left(1 + \frac{e}{-1.33}\right) = 1530 \text{ psi}$$

$$1530 + (1150)e = -300 + 225\, e$$

$$-925\, e = 1830$$

$$e = -1.98 \text{ in.}$$

From Eq. 4-4 with $e = -1.98$ in.

$$P = \frac{1530 \times 48 \times 8}{2.49} = 236{,}000 \text{ lb} \quad \text{Cover} = 4.00 - 1.98 - \frac{0.375}{2} = 1.83 \text{ in.}$$

$$= 22\text{—}\tfrac{3}{8} \text{ in. diameter tendons} \qquad \cong 1\tfrac{13}{16} \text{ in.}$$

If the tendons can be curved parabolically, the top-fiber stress will not limit the eccentricity

$$e = -4.00 + \left(1.50 + \frac{0.375}{2}\right) = -2.31 \text{ in.}$$

$$\frac{P}{384}\left(1 + \frac{-2.31}{-1.33}\right) = 1530 \text{ psi}$$

$$P = 214 \text{ k} \cong 20\text{—}\tfrac{3}{8} \text{ in. diameter}$$

ILLUSTRATIVE PROBLEM 4-9 Assuming the maximum final, top- and bottom-fiber stresses allowable are -170 psi and $+2000$ psi, respectively, determine the maximum superimposed load that can be carried by a 12 in. × 18 in. beam on a simple span of 30 ft. Determine the minimum prestressing force and corresponding eccentricity of the force, if the member is pre-tensioned with straight tendons. Determine the minimum, curved-tendon prestressing force that could be used to carry the same superimposed load if the maximum eccentricity is 6 in. (3 in. from the bottom of the beam to the center of the tendon). What is the ratio of the forces?

SOLUTION:

$$A = 216 \text{ in.}^2 \quad S = \pm \frac{bd^2}{6} = \pm \frac{12 \times (18)^2}{6} = \pm 648 \text{ in.}^3 \quad \frac{r^2}{y} = \pm \frac{d}{6} = \pm 3.00 \text{ in.}$$

Stress distribution due to final pre-tension:

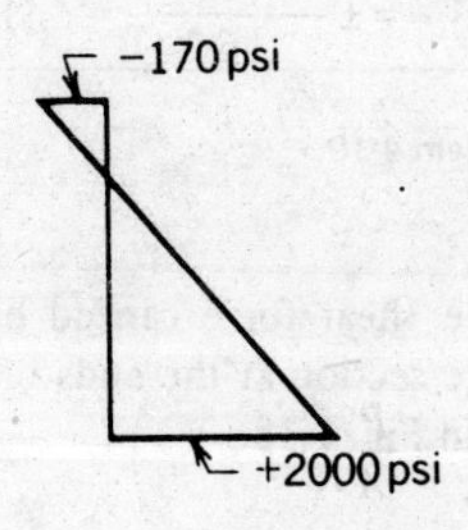

$$f_t = \frac{P}{A}\left(1 + \frac{e}{3.0}\right) = -170$$

$$f_b = \frac{P}{A}\left(1 + \frac{e}{-3.0}\right) = 2000$$

$$-170 + 56.7e = 2000 + 667\,e$$

$$-610.3e = 2170$$

$$e = -3.56 \text{ in.}$$

$$P = \frac{2000 \times 216}{1 + (-3.56/-3.00)} = 198{,}000 \text{ lb. (Straight tendons)}$$

$$M_t = 198 \text{ k} \times \frac{3.0 + 3.56}{12} = 108 \text{ k-ft}$$

$$w_t = \frac{108 \times 8}{(30)^2} = 0.960 \text{ k/ft}$$

$$w_d = 0.225 \text{ k/ft}$$

$$w_{SL} = 0.735 \text{ k/ft}$$

With curved tendon:

$$\text{Stress due to dead load} = \frac{0.225 \times (30)^2 \times 12{,}000}{8 \times \pm 648} = \pm 470 \text{ psi}$$

$$\text{Stress due to superimposed load} = \frac{0.735 \times (30)^2 \times 12{,}000}{8 \times \pm 648} = \frac{\pm 1530 \text{ psi}}{\pm 2000 \text{ psi}}$$

Stress distribution due to prestress:

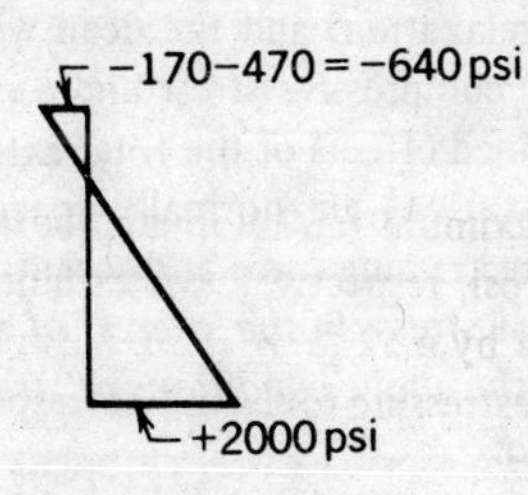

$$f_t = \frac{P}{A}\left(1 + \frac{e}{3.00}\right) = -640 \text{ psi}$$

$$f_b = \frac{P}{A}\left(1 + \frac{e}{-3.00}\right) = 2000 \text{ psi}$$

$$-640 + 213\,e = 2000 + 667\,e$$

$$e = -5.82 \text{ in.}$$

From Eq. 4-4

$$P = \frac{2000 \times 216}{2.94} = 147 \text{ k}$$

$$\text{Ratio of forces} = \frac{198}{147} = 1.35$$

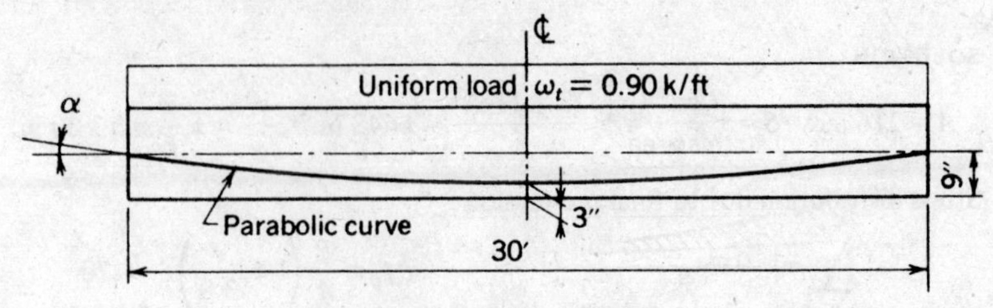

Fig. 4-18 Beam of Problem 4-10.

ILLUSTRATIVE PROBLEM 4-10 Compute the shear force carried by the prestressing tendon of 120 k and by the concrete section at the ends of the beam for the beam and condition of loading shown in Fig. 4-18.

SOLUTION:
At the ends:

$$\sin \alpha \cong \tan \alpha = \frac{4e}{L} = \frac{4 \times 6 \text{ in.}}{12 \times 30 \text{ ft}} = 0.0667$$

$$P \sin \alpha \cong 120 \text{ k} \times 0.0667 \cong 8.00 \text{ k}$$

$$V_t = 0.90 \text{ k/ft} \times 15 \text{ ft} = 13.5 \text{ k}$$

$$V_c = 13.5 \text{ k} - 8.0 \text{ k} = 5.5 \text{ k}$$

4-7 Limiting Eccentricities

It was explained in Sec. 4-6 that a greater eccentricity of the prestressing force can frequently be allowed at the midspan of a beam than at the ends, without exceeding the permissible stresses, due to the dead weight of the beam itself which acts at the time of stressing. The permissible stresses that must be satisfied in a normal, simple beam include a maximum compressive stress in the bottom fiber and a maximum tensile stress in the top fiber under the combined action of the initial prestressing force (before relaxation) and the dead weight of the beam. In addition, a maximum top-fiber compressive stress and a maximum bottom-fiber tensile stress under the combined effects of the total external load and the final prestressing force (after relaxation) are normally specified. For most beams, a number of combinations of prestressing force and eccentricity can be found that will satisfy these conditions of stress. *In the interest of economy, however, the minimum force that satisfies the above conditions of stress at the most highly stressed section is usually selected.*

For the force selected, one can compute maximum and minimum eccentricities that can be used at various locations along the length of the beam without ex-

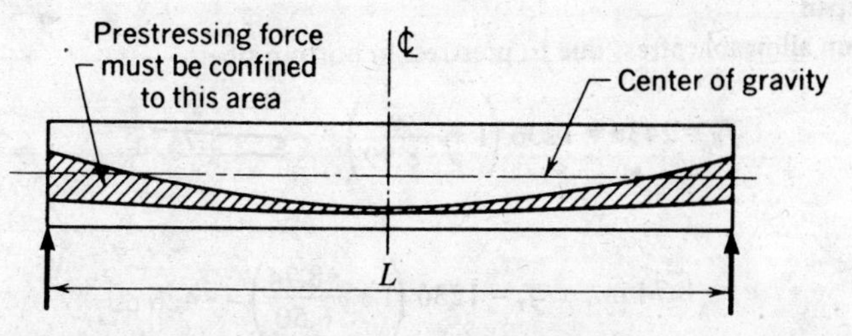

Fig. 4-19 Schematic diagram showing area in which prestressing force must be confined in order to satisfy initial and final stress requirements.

ceeding the permissible stresses enumerated above. Plotting these eccentricities in a schematic elevation of the beam, which has an exaggerated vertical scale, reveals the limiting dimensions in which the center of gravity of the prestressing force must remain in order to satisfy the conditions of allowable stress. Such a schematic diagram is shown in Fig. 4-19, where the area in which the selected prestressing force must be confined is crosshatched. It is generally not necessary to make such a diagram in designing beams subjected to normal loading, since by placing the tendons on parabolic (or near parabolic) curves, the stress conditions can generally be satisfied without difficulty. However, when non-prismatic beams, continuous beams, or beams that have acute and unusual stress conditions are encountered, such diagrams facilitate the design.

ILLUSTRATIVE PROBLEM 4-11 Compute the limits of the eccentricity of the prestressing force of 550 k at the midspan, quarter point, and end, for a simple beam, if the allowable tensile stress is −200 psi and zero in the top and bottom fibers, respectively, and if the maximum allowable compressive stress is 2000 psi and 2200 psi in the bottom and top fibers, respectively. The maximum and minimum external-load stresses (total load and beam dead load alone, respectively) are as follows:

Location	*Max. top (psi)*	*Min. top (psi)*	*Max. bottom (psi)*	*Min. bottom (psi)*
Quarter point	+1350	+453	−1530	−328
Midspan	+1800	+605	−2038	−438

The area of the beam is 445 in.^{2} and r^2/y_b and r^2/y_t are equal to −8.99 and 6.50 in., respectively.

SOLUTION:

$$\frac{P}{A} = \frac{550{,}000}{445} = 1236 \text{ psi}$$

At midspan:
Maximum allowable stress due to prestress in bottom fiber:

$$f_b = 2438 = 1236\left(1 + \frac{e}{-8.99}\right), \quad e = -8.74 \text{ in.}$$

For

$$e = -8.74 \text{ in.}, \quad f_t = 1236\left(1 + \frac{-8.74}{6.50}\right) = -426 \text{ psi}$$

Net top-fiber stress = -426 + 605 = +179 psi > -200 psi ok.

Minimum allowable stress due to prestress in bottom fiber:

$$f_b = 2038 = 1236\left(1 + \frac{e}{-8.99}\right), \quad e = -5.83 \text{ in.}$$

For

$$e = -5.83 \text{ in.}, \quad f_t = 1236\left(1 + \frac{-5.83}{6.50}\right) = +127 \text{ psi}$$

Net maximum top-fiber stress = +127 + 1800 = +1927 < +2200 psi ok.

Summary for midspan:

$$e_{max} = -5.83 \text{ in.}, e_{min} = -8.74 \text{ in.}$$

At quarter point:
Maximum allowable stress due to prestress in bottom fiber:

$$f_b = 2328 = 1236\left(1 + \frac{e}{-8.99}\right), \quad e = -7.94 \text{ in.}$$

For

$$e = -7.94 \text{ in.}, \quad f_t = 1236\left(1 + \frac{-7.94}{6.50}\right) = -274 \text{ psi}$$

Net top-fiber stress = -274 + 453 = +179 psi > -200 psi ok.

Minimum allowable stress due to prestress in bottom fiber:

$$f_b = 1530 = 1236\left(1 + \frac{e}{-8.99}\right), \quad e = -2.14 \text{ in.}$$

For

$$e = -2.14 \text{ in.}, \quad f_t = 1236\left(1 + \frac{-2.14}{6.50}\right) = +829 \text{ psi}$$

Net maximum top-fiber stress = +829 + 1350 = +2179 < +2200 psi ok.

Summary for quarter point:

$$e_{max} = -2.14 \text{ in.}, e_{min} = -7.94 \text{ in.}$$

At the support:
Maximum allowable stress due to prestress in bottom fiber:

$$f_b = 2000 = 1236\left(1 + \frac{e}{-8.99}\right), \quad e = -5.56 \text{ in.}$$

$$e = -5.56 \text{ in.}, \quad f_t = 1236\left(1 + \frac{-5.56}{6.50}\right) = +179 \text{ psi} > -200 \text{ psi ok.}$$

Minimum allowable stress due to prestress in bottom fiber:

$$f_b = 0 = 1236\left(1 + \frac{e}{-8.99}\right), \quad e = +8.99 \text{ in.}$$

For

$$e = +8.99 \text{ in.}, \quad f_t = 1236\left(1 + \frac{8.99}{6.50}\right) = +2945 \text{ psi} > +2200 \text{ psi N.G.}$$

For

$$f_t = +2200 \text{ psi} \quad e = \left(\frac{2200}{1236} - 1\right) 6.50 = +5.06 \text{ in.}$$

For

$$e = +5.06 \text{ in.}, \quad f_b = 1236\left(1 + \frac{5.06}{-8.99}\right) = +540 \text{ psi}$$

Summary at center line of support: $e_{max} = +5.06$ in., $e_{min} = -5.56$ in.

The limits in which the center of gravity of the tendons must fall are shown in Fig. 4-20.

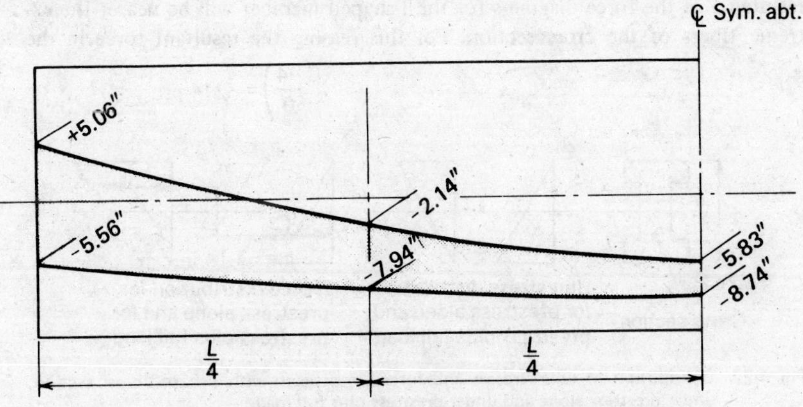

Fig. 4-20 Plot of the limits of the prestressing force for Problem 4-11.

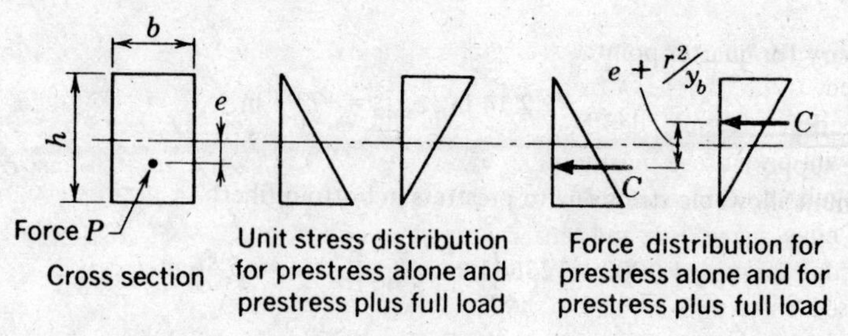

Fig. 4-21 Distribution of unit stresses and forces in a rectangular beam under prestress alone and under prestress plus full load.

4-8 Cross-Section Efficiency

In a rectangular beam the distribution of the unit flexural stresses in the concrete under prestress alone and under total load at the midspan may be as is illustrated in Fig. 4-21. The distribution of the forces in this beam will have the identical shape as the distribution of the unit stresses, and the conversion of the unit stresses to forces can be made by multiplying the unit stresses by the width of the cross section. As has been explained, the total moment to which this member is subjected can be computed by determining the distance between the points of application of the resultant forces in the concrete, under the conditions of prestressing alone and when under full load, and by multiplying this distance by the prestressing force.

Analysis of a beam with an I-shaped cross section, such as is illustrated in Fig. 4-22, will reveal that the distribution of unit stresses varies linearly, as in the case of the rectangular cross section; however due to the variable width of the cross section, the distribution of forces is variable as illustrated. It is apparent that the resultants of the force diagrams for the I-shaped member will be nearer the extreme fibers of the cross section. For this reason, the resultant force in the

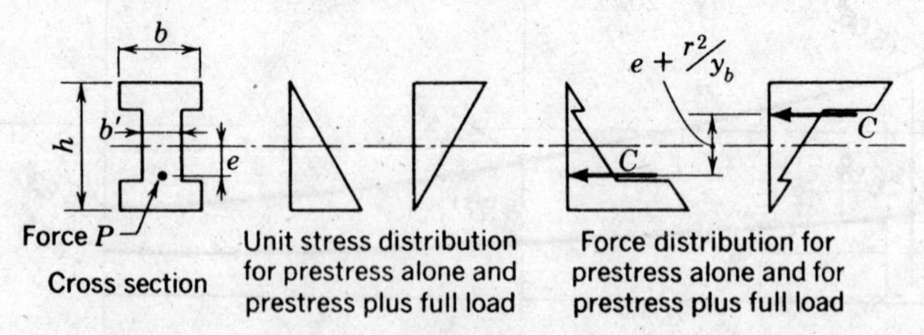

Fig. 4-22 Distribution of unit stresses and forces in a beam with I-shaped cross section, under prestress alone and under prestress plus full load.

I-shaped concrete section moves through a greater distance when external load is applied, to nullify the bottom-fiber prestress, than is the case for a rectangular cross section of equal depth. From this consideration, it is obvious that the I-shape will be more efficient and is capable of withstanding a greater load than a rectangular section of equal depth, providing each section is prestressed with a force of equal magnitude and tensile stresses are not allowed in the section.

This consideration is the primary reason for using I, T, and hollow shapes in prestressed flexural members in which major tensile stresses must be avoided and in which construction depth is of importance and must be minimized. Solid slabs and rectangular beams are economical under some conditions of span, loading, and design criteria, but the more complicated shapes generally result in the minimum quantities of prestressing steel and concrete required to carry a particular load condition, and as a result, they are frequently more economical.

The effect of allowing tensile stresses in the top and bottom fibers is discussed subsequently in Secs. 6-10 and 6-11. Selection of an efficient beam cross section for various loading conditions is discussed in Sec. 4-9.

ILLUSTRATIVE PROBLEM 4-12 Determine the maximum total moments that can be imposed upon the I-shaped and rectangular cross sections in Fig. 4-23 if each is prestressed with a straight tendon having an effective force of 200 k and if tensile stresses are not allowed under any condition of loading.

SOLUTION: Since tensile stresses are not allowed under any conditions, the maximum eccentricity of the prestressing force is r^2/y_t. (*See* Eq. 4-5.) Therefore,

For the I-shape:

$$M_T = 200\ \text{k}\left(\frac{4.22 + 4.22}{12}\right) = 140\ \text{k-ft}$$

For the rectangular shape:

$$M_T = 200\ \text{k}\left(\frac{3.00 + 3.00}{12}\right) = 100\ \text{k-ft}$$

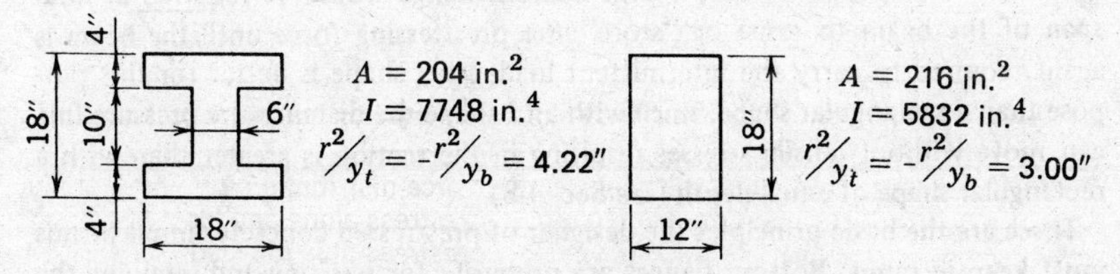

Fig. 4-23 Cross sections compared in Problem 4-12.

4-9 Selection of Beam Cross Section

It has been shown that the location of the pressure line in a prestressed-concrete flexural member changes upon the application of external load. At the end of a member where no moment exists, the pressure line in a simple prestressed-concrete beam is always coincident with the location of the center of gravity of the prestressing force. At the center of the beam, the distance from the center of gravity of the prestressing to the pressure line is equal to the total moment acting at that point divided by the prestressing force. (From Eq. 4-6.)

In order to illustrate the effect of this action on the shape of the optimum concrete section, consider a simple pre-tensioned beam that is prismatic, has straight tendons, and is subjected to a load of such magnitude that the bottom-fiber stress is zero at the midspan. At the end of the beam, the pressure line is coincident with the center of gravity of the prestressing, a condition that remains unchanged despite variations in the external load. Therefore, the optimum section at the end would be a shape that is concentric about the prestressing force, since this shape will result in minimum concrete stresses. At midspan, the pressure line acts above the center of gravity of the section, and, therefore, a top flange is necessary to resist this force. Since the stress in the bottom fibers is zero, no bottom flange is required to resist stress under this condition of loading.

The above example illustrates that, as would be expected, the optimum concrete section is materially influenced by the prestressing force and the loading. If in the above example the prestressing tendons were draped in such a manner that there was little or no eccentricity at the ends of the beam, there would be no need for any shape other than a rectangular section, which is easy to form and is efficient in resisting large, concentric, compressive forces. If the load causing zero stress in the bottom fibers at the midspan of the beam were always present, there would be no need for a large bottom flange near midspan, since the pressure line would always be acting near the top of the section and the concrete in the bottom flange would only serve to protect the prestressing steel from fire and corrosion; therefore, a T section would be efficient. On the other hand, if the load that causes zero stress in the bottom fibers at midspan is an intermittent load, and if this intermittent load is very large in comparison to the dead load of the beam itself, a large bottom flange would be required at midspan of the beam to resist or "store" the prestressing force until the beam is again required to carry the intermittent load. An I shape is better for this purpose than a rectangular shape, since with an I shape the distance the pressure line can move without tensile stresses resulting in the section is greater than with a rectangular shape of equal depth (*see* Sec. 4-8).

These are the basic principles the designer of prestressed concrete simple beams must keep in mind: Bottom flanges are primarily for resisting and retaining the prestressing force until it is needed to resist the external load, at which time the pressure line moves upward; top flanges are needed for fully loaded, flexural members, since the pressure line is in the vicinity of the top flange when the beam is fully loaded, (in addition, amply proportioned top flanges ensure that flexural failures of the brittle type cannot occur, as is discussed in Chapter 5); flanged shapes permit greater distance between the pressure line and the center of gravity of the prestressing force than is allowed by rectangular shapes, and hence smaller prestressing forces are required; and finally, the webs are primarily effective in resisting shear stresses. A complete understanding and appreciation of these functions will assist the designer in the rapid preliminary design of beams, as well as in obtaining economical and efficient designs.

Due to the fact that the dead load of a prestressed member constitutes a small portion of the total load to which it is subjected for short spans and a large portion of the total load for long spans, the use of I-shaped, hollow-rectangular, and solid-rectangular beams is more common for the short-span members, while T-shaped beams are more often used on long spans. (An exception to this is cast-in-place box girder bridge sections which are often used in long-span bridges, both simply supported and continuous. See Chapter 12.)

When straight pre-tensioned tendons are used in applications in which the dead load of the member is large in comparison with the total moment, it is often necessary to supply a large bottom flange to resist the prestressing stresses at the end. In addition, the large bottom flange may be required to assure that the concrete cover for the tendons will be adequate to protect the tendons against corrosion throughout the length of the beam. In such applications, the stress level in the bottom flange at the center of the beam, due to the combined effects of prestressing and dead load, may be relatively low. Due to the smaller area required for post-tensioned tendons, as well as the ease of placing post-tensioned tendons on curved trajectories, the size of the bottom flange of post-tensioned beams are not frequently dictated by the stresses due to prestressing at the ends or by the amount of concrete required to provide adequate concrete cover.

The designer who is experienced in field supervision as well as in the theoretical aspects of prestressed concrete will bear in mind that, although thin webs of 4 or 5 in. in thickness are often theoretically satisfactory with minimum web reinforcement, they often lead to a member in which it is difficult to place and vibrate the concrete. Therefore, honeycomb becomes a real danger. Under normal conditions, 6 in. should be regarded as the minimum web thickness for an I-shaped beam and 7 in. is the preferred minimum thickness if post-tensioning is used.

Extremely narrow, top flanges are dangerous in prestressed concrete, just as they are in structural steel. A top flange can buckle in the same manner as a column if it is of narrow dimensions, unsupported laterally, and too highly stressed. Field experience demonstrates the desirability of a reasonable width of the top flange in order to reduce the transverse flexibility of the girder during handling. This subject is treated in further detail in Sec. 10-5.

The usual ratio between depth of beam to span for simple prestressed-concrete beams varies between from 1 in 16 to 1 in 22, depending upon the conditions of loading, allowable vertical clearance and the type of construction. In lightly loaded, simple T-shaped roof members, the depth-to-span ratio may be as high as 1 in 40. Simply supported cored slabs of prestressed concrete have been successfully used with depth to span ratios of as high as 1 in 40. Solid, continuous, post-tensioned roof slabs with depth-to-span ratios as high as 1 in 45 have given good performance. Excessive deflection and vibration under transient live loads are more likely to be problems in slender members than in deeper ones.

4-10 Effective Beam Cross Section

The most commonly used procedure in prestressed-concrete design is to base the flexural computations in the elastic range upon the section properties of the gross concrete section. The gross section is defined as the concrete section from which the area of steel, or ducts in the case of post-tensioning, has not been deducted and to which the transformed area of the steel has not been added. This procedure is considered to render sufficiently accurate results in the usual applications of prestressed concrete. The change in the stresses that would result by basing the computations on the net or transformed section properties is not normally significant, in view of the fact that concrete is not a completely elastic material. Furthermore, the modulus of elasticity of concrete is not generally known precisely and a value must be assumed in computing transformed section properties. It is important, however, that the designer of prestressed concrete be aware of the nature of the section that is theoretically involved in the various types of construction, since it can be important under special conditions.*

In the case of pre-tensioning, when the prestress is applied the deformation of the concrete is a function of the net section, since the concrete is compressed by the steel, which does not assist the concrete in resiting the prestressing force. The net section is defined as the section that results when the area occupied by the tendons (or ducts in the case of post-tensioning) is deducted from the gross section. Since the pre-tensioned tendons are bonded to the concrete, when there is a change of strain in the concrete at the level of the steel, there must be a corresponding and equal change of strain in the steel. Therefore, when external loads, other than the dead load of the beam, which is acting at the time of prestressing, are applied, the deformation of the member is a function of the transformed section. The transformed section can be defined as the section that results when the area of steel is transformed into an elastically equivalent area of concrete, by multiplying the steel area by the modular ratio and adding this transformed area to the net section at the proper location. In normal pre-tensioned practice the effect of the transformed section is small and little is normally gained by taking these effects into account. The effect of the transformed section will normally be greater in large members with bundled pretensioned tendons (*see*

*Section 18.2.6 of ACI 318 is as follows: "In computing section properties prior to bonding of prestressing tendons, effect of loss of area due to open ducts shall be considered."

Sec. 6-12). However, little is to be gained under normal conditions by including these refinements in the computations.*

In the case of post-tensioning, the deformation of a member is a function of the net section under all conditions of prestressing and external load, until such time as grout is injected into the ducts and allowed to harden and bond the tendons to the concrete section. After bond is established, the deformation of the member is a function of the transformed section. As in the case of pre-tensioning, under normal conditions little is gained by including these effects in the computations.

The use of the net section for the computation of stresses that occur before the bonding of the tendons is required by ACI 318, whereas the use of the transformed section is optional for stresses that occur after bonding.

The net and transformed sections should be used in computing stresses in long-span, post-tensioned girders that have large concentrations of ducts in relatively small bottom flanges. In such a case, the ducts can have a significant influence on the compressive stresses, in the bottom flange, resulting from the prestressing, since the area occupied by the ducts may be a large portion of the total bottom flange area. Additionally, the area of the prestressing steel is generally large and has a significant effect upon the stresses due to superimposed loads under such conditions.

ILLUSTRATIVE PROBLEM 4-13 For the pre-tensioned girder illustrated in Fig. 4-24, compute the stresses in the concrete due to prestressing, based upon the gross- and net-section properties. In addition, compute the concrete stresses, based upon the gross and transformed sections at the center of a 40 ft span, when the externally applied load is 3.13 k/ft. The section properties of the gross section are:

$A = 418.5$ in.^{2}	$I = 44{,}700$ in.^{4}	$w_d = 0.44$ k/ft.
$A_{ps} = 3.20$ in.^{2}	$e = -9.40$ in.	$n = 6$
$y_t = 15.39$ in.	$r^2/y_t = 6.94$ in.	$S_t = 2904$ in.^{3}
$y_b = -14.61$ in.	$r^2/y_b = -7.31$ in.	$S_b = -3060$ in.^{3}
$P = 440$ k		

SOLUTION:

Stresses due to prestressing based upon gross section:

$$f_t = \frac{440{,}000}{418.5}\left(1 + \frac{-9.40}{6.94}\right) = -373 \text{ psi}$$

$$f_b = \frac{440{,}000}{418.5}\left(1 + \frac{-9.40}{-7.31}\right) = +2403 \text{ psi}$$

*See Section 7-2 for methods of computing section properties.

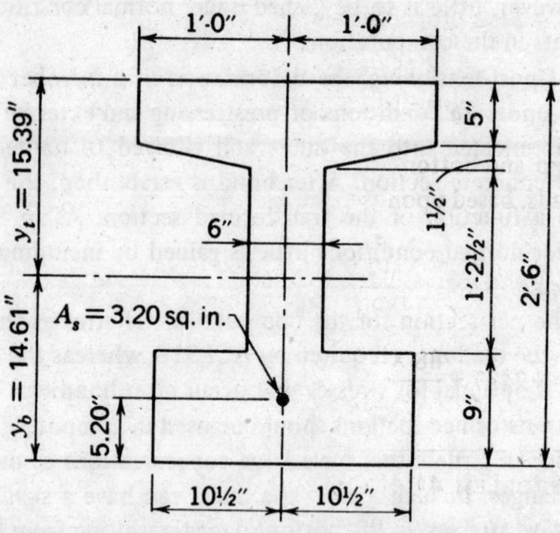

Fig. 4-24 Cross section used to demonstrate the effect of the transformed and net beam cross sections as compared to gross cross section in Problem 4-13.

The section properties for the net section are:

$$y_t = 15.32 \text{ in.} \qquad r^2/y_t = 6.98 \text{ in.} \qquad S_t = 2899 \text{ in.}^3$$

$$y_b = -14.68 \text{ in.} \qquad r^2/y_b = -7.29 \text{ in.} \qquad S_b = -3025 \text{ in.}^3$$

$$e = -14.68 + 5.20 = -9.48 \text{ in.}$$

Stresses due to prestressing based upon the net section:

$$f_t = \frac{440{,}000}{415.3}\left(1 + \frac{-9.48}{6.98}\right) = -379 \text{ psi}$$

$$f_b = \frac{440{,}000}{415.3}\left(1 + \frac{-6.98}{-7.29}\right) = +2437 \text{ psi}$$

Stresses due to prestressing, dead and superimposed loads, based upon the gross section, are computed as follows:

$$M_D + M_{SL} = 0.44 \times \frac{(40)^2}{8} + 3.13 \times \frac{(40)^2}{8} = 88 + 626 = 714 \text{ k-ft}$$

$$f_t = \frac{714 \times 12{,}000}{2904} - 373 = +2577 \text{ psi}$$

$$f_b = \frac{714 \times 12{,}000}{-3060} + 2400 = -400 \text{ psi}$$

The transformed section properties are:

$$y_t = 15.74 \text{ in.}, S_t = 2926 \text{ in.}^3$$
$$y_b = -14.26 \text{ in.}, S_b = -3230 \text{ in.}^3$$

Stresses in the top and bottom fibers, respectively, due to prestressing, dead and superimposed loads, based upon the net and transformed sections are

$$f_t = -379 + \frac{88 \times 12{,}000}{2899} + \frac{626 \times 12{,}000}{2926} = +2553 \text{ psi}$$

$$f_b = +2437 + \frac{88 \times 12{,}000}{-3025} + \frac{626 \times 12{,}000}{-3230} = -238 \text{ psi}$$

ILLUSTRATIVE PROBLEM 4-14 Compute the stresses due to prestressing in the top and bottom fibers for the post-tensioned girder of Fig. 4-25 based upon an effective prestressing force of 2380 k located 5.3 in. above the soffit based upon: (1) the gross section properties and (2) the net section properties if the area of the ducts is 39.0 sq in. Also, determine the allowable superimposed live load on the girder based upon: (3) the gross section properties and (4) the transformed section properties if $nA_{ps} = 83.5 \text{ in.}^2$. The design span is 200 ft. What is the ratio between the allowable superimposed live loads of (3) and (4)? (Note: When the net and transformed section properties are used, the net section properties are used for stresses due to prestressing and dead load because the post-tensioned tendons do not become bonded until they are stressed. The transformed section properties are used for computing stresses after bonding.)

Gross section properties

$A = 2051 \text{ in.}^2$ $\quad I = 3.735{,}950 \text{ in.}^4$

$y_t = 53.7 \text{ in.}$ $\quad S_t = 69{,}700 \text{ in.}^3$ $\quad r^2/y_t = 34.0 \text{ in.}$

$y_b = -69.3 \text{ in.}$ $\quad S_b = -54{,}000 \text{ in.}^3$ $\quad r^2/y_b = -26.3 \text{ in.}$

Net section properties

$A = 2012 \text{ in.}^2$ $\quad I = 3{,}572{,}900 \text{ in.}^4$

$y_t = 52.5 \text{ in.}$ $\quad S_t = 68{,}000 \text{ in.}^3$ $\quad r^2/y_t = 33.8 \text{ in.}$

$y_b = -70.5 \text{ in.}$ $\quad S_b = -50{,}750 \text{ in.}^3$ $\quad r^2/y_b = -25.2 \text{ in.}$

Transformed section properties

$A = 2096 \text{ in.}^2$ $\quad I = 3{,}915{,}500 \text{ in.}^4$

$y_t = 55.0 \text{ in.}$ $\quad S_t = 71{,}300 \text{ in.}^3$

$y_b = -68.0 \text{ in.}$ $\quad S_b = -57{,}600 \text{ in.}^3$

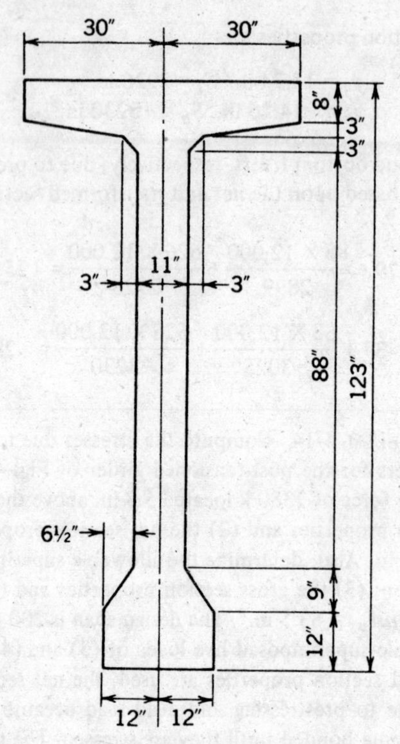

Fig. 4-25 Girder cross section for Problem 4-14.

SOLUTION:

(1) For the gross section: $f_t = \dfrac{2380}{2051}\left(1 + \dfrac{-64.0}{34.0}\right) = -1024 \text{ psi}$

$$f_b = \frac{2380}{2051}\left(1 + \frac{-64.0}{-26.3}\right) = +3984 \text{ psi}$$

(2) For the net section: $f_t = \dfrac{2380}{2012}\left(1 + \dfrac{-65.2}{33.8}\right) = -1099 \text{ psi}$

$$f_b = \frac{2380}{2012}\left(1 + \frac{-65.2}{-25.2}\right) = +4243 \text{ psi}$$

(3) For the gross section: $w_d = 2.14$ klf $\quad M_d = 10{,}700$ k-ft.

$$f_{dt} = \frac{10{,}700 \times 12{,}000}{69{,}700} = +1842 \text{ psi}$$

$$f_{db} = \frac{10{,}700 \times 12{,}000}{-54{,}000} = -2378 \text{ psi}$$

Final top fiber stress = +1842 - 1024 = +818 psi

Final bottom fiber stress = +3984 - 2378 = +1606 psi

$$w_a = \frac{1606 \times 54{,}000}{12{,}000 \times 5000} = 1.44 \text{ klf}$$

(4) For the net and transformed sections:

$$f_{dt} = \frac{10{,}700 \times 12{,}000}{68{,}000} = +1888 \text{ psi}$$

$$f_{db} = \frac{10{,}700 \times 12{,}000}{-50{,}750} = -2530 \text{ psi}$$

Final top fiber stress = +1888 - 1099 = +789 psi

Final bottom fiber stress = +4243 - 2530 = +1713 psi

$$w_a = \frac{1713 \times 57{,}600}{12{,}000 \times 5000} = 1.64 \text{ klf}$$

(5) Ratio = 1.14 (more for 4).

4-11 Variation in Steel Stress

Since the prestressing steel is never located at the extreme fiber of a prestressed beam, but is at some distance from the surface of the concrete, the maximum change in concrete stress that can normally be expected to occur at the level of the center of gravity of the steel is approximately 70 to 80% of the bottom-fiber stress that results from superimposed loads. With concrete that has a cylinder strength of 5000 psi, the stress change in the concrete at the level of the steel could be expected to be of the order of 1500 psi. The modular ratio between the prestressing steel and the concrete can be assumed to be 6 for loads of short duration. As a result, the application of the short-duration, superimposed load would cause an increase in steel stress of approximately 9000 psi, providing the steel and the concrete were adequately bonded. If the steel is not bonded to the concrete, but is anchored at the ends of the member only, the increase in steel stress resulting from the application of the superimposed load would be less than 9000 psi, since the steel can slip in the ducts. The increase in steel stress in unbonded tendons tends to be proportional to the average change in the concrete stress at the level of the steel.

It should be noted that the increase in stress of 9000 psi, which results from the application of the superimposed load, is only about 7% of the final stress normally employed in wire or strand tendons and about 11% of the final stress normally employed in bar tendons. The reduction in the stress in the prestressing steel due to the relaxation of the steel, shrinkage of the concrete, and the creep of the concrete is of the order of 10 to 30% under average conditions (*see* Sec. 6-2). Hence, the stress that exists in the tendon under the superimposed load after all of the losses of prestress have taken place is not as high as the initial stress in the steel.

The small variation in steel stress that occurs in a normal prestressed member subjected to frequent applications of the design load is responsible for the high resistance to fatigue failure that is associated with this material (*see* Sec. 9-6).

ILLUSTRATIVE PROBLEM 4-15 Compute the increase in the stress in the steel at the midspan of the beam in Problem 4-13. (Transformed section properties are to be used.)

SOLUTION: The concrete stress at the level of the steel due to the external load of 3.13 k/ft is

$$y_{c.g.s.} = -14.26 \text{ in.} + 5.20 \text{ in.} = -9.06 \text{ in.}$$

$$f_c = \frac{626 \times 12{,}000}{46{,}027} \times -9.06 = -1479 \text{ psi}$$

The increase in steel stress due to the superimposed load is

$$\Delta f_s = nf_c = 6 \times (-1485) = -8911 \text{ psi (Tension)}$$

CHAPTER 4—PROBLEMS

1. The double-tee slab shown in Fig. 4-26 has an area of 180 sq in., a moment of inertia of 2860 in.^{4}, and the distance from the top fiber to the centroid of the cross section is 4.00 in. Assume the concrete weighs 150 pcf and the

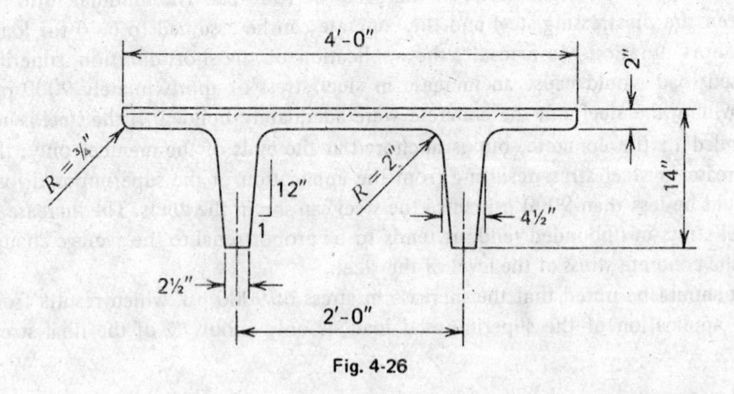

Fig. 4-26

member is to be used on a simple span of 24.0 ft. If pre-tensioned with one tendon in each stem with an initial force of 16,100 lb each, located 2 in. above the bottom fiber, determine the initial stresses due to prestressing and dead load at the support and at midspan. If the loss of prestress is 20% of the initial force, determine the maximum superimposed service load that can be imposed on the member if the allowable bottom fiber tensile stress under service load is zero, 200, 400 and 800 psi.

SOLUTION:

$$\frac{r^2}{y_t} = 3.97 \text{ in.} \qquad \frac{r^2}{y_b} = -1.59 \text{ in.}$$

Initial prestressing stresses:

$$\text{top} = \frac{2 \times 16{,}100}{180}\left(1 + \frac{-8.00}{3.97}\right) = -182 \text{ psi}$$

$$\text{bottom} = \frac{2 \times 16{,}100}{180}\left(1 + \frac{-8.00}{-1.59}\right) = +1079 \text{ psi}$$

Dead load:

$$w_d = \frac{180}{144} \times 150 = 187.5 \text{ p/f}$$

$$M_d = \frac{187.5 \times 24^2}{8} = 13{,}500 \text{ lb-ft}$$

Dead load stresses:

$$\text{top} = \frac{13{,}500 \times 12 \times 4}{2860} = 227 \text{ psi}$$

$$\text{bottom} = \frac{13{,}500 \times 12 \times -10}{2860} = -566 \text{ psi}$$

Therefore, initial stresses at support:

$$\text{top fiber} = -182 \text{ psi}$$

$$\text{bottom fiber} = +1079 \text{ psi}$$

Initial stresses at midspan:

$$\text{top fiber} = -182 + 227 = +45 \text{ psi}$$

$$\text{bottom fiber} = +1079 - 566 = +513 \text{ psi}$$

Final stresses at support:

top fiber: $0.80 \times -182 = -146$ psi

bottom fiber: $0.80 \times 1079 = +863$ psi

Final stresses at midspan:

top fiber $= -146 + 227 = +81$ psi

bottom fiber $= +863 - 566 = +297$ psi

Allowable bottom fiber stress at service load (psi)	*f_b due to superimposed service load (psi)*	*Superimposed service load (plf)*	*(psf)*	*Top fiber stress due to superimposed service load (psi)*	*Top fiber due to total service load (psi)*
0	-297	98	24	+119	+200
-200	-497	165	41	+199	+280
-400	-697	231	58	+279	+360
-800	-1097	363	91	+439	+520

Note: Design loads, rather than service loads, can be shown to govern the allowable loading on this member when the greater tensile stresses are permitted. See Chapter 5.

2. For the condition of prestressing alone, as well as for the four allowable bottom fiber stresses investigated in Problem 1, construct the locations of the pressure lines. Use a horizontal scale of 1.0 in. = 4.0 ft and a vertical scale of 1.0 in. = 0.60 ft.

SOLUTION:

$$d = \frac{M}{P} = \frac{f_b I}{y_b P} = \frac{2860 f_b}{-10 \times 12 \times 0.80 \times 32{,}200} = -0.000925 f_b \text{ feet}$$

Where f_b is the bottom fiber stress due to the applied load only.

f_b (psi)	Midspan d (ft)
-863	0.799
-1063	0.983
-1263	1.169
-1663	1.539

The results are plotted below.

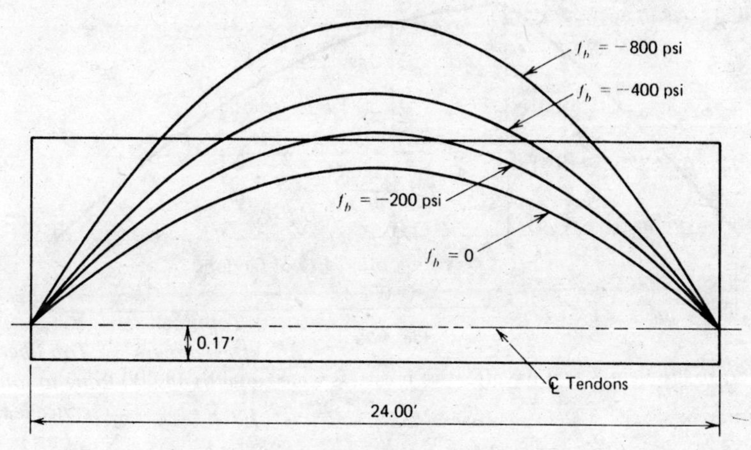

Fig. 4-27 Pressure line locations for Problem 2.

3. For the beam of Problem 1 loaded as shown in Fig. 4-28, compute and plot the location of the pressure line if the effective prestressing force is 24,000 lb. Include the dead load of the beam.

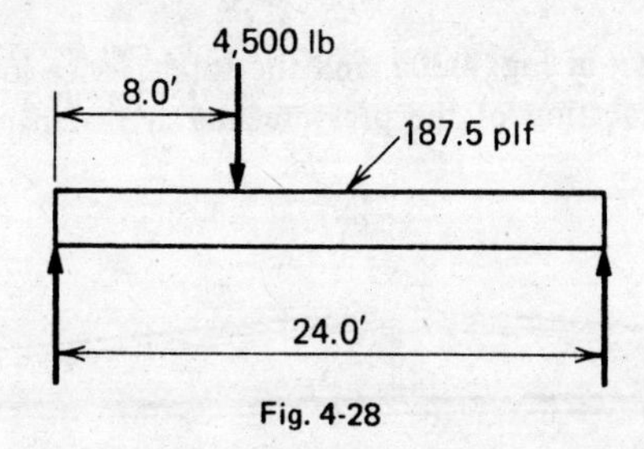

Fig. 4-28

SOLUTION:

$$M \text{ conc. max.} = \frac{4500 \times 24}{4.5} = 24{,}000 \text{ ft-lb}$$

$$M_{\text{D.L. max}} = \frac{187.5 \times 24^2}{8} = 13{,}500 \text{ ft-lb}$$

Values of $d = M/P$ (ft)

Distance from left support (ft)	0	6	8	12	18	24
For concentrated load	0	0.750	1.000	0.750	0.375	0
For uniformly distributed load	0	0.422	0.500	0.563	0.422	0
Total P.L. movement	0	1.172	1.500	1.313	0.797	0

The results are plotted in Fig. 4-29.

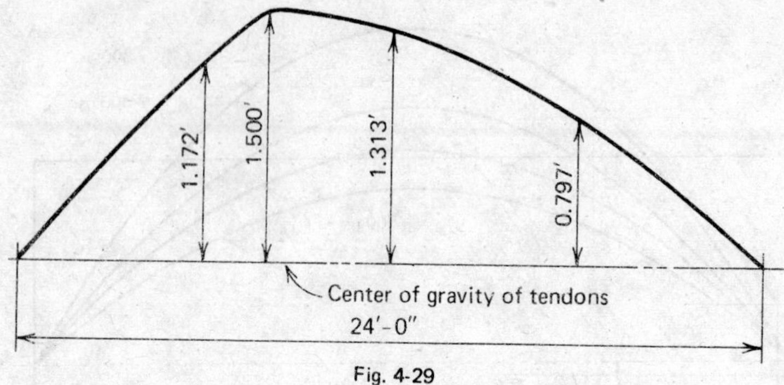

Fig. 4-29

4. Solve Problem 3 as if the effective prestress were equal to 48,000 lb.

SOLUTION:

Movement of pressure line is half as much as for an effective prestress of 24,000 lb.

Distance from left support (ft)	0	6	8	12	18	24
Total P.L. movement	0	0.586	0.750	0.657	0.399	0

5. For the beam shown in Fig. 4-30a, and the total service loads shown in Fig. 4-30(b), plot the location of the pressure line at midspan and the quarter points.

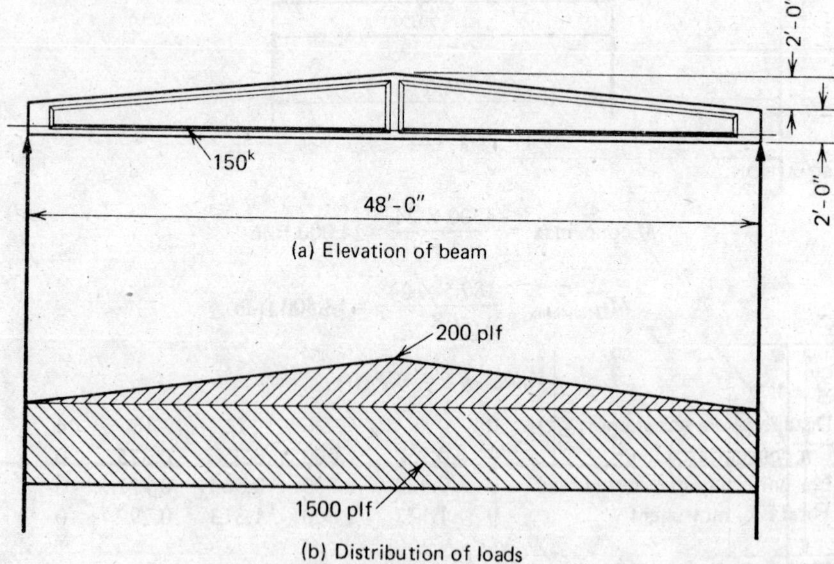

Fig. 4-30 Beam for Problem 5.

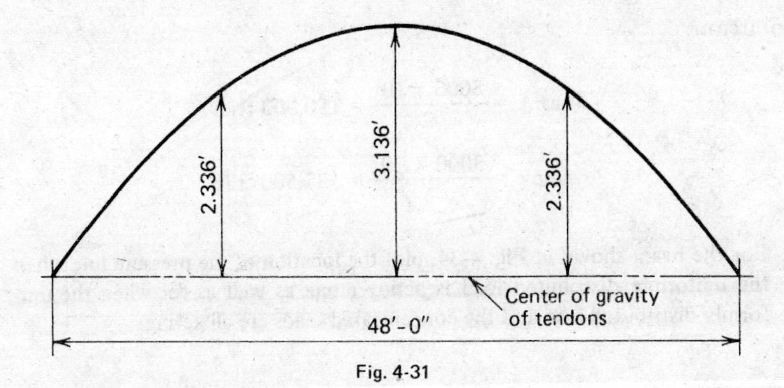

Fig. 4-31

SOLUTION: The solution is shown plotted in Fig. 4-31.

6. For the beam in Fig. 4-32, plot the location of the pressure line for a prestressing force of 100,000 lb. if the beam is subjected to a uniform service load of 3000 plf. Use scales of 1 in. = 5.00 ft. and 1 in. = 1.00 ft. horizontally and vertically, respectively.

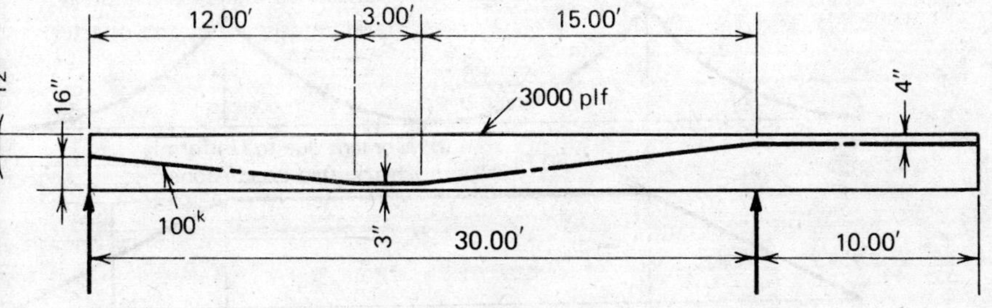

Fig. 4-32

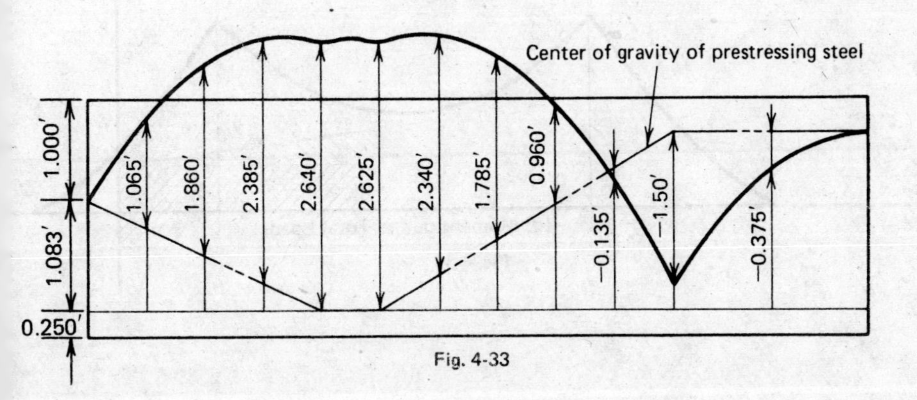

Fig. 4-33

SOLUTION:

$$M \text{ cant.} = \frac{3000 \times 10^2}{2} = 150{,}000 \text{ ft-lb}$$

$$M \text{ simple} = \frac{3000 \times 30^2}{8} = 337{,}500 \text{ ft-lb}$$

7. For the beam shown in Fig. 4-34, plot the location of the pressure line when the uniformly distributed load is acting alone as well as for when the uniformly distributed load and the concentrated loads are all acting.

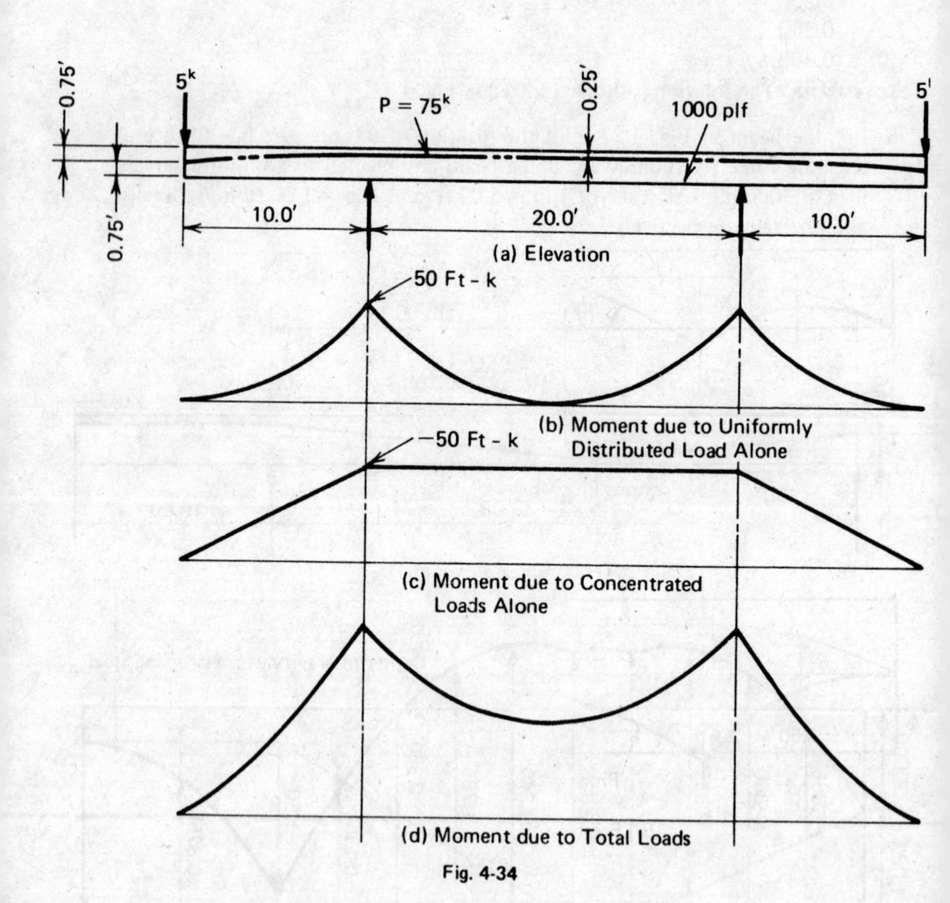

Fig. 4-34

SOLUTION:

	Eccentricity of P.L. (in.)		
Point	*Prestress only*	*Prestress and uniform load*	*Prestress and total load*
Left End	0	0	0
Center Cantilever	4.50	+2.50	-1.50
Support	6.00	-2.00	-10.00
0.05 L	6.00	-0.48	-8.48
0.10 L	6.00	0.88	-7.12
0.15 L	6.00	2.08	-5.92
0.20 L	6.00	3.12	-4.88
0.25 L	6.00	4.00	-4.00
0.30 L	6.00	4.72	-3.28
0.35 L	6.00	5.28	-2.72
0.40 L	6.00	5.68	-2.32
0.45 L	6.00	5.92	-2.08
0.50 L	6.00	6.00	-2.00

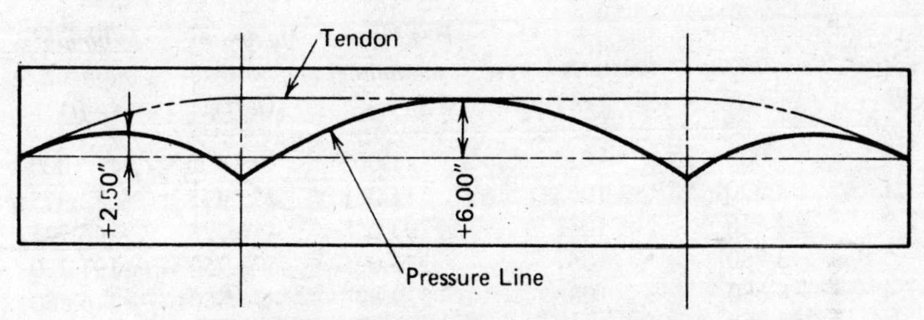

(a) Prestress Plus Uniformly Distributed Load

Fig. 4-35a

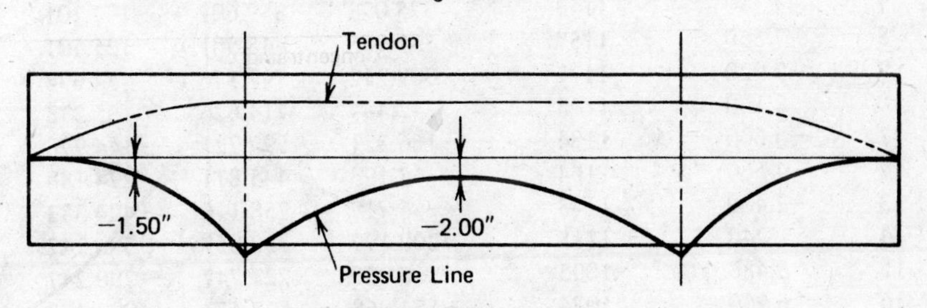

(b) Prestress Plus Total Load

Fig. 4-35b

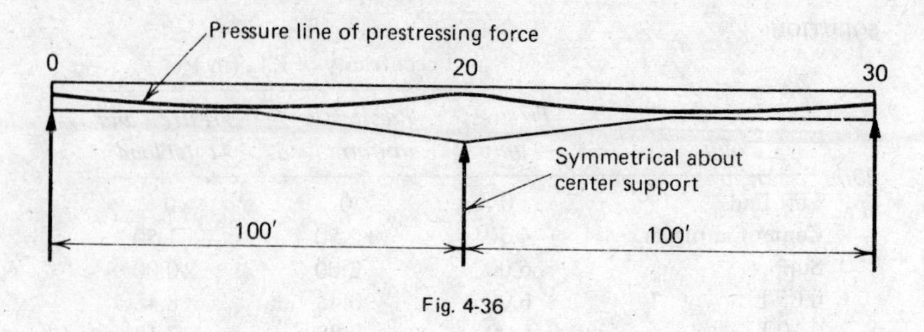

Fig. 4-36

8. The two-span continuous beam in Fig. 4-36 has a variable prestressing force. The pressure line of the prestressing force at the 20th points, as well as its eccentricity are shown in the table below. Also given in the table are the dead load and maximum and minimum live load moments. Tabulate the location of the pressure line for the beam when under dead load alone as well as when under dead plus maximum and dead plus minimum live loads.

20th Pt	*Eccentricity (In.)*	*Prestress Force (k)*	*Dead load moment (k-ft)*	*Maximum moment (k-ft)*	*Minimum moment (k-ft)*
0	.000	1012	.000	.000	.000
1	-1.320	1033	111.256	225.475	82.975
2	-2.520	1047	203.288	416.725	146.725
3	-3.780	1067	276.095	573.750	191.250
4	-4.680	1088	329.677	696.550	216.550
5	-5.700	1102	364.034	785.125	222.625
6	-6.480	1122	379.166	839.475	209.475
7	-7.140	1143	375.073	859.601	177.101
8	-7.560	1157	351.755	845.501	125.501
9	-7.920	1177	309.212	797.176	54.676
10	-8.160	1198	247.444	714.626	-35.373
11	-6.600	1184	166.321	597.721	-144.778
12	-4.560	1164	65.193	445.811	-274.188
13	-2.400	1143	-56.719	258.116	-424.383
14	.240	1115	-200.197	33.857	-596.142
15	3.480	1095	-366.020	-227.747	-790.247
16	6.960	1074	-554.968	-527.477	-1007.477
17	10.680	1047	-767.821	-866.112	-1346.903
18	14.760	1026	-1005.359	-1244.432	-1753.505
19	18.720	978	-1268.362	-1663.217	-2200.572
20	21.240	916	-1557.611	-2123.247	-2688.884

SOLUTION:

Pressure Line Eccentrically = in inches

20th Pt.	*Prestress only*	*Prestress plus dead load*	*Prestress plus dead load moment plus maximum moment*	*Prestress plus dead load moment plus minimum moment*
0	.000	.000	.000	.000
1	-1.320	-.027	1.299	-.355
2	-2.520	-.189	2.256	-.838
3	-3.780	-.673	2.673	-1.628
4	-4.680	-1.042	3.003	-2.291
5	-5.700	-1.734	2.850	-3.275
6	-6.480	-2.422	2.499	-4.238
7	-7.140	-3.199	1.885	-5.279
8	-7.560	-3.909	1.210	-6.257
9	-7.920	-4.764	.208	-7.361
10	-8.160	-5.678	-1.000	-8.513
11	-6.600	-4.910	-.540	-8.065
12	-4.560	-3.884	.037	-7.385
13	-2.400	-2.991	.311	-6.853
14	.240	-1.909	.606	-6.174
15	3.480	-.526	.986	-5.178
16	6.960	.764	1.068	-4.294
17	10.680	1.885	.755	-4.758
18	14.760	3.007	.207	-5.750
19	18.720	3.164	-1.684	-8.282
20	21.240	.842	-6.572	-13.987

9. Compare the maximum uniformly distributed total load the beam in Fig. 4-37 can withstand on a span of 70 ft if $f'_{ci} = 4000$ psi and $f'_c = 5000$ psi. Make the determination for the tendons being straight as well as curved. Use the stresses permitted by ACI 318 (*see* Sec. 3-15). Assume the loss of prestress is 20% and the curved tendons can be placed with their center of gravity as low as 3.25 in. from the bottom of the beam.

$$\text{Weight} = 488 \text{ p/f}$$
$$\text{Area} = 468 \text{ sq in.}$$
$$I = 94{,}184 \text{ in.}^4$$
$$y_t = 22.54 \text{ in.}$$
$$y_b = -19.46 \text{ in.}$$
$$S_t = 4178 \text{ in.}^3$$
$$S_b = -4840 \text{ in.}^3$$

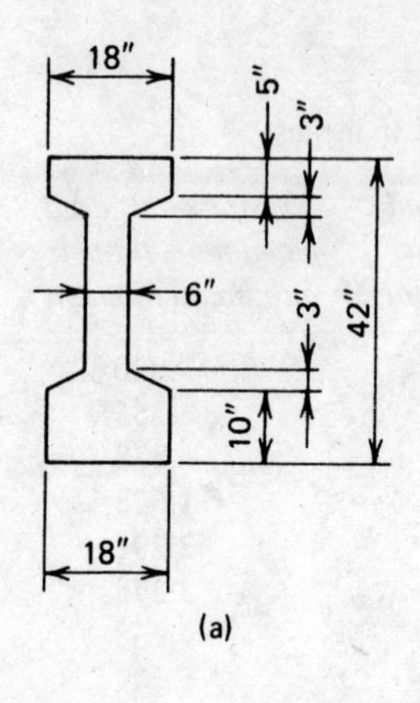

(a)

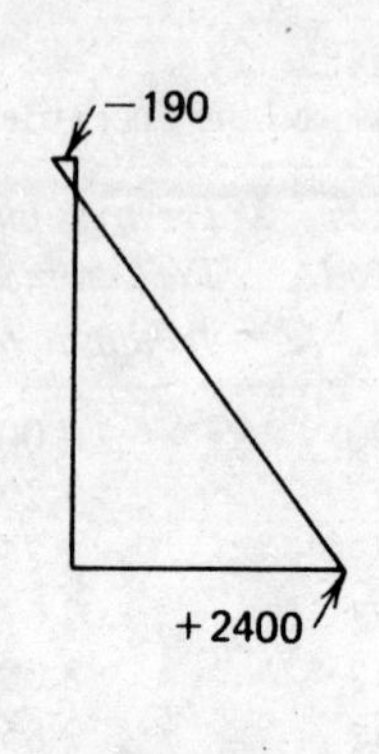

(b) Initial Stresses Straight Tendon

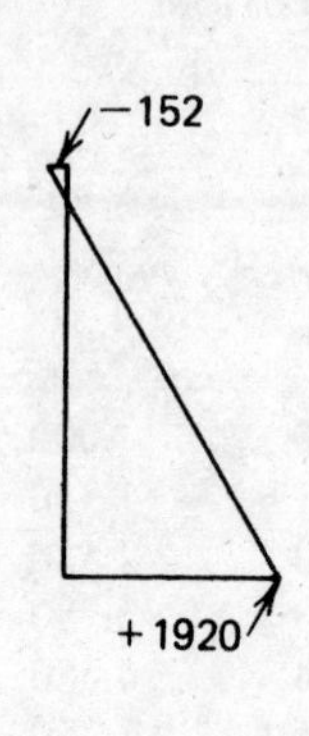

(c) Final Stresses Straight Tendon

Fig. 4-37

SOLUTION:

Allowable initial stresses = 2400 psi compression and - 190 psi tension. Allowable final stresses = 2250 psi compression and - 848 psi tension.

(a) Straight tendon

Initial stresses: (*See* Fig. 4-37(b)).

$$\frac{P}{A} = 2590 \times \frac{22.54}{42} - 190 = 1200 \text{ psi}, \qquad \frac{r^2}{y_b} = \frac{94{,}184}{468 \times -19.46} = -10.34 \text{ in.}$$

$$e = \left(\frac{2400}{1200} - 1\right)(-10.34) = -10.34 \text{ in.} < -19.46 + 3.25 \text{ in.} = -16.21 \text{ in. ok}$$

$$P = 1200 \times 468 = 561.6^{k}$$

Capacity limitations: (*See* Fig. 4-37(c)).

$$\text{By top fiber } M = \frac{(2250 + 152)}{(12{,}000)}\frac{(94{,}184)}{(22.54)} = 836 \text{ k-ft}$$

$$\text{By bottom fiber } M = \frac{(1920 + 848)}{(12{,}000)}\frac{(94{,}184)}{(19.46)} = 1116 \text{ k-ft}$$

$$w = \frac{8 \times 836}{70^2} = 1{,}364 \text{ plf}$$

(b) Curved tendon

Initial stresses:

$$M_D = 488 \times \frac{70^2}{8} = 298.9 \text{ k-ft}$$

Stresses due to dead load at midspan:

$$f_t = 858 \text{ psi}, \quad f_b = -741 \text{ psi}$$

$$d = \frac{M_D}{P} = \frac{298.9}{561.6} \times 12 = 6.39 \text{ in.}$$

$$e = -10.34 \text{ in.} - 6.39 \text{ in.} = -16.73 \text{ in.}$$

$$e_{\max} = -19.46 + 3.25 = -16.21 \text{ in. use}$$

$$2400 = \frac{P}{468}\left(1 + \frac{16.21}{10.34}\right), \quad P = 437 \text{ kips}$$

$$f_t = \frac{437{,}000}{468}\left(1 - \frac{16.21}{8.93}\right) = -761 \text{ psi}$$

$$f_b = \frac{437{,}000}{468}\left(1 + \frac{16.21}{10.34}\right) = +2398 \text{ psi}$$

Net initial top fiber stress = −761 + 858 = +97 psi

Net initial bottom fiber stress = +2398 − 741 = 1657 psi

Final prestressing stress = −609 psi and 1918 psi for top and bottom fibers, respectively.

Capacity limitations:

$$\text{By top fiber } M = \frac{(2250 + 609)\,4178}{12{,}000} = 995 \text{ k-ft}$$

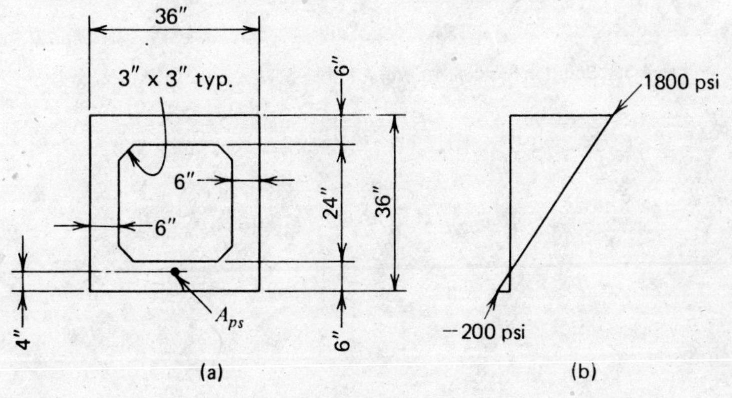

Fig. 4-38

$$\text{By bottom fiber } M = \frac{(1918 + 848)\,4840}{12{,}000} = 1116 \text{ k-ft}$$

$$w = \frac{8 \times 995}{70^2} = 1{,}624 \text{ plf}$$

$$\text{Ratio} = 1.19$$

10. For the hollow-box girder shown in Fig. 4-38(a), plot the distribution of force in the section for the stress distribution indicated in Fig. 4-38(b).

SOLUTION:

y	f	b	b'	P	P'
36	1800	36	–	64.8^k	–
30	1467	36	18	52.8^k	26.4^k
27	1300	12	–	15.6^k	–
9	300	12	–	3.6^k	–
6	133	36	18	4.79	2.39
0	-200	36	–	-7.20	–

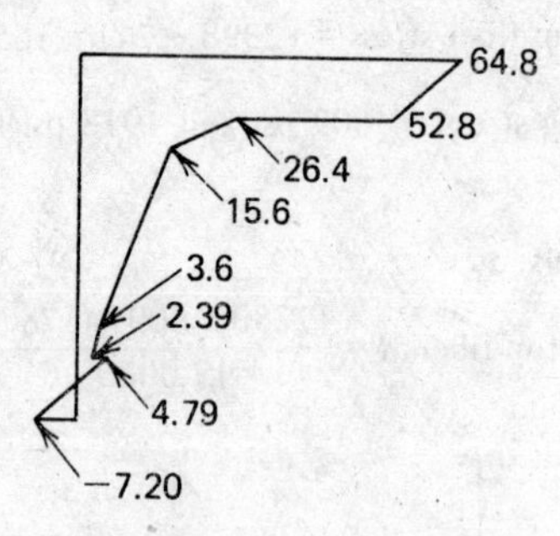

Fig. 4-39 Solution Problem 10. Note: Forces are shown in kips per inch.

5 Cracking Load, Ultimate Moment, Shear, and Bond

5-1 Action Under Overloads—Cracking Load

It has been shown that a variation in the external load acting on a prestressed beam results in a change in the location of the pressure line for beams in the elastic range. This is a fundamental principle of prestressed construction. In a normal prestressed beam, this shift in the location of the pressure line continues at a relatively uniform rate, as the external load is increased, to the point where cracks develop in the tension fiber. After the cracking load has been exceeded, the rate of movement in the pressure line decreases as additional load is applied, and a significant increase in the stress in the prestressing tendon and the resultant concrete force begins to take place. This change in the action of the internal moment continues until all movement of the pressure line ceases. The moment caused by loads that are applied thereafter is offset entirely by a corresponding and proportional change in the internal forces, just as in reinforced-concrete construction. The range of loading that is characterized by these different actions is illustrated in the load deflection curve, of Fig. 5-1. This fact, that the load in the elastic range and the plastic range is carried by actions that are fundamentally different, is very significant and renders strength computations essential for all designs in order to ensure that adequate safety factors exist. This is true

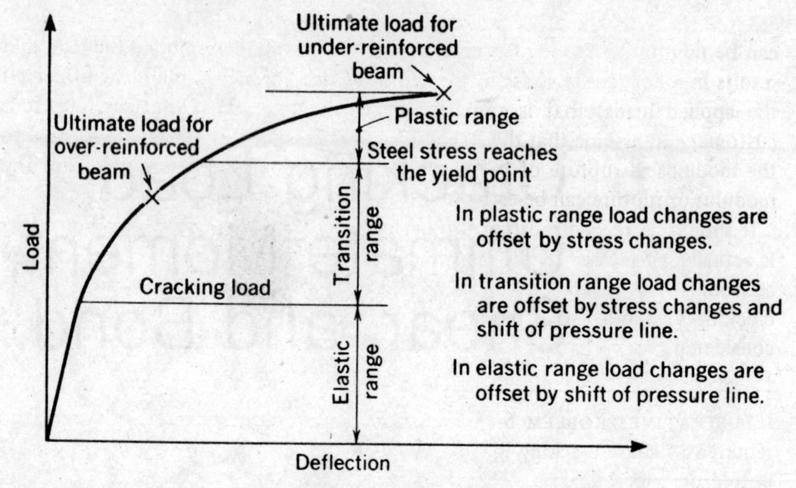

Fig. 5-1 Load-deflection curve for the prestressed beam.

even though the stresses in the elastic range may conform to a recognized elastic design criterion.

It should be noted that the load deflection curve in Fig. 5-1 is close to a straight line up to the cracking load and that the curve becomes progressively more curved as the load is increased above the cracking load.* The curvature of the load-deflection curve for loads over the cracking load is due to the change in the basic internal resisting moment action that counteracts the applied loads, as described above, as well as to plastic strains that begin to take place in the steel and the concrete when stressed to high levels.

In some structures it may be essential that the flexural members remain crack free even under significant overloads. This may be due to the structures' being exposed to exceptionally corrosive atmospheres during their useful life. In designing prestressed members to be used in special structures of this type, it may be necessary to compute the load that causes cracking of the tensile flange, in order to ensure that adequate safety against cracking is provided by the design. The computation of the moment that will cause cracking is also necessary to ensure compliance with some design criteria (see last paragraph of Sec. 5-4).

Many tests have demonstrated that the load-deflection curves of prestressed beams are approximately linear up to and slightly in excess of the load that causes the first cracks in the tensile flange. (The linearity is a function of the rate at which the load is applied.) For this reason, normal elastic-design relationships

*The presence of non-prestressed reinforcing in the tensile flange will tend to make the cracking load more difficult to detect from a load-deflection curve as well as from observations of a beam during loading.

can be used in computing the cracking load by simply determining the load that results in a net tensile stress in the tensile flange (prestress minus the effects of the applied loads) that is equal to the tensile strength of the concrete. It is customary to assume that the flexural tensile strength of the concrete is equal to the modulus of rupture of the concrete when computing the cracking load. The modulus of rupture can be estimated from Eq. 3-3.

It should be recognized that the performance of bonded prestressed members is actually a function of the transformed section rather than the gross concrete section (*see* Sec. 4-10). If it is desirable to make a precise estimate of the cracking load, such as is required in some research work, this effect should be considered.

ILLUSTRATIVE PROBLEM 5-1 Compute the total uniformly distributed load required to cause cracking in a beam that is 10 in. wide, 12 in. deep and is supported on a simple span of 25 ft, if the final prestressing force is 120,000 lb applied at an eccentricity of −2.50 in. Assume the modulus of rupture equals $7.2\sqrt{f'_c}$ psi and $f'_c = 5000$ psi.

SOLUTION:

$$A = 120 \text{ in.}^2, \qquad I = 1440 \text{ in.}^4, \qquad r^2/y = \pm 2.00 \text{ in.}$$

$$S = \frac{1440}{6} = \pm 240 \text{ in.}^3$$

$$\text{Modulus of rupture} = 7.2\sqrt{5000} = 509 \text{ psi}$$

$$f_b = \frac{120{,}000}{120}\left(1 + \frac{-2.50}{-2.00}\right) = 2250 \text{ psi}$$

Therefore, the moment that causes cracking must result in a bottom-fiber stress equal to: −509 psi − 2250 psi = −2759 psi.

$$M_{CR} = \frac{w_t L^2}{8} = f_b S = \frac{-2759 \times (-240)}{12{,}000} = 55.18 \text{ k-ft}$$

$$w_t = \frac{55.18 \times 8}{25.0^2} = 0.706 \text{ k/ft}$$

5-2 Principles of Ultimate Moment Capacity for Bonded Members

When prestressed flexural members that are stronger in shear and bond than in bending are loaded to failure, they fail in one of the following modes:

(1) *Failure at cracking load* In very lightly prestressed members, the cracking moment may be greater than the moment the member can withstand in the cracked condition and, hence, the cracking moment is the ultimate moment.

This condition is rare and is most likely to occur in members that are prestressed concentrically with small amounts of steel. It can also occur in hollow or solid prestressed concrete members that have relatively low levels of reinforcing. Determination of the possibility of this type of failure is accomplished by comparing the estimated moment that would cause cracking to the estimated ultimate moment, computed as described below. When the estimated cracking load is larger than the computed ultimate load, this type of failure would take place if the member were subjected to the required load. Because this type of failure is a brittle failure, it occurs without warning–designs that would yield this mode of failure should be avoided.

(2) *Failure due to rupture of steel* In lightly reinforced members subjected to ultimate load, the ultimate strength of the steel may be attained before the concrete has reached a highly plastic state. This type of failure is occasionally encountered in the design of structures with very large compression flanges in comparison to the amount of prestressing steel, such as a composite bridge stringer. Computation of the ultimate moment of a member subject to this type of failure can be done with a high precision. The method of computation, as well as the determination of which members are subject to this mode of failure, is described below.

(3) *Failure due to strain* The usual underreinforced, prestressed structures that are encountered in practice are of such proportions that, if loaded to ultimate, the steel would be stressed well into the plastic range and the member would evidence large deflection. Failure of the member will occur when the concrete attains the maximum strain that it is capable of withstanding. It is important to understand that research into the ultimate bending strength of reinforced and prestressed concrete has lead most investigators to the conclusion that concrete, of the quality normally encountered in prestressed work fails when the limiting strain of 0.003 is attained in the concrete. Since the ultimate bending capacity is limited by strain rather than stress in the concrete, it is a function of the elastic moduli of the concrete and steel. The magnitude of the ultimate moment for members of this category can also be predicted, as a rule, within the normal tolerances expected in structural design. The ultimate moment of underreinforced sections cannot be predicted with the same precision as the lightly reinforced members described above, since the ultimate moments of underreinforced members are a function of the elastic properties of the steel and the effective stresses in the prestressing steel, whereas the ultimate moment capacities of lightly reinforced members are not.

(4) *Failure due to crushing of the concrete* Flexural members that have relatively large amounts of prestressing steel or relatively small compressive flanges are referred to as being overreinforced. Overreinforced members, when loaded to destruction, do not attain the large deflections associated with underreinforced members–the steel stresses do not exceed the yield point and failure is the result of the concrete being crushed. Computation of the ultimate moments of overreinforced members is done by a trial and error procedure, involving assumed

strain patterns, as well as by empirical relationships. Both methods are discussed below (*see* Ref. 1).

It must be emphasized that there is no clear distinction between the different classifications of failure listed above. For convenience of design, certain parameters, which are a function of the percentage of steel, are used by different authorities to distinguish between the different types of failure that would be anticipated. These parameters, for rectangular sections, are defined as follows:

$$\text{Percentage of steel} = \rho_p = \frac{A_{ps}}{bd_p} \tag{5-1}$$

In Eq. 5-1, A_{ps} is the area of the prestressed reinforcement in the tension zone, b is the width of the compression flange of the member, and d_p is the distance from the extreme compression fiber to the centroid of the prestressing reinforcing. A dimensionless factor, called the steel index (q''), is used by some authorities. The factor is defined as

$$q'' = \frac{A_{ps}f_{pu}}{bd_p f_c'} = \rho_p \frac{f_{pu}}{f_c'} \tag{5-2}$$

where f_{pu} and f_c' are the specified tensile strength of the prestressing steel and the 28-day cylinder compressive strength of the concrete, respectively. The reinforcement index ω_p is currently used in certain building codes. It is defined as

$$\omega_p = \frac{A_{ps}f_{ps}}{bd_p f_c'} = \rho_p \frac{f_{ps}}{f_c'} \tag{5-3}$$

In Eq. 5-3, f_{ps} is the calculated stress in the prestressing steel under the load resulting in the ultimate moment.

Each of these factors is used in such a manner that the results obtained from them are virtually identical, as will be seen.

In order to simplify the explanation of the theory related to the computation of the ultimate moments, a rectangular section will be assumed throughout the derivation, in order to eliminate the variable of flange width which is frequently encountered with *I* or *T* sections (Ref. 1). In addition, the following assumptions, some of which differ slightly from those contained in ACI-318, (Ref. 2), are made:

(1) Plane sections are assumed to remain plane.
(2) The stress-strain properties of the steel are smooth curves without a definite yield point.
(3) The limiting strain of the concrete is equal to 0.0034, regardless of the strength of the concrete.
(4) The steel and concrete are completely bonded.
(5) The stress diagram of the concrete at failure is such that the average concrete stress is $0.80f_c'$ and the resultant of the stress in the concrete acts at

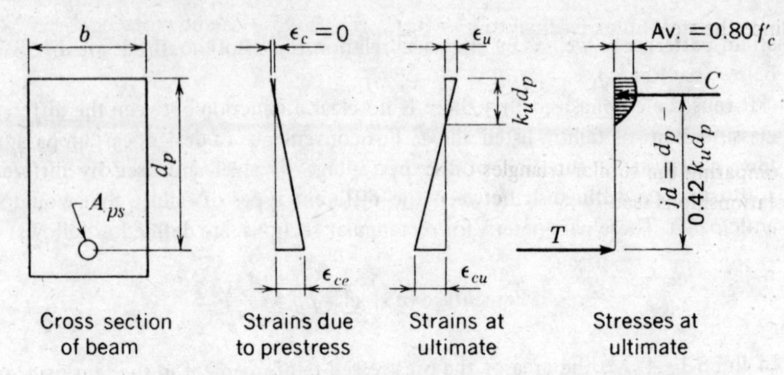

Fig. 5-2 Strain and stress distributions assumed in ultimate moment computations.

a distance from the extreme fiber equal to 0.42 of the depth of the compression block, as is illustrated in Fig. 5-2.

(6) The strain in the top fiber under prestress alone is equal to zero.
(7) The section is subject to pure bending.
(8) The analysis is for the condition of static loads of short duration.

The definition of the strains illustrated in Fig. 5-2 and used in the derivation are as follows:

ϵ_c = concrete strain at extreme fiber due to prestressing (assumed = 0)
ϵ_u = maximum concrete strain at ultimate (assumed = 0.0034)
ϵ_{ce} = concrete strain at the level of the steel due to prestressing
ϵ_{cu} = concrete strain at the level of the steel at ultimate
ϵ_{se} = steel strain due to the effective prestress
ϵ_{su} = steel strain at ultimate

Since equilibrium of the section requires that the forces in the steel and conrete be equal, we can write:

$$T = C$$

or

$$A_{ps} f_{ps} = 0.80 f'_c b k_u \, d_p$$

and

$$f_{ps} = \frac{0.80 f'_c b k_u \, d_p}{A_{ps}}$$

or

$$f_{ps} = \frac{0.80 f'_c k_u}{\rho_p} \tag{5-4}$$

Since the steel index is equal to $q'' = \rho_p f_{pu}/f'_c$, Eq. 5-4 can be written

$$f_{ps} = \frac{0.80 f_{pu} k_u}{q''} \tag{5-5}$$

Comparing the similar triangles of the concrete strains at ultimate, the following relationship is seen

$$\frac{\epsilon_{cu}}{d_p - k_u d_p} = \frac{\epsilon_u}{k_u d_p}$$

or

$$\epsilon_{cu} = \epsilon_u \left(\frac{1 - k_u}{k_u} \right) \tag{5-6}$$

Using the above value for the concrete strain at the level of the steel at ulti-ıate, the strain in the steel at ultimate (which consists of the sum of the strains ue to the effective prestress, the strain in the concrete at the level of the steel resulting from prestressing, and the strain in the concrete at the level of the steel at ultimate) can be expressed as

$$\epsilon_{su} = \epsilon_{se} + \epsilon_{ce} + \epsilon_{cu} \tag{5-7}$$

or

$$\epsilon_{su} = \epsilon_{se} + \epsilon_{ce} + \epsilon_u \left(\frac{1 - k_u}{k_u} \right) \tag{5-8}$$

Equation 5-8 can be rearranged to

$$k_u = \frac{\epsilon_u}{\epsilon_u + \epsilon_{su} - \epsilon_{se} - \epsilon_{ce}} \tag{5-9}$$

Substituting the value of k_u given in Eq. 5-9 into the relationship of Eq. 5-5, the general equation of the ultimate steel stress is obtained

$$f_{ps} = \frac{0.80 f_{pu}}{q''} \times \frac{\epsilon_u}{\epsilon_u + \epsilon_{su} - \epsilon_{se} - \epsilon_{ce}} \tag{5-10}$$

All of the terms in this relationship are known or assumed, except the steel strain at ultimate ϵ_{su} and the steel stress at ultimate f_{ps}. Therefore, by employing a trial and error procedure and solving Eq. 5-10 for various assumed values of q'' and ϵ_{su}, the values of f_{ps} can be determined and the results can be then plotted on the stress-strain diagram for the particular prestressing steel to be used. The intersection of the curves thus obtained with the stress-strain curve reveals the values of f_{ps} and ϵ_{su} for different values of q''. This is illustrated in the curve of Fig. 5-3, which is representative of one commonly used type of

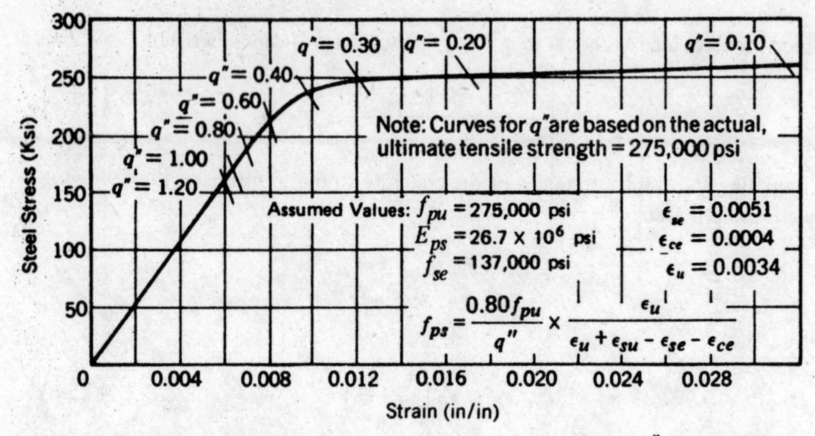

Fig. 5-3 Stress-strain diagram with f_{ps}–ϵ_{su} curves for various values of q'' superimposed.

tendon. The data used in plotting the curves are

f_{pu} = 275,000 psi
E_{ps} = 26.7 X 10^6 psi
ϵ_{se} = 0.0051
ϵ_{ce} = 0.0004, which corresponds to a concrete stress at the level of the steel due to prestressing of 2000 psi, if $E_c = 5 \times 10^6$, and 1600 psi, if $E_c = 4 \times 10^6$ psi

The values of f_{ps} obtained through the analysis of the steel stress-strain curve and the values obtained from the approximate relationship:

$$f_{ps} = f_{pu}\left(1 - 0.5\rho_p \frac{f_{pu}}{f'_c}\right) \tag{5-11}$$

are compared in Fig. 5-4. The values obtained from the approximate relationship are conservative, as would be expected. The variation of j, the ratio between the lever arm and the depth of the section, is shown in Fig. 5-5 as a function of the steel index. In a similar manner, the ratio of the ultimate moment to $f_{pu}A_{ps}d_p$ as a function of the steel index is shown in Fig. 5-6.

Tests have shown that for lightly reinforced members with very low values of q'' (0.00–0.08), the moment at ultimate can be calculated by the relationship

$$M_n = 0.95 f_{pu} A_{ps} d_p \tag{5-12}$$

Under such conditions the member fails as the result of the failure of the steel. The value of k_u is of the order of 0.10 for the very low values of q'' (*see* Fig. 5-5). The concrete is not stressed in the plastic region for these conditions because the steel fails before such concrete stresses can develop.

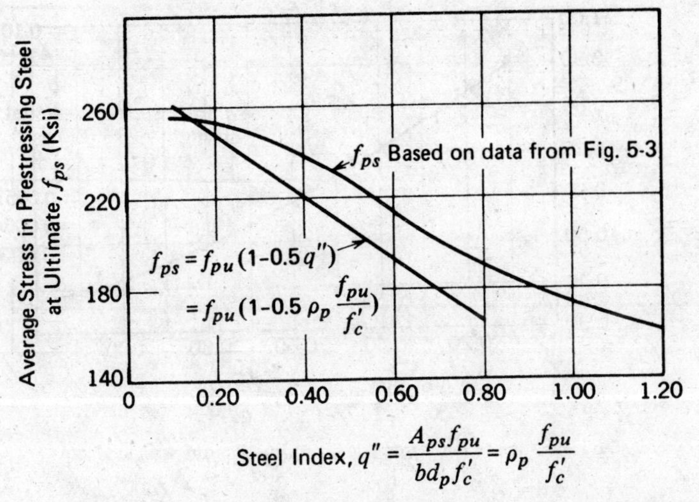

Fig. 5-4 Variation of f_{ps} with the steel index. The actual values of f_{ps} are shown based on the stress-strain curve of Fig. 5-3 and the approximate value of Eq. 5-11. (Ref. 25).

For the steel studied here, the curve of $M_n/f_{pu}A_{ps}d_p$ is nearly linear from 0.10 to 0.40 (Fig. 5-6) and is approximately equal to the following relationship

$$M_n = (1 - 0.60q'')A_{ps}f_{pu}d_p \tag{5.13}$$

In this range of steel indices, the steel would evidence large deformations, if a member were loaded to failure, and the action would be as is described above as underreinforced.

Finally, for values of q'' in excess of 0.40, the steel stress would be below the yield point at failure and the failure would be initiated in the concrete without

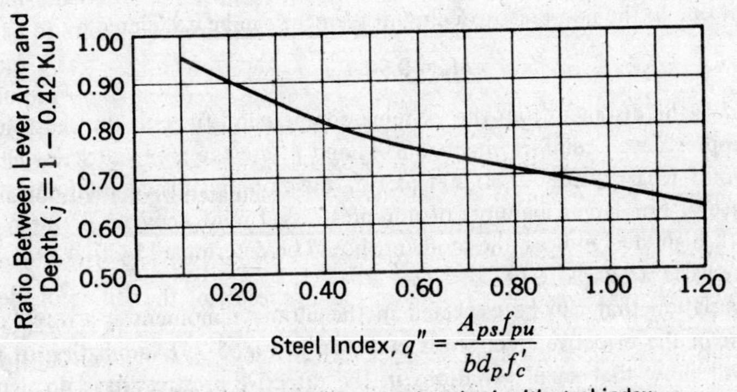

Fig. 5-5 Variation of lever-arm-depth ratio, j, with steel index.

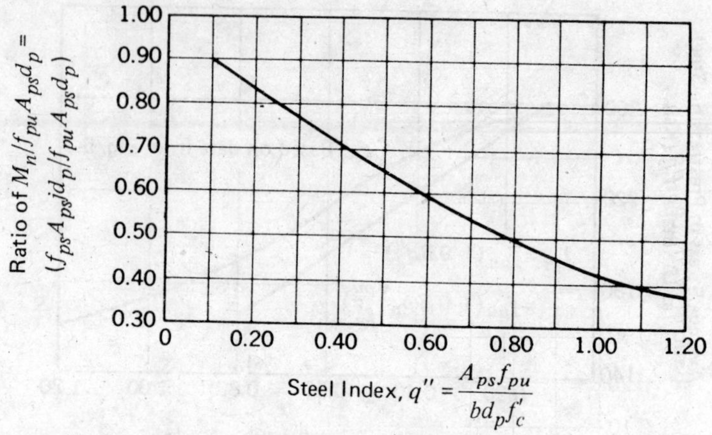

Fig. 5-6 Variation in the factor $M_n/f_{pu}A_{ps}d_p$ with the steel index.

being the result of excessive elongation of the steel and, hence, would be of the overreinforced type.

As was stated above, the relationships that were developed are applicable to rectangular sections. These relationships are equally accurate for flanged sections, provided the neutral axis of the section at ultimate is within the limits of the flange. If the neutral axis falls outside of the flange area, the same strain distribution applies as in the case of rectangular sections, but due to the variable width of the section, the distance to the resultant of the compressive block is no longer equal to $0.42k_ud_p$ and must be calculated. To facilitate the calculation of the location of the resultant, the compression block can be assumed to be rectangular rather than curved, as shown in Fig. 5-2, without introducing significant error.

When small quantities of non-prestressed reinforcement are used in combination with small quantities of prestressed reinforcement, the additional ultimate moment due to the non-prestressed reinforcement can be calculated by

$$M_n = 0.90A_sf_yd \tag{5-14}$$

where d is the distance from the extreme compression fiber to the centroid of the non-prestressed reinforcement and A_s and f_y are the area and stress at the yield point (max. value ≅ 60,000 psi) of the non-prestressed reinforcement, respectively. For larger amounts of non-prestressed reinforcement or for members with high steel indices, the moment should be determined by trial and error from the basic strain patterns.

The variation that can be expected in the ultimate moment as a result of a variation in the effective steel stress is shown in Fig. 5-7. Examination of this curve will show that small variations in the effective prestress have no significant effect on the ultimate strength of prestressed members. It is important to note that even if errors are made in estimating the losses of prestress, in

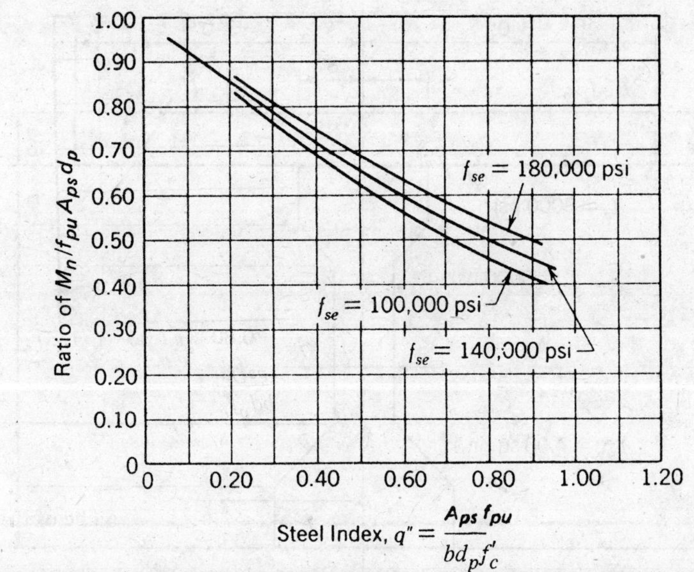

Fig. 5-7 Effect of effective prestressing stress f_{se} on the ratio $M_n/A_{ps}f_{pu}d$ for various values of steel index q''. (After J. Muller.)

estimating the stressing friction, or even if the stressing is not carried out to a high precision in the field due to poor workmanship, the effect on the ultimate moment is generally small for flexural members *with bonded tendons.*

ILLUSTRATIVE PROBLEM 5-2 Compute the ultimate moment for the composite section of Fig. 5-8, which consists of the AASHTO-PCI type III bridge stringer with a 6.50 in. cast-in-place slab. The steel area of 4.00 sq in. is located with its centroid 5.85 in. above the bottom of the beam. The steel has the characteristics indicated in Fig. 5-3. The concrete strengths are 5000 psi and 3000 psi for the stringer and the cast-in-place deck, respectively.

SOLUTION:

$$\text{Steel index} = q'' = \frac{4.00 \times 275{,}000}{72.0 \times 45.65 \times 3000} = 0.111$$

From Fig. 5-5 it will be seen that $j = 0.95$ (approximately) for $q'' = 0.11$. Therefore, the value of k_u is computed as follows

$$j = 0.95 = 1 - 0.42k_u$$

$$k_u = \frac{0.05}{0.42} = 0.119$$

$$k_ud_p = 5.43 \text{ in.}$$

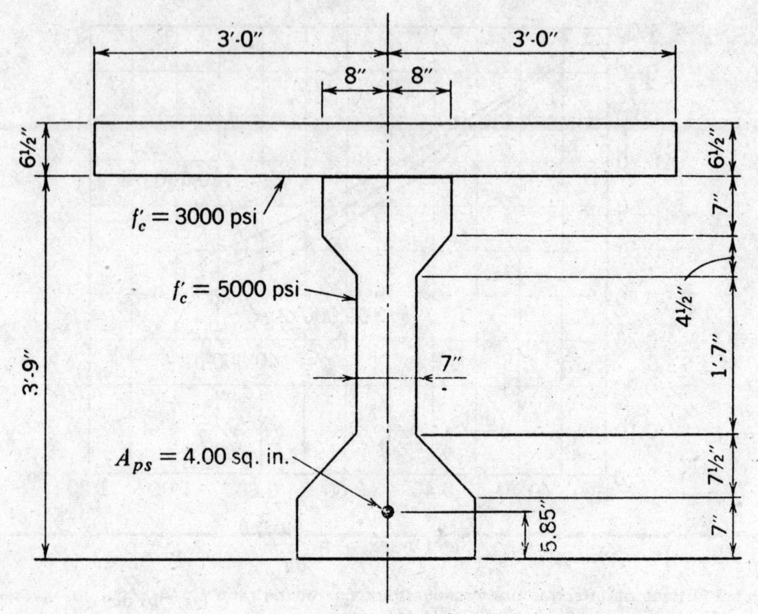

Fig. 5-8 AASHTO-PCI type III bridge stringer with composite deck.

The value of $k_u d$ is approximately $5\frac{1}{2}$ in. and the neutral axis falls within the top flange. Hence, the relationships for the ultimate moments of rectangular sections are valid

$$M_n = \frac{4.00 \times 275{,}000}{1000} \times \frac{45.65}{12} (1 - 0.6 \times 0.111) = 3906 \text{ k-ft}$$

ILLUSTRATIVE PROBLEM 5-3 Compute the ultimate moment for the stringer of Problem 5-2, neglecting the composite action of the deck.

SOLUTION:

Average concrete stress = $0.80 f'_c$ = 4000 psi

Assume

$$\epsilon_{se} = 0.0050 \qquad d_p = 45.00 \text{ in.} - 5.85 \text{ in.} = 39.15 \text{ in.}$$

$$\epsilon_{ce} = 0.0004$$

$$\epsilon_u = 0.0034 \qquad \epsilon_{su} = \epsilon_{se} + \epsilon_{ce} + \epsilon_u \left(\frac{d - k_u d_p}{k_u d_p}\right)$$

Try

$$k_u d_p = 15 \text{ in.}$$

$$C = 7 \text{ in.} \times 15 \text{ in.} \times 4 \text{ k/in.}^2 + 9 \text{ in.} \times 7 \text{ in.} \times 4 \text{ k/in.}^2 + 9 \times 4.5/2 \times 4 \text{ k/in.}^2 = 753 \text{ k}$$

$$\epsilon_{su} = 0.054 + 0.0034 \frac{39.15 - 15.00}{15.00} = 0.0109$$

From Fig. 5-3

$$f_{ps} = 245{,}000 \text{ psi} \qquad T = 980 \text{ k}$$

$$C < T \text{ try larger } k_u d$$

Assume

$$k_u d = 20 \text{ in.}$$

$$C = 753 \text{ k} + 5 \text{ in.} \times 7 \text{ in.} \times 4 \text{ k/in.}^2 = 893 \text{ k}$$

$$\epsilon_{su} = 0.0054 + 0.0034 \left(\frac{19.15}{20.00}\right) = 0.0087$$

From Fig. 5-3

$$f_{ps} = 223{,}000 \text{ psi}$$

$$T = 893 \text{ k}$$

Compute location of force C from the top of the section:

$$7 \text{ in.} \times 20 \text{ in.} \times 4 \text{ ksi} = 560 \text{ k} \times 10 \text{ in.} = 5600 \text{ k-in.}$$

$$9 \text{ in.} \times 7 \text{ in.} \times 4 \text{ ksi} = 252 \text{ k} \times 3.5 \text{ in.} = 882 \text{ k-in.}$$

$$0.5 \times 9 \text{ in.} \times 4.5 \times 4 \text{ ksi} = 81 \text{ k} \times 8.5 \text{ in.} = 688 \text{ k-in.}$$

$$893 \text{ k} \qquad 7170 \text{ k-in.}$$

$$d_t = \frac{7170}{893} = 8.03 \text{ in.}$$

$$M_n = 893 \text{ k} \left(\frac{39.15 \text{ in.} - 8.03 \text{ in.}}{12}\right) = 2315 \text{ k-ft}$$

As will be seen subsequently, strain compatibility equations can be written for T-shaped, box-shaped, and other non-rectangular sections. In addition, the effect of non-prestressed reinforcement in the compression flange can be included.

5-3 Principles of Ultimate Moment Capacity for Unbonded Members

Because the prestressing tendons can slip (with respect to the concrete) during loading of an unbonded member, the relationships for ultimate moment capacity developed in Sec. 5-2 do not apply to unbonded beams. The reader will recall that one of the basic assumptions made before the derivation of the relationships of Sec. 5-2 was that the concrete and steel are completely bonded. Because the tendons can slip with respect to the concrete, other variables affect the ultimate moment capacity of unbonded prestressed concrete members. After the "Tentative Recommendations for Prestressed Concrete" (Ref. 3) appeared, normal American practice was to consider the stress in unbonded prestressing steel under ultimate load to be as follows:

$$f_{ps} = f_{se} + 15{,}000 \tag{5-15}$$

(in psi) with the requirements that the effective stress in the prestressing steel be between 0.50 and $0.60 f_{pu}$ and that the reinforcement index not exceed 0.30.

Variables that affect the ultimate moment capacity of an unbonded beam, but which do not affect bonded beams in the same manner or not at all, include the following:

(1) Magnitude of the effective stress in the tendons.
(2) Span to depth ratio.
(3) Characteristics of the materials.
(4) Form of loading (shape of the bending moment diagram).
(5) Profile of the prestressing tendon.
(6) Friction coefficient between the prestressing steel and duct.
(7) Amount of bonded non-prestressed reinforcing.

Another relationship that has been suggested for the value of f_{ps} in unbonded members (to be used in lieu of Eq. 5-15) is as follows:

$$f_{ps} = f_{se} + \left(30{,}000 - \frac{\rho_p}{f'_c} \times 10^{10}\right) \tag{5-16}$$

in which f_{se} is limited to $0.60 f_{pu}$, ρ_p is the percentage of steel, and f_{ps}, f_{se} and f'_c are all in psi. Still another relationship that has been more recently proposed is:

$$f_{ps} = f_{se} + \frac{1.4 f'_c}{100\,\rho_p} + 10{,}000 \text{ psi} \tag{5-17}$$

The results of tests of members with unbonded tendons as well as Eqs. 5-15, 5-16 and 5-17 are shown in Fig. 5-9 (Ref. 4). Equations 5-16 and 5-17 have not been widely used, since they have not been included in any of the American codes or standards.

A method of computing the ultimate strength of prestressed members (with unbonded tendons) that takes into account the variables listed above has been proposed by Pannell (Ref. 5). This method is based upon experimental data and is considered slightly conservative. The method provides the relationship for the

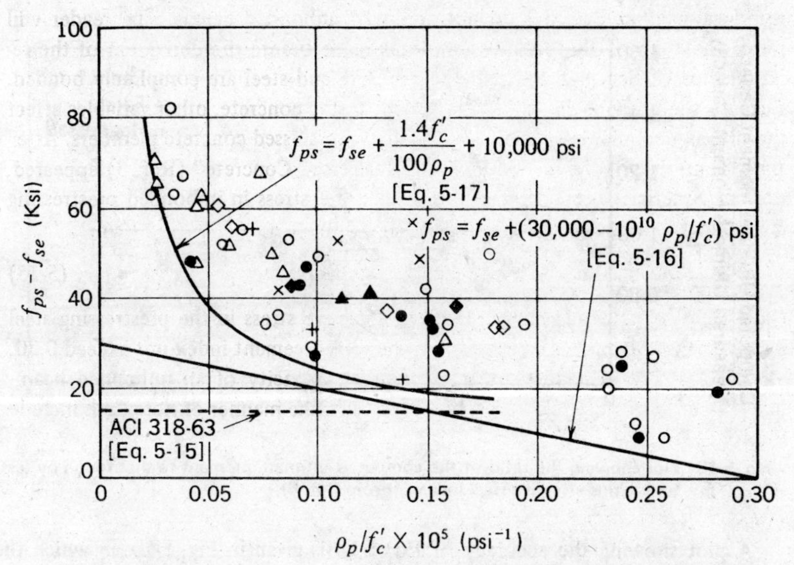

Fig. 5-9 Comparison of values of $f_{ps} - f_{se}$ for unbonded beams. Test data and suggested mathematical relationships are shown.

ultimate moment to be as follows:

$$M_n = q_u (1 - 0.80 q_u) f'_c b d_p^2 \tag{5-18}$$

in which

$$q_u = \frac{q_e + \lambda}{1 + 1.6\lambda} \tag{5-19}$$

with

$$q_e = \frac{\rho_p f_{se}}{f'_c} \tag{5-20}$$

$$\lambda = \frac{10^6\ \rho_p d_p}{f'_c L} \tag{5-21}$$

In Eqs. 5-18 through 5-21 the notation is standard and it should be recognized that the depth of the member d_p and the span length L must be in the same units.

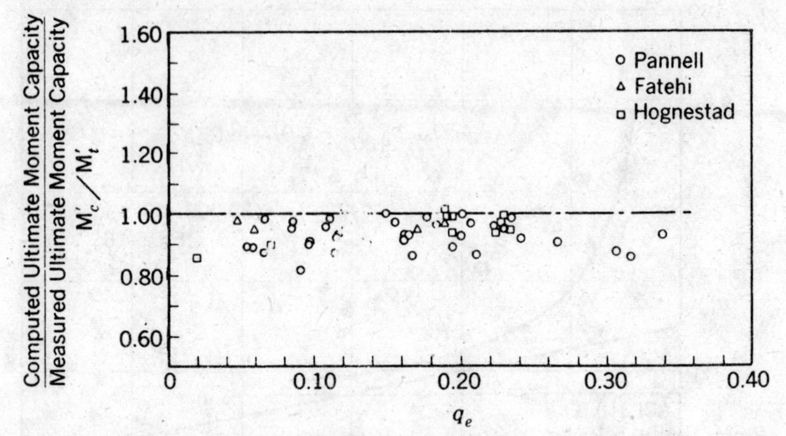

Fig. 5-10 Plot showing the ratio of the computed ultimate moment to that found by tests for various effective steel indices (From Ref. 6).

A plot showing the accuracy of Eq. 5-18 is given in Fig. 5-10, in which the ordinate is the ratio of the calculated ultimate moment to the ultimate moment measured in tests conducted by various investigators.

It should be recognized that the ultimate moment capacity of a member stressed with unbonded tendons, *unlike members with bonded tendons*, (*see* Fig. 5-7), may be adversely affected by unintentional variations in the effective prestress. Hence, it is considered prudent to exert more care in estimating the losses of prestress and in supervising the stressing of unbonded members than would be considered necessary for bonded members, in order to assure the desired results are obtained.

5-4 Ultimate Moment Code Requirements—Bonded Members

Building Code Requirements for Reinforced Concrete (ACI 318) (Ref. 2) contains basic design assumptions for the computation of ultimate moment capacities of members (preferably termed the "strength of members") in Section 10.2. The basic assumptions can be summarized as follows:

1. The principles of strain compatibility and equilibrium apply.
2. Plane sections remain plane (except for deep flexural members).
3. The maximum strain at extreme compression fibers in the concrete section is equal to 0.003.
4. Stress in non-prestressed reinforcement is equal to the product of the strain in the concrete at the level of the steel (bond between the steel and concrete is perfect) and the elastic modulus of the steel with a maximum value of f_y; see Fig. 5-11 for the assumed stress-strain curve for non-prestressed reinforcement.

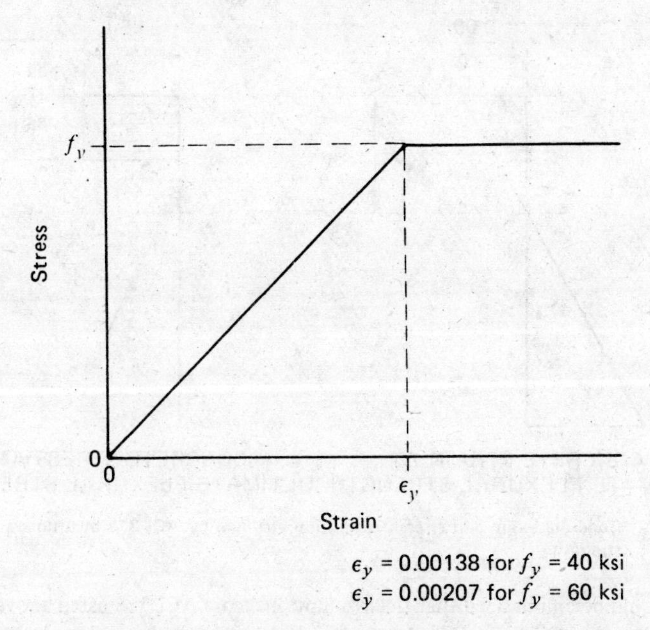

Fig. 5-11 Stress-strain curve for non-prestressed reinforcement. ($\epsilon_y = f_y \div E_s$)

5. The tensile strength of concrete shall be neglected in flexural computations (except for investigating the possibility of failure at the cracking load as described in Art. 5-2 and provided in Sec. 18.4 of ACI 318).
6. The concrete compressive stress distribution may be assumed to be of parabolic shape (as shown in Fig. 5-2), trapezoidal or rectangular (as shown in Fig. 5-12), or other shapes which can be substantiated with comprehensive test results.
7. For simplicity, a rectangular distribution of concrete compressive stress as shown in Fig. 5-12 may be assumed with the following limitations:
   a. The concrete stress shall be taken as being equal to $0.85 f'_c$.
   b. The depth of the compression block shall be taken as being equal to a distance of $a = \beta_1 c$ in which c is the distance from the extreme compression fiber to the neutral axis ($k_u d_p$ in the discussion in Art. 5-2).
   c. The factor β_1 shall be taken to be equal to 0.85 for concrete compressive strengths up to and including 4000 psi; for strengths greater than 4000 psi,

$$\beta_1 = 0.85 - \frac{(f'_c - 4000)\, 0.05}{1000} \tag{5-22}$$

   except β_1 shall not be taken to be less than $0.65 (f'_c = 8000 \text{ psi})$.

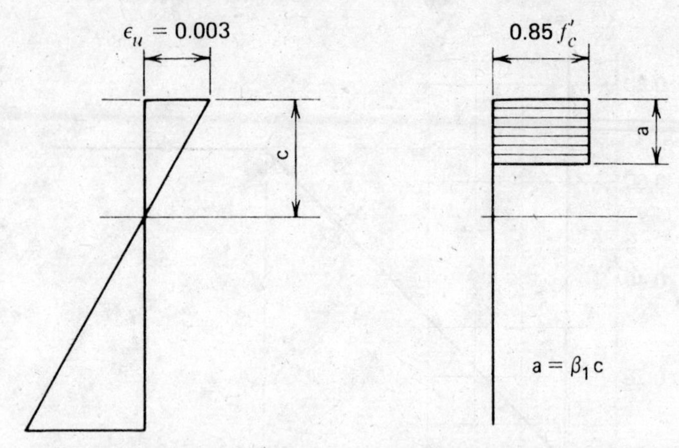

Fig. 5-12 Concrete strain and stress distribution assumed by ACI 318 at ultimate flexural strength.

It should be emphasized that the assumptions from ACI 318 listed above differ from those contained in Ref. 1 (see Art. 5-2) in that the limiting concrete strain is taken to be 0.003, rather than 0.0034, and the compression stress block is assumed to be rectangular with a uniform distribution of stress equal to $0.85f_c'$ to a depth of $\beta_1 c$, rather than a parabolic distribution having an average stress of $0.80f_c'$ acting to a depth of $k_u d_p$. It should be noted that the term β_1 is included in ACI 318 to account for the effect of concrete strength; experimental studies confirmed the need for its inclusion. Note that the term is a constant 0.85 for concrete strengths of 4000 psi and less; is a constant 0.65 for concrete strengths of 8000 psi or more; and varies linearily between 4000 psi and 8000 psi (at the rate of 0.05 per 1000 psi). This is illustrated in Fig. 5-13.

To facilitate strain compatibility computations of the strength of concrete members, the stress-strain curve for prestressed reinforcement can be assumed to be composed of three straight lines as shown in Fig. 5-14. The coordinates of points 2 and 3 can be taken as the minimum specified stresses and strains for 1% extension and ultimate tensile strength, respectively, and the coordinates of point 1 can be computed from the elastic modulus of the steel for an arbitrarily selected stress (which is less than the minimum specified stress at 1% extension) that results in the curve approximating the shape of an actual stress-strain curve as shown in Figs. 2-1, 2-2, 2-3, and 2-4. Because the stress and strain conditions at point 2 are normally adopted as the transition point between under- and over-reinforced members, the use of the latter being prohibited as will be seen below, for strength analysis the stresses and strains between points 2 and 3 alone are of interest. For stress-relieved and low-relaxation seven-wire strand, the equations

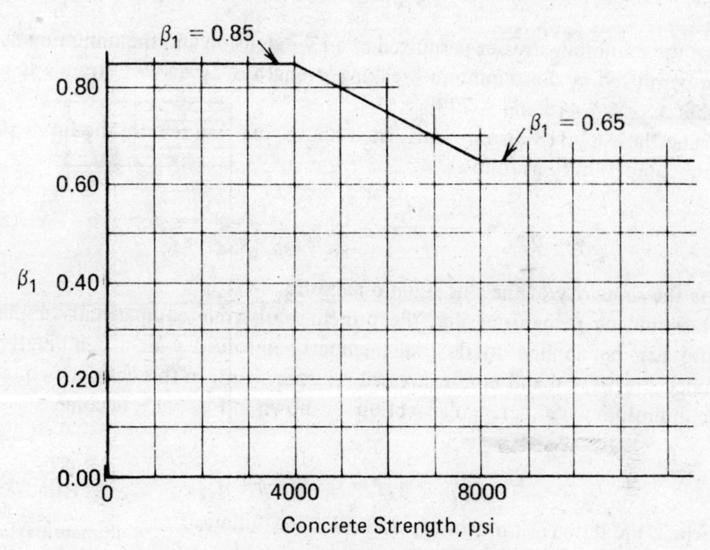

Fig. 5-13 Variation of β_1 and $f'c$.

for the stress-strain diagrams between points 2 and 3 can be taken to be

$$f_{ps} = 197.5 + 1500 f_{su} \tag{5-23}$$

and

$$f_{ps} = 232.2 + 1080 f_{su} \tag{5-24}$$

respectively, where f_{ps} and f_{su} are in *ksi.* Equations 5-23 and 5-24 are based

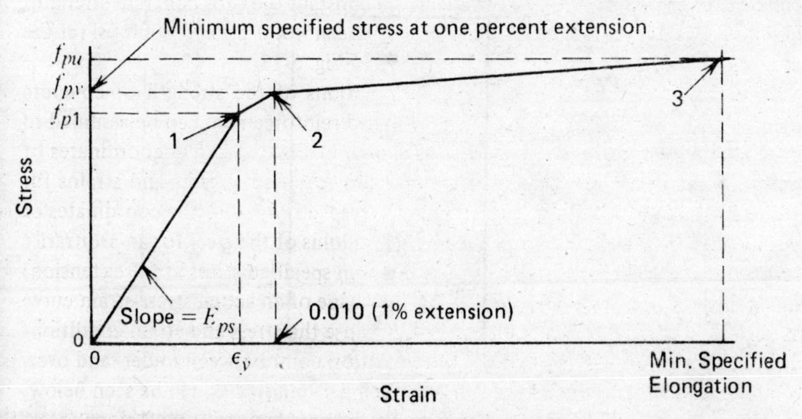

Fig. 5-14 Stress-strain curve for prestressed reinforcement. ($E_{ps} = f_{pl} \div \epsilon_y$, f_{pl} = **stress at proportional limit.)**

upon the minimum stresses permitted at a 1% extension and the minimum extension permitted at the minimum breaking strength of seven-wire strands as specified in ASTM A 416 and A 779.

Using the ACI 318 assumptions listed above, one can rewrite the Eq. 5-10 for strain compatibility as follows:

$$f_{ps} = \frac{0.85\beta_1 f_{pu}}{q''} \times \frac{\epsilon_u}{\epsilon_u + \epsilon_{su} - \epsilon_{se} - \epsilon_{ce}} \tag{5-25}$$

Note the similarity of the this equation and Eq. 5-10.

It should be recognized that the principles of strain compatibility explained above can be applied to flexural members reinforced with a combination of bonded prestressed and nonprestressed reinforcement. In this case, the equations of equilibrium for a rectangular section, as shown in Fig. 5-15, become

$$T = C$$

$$A_{ps}f_{ps} + A_s f_s = 0.85 f_c' ba + A_s' f_s' \tag{5-26}$$

in which the terms not previously defined are as follows:

A_s = area of non-prestressed tension reinforcement.
A_s' = area of non-prestressed compression reinforcement.
d = distance from extreme compression fiber to centroid of non-prestressed tension reinforcement.
d' = distance from extreme compression fiber to centroid of compression reinforcement.

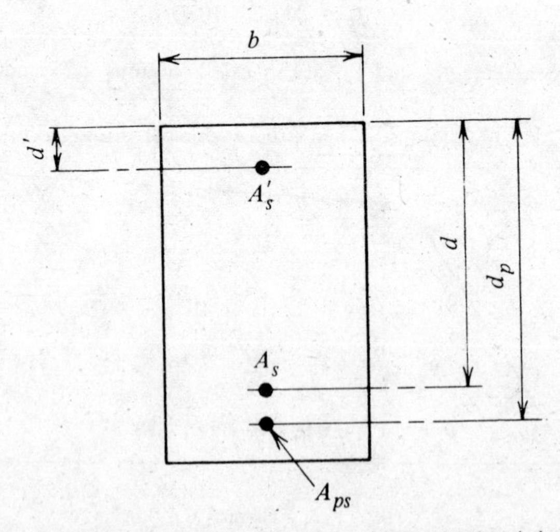

Fig. 5-15 Cross section of a rectangular flexural member having bonded non-prestresse[d] tension and compression reinforcement in addition to bonded prestressed tensio[n] reinforcement.

d_p = distance from extreme compression fiber to centroid of prestressed tension reinforcement.
f_s = stress in the non-prestressed tensile reinforcement ($\leqq fy$).
f_s' = stress in the non-prestressed compressive reinforcement ($\leqq fy$).
f_y = specified minimum yield strength of non-prestressed reinforcement.

Relationships based upon the conditions of strain in the concrete and steel, similar to those in Eqs. 5-4 through 5-10, can be written for the stresses in the non-prestressed tension and compression reinforcement. Once the stresses in all the different steels are known, the nominal moment capacity of the section can be calculated as in I.P. 5-4 and I.P. 5-5.

In a similar manner, strain compatibility equations can be written for members having other cross sections. For a T-shaped member, as shown in Fig. 5-16, the basic equation for equilibrium is

$$A_{ps}f_{ps} + A_s f_s = 0.85 f_c' \, [(b - b_w) h_f + b_w a] + A_s' f_s' \tag{5-27}$$

in which

$$b_w = \text{web width}$$

In using Eqs. 5-26 and 5-27, the stress in the non-prestressed tension reinforcement will normally be found to be equal to its yield strength. As shown in Fig. 5-11, the strain required in non-prestressed reinforcement to stress it to the yield strength is 0.00138 and 0.00207 for grades 40 and 60, respectively. Because prestressed steels, when stressed from the effective stress level to the yield stress,

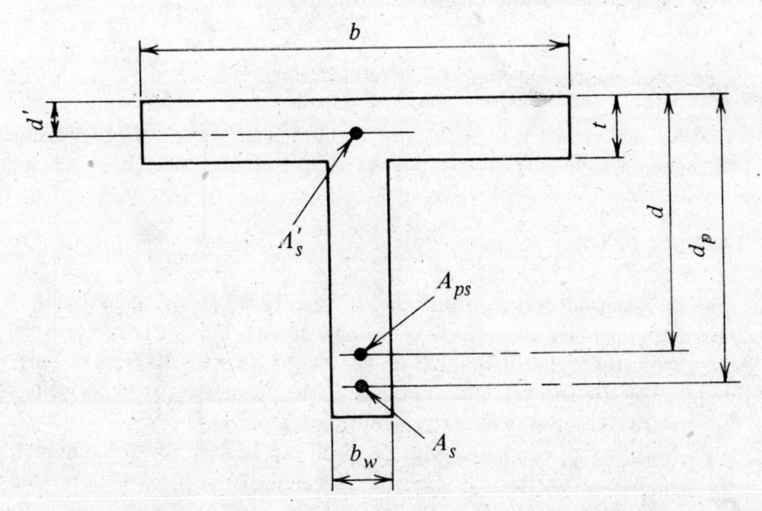

Fig. 5-16 Cross section of a T-shaped flexural member having bonded non-prestressed tension and compression reinforcement in addition to bonded prestressed reinforcement.

will normally increase in stress from 45 to 80 ksi, the stress in the non-prestressed tensile steel will normally go through a similar stress increase. As will be seen below, the stress in the non-prestressed compression reinforcement may be significantly below its yield strength at ultimate flexural strength.

Rather than require a strain compatibility analysis in computing the flexural strength of prestressed concrete members, ACI 318 (Ref. 2) permits f_{ps} to be taken as follows for bonded tendons, provided f_{se} is not less than $0.50\, f_{pu}$:

$$f_{ps} = f_{pu}\left(1 - \frac{\gamma_p}{\beta_1}\left[\rho_p \frac{f_{pu}}{f'_c} + \frac{d}{d_p}(\omega - \omega')\right]\right) \tag{5-28}$$

in which

d = distance in inches from extreme compression fiber to centroid of non-prestressed tension reinforcement

d_p = distance from extreme compression fiber to centroid of prestressed reinforcement

β_1 = factor defined above in this section

γ_p = factor for type of prestressing tendon
= 0.53 for f_{py}/f_{pu} not less than 0.80
= 0.40 for f_{py}/f_{pu} not less than 0.85
= 0.28 for f_{py}/f_{pu} not less than 0.90

ρ_p = ratio of prestressed reinforcement = A_{ps}/bd_p

ω = non-prestressed tension reinforcement index = $\dfrac{A_s f_y}{bd f'_c}$

ω' = non-prestressed compression reinforcement index = $\dfrac{A'_s f_y}{bd f'_c}$

The use of Eq. 5-28 is further restricted by ACI 318 in that if compression reinforcement is included (e.g., $\omega' > 0$) the term

$$\rho_p \frac{f_{pu}}{f'_c} + \frac{d}{d_p}(\omega - \omega')$$

shall be taken not less than 0.17 and d' shall be no greater than $0.15 d_p$. Equation 5-28 includes the effects of non-prestressed tension reinforcement, concrete strengths higher than 4000 psi ($\beta_1 = 0.85$ for f'_c of 4000 psi or less), minimum yield strength (at 1% extension) of the prestressing steel, as well as non-prestressed compression reinforcement (if any).

The value of f_{ps} computed with Eq. 5-28 can be used for the computation of the nominal design flexural strength of rectangular sections with tension reinforcement alone (Ref. 3) by

$$M_n = A_{ps} f_{ps} \left(d_p - \frac{a}{2}\right) + A_s f_y \left(d - \frac{a}{2}\right) \qquad (5\text{-}29)$$

where

$$a = \frac{A_{ps} f_{ps} + A_s f_y}{0.85 f'_c b} \qquad (5\text{-}30)$$

Equation 5-29 can also be used in flanged sections if the thickness of the flange (h_f) is not less than the depth of the compression block, a, as given in Eq. 5-30. When the depth of the compression block exceeds the flange thickness, the nominal design flexural strength of a section can be computed by (Ref. 3):

$$M_n = A_{pw} f_{ps} \left(d_p - \frac{a}{2}\right) + A_s f_y (d - d_p) + 0.85 f'_c (b - b_w) h_f \left(d_p - \frac{h_f}{2}\right) \qquad (5\text{-}31)$$

$$A_{pw} f_{ps} = A_{ps} f_{ps} + A_s f_y - 0.85 f'_c (b - b_w) h_f \qquad (5\text{-}32)$$

and

$$a = \frac{A_{pw} f_{ps}}{0.85 f'_c b_w} \qquad (5\text{-}33)$$

The tensile force represented by $A_{pw} f_{ps}$ (Eq. 5-32) is the force the web must develop; that is to say, it is the tensile force not developed by the flange.

The effect of compressive reinforcement is taken into account by basic principles. For compression reinforcement to be effective in rectangular beams, it must be positioned in such a way that it will be stressed to its yield strength under design load. For this to be the case, using the notation defined in Fig. 5-17 following relationships must be satisfied:

$$\frac{c}{d'} = \frac{\epsilon_u}{\epsilon_u - \epsilon'_s}$$

For yielding of the compression reinforcement, the strain in the concrete at the level of the compressive reinforcement (ϵ'_s) must equal or exceed the yield strain and

$$c = d' \frac{\epsilon_u}{\epsilon_u - \epsilon'_y}$$

Because $\Sigma F_y = 0$, one can write

$$A_{ps} f_{ps} + A_s f_y = 0.85 f'_c ab + A'_s f_y$$

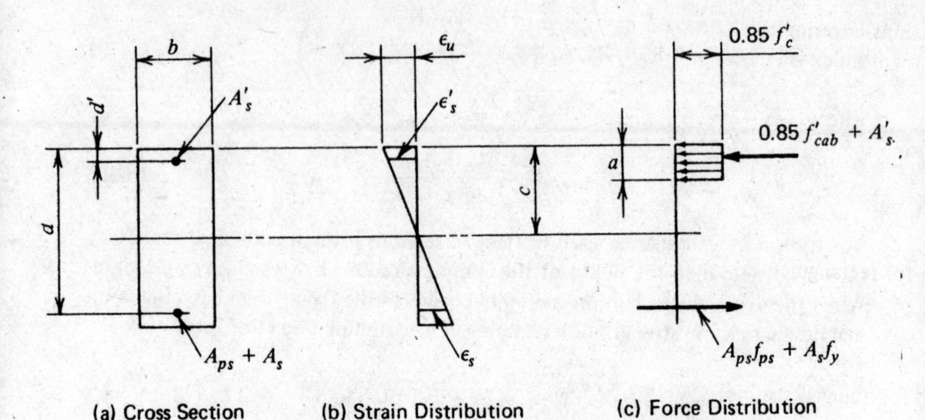

Fig. 5-17 Rectangular beam with compression reinforcement.

taking $a = \beta_1 c$, the expression becomes:

$$\omega_p + \omega - \omega' = \frac{0.85 \times \beta_1 c}{d} = \frac{0.85\beta_1 d'}{d} \cdot \frac{\epsilon_u}{\epsilon_u - \epsilon'_y}$$

Using this value for ϵ'_y and setting ϵ_u equal to 0.003, one obtains

$$\omega_p + \omega - \omega' = 0.85\beta_1 \cdot \frac{d'}{d} \cdot \frac{87000}{87000 - f_y} \tag{5-34}$$

From Eq. 5-34 one can conclude that the term on the left side must be equal to or larger than the term on the right side of the equation if the compression reinforcement is to be stressed to its yield strength. If not stressed to its yield strength the compression reinforcement should either be ignored or its effect determined from a study of the strain in the compression reinforcement. If the compression reinforcement is stressed to its yield strength as predicted from Eq. 5-34, the nominal flexural strength of a rectangular section with compression reinforcement can be computed from:

$$M_n = A_{ps} f_{ps} \left(d_p - \frac{a}{2}\right) + A_s f_y \left(d - \frac{a}{2}\right) + A'_s f_y \left(\frac{a}{2} - d'\right)$$

$$a = \frac{A_{ps} f_{ps} + A_s f_y - A'_s f_y}{0.85 f'_c b} \tag{5-35}$$

Equations 5-29, 5-31, and 5-35 are intended for use in computing the nominal flexural strength of members that are under-reinforced. Under-reinforced members are proportioned in such a way that the stress in the tension reinforcement will reach the yield strength of the reinforcement. The ACI 318 (Ref. 2) provisions covering this point are contained in Section 18.8.1 in which the reinforcement index is limited to $0.36\beta_1$. This can be expressed

$$\omega_p \leqslant 0.36\beta_1 \tag{5-36}$$

or

$$\omega_p + \frac{d}{d_p}(\omega - \omega') \leqslant 0.36\beta_1 \tag{5-37}$$

for rectangular sections and

$$\left[\omega_{pw} + \frac{d}{d_p}(\omega_w - \omega'_w)\right] \leqslant 0.36\beta_1 \tag{5-38}$$

for flanged sections. The ratio of d/d_p is to account for the difference in their magnitude and the term β_1 accounts for the effect of concrete strength as previously described.

Combining the reinforcement index limit of $0.36\,\beta_1$ with Eq. 5-34 and solving for d' gives the following:

$$d' = 0.424d\,\frac{87{,}000 - f_y}{87{,}000} \tag{5-39}$$

which can be reduced to $d' = 0.229d$ and $0.32d$ for grades 40 and 60 reinforcement, respectively. The compression reinforcement will not be stressed to its yield strength if it is placed further from the extreme compression fiber than the value of d' given by Eq. 5-39. Using these values the structural designer can rapidly check to determine the feasibility of using compression reinforcement to enhance the strength of a particular member.

Rectangular sections having reinforcement indices exceeding $0.36\beta_1$ are considered to be over-reinforced and their capacity may be computed from basic principles (*see* Problem 5-3) or from:

$$M_n = 0.25\,f'_c\,bd^2 \tag{5-40}$$

which is an approximate, (and conservative) relationship. For flange sections, neutral axis not located within the flange depth, h_f, the approximate relationship is

$$M_n = 0.25\,f'_c b_w d^2 + 0.85\,f'_c(b - b_w)\,h_f(d - 0.5h_f) \tag{5-41}$$

Note that d in Eq. 5-40 and 5-41 is the distance from the extreme compression fiber to the centroid of the tensile reinforcement.

Another important Code requirement, which is equally applicable to bonded and unbonded tendons, provides that the minimum ultimate moment a section is capable of developing must be at least 1.2 times the cracking moment, based upon a modulus of rupture of $7.5\sqrt{f'_c}$. This is to guard against failure at cracking load, which was described in Sec. 5-2. Experience has shown that some designers overlook this important provision. The author has seen load tables for standard prestressed concrete products which were prepared without consideration being given to this provision of the Code.

ILLUSTRATIVE PROBLEM 5-4 For the beam shown in the cross section in Fig. 5-18, compute the stress in the prestressing steel at ultimate (f_{ps}) and the nominal moment capacity. Use an idealized stress-strain compatibility. The following materials properties and section properties are to be used:

$f'_c = 7500$ psi
$E_c = 5000$ ksi
$A_c = 415.8$ sq. in.
$y_t = 15.3$ in.
$y_b = -14.7$ in.
$r^2/y_t = 6.97$ in.
$r^2/y_b = -7.25$ in.
$S_t = 2900$ in.^{3}
$S_b = -3020$ in.^{3}
$I = 44{,}386$ in.^{3}
$\beta_1 = 0.675$

$f_{pu} = 270$ ksi (3.5% extension)
$f_{py} = 243$ ksi (1% extension)
$E_{ps} = 28{,}000$ ksi
$A_{ps} = 2.75$ sq. in.
$P_{se} = 440$ kips
$d_p = 30.0 - 5.20 = 24.8$ in.
$e = -14.7 + 5.20 = -9.50$ in.
$r^2/e = -11.24$ in.
$w_d = 0.44$ kip per foot
span = 40.0 ft.
$M_d = 88.0$ k-ft.
$\gamma_p = 0.28$

SOLUTION:

$$q'' = \frac{2.75 \times 270}{24 \times 24.8 \times 7.5} = 0.166$$

The slope of the idealized stress-strain curve from f_{py} to f_{pu} is

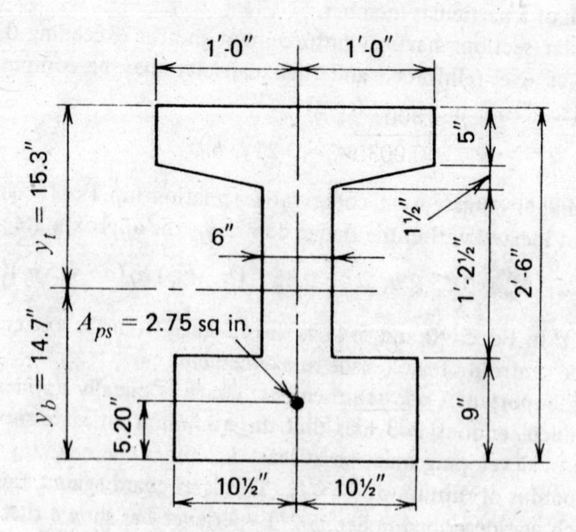

Fig. 5-18 Beam cross section for I.P. 5-4.

$$\text{slope} = \frac{(270 - 243)}{0.035 - 0.010} = 1080 \text{ ksi}$$

and the equation for the line, which is only valid between extensions of 1 and 3.5% (243 and 270 ksi), is

$$f_{ps} = 232.2 + 1080\,\epsilon_{su}$$

The stress in the concrete due to effective prestress and the dead load at the level of the steel is

$$f_e = \frac{440{,}000}{415.8}\left(1 + \frac{-9.50}{-11.24}\right) + \frac{88.0 \times 12000 \times -9.50}{44{,}386}$$

$$= 1727 \text{ psi}$$

The strains required for the analysis are

$$\epsilon_c = 0$$

$$\epsilon_u = 0.003$$

$$\epsilon_{ce} = \frac{1727}{5{,}000{,}000} = 0.00034$$

$$\epsilon_{se} = \frac{440{,}000}{28 \times 10^6 \times 2.75} = 0.0057$$

The two equations required for the determination of f_{ps} and ϵ_{su} are, from Eq. 5-25

$$f_{ps} = \frac{0.85 \times 0.65 \times 270}{0.166} \times \frac{0.003}{0.003 + \epsilon_{su} - 0.0057 - 0.00034}$$

$$= \frac{2.800}{\epsilon_{su} - 0.00304}$$

and, from above,

$$f_{ps} = 232.2 + 1080\epsilon_{su}$$

solving the two equations gives $f_{ps} = 247.69$ ksi and $\epsilon_{su} = 0.01434$

$$T = 247.69 \times 2.75 = 681.1 \text{ k}$$

$$k_u = \frac{0.003}{0.003 + 0.01434 - 0.0057 - 0.00034} = 0.266$$

$$\beta_1 = 0.85 - \frac{(7500 - 4000)\,0.05}{1000} = 0.675$$

$$a = 0.675 \times 0.266 \times 24.8 = 4.45 \text{ in.} < h_f = 5.00 \text{ in.}$$

$$C = 0.85 \times 7.5 \times 24 \times 4.45 = 680.85 \text{ k}$$

$$M_n = \frac{681.1}{12} \times \left(30.0 - 5.20 - \frac{4.45}{2}\right) = 1281.6 \text{ k-ft}$$

Using Eq. 5-28:

$$f_{ps} = 270\left(1 - \frac{0.28}{0.675}\,\frac{2.75 \times 270}{24 \times 24.8 \times 7.5}\right) = 251.4 \text{ ksi}$$

$$a = \frac{2.75 \times 251.4}{0.85 \times 7.5 \times 24} = 4.52 \text{ in.} < 5.00 \text{ in., assume a rectangular section}$$

$$M_n = \frac{2.75 \times 251.4}{12}\left(24.8 - \frac{4.52}{2}\right) = 1298.6 \text{ k-ft}$$

$$\omega_p = \frac{2.75 \times 251.4}{24 \times 24.8 \times 7.5} = 0.155 < 0.36\beta_1 = 0.243, \text{ not over-reinforced}$$

ILLUSTRATIVE PROBLEM 5-5 For the beam of I.P. 5-4, compute the nominal moment capacity of the beam if, in addition to the prestressed reinforcement, the beam is provided with non-prestressed reinforcement having a yield strength of 60.0 ksi and an area of 4.00 sq in. located with its center of gravity 4.50 inches from the soffit.

$$T = C$$

$$\frac{d}{d_p}(A_s f_y) + A_{ps} f_{ps} = 0.85 f'_c b \beta_1 k_u d_p$$

(Note that the strain in the non-prestressed steel must be at least 0.0021 for the stress to equal the yield strength.) In the above equation, d is the distance from the extreme compression fiber to the center of gravity of the non-prestressed reinforcement. Note that multiplying the term $A_s f_y$ by d/d_p converts the effect of the non-prestressed reinforcement into an equivalent amount of non-prestressed reinforcement acting at a distance of d_p from the extreme compression fiber. This relationship can be written

$$f_{ps} = \frac{0.85 f'_c b \beta_1 k_u d_p - \dfrac{d}{d_p}(A_s f_y)}{A_{ps}}$$

or

$$f_{ps} = \frac{0.85 f'_c \beta_1 k_u}{\rho_p} - \frac{d A_s f_y}{d_p A_{ps}}$$

and

$$f_{ps} = \frac{0.85\beta_1 f_{pu}}{q''} \times \frac{\epsilon_u}{\epsilon_u + \epsilon_{su} - \epsilon_{se} - \epsilon_{ce}} - \frac{dA_s f_y}{d_p A_{ps}}$$

$$f_{ps} = \frac{0.85 \times 0.675 \times 270}{0.166} \times \frac{0.003}{0.003 + \epsilon_{su} - 0.0057 - 0.00034} - \frac{25.5 \times 4.0 \times 60}{24.8 \times 2.75}$$

$$f_{ps} = \frac{2.800}{\epsilon_{su} - 0.00304} - 89.70$$

and

$$f_{ps} = 232.2 + 1080\epsilon_{su}$$

Solving the two equations simultaneously gives

$$f_{ps} = 244.53 \text{ ksi}$$

$$\epsilon_{su} = 0.011417$$

$$T = 4.0 \times 60 + 244.53 \times 2.75 = 912.4 \text{ k}$$

$$k_u = \frac{0.003}{0.011417 - 0.00304} = 0.358$$

$$a = 0.675 \times 0.358 \times 24.8 = 6.00 \text{ in.}$$

$$\text{Ave. flange thickness} = \frac{5 \times 24 + 1.5\,(9 + 6)}{24} = 5.94 \text{ in.} < 6 \text{ in.}$$

$$C = 0.85 \times 7.5 \times 24 \times 6.00 = 918.0 \text{ k}$$

$$M_n = \frac{240}{12}\left(30.0 - 4.5 - \frac{6.0}{2}\right) + \frac{672}{12}$$

$$\cdot \left(30.0 - 5.20 - \frac{6.00}{2}\right)$$

$$= 450 + 1221 = 1671 \text{ k-ft}$$

Using Eq. 5-28

$$\omega = \frac{4.00 \times 60}{24 \times 25.5 \times 7.5} = 0.523$$

$$\omega' = 0$$

$$\rho_p = \frac{2.75}{24 \times 24.8} = 0.00462$$

$$f_{ps} = 270\left(1 - \frac{0.28}{0.675}\left[\frac{0.00462 \times 270}{7.5} + \frac{25.5 \times 0.0523}{24.8}\right]\right) = 245.3 \text{ ksi}$$

$b = 60.0$ in. (flange width)	$f'_c = 4.0$ ksi
$b_w = 16.0$ in. (web width)	$f_{pu} = 270$ ksi
$h = 32.5$ in. (overall height)	$f_{py} = 229.5$ ksi
$d_p = 27.75$ in. (depth to A_{ps})	$f_y = 60$ ksi
$d = 30.0$ in. (depth to A_s)	$A_s = 4.00$ in.^{2}
$d' = 1.5$ in. (depth to A'_s)	$A'_s = 1.00$ in.^{2}
$h_f = 4.5$ in. (flange thickness)	

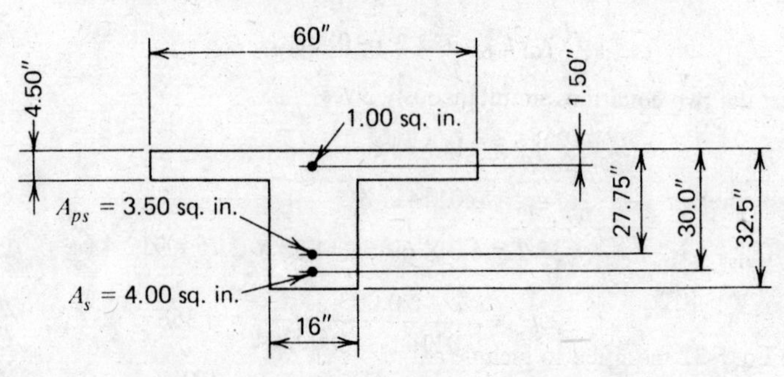

Fig. 5-19 Cross section of T-beam having compression reinforcement.

$$T = 245.3 \times 2.75 + 4 \times 60 = 914.6$$

$$a = \frac{914.6}{0.85 \times 24 \times 7.5} = 5.98 \text{ in.}$$

$$M_n = \frac{674.5}{12}\left(24.8 - \frac{5.98}{2}\right) + \frac{240}{12}\left(25.5 - \frac{5.98}{2}\right) = 1676 \text{ k-ft}$$

ILLUSTRATIVE PROBLEM 5-6 Compute the nominal moment capacity of the T-shaped beam having the non-prestressed tension and compression reinforcement shown in Fig. 5-19. The principal dimensions and properties of materials are given in the figure.

SOLUTION:

$$\rho_p = \frac{3.50}{60 \times 27.75} = 0.00210$$

$$\omega = \frac{4.0 \times 60}{60 \times 30 \times 4.0} = 0.03333$$

$$\omega' = \frac{1.0 \times 60}{60 \times 30 \times 4.0} = 0.00833$$

$$\frac{f_{py}}{f_{pu}} = \frac{229.5}{270.0} = 0.85$$

$$\beta_1 = 0.85$$

(Note that from Eq. 5-39, $0.132 \times 27.75 = 3.36$ in. $> d' = 1.50$ in.)

Using Eq. 5-28:

$$f_{ps} = 270\left(1 - \frac{0.40}{0.85}\left[0.00210 \times \frac{270}{4} + \frac{30}{27.75}(0.025)\right]\right) = 248.6 \text{ ksi}$$

$$\omega_p + \frac{d}{d_p}(\omega - \omega') = 0.158 < 0.36\beta_1 = 0.306$$

$$T = 3.50 \times 248.6 + 4.0 \times 60 = 870.1 + 240 = 1110.1 \text{ k}$$

For a rectangular member with compression reinforcement,

$$C_{max} = 0.85 \times 4.0 \times 60.0 \times 4.50 + 1.0 \times 60 = 918.0 + 60.0 = 978.0 \text{ k}$$

Because $T > C_{max}$, the member must be analyzed as a flanged section.

From Eq. 5-32 modified to include compression reinforcement (See Eq. 5-35):

$$A_{pw} f_{ps} = 870.1 + 240 - 0.85 \times 4.0(60.0 - 16.0)4.5 - 1.0 \times 60$$

$$= 870.1 + 240 - 673.2 - 60$$

$$= 376.9 \text{ k}$$

and

$$\omega_{pw} = \frac{376.9}{16 \times 27.75 \times 4.0} = 0.2121$$

$$\omega_{pw} + \frac{d}{d_p}(\omega - \omega') = 0.239 < 0.36\beta_1 = 0.306$$

and, using Eq. 5-31, modified to include compression reinforcement (See Eq. 5-35):

$$a = \frac{376.9}{0.85 \times 4.0 \times 16.0} = 6.92 \text{ in.}$$

$$M_n = \frac{376.9}{12}(27.75 - 3.46) + \frac{240}{12}(30.0 - 27.75) + \frac{673.2}{12}(27.75 - 2.25)$$

$$+ \frac{60}{12}(27.75 - 1.50)$$

$$= 762.3 + 45.0 + 1430.6 + 131.3 = 2369.2 \text{ k-ft}$$

5-5 Ultimate Moment Code Requirements—Unbonded Tendons

The Code requirements for members prestressed with unbonded tendons are the same as those for members with bonded tendons, with the exception that f_{ps} is specified to be as follows:

For members with unbonded prestressing tendons and with a span-to-depth ratio of 35 or less:

$$f_{ps} = f_{se} + 10{,}000 + \frac{f'_c}{100 p_p} \tag{5-42}$$

but f_{ps} in Eq. (5-42) shall not be taken greater than f_{py}, nor $(f_{se} + 60{,}000)$.

For members with unbonded prestressing tendons and with a span-to-depth ratio greater than 35:

$$f_{ps} = f_{se} + 10{,}000 + \frac{f'_c}{300 p_p} \tag{5-43}$$

but f_{ps} in Eq. (5-43) shall not be taken greater than $(f_{se} + 30{,}000)$.

In Eqs. 5-42 and 5-43 f_{ps}, f_{se}, and f'_c are in psi. When information is available for determining a more accurate value of f_{ps}, it may be used. Like Eq. 5-28, Eqs. 5-42 and 5-43 are limited for use in applications where f_{se} is not less than $0.50 f_{pu}$.

The relationships given in Eqs. 5-35 through 5-37 for determining if the members are to be analyzed as under- or over-reinforced are also applicable to members with unbonded tendons. Members with unbonded tendons, in general, must have ultimate moment capacities equal to or greater than 1.2 times the cracking moment with the modulus of rupture equal to $7.5\sqrt{f'_c}$, as do bonded members.

Unbonded members that do not have bonded, non-prestressed reinforcement could be subject to sudden brittle failure. For this reason the Code requires a minimum amount of non-prestressed reinforcement in the tensile zone of members stressed with unbonded tendons. The minimum amount of bonded reinforcing A_s, except for two-way flat plates (solid slabs of uniform thickness) is specified to be

$$A_s = 0.004A \tag{5-44}$$

in which A is the area of the concrete section between the flexural tension face and the center of gravity of the gross section. The reinforcing is to be placed as close as possible to the extreme tension fiber and uniformly distributed. It is required regardless of the stresses existing in the member under service loads.

It is interesting to note that the Uniform Building Code (UBC) (Ref. 6) requires one-way post-tensioned beams and slabs having unbonded reinforcement to be designed to carry the dead load plus 25% of the unreduced live load tributary to the member by some method other than the unbonded post-tensioned reinforcement. This provision applies to the design being made for strength based upon load factors and strength reduction factors of 1 (see Article 5-5). Compliance with this requirement is normally achieved through the provision of non-prestressed reinforcement detailed to comply with all requirements of the UBC (i.e., development lengths, minimum embeddments, etc.). (This provision is not included in the ACI 318 Requirements.)

In the case of flat plates, no bonded reinforcement is required in areas of positive moment when the concrete tensile stresses do not exceed $2\sqrt{f_c'}$ after all losses. If the tensile stress does exceed $2\sqrt{f_c'}$, the minimum area of bonded steel is

$$A_s = \frac{N_c}{0.5 f_y} \tag{5-45}$$

in which f_y cannot exceed 60,000 psi and N_c is the tensile force in the concrete under the sum of the service dead and live loads. Again, the steel must be uniformly distributed over the section and as close as practicable to the tension fiber. In areas of negative moment the minimum amount of bonded reinforcement required in each direction is:

$$A_s = 0.00075\,hl \tag{5-46}$$

in which l is the length of the span in the direction parallel to the reinforcement being considered and h is thickness of the member. The bonded reinforcing is to be placed in a width not exceeding the width of the supporting column plus 3 h, the maximum spacing of the bars is 12 in., and there must be at least four bars or wires in each direction. The Code contains other provisions for determining the minimum lengths of the reinforcement (Ref. 2).

ILLUSTRATIVE PROBLEM 5-7 Compute the nominal moment capacity for the member shown in Fig. 5-20. The member is stressed with an unbonded tendon, f_{se} = 144 ksi, f_{pu} = 240 ksi, f_{py} = 192 ksi, f_c' = 6.0 ksi and the span is 40.0 ft. Use the Eq. 5-42 for f_{ps} and the method proposed by Pannell.

SOLUTION:

By Eq. 5-42.

$$\rho_p = \frac{3.20}{24 \times 24.8} = 0.00538$$

$$f_{ps} = 144 + 10 + \frac{6.0}{0.538} = 165 \text{ ksi} \begin{array}{l} < f_{se} + 60 = 204 \text{ ksi} \\ < f_{py} = 192 \text{ ksi} \end{array}$$

$$A_{ps} f_{ps} = 3.20 \times 165 = 528 \text{ kips}$$

$$a = \frac{528}{0.85 \times 6 \times 24} = 4.31 \text{ in.}$$

$$M_u = \frac{528(24.8 - 2.16)}{12} = 996 \text{ ft-kips}$$

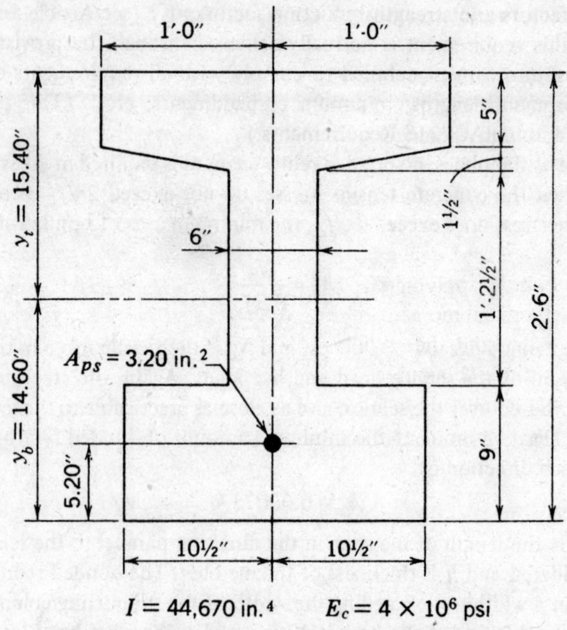

Fig. 5-20 Cross section of a beam used in Problem 5-7.

By Pannell's method:

$$\rho_p = \frac{3.20}{24 \times 24.8} = 0.00538, \qquad q_e = \frac{0.00538 \times 144}{6} = 0.129$$

$$\lambda = \frac{10^6 \times 0.00538 \times 2.5}{6000 \times 40.0} = 0.056$$

$$q_u = \frac{0.129 + 0.056}{1 + 0.0895} = 0.170$$

$$M_u = \frac{0.170(1 - 0.80 \times 0.170)6 \times 24 \times 24.8^2}{12} = 1085 \text{ ft-kips}$$

5.6 Strength Reduction Factor

The relationships given in Articles 5-2 through 5-5 have been given for use in computing the nominal moment capacity in prestressed concrete flexural members. The nominal moment capacity (or nominal strength) is the strength one would expect if the equations used in the calculations were accurate, the materials actually used in the construction had the stress and strain properties assumed in the calculations, and the members were constructed with dimensions equal to those assumed in the calculations. These conditions do not exist consistently in practice and hence a strength reduction factor, ϕ, is used to reduce the nominal strengths calculated with the equations given previously in this chapter, and these modified strengths are compared to the minimum strengths required by the applicable code or standard being used as the design criterion.

The strength reduction factors given in ACI 318 for flexure and shear are 0.90 and 0.85, respectively. The factor for shear is lower than that for flexure in view of the more brittle (less ductile) nature of shear failures as compared to flexural failures of under-reinforced members.

Hence, when one is applying the strength equations given in this chapter in actual design, the nominal moment capacity, M_n, is to be multiplied by the strength reduction factor, ϕ, and the product compared to the minimum strength permitted by the criteria being used. If ϕM_n is less than the minimum strength required, adjustments must be made in the design in order to increase the strength to equal or exceed the minimum required.

5-7 Shear and Shear Reinforcement

Two types of flexural shear cracking are currently recognized. These are illustrated in Fig. 5-21 and are described as follows:

Type I This type of cracking is associated with flexural cracking. Some authorities believe that for this type of crack to adversely affect the capacity of a member, it must be cracked in such a manner that the horizontal projection of the crack is equal in length to the depth of the member, and for this reason, it is believed a flexural crack that occurs at a distance equal to the depth of the member away (in the direction of lesser moment) from the section being investigated may lead to a critical crack (Reg. 7). In addition, principal tensile stresses along a potential crack may be aggravated by flexural cracks that may occur in the

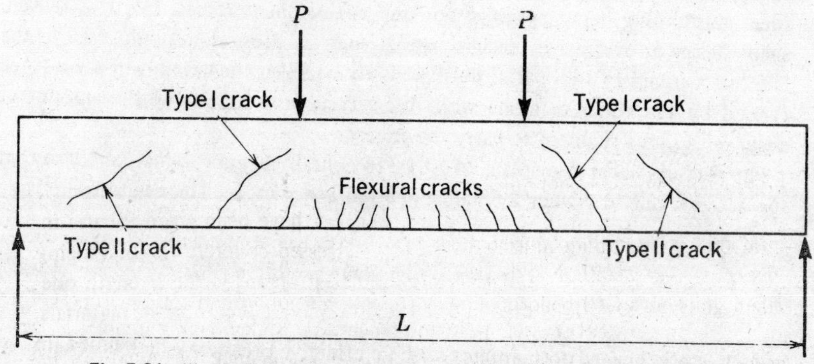

Fig. 5-21 Illustration of flexural cracks and type I and type II shear cracks.

vicinity of the potential type I crack, and since principal tensile stresses are normally maximum at the center of gravity of a beam (can be taken to be approximately equal to one-half of the depth of the beam), a flexural crack at one-half the beam depth from the section under consideration can be considered to cause a type I crack. Shear cracking in reinforced- and prestressed-concrete beams are generally type I cracks. These *flexural-shear* cracks begin as flexural cracks extending approximately vertically into the beam. When a critical combination of flexural and shear stresses develop near the top of a flexural crack, the inclined crack forms

Type II This type of cracking is associated with principal tensile stresses in areas where there are no flexural cracks, and originates in the web of the member near the centroid where the principal tensile stresses are the greatest; it subsequently extends towards the flanges (Ref. 7). Type II cracking is fairly unusual. It may appear near the supports of highly prestressed beams that have thin webs. It may also appear near inflection points and bar cut-off points of continuous, reinforced-concrete members under axial tension (Ref. 8).

Hence, in designing flexural members for shear, one must consider each type of cracking and determine the amount of shear reinforcing that is required for each of them. For members subject to moving live loads, this is sometimes confusing, and a good understanding of the fundamental differences between the types of cracking is helpful in gaining confidence and proficiency in flexural shear design. The interested reader will find a complete discussion of the shear behavior of reinforced and prestressed concrete in Ref. 10. A discussion of some of the controversial factors and provisions related to the design of members for shear is contained in Ref. 11.

In many parts of the world it is customary to assume that once a crack has formed in the concrete, all of the shear force must be carried by shear-reinforcing steel, in both reinforced and prestressed concrete beams. Traditionally, North American practice in the design of concrete flexural members for shear, has been to assume that a portion of the shear force is carried by the concrete section, shear reinforcing being provided for any excess shear forces. For this reason, shear design provisions in design criteria such as those commonly used in the U.S. are formed of two parts; the first is to establish the amount of shear force carried by the concrete alone while the second is to determine the amount of shear reinforcing required to carry the excess.

The flexural shear provisions of ACI 318 contain an approximate method that the designer may use when a detailed analysis is not made. This relationship is:

$$V_c = \left(0.6\sqrt{f_c'} + 700\,\frac{V_u d}{M_u}\right) b_w d \tag{5-47}$$

and its use is limited to members with an effective prestress which is equal to at least 40% of the tensile strength of the flexural reinforcement. In Eq. 5-47 V_u and M_u are the total applied design shear force and moment at a section, respectively. When the approximate relationship is used, the value of V_c need not be taken to be less than $2\sqrt{f'_c}\,b_w d$ and cannot be taken to be greater than $5\sqrt{f'_c}\,b_w d$. In addition, the term $V_u d/M_u$ cannot exceed unity. The term d in Eq. 5-47 is the distance from the extreme compression fiber to the centroid of the longitudinal tension reinforcement and b_w is the width of the web.

When applied to simply supported spans subjected to uniformly applied loads alone, Eq. 5-47 becomes:

$$V_c = \left(0.6\sqrt{f'_c} + 700\,\frac{d(l-2x)}{x(l-x)}\right) b_w d \qquad (5\text{-}48)$$

in which l is the span length and x is the distance from the support to the point under consideration. Equation 5-48 can be represented graphically as shown in Fig. 5-22 (from Ref. 12).

The designer may elect to make a more detailed analysis for the shear design of prestressed concrete flexural members, in conformance with the requirements of ACI 318, by determining the amount of shear reinforcing required to guard against failure as the result of types I and II cracking. Two separate analyses are required because type I cracking is a function of moment and shear while type II cracking is not a function of moment. In the case of members designed for moving loads, because maximum moment and maximum shear do not necessarily occur at any one section under the same condition of loading, more effort is required to make a complete analysis than in the case of a member subjected to non-moving live loads.

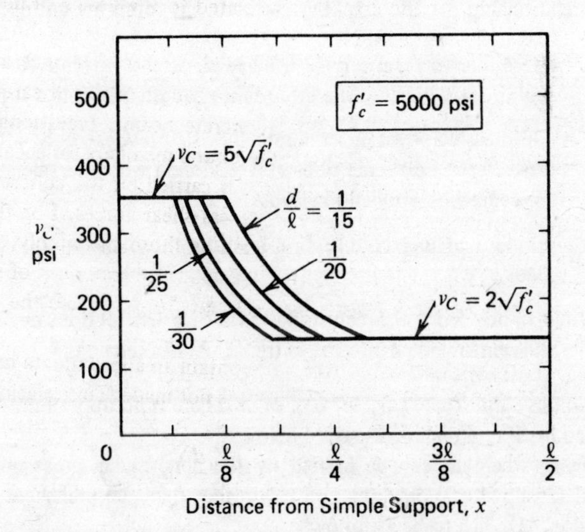

Fig. 5-22 Application of Eq. 5-48 to uniformly loaded prestressed concrete members.

The estimation of the shear force that can safely be carried by the concrete in areas subject to flexural cracking (type I), as provided in ACI 318, is by

$$V_{ci} = 0.6\sqrt{f'_c}b_w d + V_d + \frac{V_i M_{cr}}{M_{max}} \tag{5-49}$$

in which V_d is the shear force due to service dead load (unfactored) from the self-weight of the member, including any cast-in-place slabs or toppings and superimposed dead loads whether acting on a composite or non-composite member; V_i is equal to V_u minus V_d; M_{max} is equal to M_u less the moment due to service dead load (unfactored) from the self-weight of the member (M_{max} and V_i are concomitant); b_w is the width of the web; d is the effective depth; and M_{cr} is the cracking moment. In applying Eq. 5-49, one uses the loading combination which results in the greatest value of M_{max}, not V_i. V_i is the shear force concomitant with M_{max}; it is not necessarily the greatest design shear force at the section under consideration. (This fact has caused considerable difficulty among structural designers who have become accustomed to use the loading that causes the greatest shear force when designing for shear; not the loading that causes the greatest moment). The cracking moment is defined as

$$M_{cr} = (I/y_t)(6\sqrt{f'_c} + f_{pe} - f_d). \tag{5-50}$$

The equation for M_{cr} is given as it appears in ACI 318; f_{pe} and f_d are both taken to be positive. Using the sign notation followed in this book (tensile stresses are negative) this equation should be written with a plus sign between f_{pe} and f_d rather than the negative sign used in ACI 318.

In the relationship for the cracking moment, the terms are defined as follows:

f_d = stress due to total service dead load, at the extreme fiber of a section at which tensile stresses are caused by applied load, psi

f_{pe} = compressive stress in concrete due to prestress only after all losses, at the extreme fiber of a section at which tensile stresses are caused by applied loads, psi

I = moment of inertia of section resisting externally applied design loads

y_t = distance from the centroidal axis of gross section, neglecting the reinforcement, to the extreme fiber in tension

The minimum value for V_{ci} is $1.7\sqrt{f'_c}\, b_w d$. There is no upper limit for V_{ci}; this is provided by V_{cw} which is discussed below.

The shear force that can be carried by the concrete in areas where type II (principal tensile stress) cracking controls rather than flexural-shear cracking, is stipulated in ACI 318 to be as follows:

$$V_{cw} = (3.5\sqrt{f'_c} + 0.3f_{pc})b_w d + V_p \tag{5-51}$$

In lieu of using Eq. 5-51, V_{cw} can be taken as the shear corresponding to a multiple of dead load plus live load, which results in a computed principal tensile stress of $4\sqrt{f'_c}$ at the centroidal axis of the member, or at the intersection of the flange and the web when the centroidal axis is in the flange.

In Eq. 5-51, f_{pc}, V_p, b_w and d are defined as follows:

b_w = web width, or diameter of circular section, in.

d = distance from extreme compression fiber to centroid of tension reinforcement, in.

f_{pc} = compressive stress in the concrete, after all prestress losses have occurred, at the centroid of the cross section resisting the applied loads or at the junction of the web and flange when the centroid lies in the flange, psi. (In a composite member, f_{pc} will be the resultant compressive stress at the centroid of the composite section, or at the junction of the web and flange when the centroid lies within the flange, due to both prestress and to the bending moments resisted by the precast member acting alone)

V_p = vertical component of the effective prestress force at the section considered

In members which are prestressed in one direction only, the unit shear stress that will cause a principal tensile stress of $4\sqrt{f'_c}$ becomes

$$v_{cw} = \sqrt{4\sqrt{f'_c}(4\sqrt{f'_c} + f_{pc})} + \frac{V_p}{b_w d} \tag{5-52}$$

If the member is provided with prestressing in two directions, such as prestressed stirrups in addition to longitudinal prestressing, the relationship for v_{cw} becomes

$$v_{cw} = \sqrt{(4\sqrt{f'_c} + f'_{pc})(4\sqrt{f'_c} + f_{pc})} + \frac{V_p}{b_w d} \tag{5-53}$$

in which f'_{pc} is the prestressing stress acting at 90° from f_{pc}.

It should be noted that in the relationships given above for V_c, V_{ci}, and V_{cw}, it is assumed that the concrete is made of normal sand and gravel and not of lightweight aggregates. A provision of ACI 318 stipulates that $0.75\sqrt{f'_c}$ and $0.85\sqrt{f'_c}$ will be substituted for the term $\sqrt{f'_c}$ when "all-lightweight" and "sand-lightweight" concrete is used in lieu of normal weight concrete aggregates.

In applying the above equations for V_{ci} and V_{cw}, d is to be taken as the depth from the extreme compression fiber to the centroid of the longitudinal tension reinforcement or, 0.8 times the overall thickness of the member, whichever is greater. In Eq. 5.49, M_{max} and V_i are to be computed from the condition of loading causing the greatest moment at the section under consideration. In addition, the reduction in the effective prestressing force at a section, due to the length required to transfer the prestress to the concrete, must be taken into account when computing V_{cw} in pretensioned members. (*See* Sec. 5.9.)

The shear design at each section is to be based upon

$$V_u \leqslant \phi V_n \tag{5-54}$$

and

$$V_n = V_c + V_s \tag{5.55}$$

in Eqs. 5-54 and 5-55

V_c = nominal shear strength provided by the concrete; either that obtained from Eq. 5-47 or the lesser of that obtained from Eqs. 5-49 and 5-51.

V_n = nominal shear strength.

V_s = nominal shear strength provided by shear reinforcement.

V_u = total design (factored) shear force at a section. Note that for use in computing V_c and V_{cw} from Eqs. 5-47 and 5-51 the value of V_u to be used in Eq. 5-54 is the greatest value that occurs at the section. For use with V_{ci} from Eq. 5-49 the value of V_u to be used in Eq. 5-54 is a function of the shear force occurring from the load distribution causing maximum moment; *this may not be the greatest shear force that might occur at the section.*

ϕ = capacity reduction factor which for shear design = 0.85.

When the value of V_u is greater than V_c, V_{ci}, or V_{cw}, whichever is applicable, shear reinforcement must be provided for the shear stress in excess of the amount the concrete can carry. For the usual case of shear reinforcement placed perpendicular to the longitudinal reinforcement, the amount of shear the reinforcement can sustain is

$$V_s = \frac{A_v f_y d}{s} \tag{5-56}$$

in which A_v is the area of the shear reinforcement perpendicular to the flexural tension reinforcement within a distance s, and the other terms have been defined previously. Note that Eq. 5-56 can be rewritten to

$$\frac{V_s}{b_w d} = v_s = \frac{A_v f_y'}{b_w s} \tag{5-57}$$

or

$$A_v = \frac{v_s b_w s}{f_y} \tag{5-58}$$

in which v_s is the unit shear stress.

Other provisions are contained in ACI 318 for the case of shear reinforcing that is not placed perpendicular to the longitudinal reinforcing.

It should be noted that vertical stirrups are normally placed at a maximum spacing of 0.75 times the thickness of prestressed members, but when V_s exceeds $4\sqrt{f'_c}b_w d$, the maximum spacing permitted is 0.375 times the thickness. In addition, in order to guard against principal *compression* failures, the value of V_s is limited to $8\sqrt{f'_c}b_w d$.

It is interesting to note that the reduction of the shear force acting upon the section due to the vertical component of the prestressing is included in Eq. 5-51 but not in Eq. 5-49. Tests of prestressed concrete members with and without inclined tendons has shown this to be appropriate. In designing for shear in prestressed members, it is also appropriate to include the effects of variable depth of the section (see below) and shear forces resulting from prestress-induced deformations (see Chapter 8).

Beams of variable depth and normal configuration have less shear force on their webs in areas where the compression flange is inclined to the gravity axis than would be revealed from a usual analysis of the flexural shear forces. The principle is illustrated in Fig. 5-23 in which a freebody diagram of a portion of a variable depth continuous beam is shown. The portion of the beam shown is

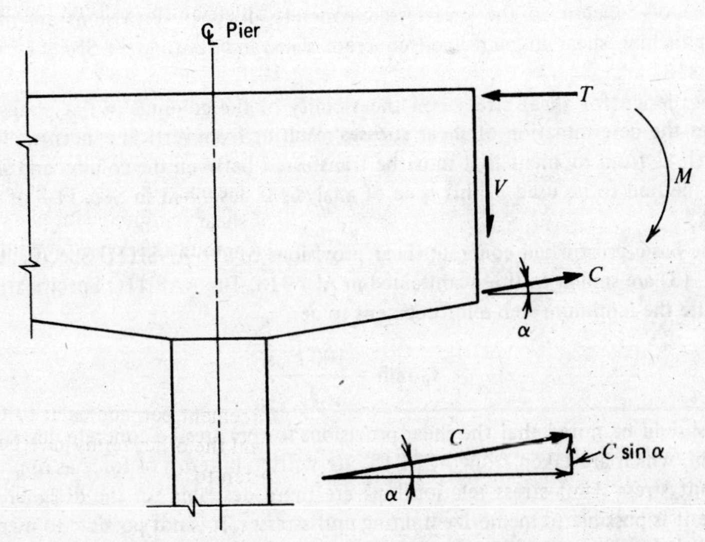

Fig. 5-23 Freebody of a portion of a bridge superstructure having variable depth. (Résal effect.)

near the support where both the shear force and the negative moment are large. If the angle of inclination of the bottom slab with respect to the gravity axis of the member is taken as α and the force in the compression flange is designated as C, there is a vertical component of the force C. The vertical component is equal to $C \sin \alpha$. The vertical component of the force C acts in a direction that reduces the shear force acting on the webs of the member. When applying this effect in an analysis under design loads, the force C should be determined based upon the design moment that is concomitant with the design shear force being considered.

Minimum shear reinforcing provisions are contained in ACI 318 for prestressed concrete, as they are for reinforced concrete. Slabs, footings, and concrete joist construction are exempt from these requirements. The minimum amount of shear reinforcing is

$$A_v = 50\, b_w s / f_y \tag{5-59}$$

for reinforced or prestressed members not subject to significant torsional moments (see ACI 318). For prestressed members the relationship

$$A_v = \frac{A_{ps}}{80} \frac{f_{pu}}{f_y} \frac{s}{d} \sqrt{\frac{d}{b_w}} \tag{5-60}$$

may be used if the effective prestress is at least 40% of the tensile strength of the flexural reinforcement. The shear stress is taken to be maximum at a distance equal to one-half the depth of the member from the support.

Punching shear stresses, as might be caused by a concentrated wheel load on a prestressed slab or by the supporting column on a flat plate structure, are a source of concern to the prestressed concrete designer. Provisions are made for punching shear in reinforced concrete slabs and footings in Sec. 11-11 of ACI 318.

The design for shear stresses in the vicinity of the columns in flat plates involves the determination of shear stresses resulting from vertical concentric load as well as from moment that must be transferred between the column and slab. The method to be used in this type of analysis is described in Sec. 11-8 of this book.

The basic prestressed concrete shear provisions of the AASHTO Specification (Ref. 13) are similar to those contained in ACI 318. The AASHTO Specifications require the minimum web reinforcement to be:

$$A_v \min = \frac{100\, b_w s}{f_y} \tag{5-61}$$

It should be noted that the shear provisions for prestressed concrete described herein, which are taken from ACI 318, are written in terms of force as opposed to unit stress. Unit stress relationships are more desirable for the designer because it is possible to memorize limiting unit stresses. It is not possible to memorize limiting forces because they are a product of limiting stresses and two dimensions of the cross section of a flexural member. Hence, it is recommended the structural designer use unit stresses in his routine design work. The forces can be shown at the completion of the calculations if necessary to show specific compliance with ACI 318.

Limiting stresses the designer should keep in mind are:

$v_c = 1.7\sqrt{f'_c}$, min value for V_{ci} in Eq. 5-49
$v_c = 2.0\sqrt{f'_c}$, min value for V_c in Eq. 5-47
$v_c = 4.0\sqrt{f'_c}$, max value for shear reinforcement to be used at normal spacings; use half spacings for higher stresses
$v_c = 5.0\sqrt{f'_c}$, max value for V_c in Eq. 5-47
$v_s = 8.0\sqrt{f'_c}$, max value of shear stress to be carried by shear reinforcement

5.8 Shear Design Expedients

As discussed in Chapter 5, Art. 5-7, ACI 318 contains several limitations for the design of prestressed concrete members for shear. In the preparation of actual design calculations, the designer should check these limitations at the outset of design for shear to confirm the following:

1. The applicability of the simplified equation for shear design (Eq. 5-47) if

$$f_{se}A_{ps} \geqslant 0.40(f_y A_s + f_{ps}A_{ps})$$

Equation 5-47 may be used; if not, Eqs. 5-49 and 5-51 must be used to determine the nominal shear strength of the concrete section (V_c). This is an important consideration in members reinforced with non-prestressed and prestressed flexural reinforcement.

2. If the calculated maximum value of the unit shear stress v_u, as modified by the strength reduction factor ϕ, does not exceed $2\sqrt{f'_c}$ and Eq. 5-47 is being used (or $1.7f'_c$ if Eq. 5-49 is being used), i.e.,

$$\frac{v_{u\,\max}}{\phi} < 2\sqrt{f'_c}$$

no further analysis is needed and the minimum shear reinforcement determined by Eq. 5-56 or 5-60 is to be provided. On the other hand, if $2v_{u\,\max}/\phi \leqslant 2\sqrt{f'_c}$ and Eq. 5-47 is being used (or $1.7\sqrt{f'_c}$ if Eq. 5-49 is being used), the minimum web reinforcement is not required.

3. The maximum shear force that can be resisted by shear reinforcing is $8\sqrt{f'_c}b_w d$. Hence, if Eq. 5-51 is being used

$$\frac{V_{u\,\max}}{\phi} \leqslant [8\sqrt{f'_c} + 2\sqrt{f'_c}]\,b_w d$$

or

$$\frac{v_{u\,\max}}{\phi} \leqslant 10\sqrt{f'_c}$$

but if Eq. 5-49 is being used

$$\frac{v_{u\,max}}{\phi} \leqslant 9.7\sqrt{f_c'}$$

4. The value of $V_{u\,max}$ or $v_{u\,max}$ should be taken as the value at $h/2$ from the support in members when the direction of the reaction induces compression in the end regions.

For simply supported members subject to uniformly distributed loads, the term $V_u d/M_u$ in Eq. 5-49 becomes

$$\frac{V_u d}{M_u} = \frac{d}{x}\,\frac{(l-2x)}{(l-x)}$$

where

x = distance from the section being investigated to the support

In computation of f_d, for use in Eq. 5-50, the extreme fiber stress for the total unfactored dead load, the appropriate section properties must be used. In the case of noncomposite members

$$f_d = \frac{y(M_{DL} + M_{SDL})}{I}$$

While for composite members

$$f_d = \frac{yM_{DL}}{I} + \frac{y'M_{SDL}}{I'}$$

in which

M_{DL} is the total dead load moment due to loads applied to the noncomposite section.

M_{SDL} is the total superimposed load moment due to loads applied to the composite section.

y and y' are the distances from the extreme fibers subject to tensile stresses from the dead loads for the noncomposite and composite sections, respectively.

I and I' are the moments of inertia for the noncomposite and composite sections, respectively.

When using Eq. 5-51 the computation of f_{pc} for noncomposite members is simply the stress at the centroidal axis of the member since in the equation

$$f_{pc} = \frac{f_{se}A_{ps}}{A}\left(1 + \frac{ey}{r^2}\right) + \frac{yM_{TDL}}{I}$$

$y = 0$; hence, the stress is equal to the effective prestressing force divided by the area of the section. In the case of composite members, however, the computation must be made for the stress at the location of the centroidal axis of the composite section or at the level of the web-flange intersection. Hence, when the centroidal axis of the composite section is not within the flange, the computation becomes

$$f_{pc} = \frac{f_{se}A_{ps}}{A}\left(1 + \frac{e\Delta y}{I}\right) + \frac{\Delta y M_{DL}}{I}$$

Where Δy is the distance between the centroidal axis of the composite and noncomposite sections, and M_{DL} is the total dead load moment acting on the noncomposite section.

For composite sections in which the centroidal axis of the composite section falls within the flange, the computation of f_{pc} becomes

$$f_{pc} = \frac{f_{se}A_{ps}}{A}\left(1 + \frac{e\Delta y'}{I}\right) + \frac{\Delta y' M_{DL}}{I}$$

where $\Delta y'$ is the distance between the centroidal axis of the noncomposite section and the intersection of the flange and web.

It should be noted that y dimensions measured above a centroidal axis are positive in the above computations for f_d and f_{pc}.

ILLUSTRATIVE PROBLEM 5-8 Using the provisions of ACI 318, investigate the double-tee slab shown in Fig. 5-24 for web reinforcing using the simplified analysis. The dead load of the double tee slab is 200 plf, the superimposed live load is 240 plf, the design span is 40 ft, f_c' is 5000 psi, and the prestressing consists of straight tendons with $A_{ps} = 0.58$ sq in., $P_{se} = 90.0$ kips, and $f_{pu} = 270$ ksi. The tendons are located 2 in. above the soffit. Plot the results. Use $b_w = 8.00$ in.

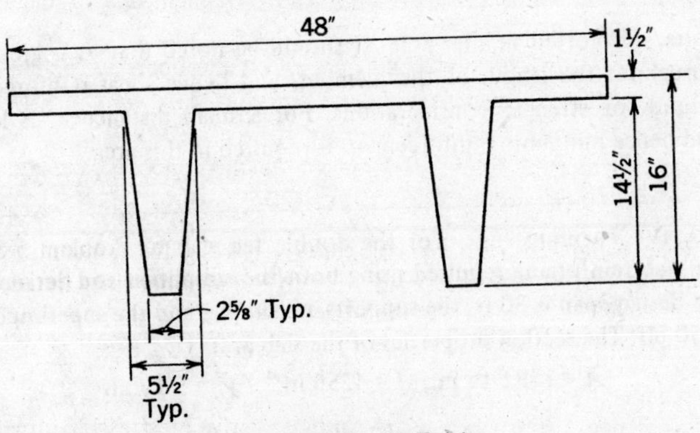

Fig. 5-24 Double T slab.

SOLUTION: Reactions:

	Service Load		Design Load
D.L. 0.200 × 20 ft =	4.00^k	× 1.4 =	5.60^k
S.L. 0.240 × 20 ft =	4.80^k	× 1.7 =	8.16^k
	8.80^k		13.76^k

$$V_u = 13{,}760 \text{ lb} \qquad v_u = \frac{13{,}760}{8 \times 14} = 123 \text{ psi}$$

$$\frac{V_u}{\phi} = 16{,}188 \text{ lb} \qquad \frac{v_u}{\phi} = 145 \text{ psi}$$

(Note: $\phi = 0.85$ for shear calculations.)

$$v_c \min = 2\sqrt{f'_c} = 141 \text{ psi}$$

$$v_c \max = 5\sqrt{f'_c} = 353 \text{ psi}$$

$$V_c = \left(0.6\sqrt{f'_c} + 700\frac{V_u d}{M_u}\right) b_w d = \left(0.6\sqrt{f'_c} + \frac{700d(l-2x)}{12x(l-x)}\right) b_w d$$

$$v_c = \frac{V_c}{b_w d} = 42 + \frac{(700)(14)(40-2x)}{12x(40-x)}$$

Solving for v_c gives

x (ft)	v_c (psi)
2	429
4	223
6	154

The results are plotted in Fig. 5-25. It should be noted that v_c is greater than v_u/ϕ throughout the length of the member, and hence shear reinforcement is not required for strength considerations. For a small distance v_c is less than $2v_u/\phi$ and hence minimum reinforcement is required in this area.

ILLUSTRATIVE PROBLEM 5-9 For the double tee slab of Problem 5-8, determine the web reinforcing required using both the simplified and detailed analyses if the design span is 30 ft, the supports are simple, and the superimposed live load is 650 plf. The section properties of the slab are:

$$A = 189.5 \text{ sq in.}, \quad I = 4256 \text{ in.}^4, \quad y_t = 5.17 \text{ in.}$$

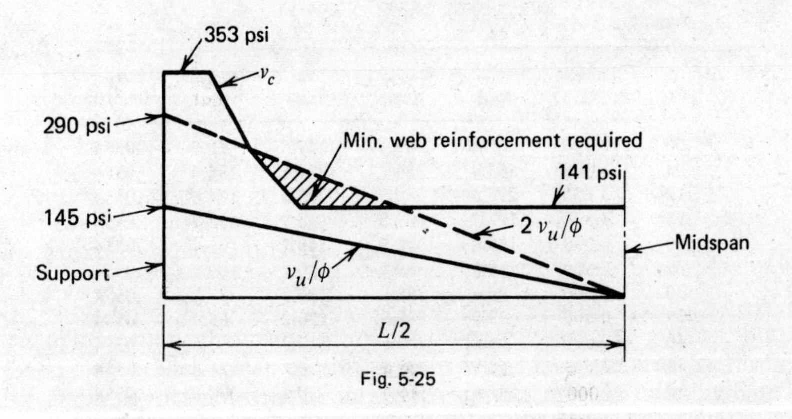

Fig. 5-25

SOLUTION:
Simplified analysis:

$$V_u = (1.4 \times 0.2 + 1.7 \times 0.65)\ 15\ \text{ft} = 20.78\ \text{kips}$$
$$v_u/\phi = 20{,}780/0.85 \times 8 \times 14 = 218\ \text{psi}$$
$$v_c\ \text{max} = 5\sqrt{f_c'} = 353\ \text{psi}$$
$$v_c\ \text{min} = 2\sqrt{f_c'} = 141\ \text{psi}$$

The computations are plotted in Fig. 5-26.

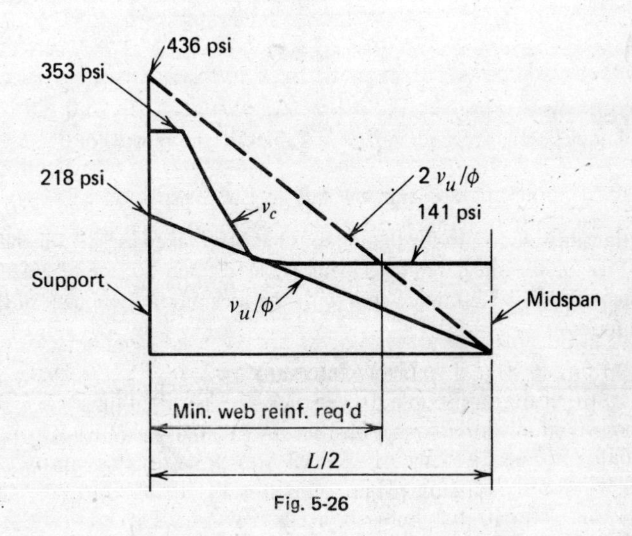

Fig. 5-26

TABLE 5-1

Pt.	Length (ft)	v_{ci} (psi)	v_{cw} (psi)	v_c (psi)	v_u (psi)	A_v (in.^{2}/ft)
.00	.000	infin	247.4	247.4	218.2	.0554
.05	1.500	563.4	384.2	384.2	196.4	.0554
.10	3.000	295.5	389.9	295.5	174.5	.0554
.15	4.500	198.7	389.9	198.7	152.7	.0554
.20	6.000	149.2	389.9	149.2	130.9	.0554
.25	7.500	118.3	389.9	120.2	109.1	.0554
.30	9.000	96.6	389.9	120.2	87.2	.0554
.35	10.500	79.9	389.9	120.2	65.4	.0554
.40	12.000	66.1	389.9	120.2	43.6	.0554
.45	13.500	53.9	389.9	120.2	21.8	.0554
.50	15.000	42.4	389.9	120.2	000.0	.0554

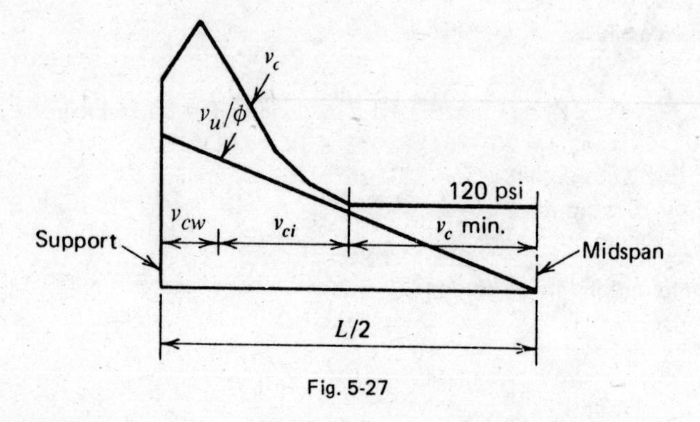

Fig. 5-27

Detailed Analysis:

The computations are summarized in Table 5-1 and plotted in Fig. 5-27.

$$v_c \text{ min} = 1.7\sqrt{f_c'} = 120.2 \text{ psi}$$

The transmission length for the tendons has been taken as 18.75 inches.

The area of the shear reinforcement is based upon the use of grade 40 reinforcement and Eq. 5-60. Note that with the area of reinforcement of 0.0554 sq in. per foot

$$V_s = \frac{0.0554 \times 40{,}000 \times 14}{12} = 2585 \text{ lb}$$

and

$$v_s = 23.1 \text{ psi}$$

COMMENT: Note that in this case the simplified analysis is only slightly conservative when compared to the detailed analysis. Note in the detailed analysis that the value of v_c increases rapidly between the end and 1.50 ft from the end; this is due to the transfer distance for the pre-tensioned tendons.

ILLUSTRATIVE PROBLEM 5-10 Investigate the post-tensioned beam shown in Fig. 5-28 for shear reinforcing if the beam has the following section properties:

$$A = 876 \text{ in.}^2 \quad y_t = 25.0 \text{ in.} \quad S_t = 17{,}300 \text{ in.}^3 \quad r^2/y_t = 19.8 \text{ in.}$$

$$I = 433{,}350 \text{ in.}^4 \quad y_b = -38.0 \text{ in.} \quad S_b = -11{,}400 \text{ in.}^3 \quad r^2/y_b = -13.0 \text{ in.}$$

The design dead loads are as follows:

Girder dead load	= 0.911 k/ft
Superimposed dead load	= 0.500 k/ft
Total Dead Load	1.411 k/ft

The design span is 80.0 ft and the design live load is 2.00 k/ft. The beam is stressed with an effective force of 670 k, $f'_c = 5000$ psi, and $f_y = 40{,}000$ psi and $f_{pu} = 270{,}000$ psi. The tendon is on a parabolic curve with $e = 0$ at the support and $e = -32.1$ in. at midspan. Use load factors of $1.4D + 1.7L$ and analyze the beam by the detailed analysis. $A_{ps} = 4.00$ sq in.

SOLUTION: The values of v_{ci}, v_{cw}, v_c, v_u and A_v are computed to be as follows:

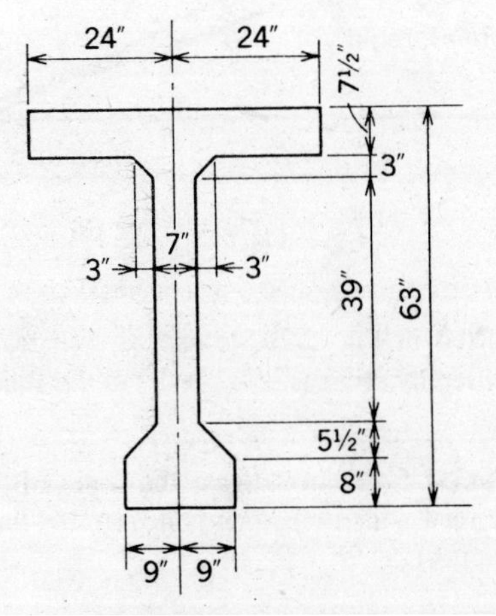

Fig. 5-28 Beam section.

TABLE 5-2

Pt.	Length (ft)	v_{ci} (psi)	v_{cw} (psi)	v_c (psi)	v_u/ϕ (psi)	A_v (in.²/ft)
.00	.000	infin	730.9	730.9	717.0	0.1050
.05	4.000	1029.6	705.5	705.5	645.3	0.1050
.10	8.000	601.5	680.1	601.5	573.6	0.1050
.15	12.000	440.0	654.7	440.0	501.9	0.1299
.20	16.000	344.9	629.3	344.9	430.2	0.1790
.25	20.000	276.1	603.9	276.1	358.5	0.1727
.30	24.000	214.9	575.8	214.9	278.1	0.1360
.35	28.000	162.3	547.7	162.3	199.9	0.1050
.40	32.000	118.4	522.8	120.2	129.4	0.1050
.45	36.000	79.3	499.4	120.2	63.6	0.1050
.50	40.000	42.4	476.9	120.2	000.0	0.1050

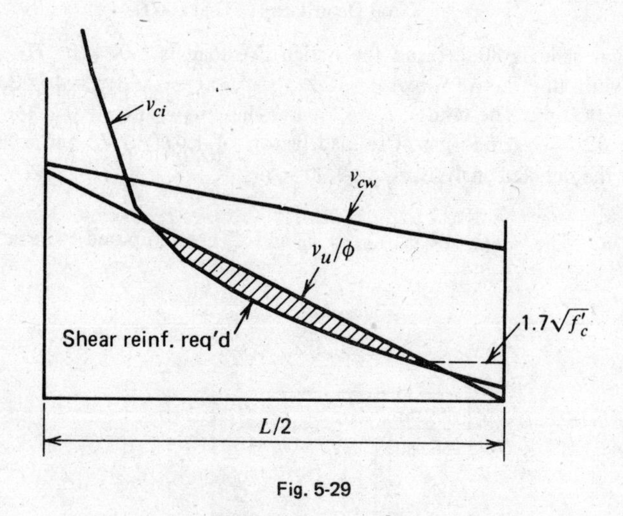

Fig. 5-29

These values are plotted in Fig. 5-29. It will be seen from the plot that shear reinforcement is required by stress (i.e., $v_u > v_c$) in the shaded area.

ILLUSTRATIVE PROBLEM 5-11 Investigate the beam of I.P. 5-10 for shear reinforcement if the live load consists of one concentrated load of 100 kips applied 20 ft from the left support.

SOLUTION: The computations are summarized in Table 5-3 and the results are plotted in Fig. 5-30.

TABLE 5-3

Pt.	Length (ft)	e (in.)	V_d (k)	M_d (k-ft)	V_i (k)	M_{max} (k-ft)	V_u (k)	V_p (k)	v_{ci} (psi)	v_{cw} (psi)	v_u (psi)	v_c (psi)	A_v (in.^{2}/ft)
.00	.000	.00	56.440	.000	150.076	.000	206.516	89.6	infin	730.9	688.6	730.9	.1050
.05	4.000	-6.10	50.796	214.472	147.818	595.788	198.614	80.7	1069.7	705.6	662.3	705.6	.1050
.10	8.000	-11.55	45.152	406.368	145.560	1182.547	190.712	71.7	647.8	680.1	635.9	647.8	.1050
.15	12.000	-16.37	39.508	575.688	143.303	1760.275	182.811	62.7	493.2	654.6	609.6	493.2	.2443
.20	16.000	-20.54	33.864	722.432	141.045	2328.972	174.909	53.8	405.2	629.4	583.2	405.2	.3738
.25	20.000	-24.08	28.220	846.600	-31.212	2888.640	-2.992	44.8	172.2	603.9	9.9	172.2	.1050
.30	24.000	-26.69	22.576	948.192	-33.469	2759.276	-10.893	35.8	160.8	575.8	35.4	160.8	.1050
.35	28.000	-29.21	16.932	1027.208	-35.727	2620.883	-18.795	26.9	149.3	547.8	58.2	149.3	.1050
.40	32.000	-30.82	11.288	1083.648	-37.984	2473.459	-26.696	17.9	140.7	522.7	80.3	140.7	.1050
.45	36.000	-31.78	5.644	1117.512	-40.242	2317.004	-34.598	9.0	134.7	499.5	102.4	134.7	.1050
.50	40.000	-32.10	.000	1128.800	-42.500	2151.520	-42.500	.0	131.0	476.9	125.0	131.0	.1050
.55	44.000	-31.78	-5.644	1117.512	-44.757	1977.004	-50.401	-9.0	158.4	499.5	149.1	158.4	.1050
.60	48.000	-30.82	-11.288	1083.648	-47.015	1793.459	-58.303	-17.9	189.8	522.7	175.5	189.8	.1050
.65	52.000	-29.21	-16.932	1027.208	-49.272	1600.883	-66.204	-26.9	227.6	547.8	205.2	227.6	.1050
.70	56.000	-26.69	-22.576	948.192	-51.530	1399.276	-74.106	-35.8	274.9	575.8	240.9	274.9	.1050
.75	60.000	-24.08	-28.220	846.600	-53.788	1188.640	-82.008	-44.8	331.2	603.9	273.4	331.2	.1050
.80	64.000	-20.54	-33.864	722.432	-56.045	968.972	-89.909	-53.8	393.2	629.4	299.8	393.2	.1050
.85	68.000	-16.37	-39.508	575.688	-58.303	740.275	-97.811	-62.7	482.2	654.6	326.1	482.2	.1050
.90	72.000	-11.55	-45.152	406.368	-60.560	502.547	-105.712	-71.7	637.8	680.1	352.5	637.8	.1050
.95	76.000	-6.10	-50.796	214.472	-62.818	255.788	-113.614	-80.7	1060.8	705.6	378.8	705.6	.1050
1.00	80.000	.00	-56.440	.000	-65.076	.000	-121.516	-89.6	infin	730.9	405.2	730.9	.1050

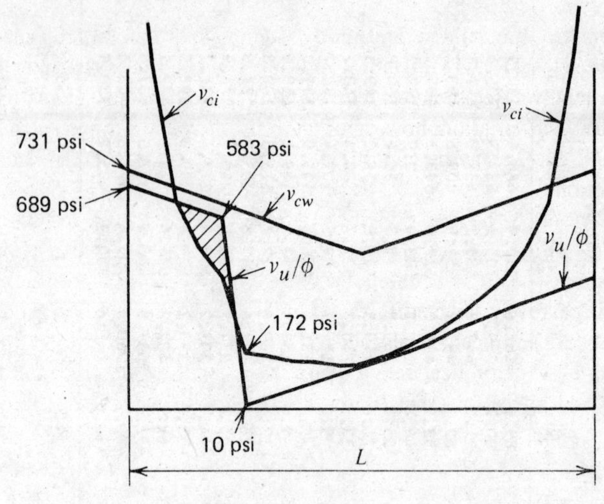

Fig. 5-30

5-9 Bond of Prestressing Tendons

Two types of bond stress must be considered in the case of prestressed concrete. The first of these is referred to as "transfer bond stress" and has the function of transferring the force in a pre-tensioned tendon to the concrete. Transfer bond stresses come into existence when the prestressing force in the tendons is transferred from the prestressing beds to the concrete section. The second type of bond is termed "flexural bond stress" and comes into existence in pre-tensioned and bonded, post-tensioned members when the members are subjected to external loads. Flexural bond stress does not exist in unbonded, post-tensioned construction, which accounts for the term "unbonded."

When a prestressing tendon is stressed, the elongation of the tendon is accompanied by a reduction in the diameter due to Poisson's effect. When the tendon is released, the diameter increases to its original diameter at the ends of the prestressed member where it is not restrained. This phenomenon is generally regarded as a primary factor that influences the bonding of pre-tensioned wires to the concrete. The stress in the wire is zero at the extreme end and is at a maximum value at some distance from the end of the member. Therefore, in the length of the tendon from the extreme end to the point where it attains maximum stress, called the "transmission length," there is a gradual decrease in the diameter of the tendon, giving the tendon a slight wedge shape over this length. This wedge shape is often referred to as the "Hoyer Effect" after the German engineer E. Hoyer, who was one of the early engineers to develop this theory. Hoyer, and others more recently, derived elastic theory to compute the transmission length as a function of Poisson's ratio for steel and concrete, the moduli of elasticity of steel and concrete, the diameter of the tendon, the coefficient of friction between the tendon and the concrete, and the initial and effective stresses in the steel (Ref. 14). Laboratory studies of the transmission lengths have indicated a relatively close agreement between the theoretical and actual values. There can be wide variation, however, due to the different properties of concrete and steel and due to surface conditions of the tendons, which affect the coefficient of friction.

There is reason to believe that the configuration of a seven-wire strand (i.e., 6 small wires twisted about a slightly larger center wire) results in very good bond characteristics. It is believed the Hoyer Effect is partially responsible for this, but the relatively large surface area and twisted configuration is believed to result in a significant mechanical bond.

Although these theoretical relationships are of academic interest, they have little practical application, due to the inability of designers and fabricators of prestressed concrete to control the several factors that influence the transmission length. Fortunately, there has been sufficient research into the magnitude of transmission lengths under both laboratory (Ref. 15) and production (Ref. 16) conditions for the following significant conclusions to be drawn.

(1) The bond of clean three and seven-wire prestressing stands and concrete is adequate for the majority of pre-tensioned concrete elements.

(2) Members that are of such a nature that high moments may occur near the ends of the members, such as short cantilevers, require special consideration.

(3) Clean smooth wires of small diameter are also adequate for use in pre-tensioning, but the transmission length for tendons of this type should be expected to be approximately double that for seven-wire strands (expressed as a multiple of the diameter).

(4) Under normal conditions, the transmission length for clean seven-wire strands can be assumed to be equal to 50 times the diameter of the strand.

(5) The transmission length of tendons can be expected to increase from 5 to 20% within one year after release as a result of relaxation.

(6) The transmission length of tendons released by flame cutting or with an abrasive wheel can be expected to be from 20 to 30% greater than tendons that are released gradually.

(7) Hard non-flaky surface rust and surface indentations effectively reduce the transmission lengths required for strand and some forms of wire tendons.

(8) Concrete compressive strengths between 1500 and 5000 psi at the time of release result in transmission lengths of the same order, except for strand tendons larger than $\frac{1}{2}$ in., in which case strengths, less than 3000 psi result in larger transmission lengths.

(9) It would seem prudent to use 3000 psi as a minimum release strength in pre-tensioned tendons, except for very unusual cases. Higher strengths may be required for tendons larger than $\frac{1}{2}$ in.

(10) Because of relaxation, a small length of tendon (3″±) at the end of a member can be expected to become completely unstressed.

(11) The degree of compaction of the concrete at the ends of pre-tensioned

members is extremely important if good bond and short transmission lengths are to be obtained. Honeycombing must be avoided at the ends of the beams.

(12) There is little if any reason to believe that the use of end blocks improves the transfer bond of pre-tensioned tendons, other than to facilitate the placing and compacting of the concrete at the ends. Hence, the use of end blocks is considered unnecessary in pre-tensioned beams, if sufficient care is given to this consideration.

(13) Tensile stresses and strains develop in the ends of pre-tensioned members along the transmission length as a result of the wedge effect of the tendons. Little if any beneficial results can be gained in attempting to reduce these stresses and strains by providing mild reinforcing steel around the ends of the tendons, since the concrete must undergo large deformations and would probably crack before such reinforcing steel could become effective.

(14) Lubricants and dirt on the surface of tendons has a detrimental effect on the bond characteristics of the tendons.

A curve showing the typical variation of stress along the length of a pre-tensioned tendon near the end of a beam is given in Fig. 5-31. It will be seen that this curve is approximately hyperbolic. The stress is zero at the extreme end and for a distance of approximately 3-4 in., as is assumed to be the case in most applications. This should be considered in the design of pretensioned members.

Bond stresses also occur between the tendons and the concrete in both pre-tensioned and bonded, post-tensioned members, as a result of changes in the external load. There are of course no transfer bond stresses in post-tensioned members, since the end anchorage device accomplishes the transfer of stress. Although it is known that flexural-bond stresses are relatively low in prestressed members for loads less than the cracking load, there is an abrupt and significant

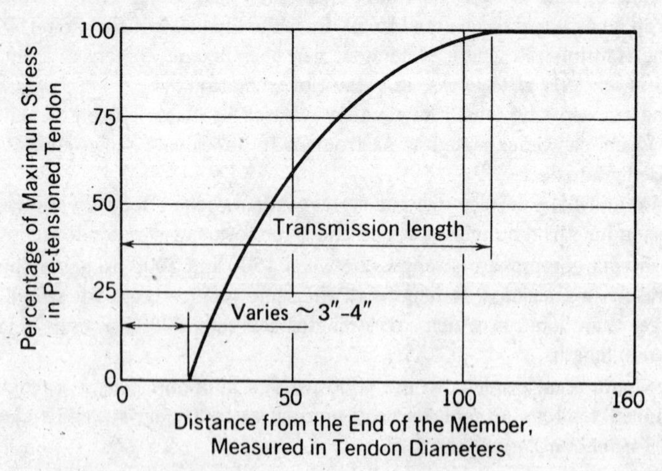

Fig. 5-31 Variation in stress in pre-tensioned wire tendon near end of beam after relaxation.

increase in these bond stresses after the cracking load is exceeded. Because of the indeterminancy which results from the plasticity of the concrete for loads exceeding the cracking load, accurate computation of the flexural-bond stresses cannot be made under such conditions. Again, tests must be relied upon as a guide for design.

The effect of flexural bond is most evident when two identical post-tensioned members, one grouted and one unbonded, are tested to destruction and the results are compared. The load-deflection curve for such tests, when plotted together, would appear as in Fig. 5-32. From these curves, it will be seen that the grouted beam does not deflect as much under a specific load as the unbonded beam. The explanation for this is that the tendon in the bonded beam must undergo changes in strain equal to the strain changes in the concrete to which it is bonded, whereas the unbonded tendon can slip in the duct and the strain changes are averaged. Hence, the grouted beam deforms and deflects as a function of a transformed section. This difference generally results in the cracking load of the grouted beam being from 10 to 15% higher than that of the unbonded beam, while the ultimate load may be as much as 50% higher. The presence of flexural bond results in many very fine cracks in a bonded member in which the cracking load is exceeded, while, in an identical unbonded member subjected to the same load, only a few wide cracks occur. This is significant because removal of the load from the bonded member will result in the fine cracks closing completely, while in the unbonded member, the large cracks do not completely close.*

It is generally believed that once a bonded prestressed member is cracked, a significant increase in flexural-bond stress occurs at the point of cracking. As the

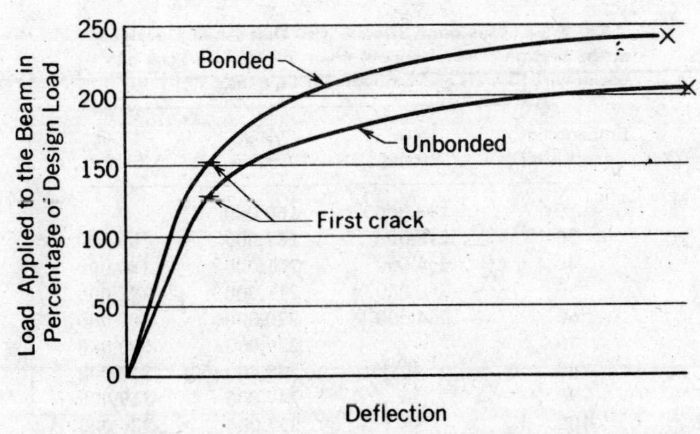

Fig. 5-32 Comparison of load-deflection curves for bonded and unbonded post-tensioned construction.

***The provision of non-prestressed reinforcement, if in sufficient quantity, will result in a non-bonded beam having deflection and cracking characteristics similar to a bonded beam.**

load on the beam is increased, the flexural bond stresses at the crack increase until slip occurs at the cracked section. Further increase in the external loads will be accompanied by additional slip in the tendon. This action will continue until the member fails either by flexure or, (in the case of a pre-tensioned member), when the flexural bond stress is destroyed over a length of a tendon that reaches the zone in which the pre-tension is developed by transfer bond, by lack of bond (Refs. 17 and 18). Research has shown that the embedment length, the length from the free end of the beam to the point of maximum steel stress, required to develop a specific steel stress for $\frac{1}{4}$, $\frac{3}{8}$, and $\frac{1}{2}$ in. strands is of the order given in Table 5-4. The data in the table are applicable to concrete with a cylinder strength of 5500 psi and steel stresses of the order of 150,000 psi. *If the distance from the section at which the critical, flexural stresses occur is less than the embedment length required to develop the computed stress in the steel at ultimate, the ultimate load may be controlled by bond rather than by flexure.* In such instances, the design should be revised, since it is generally more desirable that the controlling mode of failure be flexural rather than by bond.

Bond considerations for prestressed concrete members are treated in several different parts of ACI 318. The first of these is in Sec. 11.4.3 in which the effect of transfer bond on shear strength near the ends of pretensioned beams is considered. It is in this section that it is stated to be permissible to consider the transmission length of strand and wire tendons to be 50 and 100 diameters, respectively. Section 11.4.4 contains provisions related to shear strength computations near the ends of pretensioned members which have some tendons not bonded to the concrete all of the distance to the end of the member. (See Sec. 6-11.) The development length for prestressed three- and seven-wire strand is

TABLE 5-4 Maximum Stresses (psi) that can be Developed at the Section of Maximum Moment for Various Sizes of Seven-wire Strands and Embedment Lengths.

Embedment Length (in.)	$^1/_4$-in. Strand	$^3/_8$-in. Strand	$^1/_2$-in. Strand
20	194,000	160,000	–
30	218,000	187,000	166,000
40	234,000	201,000	180,000
50	250,000	211,000	192,000
60	264,000	220,000	200,000
70	–	229,000	206,000
80	–	238,000	213,000
90	–	247,000	219,000
100	–	257,000	226,000
120	–	–	244,000
140	–	–	272,000

From Ref. 20, p. 78.

treated in Sec. 12.10.1. In this section it is stipulated that strands of these types shall be extended a distance beyond the critical section (for moment) equal to

$$(f_{ps} - \tfrac{2}{3} f_{se})\, d_b \qquad (5\text{-}62)$$

in which d_b is the nominal diameter of the strand, and f_{ps} and f_{se} are as defined elsewhere in this chapter and have the units of ksi. The quantity within the parentheses is considered to be dimensionless. *If the bonding of the tendons does not extend to the end of the member, the length given in Eq. 5-62 must be doubled if the design allows tensile stresses in the precompressed tensile zone.* (*See* Sec. 12.10.3 of ACI 318.)

It should be noted that some designers feel the development length specified by Eq. 5-62 is *unconservative* and that the embedment lengths given in Table 5-4 more accurately reflect that which is needed. Bond is also treated in Secs. 18.7 and 18.9 of ACI 318, in which ultimate strength and minimum amounts of bonded reinforcement are treated; these have both been discussed in detail previously in this book.

ILLUSTRATIVE PROBLEM 5-12 The 4-ft wide double-tee beam in Fig. 4-26 is supported by the inverted tee beam as shown in Fig. 5-33. The span is 40 ft, the dead load of the double tee is 46 psf, the superimposed dead load is 10 psf, and the live load is 30 psf. The member is prestressed with 2 harped pre-tensioned strands in each leg. $A_{ps} = 0.4668$ in.^{2} (total for both legs), $f_{pu} = 270$ ksi, $P_{se} =$ 72.0 kips, and the center of gravity of the steel is 5.50 in. above the soffit at the ends and 2.07 in. above the soffit for a length of 4 ft at midspan. Assume the coefficient of friction between the double-tee beam and the elastomeric pad is 0.20. The effects of shrinkage and creep will cause slippage in the joint. Investigate the member for shear with the assumption that the transfer length is 22 in.

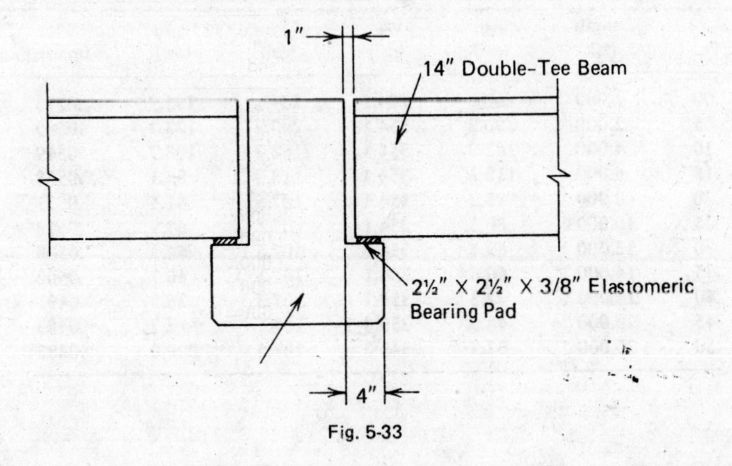

Fig. 5-33

long and the stress in the tendon varies linearly in the transfer zone. Design reinforcement for shear and support stresses taking into account the fact that the stress in the tendon is nil for the first 3 to 4 in. Use load factors of 1.4 and 1.7 for dead and live loads, respectively, use $f'_c = 4000$ psi and use $f_y = 40$ ksi.

SOLUTION:

$$A_{v\,\min} = \frac{50 \times 8 \times 12}{40{,}000} = 0.120 \text{ sq in./ft}$$

or

$$A_{v\,\min} = \frac{0.4668}{80} \cdot \frac{270}{40} \cdot \frac{12}{d} \sqrt{\frac{d}{8}} = \frac{0.167}{\sqrt{d}} \text{ controls}$$

The computations for v_{ci}, v_{cw}, v_c, v_u, and A_v are shown in Table 5-5. Note that $2\,v_u/\phi < v_c$ in the centermost 14 ft ± and hence stirrups could be omitted in this area.

The effect of the transfer length shows in the computation of v_{cw} at the support. Because $0.3 f_{pc} = 0$ at the support, from Eq. 5-51 the value of v_{cw} is only 234 psi at this point. Note that the value of v_{cw} is constant from 2.0 ft and 14.0 ft from the support. This is explained by the fact that $V_p/b_w d$ is constant in between these limits; d being taken equal to 0.80 h between the end and 14.0 from the end, d is taken as the actual value at points 16 feet and further from the end.

TABLE 5-5

Pt.	Length (ft)	v_{ci} (psi)	v_{cw} (psi)	v_c (psi)	v_u/ϕ (psi)	A_v (in.^{2}/ft)
.00	.000	infin	234.1	234.1	135.9	.0573
.05	2.000	290.9	354.1	290.9	122.3	.0560
.10	4.000	162.3	354.1	162.3	108.7	.0549
.15	6.000	118.2	354.1	118.2	95.1	.0538
.20	8.000	95.2	354.1	107.5	81.5	.0527
.25	10.000	80.3	354.1	107.5	67.9	.0518
.30	12.000	69.4	354.1	107.5	54.3	.0508
.35	14.000	60.6	354.1	107.5	40.7	.0500
.40	16.000	52.3	353.7	107.5	26.3	.0491
.45	18.000	44.9	353.3	107.5	12.7	.0483
.50	20.000	37.9	341.3	107.5	000.0	.0483

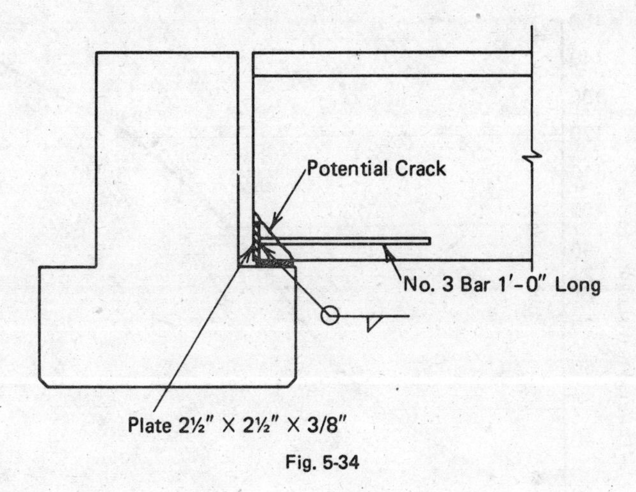

Fig. 5-34

At the end, the design reaction can be computed as follows:

$$\text{Total } R_u = [1.4(46 + 10) + 1.7(30)]\,4 \text{ ft} \times 20 \text{ ft} = 10{,}352 \text{ lb}$$

$$\text{Maximum horizontal force per leg} = \frac{10{,}352 \times 0.2}{2} = 1{,}035 \text{ lb}$$

Hence, to control cracking, steel must be provided across ***and anchored on each side*** of the potential crack. In order to control crack width, the stress in the steel should be confined to 10,000 to 20,000 psi. Using one No. 3 bar in each leg, the stress would be 9400 psi, which is adequate. A good means of anchoring the bar is shown in the Fig. 5-34.

ILLUSTRATIVE PROBLEM 5-13 For the beam in I.P. 5-12, assuming the flexural bond characteristics of the strand were the same as that given for $\frac{1}{2}$ in. strand in Table 5-4, determine the ultimate moment capacity of the member at the 20th points as controlled by bond and flexural strength considerations. Determine which controls.

SOLUTION: The stress vs. embedment length curve based upon the data in Table 5-4 is given in Fig. 5-35. The computations for the moment capacity are summarized as follows:

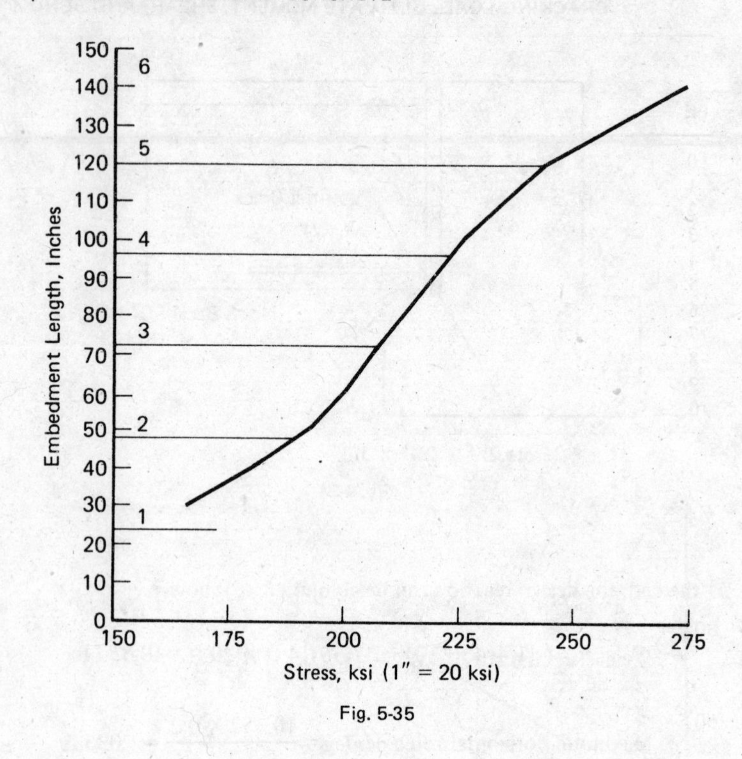

Fig. 5-35

The design moment and moment capacity are plotted in Fig. 5-36. The curve would indicate that the flexural capacity is adequate with the possible exception near the end where flexural strength as a function of embedment length is uncertain due to lack of data. This illustrates the reason why many engineers provide nominal amounts of positive moment non-prestressed reinforcement near the end of all simply supported members.

5-10 Bonded vs Unbonded Post-Tensioning

The structural advantages gained by bonding post-tensioning tendons should be apparent from the preceding article. In spite of these advantages, unbonded tendons are widely used in the United States.

Most suppliers of post-tensioning tendons, as well as some prestressed-concrete fabricators and subcontractors, report that the use of tendons coated with a bituminous rust inhibitor and wrapped with paper or plastic is significantly lower in cost than the use of tendons placed in a steel sheath or preformed hole and

Pt.	d	By Flexure f_{ps}	By Flexure $a/2$	By Flexure M_u	By Bond f_{ps}	By Bond $a/2$	By Bond M_u
0	8.50 in.	260	0.372	74	0		0
1	8.88	260	0.372	77	0		0
2	9.26	260	0.372	81	189	0.270	59
3	9.64	261	0.373	85	207	0.296	68
4	10.02	261	0.373	88	223	0.319	76
5	10.40	261	0.374	92	244	0.349	86
6	10.78	262	0.374	95	270		
7	11.17	262	0.374	99	270		
8	11.55	262	0.375	103	270		
9	11.93 in.	263	0.376	106	270		
10	11.93 in.	263	0.376	106	270		

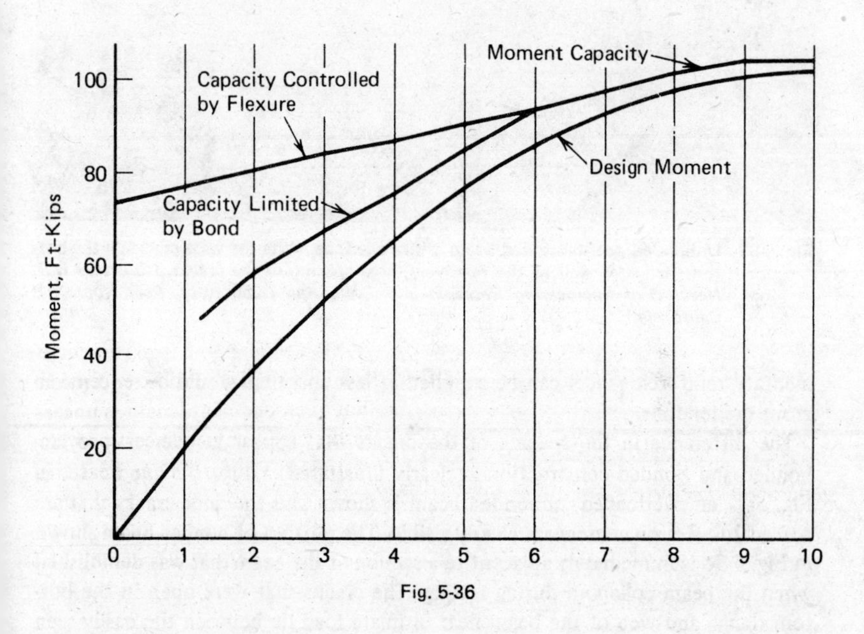

Fig. 5-36

grouted in place after stressing. The proponents of unbonded tendons point out that the lower cracking and ultimate loads that are characteristic of unbonded construction, as well as the fact that the few widely spaced cracks which would appear in the tensile flange at loads that exceed the cracking load, can be adequately controlled by including normal reinforcing steel in the tensile flanges to supplement the prestressing. It is claimed that sufficient quantity of the supple-

Fig. 5-37 Unbonded post-tensioned beam under overload. Note the wide cracks in the bottom flange as well as the relatively large spacing of the cracks. (*Courtesy U.S. Naval Civil Engineering Research and Evaluation Laboratory, Port Hueneme, California.*)

mentary reinforcing steel can be supplied at less cost than would be required to grout the tendons.

The difference in the spacing of the cracks that appear at overloads in unbonded and bonded construction is clearly illustrated in Figs. 5-37 and 5-38. In Fig. 5-37 an overloaded, unbonded beam is shown and the wide cracks that are spaced 2 to 3 ft on centers are clearly visible. The portion of bonded beam shown in Fig. 5-38 is immediately adjacent to a section of the beam that was demolished when the beam collapsed during testing. The cracks that were open in the bottom flange and web of the beam near ultimate load lie between the easily seen pencil lines. The cracks in the bonded beam were only faintly visible to the unaided eye after the failure of the beam. The effectiveness of the grouting in the beam of Fig. 5-38 is demonstrated by the fact that the cracks closed so completely after flexural failure of the beam.

During the testing of the beam shown in Fig. 5-38, the effectiveness of the grouting was also clearly evidenced by the location of the neutral axis of the beam. The location of the neutral axis was determined by measuring flexural strains and was located where it would be expected for the transformed concrete

Fig. 5-38 A portion of a bonded post-tensioned beam after testing to destruction. Note the close spacing of cracks located between the pencil lines. The effectiveness of the grouting is illustrated by the fact that the cracks are closed and virtually invisible to the unaided eye. (*Courtesy U.S. Naval Civil Engineering Research and Evaluation Laboratory, Port Hueneme, California.*)

section, which is lower than would be expected for the net or gross concrete section.

The sections of the grouted post-tensioning tendons shown in Fig. 5-39 were taken from the beam shown in Fig. 5-38. The fact that the metal sheath is very well filled with grout and virtually without voids should be noted. In addition, friction tape, which was used to seize the wires when they were being inserted in the metal sheath, is clearly seen in two sections of the tendon. The friction tape did not seriously restrict the flow of grout.

The use of unbonded tendons will certainly result in satisfactory construction if properly designed and constructed. This has been demonstrated by the large amount of building construction that has been done successfully with this method in the United States. Members that are designed with unbonded tendons should be made to conform to or exceed the minimum provisions of ACI 318-83 "Building Code Requirements for Reinforced Concrete" as well as to "Tentative Recommendations for Concrete Members Prestressed with Unbonded Tendons" (Ref. 19).

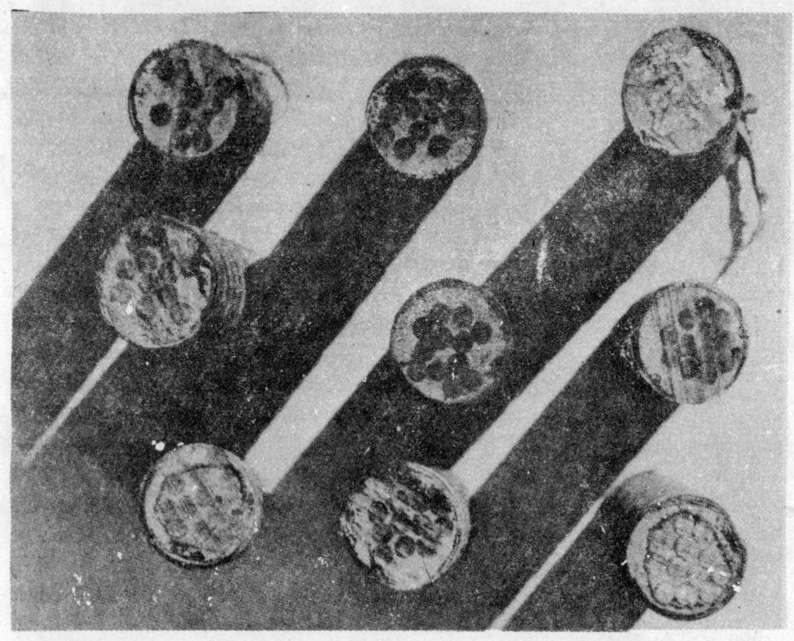

Fig. 5-39 Sections of grouted post-tensioned tendons removed from the test beam in Fig. 5-39. (*Courtesy U.S. Naval Civil Engineering Research and Evaluation Laboratory, Port Hueneme, California.*)

REFERENCES

1. Muller, Jean. "Flexural Strength of Prestressed Concrete Continuous Structures." Paper presented at the Knoxville Convention of the A.S.C.E., pp. 9–19 (Jan. 1956).
2. ACI Committee 318, Building Code Requirements for Reinforced Concrete, American Concrete Institute, Detroit, 1983.
3. ACI-ASCE Joint Committee 323, "Tentative Recommendations for Prestressed Concrete," *Journal of the American Concrete Institute*, 29, No. 7, 545–578 (Jan. 1958).
4. Yamazaki, Jun, Kattula, Basil T. and Mattock, Alan H. "A Comparison of the Behavior of Post-Tensioned Prestressed Concrete Beams With and Without Bond." Structures and Mechanics Report SM69-3. Department of Civil Engineering University of Washington, Seattle, Washington (Dec. 1969).
5. Pannell, F. N. "The Ultimate Moment Resistance of Unbonded Prestressed Concrete Beam," *Magazine of Concrete Research*, **21**, No. 66, 43–54 (Mar. 1969).
6. International Conference of Building Officials, Uniform Building Code, International Conference of Building Officials, Whittier, CA, 1982.
7. Portland Cement Association, Notes from the Building Code Seminar, Chapter 63S-28, pp. 8–31 (1963).
8. MacGregor, J. G. and Hanson, J. M. "Proposed Changes in Shear Provisions for Rein-

forced and Prestressed Concrete Beams," *ACI Journal*, **66**, No. 4, 276–288 (April 1969).

9. Kani, G. N. J., "A Rational Theory for the Function of Web Reinforcement," *ACI Journal*, **66**, No, 3, 185–197 (Mar. 1969).
10. ACI-ASCE Committee 426, "The Shear Strength of Reinforced Concrete Members," *ACI Manual of Concrete Practice*, Part 2, American Concrete Institute, Detroit, 1974.
11. Tang, Man-Chung, "Shear Design of Large Box Girders," *Shear in Reinforced Concrete*, Vol. 1, pp. 305–319, American Concrete Institute, Detroit, 1974.
12. ACI Committee 318, "Commentary on Building Code Requirements for Reinforced Concrete (ACI 318-83)," American Concrete Institute, Detroit, 1983.
13. American Association of State Highway and Transportation Officials, "Standard Specifications for Highway Bridges," Twelfth Edition, pp. 104–105, Washington, D.C., 1977.
14. Janney, Jack R. "Nature of Bond in pretensioned Concrete," *Journal of the American Concrete Institute*, **25**, 717–736 (May 1954).
15. Hanson, N. W. "Influence of Surface Roughness of Prestressing Strand on Bond Performance," *Journal of the Prestressed Concrete Institute*, **14**, No. 1, 32–45 (Feb. 1969).
16. Base, G. D. "An Investigation of Transmission Length in Pretensioned Concrete," Research Report No. 5, Cement and Concrete Association, London, 1958.
17. Nordby, G. M. "Fatigue of Concrete–A Review of Research," *Journal of the American Concrete Institute*, **31**, No. 2, 210–215 (Aug. 1958).
18. Hanson, N. W. and Kaar, P. H. "Flexural Bond Tests of Pre-tensioned Prestressed Beams," *Journal of the American Concrete Institute*, **30**, No. 7, 783–802 (Jan. 1959).
19. "Recommendations for Concrete Members Prestressed With Unbonded Tendons," *Concrete International*, July 1983.

CHAPTER 5 PROBLEMS

1. Determine ultimate moment capacity for the double-tee slab of Problem 1 of Chapter 4 using Eq. 5-27 for f_{ps}. Assume $A_{ps} = 0.171$ sq in., $f_{pu} = 270$ ksi, and $f'_c = 5000$ psi.

SOLUTION:

$$\rho_p = \frac{0.171}{48 \times 12} = 0.000297, \quad \gamma_p = 0.28, \quad \beta_1 = 0.80$$

$$f_{ps} = 270\left(1 - \frac{0.35 \times 0.000297 \times 270}{5}\right) = 268.5 \text{ ksi}$$

$$\omega_p = \rho_p \frac{f_{ps}}{f'_c} = 0.0159 < 0.30$$

$1.4\, d\omega_p = 0.267 < 2$ in. can be analyzed as a rectangular beam

$$a = \frac{0.171 \times 268.5}{0.85 \times 48 \times 5} = 0.225 \text{ in.}$$

$$\phi M_n = \frac{0.90 \times 0.171 \times 268.5}{12}\left[12.00 - \frac{0.225}{2}\right] = 40.93 \text{ k-ft}$$

$$w_{\text{design}} = \frac{40.93 \times 8}{24^2} = 568.5 \text{ plf}$$

Assuming all the superimposed service load to be live load, the maximum value permitted would be:

$$w_{\text{service}} = \frac{568.5 - 1.4 \times 187.5}{1.7} = 180 \text{ plf}$$

(Compare this to the allowable loads as controlled by service loads computed in Problem 1, Chapter 4.)

$$\text{With } f_r = 7.5\sqrt{5000} = 530 \text{ psi}$$

The bottom fiber stress due to final prestressing (from problem no. 1, Chapter 4) is +863 psi. Hence, the cracking moment for the slab is

$$M_{cr} = \frac{(530 + 863)\,2860}{12000 \times 10} = 33.20 \text{ k-ft}$$

Hence, the ultimate moment capacity can be fully utilized without the addition of more reinforcement because ϕM_n exceeds 120% of M_{cr}, or

$$\phi M_n = 40.83 > 1.20 \times 33.20 = 39.84 \text{ k-ft}$$

2. If the double-tee slab analyzed in Problem 1 had a concrete strength of 6500 psi, determine what steps would be necessary to fully utilize the flexural capacity of the member.

SOLUTION:

$$\beta_1 = 0.725$$

$$f_{ps} = 270\left[1 - \frac{0.386 \times 0.000297 \times 270}{6.5}\right] = 268.7 \text{ ksi}$$

$$a = \frac{268.7 \times 0.171}{0.85 \times 48 \times 6.5} = 0.173 \text{ in.}$$

$$\phi M_n = \frac{0.90 \times 0.171 \times 268.7}{12}\left[12 - \frac{0.173}{2}\right] = 41.06 \text{ k-ft}$$

$$f_r = 7.5\sqrt{6500} = 605 \text{ psi}$$

$$1.2\,M_{cr} = \frac{(605 + 863)\,2860}{12000 \times 10} \times 1.2 = 41.98 \text{ k-ft} > 41.06 \text{ k-ft}$$

Reinforcement must be added to increase the ultimate moment capacity at least 2.2%.

With grade 40 reinforcing steel this would require the following approximate amount:

$$A_s = \frac{0.022 \times 268.7 \times 0.171}{40} = 0.025 \text{ sq in.}$$

The provision of a number 3 bar in each leg in addition to the prestressing would result in the following (use $d = 12.5$ inches for the non-prestressed reinforcement):

$$f_{ps} = 270\left(1 - 0.386\left[\frac{0.000297 \times 270}{6.5} + \frac{12.5}{12}\left(\frac{0.22}{48 \times 12.5 \times 6.5}\right)\right]\right)$$

$$= 268.7 \text{ ksi}$$

and

$$a = \frac{0.171 \times 268.7 + 0.22 \times 40}{0.85 \times 6.5 \times 48} = 0.206$$

and

$$\phi M_n = 0.90\left(\frac{0.171 \times 268.7}{12}\left(12 - \frac{0.206}{2}\right) + \frac{0.22 \times 40}{12}\left(12.5 - \frac{0.206}{2}\right)\right)$$

$$= 49.18 \text{ k-ft} > 1.2M_{cr} = 41.98 \text{ k-ft} \qquad \text{ok}$$

3. Analyze the double-tee beam of Problem 1 for ultimate moment capacity under the conditions of $f'_c = 5000$ psi and the tendon being unbonded.

$$f_{se} = \frac{0.80 \times 2 \times 16100}{0.171} = 150.6 \text{ ksi}$$

$$\rho_p = \frac{0.171}{48 \times 12} = 0.000297$$

$$\text{Span to depth ratio} = \frac{24 \times 12}{14} = 20.6 \qquad \text{Use Eq. 5-42.}$$

$$f_{ps} = 150{,}600 + 10{,}000 + \frac{5000}{100 \times 0.000297} = 329{,}000 \text{ psi} > f_{pu}$$

$$f_{se} + 60{,}000 = 210{,}600 \text{ psi} < f_{py}$$

Therefore $f_{ps} = 210.6$ ksi, $a = 0.176$

$$\phi M_n = \frac{0.90 \times 0.171 \times 210.6}{12}\left[12 - \frac{0.176}{2}\right] = 32.17 \text{ k-ft.}$$

The cracking moment (from Problem 1) was shown to be 33.20 k-ft. Hence, additional non-prestressed reinforcement must be provided to increase the moment capacity of the section to 1.20 × 33.20 ≑ 39.84 k-ft. The tensile strength of the member must be increased approximately 24%.

4. The double-tee beam shown in Fig. 5-24 is to be used on a span of 40 ft with an overhang of 8 ft at one end. If the member is prestressed with bonded tendons having an area of 0.58 sq in., an effective prestress of 90,000 lb, and $f_{pu} = 270.0$ ksi, determine the adequacy of the member from the standpoint of negative flexural strength. The tendons are located 2.50 in. from the top of the member and $f'_c = 5000$ psi. For the double-tee section, $A = 187.5$ sq in., $I = 4256$ in.^{4}, $y_t = 5.17$ in. and $y_b = 10.83$ in.

SOLUTION:

Moment capacity

$$\rho_p = \frac{0.58}{5.25 \times 13.5} = 0.00818$$

$$f_{ps} = 270\left(1 - 0.5 \times \frac{0.00818 \times 270}{5}\right) = 210.34 \text{ ksi}$$

$$\omega_p = \frac{210.34 \times 0.00818}{5}$$

$$\omega_p = 0.344 > 0.30, \text{ over-reinforced section}$$

$$\phi M_n = \frac{0.90 \times 0.25 \times 5 \times 5.25 \times 13.5^2}{12} = 89.7 \text{ k-ft.}$$

$$\text{Effective prestress in top fiber} = \frac{90000}{187.5}\left(1 + \frac{2.67}{4.39}\right) = +772$$

$$1.2\, M_{cr} = \frac{(772 + 530)\, 823 \times 1.2}{12000} = 107.2 \text{ k-ft.}$$

Hence $1.2\, M_{cr} > M_u$ and the ultimate moment capacity must be increased about 20%. Adding compression reinforcement to a member of this shape is not a practical solution. To be effective, compression reinforcement must be tied to prevent buckling; this is not possible in a compression flange that is only 2.625 in. wide.

It should be recognized that the design load causing the moment to equal the moment capacity of the member is $(2)\,(89.7)/8^2 = 2.80$ kips per foot or 700 psf, a very sizeable load. This type of member is generally used in roof construction where the total of the dead and live loads may equal 70 psf. A strict interpretation of ACI 318 would prevent the use of this member in spite of the fact that the load factor would be of the order of 10.

One solution would be to eliminate the prestressing in the area of negative moment and design the overhang in reinforced concrete.

5. For the double-tee beam of Problem 4, study the member under the identical conditions if the tendon is unbonded.

conditions if the tendon is unbonded.

$$\text{Span-depth ratio} = \frac{2 \times 8 \times 12}{16} = 12 \qquad \text{Use Eq. 5-42.}$$

$$f_{ps} = 155 + 10 + \frac{5}{100 \times 0.00818} = 171 \text{ ksi}$$

$$f_{se} + 60 = 155 + 60 = 215 \text{ ksi} < f_y$$

Use $f_{ps} = 171$ ksi, $\omega_p = 0.280 < 0.30$ under reinforced

$$\phi M_n = \frac{0.90 \times 171 \times 0.58}{12} \left[13.5 - \frac{4.44}{2} \right] = 83.89 \text{ k-ft.}$$

The moment capacity is less than 120% of the cracking moment (107.2 k-ft) and the member does not conform to the code requirements. Addition of tensile reinforcement will not correct the situation because additional tensile reinforcement will result in the member becoming overreinforced as in Problem 4.

6. Two identical one-story buildings, having the dimensions shown in Fig. 5-40, are composed of a continuous post-tensioned roof slab that is 6 in. thick supported on concrete bearing walls. One building has bonded tendons, the other does not. The roof slab is prestressed with tendons placed on parabolic paths as shown in Fig. 5-41. The effective stress in the tendons is 15.0 kips per ft of width. The tendons have an area of 0.11 sq in./ft and $f_{pu} = 270$ ksi. If a catastrophic accident caused a downward load of 950 psf to act along the full length and width of one of the 4-ft wide overhangs of the structure, determine the effect on each of the buildings. If the catastrophic

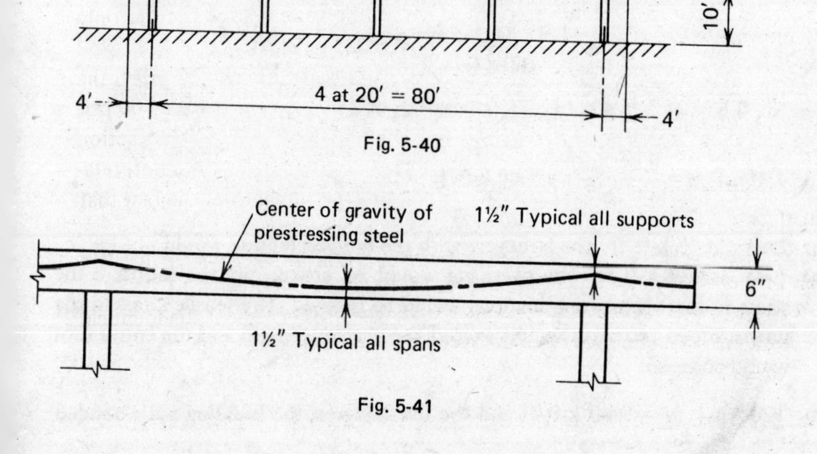

Fig. 5-40

Fig. 5-41

accident caused a downward load of 1200 psf on the overhang, determine the effect on the structures with each type of tendons. Assume $f'_c = 4500$ psi.

SOLUTION:

Bonded tendon

$$\rho_p = \frac{0.11}{12 \times 4.5} = 0.00204, f_{ps} = 253.5 \text{ ksi}$$

$$\omega_p = 0.115 \quad \phi M_n = 8.78 \text{ k-ft/ft}$$

$$f_t = \frac{15000}{72}\left(1 + \frac{1.5}{1}\right) = 521 \text{ psi}$$

$$M_{cr} = \frac{(503 + 521)\,72}{12000} = 6.14 \text{ k-ft/ft}$$

$$1.2\,M_{cr} = 7.37 \text{ k-ft/ft} < 8.78 \text{ k-ft/ft ok}$$

Unbonded tendon

$$\text{Span-depth ratio} = \frac{20 \times 12}{6} = 40$$

$\therefore$ Use Eq. 5-43

$$f_{ps} = 136 + 10 + \frac{4.5}{300 \times 0.00204} = 153.4 \text{ ksi} < 166 < f_{se} + 60 \text{ ksi}$$

$$\phi M_n = 5.46 \text{ k-ft/ft}$$

$$1.2\,M_{cr} = 7.37 > 5.46 = \phi M_n \text{ - N.G.}$$

Non-prestressed reinforcing must be added to increase the moment to 7.37 k-ft/ft.

$$A_s \cong \frac{12(7.37 - 5.46)}{(60)(4.5)} = 0.085 \text{ sq in./ft}$$

Use No. 4 bars at 2 ft-4 in. o.c. $A_s = 0.086$ sq in./ft

$$\text{For 950 psf, } M = \frac{0.95 \times 4^2}{2} = 7.60 \text{ k-ft/ft}$$

(a) 7.60 < 8.78 k-ft/ft. The structure with the bonded tendon would not fail.

(b) 7.60 > 7.37 k-ft/ft. The overhang would be broken off the roof and the stress in the prestressing tendons would be released. The tensile stress in the unreinforced roof slab would exceed the tensile strength and the entire roof would collapse.

For 1200 psf, $M = 9.60$ k-ft/ft and the overhang on the building with bonded tendons would fail. The entire roof would not fail, however, because the stress in the bonded tendon would not be released by the failure.

7. The girder used in Illustrative Problem 4-13 is designed to support precast rectangular beams and a cast-in-place slab. The girder is to be used on a span of 40 ft and support a superimposed dead load of 2.00 kips per ft and a superimposed live load of 1.13 kips/ft. Assuming $f'_c = 5000$ psi, the concrete is all–lightweight concrete, $P_{se} = 440$ kips, $A_{ps} = 3.20$ in.^{2}, and $f_{pu} = 250$ ksi, design the member for shear reinforcing using $f_y = 40{,}000$ psi and the criteria of ACI 318. Assume the bonded reinforcing is post-tensioned on a parabolic path having distances between the center of gravity of the steel and the soffit of 5.20 in. at midspan and 12.76 in. at the supports. The girder is simply supported. Use load factors of 1.5 and 1.8 for dead and live loads respectively. Plot the results. Confirm the design conforms to all of the shear requirements of ACI 318, Chapter 11.

SOLUTION: The computations are summarized below:

		Girder
y_b	(in.)	-14.61
I	(in.^{4})	44.700
h	(in.)	30.000
t	(in.)	6.500
Area	(in.^{2})	418.5
bw	(in.)	6.000
Length:		
span	(ft)	40.000
trsfr	(in.)	1.000
Soffit Distance (in.):		
end		12.760
midspan		5.200
A_{ps}	(in.^{2})	3.200
P_{se}	(k)	440.000
f_{pu}	(ksi)	250.000
f'_c	(ksi)	5.000
f_y	(ksi)	40.000
LW_{coef}		.75
LOADS (plf):		
beam		440.0
slab		.0
SDL		2000.0
LL		1130.0

$U = 1.50\text{ D} + 1.80\text{ L}$

Design End Moments (k-ft):

	Left	Right
M-SDL	.000	.000
M_{max}	.000	.000

Pt.	Length (ft)	v_{ci} (psi)	v_{cw} (psi)	v_c (psi)	v_u/ϕ (psi)	A_v (in.^{2}/ft)
.00	.000	infin	378.1	378.1	930.3	*
.05	2.000	1574.6	673.9	673.9	837.3	.2942
.10	4.000	828.3	654.6	654.6	744.3	.1613
.15	6.000	563.4	635.4	563.4	651.2	.1581
.20	8.000	418.3	616.1	418.3	558.2	.2517
.25	10.000	320.8	596.9	320.8	465.1	.2598
.30	12.000	246.4	577.6	246.4	372.1	.2261
.35	14.000	184.1	558.1	184.1	277.7	.1684
.40	16.000	128.5	538.3	128.5	182.2	.0967
.45	18.000	78.8	519.3	90.1	90.3	.0900
.50	20.000	31.8	500.6	90.1	000.0	.0900

The asterisk in the A_v column at point .00 indicates $(v_u/\phi - v_c) > 8\sqrt{f'_c}$

$$v_u/\phi - v_c = 552.2 \text{ psi}$$

$$8\sqrt{f'_c} = 565.7 \text{ psi}$$

but because the concrete is all-lightweight, the 565.7 psi (*see* ACI 318, Sec. 11.2.1.2) must be multiplied by 0.75 and hence becomes 424.3 psi. This illustrates one of the constraints of ACI 318; it does not apply here, however, because the maximum shear stress is to be computed at $d/2$ from the support (8.63 in.) and, in the case of this post-tensioned beam (parabolic tendon path), the transfer distance can be ignored. The value of 378.1 for v_{cw} shown in the above table was computed on the basis of $f_{pc} = 0$. The results are shown plotted in Fig. 5-42.

8. For the beam cross section and conditions specified in problem 7, design the central span of the girder for shear using the detailed analysis of ACI 318. The girder is to have a central span of 40 ft and overhangs of 8 ft at each end. The tendon path is composed of second-degree parabolas passing through the points indicated in Fig. 5-43. The service loads are shown in Fig. 5-44. Do not overlook the fact that the concrete is to be all-lightweight and the load factors are 1.5 and 1.8 for dead and live loads, respectively. Plot the results.

SOLUTION: The service loads cause the following service load and design load end moments:

	Service	Design
Girder = $-[5.28 \times 8 + 0.44 \times 8^2/2]$ =	-56.32 k-ft	-84.48 k-ft
S.D.L. = $-[24.00 \times 8 + 2.00 \times 8^2/2]$ =	-256.00 k-ft	-384.00 k-ft
L.L. = $-[40.8 \times 8 + 3.40 \times 8^2/2]$ =	-435.20 k-ft	-783.36 k-ft

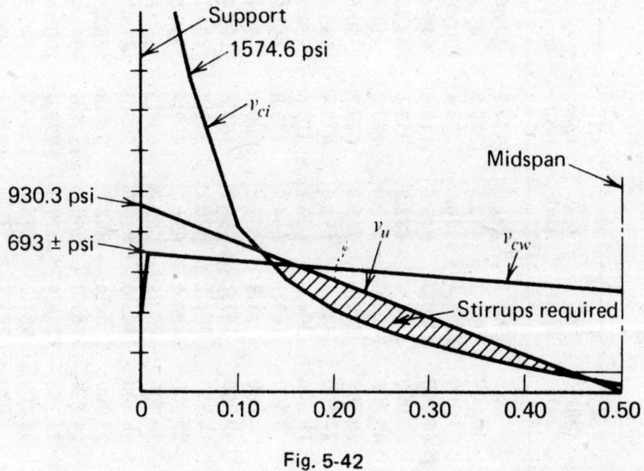

Fig. 5-42

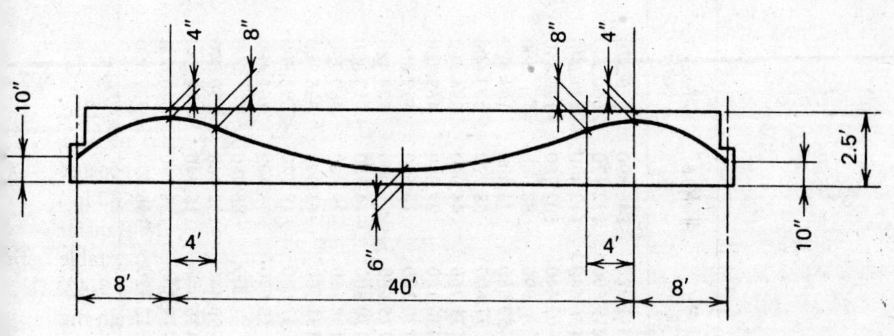

Fig. 5-43

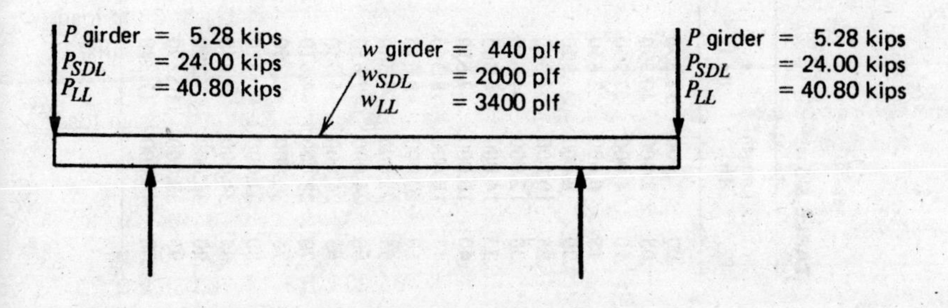

Fig. 5-44

TABLE 5-6

Pt.	Length (ft)	e (in.)	V_d (k)	M_d (k-ft)	V_i (k)	M_{max} (k-ft)	V_u (k)	V_p (k)	v_{ci} (psi)	v_{cw} (psi)	v_u/ϕ (psi)	v_c (psi)	A_v (in.^{2}/ft)
.00	.000	11.390	48.800	−312.320	146.800	−939.520	195.600	.000	781.7	591.0	1475.1	500.6	*
.05	2.000	10.390	43.920	−219.600	132.120	−660.600	176.040	36.700	981.2	745.6	1380.7	745.3	*
.10	4.000	7.390	39.040	−136.640	117.440	−411.040	156.480	73.300	1226.4	1010.0	1278.4	1009.6	.4837
.15	6.000	3.640	34.160	−63.440	102.760	−190.840	136.920	64.200	1769.0	946.8	1118.6	946.4	.3098
.20	8.000	.390	29.280	.000	88.080	.000	117.360	55.000	infin	882.9	958.8	882.5	.1372
.25	10.000	−2.360	24.400	53.680	73.400	161.480	97.800	45.800	1405.0	810.0	799.0	818.7	.0900
.30	12.000	−4.610	19.520	97.600	58.720	293.600	78.240	36.700	750.6	755.8	639.2	750.6	.0900
.35	14.000	−6.360	14.640	131.760	44.040	396.360	58.680	27.500	480.6	691.6	479.4	480.6	.0900
.40	16.000	−7.610	9.760	156.160	29.360	469.760	39.120	18.300	304.1	628.1	319.6	304.1	.0900
.45	18.000	−8.360	4.880	170.800	14.680	513.800	19.560	9.200	161.7	564.5	159.8	161.7	.0900
.50	20.000	−8.610	.000	175.680	.000	528.480	.000	.000	31.8	501.0	000.0	90.1	.0900
.55	22.000	−8.360	−4.880	170.800	−14.680	513.800	−19.560	−9.200	161.7	564.5	159.8	161.7	.0900
.60	24.000	−7.610	−9.760	156.160	−29.360	469.760	−39.120	−18.300	304.1	628.1	319.6	304.1	.0900
.65	26.000	−6.360	−14.640	131.760	−44.040	396.360	−58.680	−27.500	480.6	691.6	479.4	480.6	.0900
.70	28.000	−4.610	−19.520	97.600	−58.720	293.600	−78.240	−36.700	750.6	755.8	639.2	750.6	.0900
.75	30.000	−2.360	−24.400	53.680	−73.400	161.480	−97.800	−45.800	1405.0	810.0	799.0	818.7	.0900
.80	32.000	.390	−29.280	.000	−88.080	.000	−117.360	−55.000	infin	882.9	958.8	882.5	.1372
.85	34.000	3.640	−34.160	−63.440	−102.760	−190.840	−136.920	−64.200	1769.0	946.8	1118.6	946.4	.3098
.90	36.000	7.390	−39.040	−136.640	−117.440	−411.040	−156.480	−73.300	1226.4	1010.0	1278.4	1009.6	.4837
.95	38.000	10.390	−43.920	−219.600	−132.120	−660.600	−176.040	−36.700	981.2	745.6	1380.7	745.3	*
1.00	40.000	11.390	−48.800	−312.320	−146.800	−939.520	−195.600	.000	781.7	591.0	1475.1	500.6	*

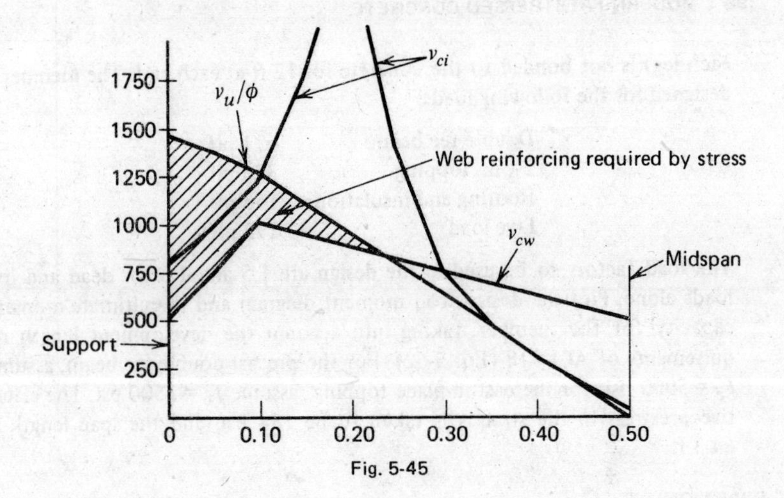

Fig. 5-45

The computations are summarized in Table 5-6. The data are given at the 20th points. The girder and loading are symmetrical.

Note that $v_u - v_c > 8 \times 0.75\sqrt{f_c'}$ near the supports.

In order to make this beam conform to the provisions of ACI 318, the web thickness would have to be increased in the vicinity of the supports. The results are shown plotted in Fig. 5-45.

9. The double-tee beam shown in Fig. 5-46 is pre-tensioned with 9 strands having a nominal diameter of 0.50 in. in each leg. The tendons have an area of 0.153 sq in. each and a minimum guaranteed ultimate tensile strength of 270,000 psi. The strands are positioned in three rows of three strands in each leg. The rows are at 2 in. on center with the center of the lowest row being 2 in. above the bottom of the legs. The bottom row of strands (3 strands in

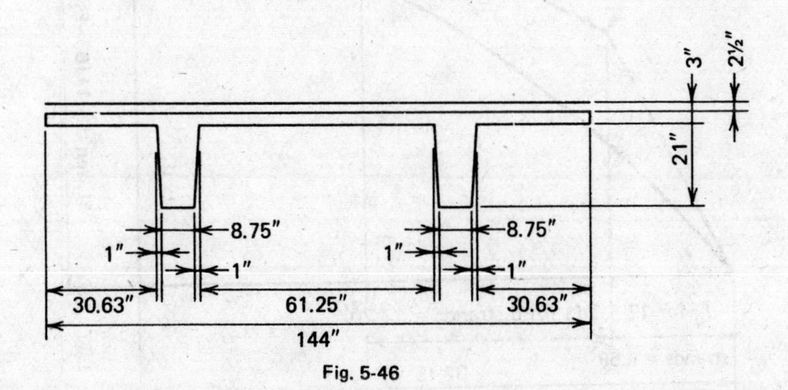

Fig. 5-46

each leg) is not bonded to the concrete for 12 ft at each end. The member is designed for the following loads:

Double tee beam	671 plf
$2\frac{1}{2}$ in. topping	375 plf
Roofing and insulation	156 plf
Live load	192 plf

The load factors to be used in the design are 1.5 and 1.8 for dead and live loads alone. Plot the design load moment diagram and the ultimate moment capacity for the member, taking into account the development length requirements of ACI 318 (Eq. 5-62). For the precast double-tee beam, assume $f'_c = 5000$ psi; for the cast-in-place topping, assume $f'_c = 3500$ psi. The effective prestress in the strands is taken to be 154 ksi, and the span length is 64.3 ft.

SOLUTION:

For 18-strands, $d = 26.50 - 4.00 = 22.50$ in.

$$\rho_p = \frac{18 \times 0.153}{144 \times 22.50} = 0.00085$$

$$f_{ps} = 261.1 \text{ ksi} \quad \omega_p = 0.0634 \quad 1.4\,\omega_p d = 2.00 \text{ in.}$$

$$\phi M_n = \frac{0.90 \times 18 \times 0.153 \times 261.1}{12}\left[22.50 - \frac{1.68}{2}\right] = 1168 \text{ k-ft.}$$

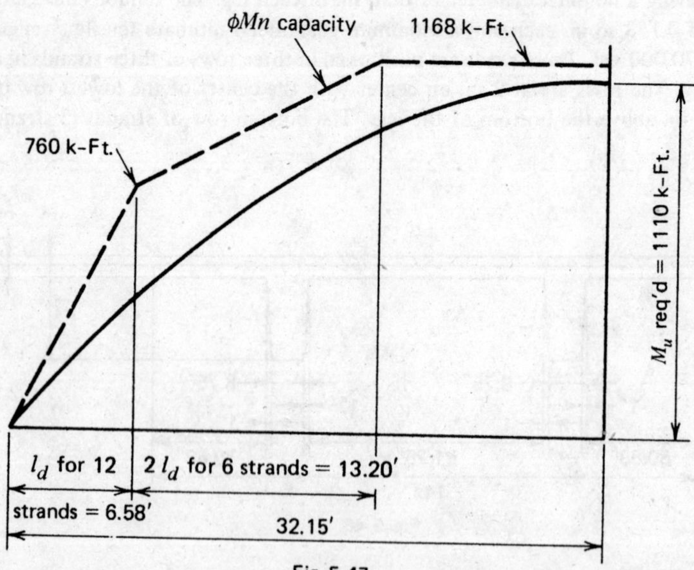

Fig. 5-47

For 12 strands, $d = 26.50 - 5.00 = 21.50$ in.

$$\rho_p = \frac{12 \times 0.153}{144 \times 21.50} = 0.000593$$

$$f_{ps} = 263.8$$

$$\phi M_n = \frac{0.90 \times 12 \times 0.153 \times 263.8}{12}\left[21.5 - \frac{1.13}{2}\right] = 760 \text{ k-ft.}$$

For the 12 strands the development length for $f_{se} = 154$ ksi is

$$\left(261.1 - \frac{2}{3} \cdot 154\right)0.50 = 79 \text{ in.}$$

For the 6 strands that are not bonded for 12 ft at each end of the member, the development length is

$$2\left(261.1 - \frac{2}{3} \cdot 154\right)0.5 = 158 \text{ in.}$$

The results are shown plotted in Fig. 5-47.

6 Additional Design Considerations

6-1 Introduction

Included in this Chapter is a discussion of a number of principles, pertaining to the elastic design of simple prestressed flexural members, which may not at first be apparent to the student of prestressed concrete. The topics treated in this chapter are often important in problems encountered in practice, and it is well for the designer and student to be familiar with these principles.

The engineer who is often engaged in the design of prestressed structures will become familiar with these relationships as a result of his design experience. On the other hand, the engineer who is only occasionally involved in the design or review of prestressed members will find that this chapter contains valuable, concise reference material presented in a manner intended to facilitate its use.

6-2 Losses of Prestress

The final stress that is required in the prestressing steel at all of the critical sections in a prestressed member should be specified by the designer. If the system of prestressing to be used is specified, the complete stressing schedule and initial stresses required should be computed and indicated on the drawings or in the specifications. In order to do this, it is necessary to either compute or assume a value for the loss of stress, in the prestressing tendons, which results from the several contributing phenomena. The losses of prestress must be included in the computation of gauge pressure and elongation for prestressing, which is discussed in Chapter 15, as is the friction loss that occurs during post-tensioning.

The various phenomena that contribute to the loss of prestress as well as the method of calculation are discussed below.

Elastic Shortening of the Concrete

When the prestress is applied to the concrete, an elastic shortening of the concrete takes place. This results in an equal and simultaneous shortening of the prestressing steel. The loss in stress in the steel from this shortening is equal to the product of the stress in the concrete at the level of the steel and the ratio of the moduli of elasticity of the steel and concrete. In a simple beam, the critial section for flexural stress, and hence the section at which the losses of prestress should be considered, is normally at the midspan of the beam. When pretensioning is used, the concrete stress that should be used in computing the reduction in prestress (due to elastic shortening) is equal to the net, initial concrete stress that results from the algebraic sum of the stress due to initial prestressing and the stress due to the dead load of the beam at the level of the steel. In post-tensioning, the first tendon that is stressed is shortened by the subsequent stressing of all other tendons, and the last tendon is not shortened by any subsequent stressing. Therefore, in post-tensioning, an average value of stress change can be computed and applied equally to all tendons.

For the case of all prestressing being applied simultaneously (as in pretensioning), the stress in the concrete at the level of the steel (after elastic shortening) f_{ci}, resulting from an initial prestressing force (before elastic shortening) P_i, can be computed from

$$f_{ci} = \frac{P_i - nf_{ci}A_{ps}}{A}\left(1 + \frac{(e_{ps})(e_{pL})}{r^2}\right) + \frac{M_d e_{ps}}{I}$$

which can be rewritten

$$f_{ci} = \frac{P_i k_s + AM_d e_{ps}/I}{A + nA_{ps}k_s} \tag{6-1}$$

in which

$$k_s = 1 + \frac{(e_{ps})(e_{pL})}{r^2}$$

In the above, e_{ps} is the eccentricity of the prestressing steel, e_{pL} is the eccentricity of the pressure line due to prestressing only (both e_{ps} and e_{pL} are negative below the centroidial axis), and dead load moments causing tension in the bottom fibers are positive. In the case of statically determinate members, the term k_s becomes

$$k_s = 1 + \frac{e_{ps}^2}{r^2}$$

because the eccentricities of the prestressing steel and pressure line due to prestressing alone are equal; as will be seen in Chapter 8, this is not the case for statically indeterminate members.

In the case of simple precast, pre-tensioned members, the value of f_{ci} can be computed at the critical section and the design adjusted accordingly. The critical section may be near the supports with members stressed with straight tendons; it may be between the supports near the point of maximum moment due to total load. If the critical section is near midspan, a higher jacking force may be permissible without exceeding 0.70 f_{pu} in the tendons immediately after transfer; in other words, the jacking stress could be increased by the amount of nf_{ci} (See Sec. 2-11).

The value of the steel stress after elastic shortening, f_{so}, is computed from

$$f_{so} = f_{si} - nf_{ci}$$

and the stress at other fibers can be computed from

$$f = \frac{f_{so} A_{ps}}{A} \left(1 + \frac{e_{ps} y}{r^2}\right)$$

As stated above, in the case of post-tensioned members having a number of tendons which are stressed sequentially, the stress in the first tendon stressed is reduced slightly by the elastic shortening of each of the tendons stressed subsequently. The stress in the last tendon stressed is not affected by the elastic shortening caused by the other tendons. Hence, the effect of elastic shortening is less than that which occurs when all tendons are stressed simultaneously. In post-tensioned members with several tendons, the effect of elastic shortening is generally taken to be 50% of that which would be computed by Eq. 6-1.

It should be recognized that the effect of elastic shortening varies along the length of the member. In simple spans where there may be one point in the span that is most critical flexurally, it is a simple matter to adjust the design for elastic shortening at the critical point. In continuous spans there may be several flexurally-critical points along the span; these result from different conditions of loading. Hence, in continuous members one may or may not be able to adjust the stress in the tendons to eliminate the reduction in the prestressing force due to elastic shortening to the same degree that is possible in simple span members.

Creep of Concrete

The loss in steel stress resulting from the creep of the concrete should also be computed on the basis of stresses that occur at the critical section for flexure rather than average values, since the greatest factor of safety against cracking is generally required at the section of maximum moment. In bonded construction, since creep is time dependent and does not take place to any significant degree until after bond has been established between the steel and the concrete, the strains along the tendons are not averaged in their effect upon the loss of steel stress. In the case of post-tensioned construction in which the tendons are not bonded to the concrete after stressing, the creep strains become averaged, since the tendon can slip in the member, and for this reason, the average concrete stress at the center of gravity of the steel should be used in the computation of this stress loss.

The magnitude of the creep of concrete, as well as the rate at which it occurs, can be estimated using the data in Sec. 3-11. When possible, the concrete stress at the level of the steel, taking into account the stress-history of the member, should be used in computing the loss of stress due to this phenomenon.

Shrinkage of the Concrete

The rate at which concrete shrinks as well as the magnitude of the ultimate shrinkage can be estimated using the data of Secs. 3-9 and 3-10. The entire shrinkage strain is effective in reducing the steel stress in pre-tensioned construction, while only the amount of shrinkage that occurs after the stressing is of significance in this respect in the case of post-tensioning.

Relaxation of the Prestressing Steel

Relaxation of the prestressing steel should be estimated on the basis of the data presented in Sec. 2-7. Careful consideration must be given in choosing the value of f_y' to be used in Eq. 2.1. The usually employed specification for stress relieved strand provides that the minimum yield strength will not be less than 0.85 f_{pu} and shall be measured as the stress at an extension of 1%. Yield strength was defined as the stress at a 0.1% offset in the research from which Eq. 2-1 evolved. Although the stress at a 0.1% offset may be slightly less than that at a 1% extension, it will normally be greater than 0.85 f_{pu}. Hence, if one uses 0.85 f_{pu} as the yield strength, he will generally be on the conservative side and over estimate the relaxation loss. The best procedure is to use the actual value of the stress at a 0.1% offset for f_y' in Eq. 2-1; this is usually not possible.

Friction Loss

Although some authors treat the friction loss due to post-tensioning in the same manner as the losses due to the *deformation of the materials* incorporated in a post-tensioned beam, this is not done in this book. Friction losses are treated in Chapter 15 because they are normally evaluated at the time shop drawings and the calculations pertinent thereto are prepared, rather than at the time a post-tensioned member is designed. An exception to this is the case of continuous post-tensioned members; some engineers evaluate friction losses during the design of structures of this type and specify minimum initial jacking forces rather than final (effective) prestressing forces as is customary with simple beams.

The computation of the losses of prestress due to the shrinkage of the concrete, elastic deformation of the concrete and relaxation of the prestressing steel is straight forward, provided the parameters governing these phenomena are known. The loss due to creep is not made accurately as easily, because the creep of concrete is a function of both time and the level of stress in the concrete. Since the stress is constantly changing, as a result of the losses in prestress that are occurring, the most accurate method of computing stress loss is to employ a numerical integration that takes into account the several variables. This procedure consists of employing a unit (or specific) creep curve, the assumed shrinkage curve, and the relaxation curve for the materials under consideration and computing the changes in stress over a large number of increments of relatively short duration. The curves are treated as step functions to facilitate the computations (Ref. 2).

The creep strain in the concrete can be computed using either the rate-of-creep method or the superposition method. These are illustrated in Figs. 6-1 and 6-2, respectively. The rate-of-creep method assumes the creep strain at time t is equal to the product of the stress and the ordinate of the specific creep curve corresponding to time t. Once the stress is removed, there is no further change in the creep strain. The superposition method predicts the same initial strains as the

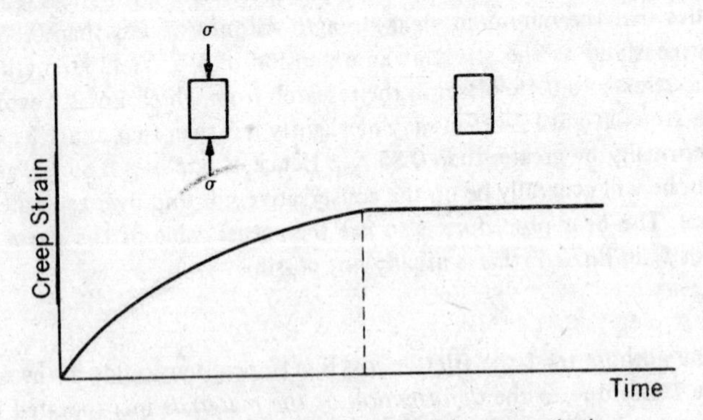

Fig. 6-1 Creep strains by the rate of creep method.

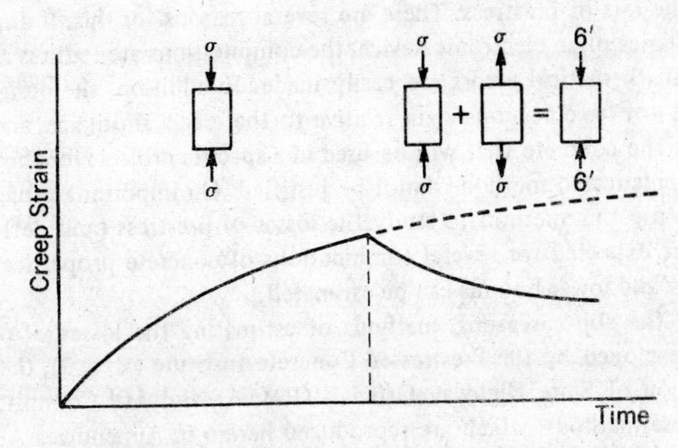

Fig. 6-2 Creep strains by the superposition method.

rate-of-creep method, but assumes the member is subjected to a tensile stress upon removal of the stress and creeps under two opposing fictitious stresses. The creep characteristics in tension and compression are assumed to be equal.

The use of the superposition method is believed to be the more accurate method, but its use is laborious and is considered to be worth using only if the creep curves and other properties of the materials are well known and high precision is required.

The following basic assumptions are made in employing the numerical integration method for predicting prestress losses (as well as in computing deflections).

(1) The initial stresses in the member under consideration are known.

(2) The specific creep vs time curve for the concrete under constant stress is known and can be considered a step function to approximate the curve.

(3) The shrinkage vs time curve is known for the concrete under consideration, and it too can be treated as a step function. The shrinkage characteristics are uniform over the section.

(4) The relaxation vs time relationship for the prestressing steel is known and can be treated as a step function.

(5) Creep strains are proportional to the concrete stress up to 50% of the concrete strength.

(6) Strains vary linearly over the depth of the section.

(7) The stress-strain relationship for the concrete is linear up to 50% of the flexural strength for loads of short duration.

(8) The stress-strain relationship for steel is linear under short duration loads.

This method lends itself to solution by programmable calculator or computer; the accuracy will be improved by using many increments of short duration.

The numerical integration procedure described above is not widely used in computing the loss of prestress. There are several reasons for this. If done without the assistance of an electronic device, the computations are tedious and time consuming; mathematical errors are easily made. In addition, the designer frequently does not have accurate data relative to the creep, shrinkage, and elastic properties of the concrete that will be used in a specific project; for this reason, the more sophisticated method cannot be justified. On important projects, the designer can use this method to study the losses of prestress (and deflections) that would be expected for several combinations of concrete properties; in this manner upper and lower bounds can be estimated.

Because of the above reasons, methods of estimating the losses of prestress have been developed by the Prestressed Concrete Institute (Ref. 3), the American Association of State Highway Officials (Ref. 4), and ACI Committee 209 (Ref. 5). These methods, which are reproduced herein in Appendices A, B, and C, with the permission of these organizations, are similar in that the loss of prestress is treated as the sum of the losses resulting from the various sources. Mathematical relationships are given for estimating each of the several losses involved.

Example calculations for losses of prestress using the several methods are given in Appendices A, B, and C.

ILLUSTRATIVE PROBLEM 6-1 Compute the loss of prestress for the composite bridge girder shown in Fig. 6-3. Use the numerical integration method in which the concrete properties are treated as follows:

	Precast	CIP
Compressive strength at 28 days (ksi)	5.500	5.000
Unit weight (pcf)	115	114
$E_{coef.}$ (33 in. Eq. 3-4)	26	26
Ultimate concrete shrinkage (inches per in. $\times 10^{-6}$)	350	300
Creep ratio	1.50	

Time dependency parameters

		Precast	CIP
Strength (Eq. 3-1)	a	2.24	2.24
	b	0.92	0.92
Shrinkage (Eq. 3-17)	e	1.00	1.00
	f	35.00	35.00
Creep (Eq. 3-18)	c	0.60	
	d	10.00	

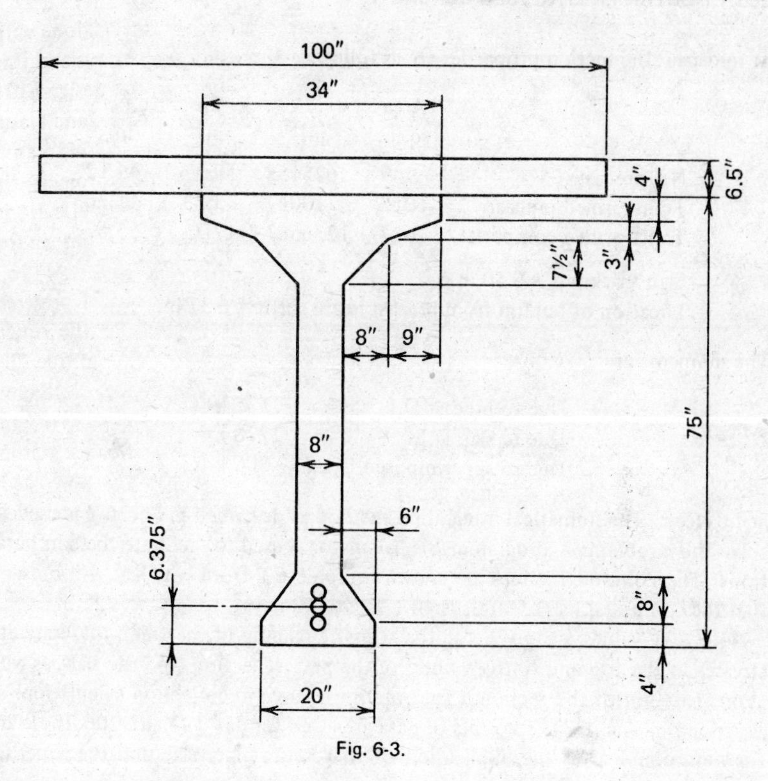

Fig. 6-3.

The steel properties are as follows:

f_{pu} = 270 ksi
f_y = 247 ksi
f_{si} = 189.5 ksi (after elastic shortening)
E_s = 27800 ksi
Relaxation coef. = 10.0 [The denominator of the second term of Eq. 2-1]
A_{ps} = 5.520 sq in.   Area of ducts = 16.23 sq. in.

Time sequence is as follows:

End precast cure = 7 days after girder cast (shrinkage begins)
Prestress = 12 days after girder cast (creep begins)
Cast slab = 197 days after girder precast
End slab cure = 204 days after girder precast
Application of SDL = 206 days after girder precast
End analysis = 600 days after girder precast

At midspan, the section properties are as follows:

	Area sq in.	I in.^{4}	y_t in.	y_b in.
Net precast	949	625158	30.87	-44.12
Transformed precast	1013	710047	33.23	-41.76
Transformed composite	1637	1226561	19.32	-55.67

Slab thickness = 6.50 in.
Location of tendon from precast beam soffit = 6.37 in.

The moments are:

Due to precast D.L.	1535 k-ft
Due to slab D.L.	1278 k-ft
Due to superimposed D.L.	300 k-ft

SOLUTION: The numerical integration method as described in Sec. 6-2 is used to solve this problem. A programmable calculator is used to facilitate the computations. The computed values are shown in Table 6-1 from which it will be seen that the computed loss of stress is 40.1 ksi at 600 days.

Note that values are given for the stress remaining in the steel, the concrete stresses in the top and bottom fibers of the precast section, and the slab, as well as for rotation of the section. (The rotation is used in deflection calculations as described in Sec. 6-3). Two sets of data are given for day 197 and 206; these are for before and after the application of the cast-in-place slab and the superimposed dead load, respectively. In this example the limiting value for loss of prestress was taken to be reached at 600 days; a more realistic time would be 1200 or 1600 days.

TABLE 6-1

		Fiber Stresses (ksi)					
		Beam			Slab		
Time (Days)	Steel Stress (ksi)	Bottom	Beam	CGS	Bottom	Top	Rotation X 10 exp 6
12	189.500	2.588	.062	2.373			14.67
197	144.449	1.825	.228	1.689			21.26
197′	152.942	.923	.946	.925			12.62
206	152.723	.919	.949	.922	-.020	.010	12.58
206′	154.330	.756	1.006	.777	.035	.086	11.41
600	149.402	.670	1.038	.701	-.159	.184	10.24

ILLUSTRATIVE PROBLEM 6-2 For the composite post-tensioned girder shown in Fig. 6-3 and the data given in I.P. 6-1, estimate the loss of prestress at 600 days using the method of Appendix C.

Assume the member is stressed when $f'_{ci} = 4000$ psi

$$E_c = (115)^{1.5}\ 26\sqrt{4000} = 2.03 \times 10^6 \text{ psi}$$

$$n = \frac{27800}{2030} = 13.7$$

SOLUTION: From Eq. 6-1, with $k_s = 3.16$,

$$f_{ci} = \frac{(189.5)(5.52)(3.16) - \dfrac{(949)(1535)(12)(37.75)}{625158}}{949 + (13.7)(5.52)(3.16)} = 1.894 \text{ ksi}$$

$$P_o = 1046 - (13.7)(1.894)(5.52) = 902.8 \text{ kips}$$

$$\text{Use } P_o = 903 \text{ kips}$$

The slab is placed at $t = 197$ days at which time from Eq. 3-1

$$f_{ct} = \frac{197}{2.24 + 0.92 \times 197}(5500) = 5900 \text{ psi}$$

$$E_c = 2.46 \times 10^6 \text{ psi}$$

$$m = 11.3$$

The stress at the level of the steel due to the application of the slab and superimposed dead load is:

$$f_{cs} = -\frac{1278 \times 12 \times 35.39}{710047} - \frac{300 \times 12 \times 49.30}{1226561}$$

$$= -0.764 - 0.144 = -0.909 \text{ ksi}$$

The relaxation loss of the prestressing steel at 600 days is

$$\Delta f_{sr} = (189.5)\frac{\log 24 \times 600}{10}\left(\frac{189.5}{247} - 0.55\right) = 17.11 \text{ ksi}$$

The value of α_s at 197 days is

$$\alpha_s = \frac{197^{0.6}}{10 + 197^{0.6}} = 0.70$$

and at 600 days, C_t is

$$C_t = \left(\frac{600^{0.6}}{10 + 600^{0.6}}\right)1.50 = 1.23$$

and β_s, from Fig. 3-12(a) is

$$\beta_s = 1.25\,(197)^{-0.118} = 0.67$$

The effect of differential shrinkage and creep strain can be estimated as follows:

For slab concrete, shrinkage from 204 days (end of cure period) to 600 days is

$$\epsilon_s = \left(\frac{396}{431}\right) 350 = 322 \text{ millionths inches per in.}$$

For the beam concrete between 197 and 600 days the shrinkage is

$$\epsilon_s = \left(\frac{600}{635} - \frac{197}{232}\right) 350 = 33 \text{ millionths inches per in.}$$

The initial top fiber stress in the beam is computed to be

$$f_t = \frac{903}{949}\left(1 - \frac{(37.75)(30.87)}{659}\right) + \frac{1535 \times 12 \times 30.87}{625158}$$

$$= -0.731 + 0.910 = 0.179 \text{ ksi}$$

At the time the superimposed dead load is placed, the top fiber stress becomes

$$f_t = (0.95)(0.179) + \frac{1278 \times 12 \times 33.23}{710047} + \frac{300 \times 12 \times 19.32}{1226561}$$

$$= 0.944 \text{ ksi}$$

and the creep of the top fiber between 197 and 600 days for a stress of 0.944 ksi is

$$\text{Creep strain} = \frac{944}{2.46}\left(\frac{600^{0.6}}{10 + 600^{0.6}} - \frac{197^{0.6}}{10 + 197^{0.6}}\right) \times 1.50$$

$$= 68 \text{ millionths inches per in.}$$

in which 2.46×10^6 is the elastic modulus of the concrete at the age of 600 days. Therefore, the differential strain is

$$322 - (33 + 68) \cong 221 \text{ millionths inches per in.}$$

Elastic interaction and relaxation of the slab concrete would reduce the effect of this strain (see Section 3.12). Assume that 150 millionths inches per in. of strain with an elastic modulus (in the slab) of 2.2×10^6 psi approximates the differential strain. Then

$$Q = \frac{2.2 \times 150 \times 8.33 \times 12 \times 6.5}{1000} = 214 \text{ kips}$$

$$f_{sds} = \frac{(11.3)(214)(-49.30)(22.57)}{1226561} = -2.19 \text{ ksi}$$

Assume $k_r = 1.0$, $\zeta = 1.25$, $\lambda = 0.88$, $\lambda' = 0.92$ and $I/I' = 0.51$. Hence,

$$\Delta f_{si} = (13.7)(1.897)\{1 + (0.70)(1.23)(0.92) + 1.23[0.88 - (0.70)(0.92)]0.51\}$$

$$+ \frac{(350)(27.5)}{(1.25)(1000)} + 17.11 - (11.3)(0.91)\left[1 + \frac{(0.67)(1.23)(0.51)}{1.25}\right] - 2.19$$

$$= 50.42 + 7.70 + 17.11 - 13.74 - 2.19 = 59.30 \text{ ksi}$$

Therefore,

$$f_{se} = \frac{903}{5.52} - 59.30 = 104.3 \text{ ksi}, \Delta P_u/P_o = 0.20$$

Assuming the initial stress (after elastic shortening) is 189.5 ksi,

$$P_0 = 189.5 \times 5.52 = 1046 \text{ kips}$$

$$f_{ci} = 2373 \text{ psi}$$

$$\Delta f_{si} = (13.7)(2.373)\{1 + (0.70)(1.23)(0.92) + 1.23[0.88 - (0.70)(0.92)]0.51\}$$

$$+ \frac{(350)(27.8)}{(1.25)(1000)} + 17.11 - (11.3)(0.91)\left[1 + \frac{(0.67)(1.23)(0.51)}{1.25}\right]$$

$$- 2.19 = 63.07 + 7.78 + 17.11 - 13.74 - 2.19 = 72.03 \text{ ksi}$$

$$f_{se} = \frac{1046}{5.52} - 72.03 = 117.46 \text{ ksi}, \Delta P_u/P_o = 0.21$$

The loss of prestress computed using the numerical integration method in I.P. 6-1 is substantially less than that obtained with the method of Appendix C. This can be explained by the fact that the effects of concrete creep and steel relaxation become less when a small time interval is used, as is the case in the numerical integration calculations. With small time intervals, one can take into account the reduction in the losses due to creep and relaxation resulting from the (generally) reducing stress levels in the concrete and prestressing steel. For example, with the numerical integration method, one can account for the fact that the prestressing steel is not subject to constant strain by substituting the computed stress in the steel at time t for the first f_{si} term appearing in Eq. 2-1. In this manner the loss of prestress due to steel relaxation will be reduced as the stress in the steel reduces. In a similar manner, the effects of reduced creep loss can be taken into account by using the "instantaneous" concrete stress levels in the creep loss computations; these levels are generally reducing. The time interval used in I.P. 6-1 was one day. The two time intervals used in I.P. 6-2 are basically 200 days and 400 days, from casting the beam to casting the slab and from casting the slab to the end of the computation. Additionally, the example used in these illustrative problems gives high losses because the sand-lightweight concrete has a relatively low elastic modulus ($n = 13.7$ and $m = 11.3$), and the initial stress in the concrete at the level of the steel is relatively high. These effects are partially offset by the relatively low creep ratio at 600 days for which the computations were made.

ILLUSTRATIVE PROBLEM 6-3 For the double-tee beam shown in Fig. 6-4, estimate the ultimate loss of prestress using the methods described in Appendix C. The dead load of the double-tee beam is 200 plf, the superimposed dead load is 40 plf, the design span is 40 ft, $A_{ps} = 0.58$ sq in., the straight tendons are pretensioned to 200 ksi, $f_{pu} = 270$ ksi, and $f_y = 250$ ksi. The gross area of the section is 189.5 sq in. and the moment of inertia is 4256 in.^{4} Assume $n = 7.3$, $m = 6.0$, $\beta_s = 0.83$, $C_u = 2.00$, $\epsilon_{sh} = 400$ millionths inches per in., and $E_s = 28{,}000$ ksi.

SOLUTION:

$$P_i = 200 \times 0.58 = 116 \text{ kips}$$

$$k_s = 1 + \frac{8.83^2 \times 189.5}{4256} = 4.47$$

$$f_{ci} = \frac{(116)(4.47) - \dfrac{(189.5)(40)(12)(8.83)}{4256}}{189.5 + (7.3)(0.58)(4.47)}$$

$$= 1.582 \text{ ksi}, \quad nf_{ci} = 11.55 \text{ ksi}$$

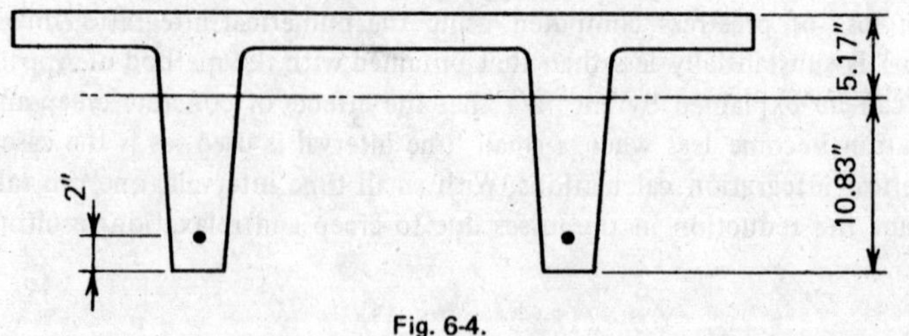

Fig. 6-4.

$$P_o = (200)(0.58) - (7.3)(1.582)(0.58) = 109.3 \text{ kips}$$

Use $P_o = 109$ kips, $nf_{ci} = 11.5$ ksi

$$\frac{A_s}{A_{ps}} = 0, \text{ hence } k_r = 1.0,$$

$$f_{cs} = \frac{0.04\,(40^2)(12)(-8.83)}{(8)(4256)} = -0.200 \text{ ksi}$$

From Eq. 2-1, @ 10^5 hours (11.4 years)

$$\Delta f_{sr} = (200)\,\frac{\log 10^5}{10}\left(\frac{200}{250} - 0.55\right) = 25.0 \text{ ksi}$$

Assume $\lambda = 0.91$ and $\zeta = 1.25$

$$\Delta f_{si} = 11.5 + (0.91)(11.5)(2.00) + \frac{400 \times 28}{1.25 \times 1000} + 25.0$$

$$- (6.0)(0.20) - \frac{(6.0)(0.20)(0.83)(2.00)}{1.25}$$

$$= 11.5 + 20.93 + 8.96 + 25 - 1.20 - 1.59 = 63.60 \text{ ksi}$$

Note that $\Delta P_u/P_o = 52.1 \times 0.58/109 = 0.28$, which is greater than the 0.18 assumed ($\lambda = 0.91$). Hence, a second iteration is indicated.

$$\Delta f_{si} = 11.5 + 0.86\,(11.5)(2.00) + \frac{400 \times 28}{1.25 \times 1000} + 25.0 - 1.20 - 1.59$$

$$= 62.45 \text{ ksi and } \Delta P_u/P_o = 0.27 \cong 0.28$$

If the relaxation loss prior to transfer is taken into account with Eq. 2-1, and the transfer is made at 24 hours,

$$\Delta f_{sr} = 200\,\frac{\log 24}{10}\left(\frac{200}{250} - 0.55\right) = 6.9 \text{ ksi}$$

$$P_i = (200 - 6.9)\,0.58 = 112 \text{ kips}$$

$$f_{ci} = \frac{(112)(4.47) - \dfrac{(189.5)(40)(12)(8.83)}{4256}}{189.5 + (7.3)(0.58)(4.47)} = 1.497 \text{ ksi}$$

$$nf_{ci} = 10.9 \text{ ksi}, \; P_o = 112 - (10.9)(0.58) = 105.7 \cong 106 \text{ kips}$$

Elastic loss + initial relaxation loss = 17.8 ksi

$$\frac{f_{st}}{182.2} = 1 - \left(\frac{182.2}{250} - 0.55\right)\left(\frac{\log 10^5 - \log 24}{10}\right) = 0.935$$

$$f_{st} = 170.3 \text{ ksi}$$

$$\Delta f_r = 6.9 + (182.2 - 170.3) = 18.8 \text{ ksi}$$

and

$$\Delta f_{si} = 11.0 + (0.91)(10.9)(2.00) + 8.96 + 18.8 - 1.20 - 1.59$$
$$= 55.99 \text{ ksi}$$

Hence, $\Delta P_u/P_o = 0.24$ which yields $\lambda = 0.88$
Using the new value for λ, $\Delta f_{si} = 55.33$ ksi and $\Delta P_u/P_o = 0.24$

ILLUSTRATIVE PROBLEM 6-4 For the double-tee beam shown in Appendix C, Fig. C-1, compute the loss of prestress at midspan at the age of 1000 days, using the numerical integration method with a time interval of one day if the values of f_{si} (after elastic loss) are 189.0, 180.8, and 174.0 ksi. Assuming 189 ksi is the steel stress immediately prior to release from the prestressing bed, 180.8 ksi would be the stress remaining in the steel after an elastic loss of 8.2 ksi. The value of 174.0 ksi is the stress one obtains if the tendons were jacked to 189 ksi and held at that level for 24 hours and then transferred to the concrete section. Under the latter condition, the loss of stress due to relaxation in the 24 hours (by Eq. 2-1) is:

$$\Delta f_{sr} = 189 \frac{\log 24}{10}\left(\frac{189}{229.5} - 0.55\right) = 7.13 \text{ ksi}$$

(Note–229.5 ksi is the minimum yield stress at 1% extension (not 0.1% offset), permitted by ASTM 416. This value is used by some designers for f_y even though the original research from which Eq. 2-1 evolved was done based upon the yield stress being defined as the stress at a 0.1% offset).

With $f_{si} = 174$ ksi

$$f_{ci} = \frac{(174)(2.14)}{615}\left(1 + \frac{18.48^2}{97.1}\right) - \frac{462.8 \times 12 \times 18.48}{59720} = 1.016 \text{ ksi}$$

and the elastic loss is $(7.3)(1.016) = 7.42$ ksi

$$f_{si} = 189 - 7.13 - 7.42 = 174.45 \text{ ksi} \cong 174 \text{ ksi ok}$$

Other parameters used in the three computations are summarized below:

CONSTANT DATA

Times (Days)		Prestressing Steel	
End precast cure	7	Yield strength (ksi)	229.5
Prestressing	12	Es (ksi)	28000
Application of slab	None	Relaxation coef.	10.0
		Area (in.^{2})	2.140
Application of SDL 1	None		
Application of SDL 2	None		
End analysis	1000		

Concrete	Precast
Compressive strength, 28 days (ksi)	4.400
Weight (pcf)	145
E Coef.	33
Ultimate shrinkage	.000546
Creep ratio	1.88
Time dependency parameters	
c-Strength	2.24
d-Strength	.92
a-Shrinkage	1.00
b-Shrinkage	35.00
a′-Creep	.60
b′-Creep	10.00

Section Properties

	Area (in.^{2})	I (in.^{4})	Distance (in.) from centroid to extreme precast fibers: Top	Bottom
Net precast	615	59720	10.020	21.980
Transformed precast	615	59720	10.020	21.980

Moment (ft-k)		Tendon	
Precast DL	462.800	Initial prestress (ksi)	189
		Location, from bottom (in.)	3.500

SOLUTION: The computations were made with the aid of an electronic programmable calculator. The results are as follows:

f_{si}* at 12 days from casting	f_s at 1000 days from casting (ksi)
189.0	145.3
180.8	141.5
174.0	138.2

*After elastic shortening.

6-3 Deflection

The computations of short term deflections in prestressed-concrete flexural members are made with the assumption that the concrete section acts as an elastic and homogeneous material. This assumption is only approximately correct, because the elastic modulus for concrete is not a constant value for all stress levels, and in addition, the elastic modulus varies with the age of the concrete and is influenced by other factors. As a result, deflection computations for prestressed concrete are approximations.

The deflections for dead and live loads are calculated using the fundamental principles of the mechanics of materials. Normally, the moment of inertia of the

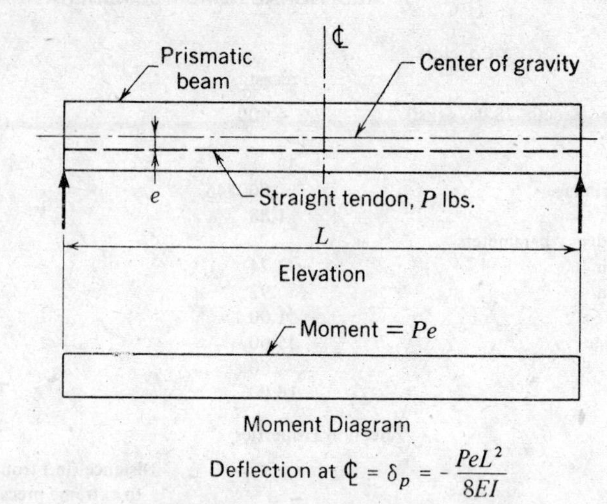

Deflection at ℄ $= \delta_p = -\dfrac{PeL^2}{8EI}$

Fig. 6-5 Layout and prestressing-moment diagram for a beam and a straight tendon.

gross section is used in the computations. In members that have a large amount of reinforcing steel, the moment of inertia of the transformed section can be used. The deflection resulting from the prestressing can be readily calculated for prismatic members with known prestressing force and eccentricity by use of the area-moment principle. The results of such a calculation for a prismatic member with straight tendons (*see* Fig. 6-5) is

$$\delta_p = -\frac{PeL^2}{8\,EI}$$

where P is the prestressing force in pounds, e is the eccentricity in inches, L is the span in inches, E is the modulus of elasticity of the concrete in psi, and I is the moment of inertia of the gross section in inches to the fourth power. The negative sign results in the upward deflection of the beam being positive for negative eccentricities.*

In Fig. 6-6, the moment diagram and corresponding deflection due to prestressing is shown for a member prestressed with a tendon that is on a parabolic curve having zero eccentricity at the ends. Finally, in Fig. 6-7, the moment diagram and corresponding deflection is indicated for a prismatic member prestressed with a tendon that has a path composed of three straight lines that are symmetrical about midspan and has zero eccentricity at each end.

***In this book, upward deflections are positive, downward deflections are negative. The word "camber" is reserved for deflection built into a member, other than by prestressing, in order to achieve a desired shape.**

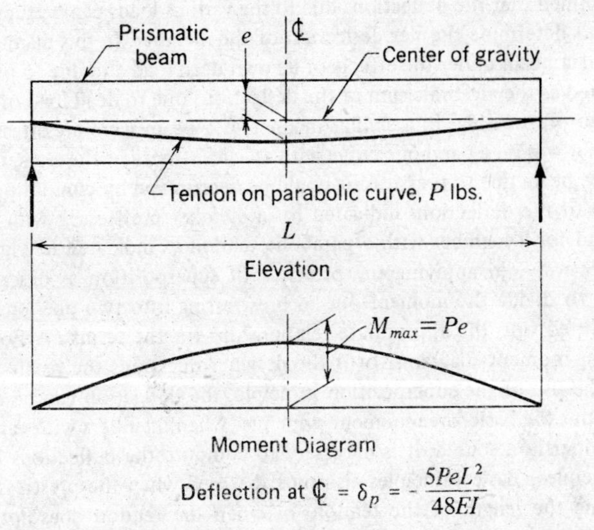

$$\text{Deflection at } ℄ = \delta_p = -\frac{5PeL^2}{48EI}$$

Fig. 6-6 Layout and prestressing-moment diagram for a beam with a parabolic tendon having no eccentricity at the supports.

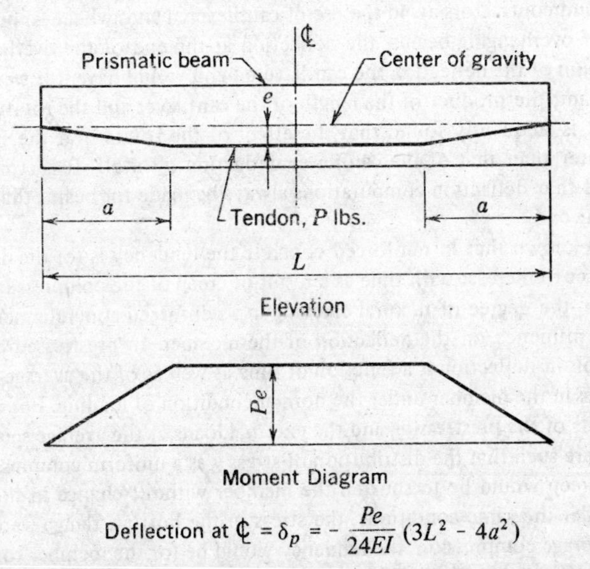

$$\text{Deflection at } ℄ = \delta_p = -\frac{Pe}{24EI}(3L^2 - 4a^2)$$

Fig. 6-7 Layout and prestressed-moment diagram for a beam with a deflected tendon having constant eccentricity in the center portion and zero eccentricity at the ends.

It is assumed that the deflections due to the various loads can be superimposed in order to determine the net deflection of the member. In this manner, the deflection of a beam under the effects of its own dead load and due to prestressing is computed as the algebraic sum of the deflections due to dead load of the beam and due to prestressing. In a similar manner, if a beam were presstressed with a tendon that was on a parabolic curve with an eccentricity at the ends, the deflection of the beam due to prestressing could be determined by computing the algebraic sum of the deflections indicated for a member prestressed with a straight tendon and for a member with a parabolic tendon as indicated in Figs. 6-5 and 6-6, respectively. In applying the principle of superposition as described, it is necessary to divide the moment due to prestressing into two portions that can be substituted into the appropriate relationships for the terms Pe. For unusual prestressing-moment diagrams, or if the designer questions the results obtained through the use of the superposition principle, the deflection can be easily calculated from the basic, area-moment principle. When members with variable moments of inertia are used, it is necessary to compute the deflections by use of basic principles. Basic principles also must be used when the prestressing force varies along the length of the tendon, or when the tendon does not follow a mathematical curve, as is frequently the case in cast-in-place post-tensioned bridges.

The deflections at the ends of members that have overhanging ends are frequently large and often result in an undesired appearance. Because of this, many engineers and contractors avoid the use of cantilevered ends whenever possible. In the case of overhanging beams, the deflection at the end of the overhang is the algebraic sum of the deflection the cantilevered end would have if it were a fixed cantilever and the product of the length of the cantilever and the rotation at the support. It is frequently found that the effect of the rotation at the support is much greater than that of the cantilever deflection by itself. It is strongly recommended that deflection computations always be made for beams that have an overhanging end.

It is well known that in reinforced concrete the tendency is for the deflection of a member to increase with time as a result of creep of the compression flange. In addition, the degree of flexural cracking in a reinforced concrete member has significant influence on the deflection of the member. In prestressed concrete, the variation in deflection is a function of time as well as of the average distribution of stress in the member under the normal condition of loading. For example, if the effects of the prestressing and the external loads at the average section of a member were such that the distribution of stress was a uniform compression, the effect of creep would be to shorten the member without change in the deflection. If under the same conditions, the stress in the bottom flange were greater than the average compression, the tendency would be for the member to increase in upward deflection with the passage of time. If the top-fiber compressive stress under the normal loading were higher than the average compression, the tendency would be for the deflection to increase downward as a result of the creep.

It is interesting to note that for the deflection due to prestressing alone, the effects of concrete shrinkage and steel relaxation are to reduce the deflection due to prestressing, because these two effects tend to reduce the prestressing force. The effect of creep is to alter the deflection for cases where the resultant force in the concrete is significantly eccentric, because the rotational changes due to creep are normally greater than is the shortening effect on reducing the prestress.

It should be recognized that the rotation at any section of a beam is equal to

$$\phi = \frac{M}{EI}$$

This relationship is useful in estimating the time-dependent deflection changes that take place in a prestressed concrete member using the numerical integration method. The relationships for deflection due to prestressing of different types that are given in Figs. 6-5, 6-6 and 6-7 can be rewritten in the form

$$\delta = -\frac{\phi L^2}{K}$$

in which K is a constant depending upon the path of the tendon. From Fig. 6-8 it will be seen that the rotation at any section can be computed if the strain distribution is known. The rotation is

$$\phi = \frac{\epsilon_b - \epsilon_t}{h}$$

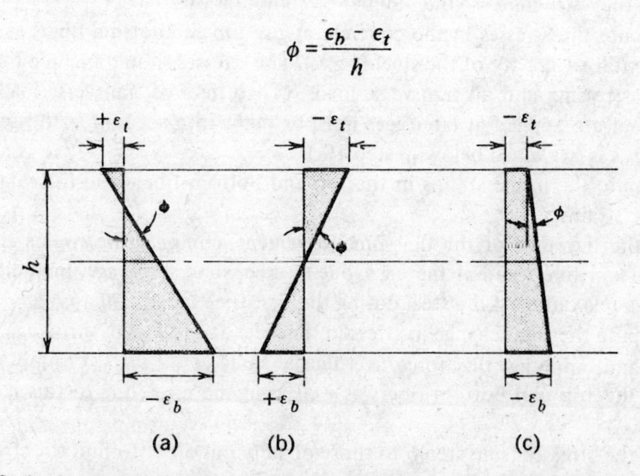

Fig. 6-8 Strain distributions due to: (a) prestressing only; (b) effects of a transverse load; and (c) the algebraic sum of (a) and (b).

where compressive strains are negative and tensile strains and rotations that cause downward deflections are positive. This relationship applies equally to short and long term rotations.

In computing long term deflections, using the numerical integration method, the principles used in computing the loss of prestress described in Sec. 6-2 can be used. All of the computations for the loss of prestress must be made in the deflection computations, because the variation in prestressing force must be taken into account. In addition, at each time increment, the strain in the top and bottom fibers must be determined so that the rotation can be computed. The computations must be made at several sections along the length of the beam and not just at the location of maximum moment, as in the computation for loss of prestress, because the conditions of stress and hence creep, loss of prestress, and rotation vary along the length of the beam. After the rotations have been determined at a number of locations along the beam length, the deflection can be determined by the classical methods.

The step by step procedure for computing the deflection of a precast beam with a composite cast-in-place slab (such as shown in Fig. 6-3) is as follows:

(1) Determine shrinkage and creep characteristics of the concrete to be used as a function of time. In addition, determine the relaxation characteristics of the prestressing steel as a function of time.

(2) Divide the beam into a number of incremental lengths. The computations that follow must be done for each increment. (Symmetry of beam and loading conditions will reduce the calculations required.) Determine the time increments to be used. Small increments are desirable from the standpoint of improving accuracy, but they also increase the number of computations.

(3) Compute the stresses in the concrete at the top and bottom fibers as well as at the center of gravity of the steel (c.g.s.). The stresses should include the effects of prestressing and all transverse loads. (The effect of transverse loads or restraints that are applied at later ages must be taken into account at the appropriate time, as is explained below in step 10.)

(4) Compute the initial strains in the top and bottom fibers and the rotation due to these strains.

(5) For the duration of the first time increment, compute the strains at the top and bottom fibers and at the c.g.s. due to creep, and shrinkage. In addition, compute the relaxation of the steel during the first time increment.

(6) Compute the total change in stress in the steel due to creep, shrinkage, and relaxation and, applying this force as a tensile force at the c.g.s. Compute the stresses in the top and bottom fibers as well as at the c.g.s. due to this tensile force.

(7) Add the stresses from step 6 to those of step 3 in order to find the stresses at the end of the time increment.

(8) Compute the strains in the top and bottom fibers at the end of the time terval as well as the rotation.

(9) Using the stresses from step 7, repeat the procedure (steps 5 through 8). his procedure is repeated until the total time being considered is covered.

(10) At the appropriate time, the effect of a superimposed load is taken into ccount by computing the changes in stress at the top and bottom fibers and at ne c.g.s. due to the load. The stress at the c.g.s. should be multiplied by the odular ratio and the resulting steel stress applied as an incremental change in ne prestressing force. The effect of the incremental change in the prestressing orce on the stresses in the top and bottom fibers and at the c.g.s. should be omputed. From these computations, the resulting stresses in the concrete can e determined and the rotation computed.

(11) The procedure continues (steps 5 through 8) until the total time under onsideration is covered.

(12) The effect of the differential strains in the cast-in-place concrete and the recast concrete are taken into account by first computing the difference in the nrestrained changes in strain in the cast-in-place concrete and the precast concrete at the interface between the two. Strain compatibility is then forced by pplying equal and opposite forces to the cast-in-place concrete and the beam.

(13) The stresses from the forces computed in step 12 are to be computed in ach subsequent time-interval computation and taken into account in the roune of steps 5 through 8.

(14) The deflection at the end of any time interval can be computed by the rea-moment principle, since the rotation ϕ is equal to M/EI.

It should be apparent that a large amount of tedious computations must be nade in applying this method. Because of this, the method can best be applied ith the aid of programmable calculators or computers.

Special structures, such as prestressed concrete bridges that are constructed egmentally in cantilever (See Fig. 8-3), may involve many steps in the construcon, at which times new increments of load or prestressing are added to the tructure. The numerical integration method of estimating losses of prestress nd deflection is the logical method to be used in such a case. This method ermits the effect of time to be computed independently for each of the omponents.

When the numerical integration method of computing the deflection of prestressed concrete members is not used, one can estimate the deflection of prestressed concrete members using the modified step function method recomnended by ACI Committee 209 (Ref. 5) and Branson (Ref. 6). In this method, he principle of superposition is used with the final deflection taken as the algeraic sum of the effects of dead and live loads combined with the effects of creep nd loss of prestress. Two basic relationships are used.

These are for:

1. Prestressed Concrete beam with or without superimposed loads.
2. Composite prestressed concrete beam constructed without or with shoring supporting the beam at the time the composite slab is placed.

The following notation and definitions are used in these relationships:

C_u = Ultimate creep ratio
I = Moment of inertia of precast beam
I' = Moment of inertia of composite beam

$k_r = \dfrac{1}{1 + b_{12}}$, or when $\dfrac{A_s}{A_{ps}} \leqslant 2, k_r \cong 1 + \dfrac{A_s}{A_{ps}}$

(See Appendix C. k_r is frequently taken to be unity)

P_o = Prestressing force after transfer (after elastic loss)
ΔP_s = Loss of prestress at time slab is cast (excluding the initial elastic loss)
ΔP_u = Total loss of prestress (excluding the initial elastic loss)
α_s = Ratio of the creep ratio for the concrete of the beam at the time the slab is cast to the ultimate creep ratio ($\alpha_s = t^c / d + t^c$ from Eq. 3-18)
β_s = Creep correction factor for beam concrete age when loaded (see Fig. 3-12)
δ_d = Deflection due to beam dead load
δ_L = Deflection due to live load
δ_{DS} = Deflection due to differential shrinkage and creep between beam and slab concretes

$$\delta_{DS} = \frac{\Delta Q y_{cs} L^2}{8\, EI'}$$

(See Eq. 6-4 below.)

δ_S = Deflection due to slab dead load
δ_{SDL} = Deflection due to superimposed dead load
δ_p = Deflection due to prestressing
δ_u = Ultimate deflection

$$\lambda = 1 - \frac{\Delta P_u}{2 P_o}$$

$$\lambda' = 1 - \frac{\Delta P_s}{2\, P_o}$$

For a non-composite beam, the ultimate total load deflection is computed by

$$\delta_u = \delta_p \left[1 - \frac{\Delta P_u}{P_o} + \lambda (k_r C_u) \right] + \delta_d\, [1 + k_r C_u] + \delta_{SDL}\, [1 + \beta_s k_r C_u] + \delta_L \tag{6-2}$$

For a composite beam without shoring at the time the slab is placed, the ultimate total load deflection is

$$\delta_u = \delta_p \left[1 - \frac{\Delta P_s}{P_o} + \alpha_s k_r C_u \lambda'\right] + \delta_d \left[1 + \alpha_s k_r C_u\right]$$
$$+ \delta_p \frac{I}{I'}\left[-\frac{\Delta P_u - \Delta P_s}{P_o} + k_r C_u (\lambda - \alpha_s \lambda')\right] \tag{6-3}$$
$$+ (1 - \alpha_s) k_r C_u \delta_d \frac{I}{I'} + \delta_s \left[1 + \alpha_s k_r C_u \frac{I}{I'}\right]$$
$$+ \delta_{DS} + \delta_L$$

For a beam which is shored at the time the composite slab is placed, the deflection due to the slab and due to live load are computed using the moment of inertia of the composite section, and the ratio I/I' is deleted from the third and from last term of Eq. 6-3.

For a simple beam, the deflection due to differential shrinkage and creep between the slab and beam concretes ($\Delta\epsilon_s$) is computed from

$$\delta_{DS} = \frac{Q y_{cs} L^2}{8 E_c I'} \tag{6-4}$$

in which

E_c = Elastic modulus of the composite beam
L = Span length of the beam
Q = The force generated by the differential shrinkage and creep
y_{cs} = Distance from the centroid of the composite section to centroid of the slab

The value of δ_{DS} must be computed from basic relationships for beams which are not simply supported.

Deflections at times between the initial and ultimate values can be estimated using relationships similar to Eqs. 6-2 and 6-3. One can either estimate the deflection at an intermediate time based upon the initial and final values computed as described above, or employ the more detailed procedures suggested by Branson (Ref. 6).

In using Eqs. 6-2 and 6-3, the losses of prestress can be estimated by using the methods described in Appendices A, B, and C.

Allowable tensile stresses as high as $12\sqrt{f_c'}$ are permitted by ACI-318 (Sec. 18.4.2(c)), provided computations are made to demonstrate that the immediate and long-term deflections comply with the requirements of Sec. 9.5 of ACI 318. This requires that the effects of creep, shrinkage, relaxation, and cracking all be accounted for in the computations. With a significant degree of cracking, the

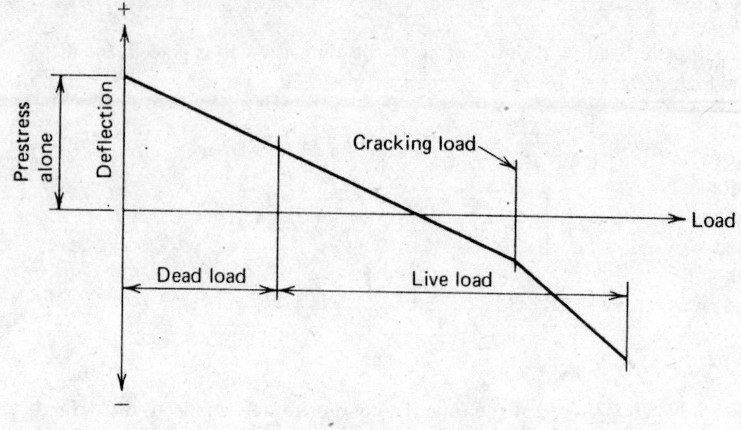

Fig. 6-9

load-deflection curve of a prestressed concrete member can be represented as shown in Fig. 6-9. This curve does not accurately depict the deflection characteristics of this type of member because in reality the components of the figure should be curved lines rather than straight ones. For loads below the cracking loads, the deflections are computed as described above. For loads exceeding the cracking load the "effective moment of inertia", I_e, is to be used in the computations. This term is defined in ACI-318, Sec. 9.5.2.3 as follows:

$$I_e = \left(\frac{M_{cr}}{M_a}\right)^3 I_g + \left[1 - \left(\frac{M_{cr}}{M_a}\right)^3\right] I_{cr} \qquad (6\text{-}5)$$

The terms in the above are defined as follows:

I_{cr} = Moment of inertia of the cracked transformed concrete section about the centroidal axis

I_g = Moment of inertia of the gross concrete section about the centroidal axis, neglecting the reinforcement

M_a = Maximum moment in member at stage for which deflection is being computed

M_{cr} = Cracking moment (based upon the modulus of rupture equal to $7.5\sqrt{f_c'}$ for normal concrete. See ACI 318, Sec. 9.5.2.3 for lightweight concrete) concrete)

Experience has shown that the deflection of uncracked prestressed concrete flexural members cannot be predicted with precision. This is often found to be true for members where creep and shrinkage would not be expected to have had a major influence on the deflection of the member. It is well known that cracking, which may be confined to small portions of a flexural member or which may be more general, adds another variable that is difficult to accurately evaluate. It is doubtful that reliable methods currently exist that will permit the average practicing engineer to conform to the requirements of Sec. 18.4.2(c) of ACI 318.

ILLUSTRATIVE PROBLEM 6-5 Determine the midspan deflections of the composite beam having the cross-section shown in Fig. 6-3 if the span is 126.3 ft. The beam is the end span of a three span beam that is continuous for superimposed dead load and live load but simply supported for other dead loads. The 10th point section properties are as shown in Table 6-2. Plot the midspan vs. time curve.

TABLE 6-2

Net Precast Section			
Pt.	Area (sq in.)	Moment of Inertia, (in.^{4})	y_b (in.)
0 & 10	949	650070	43.41
1 & 9	949	647179	43.67
2 & 8	949	640206	43.87
3 & 7	949	632685	44.01
4 & 6	949	627168	44.10
5	949	625158	44.12
Precast Transformed Section			
Pt.	Area (sq in.)	Moment of Inertia, (in.^{4})	y_b (in.)
0 & 10	1013	650140	43.48
1 & 9	1013	657091	42.86
2 & 8	1013	673860	42.37
3 & 7	1013	691947	42.03
4 & 6	1013	705214	41.82
5	1013	710047	41.76
Composite Transformed Section			
Pt.	Area (sq in.)	Moment of Inertia, (in.^{4})	y_b (in.)
0 & 10	1637	1119276	56.74
1 & 9	1637	1143025	56.35
2 & 8	1637	1173053	56.06
3 & 7	1637	1200732	55.84
4 & 6	1637	1219788	55.72
5	1637	1226561	55.67

The dead load moments, tendon jacking stress, and soffit distance for the prestressing at the 10th points are shown in Table 6-3.

TABLE 6-3

Pt.	Precast DLM (ft-k)	CIP Slab DLM (ft-k)	SDL Moment (ft-k)	f_{si} (ksi)	Soffit Dist. (in.)
0	0	0	0	189.5	44.50
1	553	460	125	189.5	30.77
2	982	818	200	189.5	20.10
3	1289	1074	250	189.5	12.47
4	1473	1227	300	189.5	7.90
5	1535	1278	300	189.5	6.38
6	1473	1227	250	189.5	7.90
7	1289	1074	150	189.5	12.47
8	982	818	0	189.5	20.10
9	553	460	-175	189.5	30.77
10	0	0	-445	189.5	44.50

The materials properties and time sequence are as given in Illustrative Problem 6-1.

SOLUTION: The 10th point rotations, using the numerical integration method of analysis with the aid of a programmable calculator, are found to be as follows:

TABLE 6-4

	Rotation × 10^6					
Pt.	12 Days	197 Days	197′ Days	206 Days	206′ Days	600 Days
0	-0.76	-1.33	-1.33	-1.37	-1.37	-2.06
1	4.61	6.93	3.58	3.54	3.01	2.23
2	8.90	13.29	7.46	7.42	6.61	5.72
3	12.06	17.75	10.30	10.26	9.26	8.13
4	14.02	20.40	12.05	12.01	10.83	9.66
5	14.67	21.26	12.62	12.58	11.41	10.24
6	14.02	20.40	12.05	12.01	11.03	9.90
7	12.06	17.75	10.30	10.26	9.66	8.63
8	8.90	13.29	7.46	7.42	7.42	6.57
9	4.61	6.93	3.58	3.54	4.27	3.69
10	-0.76	-1.33	-1.33	-1.37	0.53	0.42

The deflections at the 10th points for the significant days in the time sequence are summarized as follows:

TABLE 6-5

	Deflections (in.)					
Pt.	12 Days	197 Days	197′ Days	206 Days	206′ Days	600 Days
0	0	0	0	0	0	0
1	1.05	1.54	0.88	0.88	0.80	0.70
2	1.99	2.93	1.69	1.68	1.54	1.35
3	2.74	4.02	2.33	2.32	2.13	1.88
4	3.22	4.72	2.74	2.73	2.51	2.22
5	3.39	4.96	2.89	2.88	2.65	2.35
6	3.22	4.72	2.74	2.73	2.53	2.25
7	2.74	4.02	2.33	2.32	2.17	1.93
8	1.99	2.93	1.69	1.68	1.59	1.41
9	1.05	1.54	0.88	0.88	0.84	0.75
10	0	0	0	0	0	0

The midspan deflections, plotted as a function of time, are shown in Fig. 6-10.

ILLUSTRATIVE PROBLEM 6-6 For the composite post-tensioned beam of I.P. 6-1, 6-2, and 6-5, compute the dead load deflection at 600 days based upon the data computed in I.P. 6-2. Assume the tendon to be on a second-degree parabolic path with no eccentricity at the supports and an eccentricity at midspan of −37.75 in. The span length is 126.3 ft. The superimposed dead load results in a negative moment of 445 k/ft at one end of the beam only. Use $P_0 = 1046$ kips and $\Delta P_u/P_0 = 0.18$.

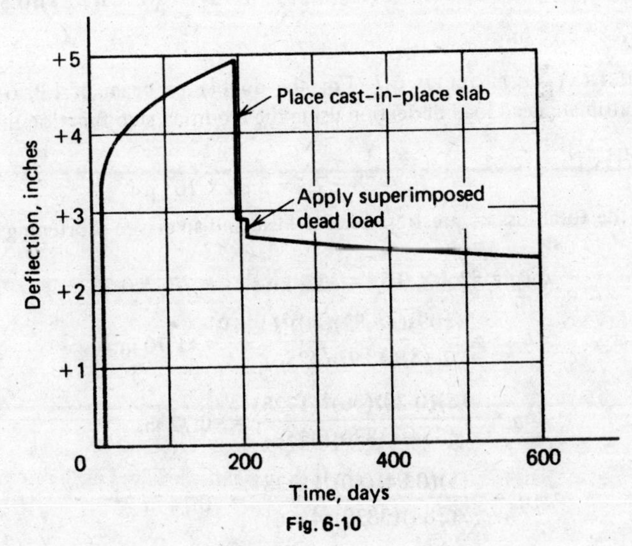

Fig. 6-10

SOLUTION:

$$\delta_p = -\frac{(5)(1046)(-37.75)(126.3)^2 \times 144}{48 \times 2030 \times 625158} = +7.44 \text{ in.}$$

$$\delta_d = -\frac{(5)(1535)(126.3)^2 \times 1728}{48 \times 2030 \times 625158} = -3.47 \text{ in.}$$

$$\delta_s = -\frac{(5)(126.3)^2(1728)}{(48)(2460)}\left[\frac{1278}{710047} + \frac{300}{1226561}\right]$$

$$+ \frac{(3)(445)(126.3)^2(1728)}{(48)(2460)(1226561)} = -2.13 \text{ in.}$$

$$\Delta P_u = 34.90 \times 5.52 = 193 \text{ kips} \quad \text{(at 600 days)}$$

$$\Delta P_u/P_o = 0.18 \qquad \text{(at 600 days)}$$

$$\text{Assume } \Delta P_s/P_o = 0.10 \quad \text{(at 200 days)}$$

Assume $\lambda = 0.91$, $\lambda' = 0.95$, $\alpha_s = 0.70$, $\beta_s = 0.67$, $k_r = 1.0$, $C_u = 1.23$ (at 600 days), $I/I' = 0.51$

$$\delta_{DS} = -\frac{214(22.57)(126.3)^2(144)}{(8)(2460)(1226561)} = -0.46 \text{ in.}$$

$$\delta_{600} = 7.44[1 - 0.10 + (0.70)(1.23)(0.95)] - 3.47[1 + 0.70)(1.23)]$$

$$+ (7.44)(0.51)\{-0.08 + 1.23[0.91 - (0.70)(0.95)]\}$$

$$- (1 - 0.70)(1.23)(3.47)(0.51) - 2.13[1 + (0.70)(1.23)(0.51)] - 0.46$$

$$= 2.98 \text{ in.}$$

ILLUSTRATIVE PROBLEM 6-7 For the double-tee beam of I.P. 6-3, compute the ultimate dead load deflection using the modified step function method.

SOLUTION:

$$E_c = 28 \div 7.3 = 3.83 \times 10^6 \text{ psi}$$

For the total loss of prestress of 63.60 ksi, and an elastic shortening loss of 11.5 ksi,

$$\Delta P_u = 52.1 \times 0.58 = 30.2 \text{ kips} \qquad P_u/P_o = 0.28$$

$$\delta_p = -\frac{(109)(-8.83)(40)^2(144)}{(8)(3830)(4256)} = +1.70 \text{ in.}$$

$$\delta_d = -\frac{(5)(0.20)(40)^4(1728)}{(384)(3830)(4256)} = -0.72 \text{ in.}$$

$$\delta_{SDL} = -\frac{(5)(0.04)(40)^4(1728)}{(384)(3830)(4256)} = -0.14 \text{ in.}$$

$$\delta_u = +1.70[1 - 0.28 + (0.86)(2.00)] - 0.72[1 + 2.00]$$

$$- 0.14[1 + (0.83)(2.00)] = 1.61 \text{ in.}$$

6-4 Composite Beams

Flexural members that are formed of precast and cast-in-place elements are frequently employed in bridge and building construction. Beams of this type are referred to as composite beams. An illustration of a typical, composite, bridge beam is given in Fig. 6-11.

Composite beams are used to permit the precasting of the elements that are difficult to form and that have the bulk of the reinforcing. Falsework is not normally needed with composite construction, since the precast elements are usually designed to carry the dead load of the precast beams, the diaphragms as well as the composite slab, without the composite top flange. Dead load and live load, which are applied after the cast-in-place deck has hardened, are carried by the composite beam.

The use of large, composite top flanges results in high, ultimate resisting moments and greater flexural strength in the elastic range, but does not improve the shear strength of a prestressed beam to a significant degree. For this reason, there is little if any advantage to be gained by using composite construction for short-span members in which shear stresses are more significant than flexural stresses.

In designing composite beams, it is necessary to compute the section properties of the precast and composite sections. The flexural stresses due to the various causes at the critical fibers are then computed using the appropriate section properties. In addition, the designer must provide adequate means of transferring the shear stress from the precast to the cast-in-place elements. Nominal shear

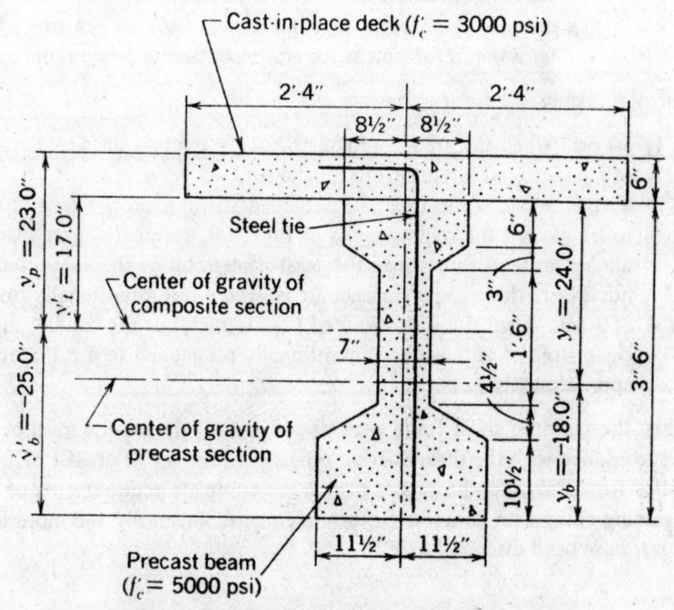

Fig. 6-11 Typical cross section of a composite beam.

stresses can be transferred by bond alone, if the concrete surface receiving the composite topping is clean and rough. Higher shear stresses can be transferred if, in addition to the above, steel dowels are extended from the precast section into the cast-in-place section. Shear keys are not required to transfer shear in composite members.

The provisions of ACI 318 permit the full transfer of horizontal shear forces to be assumed without calculations if: (1) the contact surfaces are clean and intentionally roughened to a full amplitude of approximately $\frac{1}{4}$ inch; (2) the minimum amount of reinforcing connecting the components is not less than that calculated with Eq. 5-59 with a spacing not exceeding four times the least dimension of the supported element nor 24 in.; (3) web members are designed to resist the entire vertical shear; and (4) all stirrups are fully anchored into all intersecting components. If this is not done, the horizontal shear stresses must be fully investigated.

ACI 318 permits composite sections to be designed for shear based upon

$$V_u \leqslant \phi V_{nh} \tag{6-6}$$

in which V_u is the factored (design) shear force at the section under consideration, ϕ is 0.85 and V_{nh} is the nominal horizontal shear strength. The nominal horizontal shear is computed as

$$V_{nh} = v_{nh}\, b_v\, d \tag{6-7}$$

in which

b_v = width of cross section at contact surface being investigated for horizontal shear

d = distance from extreme compression fiber to centroid of tension reinforcement for entire composite section, in.

The permissible values for v_{nh} are:

(1) 80 psi When ties are not provided, but the contact surfaces are clean, free of laitance, and intentionally roughened.

(2) 80 psi When vertical bars or extended stirrups, proportioned to be equal to or exceed the requirements of Eq. 5-59, are provided at spacings that do not exceed four times the least dimension of the supported element nor 24 in., the contact surfaces are clean but not intentionally roughened.

(3) 350 psi When the conditions of (2) are met, but the contact surfaces are clean, free of laitance and intentionally roughened to a full amplitude of approximately $\frac{1}{4}$ in.

When the factored shear force exceeds $\phi(350\, b_v d)$ the section must be designed in accordance with the shear friction provisions (Sec. 11.7) of ACI 318.

It is recommended the reader review the complete requirements of ACI 318 regarding composite concrete flexural members, since only the more important points have been discussed here.

Differential-shrinkage stresses in composite construction can result in tensile stresses being developed in the cast-in-place concrete and a reduction in the precompression of the tensile flange of the precast element. The differential shrinkage has no effect on the flexural strength of the composite beam, but it does slightly reduce the load required to crack the tensile flange of the precast element. This effect should be considered in structures in which the cracking load is significant. The effect is normally ignored.

When computing the properties of the transformed section, the difference in the elastic properties of the cast-in-place concrete and the concrete in the precast element should be taken into account by adjusting the width of the composite flange in proportion to the modular ratio of the two concretes. This is exemplified in Problem 6-8.

ILLUSTRATIVE PROBLEM 6-8 Compute the flexural stresses in the precast and cast-in-place concrete for the composite section shown in Fig. 6-11, if the moment due to the dead load of the beam, slab, and diaphragms is 673 k-ft and the moment due to the future wearing surface, live load, and impact is 830 k-ft. The section properties for the precast and composite sections are as follows:

Precast section

$y_t = 24.0$ in. $\quad S_t = 4450$ in.^{3} $\quad y_b = -18.0$ in. $\quad S_b = -5950$ in.^{3}

Composite section properties are based upon the transformed cast-in-place flange width of 0.6 × 56 in. = 33.6 in. It is assumed that the ratio of the elastic modulus of the slab and girder concrete is 0.60.

$y_p = 23.0$ in. $\quad S_p = 9400$ in.^{3} $\quad I = 216{,}000$ in.^{4}
$y_t = 17.0$ in. $\quad S_t = 12{,}700$ in.^{3}
$y_b = -25.0$ in. $\quad S_b = -8650$ in.^{3}

SOLUTION:

	Top Fiber Stress in Cast-in-Place Slab	Stresses in Precast Section	
		Top	Bottom
Dead load: $\dfrac{673 \times 12{,}000}{4450/-5950}$		+1810 psi	-1360 psi
$L + I$: $\dfrac{830 \times 12{,}000}{9400/12{,}700/-8650}$	+1060 psi	+ 785 psi	-1150 psi
Totals:	+1060 psi	+2595 psi	-2510 psi

ILLUSTRATIVE PROBLEM 6-9 Compute the section properties of the precast and composite sections for the cross section shown in Fig. 6-11, if the beam is made of sand-light weight concrete having a unit weight of 112 pcf and the deck slab is normal weight concrete (145 pcf). Assume the 28-day compressive strengths to be 4,500 and 3,500 psi for the beam and deck concretes respectively. Assume Eq. 3-4 is sufficiently accurate for both concretes.

SOLUTION:

For the beam concrete

$$E_{cb} = 33(112)^{1.5}\sqrt{4500} = 2{,}620{,}000 \text{ psi}$$

For the slab concrete

$$E_{cs} = 33(145)^{1.5}\sqrt{3{,}500} = 3{,}400{,}000 \text{ psi}$$

Ratio $E_{cs}/E_{cb} = 1.30$

Beam section properties:

Area		y	Ay	d^2	$\left(\frac{I_{cg}}{A}\right)$	xA	Io
7×42	$= 294$	$\times 21$	$= 6174$	9	$+ 147$	$= 156$	45864
10×6	$= 60$	$\times 3$	$= 180$	441	$+ 3$	$= 444$	26640
$10 \times \frac{3}{2}$	$= 15$	$\times 7$	$= 105$	289	$+ 0.5$	$= 289.5$	4342
$16 \times \frac{4.5}{2}$	$= 36$	$\times 30$	$= 1080$	36	$+ 1.13$	$= 37.13$	1337
16×10.5	$= 168$	$\times 36.75$	$= 6174$	162.6	$+ 9.19$	$= 171.8$	28862
Σ Area	573	ΣAy	13713				107045

$$y_t = 24.0 \text{ in.}$$
$$y_b = -18.0 \text{ in.}$$

Composite section properties:

$1.30 \times 56 \times 6 =$	437	$\times 3 =$	1310	$234 + 3 = 237$		103569
	573	$\times 30$	17190	$137 + 0 = 137$		78501
	1010		18500			107045
						289,115

$$y_p = 18.3 \text{ in.}$$
$$y_t = 12.3 \text{ in.}$$
$$y_b = -29.7 \text{ in.}$$

6-5 Beams with Variable Moments of Inertia

The magnitude of the prestressing moment in a simple beam can be made t vary along the length of the beam by varying the depth of the member. Mem bers of variable depth may be prestressed with either a straight or curved tendo

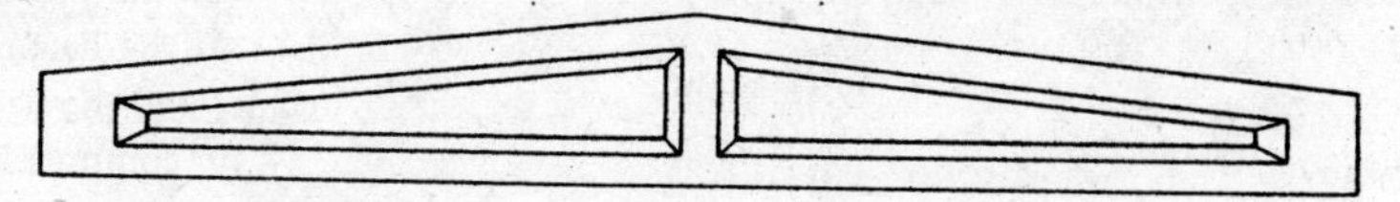

Fig. 6-12 Elevation of beam with sloping top flange.

This is illustrated by the sloped beam of Fig. 6-12. This type of beam is obviously adaptable to roof construction where the pitch of the roof required for drainage will be provided by the sloping top flange. Although beams of this shape have been produced in this country by a few prestressing plants and on a few special jobs, they are not used to a great extent. The following disadvantages, characteristic of this type of beam, account for the limited use of this mode of framing:

(1) The design of such beams must be done with care, since the maximum moment and maximum flexural stresses may not (probably do not) occur at the same section. Therefore, in order to be certain that the most severe conditions are investigated, the stresses must be determined at several points along the span and plotted to determine the most severe cases. This is a refinement not normally required in simple beam design and is a not always recognized characteristic of this type of beam.

(2) The sloping top flanges intersect at the center of the beam, which is the point of maximum moment. The large inclined compressive forces that are carried by these flanges intersect at an angle as shown in the vector diagram of Fig. 6-13. Provision must be made for the vertical component of these forces; otherwise, there is danger the top flange may buckle upward under load. The danger of this occurring is even greater if a Vierendeel truss is used, rather than a beam, or if an opening is made in the web of the beam at midspan for the passage of utilities.

(3) Forms for members with sloping flanges are expensive and are not easily converted for manufacturing the many different span lengths encountered in modern commercial and industrial building construction.

Another type of beam with a variable moment of inertia and depth that can be used to advantage in roof as well as bridge construction is illustrated in Fig. 6-14.

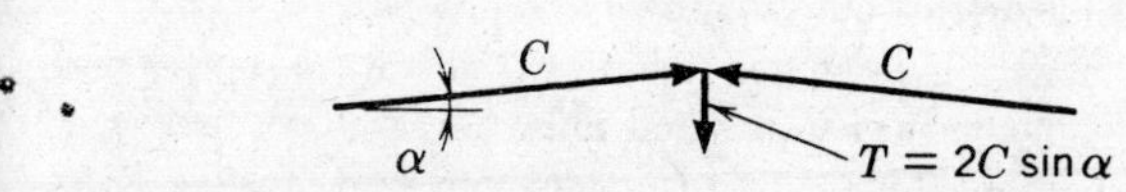

Fig. 6-13 Vector diagram of forces at ridge of beam with sloping top flange.

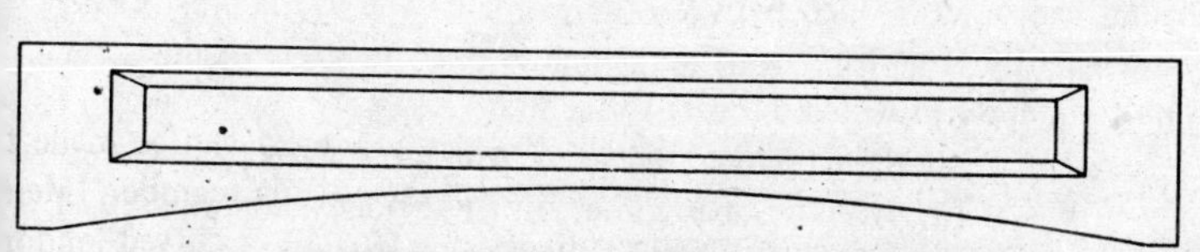

Fig. 6-14 Elevation of beam with variable bottom-flange thickness.

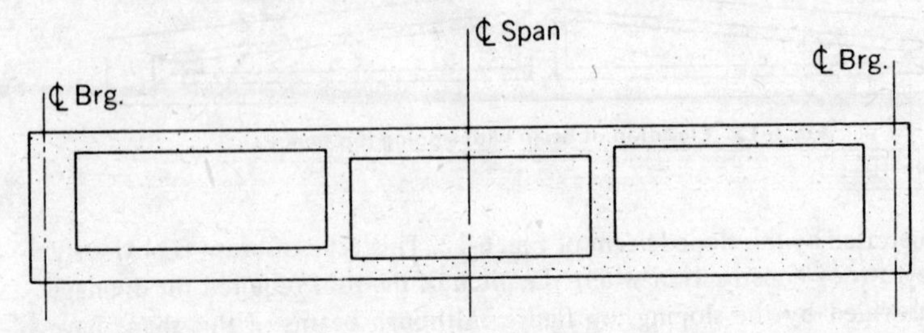

Fig. 6-15 Longitudinal cross section of a hollow box beam showing a method of varying the moment of inertia.

This beam can be stressed with straight tendons, and since the depth of the section is greater at the ends, the relative eccentricity at the ends will be less at the ends than at the center. As a result, the stresses due to prestressing will not be as high at the ends. In this manner, an effect similar to curving the tendons in a prismatic beam is obtained.

Another economical method of forming a beam with a variable moment of inertia is to use a box section, as illustrated in Fig. 6-15. In such a beam, the hollow core is frequently made with inexpensive plywood or paper forms that can be placed lower at the center than at the ends. In this manner, the larger concrete flanges are placed where needed to resist the large compressive stresses due to pre-tensioning at the ends and to the flexural stresses at midspan.

Beams with variable moments of inertia and depth are frequently used in continuous prestressed-concrete structures, for the same reasons variable depths are employed in continuous members made of other materials. Continuity in prestressed concrete construction is discussed in detail in Chapter 8.

6-6 Segmental Beams

Post-tensioned beams consisting of two or more elements or components held together by prestressing are used occasionally to facilitate fabrication, transportation, erection, or for other economic considerations. An example of a multi-element beam formed of three precast units is shown in Fig. 6-16.

In beams on which this method is employed, the prestressing force is generally very large in comparison with the shear force that must be developed between the elements. This is true because this method is most often used on large, long beams in which shear forces are not as important as in short beams. As a result, the friction that can be developed between the elements due to the prestressing is normally sufficiently large to provide very high factors of safety against slipping. Keys are frequently provided in order to facilitate assembly of the units.

A precast segment of the Downstream Bridge at Autevil in Paris during erection is shown in Fig. 6-17. The long-span girders of this bridge are composed of many precast segments held together by longitudinal prestressing.

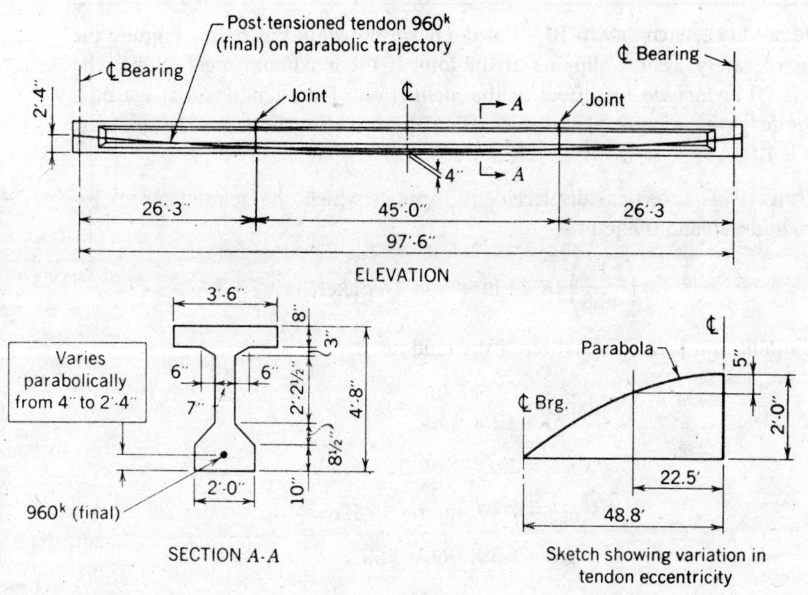

Fig. 6-16 Segmental post-tensioned beam. Adapted from bridge over Naugatuck River, Route 68 in Connecticut.

Fig. 6-17 A precast segment of the Downstream Bridge at Autevil, Paris. (*Courtesy of the Freyssinet Co., New York.*)

ILLUSTRATIVE PROBLEM 6-10 For the beam shown in Fig. 6-16, compute the factor of safety against slipping at the joint if the maximum shear load at the joint is 70 k. Include the effect of the inclination of the tendons that are on a parabolic path. Assume that the coefficient of friction between the concrete units is 1.0.

SOLUTION: The vertical displacement through which the tendon moves between midspan and the joint is

$$\left(\frac{22.5}{48.8}\right)^2 \times 24 \text{ in.} = 5 \text{ in. } (\textit{see} \text{ sketch})$$

At the joint, if α is the inclination of the tendons

$$\tan \alpha = \frac{2 \times 5 \text{ in.}}{12 \times 22.5} = 0.037$$

$$P \sin \alpha = 0.037 \times 960 = 35.5 \text{ k}$$

$$V_{\text{conc}} = V - P \sin \alpha = 34.5 \text{ k}$$

$$\mu P = 1.0 \times 960 = 960 \text{ k}$$

$$F/S \text{ slipping} = \frac{960}{34.5} \cong 27.8$$

6-7 Partial Prestressing

When tensile stresses are permitted in the tensile flange of a prestressed concrete flexural member under the condition of service load, a member is said to be partially prestressed. Partial prestressing was first suggested as a means of permitting the use of higher stresses in non-prestressed reinforcing bars together with supplementary pretensioned tendons as a means of reducing the cracking of the concrete. Later, Abeles suggested the use of high-tensile steel for the entire tensile reinforcement, but with only a portion of the steel being prestressed (Ref. 7). In this manner, economy would result from the reduction in labor required to stress and grout the tendons. In addition, the use of high-tensile steel for the entire tensile reinforcement is the most economical, since the cost per pound of ultimate tensile strength is less for high-tensile steel than for lower quality steels.

Currently there are three motivations for the use of partial prestressing. The first is economy of labor and steel costs, as was explained above. Partial prestressed beams made in this manner will have somewhat lower flexural strengths than fully-prestressed members, since the average effective stress in the tendon will be lower than in fully-prestressed members. Therefore, the flexural strength must be computed from basic strain computations. The fundamental principles of ultimate moment analysis developed in Sec. 5-2 are applicable to this condition of reinforcing.

If partial prestressing is used in a member because the concrete cross section to be used is inefficient, but all of the steel is stressed to the normal values, the ultimate moment will not be affected as a result of the use of partial prestressing. This can be better understood when the basic reason for using I- and T-shaped members (Secs. 4-8 and 4-9) is analyzed and compared to a rectangular section. The use of I and T shapes is based upon elastic-design considerations and not upon strength considerations. If a rectangular section, which is easier to manufacture and more resistant to large shear stresses, can be found that will work satisfactorily at service loads with moderate tensile stresses in the tensile flange, the flexural strength may very well be as high as would be found for an I or T beam designed for the same loads, but without tensile stresses in the bottom flange under full load. This is particularly true for short-span members in which the dead weight of the member itself is not important in comparison to the total moment. The motivations for using partial prestressing of this second type is the reduction of form costs and labor that can be derived through the use of simple rectangular or tapered sections.

Deflection due to prestressing and differentials in initial deflections between members have been a significant problem in the manufacture of prestressed-concrete members. The total deflection due to prestressing as well as the variation of deflection between members, which is assumed to be a function of the total deflection, can be reduced by not fully prestressing the members. Assuming the flexural strength is still adequate, this procedure will result in satisfactory construction for many types of applications. Hence, the third motivation, which is the principal one, is the desire to achieve better performance at service loads without sacrificing the minimum safety requirements of the codes.

In most roof and floor members, partial prestressing can be used without risk, since the loads can generally be predicted with reasonable accuracy, and fatigue or frequent overloading is not normally a possibility. In bridge construction, due to the uncertainty of impact loads, vibration and fatigue, or repetition of loads, partial prestressing has not been allowed to the same extent it has been in building construction. Nominal tensile stresses are currently permitted in the design of prestressed-concrete highway bridges. There is reason to believe that more use will be made of partial prestressing in bridge construction in the future.

6-8 End Blocks

In post-tensioned beams, it is customary and often necessary to curve the tendons up at the ends of the beams as a means of reducing the eccentricity of the prestressing force. In order to have sufficient concrete section in which to slope the tendons to the end anchorages and in which to embed the anchorages, a

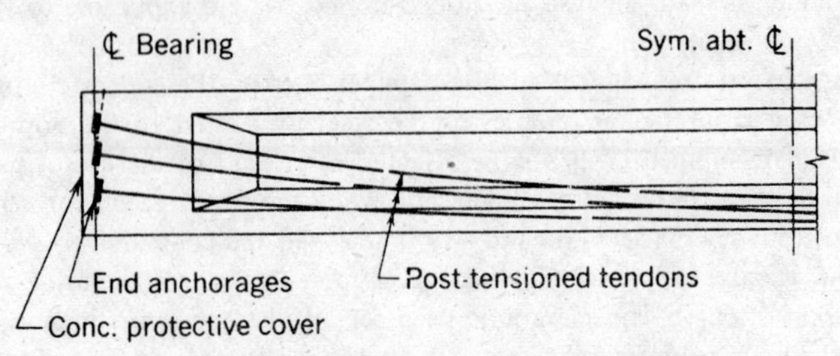

Fig. 6-18 Half-elevation of a post-tensioned beam.

short section at the end of the beam is often enlarged and made rectangular in cross section. This rectangular section, which is called an end block, is illustrated in Fig. 6-18. End blocks are occasionally used on pre-tensioned members as well, but experience has shown that they are not normally needed with pre-tensioned tendons. End blocks do facilitate the placing and compacting of the concrete at the ends of the beams, an important consideration in pre-tensioned and post-tensioned members.

Cracking has occurred along the path of the tendons or near the tendons at the ends of prestressed beams, as illustrated in Fig. 6-19. The first of these types of cracks is called "bursting cracks" and the second is referred to as "spalling cracks." The cracking is due to tensile stresses that result from the distribution of the highly concentrated compressive stresses at the end of the member. This can be visualized by considering the schematic diagram of stress paths that result from a large, concentrated load acting upon a short prism over a small bearing area that extends across the width of the beam, as is shown in Fig. 6-20. From this diagram it will be seen that the stress paths are closely spaced near the ends, and spread to the distribution that results from the normal elastic analysis at a distance of approximately one times the depth of the beam from its end.

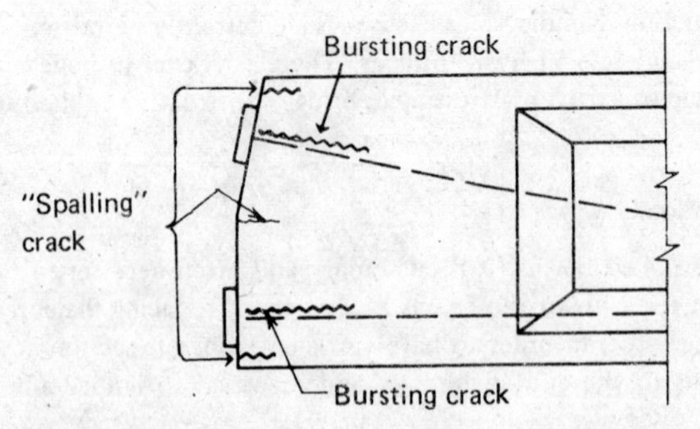

Fig. 6-19 Elevation of an end block showing cracks due to secondary tensile stresses.

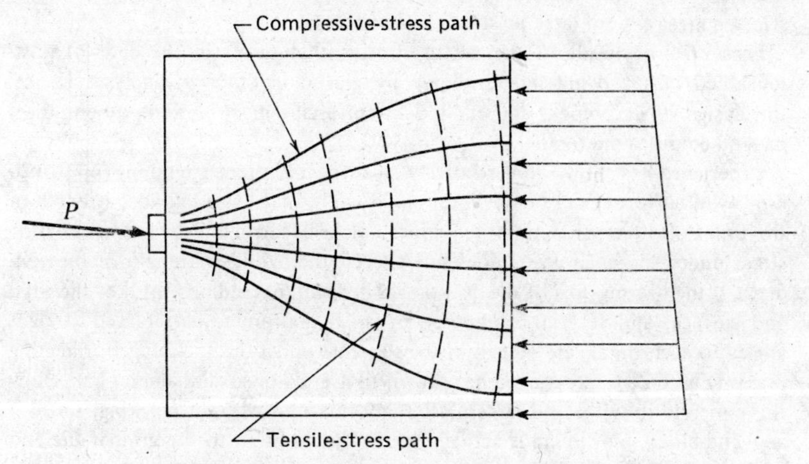

Fig. 6-20 Idealized stress paths in end block with a single load.

If the same total prestressing force is applied to the beam by a number of smaller tendons distributed over the end of the beam rather than by one large tendon, the condition of stress can be approximated by the diagram of Fig. 6-21, in which the load is represented as several small forces acting on a common bearing plate. Under this condition of loading, the stress paths are seen to be further apart, and the same distribution of stress is obtained at one times the depth of the beam from the end of the beam.

The above examples are an oversimplification of the problem because: (1) The bearing plates or end anchorages do not normally extend across the entire width of a post-tensioned beam, but are of small width; (2) Bearing plates are not provided for pre-tensioned tendons; (3) Highly stressed concrete acts as a plastic

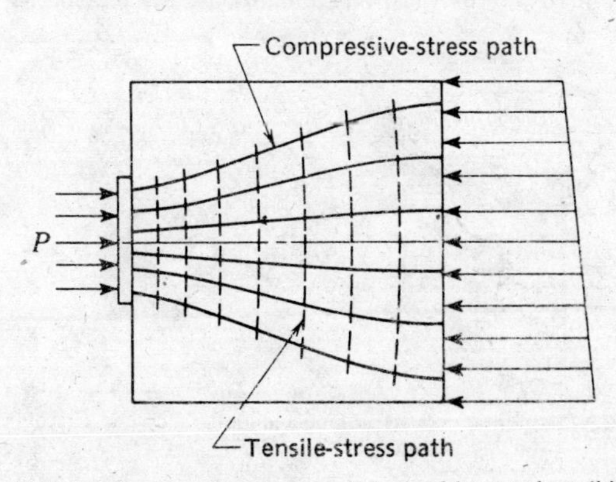

Fig. 6-21 Idealized stress paths in end block with several small loads.

material rather than an elastic material, with the result that the more highly stressed areas exhibit large plastic deformations.

These factors result in the actual stresses that occur in the ends of post-tensioned beams being indeterminate by elastic methods of analysis. Hence, the design of end blocks and end-block reinforcing must be done by empirical or semi-empirical methods.

Experience has shown that relatively small post-tensioned tendons (up to 120 kips ± initial force) can be anchored in concrete that has a cylinder strength of the order of 4000 psi, without end-block cracks occurring, if the average bearing stress under the anchorages does not exceed the cylinder strength of the concrete, if the bearing area of the anchorages does not exceed one-third of the total end surface, and if grids, composed of small reinforcing bars placed at right angles to each other, are placed in the concrete under the anchorages or bearing plates (Fig. 6-22). Experience has shown that grids of small-diameter bars allow the concrete to withstand large plastic deformations without cracking. Little if any end-block reinforcing is actually needed in addition to the grids if the end anchorages can be distributed uniformly over the ends of the beam.

When large tendons are used, the stress conditions at the ends are more severe, and it is customary to provide vertical and horizontal reinforcing in the end blocks to restrict cracking. The amount of reinforcing required can be estimated by computing the area of reinforcing steel required to control the cracking based upon the following assumptions:

(1) As shown in Fig. 6-23, the end of the beam can be represented by a free-body subjected to the several forces shown. To simplify the analysis, the vertical forces can be ignored.
(2) With the vertical components of the forces neglected, and with the variable width of the cross section being taken into account, one obtains the free-body shown in Fig. 6-24.
(3) Moments acting on horizontal planes at y above the beam soffit can be computed at various locations and the results plotted as shown in Fig. 6-25(a).

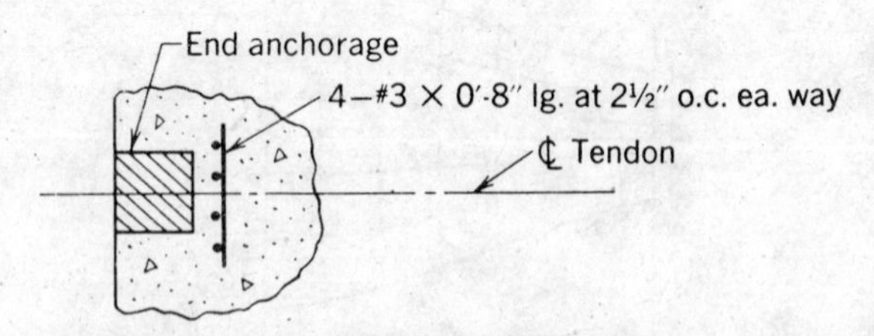

Fig. 6-22 Section at post-tensioning anchorage showing grid reinforcing.

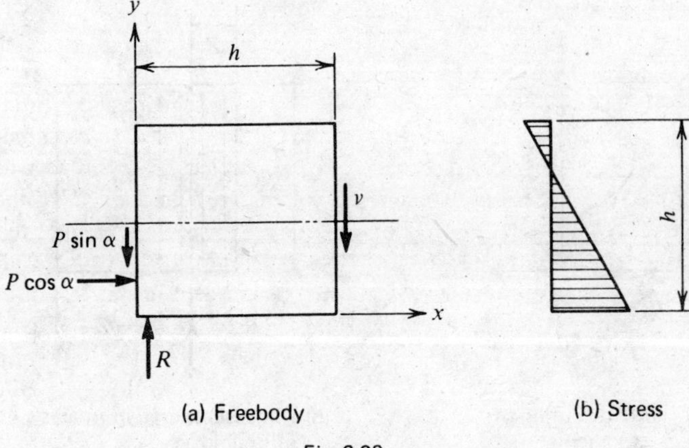

(a) Freebody

(b) Stress

Fig. 6-23

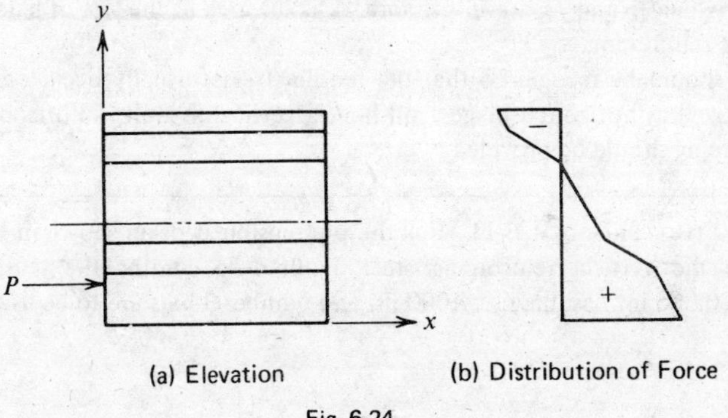

(a) Elevation

(b) Distribution of Force

Fig. 6-24

(4) Assuming a resisting couple as shown in Fig. 6-25(b), one can compute the maximum tensile force for which reinforcing must be supplied from

$$T_{max} = \frac{M_{max}}{h - z} \tag{6-8}$$

(5) In order to restrict the crack width resulting from the tensile force to 0.005 in., the stress in the reinforcing steel provided for T_{max} should be restricted to

$$f_s = 775\left(\frac{\sqrt{f'_c}}{A_b}\right)^{1/2} \tag{6-9}$$

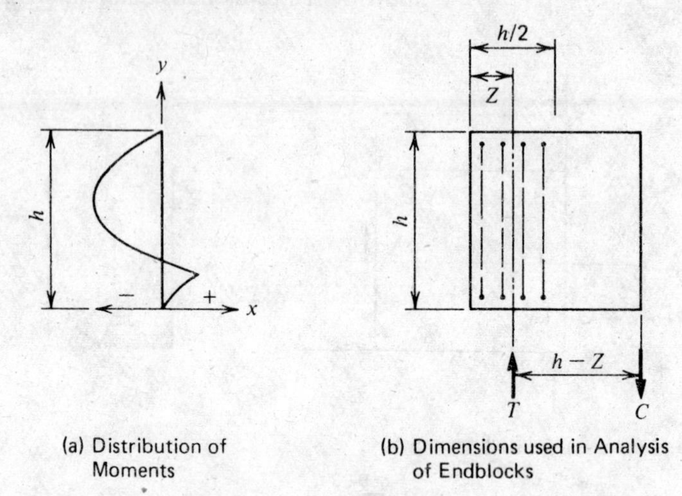

(a) Distribution of Moments

(b) Dimensions used in Analysis of Endblocks

Fig. 6-25

in which f_s and f'_c are in psi and A_b is the area of the size of bars used in the reinforcing.

(6) It should be recognized that the tensile stresses usually occur on vertical as well as horizontal planes and hence, vertical as well as horizontal reinforcing should be provided.

ILLUSTRATIVE PROBLEM 6-11 For the pre-tensioned beam shown in Fig. 6-26 compute the vertical reinforcing steel required to confine the tensile-crack width to 0.005 in. Assume f'_c = 4000 psi and number 3 bars are to be used.

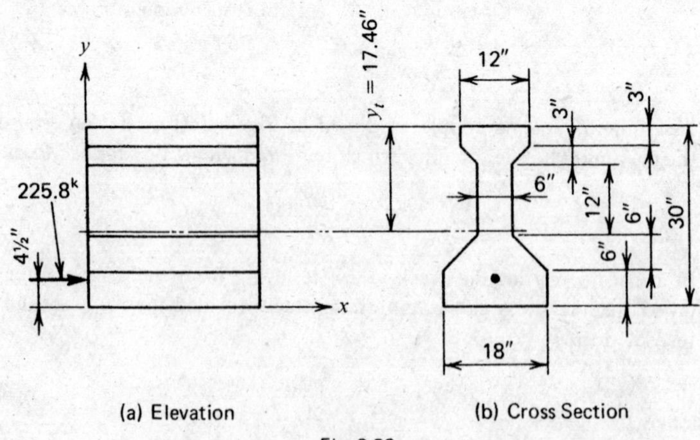

(a) Elevation

(b) Cross Section

Fig. 6-26

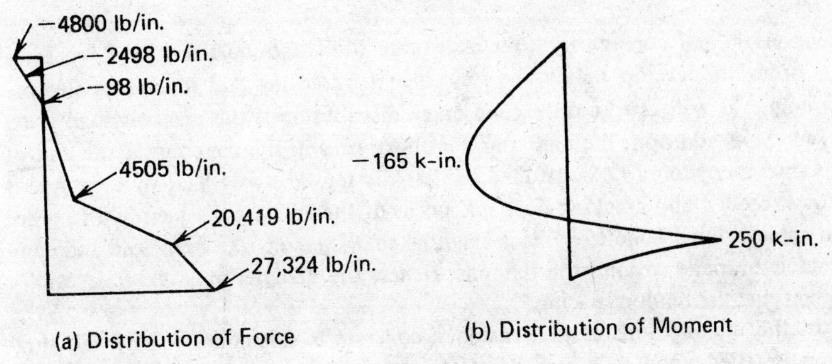

Fig. 6-27

SOLUTION: The distribution of force is plotted in Fig. 6-27(a). The unit stress and force for various values of y are:

y (in.)	f (psi)	Width (in.)	Force (lb/in.)
0	1518.0	18	27,324
6	1134.4	18	20,419
12	750.8	6	4,505
24	-16.4	6	-98
27	-208.2	12	-2,498
30	-400.0	12	-4,800

The distribution of moments shown in Fig. 6-27(b) are computed from the forces given in Fig. 6-27(a).

The allowable steel stress is:

$$f_s = 755\left(\frac{\sqrt{4000}}{0.11}\right)^{1/2} = 18{,}100 \text{ psi}$$

Assuming $Z = 6$ in. and $A_s = 6 \times 0.11 = 0.66$ sq in.,

$$f_s = \frac{165\ k\text{ - in.}}{(30-6)(0.66)} = 10.4 \text{ ksi} < 18{,}100 \text{ psi}$$

Use No. 3 U-shaped stirrups 4 in. on center, with the first stirrup 2 in. from the end of the beam.

6-9 Spacing of Pre-tensioning Tendons

The bond between the tendons and the concrete section at the ends of pre-tensioned members is relied upon to transfer the pre-tensioning force from the tendons to the concrete section. Adequate flexural bond strength is necessary in order to develop maximum resistance to cracking and maximum flexural strength, as well as to ensure good crack distribution if the cracking load is exceeded. In addition, the end zone tensile stresses in the concrete at the end of the member should be controlled so that the tensile strength of the concrete is not exceeded and cracking does not occur at the ends of the member. In order to achieve these conditions, it is essential that the concrete be placed and compacted properly around the tendons. Hence, the designer must exercise care in specifying the tendon spacing.

In the manufacture of pre-tensioned concrete in this country, internal vibration is relied upon to a high degree to ensure the concrete is adequately compacted. For this reason, particularly in deep beams, it is important that the pre-tensioning tendons be spaced in such a manner that the head of the internal vibrator can penetrate the extreme bottom of the forms. In addition, the tendons should not be placed in such a manner that the flow and compaction of the plastic concrete is unduly restricted in areas not directly accessible to the vibrator head.

In order to save labor in handling and stressing the strands in the manufacture of pre-tensioned concrete, the trend has been toward the use of a few of the larger seven-wire strands in lieu of many lower strength, solid wires commonly used abroad. As progress has been made in prestressed-concrete manufacturing techniques and as more has been learned about the action of transfer bond and fatigue on flexural bond stresses, the size of tendons commonly used has been increased from $\frac{1}{4}$ to $\frac{1}{2}$ in., with strands as large as 0.60 in. in diameter being used occasionally. It should be recognized that the head of the internal vibrator used in compacting the concrete in beams with 0.50 in. tendons at 2 in. on centers, which is a common spacing with tendons of this size, is restricted to something less than $1\frac{1}{2}$ in. in diameter. Because large, internal vibrators are generally much more effective than smaller vibrators, placing of the concrete is often materially facilitated if at least one, wide, vertical opening (through which a large internal vibrator can be inserted) is left down the center of deep members. This is illustrated in Fig. 6-28, in which it will be seen that omission of the strands in the center row allows the use of a 3-in. vibrator head.

Heavy concentrations of two groups of tendons at the ends of pre-tensioned members, when deflected pre-tensioned tendons are used, can result in cracking at the ends of the members, just as such concentrations of load may cause cracking in the ends of post-tensioned members.

The distance from the edge or center line of the tendons to the surface of the concrete in the area in which transfer bond is developed must also be chosen with care. Placing the tendons too close to the surface can cause cracking or splitting along the tendon. In general, the smaller tendons can be placed as close as 1 in. to the surface without ill effects, while the larger tendons should

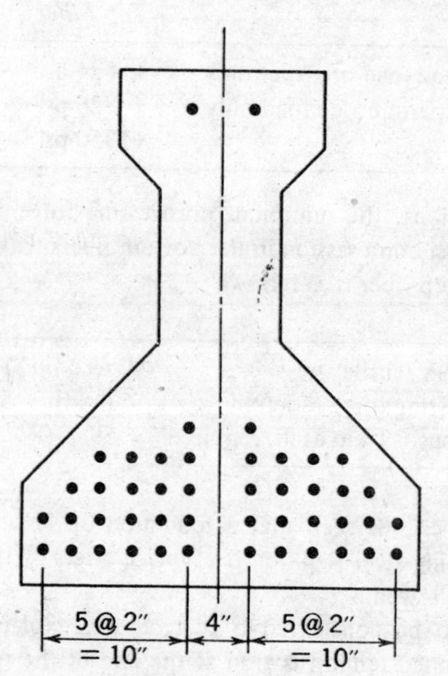

Fig. 6-28 Section through a pre-tensioned beam showing tendon spacing to facilitate placing and vibrating concrete.

not be closer than 2 in. to the surface. The center-to-center spacing of seven-wire strands is frequently restricted to four times the nominal strand diameter, for example, $1\frac{1}{2}$ in. for a $\frac{3}{8}$-in. strand.

6-10 Pre-tensioning Stresses at Ends of Beams

In simple prismatic beams, pre-tensioned with straight tendons, the eccentric pre-tensioning force results in a high compressive stress in the bottom flange and a tendency for tensile stresses in the top flange. Furthermore, in well-proportioned beams, the service-load compressive stresses in the top flange at midspan are not normally a problem. The amount of prestressing required, as a rule, is controlled by the flexural, tensile stress in the bottom flange, which the prestressing force must fully or partially nullify. The usual specifications permit some tensile stresses resulting from prestressing in the top flange without reinforcing, and higher-tensile stresses if unstressed reinforcing steel is provided to carry the entire tensile force. As a means of illustrating the effect of top-flange tensile stresses on the quantities required for a given *elastic* design, consider the beam shown in Fig. 6-29. Assume this beam is to be used on a span of 70 ft, no tensile stress is to be allowed in the bottom fibers under full load, and the flexural stresses due to dead load of the girder and due to the superimposed load are as follows:

	Top Fiber	Bottom Fiber
Stress due to dead load of girder only	+ 524 psi	- 780 psi
Stress due to superimposed load only	+ 826 psi	-1220 psi
Total Stress	+1350 psi	-2000 psi

Under these conditions, the minimum prestressing force required to produce the required 2000 psi compression in the bottom fibers with various amounts of tensile stress in the top fiber is as follows:

Allowable top-fiber tensile stress	zero	160 psi	320 psi
Minimum prestressing force required	644 kip	570 kip	491 kip
Number of tendons, 11 kip each, required	59	52	45

From this table, it can be seen that a reduction of 12% can be made in the amount of prestressing steel required if a tensile stress of 160 psi is allowed and 23.7% if 320 psi is allowed.

Another factor to be considered is that, as was explained in Sec. 5-9, the stress in a pre-tensioned tendon is zero at the end of the tendon and the stress increases to a maximum value at a distance, from the end of the beam, that is referred to as the transmission length. The transmission length is primarily a function of the type and size of the tendon. Values of 50 and 100 diameters are recommended for use in estimating the transmission length for strand or wire

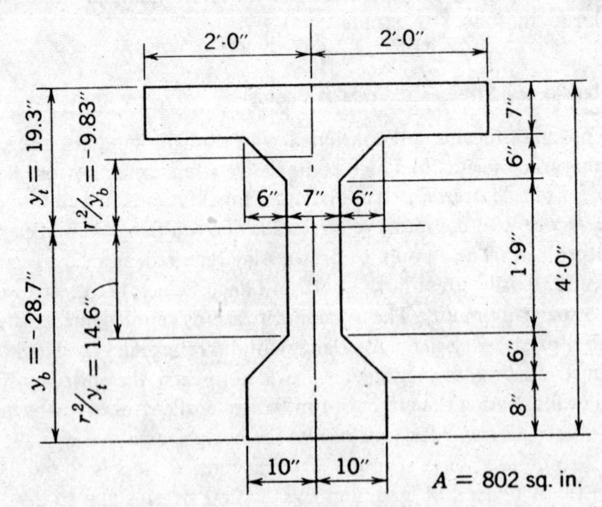

Fig. 6-29 Beam cross-section used in illustrating the effect of top-fiber tensile stresses on the prestressing force.

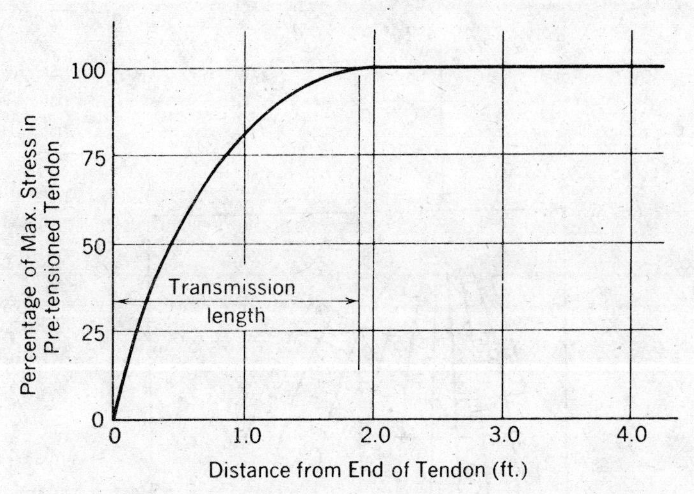

Fig. 6-30 Variation in initial stress in a pre-tensioned tendon near the end of a beam.

tendons, respectively. The initial stress in the tendon varies along the transmission length approximately as shown in Fig. 6-30.

An additional consideration is that a force applied to an elastic body causes stresses in the body which flow out along smooth curves or stress paths. A large force applied to the end of a prism, such as shown in Fig. 6-31, results in principal compressive stresses that follow a pattern similar to the solid lines and principal tensile stresses that follow along lines similar to the dashed lines. At a distance of about one times the depth of the block from the end, the stresses are approximately equal to the values that would be computed from the usual combined stress relationship used in prestressed-concrete design. In other words, the effect of the concentration of the load is virtually reduced to zero at a distance of one times the depth of the prism from the end, under normal conditions.

As a result of the combined effects of the transmission length required to develop full bond and the distance required for the concentrated prestressing force to distribute fully, the maximum tensile stress in the top fibers does not occur at the immediate end of the beam. This is a significant fact that can affect the economy of a design.

Returning to the example used above, if the effects of transmission length and distribution of the concentrated force are taken into account on the beam acting on a span of 70 ft, the 320-psi tension resulting from the 45 tendons would not be acting at the immediate end of the beam, but at a distance of from 4 to 8 ft from the end of the beam. The actual tensile stress in the top fibers of the beam would be less than 320 psi, due to the effect of the dead load of the beam. The effect of the stress in the top fiber of the beam resulting from

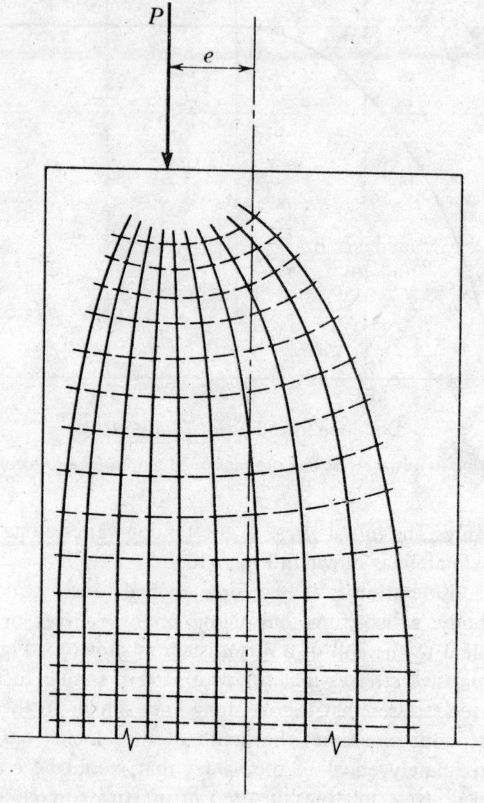

Fig. 6-31 Approximate paths of principal tensile and compressive stresses in a prism that is eccentrically loaded.

the dead weight of the beam on the net, final tensile stress in the top fibers as well as the amount of unstressed reinforcing that would be required is summarized as follows:

Distance from the end to the point under consideration (ft)	zero	4	6	8
Theoretical top-fiber tensile stress (psi)	−320	−320	−320	−320
Top-fiber stress due to girder dead load (psi)	0	+113	+164	+213
Net tension under dead load of girder plus prestress (psi)	−320	−207	−156	−107
Area of unstressed reinforcing steel required (sq in.)	3.00	1.42	0	0

From this study it is quite apparent that, in taking all of these factors into consideration, the designer may be able to reduce the pre-tensioning steel required in a specific elastic design as much as 25% without adding any mild steel in the top flange.

6-11 Bond Prevention in Pre-tensioned Construction

In Sec. 4-6, it was shown that the stresses at the ends of a member prestressed with straight tendons may limit the capacity of the member and that, by varying the eccentricity of the prestress, these stresses at the ends can be reduced and the capacity of the member thereby increased. The prestressing moment and stresses at the ends can also be reduced in a pre-tensioned member by varying the prestressing force. This can be done by preventing a portion of the tendons from bonding to the concrete at the immediate ends of the member and, in so doing, preventing these tendons from stressing the concrete at the ends.

This principle can best be explained by considering an example such as the beam shown in Fig. 6-32. The pre-tensioning tendons, as located in the figure, result in an initial, top-fiber tensile stress due to prestressing of 384 psi. It can be shown that by preventing bond on five tendons in the bottom row and four

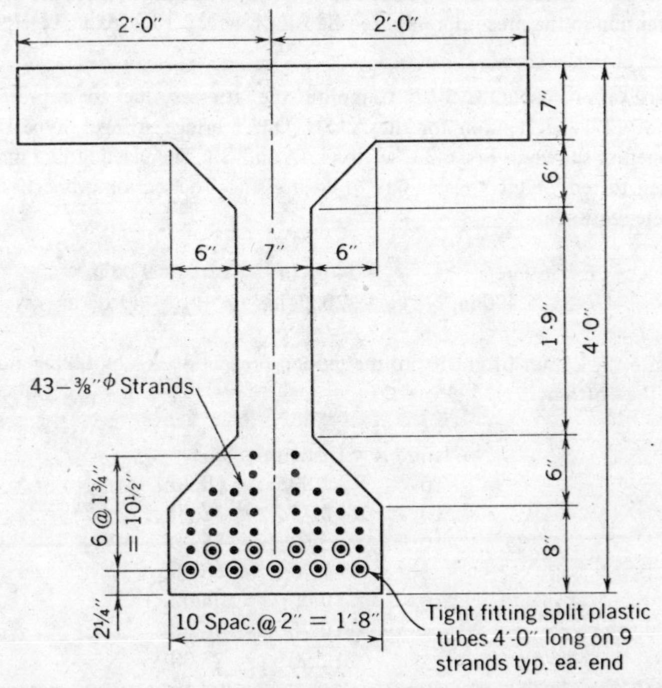

Fig. 6-32 Beam section indicating method of preventing bond on pre-tensioned tendons.

tendons in the second row, as indicated in the figure, the initial tensile stress in the top fibers at the ends can be reduced to 270 psi.

The length over which the bond must be prevented is a function of the beam dead-load stresses. In most cases, the beam dead-load stresses reduce the initial prestressing stresses to permissible values at only a few feet from the end of the beam. The transmission length required for the tendons to develop full tension, as well as the distance required for the pretensioning force to distribute, which are discussed in Sec. 6-10, should also be taken into account when calculating the maximum tensile stresses at the ends of members.

It is believed that bond prevention can be used to advantage with complete safety if a split, plastic tube, or a heavy paper or cloth tape with a waterproof adhesive is used as the bond-preventing media. Grease and chemicals that retard the concrete set have been used in lieu of plastic tubes or tape as a means of preventing bond. Because of the obvious danger of the workmen inadvertently or carelessly applying the grease or retarder to the incorrect tendons or number of tendons, this procedure should only be permitted when adequate supervision and inspection will be provided to prevent errors.

As pointed out in Sec. 5-9, if bond is prevented at the end of a strand tendon, the length required to develop the strength of the tendon is taken to be *twice as great* as for a tendon that is bonded to the end of the member; if the design allows tension in the precompressed tensile zone (Sec. 12.10.3, ACI 318).

ILLUSTRATIVE PROBLEM 6-12 Compute the stresses due to a prestressing force of 11 k per tendon for the AASHTO-PCI bridge stringer, type III, pretensioned as shown in Fig. 6-33, sections AA and BB. The plastic tubes indicated are used to reduce the stresses due to prestressing. The section properties of the concrete section are:

$$A = 560 \text{ in.}^2 \qquad y_t = 24.63 \text{ in.} \qquad r^2/y_t = 9.06 \text{ in.}$$
$$I = 125{,}400 \text{ in.}^4 \qquad y_b = -20.27 \text{ in.} \qquad r^2/y_b = -11.03 \text{ in.}$$

Compute the center of gravity of the tendon for section AA by taking moments about the bottom

No. tendons	× Distance	= N.D.
10	× 2.00	= 20.00
10	× 3.75	= 37.50
10	× 5.50	= 55.00
8	× 7.25	= 58.00
6	× 9.00	= 54.00
4	× 10.75	= 43.00
2	× 12.50	= 25.00
50		292.50

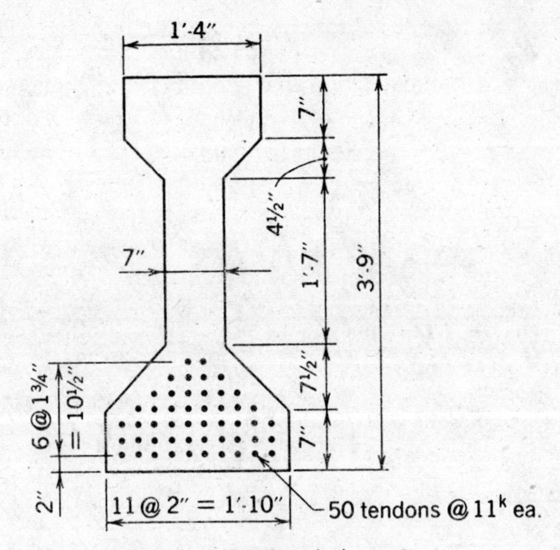

Section A-A

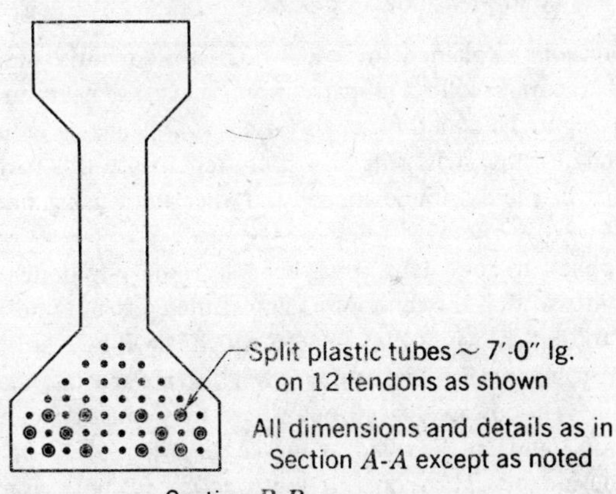

Section B-B

Fig. 6-33 AASHTO-PCI type III bridge stringer pre-tensioned with 50 tendons. Section *A-A* shows details in typical section. Section *B-B* shows details near end where plastic tubes are used to prevent 12 tendons from bonding.

$$e = -20.27 + \frac{292.5}{50} = -14.43 \text{ in.}$$

$$f_t = \frac{50 \times 11{,}000}{560}\left(1 + \frac{-14.43}{9.06}\right) = -580 \text{ psi}$$

$$f_b = \frac{50 \times 11{,}000}{560}\left(1 + \frac{-14.43}{-11.03}\right) = +2270 \text{ psi}$$

At section B-B:

$$\begin{array}{rr}
50 & 292.50 \\
-\ 4 \times 2.00 = - & 8.00 \\
-\ 4 \times 3.75 = - & 15.00 \\
-\ \underline{4} \times 5.50 = - & \underline{22.00} \\
38 & 247.50
\end{array}$$

$$e = -20.27 + \frac{247.5}{38} = -13.75 \text{ in.}$$

$$f_t = \frac{38 \times 11{,}000}{560}\left(1 + \frac{-13.75}{9.06}\right) = -386 \text{ psi}$$

$$f_b = \frac{38 \times 11{,}000}{560}\left(1 + \frac{-13.75}{-11.03}\right) = +1680 \text{ psi}$$

6-12 Deflected Pre-tensioned Tendons

For the reasons explained in Sec. 4-6, it is frequently desirable to have pre-tensioned tendons follow a path more eccentric near midspan than at the ends of a beam. This method is also used as a means of reducing the deflection due to prestressing. It is generally preferred to use this method of controlling the stresses in pre-tensioned members rather than using bond prevention (see Sec. 6-11).

When applied to roof slabs, it has been customary to deflect the tendon at one or two points so that the tendon spacing at the ends and midspan of the member is similar to that shown in Fig. 6-34, in which it will be seen that the tendons are allowed to touch each other at the center and are spaced out at the ends. In this manner, the tendons are spaced out where they must develop the all important transfer bond and are bundled at the center where only flexural bond stresses must be developed. This construction practice has been used a great deal with satisfactory results. The flexural bond strength at the center of such members is considered as good or better than that which is achieved in grouted, post-tensioned construction.

In recent years, deflected tendons have been used extensively in the construction of bridge beams. The theoretical principles involved in the use of deflected tendons in bridge construction is the same as in roof slabs. It can be shown that the same flexural strength that is obtained with spaced, deflected tendons can be achieved by using bond prevention, although some unstressed reinforcing may be required at the ends of the members, and by so doing, the large capital investment required for deflecting equipment, as well as the labor required in the de-

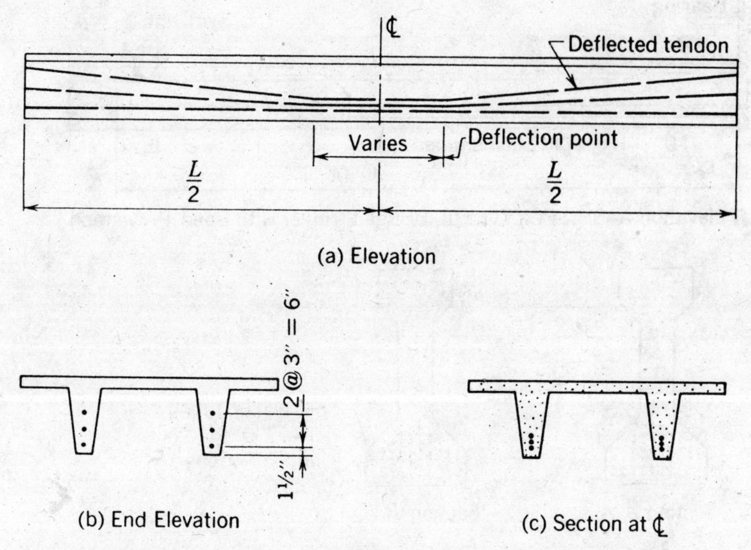

Fig. 6-34 Pre-tensioned double-T roof slab with deflected tendons.

flecting operation, can be avoided. This is illustrated in Fig. 6-35, in which two AASHTO-PCI, type III, bridge stringers of equivalent flexural strength, but detailed for deflected tendons and bond prevention, are shown.

If the deflected, pre-tensioned tendons in the AASHTO-PCI, type III, stringer were bundled at the center rather than being spaced out, the stress in the bottom flange at the midspan due to prestressing alone, and therefore the capacity of the stringer, could be increased 4% without additional materials or labor being required. If the number of bundled, deflected tendons were increased to 20 and each tendon had an initial prestressing force of 13,000 lb, the initial net bottom-fiber compressive stress (prestress-dead load) at the center of the girder would be of the order of 2425 psi if the girder were to have a span of 70 ft. This latter tendon layout would develop the maximum practical capacity of this concrete section for the span of 70 ft and would not be possible with spaced tendons of the same size.

6-13 Combined Pre-tensioned and Post-tensioned Tendons

The structural advantages of draped tendons can be obtained without materially reducing the economy of pre-tensioned construction with straight tendons by using a combination of pre-tensioned and post-tensioned tendons. This is illustrated in Fig. 6-36, in which the details of the AASHTO-PCI, type III, bridge

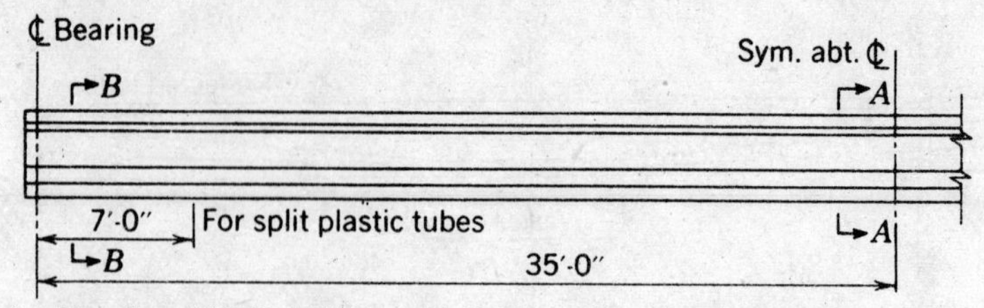

(a) Elevation AASHO-PCI, Type III, Bridge Stringer with Bond Prevention

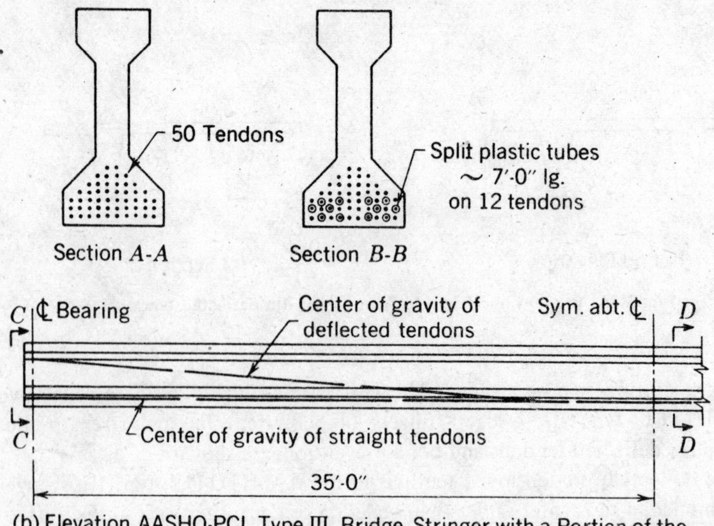

(b) Elevation AASHO-PCI, Type III, Bridge Stringer with a Portion of the Tendons Deflected.

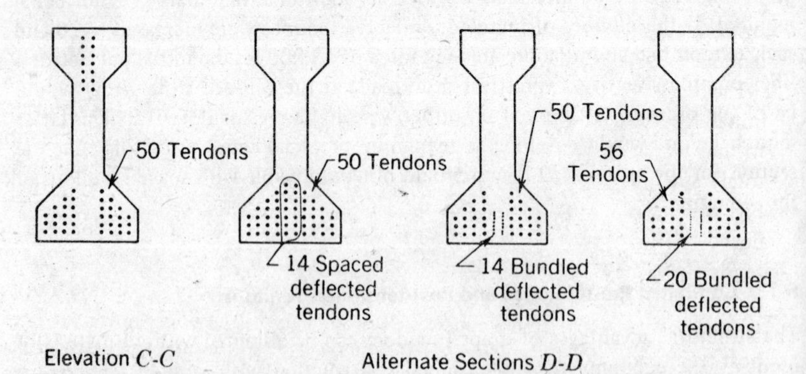

Fig. 6-35 Elevations and sections of AASHTO-PCI type III bridge stringer shown (a) prestressed with pre-tensioned tendons utilizing bond prevention, and (b) pretensioned with a portion of the tendons deflected.

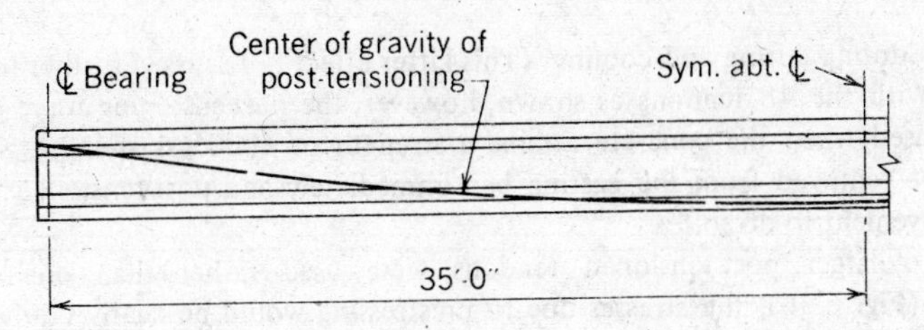

(a) Half Elevation AASHO-PCI, Type III, Bridge Stringer with Combined Pre- and Post-tensioned tendons

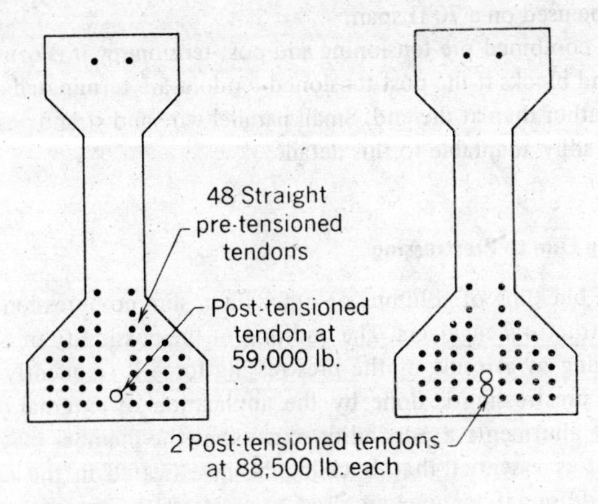

(b) Alternate Sections at Center Line

Fig. 6-36 Half-elevation and section of beam prestressed with pre-tensioned and post-tensioned tendons.

stringer are shown with two combinations of pre-tensioned and post-tensioned tendons.

The use of the 48 pre-tensioned and one small post-tensioned tendon (59 k initial force) results in a distribution of prestressing stresses that is equivalent to that which is obtained with 36 straight and 14 deflected tendons (50 tendons total) and with bond prevention in combination with 50 tendons, as shown in Fig. 6-35. The number of pre-tensioned tendons in this solution could be reduced to 42 tendons if the post-tensioned tendons were stressed before the pre-tensioning force were completely released on the concrete section. In such a procedure, if steam curing were used, it would be necessary to release that pre-tensioned tendons partially and to allow the girders to cool somewhat before post-tensioning and the complete release of the pre-tensioning force; this is done to eliminate the possibility of vertical cracks forming in the girder as a result of the strain changes that take place in the concrete and in the pre-tensioning tendons during curing and cooling. (This latter effect is discussed further in Sec. 10-2.) With the 48 tendons as shown, however, the pre-tensioning force could be released when the concrete attains a strength of 4000 psi and the girders could be removed from the casting bed immediately and post-tensioned when it is convenient to do so.

If two larger post-tensioned tendons were used rather than one small tendon (Fig. 6-36), the stresses due to prestressing would be nearly equivalent to those in the same beam section that would result from 56 tendons of which 20 are deflected, as shown in Fig. 6-35, and which, as was explained previously, would be the maximum stresses that could normally be imposed on this section when it is to be used on a 70-ft span.

When using combined pre-tensioning and post-tensioning, it is often not necessary to use end blocks if the post-tensioned tendons are terminated at the top of the member rather than at the end. Small parallel wire and strand post-tensioning tendons are readily adaptable to this detail.

6-14 Buckling Due to Prestressing

The danger of buckling of columns or other long, slim compression members is known to all structural engineers. The question of the possibility of a pre-stressed member buckling as a result of the prestressing force is frequently raised. Obviously, when prestressing is done by the application of external load such as jacking against abutments a possibility exists that the member may buckle. In such a case, it is essential that buckling be investigated in the conventional manner. In addition, if tendons are used to prestress the member and the tendons are placed externally in such a manner that they are in contact with the member at the ends only, there would be a possibility of buckling.

When the tendons are placed internally and they are in contact with the member at points between the ends of the member, the tendency to buckle is reduced a significant degree. When the tendons are in intimate contact with the member throughout the length of the member, as is the normal case, in post-tensioning and in pre-tensioning, there is no possibility of buckling. This fact has been demonstrated experimentally and mathematically and can be understood by considering the difference between the action of prestressing and column action.

Column action is characterized by an increase in eccentricity of the load as the load is increased over a critical value. This is illustrated in Fig. 6-37, in which it is seen that the column load has an eccentricity of e at load P, and if the load is increased to ΔP, the member deflects an additional amount, Δe. This action continues until the critical value of $P + \Delta P$ is reached, and the column buckles. If there were no eccentricity of the load, the column would fail by crushing, as is indeed the case for short columns.

Prestressing action results in a specific distribution of stresses in a member.

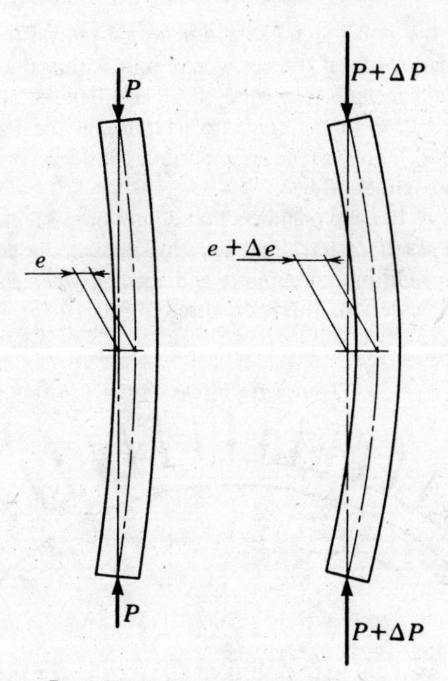

Fig. 6-37 Illustration of column action.

The eccentricity of the prestressing force remains constant, even if the member is deflected laterally, providing, as was mentioned above, the tendons and concrete are in intimate contact with each other. If the concrete section were cast slightly curved or crooked, as is often the case, the effect of the pre-stressing alone would be to straighten the concrete member (opposite to column action), since the taut tendon would attempt to assume a straight path.

Prestressed columns and piles, which are pre-tensioned or post-tensioned with the tendons in ducts through the members in the normal manner, can of course buckle, due to externally applied loads, and these members must be designed with care. Prestressed columns and prestressed piles are treated in Sec. 9-3.

Consider a square, prismatic concrete member cast as a segment of a circular arc and having a single post-tensioned tendon located at the center of gravity of the member throughout its length. When the tendon is stressed, the concrete becomes subjected to a compressive stress uniformly distributed over the square cross section. In addition, a transverse force exists between the tendon and the concrete; this force, which can be calculated using the methods in the following section, has to exist because if it did not, the tendon would not retain its curved shape. If one draws a freebody of the concrete section alone, it will be apparent it is stressed in an arch-like manner as shown in Fig. 6-38(a). If the member were to have an I-shaped cross section as shown in Fig. 6-38(b), rather than being solid and square, the global arch-like action would remain for the member as a whole but secondary or local stresses would exist within the cross section. The local stresses would include transverse shear and flexural stresses because the radial force distribution in the concrete section and the tendon would be as shown in Fig. 6-38(c). This can be an important consideration in curved, flanged sections such as box girder bridges.

The top flanges of flexural members that do not have adequate lateral support can also fail as a result of buckling. For this reason, the designer should give attention to the conditions of support and loading when selecting the dimensions of the concrete section. This is discussed in Sec. 10-5.

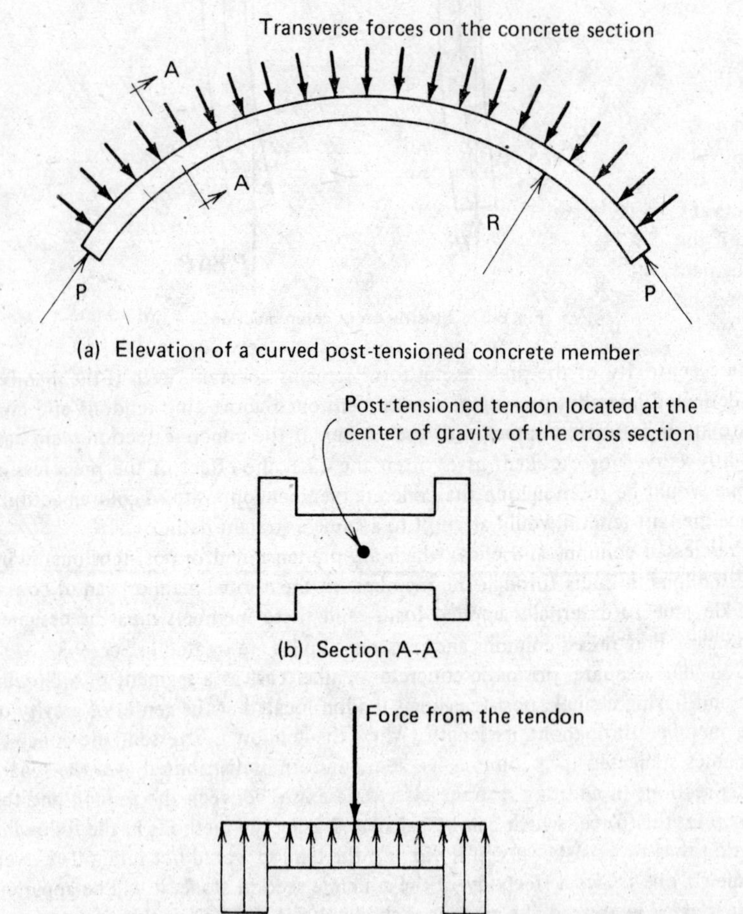

Fig. 6-38 Global and local force distributions on a curved, prismatic post-tensioned member.

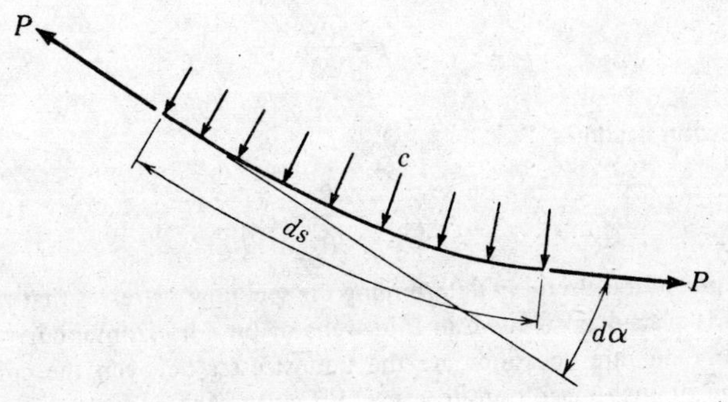

Fig. 6-39 Free-body diagram of an infinitesimal length of a curved tendon.

6-15 Secondary Stresses Due to Tendon Curvature

In considering a short segment of a curved post-tensioned tendon, such as is shown in Fig. 6-39, neglecting friction between the tendon and the concrete, it will be seen that the forces acting upon the tendon are the tension P, which acts at each end of the tendon, and the unit stress c, which is between the tendon and the concrete and which holds the tendon in the curved trajectory. If the segment under consideration is infinitesimal, the length of the segment is ds, the angular change in length ds is $d\alpha$, and the radius of curvature of the tendon is ρ. Since a very small angle is equal to the tangent of the angle, one can write

$$\tan \alpha = d\alpha = \frac{ds}{\rho}$$

and

$$\rho = \frac{ds}{d\alpha}$$

It is evident from the vector diagram, Fig. 6-40, that the unit stress exerted by the steel on the concrete is

$$c\,ds = P\,d\alpha$$

which can be rewritten

$$c = \frac{P\,d\alpha}{ds}$$

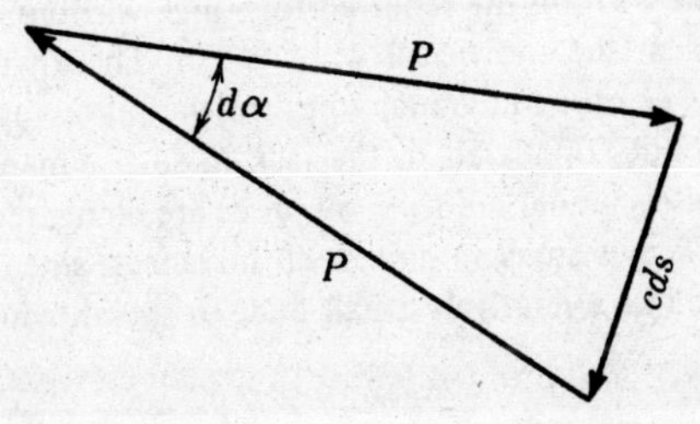

Fig. 6-40 Vector diagram of forces acting on curved tendon.

and since

$$\rho = \frac{ds}{d\alpha}$$

the expression becomes

$$c = \frac{P}{\rho}$$

This expression is useful in determining the secondary stresses that result when a tendon is placed on a curve in thin webs or on a horizontal curve in an end block. Only on rare occasions are the unit stresses between the concrete and the tendon of such magnitude to cause difficulty. The curvatures must be high and the concrete cover must be small in order to produce critically high stresses; conditions which are normally avoided.

ILLUSTRATIVE PROBLEM 6-13 Compute the secondary stress between a curved tendon and the duct if the radius of curvature is 25 ft and the force in the tendon is 500 k.

SOLUTION:

$$c = \frac{P}{\rho} = \frac{500 \text{ k}}{25 \text{ ft}} = 20 \text{ k/ft}$$

6-16 Variation in Tendon Stress

The question of the effect of the variation in the stress in the individual wires of a parallel-wire, post-tensioning tendon or in the individual tendons in a pre-tensioned member is often raised. It can be stated that the normal variations in stress encountered in practice do not exceed the normal tolerances expected in structural design.

Consider the case of a parallel-wire, post-tensioned tendon composed of n wires stressed to a total force, P. The force P is measured during construction by determining the elongation of the tendon during stressing as well as by observing the hydraulic pressure required to stress the tendon. The value of P can normally be controlled within the required tolerance without difficulty.

The average stress in the individual wires is P/n. There will be a variation in the unit stresses between the individual wires for the following reasons:

(1) The wires are not connected to the jack in such a manner that the length of each wire is precisely equal, and yet all wires are elongated the same amount by the jack. This effect is small in almost all instances, since the wires are very stiff and are confined in a relatively small duct or sheath which renders it phys-

ically impossible for one wire to have significantly more curvature than the average wire, as would be required if the wire were to be materially longer than the average length.

(2) The difference in the length between individual wires is only important with respect to the magnitude of the elongation of the tendon that is obtained during stressing. If, for example, the elongation of a tendon that is 100 ft long is 7 in., a difference in length of $\frac{1}{4}$ in. between an individual wire and the average wire length will only result in a stress variation of $\pm 3\frac{1}{2}\%$ from the average stress in the tendon. The total force in the tendon will not be affected, since the average elongation is not affected.

(3) Variation in the modulus of elasticity of prestressing wire, along the length of one wire and from coil to coil, as much as ±4% is not uncommon.

(4) Although the relaxation loss of a wire that is stressed higher than the average will be higher than the average relaxation loss, this will be offset by the wires stressed less than the average.

(5) The estimate of losses of stress in the tendons is generally not as precise as the initial prestressing stresses in the tendon.

For a relatively short, large, multi-wire or multi-strand tendon placed on small radii and through considerable curvature (such as in nuclear reactor vessels), the difference in length between individual wires or strands on the inside and outside of the curvature can be significant and hence cannot be permitted. In instances such as this, the tendons are frequently twisted to equalize the lengths of the individual wires or strands in the tendon.

In general, the same factors that affect parallel-wire, post-tensioned tendons affect pre-tensioned tendons. The exception is that the pre-tensioned tendons are not confined in a sheath or duct, and therefore, the variation in length could be significant if the tendons were not laid out approximately parallel, prior to stressing. Although the wires are usually sufficiently parallel before stressing, a small force is normally applied to each tendon before the tendons are stressed to their final value. This force straightens the tendons and equalizes the lengths. This procedure is not necessary when the tendons are stressed individually, as is discussed in Chapter 14.

6-17 Standard vs Custom Prestressed Members

In recent years, there has been a trend in the prestressing industry toward the adoption of standard beams and slab cross sections. There are several reasons why this is taking place, but the primary reasons are that the manufacturers of prestressed concrete prefer to use the same forms many times in order to reduce the amount of form cost that must be charged to each unit. In addition, the labor required to produce the prestressed concrete can be reduced to a minimum if the workmen perform the same duties each day and are not confronted with variable duties and operations. Furthermore, when standard products are used, load tables and advertising literature can be prepared for distribution to architects and engineers, and the manufacturers can often operate with a smaller sales-engineering force than would be required if custom products were used exclusively.

There are many instances where standard prestressed-concrete members have been used on small structures on which the use of custom-made members would have been out of the question, due to the high cost of the special forms required. Since all structural methods and framing schemes have their limitations, and because many large structures have peculiar framing or loading requirements, the designer should carefully consider the economy that could result from the use of custom-made members on large projects.

6-18 Precision of Elastic Design Computations

Prestressed-concrete flexural members are normally designed with the assumption that the concrete is an elastic material under the service loads and the stresses under such conditions of loading are made to conform to a standard or criteria. In addition, as has been pointed out, it is essential that the flexural strength be computed, in order to be sure that the elastic design has resulted in adequate safety factors.

It is well known that concrete is not an elastic material and that stresses computed on the basis of elastic assumptions can only be considered as approximations. Furthermore, in order to facilitate the design of prestressed members, most engineers base their computations upon the gross concrete section rather than using the net and transformed sections. Errors in the elastic computations are introduced as a result of this simplification, as is apparent from the discussion of Sec. 4-10. These considerations can only lead to the conclusion that normal elastic-design computations can only be considered approximate and nothing is gained by using more than three significant figures in such computations.

It is significant that strength computations of bonded construction can be made with good precision if the characteristics of the steel are known. Ultimate-moment computations are virtually independent of the elastic properties of the concrete and are not materially influenced by variations in the effective prestress. For these reasons, the flexural strength computations are usually more important and precise than the elastic design computations.

6-19 Load Balancing

Consider a tendon that is placed on a parabolic path in a simple beam in such a manner that the sag of the tendon, as measured vertically from a straight line which connects the ends of the tendon, is equal to e as shown in Fig. 6-41. If the total uniform load on the beam is equal to w, the load will be exactly

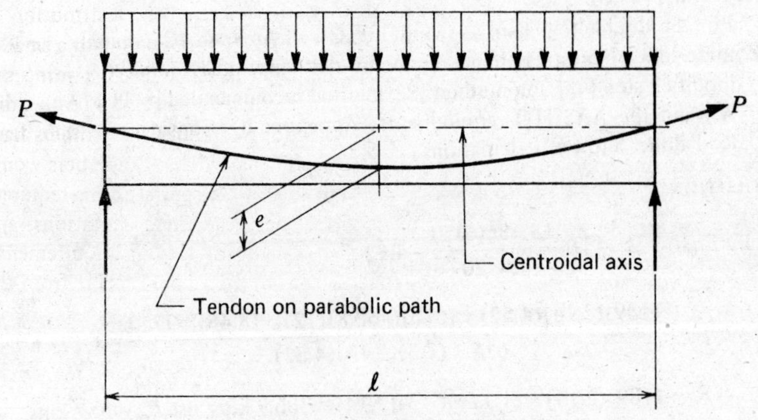

Fig. 6-41 The principle of load balancing.

balanced if

$$Pe = \frac{wl^2}{8} \tag{6-10}$$

because the parabolic tendon results in a uniform upward force (neglecting friction) as has been explained. In such circumstances, the pressure line acts along the centroidal axis of the members if the end eccentricities are equal to zero. This is the principle of load balancing and is a useful design aid in certain circumstances.

REFERENCES

1. ACI Committee 318, *Building Code Requirements for Reinforced Concrete* (ACI 318-83). American Concrete Institute, Detroit, 1983.

2. Subcommittee 5, ACI Committee 435, "Deflection of Prestressed Concrete Members," *Journal of the American Concrete Institute*, **60**, No. 12 (Dec. 1963).

3. PCI Committee on Prestress Losses, "Recommendations for Estimating Prestress Losses" *Journal, Prestressed Concrete Institute*, **20**, No. 4, 43–75, Chicago (July/Aug. 1975).

4. American Association of State Highway and Transportation Officials, "Interim Specifications Bridges 1975," pp. 41–79, Washington D.C., 1975.

5. ACI Committee 209, "Prediction of Creep, Shrinkage, and Temperature Effects in Concrete Structures," Special Publication 27, pp. 51–93, American Concrete Institute, Detroit, 1971.

6. Branson, D. E., "The Deformation of Non Composite and Composite Prestressed Concrete Members," *Deflections of Concrete Structures*, pp. 83–127, American Concrete Institute, Detroit, 1974.

7. Ables, P. W. "Principles and Practice of Prestressed Concrete," p. 46, Crosby Lockwood and Son, Ltd., London, 1949.

CHAPTER 6–PROBLEMS

1. For the double-tee beam shown in Fig. C-1 of Appendix C, compute the ultimate loss of prestress based upon the data given in Example C-1 using the modified step function method, the method recommended by PCI (Appendix A), and the AASHTO requirements (Appendix B). Consider the slab dead load alone. Show all calculations.

SOLUTION:

By Eq. 6-1, $k_s = 1 + \dfrac{18.48^2(615)}{59720} = 4.52$

$$f_{ci} = \frac{(189)(2.14)(4.52) + (615)(462.8)(12)(-18.48)/59720}{615 + (7.3)(2.14)(4.52)} = 1.124 \text{ ksi}$$

$$P_o = (189)(2.14) - (1.124)(7.3)(2.14) = 386.9 \text{ kips, use } P_o = 387 \text{ kips}$$

By modified step function method ($k_r = 1$)

$$\Delta f_{si} = (7.3)(1.124) + (0.91)(7.3)(1.124)(1.88) + \frac{(546)(28)}{(1.25)(1000)} + (0.075)(189)$$

$$\Delta f_{si} = 8.21 + 14.04 + 12.23 + 14.18 = 48.66 \text{ ksi}$$

By PCI recommendations: General method

Elastic shortening

$$f_{cr} = f_{ci} = 1.124 \text{ ksi}$$
$$ES = (1.124)(7.3) = 8.21 \text{ ksi}$$

Creep loss ($E_c = 28 \times 10^6 \div 7.3 = 3.83 \times 10^6$ psi)

$$UCR = 63 - (20)(3.83) = -13.6 \text{ use } 11.0$$

$$\text{Vol. to surface ratio} \cong \frac{615}{360} \cong 1.70 \text{ in.}$$

$$SCF = 1.05 - 0.70\,(1.05 - 0.96) = 0.99$$
$$MCF = 1.0$$
$$CR = (11)(0.99)(1.0)(1.124) = 12.24 \text{ ksi}$$

Shrinkage loss

$$USH = \frac{27{,}000 - (3000)(3.83)}{1000} = 15.51 \text{ ksi}$$

$$SSF = 1.04 - 0.70\,(1.04 - 0.96) = 0.98$$
$$SH = (15.5)(0.98) = 15.2 \text{ ksi}$$

Relaxation loss at $t = 10^5$ hours, $f'_y = (0.90 \times 270) = 243$ ksi

$$RET = (189)\frac{\text{Log } 10^5}{10}\left(\frac{189}{243} - 0.55\right) = 21.52 \text{ ksi}$$

$$\Delta f_{si} = 8.21 + 12.24 + 15.21 + 21.52 = 57.18 \text{ ksi}$$

By PCI simplified method

$$\Delta f_{si} = 33.0 + (13.8)(1.124) = 48.5 \text{ ksi}$$

By AASHTO

$$SH = \frac{17000 - (150)(70)}{1000} = 6.50 \text{ ksi}$$

$$ES = (7.3)(1.124) = 8.21 \text{ ksi}$$
$$CR_c = (12)(1.24) = 14.88 \text{ ksi}$$

$$\text{Relaxation} = 20 - (0.4)(8.21) - 0.2\,(6.50 + 14.88) = 12.44 \text{ ksi}$$

$$\Delta f_{si} = 42.03 \text{ ksi}$$

SUMMARY

	Total Loss (ksi)
Mod. step function	48.66
PCI–General	57.18
–Simplified	48.50
AASHTO	42.03

Note that the PCI methods did not include an adjustment for relative humidity; this was a consideration for the other two methods.

2. A particular concrete has an elastic modulus of 3000 ksi and is under an initial axial stress of 600 psi. The stress is caused by a prestressing tendon stressed at 200 ksi and having an elastic modulus of 30,000 ksi. The tendon is perfectly elastic and not subject to relaxation. The concrete is free of shrinkage. If the creep characteristics of the concrete are defined by Eq. 3-18 with $c = 0.60$ and $d = 10.0$, and if the creep ratio is 3.00, determine the stress remaining in the concrete and in the steel after ten days using the numerical integration procedure with a time interval of one day and with ten days.

SOLUTION:

$$\text{Specific elastic strain} = \frac{1}{3 \times 10^6} = 0.333 \times 10^{-6} \text{ in./in./psi}$$

$$\text{Specific creep strain} = 1.000 \times 10^{-6} \text{ in./in./psi}$$

The computations with a one-day interval are summarized in Table 6-6. For an interval of ten days the computation becomes:

$$\text{Creep strain} = -\left(\frac{10^{0.6}}{10 + 10^{0.6}}\right)600 \times 10^{-6} = -170.85 \times 10^{-6} \text{ in./in.}$$

$$f_s = 200 - \frac{170 \times 30}{1000} = 194.9 \text{ ksi}$$

$$f_c = \frac{194.9}{200.00} \times 600 = 584.7 \text{ psi}$$

TABLE 6-6

Day	C_t/C_u	$\Delta C_t/C_u$	$\Delta\epsilon_{creep}$ millionths-inches/ in.	f_s (ksi)	$\epsilon_{conc.}$ millionths-inches/ in.	f_c (psi)
0	0			200	-200	600
		0.0909	-54.54			
1	0.0909			198.36	-254.54	595.08
		0 0407	-24.22			
2	0.1316			197.64	-278.76	592.92
		0.0304	-18.03			
3	0.1620			197.09	-296.79	591.27
		0.0248	-14.66			
4	0.1868			196.66	-311.45	589.97
		0.0212	-12.507			
5	0.2080			196.28	-323.96	588.84
		0.0186	-10.952			
6	0.2266			195.95	-334.91	587.86
		0.0166	-9.758			
7	0.2432			195.66	-344.67	586.98
		0.0151	-8.863			
8	0.2583			195.39	-353.53	586.18
		0.0137	-8.03			
9	0.2720			195.15	-361.56	585.46
		0.0127	-7.435			
10	0.2847			194.93	-368.99	584.79

3. For the conditions of Problem 2, compute the steel stress and concrete stress remaining at the age of ten days if, in addition to the creep strain, the concrete shrinks as described by Eq. 3-17 with $e = 1$ and $f = 35$. Use a one-day time interval and a ten-day interval. Ultimate shrinkage can be assumed to be 400 millionths inches/in. and the shrinkage starts at day 0.

SOLUTION: The computations for a one-day interval are summarized in Table 6-7.

TABLE 6-7

Day	$\epsilon_{shrink.}$ millionths-inches in.	$\Delta\epsilon_{shrink.}$ millionths-inches in.	$\Delta\epsilon_{creep}$ millionths-inches in.	f_s (ksi)	$\epsilon_{conc.}$ millionths-inches in.	f_c (psi)
0	0			200.00	-200.00	600
		-11.11	-54.54			
1	-11.11			198.03	-265.65	594.09
		-10.51	-24.18			
2	-21.62			196.99	-300.34	590.97
		-9.96	-17.97			
3	-31.58			196.15	-328.27	588.46
		-9.45	-14.59			
4	-41.03			195.43	-352.31	586.29
		-8.97	-12.43			
5	-50.00			194.79	-373.71	584.36
		-8.54	-10.87			
6	-58.54			194.20	-393.12	582.61
		-8.12	-9.67			
7	-66.66			193.67	-410.91	581.01
		-7.76	-8.77			
8	-74.42			193.17	-427.44	579.52
		-7.40	-7.93			
9	-81.82			192.71	-442.78	578.14
		-7.07	-7.34			
10	-88.89			192.28	-457.19	576.85

For 10-day interval, $\Delta\epsilon = -170.85 - 88.89 = -259.74$ millionths-inches/in.

$$\Delta f_s = -7.79 \text{ ksi}, \qquad f_s = 192.21 \text{ ksi}, \qquad f_c = 576.62 \text{ psi}$$

4. For the double-tee beam shown in Fig. C-1, determine the deflection assuming $E_{ci} = 3.64 \times 10^6$ psi, $E_c = 3.89 \times 10^6$ psi, $\Delta P_u/P_o = 0.22$ and the remaining data as given in Example C-1.

SOLUTION: From Eq. 6-1, $f_{ci} = 1.124$ ksi, $nf_{ci} = 8.20$ ksi

$$f_{so} = 189 - 8.2 = 180.8 \text{ ksi}, \quad P_o = 387 \text{ kip}$$

$$\delta_p = -\frac{(387)(-11.12)(76)^2(144)}{(8)(3890)(59{,}720)} - \frac{(387)(-7.36)[3(76)^2 - 4(38)^2](144)}{(24)(3890)(59{,}720)}$$

$$= +1.93 + 0.85 = +2.78 \text{ in.}$$

$$\delta_d = -\frac{(5)(0.641)(76)^4(1728)}{(384)(3890)(59720)} = -2.07 \text{ in.}$$

$$\delta_u = 2.78\ [1 - 0.22 + (0.91)(1.88)] - 2.07\ [1 + 1.88]$$
$$= 6.92 - 5.96 = +0.96 \text{ in.}$$

5. Compute the gross precast, precast transformed, and composite transformed section properties for the girder shown in Fig. C-3, assuming $f'_c = 5000$ psi and 3500 psi for the precast girder and the cast-in-place slab, respectively. The steel is located 3.50 in. from the soffit at midspan, $A_{ps} = 4.90$ sq. in., and $E_s = 28{,}000$ ksi. The dimensions of the precast section are given in Fig. 7-3. The concrete is normal weight (145 pcf). The elastic moduli are 4074 ksi and 3409 ksi for the precast beam and cast-in-place slab, respectively.

SOLUTION:

Gross Section:

Area	789.000				
I_{xx}	260740				
y_t	29.266	y_b	-24.733		
S_t	8909	S_b	- 10541		
r^2/y_t	11.291	r^2/y_b	-13.361		

Net Section:

Area	784.100				
I_{xx}	258517				
y_t	29.133	y_b	-24.866		
S_t	8873	S_b	- 10396		
r^2/y_t	11.316	r^2/y_b	-13.258		

Transformed Section:

Area	817.777				
I_{xx}	273259				
y_t	30.013	y_b	-23.986		
S_t	9104	S_b	- 11392		

Composite Section:

Area	1356.805				
I_{xx}	640353				
y_t	16.699	y_b	-37.300	y_p	23.699
S_t	38346	S_b	- 17167	S_p	27019

6. Determine the section properties for the girder of Problem 5 if the elastic moduli are 2200 ksi and 4200 ksi for the precast beam and cast-in-place, respectively.

SOLUTION:

Gross Section:

Area	789.000		
I_{xx}	260740		
y_t	29.266	y_b	-24.733
S_t	8909	S_b	- 10541
r^2/y_t	11.291	r^2/y_b	-13.361

Net Section:

Area	784.100				
I_{xx}	258517				
y_t	29.133	y_b	-24.866		
S_t	8873	S_b	- 10396		
r^2/y_t	11.316	r^2/y_b	-13.258		

Transformed Section:

Area	846.462				
I_{xx}	284890				
y_t	30.707	y_b	-23.292		
S_t	9277	S_b	- 12231		

Composite Section:

Area	2075.858				
I_{xx}	876517				
y_t	10.448	y_b	-43.551	y_p	17.448
S_t	83887	S_b	- 20126	S_p	50234

7. A 12 in. wide beam of rectangular cross section varies in depth from 12 in. at the supports to 42 in. at midspan of its span of 60 ft. The member is post-tensioned with a straight bonded tendon located 4 in. from the soffit. The properties and dimensions of the tendon are A_{ps} = 1.224 sq in., f_{pu} = 270 ksi and f_{se} = 140 ksi. If f_c' = 5000 psi and the beam is to support a superimposed dead load of 288 plf and a superimposed live load of 200 plf, determine if the beam conforms to the service load and strength requirements of ACI 318-71. Using Grade 40 reinforcing steel, design the member for shear.

SOLUTION: The service load stresses under prestress and dead load as well as for prestress and total load are shown in the table. It will be seen that the greatest compressive stress is 2379 psi, which exceeds the allowable value of 2250 psi (6%); this is considered acceptable at this location. The largest tensile stress is -794 psi, which is less than the maximum allowable value of -848 psi. The design moment and ultimate moment capacity are also shown in the table as is 1.20 M_{cr}. It will be seen that these conform to the requirements of ACI 318-71, except 1.2 $M_{cr} > M_u$ at point 0. This is considered acceptable because there is no moment at point 0. The values of v_{ci}, v_{cw}, v_u, v_c and A_v are also summarized in the table on p. 246. By observation, one can see that minimum web reinforcement is required in the end few feet (less than 6 ft at each end) only; no stirrups are required by stress considerations.

8. For the beam of Problem 7, determine the amount of reinforcing steel, Grade 40, located 2 in. above the soffit, if the tendon is not bonded and the strength requirements of ACI 318-71 are to be met. Use f_{py} = 229.5 ksi.

SOLUTION: The ultimate moment computations are summarized in the table. The non-prestressed reinforcement has been proportioned for No. 6 bars. The tabulated values are for the 20th points with the support section at the top and

TABLE FOR PROBLEM 7

20th Pt.	Service Load Stresses (ksi) W/O Live Load Top	Bottom	With Live Load Top	Bottom	Moment (k-ft) Design	Capacity	1.2 M_{cr}
0	.000	2.379	.000	2.379	.000	72.000	83.78
1	.409	1.494	1.522	.381	105.027	136.124	126.63
2	.537	1.048	2.001	-.415	201.006	220.499	171.40
3	.553	.806	2.077	-.717	287.462	296.706	219.51
4	.520	.669	1.984	-.794	363.924	367.746	266.57
5	.465	.592	1.821	-.763	429.918	439.591	317.11
6	.400	.551	1.630	-.678	484.974	511.980	369.54
7	.332	.533	1.433	-.567	528.617	584.755	423.84
8	.262	.530	1.238	-.445	560.376	657.812	480.04
9	.193	.538	1.051	-.319	579.777	731.083	538.13
10	.126	.553	.873	-193	586.350	804.519	598.00

Shear Stresses and Reinforcing

20th Pt.	v_{ci} (psi)	v_{cw} (psi)	v_u (psi)	v_c (psi)	A_v (in.^{2}/ft)
0	infin	604.4	372.4	604.4	.1154
1	2068.3	533.0	273.9	533.0	.1032
2	963.0	485.3	207.2	485.3	.0942
3	612.3	451.4	156.7	451.4	.0867
4	431.7	425.9	116.5	425.9	.0799
5	328.2	406.1	86.1	328.2	.0745
6	256.9	390.2	62.1	256.9	.0701
7	200.3	377.3	42.5	200.3	.0664
8	149.3	366.5	26.2	149.3	.0632
9	98.3	357.2	12.1	120.2	.0604
10	42.4	349.4	000.0	120.2	.0580

midspan at the bottom of the listing. (The values are symmetrical about the midspan.)

9. A rectangular beam 12 in. wide and 42 in. deep is prestressed with a single bonded tendon having a steel area of 0.918 sq in. The beam is to be used on a span of 60.0 ft. The stress in the steel upon stressing (after elastic shortening) will be 189 ksi and immediately after the stressing, a superimposed load of 825 plf is to be applied to the beam. The parabolic tendon path has a distance from the soffit of 3 in. at midspan and 21 in. at each end. Compute the deflection of the beams using a bilinear deflection analysis. Use E_c = 4000 ksi, E_s = 2800 ksi, and f_c' = 5000 psi.

SOLUTION: For gross section, A = 504 sq in., I = 74088 in.^{4}, r^2 = 147 sq in. and the section weighs 525 plf.

TABLE FOR PROBLEM 8

Pt.	h (in.)	M_u req'd (k-ft)	A_{ps} (in.^{2})	A_s (in.^{2})	f_{su} = (ksi)	Steel index	t min (in.)	M_u (k-ft)	M_u ratio
0	12.000	.000	1.224	.000	153.921	.3925	4.396	72.000	infin
1	15.000	105.027	1.224	.000	155.392	.2881	4.438	130.314	1.2407
2	18.000	201.006	1.224	.880	156.862	.2652	5.301	205.884	1.0242
3	21.000	287.462	1.224	1.760	158.333	.2517	6.164	296.090	1.0300
4	24.000	363.924	1.224	2.200	159.803	.2296	6.617	379.461	1.0426
5	27.000	429.918	1.224	2.200	161.274	.2017	6.659	445.622	1.0365
6	30.000	484.974	1.224	2.200	162.745	.1800	6.701	512.590	1.0569
7	33.000	528.617	1.224	1.760	164.215	.1533	6.332	546.694	1.0341
8	36.000	560.376	1.224	1.320	165.686	.1315	5.963	573.322	1.0231
9	39.000	579.777	1.224	.880	167.156	.1132	5.595	592.472	1.0218
10	42.000	586.350	1.224	.440	168.627	.0978	5.226	604.145	1.0303

Stresses due to the causes listed are:

	Top Fiber (psi)	Bottom Fiber (psi)
Initial prestressing	-540	+1230
Beam dead load	+804	-804
Superimposed load	+1263	-1263
Total	+1527	-837

The bottom fiber stress under full load exceeds $6\sqrt{f_c'}$ but is less than $12\sqrt{f_c'}$. Hence, a bilinear analysis is required. The cracking moment is:

$$M_{cr} = \frac{74088}{21} \frac{(1230 + 7.5\sqrt{5000})}{12000} = 517 \text{ k-ft}$$

$$w_{cr} = \frac{8 \times 517}{(60)^2} = 1.149 \text{ klf}$$

Superimposed load required to crack the section = 1149 - 525 = 624 plf.

With nA_s = 6.426 sq in., y, the depth to the neutral axis of the transformed section, is

$$\frac{12\,y^2}{2} = 6.426\,(39.0 - y)$$

$$y = 5.95 \text{ in.}$$

$$I = \frac{12 \times 5.95^3}{3} + 6.426\,(39.0 - 5.95)^2 = 7862 \text{ in.}^4$$

$$I_e = \left(\frac{1149}{1350}\right)^3 (74088) + \left[1 - \left(\frac{1149}{1350}\right)^3\right] 7862 = 48693 \text{ in.}^4$$

The instantaneous deflections are

$$\delta_p = -\frac{(5)(0.918)(189)(-18)(60)^2(144)}{(48)(4000)(74088)} = 0.569 \text{ in.}$$

$$\delta_{sL\,1} = -\frac{(5)(1.149)(60)^4(1728)}{(384)(4000)(74088)} = -1.131 \text{ in.}$$

$$\delta_{sL\,2} = -\frac{(5)(0.201)(60^4)(1728)}{(384)(4000)(48693)} = -0.301 \text{ in.}$$

The deflections are plotted in Fig. 6-42.

10. For a rectangular endblock 75 in. deep and 20 in. thick, determine the reinforcing steel required to confine the width of the bursting cracks to 0.005 in. if it is subjected to 990 kips applied uniformly over a length of 36 in. commencing 26 in. from the bottom.

SOLUTION:

For b = 8 in. and h = 75 in.

$$A = 1500 \text{ sq in.}, \quad e = 44.0 - 37.5 = 6.5 \text{ in.}$$

$$f_t = \frac{990}{1500}\left(1 + \frac{6.5}{12.5}\right) = 1.003 \text{ ksi}$$

$$f_s = \frac{990}{1500}\left(1 + \frac{6.5}{-12.5}\right) = +0.317 \text{ ksi}$$

Hence, at 75 in. from the end of the member, the force distribution can be taken as a linear variation from 20.06 kips per in. at the top to 6.34 kips per in. at the bottom. The force on the end is 990 ÷ 36 = 27.5 kips per in.

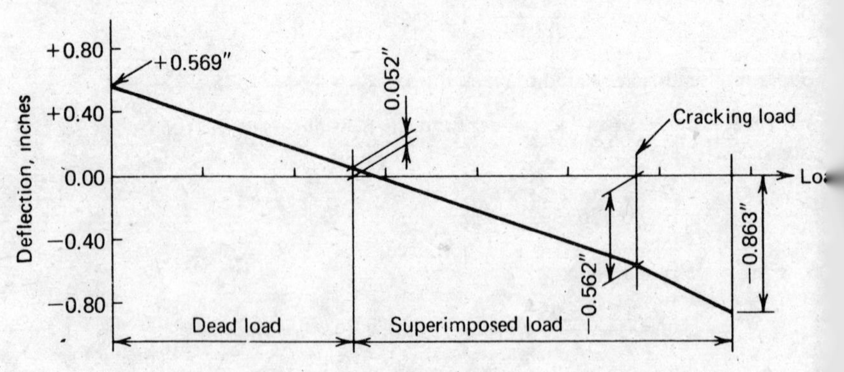

Fig. 6-42

Equations for the bending moment can be written as follows:
From the bottom of the member up to $y = 26$ in.,

$$M = \frac{6.34\,y^2}{2} + \frac{(20.06 - 6.34)}{75}\frac{y^3}{6} = \frac{y^2}{2}(6.34 + 0.06098\,y)$$

Between $y = 26$ in. and $y = 62$ in.,

$$M = \frac{y^2}{2}(6.34 + 0.06098\,y) - 13.75\,(y - 26)^2$$

Between $y = 62$ in. and $y = 75$ in.,

$$M = \frac{y^2}{2}(6.34 + 0.06098\,y) - 990\,(y - 44)$$

The results are shown in Fig. 6-43 from which it will be seen the maximum moment is 4377 k-in.

Taking z to be $75 \div 4 = 18.75$ in.,

$$T_{max} = \frac{4377}{75 - 18.75} = 77.81 \text{ kips}$$

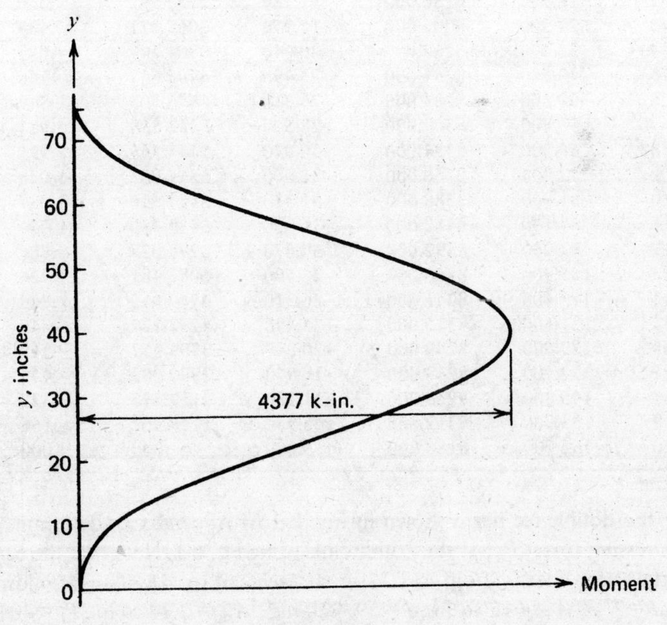

Fig. 6-43

Using No. 5 bars for reinforcing

$$f_s = 775 \left(\frac{\sqrt{5000}}{0.31}\right)^{1/2} = 11{,}700 \text{ psi}$$

$$A_s = \frac{77810}{11700} = 6.65 \text{ sq in.}$$

This would require 11 No. 5 ties spaced over a distance of about 35 in. ($h \div 2$) and represents a reasonable solution.

11. Compute the initial deflection of the simply supported box girder bridge of Example IV in Appendix B using the initial stress distribution given in the example. $A_{ps} = 290 \times 0.153 = 44.37$ sq in. Assume the loss due to elastic shortening is accounted for in the stressing procedure (i.e. slight over stressing to offset the loss due to elastic shortening). Use $w_d = 8.36\ k/f$.

SOLUTION: The computations are summarized in the following table.

20th Pt.	Dist. (ft)	Prestress Force (k)	e (in.)	Moment (k-ft)	Defl. (in.)
0	.000	8129.000	.000	.000	.000
1	8.100	8154.000	-5.950	1167.721	-.167
2	16.200	8180.000	-11.280	2183.792	-.330
3	24.300	8206.000	-15.970	3065.921	-.483
4	32.400	8232.000	-20.040	3804.547	-.623
5	40.500	8257.000	-23.490	4405.657	-.746
6	48.600	8283.000	-26.310	4876.505	-.849
7	56.700	8309.000	-28.500	5222.856	-.931
8	64.800	8334.000	-30.070	5444.365	-.991
9	72.900	8360.000	-31.000	5554.063	-1.026
10	81.000	8386.000	-31.320	5537.520	-1.037
11	89.100	8412.000	-31.000	5419.730	-1.024
12	97.200	8392.000	-30.070	5299.027	-.986
13	105.300	8366.000	-28.500	5087.481	-.926
14	113.400	8341.000	-26.310	4749.340	-.843
15	121.500	8315.000	-23.490	4292.122	-.740
16	129.600	8289.000	-20.040	3709.357	-.617
17	137.700	8263.000	-15.970	2990.063	-.479
18	145.800	8238.000	-11.280	2129.272	-.327
19	153.900	8212.000	-5.950	1138.962	-.166
20	162.000	8186.000	.000	.000	.000

12. For the double-tee beam shown in Fig. C-1 of Appendix C, determine the prestressing force, P_o, at the 20th points using Eq. 6-1. Note that the eccentricity varies linearly from -11.12 in. at each end to -18.48 in. at midspan. Use $n = 7.3$, $A = 615$ sq in., $I = 59{,}720$ in.^{4}, $A_{ps} = 2.14$ sq in., $P_i = 404.46$ kips, and $M_{d\,max} = 462.8$ k-ft.

SOLUTION:

$$f_{ci} = \frac{(404.46)\,k_s + \dfrac{(615)(12)(M_d)(e)}{I}}{615 + (7.3)(2.14)\,k_s} = \frac{404.46\,k_s + 0.12357\,M_d e}{615 + 15.622\,k_s}$$

$$P_o = 404.46 - nf_{ci}A_{ps} = 404.46 - 15.622\,f_{ci}$$

The computations are summarized in the table.

Pt.	e (in.)	k_s	M_d (k-ft)	f_{ci} (ksi)	P_o (kips)
0	-11.12	2.273	0	1.489	381.21
1	-11.86	2.449	87.9	1.319	383.85
2	-12.59	2.632	166.6	1.227	385.29
3	-13.33	2.830	236.0	1.147	386.55
4	-14.06	3.036	296.2	1.077	387.64
5	-14.80	3.256	347.1	1.024	388.46
6	-15.54	3.487	388.8	0.991	388.97
7	-16.27	3.726	421.1	0.981	389.14
8	-17.01	3.980	444.3	0.998	388.87
9	-17.74	4.241	458.2	1.043	388.16
10	-18.48	4.517	462.8	1.123	386.91

13. For the double-tee of Problem 12, determine the instantaneous deflection due to prestressing alone, using the values of P_o computed in Problem 12 as well as with P_o being a constant value of 386.91 kips. $E_c = E_s/n = 3836$ ksi.

SOLUTION: The deflections at the 20th points are summarized in the following two tables, the first for a variable P_o and the second for a constant P_o. The slight variation in P_o does not have significant effect on the deflections in this instance.

20th Pt.	Dist. (ft)	Prestress Force (k)	e (in.)	Moment (k-ft)	Defl. (in.)
0	.000	381.210	-11.120	-353.254	.000
1	3.800	383.850	-11.860	-379.371	-.501
2	7.600	385.290	-12.590	-404.233	-.960
3	11.400	386.550	-13.330	-429.392	-1.376
4	15.200	387.640	-14.060	-454.184	-1.745
5	19.000	388.460	-14.800	-479.100	-2.064
6	22.800	388.970	-15.540	-503.716	-2.332
7	26.600	389.140	-16.270	-527.608	-2.544
8	30.400	388.870	-17.010	-551.223	-2.699
9	34.200	388.160	-17.740	-573.829	-2.794
10	38.000	386.160	-18.480	-594.686	-2.826

20th Pt.	Dist. (ft)	Prestress Force (k)	e (in.)	Moment (k-ft)	Defl. (in.)
0	.000	386.910	-11.120	-358.536	.000
1	3.800	386.910	-11.860	-382.396	-.500
2	7.600	386.910	-12.590	-405.933	-.958
3	11.400	386.910	-13.330	-429.792	-1.373
4	15.200	386.910	-14.060	-453.329	-1.740
5	19.000	386.910	-14.800	-477.189	-2.058
6	22.800	386.910	-15.540	-501.048	-2.325
7	26.600	386.910	-16.270	-524.585	-2.536
8	30.400	386.910	-17.010	-548.444	-2.691
9	34.200	386.910	-17.740	-571.981	-2.785
10	38.000	386.910	-18.480	-595.841	-2.818

14. The initial midspan deflection computed for the beam in I. P. 6-4 is 3.39 in. For the same beam in I. P. 6-5, the initial deflection at midspan is 3.97 in. Explain the difference.

SOLUTION: In I. P. 6-4, the eccentricity of the tendon at the end is 44.50 - 43.41 = 1.09 in. In I. P. 6-5, the eccentricity at the supports was taken to be nil. The effect of this eccentricity is:

$$\delta_p = -\frac{(1046)(1.09)(126.3)^2(144)}{(8)(2030)(625158)} = -0.26 \text{ in.}$$

The elastic modulus used in I. P. 6-4 was computed for the age of 12 days from Equs. 3-1 and 3-4 modified as follows:

$$f_{ct} = \left[\frac{12}{2.24 + (0.92)(12)}\right] 5500 = 4970 \text{ psi}$$

and

$$E_{ct} = \frac{26\,(115)^{1.5}\sqrt{4970}}{1000} = 2260 \text{ ksi}$$

Note that 26 was used rather than 33 inches in Eq. 3-4. This was done to "fit" Eq. 3-4 to test data for known properties of a particular concrete.

Adjusting the results of I. P. 6-5 to fit the data used in I. P. 6-4 gives the following

$$\delta_i = (3.97 - 0.26)\frac{2030}{2260} = 3.33 \text{ in.}$$

The agreement is reasonable; the difference of 0.06 in. is not significant and is the result of the electronically made calculations in I. P. 6-4 not being rounded off.

7 Design Expedients and Computation Methods

7-1 Introduction

In the past, one of the major deterrents to the use of prestressed concrete in all forms of construction has been the relatively great amount of time required to design prestressed structures in comparison to the time required to design reinforced-concrete or structural-steel structures. This has been due to the fact that the average designer has not been familiar with the basic design principles of prestressed concrete, nor has the average designer had sufficient experience in prestressed design to develop short cuts and design expedients. The fact that prestressed concrete design is now taught in most of our universities has helped alleviate this problem.

This chapter is intended to bridge the gap between theoretical considerations and practical design methods. The theorems and methods explained here can be applied in many different ways and can be modified by the individual designer for any special conditions. Proper use of these design methods and expedients will greatly reduce the time and labor required to prepare economical prestressed designs.

These design methods were developed to facilitate design calculations being made with a slide rule. The use of modern calculators and computers will render some of these methods unnecessary.

7-2 Computation of Section Properties

In order to determine the elastic stresses that result in the concrete section from the prestressing force and from the external loads, the physical properties of the concrete section must be known. The basic dimensions and properties to be determined are the location of the center of gravity, the area, and the moment of inertia of the section. The other properties used to facilitate the computation of stresses are determined from these basic properties.

The computation of the basic properties of a section can be done by several methods, all of which produce the same results. These methods differ only in the organization of the computations, the datum for taking the static moments, and the procedure used in computing the moment of inertia. One convenient method is to use the top of the section as the datum for taking moments in computing the location of the center of gravity of the section. The moment of inertia of the section can then be computed about the center of gravity of the gross section. Whichever datum is used in any design office or by any individual, it should be used consistently in order to facilitate the checking and reviewing of the computations.

All moment-of-inertia computations are made using one or more variations of the basic relationship

$$I_{xx} = I_o + Ay^2 \tag{7-1}$$

which can be expressed verbally as: The moment of inertia of any section or shape about the axis, $x - x$, is equal to the moment of inertia of that shape or section about an axis parallel to axis $x - x$ and passing through the center of gravity of the section (I_o), plus the product obtained from multiplying the area of the shape or section by the square of the distance between the axis $x - x$ and the center of gravity of the section. This is illustrated in Fig. 7-1.

The location of the center of gravity and the moment of inertia, about the center of gravity, of various shapes frequently encountered in prestressed concrete design are given in Fig. 7-2. The location of the center of gravity and moments of inertia of other, less-common sections that may be encountered can be found in standard engineering references or calculated by fundamental mathematical relationships. It should be noted that the moments of inertia given in Fig. 7-2 are expressed in terms of the dimensions of the section as well as a function of the area of the section and the height or diameter of the section. The expression that gives the moment of inertia in terms of the area of the section is a device used to facilitate the moment-of-inertia computation for complex shapes when done in tabular form. This is illustrated in the following discussion.

This method of computation can best be explained with an example, and the

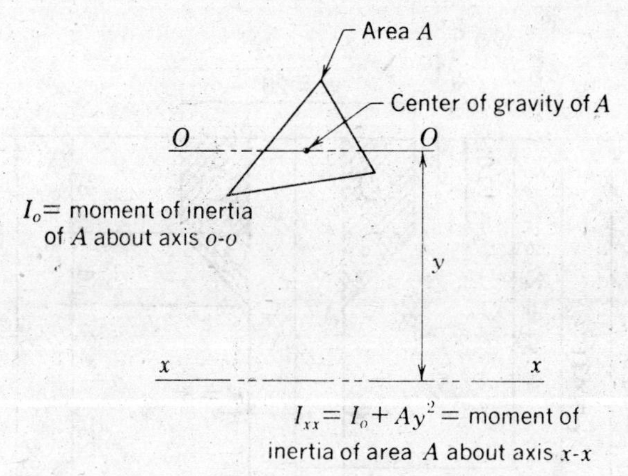

Fig. 7-1 Notation for moment-of-inertia computations.

location of the center of gravity, area, moment of inertia, and other section properties for the AASHTO-PCI standard bridge beam, type IV, is computed as an illustration of the recommended procedure. Referring to Fig. 7-3 and Table 7-1, the procedure is as follows:

(1) Divide the cross section into shapes of known area, centers of gravity, and moments of inertia, such as the rectangles and triangles numbered 1 through 5 in

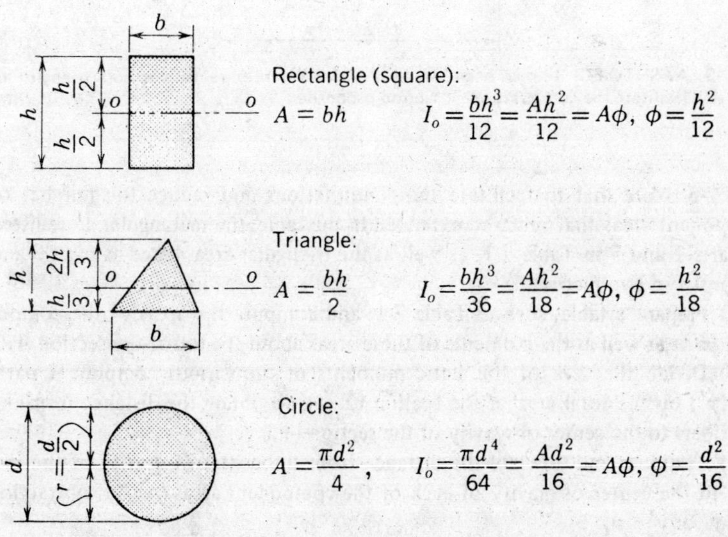

Fig. 7-2 Area and moment of inertia for common geometric shapes.

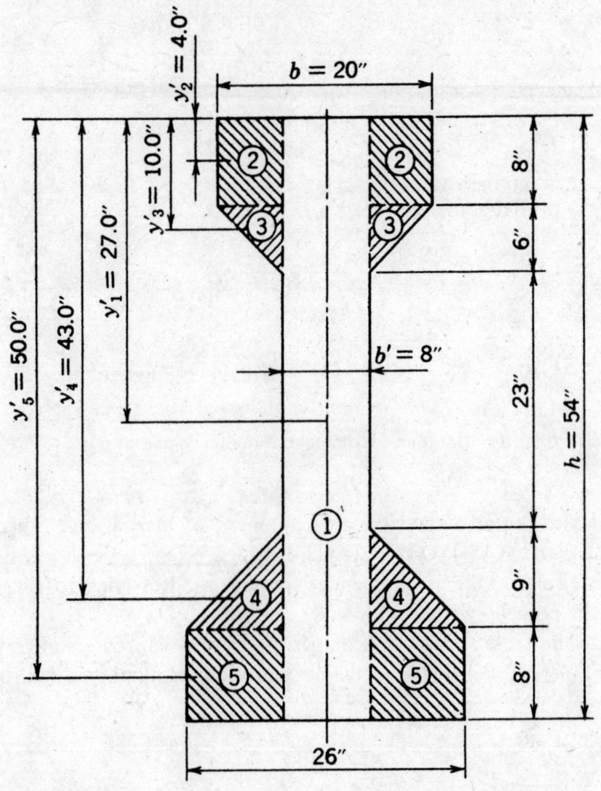

Fig. 7-3 AASHTO-PCI bridge beam, type IV, divided into rectangles and triangles to facilitate the computation of section properties.

Fig. 7-3. Note that to facilitate the computations and reduce the number of component areas that must be extended in the table, the rectangular areas listed as parts 2 and 5 in Table 7-1, as well as the triangular areas listed as parts 3 and 4, consist of two pieces.

(2) Prepare a table, such as Table 7-1, and compute the areas of the component parts as well as the moments of these areas about the top of the section Ay'.

(3) Divide the sum of the static moments of the various component parts ($\Sigma\ Ay'$) by the total area of the section ($\Sigma\ A$), to obtain the distance from the top fibers to the center of gravity of the section (y_t).

(4) Compute and tabulate the distance from the center of gravity of the section to the center of gravity of each of the component areas (y) by subtracting $y_t - y'$ or $y' - y_t$.

(5) Square and tabulate the distances obtained (y^2).

TABLE 7-1 Computation of Section Properties of the AASHO Type IV Bridge Beam, in Tabular Form.

Part	Area (A) Computation (in. X in. = in.^{2})	y' (in.)	Ay' (in.^{3})	y (in.)	y^2 (in.^{2})	ϕ (in.^{2})	$y^2+\phi$ (in.^{2})	$A(y^2+\phi)$ (in.^{4})
1	8 X 54 = 432	27.0	11,700	2.30	5.3	243	248	107,000
2	12 X 8 = 96	4.0	384	25.3	640	5.33	645	62,000
3	12 X 6/2 = 36	10.0	360	19.3	373	2.0	375	13,500
4	18 X 9/2 = 81	43.0	3,480	13.7	188	4.5	193	15,600
5	18 X 8 = 144	50.0	7,200	20.7	428	5.33	433	62,300

$\Sigma A = 789 \quad \Sigma Ay' = 23{,}124 \qquad I = 260{,}400$

$y_t = 23{,}124/789 = 29.3$ in. $\qquad S_t = 260{,}400/29.3 = 8880$ in.^{3}

$$r^2/y_t = \frac{260{,}400}{789 \times 29.3} = 11.3 \text{ in.}$$

$y_b = -54.0 + 29.3 = -24.7$ in. $\qquad S_b = 260{,}400/-24.7 = -10{,}500$ in.^{3}

$$r^2/y_b = \frac{260{,}400}{789 \times -24.7} = -13.4 \text{ in.}$$

(6) Compute and tabulate the factors by which the areas of the component parts can be multiplied in order to obtain their respective moments of inertia about their centers of gravity (ϕ) (*see* Fig. 7-2).

(7) Tabulate the sums of $(y^2 + \phi)$.

(8) Multiply and tabulate the terms $A(y^2 + \phi)$ for each of the component areas. The summation of the $A(y^2 + \phi)$ is the moment of inertia of the section.

(9) The remaining properties used in the design computations are computed as follows:

$$y_b = -h + y_t, \quad S_t = I/y_t, \quad S_b = I/y_b, \quad r^2/y_t = I/Ay_t, \quad r^2/y_b = I/Ay_b$$

Bridge stringers, such as the AASHTO-PCI standard, prestressed-concrete beams for highway bridges, are frequently used with a cast-in-place deck slab that acts compositely with the stringers, as a result of shear stresses that develop between the slab and the top of the stringers (*see* Sec. 6-4). The computation of the composite section properties for the AASHTO-PCI bridge beam, type IV, with a 6 X 36 in. cast-in-place top flange, as is illustrated in Fig. 7-4, can be done using the same fundamental procedure described above. This is illustrated in Table 7-2.

The concrete in the deck slab does not have the same elastic properties as the prestressed stringers under normal conditions, since the quality of the concrete normally employed in cast-in-place bridge decks is not as high as that used in prestressed stringers. This effect is taken into consideration by using a transformed cross section when computing the properties of the composite section. The transformed cross-section to be used consists of the gross section of the pre-

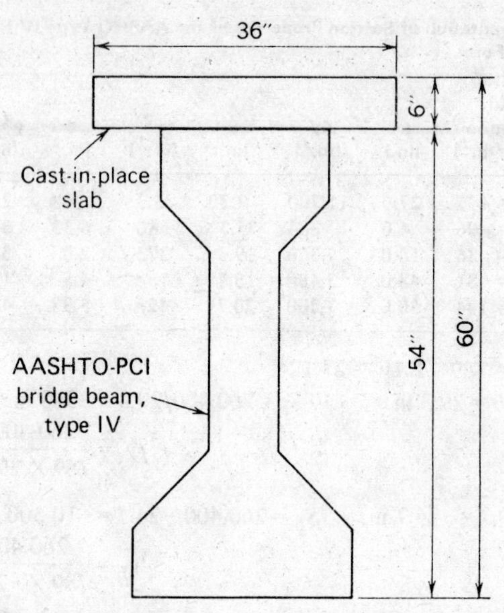

Fig. 7-4 Composite section composed of AASHTO-PCI bridge beam, type IV, and 6 X 36 in. cast-in-place slab.

TABLE 7-2 Computation of the Section Properties of the Composite Beam Illustrated in Fig. 7-4.

Part	Area Computation (in. X in. = in.^{2})	y' (in.)	Ay' (in.^{3})	y (in.)	y^2 (in.^{2})	ϕ (in.^{2})	$y^2 + \phi$ (in.^{2})	$A(y + \phi^2)$ (in.^{4})
Slab	36 X 6 in. = 216	3.0	648	25.3	640	3.0	643	139,000
Beam	= 789	35.3	27,800	7.0	49	*	49	38,600
								260,400*
	$\Sigma A = 1005$ in.^{2}		$\Sigma Ay' = 28{,}448$ in.^{3}					$I = 438{,}000$ in.^{4}

$y_p = 28.3$ in. $\quad S_p = 15{,}500$ in.^{3}
$y_t = 22.3$ in. $\quad S_t = 19{,}600$ in.^{3}
$y_b = -31.7$ in. $\quad S_b = -13{,}800$ in.^{3}

*The value of I_0 is known to equal 260,400 in.^{4} for the precast section. Hence, the value of ϕ is not computed for this portion of the composite section and the value of I_0 for the precast section is simply added in the $A(y^2 + \phi)$ column.

stressed beam and a slab section having a depth equal to that of the actual slab (less any allowance for wearing surface) and the effective slab width, equal to the actual width normally assumed to be effective, multiplied by the ratio of the elastic modulus of the slab concrete to the elastic modulus of the concrete in the beam. If the slab and beam concretes are assumed to have moduli of 3.5×10^6 psi and 5×10^6 psi, respectively, the width of the slab that should be used in the composite section would be 3.5/5.0 or 0.70 of the actual effective width.

It should be noted that y_p and y_t are used to denote the distances from the center of gravity of the composite section to the top fibers of the cast-in-place deck and the precast stringer, respectively. This procedure is recommended in order to avoid confusion; the subscript t is used to denote the distance to the top fibers of the precast section during the computation of the section properties of the precast section.

Finally, in Table 7-2, it will be noted that there is no entry for the second part in the ϕ column. The moment of inertia of the precast section about its center of gravity is known, and hence, is simply added to the last column.

After the designer becomes familiar with this tabular form, the computation of section properties becomes rapid and routine. The effects of minor adjustments in the concrete sections are often determined by subtracting or adding areas to the section which is being modified, rather than by entirely recomputing the section properties of the modified section. In addition, the column marked $(y^2 + \phi)$ is often eliminated in actual calculations and the terms y^2 and ϕ are added mentally as their sum is multiplied by the area. When electronic calculators are used, the columns marked y, y^2, ϕ, and $y^2 + \phi$ can frequently be eliminated from the table, thereby reducing the amount of writing required.

7-3 Allowable Concrete Stresses to be Used in Design Computations

Most prestressed-concrete design criteria specify maximum allowable initial, or temporary, compressive and tensile stresses, as well as maximum allowable final, or permanent, compressive and tensile stresses. This procedure is generally considered necessary or justified for the following reasons:

(1) In order to obtain an economical and realistic production schedule under many conditions, it is essential that the prestress be applied to the member before the concrete attains the minimum required 28-day cylinder strength. Hence, it is normal practice to apply the prestress to the concrete when the strength of the concrete is of the order of 4000 psi (or less), although the required minimum cylinder strength of the concrete at the age of 28 days is generally of the order of 5000 psi (or more). Therefore, the temporary, or initial, allowable stresses are reckoned from a cylinder strength lower than that used in determining the final, or permanent, allowable concrete stresses.

(2) The initial prestressing force is the maximum prestressing force that will ever be imposed on the member. This force is subject to a reduction in the amount of 10 to 30%. The reduction or relaxation of the prestressing force starts to take place immediately after stressing and requires 3 years or more to reach its practical maximum value.

(3) The stresses imposed on the member due to prestressing are of opposite direction to those imposed by the service loads, i.e., the prestressing normally causes small tensile stresses in the top fibers and large compressive stresses in the bottom fibers of simple beams, while the superimposed loads that are to be carried by the beams cause tensile stresses in the bottom fibers and compressive stresses in the top fibers.

(4) The stresses resulting from prestressing can be controlled by the fabricator to a relatively high precision, but usually neither the designer nor the fabricator can control or predict with high precision the loads that will be imposed on the structure while it is in service. For this reason, and in view of the reasons listed under 2 and 3, the safety factor required to guard against failure of the concrete during stressing does not need to be as high as that required for the design loads.

In a beam pre-tensioned with straight tendons, the highest initial stresses occur near the ends where there is no moment, due to dead load, to counteract the the effects of the prestressing (*see* Sec. 4-6). Therefore, the restrictions of the temporary allowable stresses for a pre-tensioned member can be expressed mathematically as follows:

$$\frac{f_{si}A_{ps}}{A}\left(1+\frac{ey_t}{r^2}\right) \leqq -3\sqrt{f'_{ci}} \tag{7-2}$$

and

$$\frac{f_{si}A_{ps}}{A}\left(1+\frac{ey_b}{r^2}\right) \leqq 0.60\, f'_{ci} \tag{7-3}$$

in which the minus sign denotes a tensile stress, f_{si} is the initial stress in the prestressing steel, A_{ps} is the area of the tendons, A is the area of the concrete, e is the eccentricity of the tendon, r is the radius of gyration, f'_{ci} is the concrete cylinder strength at the time of stressing, and y_t and y_b are the distances from the centroidal axis to the top and bottom fibers, respectively.* If f_{se} is the effective stress in the tendons, f'_c is the cylinder strength at 28 days, f_{tt} and f_{tb} designate the total stresses due to dead and live loads in the top and bottom fibers at the section of maximum moment, respectively, the restrictions of the final allowable stresses can be expressed mathematically as follows:

$$\frac{f_{se}A_{ps}}{A}\left(1+\frac{ey_t}{r^2}\right)+f_{tt} \leqq 0.40\, f'_c \tag{7-4}$$

and

$$\frac{f_{se}A_{ps}}{A}\left(1+\frac{ey_b}{r^2}\right)+f_{tb} \geqq -3\sqrt{f'_c} \tag{7-5}$$

*Values of e and y are positive when above the center of gravity of the section, negative when below. The terms on the right side of Eqs. 7-2 through 7-5 are based upon stresses allowed by the AASHTO Specifications for Highway Bridges.

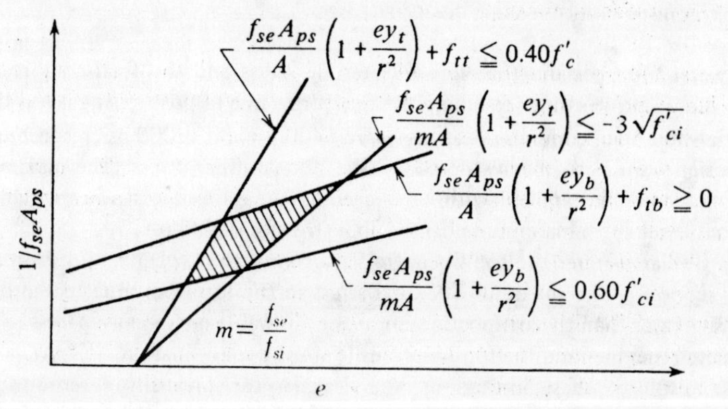

Fig. 7-5 Graphical solution of four equations in solving for the prestressing force and eccentricity.

It should be noted that f_{tb} is a negative stress and the negative sign is included in this symbol.

For an assumed concrete section and an assumed ratio between the effective steel stress and the initial steel stress $m = f_{se}/f_{si}$, the values of f_{tt} and f_{tb} can be computed and substituted in Eqs. 7-2 through 7-5, in which case, all of the terms that appear in the equations will be known or assumed, except the values of $f_{se}A_{ps}$ and e. Since a number of combinations of these terms will normally satisfy each of the four equations, the combinations that will satisfy all of the equations can be determined by plotting each of the four relationships, as shown in Fig. 7-5. The shaded area of Fig. 7-5 indicates the combinations of $f_{se}A_{ps}$ and e that satisfy the conditions of the allowable stresses for the assumed section.

Although the procedure of plotting a figure similar to that shown in Fig. 7-5, first suggested by Magnel (Ref. 1), will yield accurate results, it is obviously too cumbersome and time consuming to be used as a general design procedure. It does, however, illustrate the fact that there is frequently a combination of prestressing forces and eccentricities that will yield a satisfactory solution with a specific beam section.

7-4 Limitations of Sections Prestressed with Straight Tendons

It should be apparent that fully bonded, straight pretensioning tendons can only be used in prismatic beams in which the maximum flexural stress in the bottom fibers, due to the total load, does not exceed the arithmetic sum of either the allowable tensile stress and the final bottom fiber stress due to prestressing or the sum of the allowable tensile stress and the allowable compressive stress. If, for example the maximum stress in the bottom fiber due to the total external load were 2300 psi and the allowable tensile stress and the final compressive stress due to prestressing (assuming the final stress due to prestressing is less than the allowable final compressive stress) were − 400 psi and +2000 psi, respectively, the design would not be restricted by the bottom-fiber stress. The maximum load that could be applied without exceeding the allowable stresses would be one that results in a maximum bottom-fiber stress of −2400 psi.

In a similar manner the top-fiber stress may limit the capacity of a prismatic beam section if the maximum flexural stress in the top fiber, due to the total load, is greater than the arithmetic sum of the allowable compressive stress in the completed member and the top fiber tensile stress due to final prestressing.

As a result of these limitations, the designer can normally determine if a specific concrete section can be used with straight tendons without calculating the magnitude and eccentricity of the prestressing force. It is only necessary to determine the stresses in the section due to the total load and compare these values with the sum of the appropriate, allowable stresses.

ILLUSTRATIVE PROBLEM 7-1 Determine the maximum moment the section in Fig. 7-6 can withstand if pretensioned with straight tendons having an initial and final stress of 180 and 154 ksi, respectively, if f'_{ci} and f'_c are 4000 and 5000 psi, respectively and the initial and final concrete stresses cannot exceed the following:

Top fiber:

$$\text{Initial tension} = -3\sqrt{4000} = -190 \text{ psi}$$
$$\text{Final compression} = 0.40 \times 5000 = 2000 \text{ psi}$$

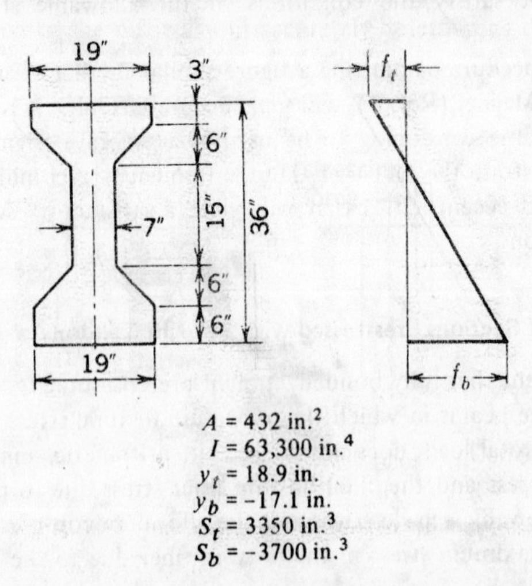

Fig. 7-6 Beam section and stress distribution for Problem 7-1.

Bottom fiber:

$$\text{Initial compression} = 0.60 \times 4000 = 2400 \text{ psi}$$
$$\text{Final tension} = -3\sqrt{5000} = -212 \text{ psi}$$
$$\text{Final compression} = 0.40 \times 5000 = 2000 \text{ psi}$$

SOLUTION: The stresses at the time of stressing will reduce to 154/180 = 0.856 of their initial value as a result of the prestressing losses. Therefore, the maximum stress due to all loads will be limited as follows:

$$\text{Top fiber} = 2000 + 0.856 \times 190 = 2163 \text{ psi}$$
$$\text{Bottom fiber} = 212 + 0.856 \times 2400 = 2266 \text{ psi}$$
$$\text{Bottom fiber} = 212 + 0.40 \times 5000 = 2212 \text{ psi}$$

The allowable moments as controlled by the top and bottom fibers are

$$M_T = \frac{2163 \times 63{,}300}{18.9 \times 12000} = 604 \text{ 'k}$$

$$M_B = \frac{-2212 \times 63{,}300}{-17.1 \times 12000} = 682 \text{ 'k}$$

The stresses in the top fiber control the capacity of this section with these design conditions.

7-5 Limitations of Sections Prestressed with Curved Tendons

In considering the stress in the bottom fiber at any specific section of a simple beam prestressed with a curved tendon, it should be apparent that the maximum stress due to external loads must not exceed the arithmetic sum of the stress due to the effective prestressing force and the allowable tensile stress in the completed structure. In addition, the algebraic sum of the stress due to the initial prestressing force and the stress due to the minimum loading condition must not exceed the allowable, initial compressive stress.

ILLUSTRATIVE PROBLEM 7-2 Determine the maximum possible allowable total moment that could be permitted on the beam section of Fig. 7-6 if a curved tendon is used and the initial and final prestressing steel stresses are 187 and 162 ksi, respectively. The design span is 50.0 ft. The allowable stresses are those used in I.P. 7-1, and $f'_{ci} = 4000$ psi, $f'_c = 5000$ psi.

SOLUTION:

$$w_d = \frac{432 \times 0.15}{144} = 0.450 \text{ kip/ft} \qquad M_d = \frac{0.450 \times 50^2}{8} = 141 \text{ k-ft}$$

Dead load flexural stresses:

$$f_t = \frac{141 \times 12}{3350} = +0.505 \text{ ksi}$$

$$f_b = \frac{141 \times 12}{-3700} = -0.457 \text{ ksi}$$

Maximum allowable top fiber stress and moment:

$$f_t = 0.40 \times 5000 + 505 + \frac{162}{187}(3\sqrt{4000}) = 2669 \text{ psi}$$

$$M_T = \frac{2669 \times 3350}{12000} = 745 \text{ k-ft}$$

Maximum allowable bottom fiber stress and moments:

$$(0.60 \times 4000)\frac{162}{187} + 3\sqrt{4000} = 2269 \text{ psi}$$

$$(0.40 \times 5000) + 3\sqrt{5000} = 2212 \text{ psi}$$

$$f_b = 2212 + 457 = 2669 \text{ psi}$$

$$M_T = \frac{-2669 \times -3700}{12000} = 823 \text{ k-ft}$$

It should be recognized that the condition of prestress assumed by this type of analysis may not be attainable due to practical considerations. (*See* Sec. 7-7.)

The initial tensile stresses in the top fibers of beams prestressed with curved tendons are not normally critical at the section of maximum moment in beams of good proportions. If the top-fiber stresses limit the design of beams with curved tendons, it is usually due to excessive, compressive stress under maximum loading conditions. Top fiber stresses are much more apt to be a problem in a beam with a narrow top flange than in a beam with a wide top flange.

7-6 Determination of Minimum Prestressing Force for Straight Tendons

In the trial and error procedure used in designing prestressed flexural members, a beam with known cross sectional properties is tentatively adopted and reviewed to determine the stresses due to the external loads. If the external loads result in stresses within practical limits (*see* Sec. 7-4), the magnitude and eccentricity of the prestressing force required to develop the desired net concrete stresses must then be determined. When straight tendons are used in prismatic simple beams subjected to usual loading conditions, the maximum stresses under minimum loading conditions (dead load of the beam alone, usually) occur at the ends of the beam where there is no moment due to external loads. The maximum stresses under the maximum loading conditions occur near midspan. The procedures recommended for determining the magnitude of the minimum prestressing force and the required eccentricity for different specific conditions are illustrated with explanation in Problems 7-3 through 7-5.

ILLUSTRATIVE PROBLEM 7-3 Determine the prestressing force and eccentricity required to prestress a slab, 4 ft wide and 8 in. deep, with straight tendons. The slab is to be used, simply supported, on a span of 30 ft and is to be composed of normal concrete (150 pcf) with $f'_{ci} = 4000$ psi and $f'_c = 5000$ psi. The superimposed load is 45 psf and the member will be exposed to a corrosive atmosphere in service.

SOLUTION:

Loads and moments:

$$
\begin{aligned}
\text{Slab dead load} &= 4 \times 100 = 400 \text{ plf} \\
\text{Superimposed load} &= 4 \times 45 = \underline{180} \text{ plf} \\
\text{Total load} &= 580 \text{ plf}
\end{aligned}
$$

$$\text{Total bending moment} = 580 \times \frac{30^2}{8} = 65{,}250 \text{ ft-lb}$$

Section modulus $S = bd^2/6 = 48 \times 8^2/6 = 512$ in.^{3}

Top and bottom fiber stresses due to the total bending moment

$$= \frac{65{,}250 \times 12}{\pm 512} = \pm 1530 \text{ psi}$$

Assume non-prestressed reinforcement is not to be used in the top flange to resist tensile stresses in the concrete. The final prestressing stress in the bottom fiber must be equal to +1530 psi. (Since the slabs are to be exposed to a corrosive atmosphere, the net bottom-fiber stress should not be tensile under full load.) If non-prestressed reinforcement is not to be provided at the ends, the top-fiber tensile stress at the ends, due to initial prestress, should be equal to or less than $3\sqrt{f'_{ci}} = 190$ psi. Assuming the ratio of $f_{se}/f_{si} = 0.85$, the tensile stress in the top fiber resulting from the effective prestress must be less than $0.85 \times -190 = -160$ psi. Therefore, the stress distribution due to the final prestressing force should be as shown in Fig. 7-7.

The prestressing stress at the center of gravity of a section is equal to P/A, since the familiar equation for stress due to prestressing

$$f = \frac{P}{A}\left(1 + \frac{ey}{r^2}\right)$$

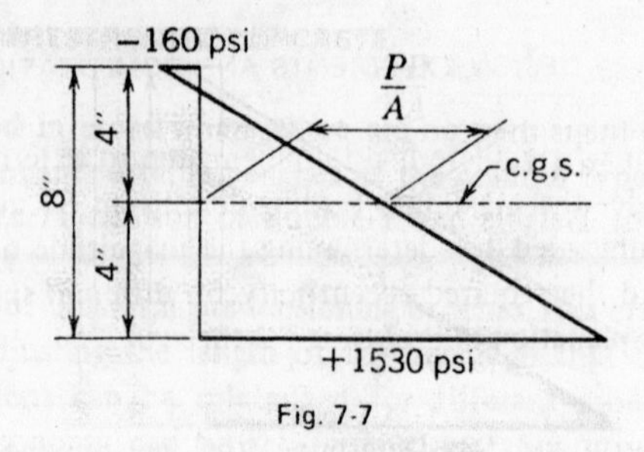

Fig. 7-7

becomes

$$f = \frac{P}{A}$$

for the fibers located at the center of gravity of the section, where $y = 0$. This is a fundamental principle that holds true for sections that are symmetrical or asymmetrical about an axis which passes through the center of gravity of the section. The average stress, which is at the center of gravity of the section (in this case, mid-depth of the section), can be rapidly computed by use of the proportional triangles indicated in Fig. 7-8, and from which it will be seen that

$$\frac{P}{A} = (f_b - f_t)\frac{y_t}{d} + f_t = 1690\left(\frac{4.0}{8.0}\right) - 160 = 685 \text{ psi}$$

Therefore, the final prestressing force P can be computed by

$$P = 685 \times 48 \times 8 = 263{,}000 \text{ lb}$$

Since this force must develop +1530 psi in the bottom fiber, the familiar relationship for the bottom-fiber stress due to prestressing

$$f_b = \frac{P}{A}\left(1 + \frac{ey_b}{r^2}\right)$$

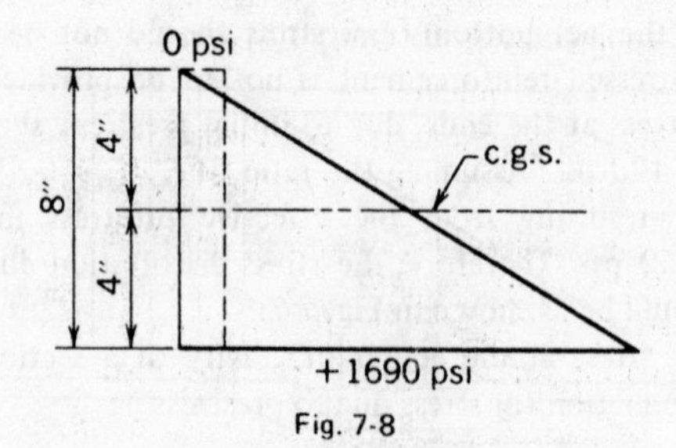

Fig. 7-8

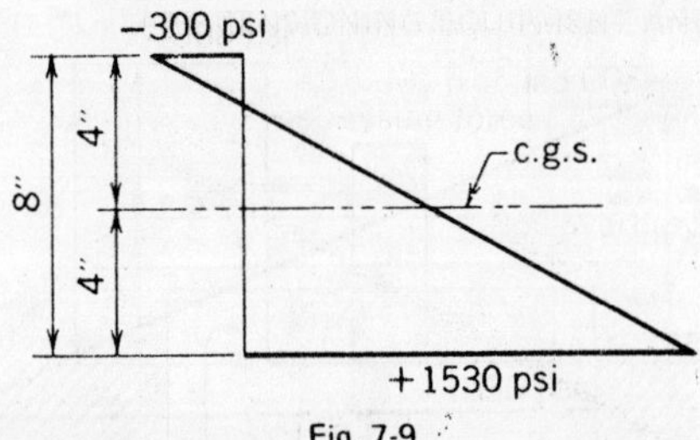

Fig. 7-9

can be rewritten

$$e = \left(\frac{f_b}{P/A} - 1\right)\frac{r^2}{y_b} = \left(\frac{1530}{685} - 1\right) - \frac{8}{6} = -1.65 \text{ in.}$$

Note that $r^2/y = d/6$ for a rectangular section.

If it is decided to use non-prestressed reinforcement in the top fibers to resist the tensile stresses in the concrete due to prestressing, the initial, top-fiber tensile stress might be as high as $6\sqrt{f'_{ci}} = -380$ psi and the top-fiber tensile stress due to the effective prestressing force could be as high as $0.85 \times -380 = -320$ psi. Assuming it is desired to limit the top-fiber stress to -300 psi, the required distribution of prestress would be as shown in Fig. 7-9, and the computation of P and e becomes

$$\frac{P}{A} = 1830 \times \frac{4.0}{8.0} - 300 = 615 \text{ psi}$$

$$P = 236{,}000 \text{ lb}$$

$$e = \left(\frac{1530}{615} - 1\right)\left(-\frac{8}{6}\right) = -1.98 \text{ in.}$$

Compare the simplicity of this computation to that required using the classical relationship demonstrated in Problem 4-8.

ILLUSTRATIVE PROBLEM 7-4 For the slab of Problem 7-3 assume the superimposed load is to be 100 psf. Compute the required prestressing force and eccentricity assuming: (1) The final top-fiber stress due to prestressing must not exceed -300 psi and no tension is to be allowed in the bottom fibers, and (2) the final top-fiber stress due to prestressing must not exceed -300 psi and the net stress in the bottom fiber under full load must not exceed -400 psi.

SOLUTION:

Loads and moments:

Slab dead load = 4 × 100 = 400 plf
Superimposed load = 4 × 100 = 400 plf
Total load = 800 plf

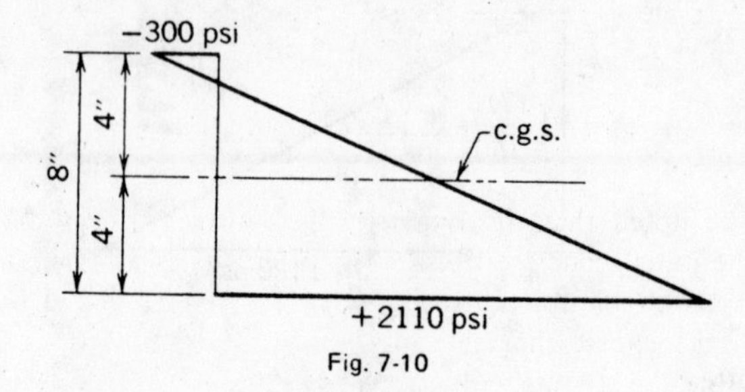

Fig. 7-10

Total moment = 800 × 30²/8 = 90,000 ft-lb

$$\text{Top and bottom fiber stresses} = \frac{90{,}000 \times 12}{\pm 512} = \pm 2110 \text{ psi.}$$

Part (1): The bottom fiber stress of +2110 is too high for $f'_{ci} = 4000$ psi, since 0.85 × 0.60 × 4000 psi = 2040 psi. If this solution is to be used and if the design is to conform to the allowable stresses specified above, the value of f'_{ci} must be 2110/(0.85 × 0.60), which is equal to 4150 psi. It should be pointed out that the net compression in the top fiber will be 2110 − 300 psi = 1810 psi. If the final net compressive stress is to be 0.40 f'_c or less, the minimum value of f'_c is 4500 psi. Assuming these values of f'_{ci} and f'_c are to be used, the required values of *P* and *e* are (*see* Fig. 7-10)

$$\frac{P}{A} = 2410 \times \frac{4.0}{8.0} - 300 = 905 \text{ psi}$$

$$P = 348{,}000 \text{ lb}$$

$$e = \left(\frac{2110}{905} - 1\right)\left(-\frac{8}{6}\right) = -1.77 \text{ in.}$$

Part (2): The desired distribution of stresses due to prestressing are as shown in Fig. 7-11, and the values of *P* and *e* become

$$\frac{P}{A} = 2010 \times \frac{4.0}{8.0} - 300 = 705 \text{ psi}$$

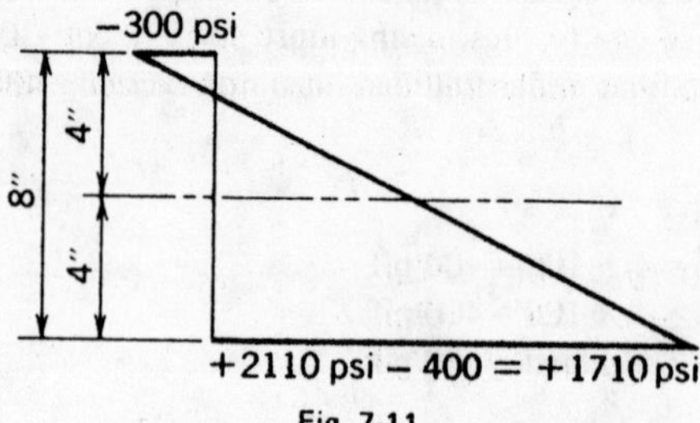

Fig. 7-11

$$P = 271{,}000 \text{ lb}$$

$$e = \left(\frac{1710}{750} - 1\right)\left(-\frac{8}{6}\right) = -1.90 \text{ in.}$$

It should be noted that this example illustrates a procedure that can be adopted if the initial concrete stresses (or final stresses) are nominally higher than would be allowable for the quality of concrete which was at first assumed; the designer can increase the value of f'_{ci} and f'_c (within reasonable limits) in order to confine the stresses within the allowable limits. Also illustrated is the procedure used in the calculation of the prestressing required for members in which tensile stresses are permitted in the bottom fibers of the members under full load.

ILLUSTRATIVE PROBLEM 7-5 Compute the prestressing force and eccentricity required to produce a final stress of −300 psi in the top fibers and +2000 psi in the bottom fibers of the AASHTO-PCI type III bridge beam as shown in Fig. 7-12. The required section properties are: $A = 560 \text{ in.}^2$, $y_t = 24.7$ in., and $r^2/y_b = -11.03$ in.

SOLUTION: The desired distribution of stresses is illustrated in Fig. 7-12 and P and e are computed as follows:

$$\frac{P}{A} = 2300 \times \frac{24.7}{45.0} - 300 = 962 \text{ psi}$$

$$P = 539{,}000 \text{ lb}$$

$$e = \left(\frac{2000}{962} - 1\right)(-11.03) = -11.90 \text{ in.}$$

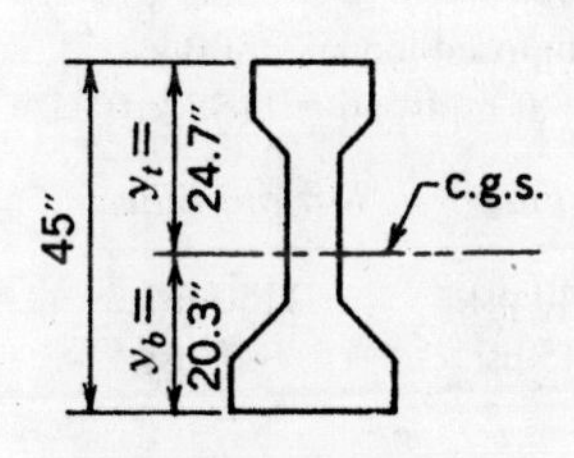

(a) AASHTO-PCI Type III

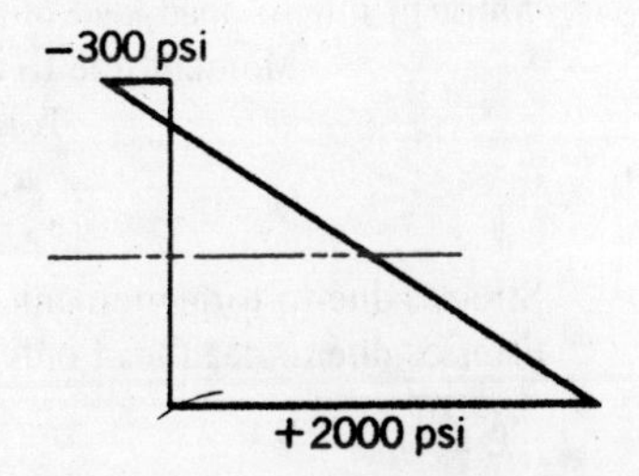

(b) Effective Prestress Distribution

Fig. 7-12 (a) AASHTO-PCI type III bridge beam. (b) Required effective prestress distribution.

7-7 Determination of Minimum Prestressing Force for Curved Tendons

Since the dead weight of a simple prismatic beam is normally acting at the time of stressing, the eccentricity of prestressing can be greater near midspan of the beam than at the ends, without the net concrete stresses exceeding the allowable values, as was explained in Sec. 4-6. This is the principle reason for draping or curving the tendons, and because this method results in a variable prestressing moment along the length of the beam, the concrete stresses resulting from curved prestressing tendons must be considered at several locations. For prismatic members, the magnitude of the prestressing force required with curved tendons is determined from the condition of stress at the position of maximum moment. *For members of variable depth, the magnitude of the prestressing force required may be controlled by conditions at a section which is not the section of maximum moment.*

In detailing an actual design, the prestressing force must be developed by a specific number of tendons. For reasons of economy, it is desirable to use tendons stressed near their maximum allowable values.

ILLUSTRATIVE PROBLEM 7-6 For the AASHTO-PCI bridge beam, type III, which is to be used on a span of 70 ft, compute the minimum prestressing force and eccentricity that can be used if the member must withstand a superimposed moment of 800 k-ft at midspan. The superimposed moment varies parabolically from a maximum value at midspan to zero at the support. Assume $f'_{ci} = 4000$ psi and $f'_c = 5000$ psi. The section properties of the AASHTO-PCI bridge beam, type III, are

Area = 560 in.^{2}	Moment of Inertia = 125,400 in.^{4}	
$y_t = 24.7$ in.	$S_t = 5080$ in.^{3}	$r^2/y_t = 9.06$ in.
$y_b = -20.3$ in.	$S_b = -6180$ in.^{3}	$r^2/y_b = -11.03$ in.

Dead weight of the beam = 0.585 k/ft

SOLUTION:

Moment due to dead load of beam = $0.585 \times 70^2/8$ = 358 k-ft
Moment due to the superimposed load = 800 k-ft
Total moment at midspan = 1158 k-ft

	Top Fiber	*Bottom Fiber*
Stresses due to total moment	+2740 psi	-2250 psi
Stresses due to dead load only	+845 psi	-695 psi

The distribution of concrete stresses due to the effective prestress at midspan must be as shown in Fig. 7-13(a), if the net top-fiber stress due to total load plus effective prestress is to be held to $0.40 f'_c = 0.40 \times 5000 = 2000$ psi and if the tensile stresses in the bottom fiber due to the total load are to be exactly

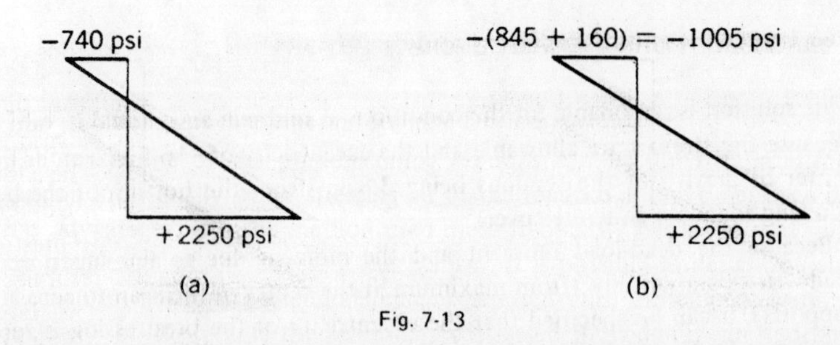

Fig. 7-13

nullified by the effective prestress. Non-prestressed reinforcement is not to be used and the top-fiber concrete stress due to initial prestressing plus dead load of the beam is to be limited to −190 psi. Assuming $f_{se}/f_{si} = 0.85$, the prestressing distribution shown in Fig. 7-13(b) limits the top fiber stress to the allowable value and exactly nullifies the total load stress in the bottom fiber. The bottom-fiber stress, due to the effective prestress, could be as high as $0.40\ f_c' + 695 = 2695$ psi. The most economical design will result from a prestressing force that can develop the required minimum effective prestress in the bottom fibers (+2250 psi), without exceeding the allowable, initial tensile stress in the top fibers, such as is shown in Fig. 7-13(b), if such a stress distribution can be obtained with a practical eccentricity.

For the distribution of stress shown in Fig. 7-13(b) the values of P and e are

$$\frac{P}{A} = 3255 \times \frac{24.7}{45.0} - 1005 = 782 \text{ psi}$$

$$P = 438 \text{ k}$$

$$e = \left(\frac{2250}{782} - 1\right)(-11.03) = -20.70 \text{ in.}$$

It is apparent that this is not a practical solution for this case, because the required eccentricity is greater than the distance from the center of gravity of the section to the bottom fibers (y_b); hence, if this solution were used, the center of gravity of the tendons would be below the bottom of the beam. Therefore, the distribution of stress due to the effective prestress must be revised in such a manner that the eccentricity is reduced.

Using the distribution of stress indicated in Fig. 7-13(a), the values of P and e are

$$\frac{P}{A} = 2990 \times \frac{24.7}{45.0} - 740 = 900 \text{ psi}$$

$$P = 504 \text{ k}$$

$$e = \left(\frac{2250}{900} - 1\right)(-11.03) = -16.5 \text{ in.}$$

This solution is reasonable for the conditions at midspan and should be adopted, because the stresses are allowable and the eccentricity of -16.5 in. results in the center of gravity of the tendons being 3.8 in. from the bottom of the beam, allowing adequate concrete cover.

Because the dead-load moment and the moment due to the superimposed loads vary parabolically (from maximum at the center of the span to zero at the supports), it can be specified that the eccentricity of the prestressing is zero at the supports. Nominal eccentricities above or below the center of gravity of the section could be allowed at the supports without exceeding the allowable stresses.

At the quarter point, the stresses due to the dead and superimposed loads are only 75% of the stresses due to these loads at midspan, or +2060 and -1690 psi in the top and bottom fibers, respectively. Assuming the prestressing force is 504 k, the eccentricity is -16.5 in. at midspan, and, assuming the eccentricity varies parabolically to zero at the supports, the eccentricity would be 75% of -16.5 in. at the quarter point and the stresses due to the effective prestress would be

$$f_t = \frac{504{,}000}{560}\left(1 - \frac{12.4}{9.06}\right) = -332 \text{ psi}$$

$$f_b = \frac{504{,}000}{560}\left(1 + \frac{12.4}{11.03}\right) = +1912 \text{ psi}$$

It can be shown that these stresses will result in net concrete stresses under minimum and maximum loading conditions within the allowables, as follows:

	Top Fiber	*Bottom Fiber*
Stress due to dead load of beam	+634	-521
Effective prestress	-332	+1912
Net, minimum loading	+302	+1391
Stress due to superimposed load	+1420	-1165
Net, maximum loading	+1722	+226

ILLUSTRATIVE PROBLEM 7-7 For the beam and the conditions specified in Problem 7-6, determine the number of high-tensile, steel rods, in 1.5 in. dia. sheaths, that could be used, if the rods were to be used with an effective prestress of 82 k each and $f_{se}/f_{si} = 0.85$.

SOLUTION:

Assume six rods.

$$P = 6 \times 82 = 492 \text{ k}$$

$$M_T = -P(e + r^2/y_b) = 1158 \text{ k-ft} = -492\,(e + 11.03)$$

$$e = -17.3 \text{ in.}$$

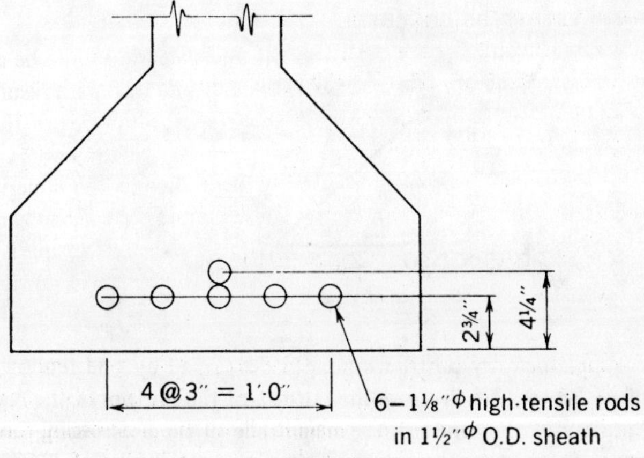

Fig. 7-14

Check:

$$f_t = \frac{492{,}000}{560}\left(1 - \frac{17.3}{9.06}\right) = -800 \text{ psi}$$

$$f_b = \frac{492{,}000}{560}\left(1 + \frac{17.3}{11.03}\right) = +2250 \text{ psi}$$

This solution is satisfactory, since the stress distribution is between those of Fig. 7-13(a) and (b) and the distance from the bottom of the beam to the center of gravity of the tendon is 3.0 in., which will allow a clear concrete cover of 2 in. for the $1\frac{1}{2}$ in. sheaths if placed as shown in Fig. 7-14.

ILLUSTRATIVE PROBLEM 7-8 Assume the beam for Problem 7-6 is to be stressed with a combination of pre-tensioning and post-tensioning. Determine the amount and eccentricity of the prestressing required for each of these methods, if it is assumed the maximum, initial, tensile and compressive concrete stresses are –350 psi and 2400 psi, respectively, and $f_{se}/f_{si} = 0.85$. The magnitude of the post-tensioning force is to be kept as small as possible, in the interest of economy. Assume the post-tensioned tendon is not to be stressed until the beam is pre-tensioned and removed from the casting bed.

SOLUTION: The maximum distribution of stress that can be allowed by the effective prestress in the straight pretensioned tendons is as shown in Fig. 7-15, for which the values of P and e are

$$\frac{P}{A} = 2340 \times \frac{24.7}{45.0} - 300 = 985 \text{ psi}, \qquad P = 551 \text{ k}$$

$$e = \left(\frac{2040}{985} - 1\right) - 11.03 = -11.8 \text{ in.}$$

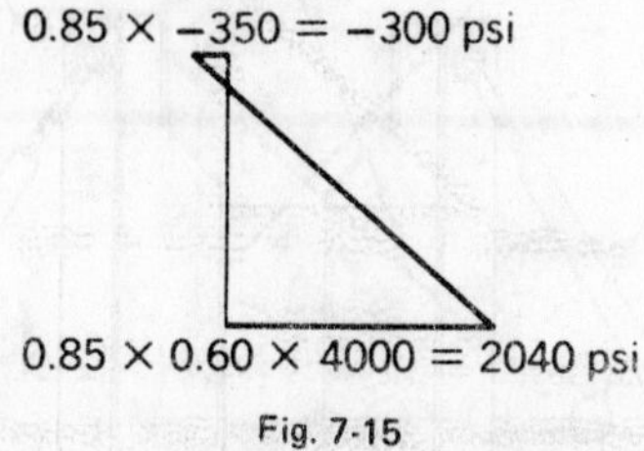

Fig. 7-15

Assuming the supplementary post-tensioning is developed by one tendon at an eccentricity of −17.5 in. (2.8 in. from the bottom of the bottom of the beam to the center of gravity of the tendon), the magnitude of the prestressing force required to increase the bottom-fiber stress, 210 psi to a total value of 2250 psi, is calculated as follows:

$$\frac{P}{A} = \frac{210}{1 + \left(\dfrac{17.5}{11.03}\right)} = 81.2 \text{ psi} \qquad f_t = \frac{45{,}500}{560}\left(1 - \frac{17.5}{9.06}\right) = -76 \text{ psi}$$

$$P = 45.5 \text{ k} \qquad f_b = \frac{45{,}500}{560}\left(1 + \frac{17.5}{11.03}\right) = +210 \text{ psi}$$

Summarizing the net concrete stresses for this solution reveals the following:

	Top Fiber	*Bottom Fiber*
Stress due to total moment	+2740 psi	−2250 psi
Effective prestress: pre-tension	−300 psi	+2040 psi
Effective prestress: post-tension	−76 psi	+210 psi
Net concrete stresses	+2364 psi	0 psi

If this combination of prestressing is allowed, the value of f_c' required for conformance to the design criteria is 2364/0.40 = 5900 psi. This value is too high for general use and the member must be redesigned with a larger top flange, in order to resist the compressive stresses or the tensile stresses in the top flange due to prestressing must be increased.

Since the top-fiber tensile stress resulting from the effective pre-tensioning is confined to −300 psi, in order to satisfactorily revise the prestressing so that the value of f_c' can be held to 5000 psi, the post-tensioning must develop −440 psi. In this manner, the top-fiber compressive stress can be confined to 2000 psi when the member is subjected to maximum loading conditions. Assuming the center of gravity of the post-tensioning tendons is 4.40 in. above the bottom of the beam ($e = -15.9$ in.), the force required to develop the required tension of

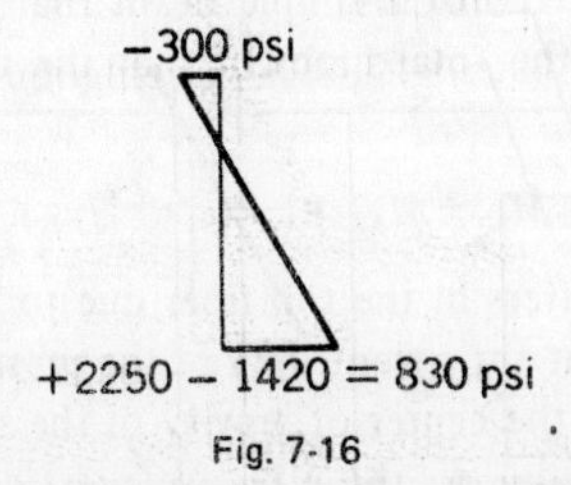

Fig. 7-16

-440 psi in the top fibers and the corresponding compressive stress in the bottom fibers are

$$f_t = -440 = \frac{P}{560}\left(1 - \frac{15.9}{9.06}\right)$$

$$P = 326 \text{ k}$$

$$f_b = \frac{326{,}000}{560}\left(1 + \frac{15.9}{11.03}\right) = +1420 \text{ psi}$$

Hence, the required stress distribution due to the supplementary effective pretensioning is as shown in Fig. 7-16 and the P and e required are as follows:

$$\frac{P}{A} = 1130 \times \frac{24.7}{45.0} - 300 = 320 \text{ psi}, \qquad P = 179 \text{ k}$$

$$e = \left(\frac{830}{320} - 1\right) - 11.03 = -17.6 \text{ in.}$$

The net concrete stresses resulting from this combination of prestress are summarized as follows:

	Top Fiber	*Bottom Fiber*
Stresses due to M_T	+2740 psi	−2250 psi
Stresses due to post-tensioning	−440 psi	+1420 psi
Stresses due to pre-tensioning	−300 psi	+830 psi
Net concrete stresses	+2000 psi	−0 psi

This problem illustrates the fact that the capacity and method of prestressing used in a beam may be limited by compressive stresses in the top fiber of the beam. This condition is particularly acute in beams of long span with narrow top flanges.

7-8 Estimating Prestressing Force and Cross-Sectional Characteristics

In Sec. 4-4, it was shown that for a simple beam that has zero bottom-fiber stress in the loaded state, the total moment that the beam can withstand is expressed by

$$M_T = M_D + M_{SL} = -P(e + r^2/y_b) \tag{7-6}$$

In a similar manner, if the stress in the top fiber due to prestressing alone is to be zero, it can be shown that the eccentricity of the prestressing force must not be greater than r^2/y_t below the center of gravity of the section. For this limitation of the concrete stress due to the effective prestress, the relationship of Eq. 7-6 becomes

$$M_T = M_D + M_{SL} = -P(r^2/y_t + r^2/y_b) \tag{7-7}$$

These relationships are useful in making preliminary designs since, by assuming values for M_D and P, the required value of the quantity $(e + r^2/y_b)$ can be computed. In employing these relationships, the designer must bear in mind the following fundamental factors:

(1) Most economical designs of simple beams result in values of P/A (the average compressive stress) between 500 and 900 psi. This gives a means of making a rough check on the estimated dead weight of the beam without assuming a specific cross section. (Considerably lower values of average prestress (P/A) are generally used in precast members with large top flanges, such as T-shaped beams, as well as in solid slabs).

(2) The dead weight of the beam itself is a small portion of the total load for short-span beams, whereas for long-span beams, the dead load of the beam itself may be of great importance.

(3) When straight tendons are used in short-span members, the value of the quantity $(e + r^2/y_b)$ approaches the lower limit given in Eq. 7-7 as $(r^2/y_t + r^2/y_b)$.

(4) When curved or draped tendons are used in beams of moderate to long spans, the value of e is frequently limited by the dimensions of the concrete section, rather than by top- or bottom-fiber stresses, and Eq. 7-6 approaches

$$M_T = -P(y_b + r^2/y_b) \tag{7-8}$$

(5) When tensile stresses are allowed in the bottom fibers in the fully loaded state, the pressure line goes higher than r^2/y_b above the center of gravity of the section and the relationship given by Eq. 7-6 can be rewritten

$$M_T = -\psi P(e + r^2/y_b) \tag{7-9}$$

The value of ψ to be assumed in the above relationship must be estimated by considering the magnitude of the allowable bottom-fiber tensile stress with respect to the bottom-fiber stress resulting from prestressing alone. For example, if the bottom fiber stress due to the effective prestress must be confined to 2000 psi and the allowable tensile stress in the bottom fiber is 400 psi, the value of $(e + r^2/y_b)$ must be increased by the ratio $\psi = 2400/2000 = 1.20$ to give an accurate estimate of the movement of the pressure line that will take place when the beam is loaded from the condition of zero bottom-fiber stress (due to external loads) to the point where the stress in the bottom fiber is -400 psi. If the bottom-fiber stress due to the effective prestress alone is as high as 3000 psi, as it frequently is in long-span, post-tensioned members, and allowable tensile stress of 400 psi in the bottom fiber would result in a ratio for ψ of only 3400/3000, which is equal to 1.13.

(6) The value of the term $(e + r^2/y_b)$ varies from 33 to 80% of the depth of the beam, depending upon the efficiency of the cross section, the allowable stresses, and the dead-load moment. Average values of this factor for use in estimating the preliminary design of roof and bridge girders are between 60 and 75% of the depth of the member, with the larger values being applicable to the longer spans and to members with relatively large flanges.

(7) The average value for depth-to-span ratio for most simple beams can be assumed to be $\frac{1}{20}$. This ratio does vary between relatively wide limits, but for simple beams it is rarely greater than $\frac{1}{16}$ or less than $\frac{1}{24}$, except for solid and cored slabs.

ILLUSTRATIVE PROBLEM 7-9 Estimate the depth of a beam required to carry a superimposed moment of 800 k-ft on a span of 70 ft. using curved tendons.

SOLUTION: Assume the weight of the girder will be 400 plf.

$$\text{Total moment} = 0.40 \times \frac{70^2}{8} + 800 = 1045 \text{ k-ft}$$

f $(e + r^2/y_b) = 0.70d$ and $P = 400$ k,

$$d = \frac{1045}{0.70 \times 400} = 3.73 \text{ ft}$$

his amounts to a depth-to-span ratio of 1 to 18.8. If $P/A = 1000$ psi, $P = 400$ k, nd $A = 400$ in.^{2}, then the weight of the girder will be 415 plf. The estimated ead weight of the beam is reasonably close to the assumed value, but the depth somewhat greater than normal for this span. Therefore, try $P = 450$ k, $A = 450$.^{2} and $w_d = 470$ plf.

$$\text{Total moment} = 0.47 \times \frac{70^2}{8} + 800 = 1088 \text{ k-ft}$$

$$d = \frac{1088}{0.70 \times 450} = 3.46 \text{ ft}$$

e depth-to-span ratio for this prestressing force is 1 to 20.2, which is reasonle and slightly more slender than average. It is apparent that a preliminary estimate can be made with the data developed here, since the magnitude of the prestressing force and the concrete quantity are known approximately.

ILLUSTRATIVE PROBLEM 7-10 For the conditions stated for Problem 7-9, assume that the depth of the beam cannot exceed 3.5 ft, due to headroom restrictions. Estimate the prestressing force required as well as determine a preliminary cross-sectional shape. Assume that the member is to be post-tensioned, $f_c' = 5000$ psi, and that no tensile stresses are to be allowed in the bottom fibers. Check the estimate.

SOLUTION:

Assume

$$w_d = 0.50 \text{ and } (e + r^2/y_b) = 0.70\,d$$

$$M_T = 0.50 \times \frac{70^2}{8} + 800 = 1106 \text{ k-ft}$$

$$P = \frac{1106}{0.70 \times 3.5} = 451 \text{ k}$$

For a beam 42 in. deep, the web thickness is normally about 7 in. The average stresses in the flanges can be estimated at 2000 psi for the purpose of selecting dimensions of the trial section. For the top flange, the total force to be resisted is 451 k, since the pressure line will be quite high when the beam is fully loaded. Furthermore, the width of the top flange of a member that is 70 ft long should be about $\frac{70}{35}$ or 2 ft wide. Therefore, the thickness of the top flange can be computed by

$$t = \frac{451{,}000}{2000 \times 24} \cong 9.40 \text{ in.}$$

The bottom flange must resist a smaller force than the top flange, since the dead load of the beam is acting at the time of stressing. The force the bottom flange must resist can be approximated by multiplying the estimated prestressing force by the ratio of the moment due to the superimposed load and the moment due to the total load, or

$$451 \text{ k} \times \frac{800}{1106} \cong 330 \text{ k}$$

Assuming the width of the bottom flange is to be 18 in., the thickness of the bottom flange can be computed by

$$t = \frac{330{,}000}{2000 \times 18} \cong 9.20 \text{ in.}$$

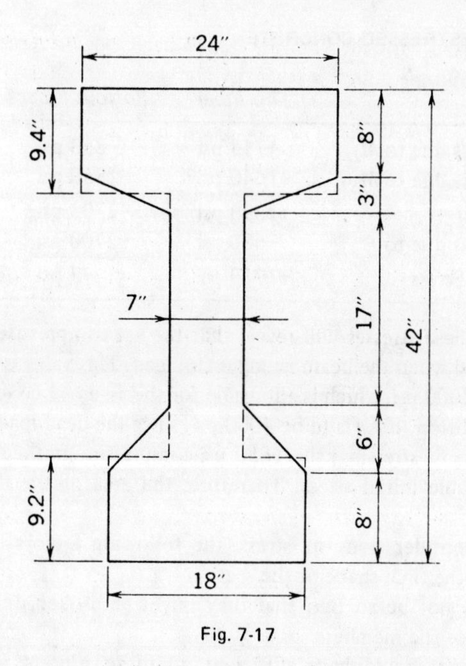

Fig. 7-17

The assumed trial section is shown in Fig. 7-17, where the estimated values of the thicknesses for the top and bottom flanges calculated above are shown superimposed.

To check the estimated prestressing force and concrete area, the section properties of the trial section are computed to be as follows:

$$A = 576.5 \text{ in.}^2 \qquad I = 115{,}680 \text{ in.}^4$$

$$y_t = 19.7 \text{ in.} \qquad S_t = 5860 \text{ in.}^3 \qquad r^2/y_t = 10.2 \text{ in.}$$

$$y_b = -22.3 \text{ in.} \qquad S_b = -5180 \text{ in.}^3 \qquad r^2/y_b = -9.00 \text{ in.}$$

$$M_D = 0.60 \times \frac{(70)^2}{8} = 368 \text{ k-ft}$$

$$M_{SL} = \underline{800 \text{ k-ft}}$$

$$M_T = 1168 \text{ k-ft}$$

The section is reviewed as follows:

Assume $e = -22.3 \text{ in.} + 4.5 \text{ in.} = -17.8 \text{ in.}$

$$P = \frac{1168 \text{ k-ft} \times 12}{17.8 + 9.0} = 523 \text{ k}$$

	Top Fiber	Bottom Fiber
Stress due to M_D	+753 psi	-853 psi
Stress due to M_{SL}	+1640 psi	-1850 psi
Total	+2393 psi	-2703 psi
Stress due to P	-680 psi	+2700 psi
Net Stress	+1713 psi	-3 psi

Examination of these stresses will reveal that the net compressive stress in the top fiber is 1713 psi when the beam is under full load. This value is substantially below the value of 2000 psi, which is allowable for the assumed concrete strength of 5000 psi. In addition, if f'_{ci} is to be 4000 psi, since the dead load of the beam is acting at the time of stressing, the 1850 psi compression in the bottom fibers is below the allowable initial stress. Therefore, the area of the flanges can be reduced.

In addition to considerations of stress, the following factors must be considered in selecting the final shape of the section.

(1) Flanges must not be so thin that they might be broken during handling and transportation of the members.

(2) The top flange should have sufficient width to protect against undue lateral flexibility during transportation and erection, as well as to ensure that the flange will not buckle under load if it is to be used in the completed structure without supplementary lateral support.

(3) The bottom flange must be of such shape that the prestressing tendons can be positioned with adequate cover and spacing to protect them against corrosion and to facilitate placing of the concrete, and, when post-tensioning is used, the shape of the bottom flange must allow curving of the tendons up into the web while the minimum cover is maintained.

(4) For reasons of economy, the shape should be as simple as possible to facilitate the fabrication.

(5) The slopes provided as transitions between the flanges and the webs should be of such size and shape that danger of honeycomb in the bottom flange is minimized. In addition, stripping of the form is facilitated by large slopes on the flanges.

In this example, it is assumed that the flanges can be reduced to 6 in. in thickness at their extremeties, as shown in Fig. 7-18, and the revised section properties and stresses are computed as follows:

$$A = 520.5 \text{ in.}^2 \qquad I = 108{,}090 \text{ in.}^4$$

$$y_t = 19.9 \text{ in.} \qquad S_t = 5430 \text{ in.}^3 \qquad r^2/y_t = 10.4 \text{ in.}$$

$$y_b = -22.1 \text{ in.} \qquad S_b = -4900 \text{ in.}^3 \qquad r^2/y_b = -9.39 \text{ in.}$$

$$M_D = 0.542 \times \frac{(70)^2}{8} = 332 \text{ k-ft}$$

$$M_{SL} = 800 \text{ k-ft}$$

$$M_T = 1132 \text{ k-ft}$$

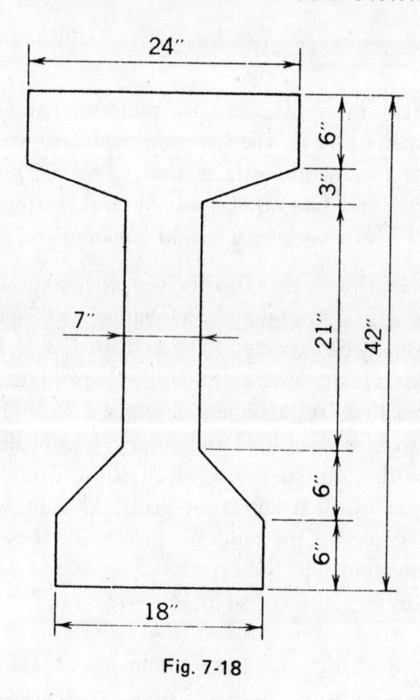

Fig. 7-18

The section is evaluated as follows:

Assume $e = -22.1$ in. + 4.5 in. = -17.6 in.

$$P = \frac{1132 \times 12}{9.39 + 17.6} = 503 \text{ k}$$

	Top Fiber	Bottom Fiber
Stress due to M_D	+735 psi	-814 psi
Stresses due to M_{SL}	+1770 psi	-1960 psi
	+2505 psi	-2774 psi
Stresses due to P	-665 psi	+2770 psi
Net stresses	+1840 psi	-4 psi

It should be noted that the prestressing force required in the final design is about 11% higher than the preliminary estimate and the concrete quantity is about 4% higher in the final design. These errors are the result of an error in the assumed value of $(e + r^2/y_b)$, which was assumed to be $0.70d$ and which is only $0.644d$ in the final design.

The initial stress in the bottom fiber should be checked. Assuming $f_{se}/f_{si} = 0.85$, the initial bottom-fiber stress is approximately +2445 psi, in which case, the value of f'_{ci} should be 4450 psi, if the initial compression is confined to $0.55\, f'_{ci}$, and 4100 psi and if the maximum allowable initial compression is $0.60\, f'_{ci}$.

ILLUSTRATIVE PROBLEM 7-11 Design a pretensioned T-beam to be used in a roof on a simple span of 60 ft. The superimposed dead load is 16 psf and the live load is 30 psf. Use a non-composite section, f'_c = 4000 psi, f_{pu} = 270 ksi, loss of prestress 40,000 psi, and f_y = 60,000 psi. Allowable stresses and load factors are to conform to ACI 318. Use normal weight concrete and a width of 8 ft.

SOLUTION: The depth-to-span ratio for simple prestressed beams is usually between 1 in 16 and 1 in 22 but may be as shallow as 1 in 40. For a span of 60 ft, the depth would usually be expected to be as deep as 3.75 ft or as shallow as 2.75 ft. Because the span is relatively long, a depth-to-span ratio of 1 in 40 is probably not practical. For a first try, assume a depth of 2.75 ft. The top flange can have a variable depth to save dead load and to give greater depth at the face of the web where the cantilever moment is greatest. Shear would not be expected to be a problem on a span of 60 ft with roof loads. An 8 in. wide web will probably provide sufficient space for the tendons. In view of these considerations, adopt the trial section shown in Fig. 7-19(a).

Section properties are: A = 572 sq in., I = 54,863 in.^{4}

$$y_t = 8.62 \text{ in.} \qquad S_t = 6365 \text{ in.}^3 \qquad r^2/y_t = 11.13 \text{ in.}$$
$$y_b = -24.38 \text{ in.} \qquad S_b = -2250 \text{ in.}^3 \qquad r^2/y_b = -3.93 \text{ in.}$$

Loads and maximum moments are

$$\text{T.D.L.} \ \frac{572 \times 150}{144} = 596 \text{ plf} \quad 268.2 \text{ k-ft}$$

$$\text{S.D.L.} \ 8 \times 16 = 128 \text{ plf} \quad 57.6 \text{ k-ft}$$

$$\text{S.L.L.} \ 8 \times 30 = 240 \text{ plf} \quad 108.0 \text{ k-ft}$$

Stresses at midspan are

	Top (psi)	Bottom (psi)
T.D.L.	+506	-1430
S.D.L.	+109	- 307
S.L.L.	+204	- 576
Total	+819	-2313

The stresses are relatively low and it is apparent a solution can be found with this depth. Allowing a bottom fiber tensile stress of $8\sqrt{f'_c}$ = -506 psi, the prestress must equal 1807 psi in the bottom fibers.

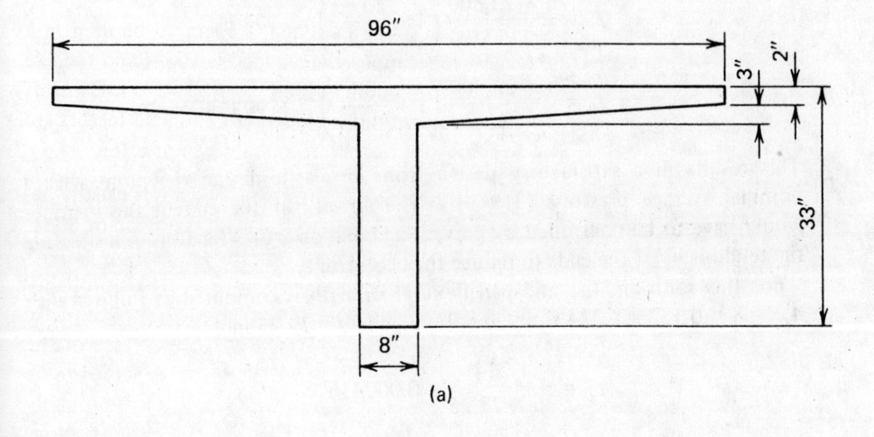

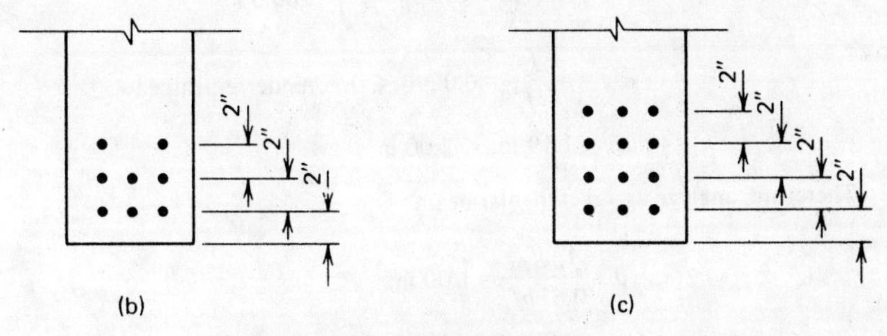

Fig. 7-19 T-beam for I.P. 7-11.

$$\text{Assume } e = -24.38 + 4.0 = -20.38 \text{ in.}$$

for

$$f_b = 1807 \text{ psi}, \quad \frac{P}{A} = \frac{1807}{1 + 20.38/3.93} = 292 \text{ psi}$$

$$P = (292)(572) = 167 \text{ kips}$$

Assuming the strands have a stress of 0.70 f_{pu} = 189 ksi after transfer and the loss of prestress is 40 ksi, each strand will have an effective force of (0.153)(149) = 22.8 k and 7.32 would be required to provide a force of 167 kips. Using the 8-tendon layout shown in Fig. 7-19(b),

$$e = -24.38 + \frac{3 \times 2 + 3 \times 4 + 2 \times 6}{8} = -20.63 \text{ in.}$$

and the top and bottom fiber stresses are

$$f_t = \frac{8 \times 22{,}800}{572}\left(1 - \frac{20.63}{11.13}\right) = -272 \text{ psi}$$

$$f_b = \frac{8 \times 22{,}800}{572}\left(1 + \frac{20.63}{3.93}\right) = +1993 \text{ psi}$$

This results in a satisfactory design from an elastic design viewpoint with a nominal average prestress (319 psi). The stresses at the end of the member would have to be controlled by preventing bond on 2 or 3 tendons or deflecting the tendons up at the ends to reduce the eccentricity.

For this solution, the midspan flexural strength is computed as follows with $A_{ps} = 8 \times 0.153 = 1.224$ sq in., $b = 96$ in., and $d = 29.25$ in.

$$\rho_p = \frac{1.224}{96 \times 29.25} = 0.000436$$

$$f_{ps} = 270\left(1 - 0.5\,\rho_p \frac{f_{pu}}{f_c'}\right) = 266.0 \text{ ksi}$$

$$\omega_p = \rho_p \frac{f_{ps}}{f_c'} = 0.0290 < 0.30 \text{ underreinforced}$$

$$1.4\,d\omega_p = 1.19 \text{ in.} < 2.00 \text{ in.}$$

Therefore, analyze as a rectangular beam

$$a = \frac{A_{ps}f_{ps}}{0.85\,bf_c'} = 1.00 \text{ in.}$$

$$\phi M_n = \frac{(0.90)A_{ps}f_{ps}}{12}\left(d - \frac{a}{2}\right) = 702 \text{ k-ft}$$

$$\text{Design } M_u = 1.4\,[268.2 + 57.6] + 1.7(108.0) = 639.7 \text{ k-ft}$$

The flexural capacity is about 10% greater than the minimum required.

The design shear at each reaction is:

$$V_u = [1.4\,(724) + 1.7\,(240)]\,\frac{60'}{2} = 42{,}650 \text{ lb}$$

$$v_u = \frac{42650}{(8)(29.25)} = 182 \text{ psi} = 2.88\sqrt{f_c'}$$

$$A_{v\,\text{min.}} = \frac{1.224}{80} \cdot \frac{270}{60} \cdot \frac{12}{29.25}\sqrt{\frac{29.25}{8}} = 0.054 \text{ sq in./ft}$$

$$v_s = \frac{A_v f_y}{b_w s} = \frac{0.054 \times 60{,}000}{8 \times 12} = 34 \text{ psi}$$

$$v_{c\,\text{min.}} = 2\sqrt{f_c'} = 2\sqrt{4000} = 126 \text{ psi}$$

Therefore, with minimum reinforcement:

$$\phi v_n = 0.85(126 + 34) = 136 \text{ psi}$$

Using Eq. 5-48 converted to unit stress:

$$v_c = 0.6\sqrt{4000} + (700)\frac{29.25\,(60 - 2x)}{12x\,(60 - x)} = 37.9 + \frac{1706\,(60 - 2x)}{x(60 - x)}$$

The computation for v_c can be summarized as follows:

x (ft)	v_c (psi)	$v_{s\,(\text{min})}$ (psi)	$\phi(v_c + v_s)$ (psi)	v_u (psi)
0	316	34	298	182
6	316	34	298	146
9	194	34	194	127
12	145	34	152	109
15	126	34	136	91

Note that $5\sqrt{f_c'} = 316$ psi and $2\sqrt{f_c'} = 126$ psi and stirrups are not required by stress considerations.

The design should be completed by making a short- and long-term deflection study and comparing the results to the design criteria being used, or to other performance standards the designer may wish to adopt.

If the designer were to elect to use a depth-to-span ratio of 1 in 30, the depth would be 24 in. rather than 33 in. and the section properties become: $A = 500$ sq in. and $I = 21810$ in.^{4}

$$y_t = 5.75 \text{ in.} \qquad S_t = 3793 \text{ in.}^3 \qquad r^2/y_t = 7.59 \text{ in.}$$
$$y_b = -18.25 \text{ in.} \qquad S_b = -1195 \text{ in.}^3 \qquad r^2/y_b = 2.39 \text{ in.}$$

Loads and maximum moments become

$$\text{T.D.L.} \ \frac{500 \times 150}{144} = 521 \text{ plf} \quad 234.4 \text{ k-ft}$$
$$\text{S.D.L.} = 128 \text{ plf} \quad 57.6 \text{ k-ft}$$
$$\text{S.L.L.} = 240 \text{ plf} \quad 108.0 \text{ k-ft}$$

Flexural stresses at midspan are

	Top (psi)	Bottom (psi)
T.D.L.	742	-2354
S.D.L.	182	-578
S.L.L.	342	-1085
Total	1266	-4017

Using $8\sqrt{f_c'}$ tension in the bottom fiber, the required bottom fiber prestress is 3511 psi. Assuming $e = -18.25 + 4.0 = -14.25$ in.

$$P = \frac{500 \times 3511}{1 + 5.96} = 252 \text{ kips}$$

This requires 11 strands and the average prestress is 504 psi. Using 12 strands spaced 2 in. on center, as shown in Fig. 7-19(c), $e = -18.25 + 5.00 = -13.25$ in.

$$f_t = \frac{22800 \times 12}{500}\left(1 - \frac{13.25}{7.59}\right) = -408 \text{ psi}$$

$$f_b = \frac{22800 \times 12}{500}\left(1 + \frac{13.25}{2.39}\right) = +3581 \text{ psi}$$

This too, is a satisfactory solution for elastic flexural stresses. The average prestress of 547 psi is greater than the 319 psi required with a depth of 33 in.; hence, more creep deformation must be accommodated in the structure if the 24 in. depth is used. Bond prevention or tendon deflection must be used to control the stresses the latter would be preferred for the 24 in. depth due to the high bottom fiber prestress.

The flexural strength computations become:

$$\rho_p = \frac{(12)(0.153)}{(96)(19)} = 0.00101$$

$$f_{ps} = 260.8 \text{ ksi}, \; \omega_p = 0.0656 < 0.30$$

$$1.4\, d\omega_p = 1.75 \text{ in.} < 2.00 \text{ in.}, \; a = 1.47 \text{ in.}$$

$$\phi M_n = \frac{(0.90)(12 \times 0.153)(260.8)}{12}\left[19.00 - \frac{1.47}{2}\right] = 656 \text{ k-ft}$$

Required $M_u = 592.4$ k-ft.

The flexural strength is about 11% greater than the minimum required.

The design shear at each reaction is 39,500 lb and

$$v_u = \frac{39{,}500}{(8)(19.0)} = 260 \text{ psi}$$

$$A_{u\,\text{min.}} = \frac{1.836}{80} \cdot \frac{270}{60} \cdot \frac{12}{19} \cdot \sqrt{\frac{19}{8}} = 0.100 \text{ sq in./ft}$$

Using Eq. 5-48 converted to unit stress:

$$v_c = 37.9 + \frac{1108\,(60 - 2x)}{x(60 - x)}$$

and the computations can be summarized as follows:

x (ft)	v_c (psi)	$v_{s\,\text{min}}$ (psi)	$\phi(v_c + v_s)$ (psi)	v_u (psi)
0	316	63	322	260
3	316	63	322	234
6	202	63	225	208
9	139	63	172	182
12	126	63	160	156
15	126	63	160	130
18	126	63	160	104

As will be seen, shear reinforcing greater than the minimum permitted is required by stress considerations between 6 and 12 ft from the ends of the beam.

To complete the design, a deflection study should be made.

This example demonstrates that there is a family of acceptable designs. One designer may prefer the deeper T-beam over the more shallow one because of the lower average prestress, and hence lower deferred strain, as well as because of the smaller deflections associated with deeper members. *Concrete stresses are only one design parameter that must be considered*; frequently, the design selected has concrete stresses lower than the maximum permitted under service loads.

ILLUSTRATIVE PROBLEM 7-12 Prepare the preliminary design for a simple post-tensioned beam that is to be used on a span of 32 ft. The beam is to have a composite concrete slab that is 5 in. thick ($f_c' = 4000$ psi). Superimposed dead and live loads are 20 psf and 125 psf, respectively. The width tributary to the beam is 30 ft. Assume the beam concrete is to have a strength at 28 days of 4000 psi and that the beam and slab are cast-in-place monolithically. Use the allowable stresses of ACI 318.

SOLUTION: The relatively short span and high superimposed loads will render shear stresses an important design consideration. The loads without the beam stem are.

$$\text{Slab } 30 \times 5 \times 150/12 = 1875 \text{ plf}$$
$$\text{S.D.L. } 30 \times 20 = 600 \text{ plf}$$
$$\text{S.L.L. } 30 \times 125 = 3750 \text{ plf}$$

The live load is greater than the dead loads and hence, a bottom flange may be required to resist the prestressing force when the live load is not applied.

The span-depth ratio for a heavily loaded beam is generally lower than for a lightly loaded beam. In view of these considerations, for a first trial section, adopt a beam that has an overall depth of 2 ft (span-depth ratio of 16) and a width of 12 in. with no bottom flange. The trial section is shown in Fig. 7-20(a). The top flange width is taken to be 16 times the flange thickness plus the width of the web. For the assumed section the section properties are

$$A = 688 \text{ sq in.} \quad \text{and} \quad I = 29768 \text{ in.}^4$$

$$y_t = 6.48 \text{ in.} \qquad S_t = 4594 \text{ in.}^3 \qquad r^2/y_t = 6.68 \text{ in.}$$

$$y_b = -17.52 \text{ in.} \qquad S_b = -1699 \text{ in.}^3 \qquad r^2/y_b = -2.47 \text{ in.}$$

The weight of the beam stem is 238 plf and the midspan moments are

$$\text{D.L.} \ (238 + 1875)(128) = 270.5 \text{ k-ft}$$

$$\text{S.D.L.} \ (600)(128) = 76.8 \text{ k-ft}$$

$$\text{S.L.L.} \ (3750)(128) = \underline{480.0} \text{ k-ft}$$

$$827.3 \text{ k-ft}$$

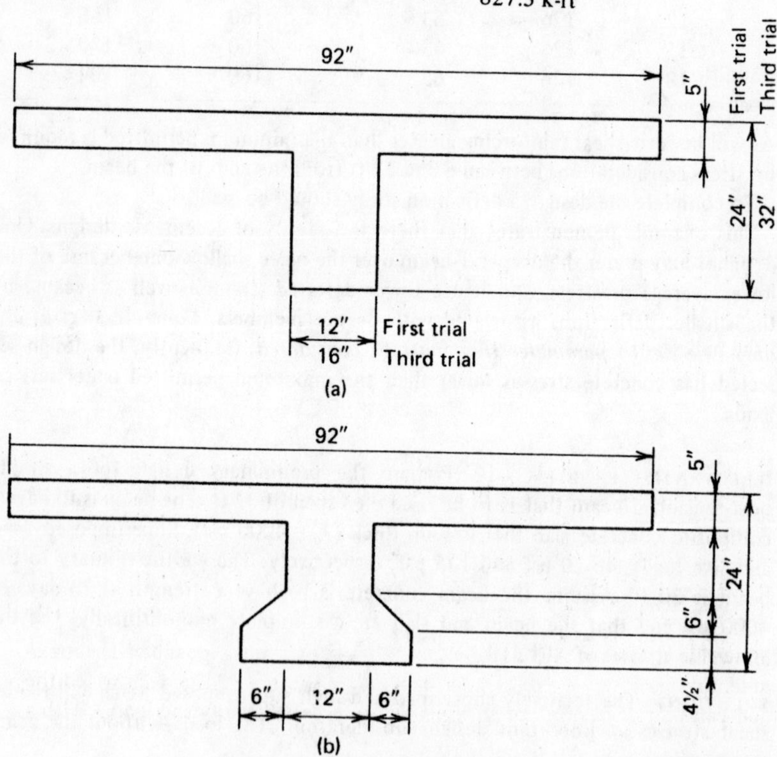

Fig. 7-20 Cross sections used in I.P. 7-12.

Midspan flexural stresses are

	Top (psi)	Bottom (psi)
D.L.	+707	-1911
S.D.L.	+201	-542
S.L.L.	+1254	-3390
T.L.	+2162	-5843

A review of the stresses reveals that the section will not be satisfactory because the bottom fiber stresses are too high. If the bottom flange were increased in width to 24 in., the resulting bottom fiber stress should be reduced and effect a workable solution.

Adding a bottom flange to the section, as shown in Fig. 7-20(b), the section properties become:

$$A = 778 \text{ sq in.} \qquad I = 44985 \text{ in.}^4$$

$$y_t = 8.05 \text{ in.} \qquad S_t = 5588 \text{ in.}^3 \qquad r^2/y_t = 7.18 \text{ in.}$$

$$y_b = -15.95 \text{ in.} \qquad S_b = -2820 \text{ in.}^3 \qquad r^2/y_b = -3.63 \text{ in.}$$

The additional dead load due to the bottom flange is 94 plf and the loads and midspan moments are

	Loads	Moment
D.L.	2207 plf	282.5 k-ft
S.D.L.	600 plf	76.8 k-ft
S.L.L.	3750 plf	480.0 k-ft

and the midspan flexural stresses are

	Top (psi)	Bottom (psi)
D.L.	+607	-1202
S.D.L.	+165	-327
S.L.L.	+1031	-2042
T.L.	+1803	-3571

Using tensile stresses in this member, a solution is obviously possible. The maximum tension necessary is -3571 + (1202 + 327 + 1800) = -242 psi, provided the tendon can be sufficiently eccentric to nullify the effects of dead load.* With a

*Note that 1800 psi = 0.45 f'_c for these design conditions.

TABLE 7-3

Pt.	Length (ft)	v_{ci} (psi)	v_{cw} (psi)	v_c (psi)	v_u (psi)	A_v (in.^{2}/ft)
.00	.000	infin	779.0	779.0	841.8	0.2419
.05	1.600	1303.7	747.0	747.0	757.7	.1200
.10	3.200	870.8	714.3	714.3	673.5	.1200
.15	4.800	682.6	681.7	681.7	589.3	.1200
.20	6.400	555.4	649.1	555.4	505.1	.1200
.25	8.000	452.3	616.4	452.3	420.9	.1200
.30	9.600	361.0	583.8	361.0	336.7	.1200
.35	11.200	276.2	551.2	276.2	252.5	.1200
.40	12.800	192.1	517.3	192.1	165.1	.1200
.45	14.400	113.2	484.7	113.2	81.1	.1200
.50	16.000	37.9	453.3	107.5	000.0	.1200

tension of −242 psi and an assumed eccentricity of −12.00 in., the prestressing force required is

$$P = \frac{(3329)(778)}{1 + \dfrac{12.00}{3.63}} = 601.5 \text{ kips}$$

and the average prestress is 773 psi. This value of average prestress is not unrealistic but will cause significant creep deformation.

A shear analysis for the member, based upon $f_y = 60{,}000$ psi and a parabolic trajectory for the tendon ($e = 0$ at support, $e = -12.00$ in. at midspan) is summarized in Table 7-3. A review of this table will show that the 12-in. thick web results in only minimum shear reinforcement being required. Hence, the web thickness could be reduced if the designer so desired.

The cost of forming the bottom flange, as well as the added costs of placing reinforcing steel stirrups in a beam of this shape, are objections to adopting this section as a final one.

Rather than adding the bottom flange, another solution would be to increase the depth of the beam.

Still another solution would be to increase the depth as well as the stem width. Increasing the depth to 32 in. and increasing the stem width to 16 in. will be adopted for another trial. Using a top flange width of 8 ft ($L \div 4$), the section properties are:

$$16 \times 32 = 512 \times 16 = 8192$$

$$80 \times 5 = \frac{400}{912} \times 2.5 = \frac{1000}{9192}$$

$$A = 912 \text{ sq in.,} \qquad I = 85{,}450 \text{ in.}^4$$

$$y_t = 10.08 \text{ in.} \qquad S_t = 8477 \text{ in.}^3 \qquad r^2/y_t = 9.30 \text{ in.}$$

$$y_b = 21.92 \text{ in.} \qquad S_b = 3898 \text{ in.}^3 \qquad r^2/y_b = 4.27 \text{ in.}$$

The midspan stresses become

	Top (psi)	Bottom (psi)
D.L.	421	-916
S.D.L.	109	-236
S.L.L.	679	-1477
T.L.	1209	-2629

An examination of these stresses will show that the superimposed dead and live loads cause a bottom fiber stress of -1713 psi. Hence, tensile stresses can be avoided with this solution, if so desired. It should also be apparent that the height or thickness of the stem could be reduced if so desired.

The prestressing force required can be determined for the case of zero tension by

$$P_{se} = \frac{M_T}{(e + r^2/y_b)} = \frac{854.4 \times 12}{(17.92 + 4.27)} = 462.0 \text{ kips}$$

in which $e = -21.92 + 4.00 = -17.92$ in.

This solution can be checked as follows

$$f_t = \frac{462{,}000}{912}\left(1 - \frac{17.92}{9.30}\right) = -470 \text{ psi}$$

$$f_s = \frac{462{,}000}{912}\left(1 + \frac{17.92}{4.27}\right) = +2633 \text{ psi}$$

If one wishes to permit tensile stresses under full load, the prestressing force could be reduced.

For a tensile stress of $6\sqrt{f_c'}$, the prestressing force can be determined from Eq. 4-4 as follows:

$$\text{Required } f_b = 2629 - 6\sqrt{4000} = 2250 \text{ psi}$$

$$P = \frac{(912)(2250)}{1 + \dfrac{17.92}{4.27}} = 394.9 \text{ kips}$$

To complete the design, one must investigate short- and long-term deflections, design the shear reinforcement, and confirm the adequacy of the flexural strength.

7-9 Reduction in Shear Force Due to Curvature of Parabolic Tendons

In Sec. 4-6, it was shown that the curvature of prestressing tendons results in a reduction in the shear force the concrete must withstand. Furthermore, it was shown that this reduction is equal to the vertical component of the prestressing force at the point under consideration. The vertical component of the prestressing force is equal to $P \sin \alpha$, in which α is the angle of inclination of the tangent to the prestressing tendon at the point under consideration.

Because the angle α is small in almost all instances, the sine and tangent are practically equal. Hence, the tangent can be used in computing the vertical component of the prestressing force without introducing significant error.

The computation of the tangent of the angle of inclination for tendons placed on a series of chords is basic and requires no explanation. For tendons on parabolic curves, the computation of the tangent of the angle of inclination is equally simple, when the properties of a parabola are understood.

A parabola is shown in Fig. 7-21 with the dimensions and tangents that are most important in the analysis of prestressing shear forces. It will be seen from the figure that the tangent to the parabola at the end is at an angle of α to the horizontal and the tangent of the angle α is equal to

$$\tan \alpha = \frac{2E}{L/2} = \frac{4E}{L} \tag{7-10}$$

The dimension E is the total displacement of the prestressing force and is equal to the normal eccentricity of the force only when the eccentricity of the prestressing force is zero at the ends. The units of E and L must be the same.

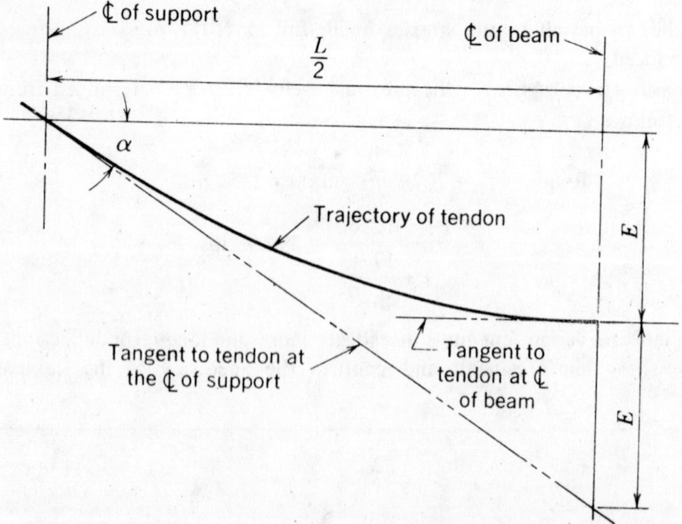

Fig. 7-21 Diagram illustrating fundamental properties of a parabola.

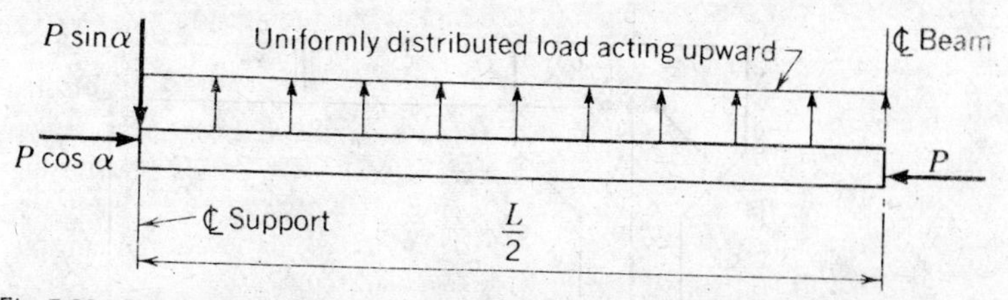

Fig. 7-22 Free-body diagram of the forces on a beam prestressed by a parabolic tendon.

It is apparent from the free-body diagram of Fig. 7-22, in which the forces that act on the concrete as a result of stressing with a parabolic tendon are shown, that the shear stresses resulting from the tendon vary uniformly from a maximum value at the end to zero at midspan, just as the shear stresses in a simple beam subjected to a uniform load vary from zero at midspan to a maximum value at the support. Hence, the vertical component of the shear stress carried by the tendon at points between the end and the midspan of the beam can be determined by the following relationship:

$$P \sin \alpha_x = P \sin \alpha \times \frac{(L/2 - x)}{L/2} \tag{7-11}$$

in which x is the distance from the support to the point under consideration.

Although the relationships presented here are derived for tendons on parabolic trajectories, they can normally be applied to tendons on other curved trajectories encountered in prestressed concrete, without introducing significant error. If the displacement of a tendon is very large in comparison to the span, as is sometimes the case in post-tensioned folded plates or shells, it is advisable to compute the reduction in shear using the sine of the actual angle as determined from the tendon layout.

An example of the computation of the shear component carried by a tendon on a parabolic curve is given in Problem 4-10.

7-10 Computing the Location of Pre-tensioning Tendons

The selection of the location or pattern of the pre-tensioning tendons must be made after the cross-sectional shape, the prestressing force, and the eccentricity have been determined. This is done by trial and error and can generally be accomplished quickly if the trial and error computations are made according to a specific procedure. This procedure consists of first determining the number of tendons required by dividing the required effective prestressing force by the maximum allowable effective prestressing force that one tendon can withstand. Secondly, locate the position of the required number of tendons, by computing moments of the tendons at assumed locations, that are adjusted by trial and error, until the center of gravity of the tendons is at the desired location. The procedure can be illustrated by the sketch of Fig. 7-23 which represents the

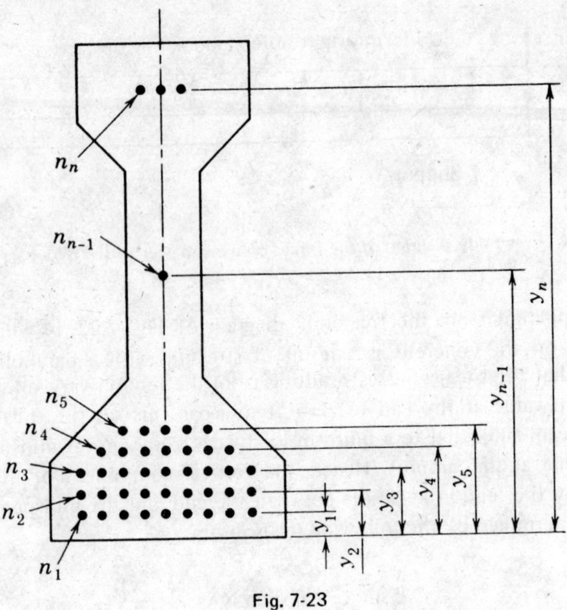

Fig. 7-23

cross section of a beam that requires N tendons placed with their center of gravity at a distance d' from the bottom of the beam. If $n_1, n_2, n_3 \ldots n_n$ represent the number of tendons in the first, second, third, and the nth rows from the bottom of the beam and $y_1, y_2, y_3, \ldots y_n$ represent the distances from these rows to the bottom of the beam, respectively, it is apparent that in order to obtain the desired location of the center of gravity, the following relationship must be satisfied

$$Nd' = \sum (y_1 n_1 + y_2 n_2 + y_3 n_3 + \cdots y_n n_n) \tag{7-12}$$

Since the values of N and d' are known, the majority of the tendons can be located and the value of $\Sigma\ (y_1 n_1 + y_2 n_2 + y_3 n_3, \ldots y_n n_n)$ adjusted to the desired value with the remaining tendons.

It is apparent that the majority of the tendons will be near the bottom of the member in order to achieve the required eccentricity. It is desirable that some tendons be supplied near the top of most members for the purpose of supporting web reinforcing, inserts, and other accessories. This can frequently be done with the required number of tendons and without supplying additional tendons specifically for this purpose.

In tendon patterns which have some tendons high in the section, the upper tendons should be disregarded when computing the flexural strength of the section. The distance to the center of gravity of the lower group of tendons, which would be very highly stressed at ultimate load, must be determined for use in calculating the ultimate moment capacity.

ILLUSTRATIVE PROBLEM 7-13 Compute the location of the tendons required to produce a prestressing force of 538 k at an eccentricity of −11.9 in. in the AASHTO-PCI bridge beam, type III, if the tendons to be used have an effective force of 11 k each and $y_b = -20.3$ in.

SOLUTION: The number of tendons required is: 538/11 = 49 each. The distance from the bottom of the beam to the center of gravity of the steel is: $d' = 20.3 - 11.9 = 8.4$ in. $Nd' = 49 \times 8.4 = 412$. The summation of the moments of the tendons in their final location should equal 412. Forty-five of the tendons are tentatively positioned as shown in Fig. 7-24 and the moment of the tendons computed about the bottom of the section is equal to 278. Therefore, the remaining four tendons must have an average distance from the bottom of the beam equal

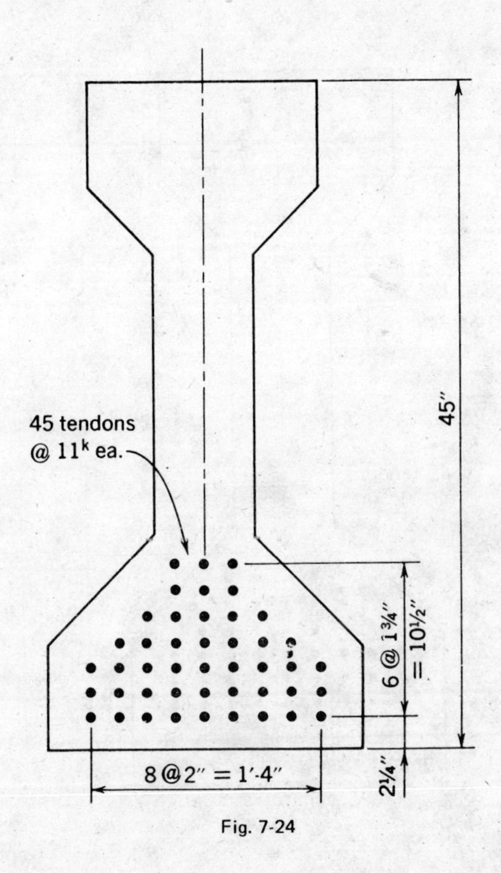

Fig. 7-24

to:

$$\text{Average } y = \frac{412 - 278}{4} = 33.5 \text{ in.}$$

This average distance is of the order of 75% of the depth of the beam and it appears that the tendon layout should be adjusted in order to include one more tendon in the bottom group. Therefore, the pattern is revised by increasing the number of tendons in the sixth row, from the bottom of the section, to 5 and reducing the number of tendons in the seventh row, to 2. The moment of the 45 tendons in the revised tendon pattern is 287.25. The average distance required for the 3 remaining tendons is computed as follows:

$$\frac{412 - 287.25}{3} = 41.5$$

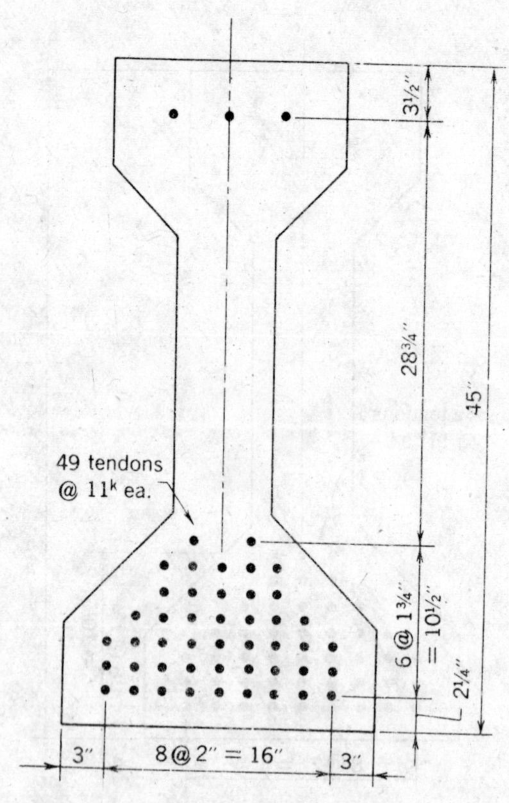

Fig. 7-25

The final tendon layout is illustrated in Fig. 7-25. The values of d' and d'_u are computed as follows:

$$d' = \frac{411.75}{49} = 8.40 \text{ in.}$$

$$d'_u = \frac{287.25}{46} = 6.25 \text{ in.}$$

The flexural strength should be computed on the basis of 46 tendons, 6.25 in. from the bottom of the beam.

7-11 Fiber Stresses at Ends of Prismatic Beams

In employing bond prevention or in using non-prestressed reinforcing as a means of controlling the stresses resulting from the initial prestress at the end of a beam, it is necessary to analyze the flexural stresses resulting from the dead weight of the beam near the end of the beam, in order to determine the limits over which the bond must be prevented or over which the special end reinforcement should be provided. This can be done using the fundamental principles of strength of materials, through the use of factors selected from unit parabolic curves, or by computing the location of the required dead load stresses through the use of the known properties of parabolas. Each of these methods should yield identical results. The use of the latter method is shown in the following problem.

ILLUSTRATIVE PROBLEM 7-14 Compute the length over which non-prestressed reinforcing is required at the ends of a simple prismatic beam in which the top-fiber stresses due to initial prestressing are equal to −360 psi and the maximum, allowable tensile stresses without non-prestressed reinforcing is −190 psi. The stress in the top fibers at midspan of the beam due to dead load alone is +730 psi. The beam is 70 ft long.

SOLUTION: The stress due to dead load in the top fiber varies parabolically as shown in Fig. 7-26. It is required to determine the distance from the end of the beam, where the top-fiber stress is 360 − 190 = 170 psi, since at this location, the net concrete stress will be −190 psi, which can be allowed without non-prestressed reinforcement. The ordinates of the parabola vary according to the relationship:

$$f_{\text{℄}} - f_x = \left(\frac{L/2 - x}{L/2}\right)^2 f_{\text{℄}}$$

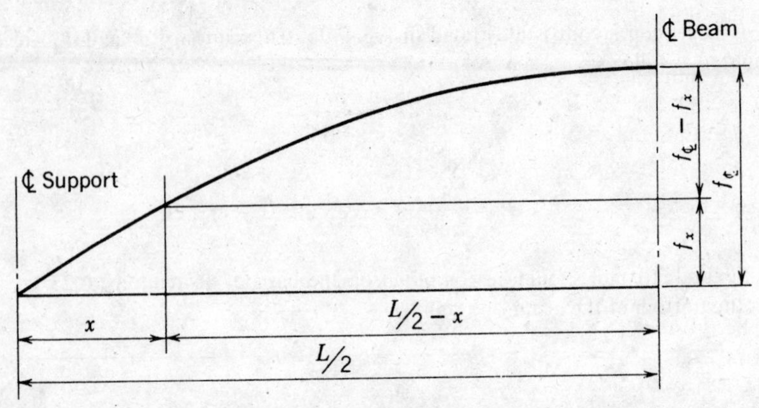

Fig. 7-26 Diagram used to compute stresses at different locations along the length of a beam having a parabolic moment diagram.

or

$$x = \frac{L}{2}\left[1 - \left(\frac{f_L - f_x}{f_{℄}}\right)^{1/2}\right]$$

Using the values given in the example

$$x = 35 \text{ ft}\left[1 - \left(\frac{730 - 170}{730}\right)^{1/2}\right] = 35 \text{ ft} \times 0.125 = 4.37 \text{ ft}$$

7-12 Computing the Effects of Bond Prevention

When the initial prestressing stresses at the ends of a simple pre-tensioned beam exceed the allowable stresses, the effect of the prestressing can be reduced by preventing bond on a specific number of tendons over a specific length, as is explained in detail in Sec. 6-11. The length over which the bond must be prevented can be computed according to the methods suggested in Sec. 7-11. The number and location of tendons that should be prevented from bonding to the concrete can be determined by computing the effect of one tendon in each of the lower rows of the tendon pattern and then, by trial and error, determining the number of tendons in each row that should be prevented from bonding to the concrete.

REFERENCES

1. Magnel, G. "Prestressed Concrete," pp. 20–25, Concrete Publications, Ltd., London, 1948.

CHAPTER 7—PROBLEMS

1. For the lightweight concrete (115 pcf) double-t-beam in Problem 9, Chapter 5, (Fig. 5-46), compute the distance over which bond must be prevented if the initial tensile stress in the straight tendons immediately before transfer is 189 ksi, and the allowable initial compressive and tensile stresses are 2100 psi and −177 psi, respectively. Assume $n = 14.0$; take the elastic shortening into account in computing initial concrete stresses.

   Section properties for the gross precast section are

$A = 840$ sq in.	$I = 43759$ in.^{4}	$r^2/y_t = 7.26$ in
$y_t = 7.175$ in.	$s_t = 6098$ in.^{3}	$r^2/y_b = -3.10$ in.
$y_b = -16.825$ in.	$s_b = -2600$ in.^{3}	

SOLUTION:

For 18 strands, $A_{ps} = 2.754$ sq in., $e = -12.825$ in.

$$k_s = 1 + \frac{(12.825)^2(840)}{43759} = 4.16$$

From Eq. 6-1 with M_d expressed in kip-feet,

$$f_{\text{c.g.s.}} = \frac{2165 - 2.954\,M_d}{1000.4}$$

For 12 strands, $A_{ps} = 1.836$ sq in., $e = -11.825$ in.

$$k_s = 3.68$$

$$f_{\text{c.g.s.}} = \frac{1277 - 2.724\,M_d}{934.6}$$

f_{so}, the stress in the prestressing steel after elastic shortening, is

$$f_{so} = f_{si} - nf_{\text{c.g.s.}}$$

The initial stresses in the top and bottom fibers are

$$f_t = \frac{f_{so}A_{ps}}{A}\left(1 + \frac{e_{p1}y_t}{r^2}\right) + \frac{12M_d}{S_t}$$

and

$$f_b = \frac{f_{so}A_{ps}}{A}\left(1 + \frac{e_{p1}y_b}{r^2}\right) + \frac{12M_d}{S_b}$$

respectively.

The computations are summarized in the following table. The results at the 20th points (from the support to midspan) as well as at 6 and 7 ft from the

	Distance from End (ft)	M_d (k-ft)	e_{ps} (in.)	e_{pl} (in.)	$f_{c.g.s.}$ (ksi)	f_t (ksi)	f_b (ksi)	f_{so} (ksi)	P_o (k)
18 Strands A_{ps} = 2.754 sq in.	0.000	.000	−12.825	−12.825	2.163	−.398	2.675	158.713	437.097
	3.215	65.868	−12.825	−12.825	1.968	−.276	2.417	161.438	444.600
	6.430	124.840	−12.825	−12.825	1.794	−.166	2.186	163.875	451.313
	9.645	176.857	−12.825	−12.825	1.640	−.069	1.983	166.026	457.236
	12.860	221.939	−12.825	−12.825	1.507	.014	1.806	167.890	462.370
	16.075	260.085	−12.825	−12.825	1.395	.085	1.657	169.467	466.713
	19.290	291.295	−12.825	−12.825	1.302	.144	1.534	170.758	470.267
	22.505	315.570	−12.825	−12.825	1.231	.189	1.439	171.761	473.031
	25.720	332.909	−12.825	−12.825	1.180	.221	1.371	172.478	475.006
	28.935	343.312	−12.825	−12.825	1.149	.241	1.331	172.908	476.191
	32.150	346.780	−12.825	−12.825	1.139	.247	1.317	173.052	476.585
	0.000	.000	−12.825	−12.825	2.163	−.398	2.675	158.713	437.097
	6.000	117.400	−12.825	−12.825	1.816	−.180	2.215	163.567	450.465
	7.000	134.600	−12.825	−12.825	1.765	−.147	2.148	164.279	452.424
12 Strands A_{ps} = 1.836 sq in.	0.000	.000	−11.825	−11.825	1.367	−.233	1.789	169.851	311.847
	2.000	41.800	−11.825	−11.825	1.245	−.153	1.614	171.556	314.978

support are given for 18 strands; for 12 strands, the results are given at the support and at 2 ft from the support. Because the transmission length for $\frac{1}{2}$ in. strands is of the order of 25 in., the 12 strands can be bonded full length without exceeding the permissible stresses. All 18 strands can be fully bonded at 7 ft from the end without the stresses being excessive; in consideration of the transmission length the bottom 6 strands can be prevented from bonding to the concrete for 5 ft at each end.

2. A continuous cast-in-place post-tensioned bridge is to have the cross section shown in Example IV of Appendix B. Assuming the bottom slab thickness varies linearly from 5.75 in. to 12 in. over a length equal to 0.20 times the span length, compute the gross section properties for the cross-section at the 20th points between 0 and 0.20L.

SOLUTION: The results of the computations are summarized in the table.

20th Pt.	Bottom Slab Thickness (in.)	Area (sq in.)	Moment of Inertia (in.^{4})	y_t (in.)	y_b (in.)	Weight (plf)
0.00	12.000	9744.5	8847614	40.7	37.3	10,150
0.05	10.438	9313.4	8543251	39.5	38.5	9,700
0.10	8.875	8882.0	8169629	38.1	39.9	9,250
0.15	7.313	8450.1	7714397	36.5	41.5	8,800
0.20	5.750	8019.5	7161480	34.6	43.4	8,350

8 Continuity in Prestressed Concrete Flexural Members

8-1 Introduction

The theoretical reduction in moments, stresses and hence cost of materials, achieved through the use of continuity in prestressed concrete, is comparable to that which can theoretically be made with other structural materials. The actual economy of materials and cost of construction resulting from the use of continuity is greatly influenced by the design criteria used, the magnitude of the spans involved, the type of structure under consideration, the type of loading to be carried by the structure in question, and the methods of prestressing available.

The economy associated with the majority of prestressed structural elements would be non-existent if the elements were not precast. Although precast elements can be field-connected in order to form fully continuous prestressed members, this procedure has proved economical only under special conditions.

One of the important uses of cast-in-place, continuous prestressed concrete in the United States has been in the construction of roof and floor slabs. The slabs may be flat plates which are continuous in both directions or may be "one-way" slabs supported by beams or walls. Another important use of cast-in-place continuous prestressed concrete members has been in the construction of highway and railroad bridges. Continuity has been used extensively in cast-in-place long span bridge construction in the Western United States. The use of cast-in-place post-tensioned bridges is increasing in other parts of the country.

Precast, prestressed elements, which are rendered continuous by cast-in-place concrete and normal reinforcing steel, have been widely used. In this type of construction, the precast, prestressed elements act as simple beams resisting a portion of the dead load, but the live load, as well as the dead load which is applied after the hardening of the cast-in-place concrete, is carried by the continuous beam. This type of construction has proved to be economical in the American market, and it is expected that the use of this method will continue. The method results in a structure with the fundamental advantages of precast, prestressed concrete, as well as those resulting from the use of continuity. The basic structural analysis of this type of construction does not involve principles unfamiliar to structural engineers.

Continuous prestressed concrete spans have wide variation in their depth-to-span ratios. Continuous flat plate roof and floor slabs with depth-to-span ratios of 1 to 45 and 1 to 40, respectively, are not uncommon. Continuous cast-in-place box girder bridges are frequently constructed with depth-to-span ratios of the order of 1 to 25, while similar structures erected with the cantilever technique (*see* Sec. 8-3) frequently have depth-to-span ratios from 1 in 18 to 1 in 25.

The greater rigidity of continuous prestressed members also results in less vibration from moving or alternating loads. A significant advantage gained through the use of continuity in prestressed concrete construction, as in the case with other materials, is the reduction of deflections. Over-all structual stability and resistance to longitudinal and lateral loads is normally improved through the introduction of continuity, in any structural system.

8-2 Disadvantages of Continuity

The construction of cast-in-place, continuous, prestressed-concrete box girder bridges of conventional design do not present any unusual or serious construction problems. The more sophisticated methods of constructing continuous prestressed bridges, such as cantilevered construction (*see* Sec. 8-13), require a great deal of technical skill on the part of the contractor. Additionally, these methods would appear to require more construction labor than is required with conventional construction methods. These factors have deterred the use of the more sophisticated methods of constructing continuous prestressed concrete members in the United States. Except for very special cases, it is expected this situation will continue as the industry seeks methods of reducing the amount of labor, as well as the degree of skill required in construction.

Many engineers are under the impression that continuous prestressed structures are difficult to design and analyze, because of the moments that result from the deformation of the structure during prestressing. As will be seen in the following discussion, the analysis of continuous prestressed structures is not particularly complex and involves only the familiar principles used in the analysis of ordinary statically indeterminate structures.

As will be explained in detail in Secs. 8-5 and 8-6, the prestressing of a continuous structure induces moments and reactions in the structure. For simplicity in this book, these moments and reactions are referred to as the "secondary moment (or reaction) due to prestressing" or simply the secondary moment. Other authors use other terms for these moments and forces.

8-3 Methods of Framing Continuous Beams

There are a number of different methods that have been used or suggested for framing continuous beams. The possible variations are infinite and are limited only by the imagination of the designer. In general, these methods can be divided into three categories, cast-in-place structures, structures formed of precast elements, which may or may not have cast-in-place joints, and supplementary elements and members constructed of precast or cast-in-place segments utilizing the cantilever method of erection.

The cast-in-place, structure is generally used on the longer spans where the use of falsework is feasible and where precast beams cannot be used with ease, because of the dead weight of the beam and resulting difficulties in transportation and erection. Such construction is generally confined to one of the five types illustrated in Fig. 8-1. The construction can be described briefly by the following characteristics.

(1) Prismatic beam with curved tendon (Fig. 8-1(a)). This type of beam is simple to analyze, since the moments can be computed without the mathematical complications that result from a variable moment of inertia. This type of beam requires relatively simple formwork and has a relatively high friction loss. In this mode of framing, the concrete is not used as efficiently as it might be, but the over-all economy may remain good in spite of this, due to the simplicity of the member. Designs of this type normally have good appearance.

(2) Beams with variable depth and straight tendons (Fig. 8-1(b)). The structural analysis of such a beam is complicated by the variable moment of inertia and difficulty is frequently experienced in determining the beam proportions that will result in the desired balance of moments, due to dead and live loads, in combination with the moments, due to prestressing. Although this shape of beam results in minimum friction loss in the tendons for a specific length of structure, the formwork required to construct beams of this type is somewhat more complicated than that of prismatic beams. In addition, it is frequently difficult to achieve the required eccentricity of prestress at the several critical sections in this type of framing without resorting to impractical amounts of haunching.

(3) Beams with variable depth and curved tendons (Fig. 8-1(c)). This type of beam, which is among the more complex to analyze, frequently results in the

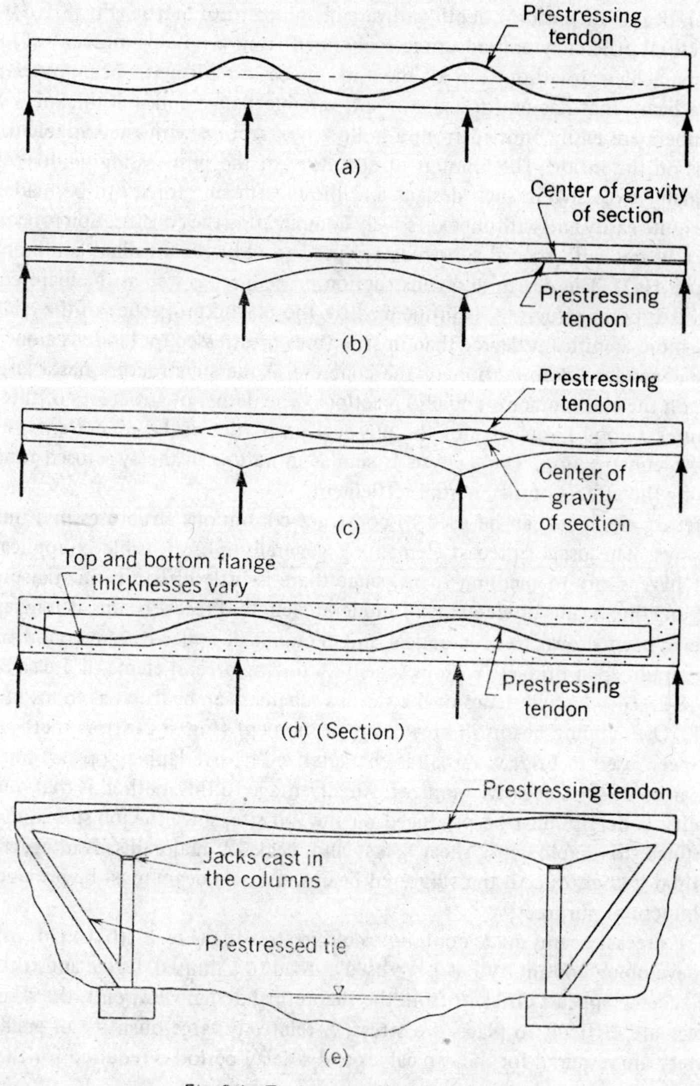

Fig. 8-1 Typical monolithic framing.

best solution, since the curvature (and hence friction loss) required is generally not as severe as in the case of the prismatic beam, and the proportions of the beam and tendon trajectory can be adjusted during the design, in order to obtain an efficient use of materials.

(4) Beams of uniform depth and variable moment of inertia (Fig. 8-1(d)). The structural efficiency and advantages obtained from a variable moment of inertia can be achieved without loss of the appearance of a prismatic beam by employing a beam that has uniform depth and variable flange and web thickness. Such members are easily formed from a hollow box section with the variable dimensions on the inside. The amount of curvature of the prestressing tendons is not normally excessive in such designs and the prestressing force can be made variable quite easily and without excessively complicating the construction procedure.

(5) Beams with special substructures used to induce controlled end moments (Fig. 8-1(e)). The design and construction procedure, as well as the maintenance of this type of structure, is influenced by the plastic properties of the concrete to a more significant degree than in structures prestressed by tendons alone. The time-dependent deformation of the concrete in the substructure has a large effect on the end moments and end reactions. The design of structures of this type requires a good knowledge of the physical properties of the concrete to be used in the construction. The analysis is similar in nature to the type used in determining the effects of differential settlement.

Precast elements can be used to construct continuous structures in a number of ways. The use of precast elements is generally more feasible in applications that have short to medium spans, since there is little difficulty in transporting and erecting beams or elements of moderate length. Precast beams as long as 250 ft have been shown to be practical and economical under certain conditions on large multi-span bridges. Various schemes utilizing precast elements are shown in Fig. 8-2. The characteristics of the various schemes can be listed as follows:

(1) Overlapping beams in areas of high moment (Fig. 8-2(a)). A method that has been used in bridges abroad is characterized by overlapping precast beams in the areas of high negative moment. An advantage to this method is that no high-quality concrete must be produced on the job site, since the job site concrete is confined to diaphragms, shear keys, and deck. A major disadvantage to the method is that, due to the staggered beams, a fascia beam must be provided for architectural purposes.

(2) Precast beams made continuous by prestressing (Fig. 8-2(b)). This method of developing continuity has been used abroad to a limited degree and, although the scheme appears efficient from the theoretical design viewpoint, the short cap cables are difficult to place and stress. A relatively large quantity of anchorage devices are required for the cap cables and a delay period is required for the cast-in-place concrete to gain sufficient strength for stressing.

(3) Continuity developed by ordinary reinforcing steel (Fig. 8-2(c)). This method is equally applicable to prismatic beams and beams with variable depths, and it has been used extensively in the United States in bridge construction. The cast-in-place joint is required to assure uniform stress transfer at the section of negative moment. In order to attain economy of materials with this method, when used with prismatic precast beams, it is necessary to use curved tendons or

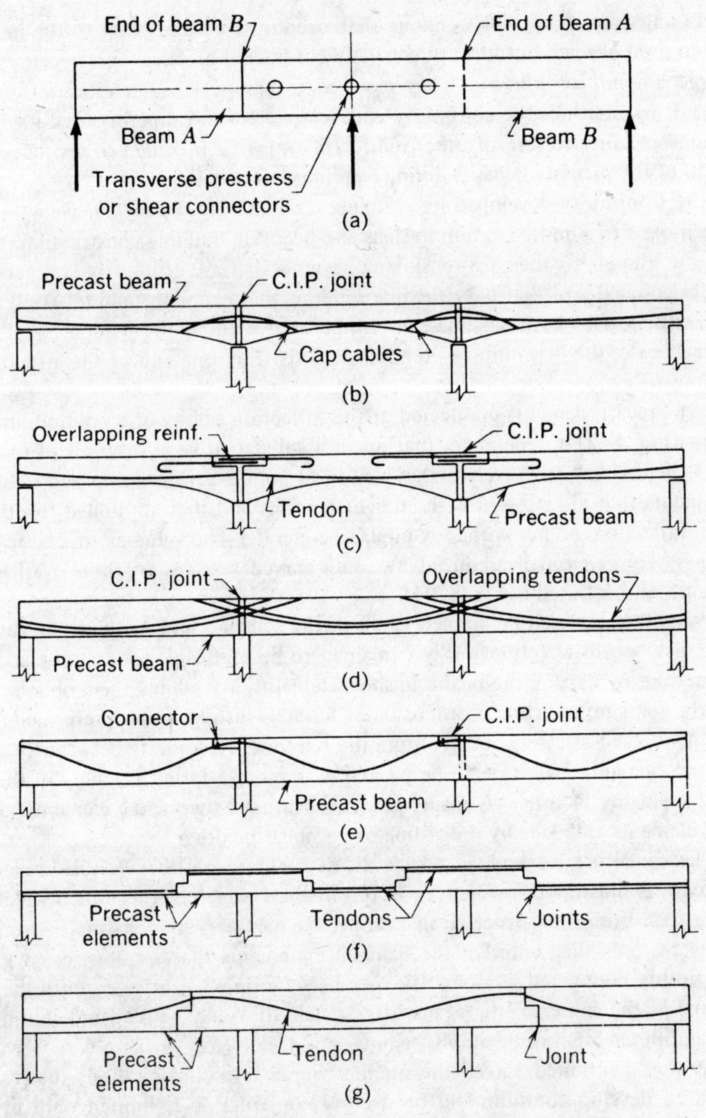

Fig. 8-2 Continuous beams formed of precast elements.

bond prevention as a means of reducing the bottomfiber stresses, due to prestressing at the ends of the precast member. No job site post-tensioning is required after erection of the precast elements with this method of framing.

(4) Continuity due to overlapping tendons (Fig. 8-2(d)). The method of de-

veloping continuity with tendons that overlap from one beam to the next has been used abroad, but it is not generally considered practical here because of the large amount of job site labor required to erect, thread tendons, stress, and grout. In addition, the continuity is developed for the superimposed load only, and some prestressing or other reinforcing must be provided to resist the dead load of the precast elements during handling and erection.

(5) Continuity developed by coupling tendons (Fig. 8-2(e)). This method has been used to a limited extent in slabs and beams in building construction, but its use is limited to the post-tensioning systems that are efficiently and economically coupled and that have flexible tendons that can withstand relatively sharp curvatures. This basic scheme can also be used without the tendon couplers, in which case, the "continuity" tendon extends from one end of the structure to the other.

(6) Precast elements connected at the inflection points of a continuous stucture (Fig. 8-2(f)). Members that are not subject to large reversals of moment and, hence, have relatively stable inflection points, can be composed of simple elements that are stressed with straight tendons and that are joined together at the inflection points with cast-in-place concrete. The number of cast-in-place joints required can be reduced by using curved tendons to form overhanging beams, as illustrated in Fig. 8-2(g).

Precasting can also be applied to all of the schemes and methods indicated in Fig. 8-1, which were previously explained to be praticable for cast-in-place construction, by casting the beams in short elements, assembling them on falsework at the job site, inserting post-tensioning tendons through ducts preformed in the elements, and stressing and grouting the tendons. The joints between the individual elements may have to be packed with mortar during assembly of the precast elements in order to ensure uniform bearing between the elements, unless the elements are made by precasting one against the other.

The cantilever method, in which the segments can either be cast-in-place or precast, is illustrated in Fig. 8-3. In this method, which has been used widely on long span bridge construction in Europe, the members are constructed from the piers towards the center of the spans in increments that are prestressed to the previously completed sections. The members are simple cantilevers until they are joined at the center of the spans, after which, they may be rendered continuous by additional tendons or simply be hinge-connected (Ref. 1).

As was mentioned above, the number of schemes and methods that can be used to develop continuity in prestressed construction is limited only by the imagination of the designer. This is the primary advantage of concrete construction, since the designer need not confine his thinking to standard shapes. A complete discussion of all the schemes that have been used in developing continuity is beyond the scope of this book. The interested reader will find considerable detailed descriptions of various other methods in the literature pertaining to prestressed concrete.

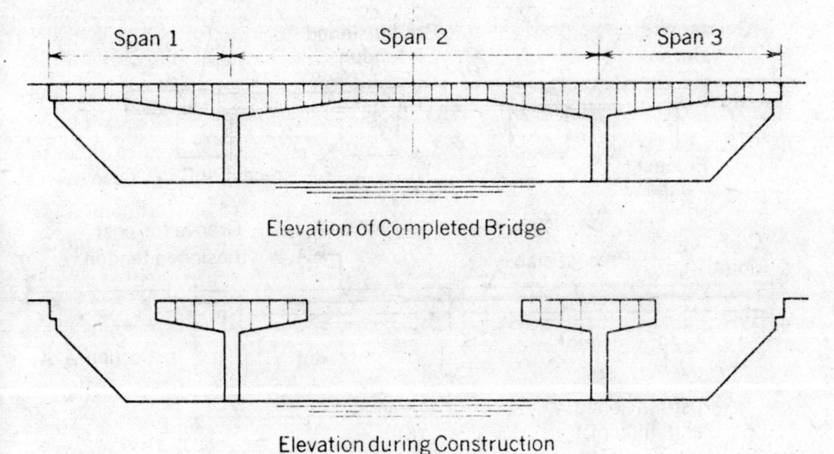

Fig. 8-3 Cantilever bridge construction.

8-4 Continuous Prestressed Slabs

The methods of fabricating continuous prestressed slabs can also be divided into cast-in-place monolithic slabs and precast slabs, in order to simplify considera-tion of these structural elements. The cast-in-place monolithic slabs, which may be cast-in-place in the conventional manner or which may utilize the lift-slab technique, are most frequently one of the following:

(1) Solid slabs of uniform thickness prestressed in one or in each direction Slabs of this type are normally from 6 to 10 in. thick and are post-tensioned with curved tendons.

(2) Monolithic joist and slab construction.

(3) Two-way, monolithic joist and slab construction (waffle slabs).

(4) Hollow slabs of variable or uniform dimensions.

Monolithic slabs are generally prestressed with the tendons placed in the same general shape that is used in monolithic beams and, from the designer's view-point, can be considered a special case of a continuous beam.

Precast continuous slabs have been used extensively. The slabs frequently con-tain prestressed and ordinary reinforcing and can be divided into the following general types:

(1) Solid and hollow slabs, reinforced for positive moments with ordinary reinforcing or, in some instances when limit design is used, in which the ends of the members are partially restrained by post-tensioned tendons, which pass through ducts or grooves in the members. These are illustrated in Fig. 8.4(a), (b), and (c).

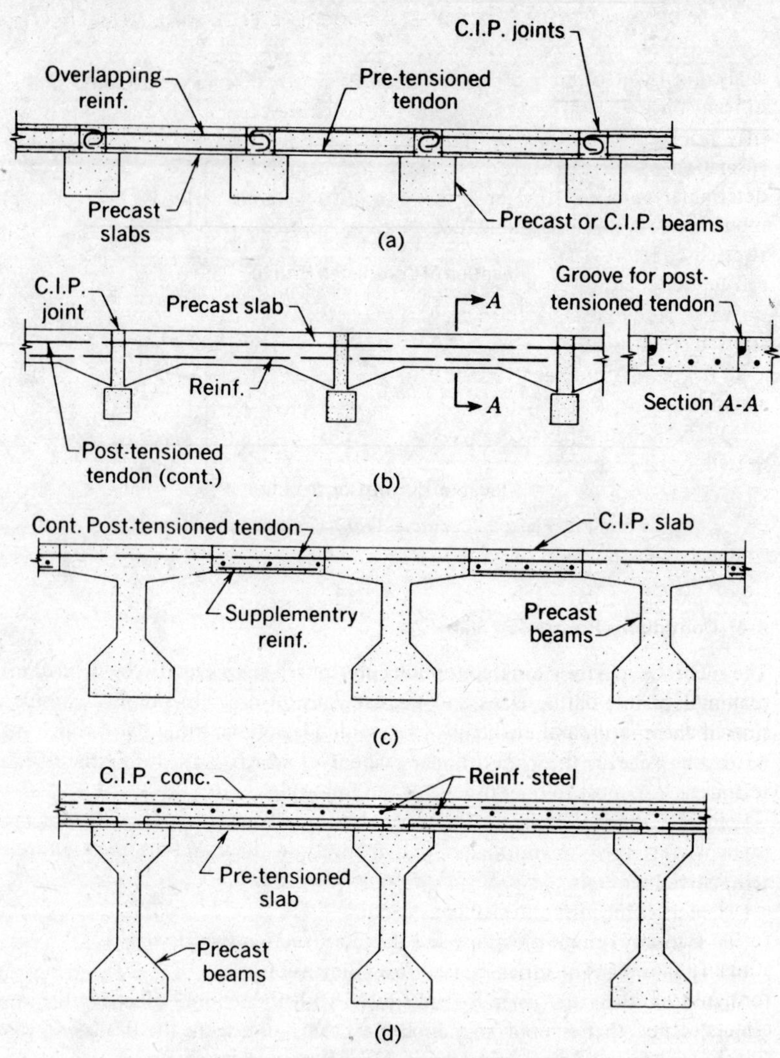

Fig. 8-4 Various types of continuous prestressed slabs.

(2) Slabs, such as bridge decks, which are formed of precast units that carry the positive moments, and cast-in-place elements, which transfer shear and carry negative moment. This is illustrated in Fig. 8-4(d).

8-5 Elastic Analysis of Beams with Straight Tendons

The moment due to dead and live loads in an indeterminate prestressed-concrete structure are calculated using the same classical methods employed in analyzing indeterminate structures composed of other materials. The one significant difference in a prestressed structure is that secondary moments may or may not result from the prestressing. These moments are the result of the deformation of the structure and are also calculated by the usual methods of indeterminate analysis. In most areas of structural design, the term secondary moments denotes undesirable moments that are to be avoided if possible. In prestressed concrete design, the secondary moments are not always undesirable and can be quite helpful. It is essential that the designer be aware that such moments do exist and that they must be included in the design of indeterminate prestressed structures.

In the design and analysis of continuous prestressed beams, the following assumptions are generally made:

(1) The concrete acts as an elastic material within the range of stresses permitted in the design.

(2) Plane sections remain plane.

(3) The effects of each cause of moments can be calculated independently and superimposed to attain the result of the combined effect of the several causes (the principle of superposition).

(4) The effect of friction on the prestressing force is small and can be neglected.

(5) The eccentricity of the prestressing force is small in comparison to the span and, hence, the horizontal component of the prestressing force can be considered uniform throughout the length of the member.

(6) Axial deformation of the member is assumed to take place without restraint.

Research into the performance of continuous prestressed-concrete beams has revealed that these assumptions do not introduce significant errors in normal applications. If the cracking load of a beam is exceeded, and in cases where the effect of friction during stressing is significant, special attention should be given to the effects of these conditions. The axial deformation that results from prestressing can have a significant effect on the moments and stresses when such deformations are restrained, as in rigid frames; hence, special investigation into the effects of this phenomenon may be required. Some of these effects will be considered subsequently, but for the general discussion which follows, the above assumptions will be assumed to be valid.

The magnitude and nature of secondary moments can be illustrated by considering a two-span, continuous, prismatic beam that is not restrained by its supports, but which must remain in contact with them, as is illustrated in Fig. 8-5(a). This beam is prestressed with a straight tendon that has a force of P and eccentricity of e. This would tend to make the beam deflect away from the center support by the amount

$$\delta = \frac{Pe(2l)^2}{8EI} = \frac{Pel^2}{2EI} \tag{8-1}$$

Because the beam must remain in contact with the center support, a downward reaction must exist at the location of the center support to cause an equal but

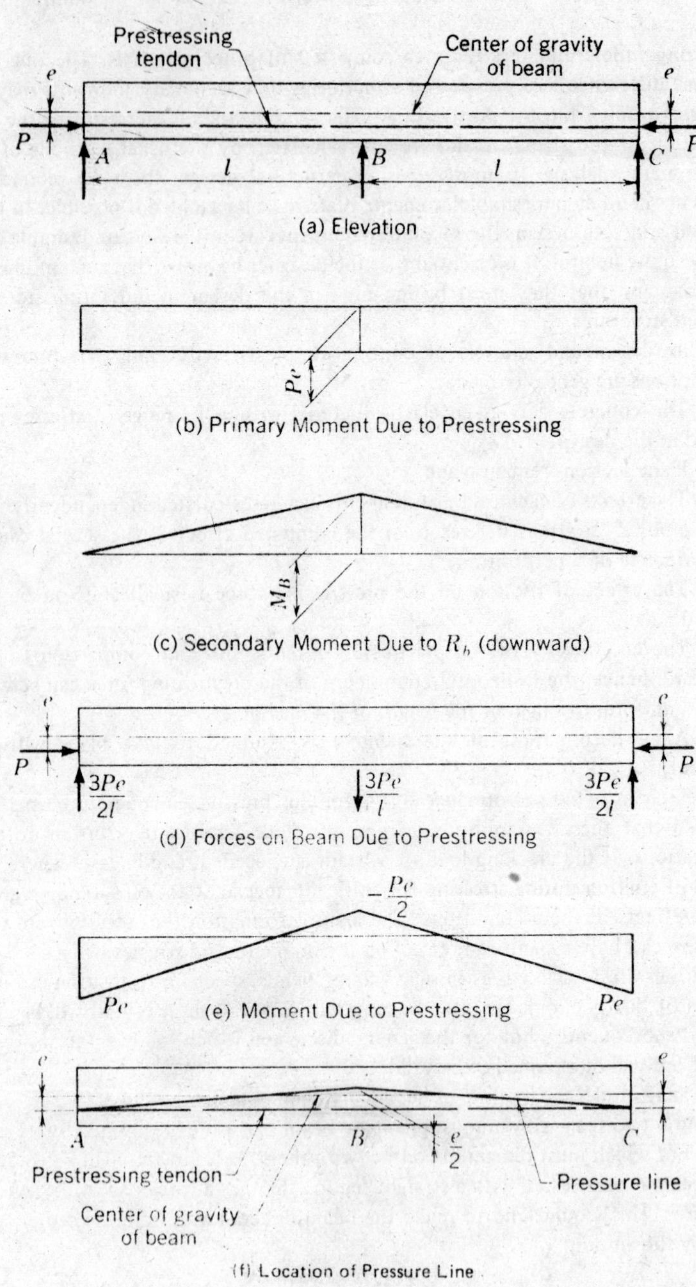
Prestressing
tendon
Center of gravity
of beam
e
e
P
P
A
B
C
l
l
(a) Elevation
Pe
(b) Primary Moment Due to Prestressing
M_B
(c) Secondary Moment Due to R_b (downward)
e
e
P
P
3Pe/2l
3Pe/l
3Pe/2l
(d) Forces on Beam Due to Prestressing
-Pe/2
Pe
Pe
(e) Moment Due to Prestressing
e
e
A
B
C
e/2
Prestressing tendon
Pressure line
Center of gravity
of beam
(f) Location of Pressure Line

Fig. 8-5

opposite deflection. The deflection at the center of a beam which has a span of $2l$, due to a concentrated load (R_b) applied at the center, is equal to

$$\delta = \frac{R_b l^3}{6EI} \tag{8-2}$$

Since the deflections must be equal in magnitude, by equating Eq. (8-1) and (8-2), the value of R_b is found to be

$$R_b = \frac{3Pe}{l} \tag{8-3}$$

Therefore, by applying the rules of statics, it can be shown that the forces that are acting upon the beam must be as shown in Fig. 8-5(d) in order to maintain equilibrium, and the moment diagram due to prestressing alone is as shown in Fig. 8-5(e) (Ref. 2).

By dividing the moment to prestressing at each section by the prestressing force P, the eccentricity of the pressure line, e, can be found and plotted (Fig. 8-5(f)). At the ends, as would be expected, the pressure line is seen to be coincident with the location of the prestressing force, while at the center support, the pressure line is at $e/2$ above the center of gravity of the section. The effect of the secondary reaction, R_b, at the center support has been to move the pressure line from an eccentricity of e below the center of gravity of the section to an eccentricity of $e/2$ above the center of gravity. Since there are no additional loads between the end support and the center support, the pressure line is a straight line, as shown in Fig. 8-5(f). It will be seen from this example that the secondary moment is secondary in nature, *but not in magnitude.*

As the above example illustrates, if the prestressing results in a deformation of the structure, (there are cases where this does not occur) secondary reactions as well as secondary moments are the result. Because the secondary reactions and secondary moments *are functions of each other, the secondary moments vary linearly between the supports*. The addition of the secondary moment, as shown in Fig. 8-5(c), to the prestressing moment diagram, Fig. 8-5(b), results in the *total moment diagram* resulting from the prestressing Fig. 8-5(e). Because the secondary reactions can only cause a moment which varies linearly, the effect of the secondary moment is to displace the pressure line linearly from the center of gravity of the steel, in direct proportion to the distance from the supports. It is also apparent from this example that, if the reactions which result from the prestressing are known, the location of the pressure line can be determined and the stresses due to prestressing at any point can be determined thereby.

The stresses due to prestressing in continuous beams are calculated from the basic relationship

$$f = \frac{P}{A}\left(1 + \frac{ey}{r^2}\right) \tag{8-4}$$

in which e is the *eccentricity of the pressure line and not the eccentricity of the tendon* (although it may be both, as will be seen). An important axiom illustrated by the above is that *the pressure line due to prestressing is not necessarily coincident with the center of gravity of the steel in indeterminate prestressed structures.*

If the position of the tendon in the above example is revised so that it is coincident with the location of the pressure line, which was computed above, the resulting tendon location and moment diagram due to prestressing alone would be as shown in Fig. 8-6(a) and (b), respectively. Removing the reaction at B, in order to render the structure statically determinate, and using the principle of elastic weights, the reactions and forces acting on the beam are as shown in Fig.

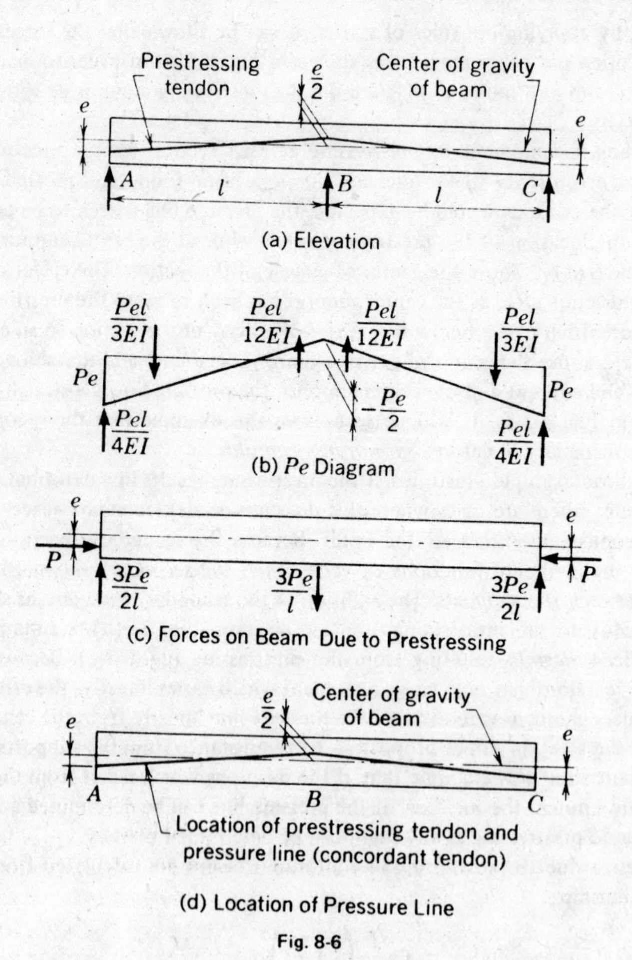

Fig. 8-6

8-6(c). The deflection of the beam at B due to the prestressing is equal to the moment at B resulting from the elastic weights or

$$\delta = \frac{Pel}{4EI} \times l - \frac{Pel}{3EI} \times \frac{7l}{9} + \frac{Pel}{12EI} \times \frac{l}{9}$$

$$= \frac{27Pel^2}{108EI} - \frac{28Pel^2}{108EI} + \frac{Pel^2}{108EI} = 0$$

Since the deflection due to prestressing at B is equal to zero, no secondary reaction is required at B to keep the center support in contact with the beam, and there are no secondary moments. In this example, the pressure line and the center of gravity of the steel are coincident and the tendon is said to be "concordant." In the first example, the pressure line and the center of gravity of the prestressing were not coincident and the tendon is said to be "nonconcordant."

If the tendon is placed in the trajectory shown in Fig. 8-7(a), the elastic weights are as shown in Fig. 8-7(b) and the reactions and moments due to prestressing are found to be as shown in Fig. 8-7(c) and (d). It will be seen that in this case the moment diagram due to prestressing, and hence the location of the pressure line, is identical to that in the above two examples. This tendon is also "non-concordant."

This series of three examples illustrates several principles that are extremely useful in the elastic design and analysis of statically indeterminate prestressed structures. In the three examples, the only variable is the eccentricity of the prestressing tendon at the center support. The force in the tendon, P, as well as the eccentricity of the tendon, e, at the end supports was held constant. The inspection of the three solutions reveals that the moments on the concrete section which resulted from each layout of the tendon are identical, although the secondary reactions and secondary moments are not equal. It is apparent, from inspection of the moment diagrams used in the three examples, that if the eccentricity at the end of the member were changed to another value, e', the moment due to prestressing and, therefore, the location of the pressure line, would be changed. Free-body diagrams for each of the three examples (Fig. 8-8) reveal that the net forces that act on the members (combination of components of the prestressing force and the secondary reactions) are identical for the three conditions of prestressing, as of course would be expected if the moments are equal.

The above examples also illustrate the very important principle of linear transformation, which can be defined as follows: *The path of the prestressing force in any continuous prestressed beam is said to be linearly transformed when the location of the path at the interior supports is altered without altering the position of the path at the end supports and without changing the basic shape (straight, curved, or series of chords) of the path between any supports.* Linear transformation of any tendon can be made without altering the location of the pressure line.

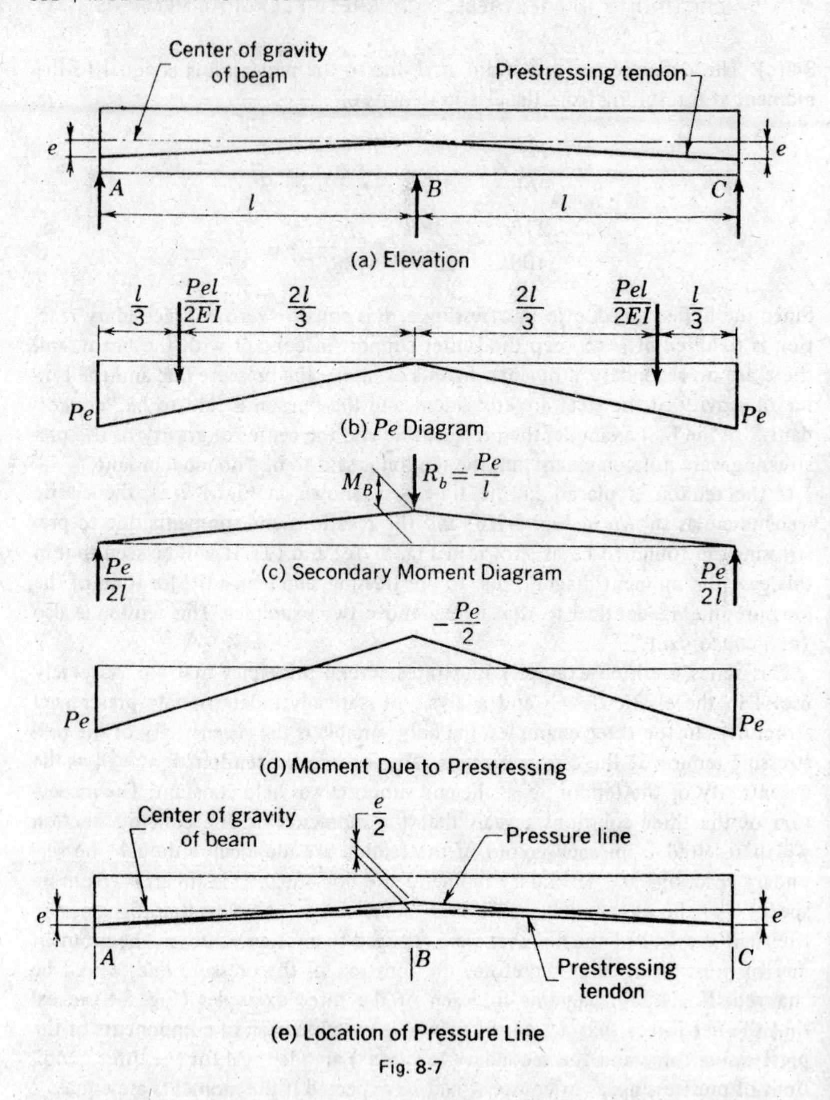

Fig. 8-7

The only difference between the three tendon paths in the above three examples is that they are displaced from each other, at every section, by an amount which is in direct (linear) proportion to the distance of the section from the end of the member. The eccentricity of the tendon at the end supports and the shape of the tendon between the supports were not changed. It will be subsequently shown that this principle of linear transformation is equally applicable to tendons that are curved and, from the above examples, it is apparent that the principle applies to concordant tendons as well as to non-concordant tendons.

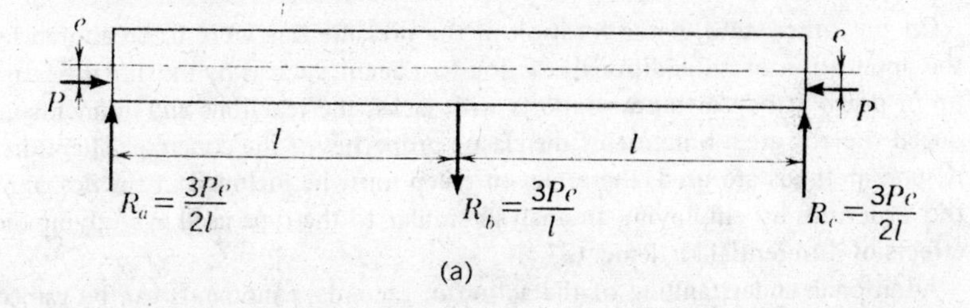

(a)

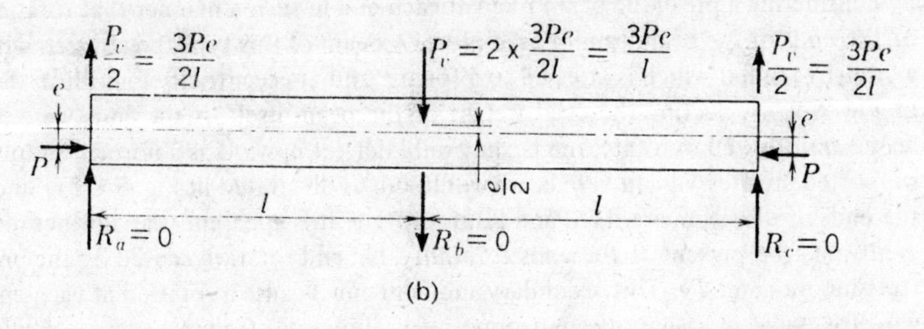

(b)

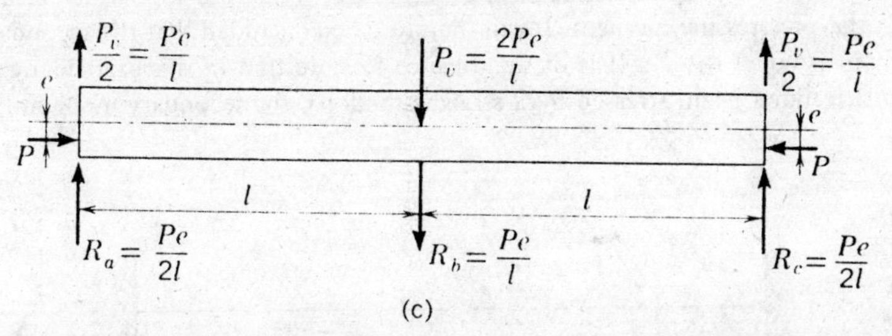

(c)

Fig. 8-8 Free-body diagrams with (a) straight tendon of eccentricity e, (b) tendon sloping from eccentricity of e at each end to $e/2$ above the center support, and (c) tendon sloping from eccentricity of e at the ends to 0 at the center support.

The principle of linear transformation is particularly useful in designing continuous beams when it may be desirable to adjust the location of the tendon in order to provide more protective cover over the prestressing tendons, without altering the location of the pressure line.

Another significant principle, apparent from these examples, is that the location of the pressure line in beams stressed by tendons alone is not a function of the elastic properties of the concrete. The elastic modulus of the concrete did not appear in the values of the secondary reactions. Therefore, the only effect of changes in the elastic properties of the concrete (i.e., creep) is a reduction in the magnitude of the prestressing force, just as it is in statically determinate structures. This has been proved by tests (Ref. 3).*

*This does not apply to structures built in a structural configuration that is different from the final one (*see* Sec. 8-13).

On the other hand, if the location of the pressure line were to be altered by the application of an additional reaction to a beam, such as by moving the beam up or down at one or more supports with jacks, the reactions and moments induced thereby are a function of the elastic properties of the concrete. Therefore, if such methods are used, the effect of creep must be included in the design of the structures by employing an analysis similar to the type used in studying the effects of differential settlement.

Additional understanding of the action of secondary moments can be gained by considering a prismatic beam fixed at each end in such a manner that rotation of the ends of the beam cannot take place. A beam of this type, prestressed with a straight tendon which is stressed to a force P and at eccentricity e, is illustrated in Fig. 8-9. Neglecting the dead weight of the beam itself, if the ends were released and allowed to rotate, the beam would deflect upward as a normal, simply-supported prestressed-concrete beam would do, as illustrated in Fig. 8-9(b). Since the ends of the beam are fixed and cannot rotate, it is apparent that another moment must be present at the ends to nullify the end rotation caused by the prestressing moment Pe. This secondary moment must cause a rotation at each end which is equal in magnitude, but opposite in direction, from that which results from the prestressing moment. It can therefore be concluded that the secondary moment is equal to $-Pe$. It is of significance to note that in this particular case (primatic fixed beam stressed by a straight tendon), the secondary moment has

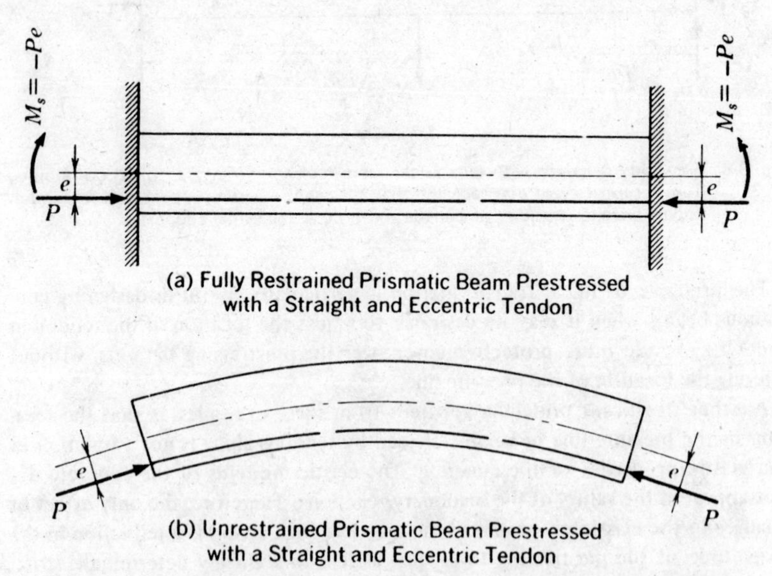

Fig. 8-9 Effect of end restraint on prismatic prestressed beams. (a) Fully restrained prismatic beam with a straight eccentric tendon. (b) Unrestrained prismatic beam with a straight eccentric tendon.

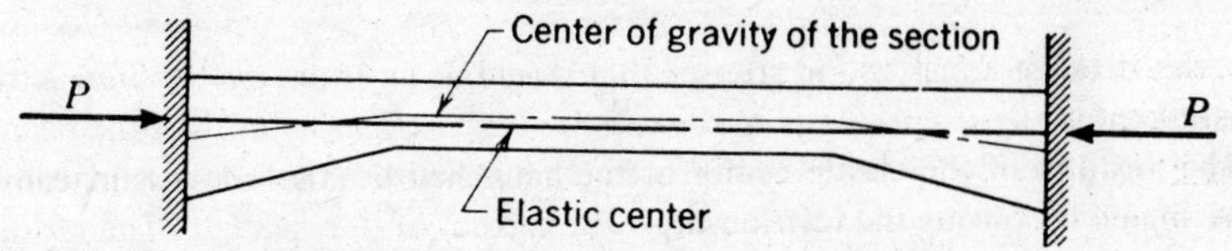

Fig. 8-10 Haunched beam of rectangular cross section prestressed with a straight tendon.

the effect of nullifying the eccentricity of the prestressing force and results in a uniform compressive stress due to prestressing at every cross section. Again in this example, as has been noted previously, the secondary moment is secondary in nature, but not in magnitude.

If a haunched beam of rectangular cross section, as shown in Fig. 8-10, is prestressed with a straight tendon located at the elastic center of the member, the prestressing will result in stresses of the proper direction for resisting negative moments at the ends of the beam and positive moments at the center of the beam, and there will be no secondary moments. *The elastic center is defined as the location through which a force may be applied without causing rotations at the ends of the beam.* Applying the same principles used in the above discussion of the prismatic beam, it can be shown that if the straight, prestressing tendon is not located at the elastic center, secondary moments will result in combined

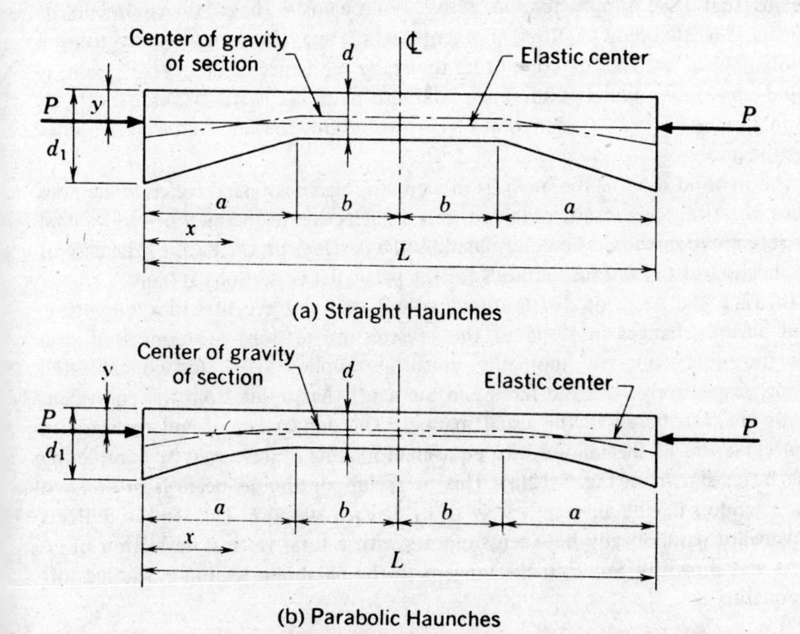

Fig. 8-11 Beams with (a) straight haunches and (b) parabolic haunches.

stresses that are equal to the stresses that would occur if the tendon were at the elastic center.

The location of the elastic center of the haunched beams of Fig. 8-11 can be determined by solving the relationship

$$\int_0^{L/2} \frac{e\,dx}{I} = 0 \tag{8-5}$$

The value of y, which satisfies Eq. 8-5, is the location of the elastic center.

The example given above for straight tendons are useful in all design problems, but are of particular use in short-span continuous slabs, such as those which occur in bridge decks, viaducts, and waterfront structures.

8-6 Elastic Analysis of Beams with Curved Tendons

The introduction of curvature to the tendon does not affect the basic methods or principles of analysis in any way; the calculation of the secondary reactions or moments can be made using the principle of elastic weights, the theorem of three moments, or other classical methods, if desired. The familiar moment distribution method is considered among the easier methods for analyzing prismatic beams that have simple tendon paths. For beams with variable moments of inertia, the theorem of three moments, which may be simplified by using a semigraphical method of computing the static moments of the M/I diagram, is rapid and easily understood. Each of these methods is used subsequently in problems, and a brief explanation of the procedures that are followed in applying the methods is given.

The method used in the analysis of a continuous prestressed-concrete member does not affect the results obtained, and the selection of the methods to be used in actual design should be determined by the designer on the basis of the ease of application of the various methods for the particular conditions at hand.

In using the moment distribution method, the end eccentricities, curvature, and abrupt changes in slope of the prestressing tendons are converted into specific, equivalent end moments, uniformly applied loads, and concentrated loads, respectively, and the fixed end moments that result from the equivalent loads are distributed in the usual manner. The conversion of end eccentricity and curvature of the tendon into equivalent loading is illustrated by considering the beam shown in Fig. 8-12(a). This two-span continuous beam is prestressed by a tendon having an eccentricity of e_1 at each support. The tendon deflects downward parabolically between supports with a total vertical deflection of e_t.

As was shown in Sec. 7-9, the tangent to the parabolic tendon at the support is equal to

$$\tan \theta = \frac{4e_t}{l}$$

Since the curvatures are small (*see* design assumptions in Sec. 8-5), $\tan\theta = \sin\theta = \theta$ and the vertical component of the tendon at each end of each span is:

$$V_p = P\sin\theta = \frac{4Pe_t}{l}$$

Therefore, the total vertical component of the prestressing force that acts on each span of the beam is equal to twice the force that acts at each end, and the equivalent uniformly applied load is found to be

$$wl = \frac{2 \times 4Pe_t}{l}$$

$$w = \frac{8Pe_t}{l^2} \tag{8-6}$$

The vertical components of the prestressing force that occur at the supports do not cause moments in the beam, but pass directly through to the supports, and, for this reason, these forces are disregarded in the equivalent loading. The horizontal component of the prestressing tendon is eccentric by an amount equal to e_1 at each end of the beam and the equivalent loading must, therefore, include end moments in the amount of Pe_1. The end moment Pe_1 is treated just as the moment due to a cantilevered end would be treated in the analysis. The equivalent loading is shown in Fig. 8-12(b).

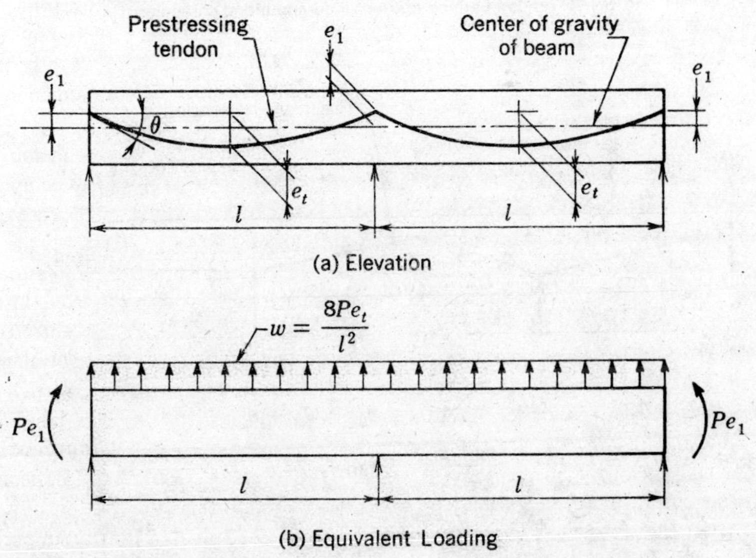

Fig. 8-12 (a) Beam continuous over two spans—prestressed with tendon on parabolic path. (b) Equivalent loading for beam in (a).

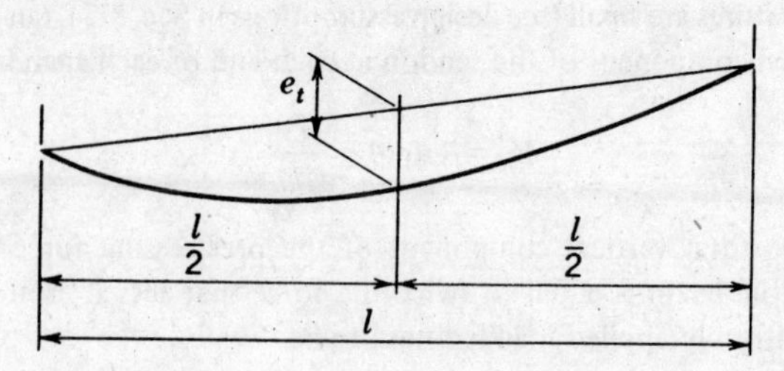

Fig. 8-13 Value of e_t to be used for parabolic tendon path terminated at different elevations. elevations.

When the tendon curves parabolically over the length of the span but with the ends of the curve at different elevations, the value of e_t and l to be used in Eq. 8-6 are as shown in Fig. 8-13. If the tendon trajectory is formed of compounded parabolas, as shown in Fig. 8-14, the equivalent loads due to tendon

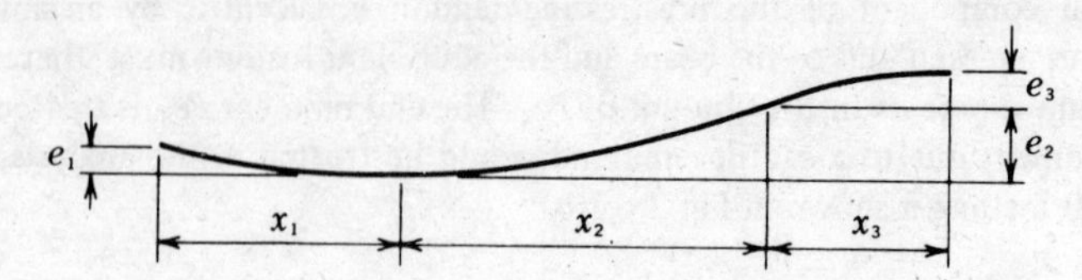

Fig. 8-14 Tendon path composed of compounded parabolas.

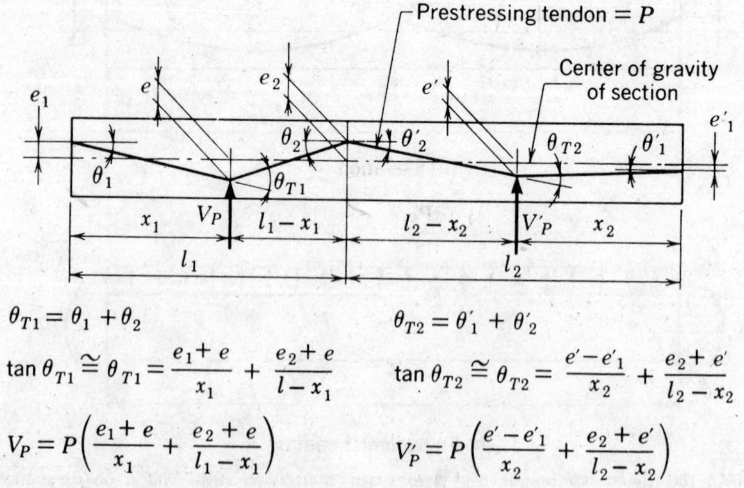

$$\theta_{T1} = \theta_1 + \theta_2 \qquad \theta_{T2} = \theta'_1 + \theta'_2$$

$$\tan\theta_{T1} \cong \theta_{T1} = \frac{e_1 + e}{x_1} + \frac{e_2 + e}{l - x_1} \qquad \tan\theta_{T2} \cong \theta_{T2} = \frac{e' - e'_1}{x_2} + \frac{e_2 + e'}{l_2 - x_2}$$

$$V_P = P\left(\frac{e_1 + e}{x_1} + \frac{e_2 + e}{l_1 - x_1}\right) \qquad V'_P = P\left(\frac{e' - e'_1}{x_2} + \frac{e_2 + e'}{l_2 - x_2}\right)$$

Fig. 8-15 Equivalent loads for tendons placed on a series of straight slopes.

curvature is computed by

$$w_n = \frac{2Pe_n}{x_n^2} \tag{8-7}$$

It is usually sufficiently accurate to assume all curves are parabolic even though they may be circular or of other shape. Since the eccentricity is normally small in comparison to the span, the error introduced by this assumption is small.

The vertical load resulting from an abrupt change in slope of the tendons is computed as follows:

$$V_p = P(\sin \theta_t) \cong P(\tan \theta_t)$$

The value of $\tan \theta$ is determined by the dimensions of the tendon trajectory, as is illustrated in Fig. 8-15.

ILLUSTRATIVE PROBLEM 8-1 Compute the moments due to prestressing and draw the pressure line for the prismatic beam shown in Fig. 8-16. Use moment distribution and the theorem of three moments.

SOLUTION:

$$\text{Equivalent uniform load} = w = \frac{8P_e}{l^2}$$

$$e = 0.80 + \frac{0.50 + 1.20}{2} = 1.65 \text{ ft} \qquad w = \frac{8 \times 500 \times 1.65}{(100)^2} = 0.66 \text{ k/ft}$$

Fixed end moments:

$$M_{AB}^F = -\frac{0.66 \times (100)^2}{12} = -550 \text{ k-ft} = M_{BC}^F$$

$$M_{BA}^F = +\frac{0.66 \times (100)^2}{12} = +550 \text{ k-ft} = M_{CB}^F$$

Overhang moments:

$$M_A = 0.5 \text{ ft} \times 500 = -250 \text{ k-ft}$$

$$M_C = +0.5 \text{ ft} \times 500 = +250 \text{ k-ft}$$

The distribution of moments is performed in Fig. 8-16 and the moment diagram due to prestressing is plotted in Fig. 8-16(d). The computed moment at B is 700 k-ft and the eccentricity of the pressure line is computed as follows:

$$e = \frac{700}{500} = 1.40 \text{ ft}$$

The pressure line is then 1.40 ft - 1.20 ft = 0.20 ft above the tendon at B, since the pressure line is linearly transformed from the tendon trajectory. At the

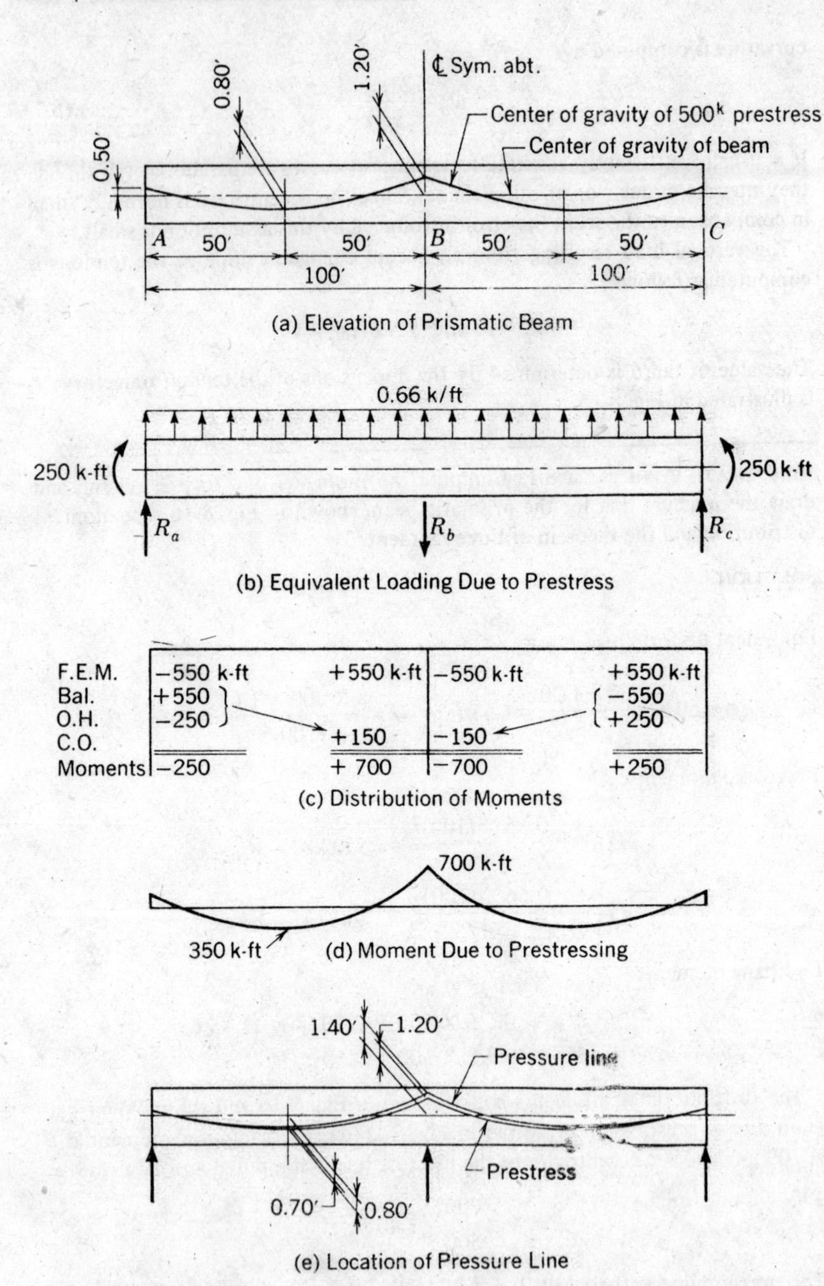

Fig. 8-16 Analysis of two-span continuous beam with moment distribution of equivalent loading.

center of the span, the pressure line is 0.10 ft higher than the tendon trajectory since $(50 \times 0.20)/100 = 0.10$ ft. This is shown in Fig. 8-16(e). Due to the symmetry of the beam and the loading, the three-moment equation for this example is reduced to

$$M_A + 4M_B + M_C = -\frac{wL^2}{2}$$

$$-250 + 4M_B - 250 = -\frac{0.66 \times (100)^2}{2}$$

$$M_B = -700 \text{ k-ft}$$

ILLUSTRATIVE PROBLEM 8-2 Compute the moments due to prestressing for the prismatic beam and condition of loading illustrated in Fig. 8-17. Use moment distribution and the theorem of three moments.

SOLUTION:

$$\text{Concentrated load in span } AB = 500\left(\frac{1.50}{60} + \frac{2.30}{40}\right) = 41.2 \text{ k}$$

$$\text{Concentrated load in span } BC = 500\left(\frac{0.80 + 1.60}{50}\right) = 24.0 \text{ k}$$

Fixed end moments:

$$M_{AB}^F = \frac{41.2 \times 60 \times (40)^2}{(100)^2} = -395 \text{ k-ft}$$

$$M_{BA}^F = \frac{41.2 \times (60)^2 \times 40}{(100)^2} = +593 \text{ k-ft}$$

$$M_{BC}^F = \frac{24 \times 50 \times (50)^2}{(100)^2} = -300 \text{ k-ft}$$

$$M_{CB}^F = +300 \text{ k-ft}$$

The moments are distributed in Fig. 8-17, and the moment at B is found to be 621 k-ft. The eccentricity of the pressure line at B is then

$$e = \frac{621}{500} = 1.24 \text{ ft}$$

Therefore, the pressure line is 1.24 ft - 0.80 ft = 0.44 ft above the location of the tendon at B. The distance from the pressure line to the tendon at points D and E are found from the principle of linear transformation as

$$\text{At point } D \qquad \frac{0.44 \text{ ft} \times 60}{100} = 0.264 \text{ ft}$$

$$\text{At point } E \qquad \frac{0.44 \text{ ft} \times 50}{100} = 0.220 \text{ ft}$$

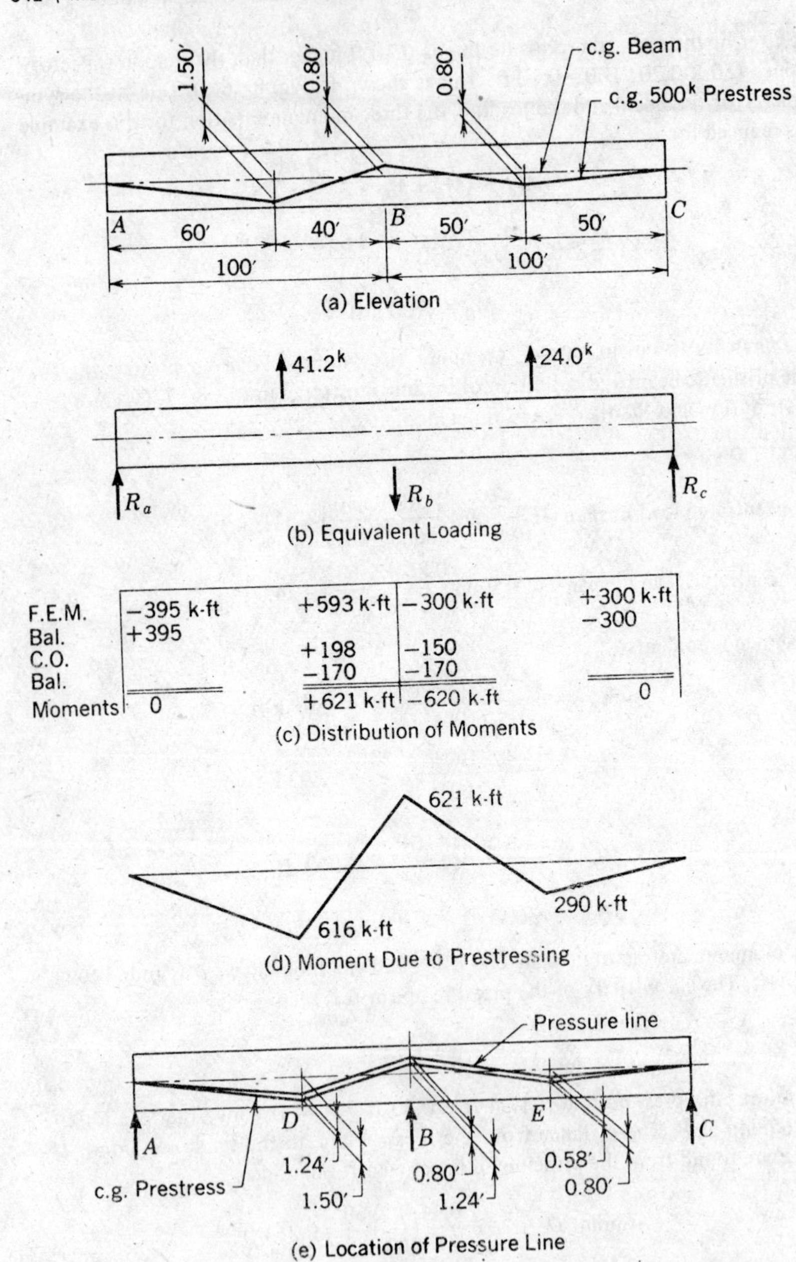

Fig. 8-17 Analysis of two-span continuous beam with moment distribution of equivalent loading.

Using the theorem of three moments, the computation of the moment at B for the above example is as follows:

$$M_A L_1 + 2M_B(L_1 + L_2) + M_C L_2 = -\frac{41.2 \times 60}{100}[(100)^2 - (60)^2]$$

$$-\frac{24 \times 50}{100}[(100)^2 - (50)^2]$$

$$400\, M_B = -158{,}000 - 90{,}000$$

$$M_B = -620 \text{ k-ft}$$

ILLUSTRATIVE PROBLEM 8-3 Compute the moments due to prestressing for the beam illustrated in Fig. 8-18. Note that the relative moment of inertia is 1.00 for the outermost 60 ft of each span and 1.15 for the center 40 ft of the beam. The center of gravity of the section is a straight line (the variable moment of inertia is the result of an abrupt change in web thickness or abrupt, symmetrical change in top and bottom flange thicknesses, or both).

SOLUTION: In Fig. 8-18(b), (c), (d), and (e) are plotted the Pe diagram, the Pe/I diagram, the assumed M_b diagram, and the M_b/I diagram, respectively. The magnitude of M_b can be determined rapidly by employing the principle of elastic weights. The former is used here by computing and equating the moments (deflections) at the center of the span AC of the beam loaded with the M/I diagrams for Pe and M_b.

The upward deflection due to Pe/I diagram (δpe) is

$$\begin{aligned}
-800 \times 50/2 &= -20{,}000 \times 46.70 = -933{,}000\\
-300 \times 10 &= -3{,}000 \times 25.00 = -75{,}000\\
-500 \times 10/2 &= -2{,}500 \times 26.70 = -66{,}700\\
-261 \times 6/2 &= -783 \times 18.00 = -14{,}100\\
+608 \times 14/2 &= +4{,}256 \times 4.67 = +19{,}800\\
& -22{,}027 \qquad\qquad -1{,}069{,}000
\end{aligned}$$

$$\delta_{pe} = +22{,}027 \times 80 \text{ ft} - 1{,}069{,}000 = 691{,}100 \text{ k-ft}^3/\text{in.}^4$$

The downward deflection due to M_b/I diagram (δM_b) is

$$\begin{aligned}
+0.75\, M_b \times 60/2 &= 22.5M_b \times 40.00 = +900.0M_b\\
+0.652M_b \times 20 &= 13.0M_b \times 10.00 = +130.0M_b\\
+0.218M_b \times 20/2 &= 2.2M_b \times 6.67 = +14.6M_b\\
& +37.7M_b \qquad\qquad +1044.6M_b
\end{aligned}$$

$$\delta_{Mb} = -37.7M_b \times 80 + 1044.6M_b = -1965M_b$$

Equating the two deflections

$$\delta_{pe} = \delta_{Mb}$$

$$M_b = \frac{691{,}000}{-1965} = -352 \text{ k-ft}$$

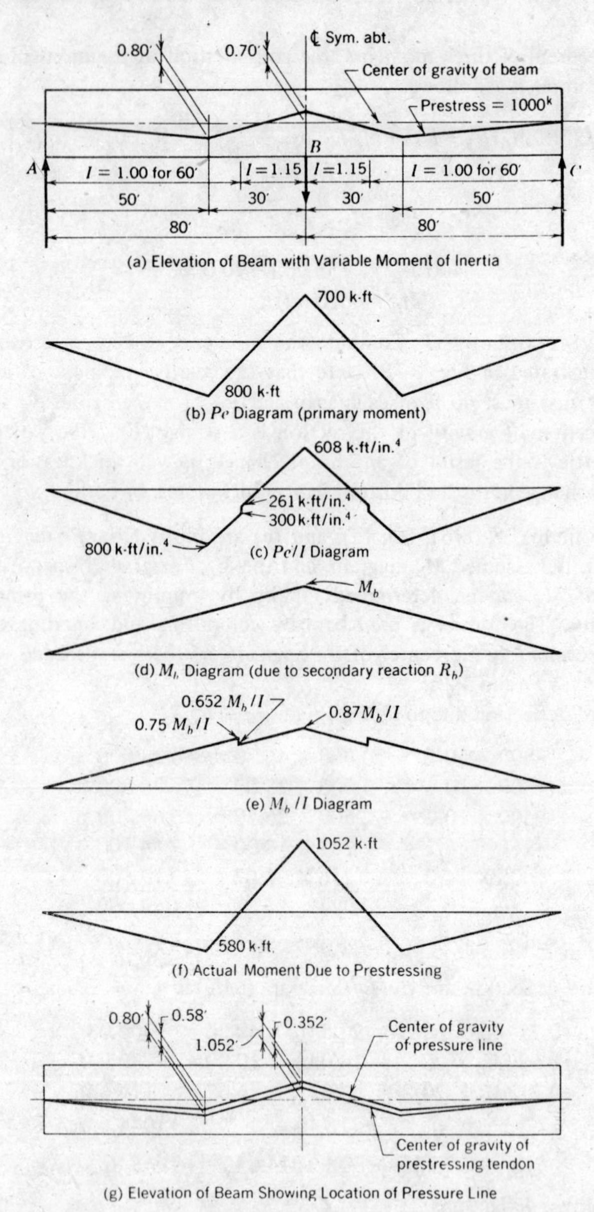

Fig. 8-18 Analysis of two-span continuous beam with a variable moment of inertia, using the principle of elastic weights.

The combined or actual moment diagram due to prestressing and the location of the pressure line are as shown in Fig. 8-18(f) and (g) respectively.

It should be noted that the total moment due to prestressing and not the secondary moment was computed in Illustrative Problems 8-1 and 8-2. When equivalent loads are used for the analysis of moment due to prestressing, the total moment due to prestressing is computed. In Illustrative Problem 8-3, the secondary moment and not the total moment due to prestressing was computed. The secondary moment due to prestressing is computed when the effects of prestressing are analyzed, using the basic principles of indeterminate structural analysis with the primary moment due to prestressing being considered as the initial loading condition.

With the general availability of electronic computers and programmable calculators, the solution of continuous prestressed concrete structures is greatly facilitated. These devices particularly facilitate the solution of members with variable prestressing forces, variable moments of inertia, and variable eccentricity. There is no longer a need for engineers to avoid the inclusion in their analyses affects such as the variation in the prestressing force (due to friction) or other causes.

8-7 Additional Elastic-Design Considerations

From the previous discussions, it should be recognized that in applying the moment distribution method in the analysis of the moments due to prestressing, the effect of the prestressing is analyzed by determining an equivalent loading that is a function of the eccentricity of the tendon at the end supports and of the intrinsic shape of the tendon between supports, but is independent of the eccentricity of the tendon at the interior supports. There is a family of curves or trajectories that will result in this equivalent loading but, for this loading, there will be only one pressure line. It should also be apparent that the pressure line that results from the distribution of the moments, due to the equivalent loading, is the location of the concordant tendon for the particular condition of end eccentricities and intrinsic shape of the tendon. Finally, it should be recognized that the location of the pressure line is determined by dividing the moment diagram, resulting from the equivalent loading, by the effective prestressing force. From these considerations, the following corollary should be apparent: *A moment diagram for a particular beam, due to any condition of loading on the beam, defines a location for a concordant tendon and, if a tendon is placed on a trajectory that is proportional to any moment diagram for the beam, the tendon will be concordant.* This principle is useful in selecting a trial trajectory for a tendon for which it is not necessary to compute the effects of secondary moments.

From the discussion of Sec. 8-5, it should be apparent that revision of the eccentricities of the tendon at the end support, without revision of the shape of the tendon between supports, results in a linear change in the pressure line. This is evident when the pressure lines of the example problems are studied. From this, it is seen that if the end eccentricity is changed from e to e' below the center of gravity, the location of the pressure line at the center support is changed from $e/2$ to $e'/2$ above the center of gravity; moreover, the location of the pressure line is independent of the eccentricity of the tendon at the interior support. This principle is useful in the design of continuous beams, when it is desired to modify the location of the tendon and the pressure line. *The revision of the end eccentricity can be treated as a linear adjustment in the pressure line.*

From the discussions of linear transformation in Sec. 8-5, from the principles of elastic design that are the basis for the analysis of statically indeterminate structures, and from the discussion at the beginning of this article, the following axioms and corollaries applicable to most continuous beams can be stated (*see* Sec. 8-13 for an exception):

(1) For any particular beam, a specific shape or path of the prestressing tendon between the supports will result in a specific pressure-line location for each condition of eccentricity at the end supports.

(2) Alteration of the eccentricity at one or both end supports will result in a shift in the location of the pressure line for any particular continuous beam.

(3) Alteration of the eccentricity of the tendon path at the interior supports will not affect the location of the pressure line, if the eccentricities at the ends of the beam are not changed and if the intrinsic shape of the tendon path is maintained.

(4) The conditions of continuity require that the deflection of a beam at the supports is equal to zero and that the end rotations of each span that adjoin at a common interior support are equal. These conditions apply equally to the effect of the prestressing force (pressure line) and to the effect of external loads.

(5) The moments and the location of the pressure line due to prestressing can be determined by resolving the primary prestressing moment into equivalent loads and end moments, then calculating the resulting moment in the continuous structure.

(6) The location of a pressure line is directly proportional to the moment in the continuous beam which results from a specific condition of (equivalent) loading.

(7) Since the pressure line defines the location of a concordant tendon, which has specific end eccentricities and shape between supports, a tendon path can be made to be concordant, if it is located along a path proportional to the moment diagram that results from any specific loading for the beam under consideration.

(8) The location of the pressure line determined for any non-concordant tendon path is the pressure-line location that will result from all non-concordant

tendons in the particular beam, if the eccentricity of the tendons are equal at the end supports and the basic shape of the tendon is not changed between supports.

(9) The location of the pressure line determined for any non-concordant tendon path is the only location for a concordant tendon for the particular beam, basic shape of the tendon path between supports, and eccentricities at the end supports.

(10) The combination of two concordant tendon paths will result in a concordant tendon, since the principle of superposition applies and each concordant tendon satisfies the requirements of continuous structures, i.e., the deflection at the supports is zero and the end rotations of adjacent spans are equal at common supports.

(11) Superposition of a non-concordant tendon with a concordant tendon will result in a non-concordant tendon that has a pressure-line location which can be determined by superimposing the path of the concordant tendon on the pressure line resulting from the non-concordant tendon.

It should be pointed out that concordant tendon locations are not more desirable than non-concordant tendon locations. It is a fact that non-concordant tendon locations often result in a more efficient design, since the protective cover over the prestressing steel may be greater for non-concordant tendon locations than for equivalent concordant tendon locations. It is often desirable to start a design with an assumed tendon trajectory that is concordant, in order to avoid the necessity of determining secondary moments. The tendon location can then be linearly transformed, or otherwise altered, into a non-concordant tendon that may result in a better design.

8-8 Elastic Design Procedure

It is difficult to generalize on the best procedure to be followed in the complete design of a structure, since there are so many considerations that must be taken into account. These include the theoretical elastic-design considerations, as well as practical construction and economic considerations, such as the methods of construction that are feasible on the site, clearance requirements, etc. Therefore, the procedure outlined here must be considered for the general case of "design," by reviewing a concrete section which has been selected after due consideration of all governing factors.

The computation of the maximum and minimum live-load moments that act at various sections along a continuous prestressed beam is no different than it is for other types of construction. The unique procedures in the design of continuous prestressed beams are the determination of the minimum prestressing force that will perform satisfactorily with the assumed concrete section and the determination of a satisfactory layout for the tendon. The normal design procedure is as follows:

(1) Compute the moments due to dead load, superimposed dead load, and live loads. For structures subject to moving live loads, the maxima and minima moments must be determined.

(2) Adopt a trial layout for the prestressing tendon. The shape and location of the trial layout should consider practical controls, such as the space required to place the tendons with adequate cover and the position of other embedded items.

(3) Determine the secondary moments due to prestressing.

(4) Compute the flexural stresses at the critical points along the span under the combinations of dead loads and prestressing alone, as well as with maximum or minimum live load.

(5) If the elastic analysis reveals stresses that exceed those permitted by the applicable design criteria, or if it reveals the design of some portions of the member to be overly conservative, the dimensions of the concrete section, the amount of prestressing, or the layout of the prestressing, or all three, may be revised and the procedure repeated until an acceptable solution is obtained.

(6) After the designer is satisfied with the flexural design under service loads, he must design the member for shear and for flexural strength (*see* Secs. 8-9 and 8-10).

ILLUSTRATIVE PROBLEM 8-4 Determine a satisfactory means of prestressing the beam shown in Fig. 8-19. The beam is a hollow box and varies in depth from 24 to 49 in. The superimposed live load is 600 plf. Each span may be loaded independently of the other span.

SOLUTION: The section properties at the 20th points (intervals of 5 ft.) of span *AB* are summarized in Table 8.1 together with the following moments:

(1) Dead load on spans *AB* and *BC*

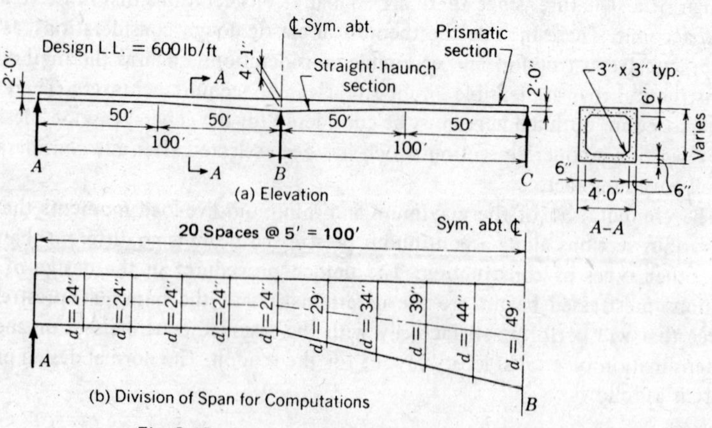

Fig. 8-19 Continuous beam analyzed in I.P. 8-4.

…MOMENTS FOR I.P. 8-4.

20th Pt.	Location (ft)	Section: I (in.^{4})	Area (in.^{2})	Location (in.) CG Top	Location (in.) CG Bottom	(1) Moment (k-ft)	(2) Moment (k-ft)	(3) Moment (k-ft)	(4) Moment (k-ft)	(5) Moment (k-ft)	(6) Moment (k-ft)
0	.000	50571	738	12.000	12.000	.000	.000	.000	.000	.000	.000
1	5.000	50571	738	12.000	12.000	111.256	197.193	225.475	82.975	225.475	82.975
2	10.000	50571	738	12.000	12.000	203.288	360.161	416.725	146.725	416.725	146.725
3	15.000	50571	738	12.000	12.000	276.095	488.904	573.750	191.250	573.750	191.250
4	20.000	50571	738	12.000	12.000	329.677	583.423	696.550	216.550	696.550	216.550
5	25.000	50571	738	12.000	12.000	364.034	643.716	785.125	222.625	785.125	222.625
6	30.000	50571	738	12.000	12.000	379.166	669.784	839.475	209.475	839.475	209.475
7	35.000	50571	738	12.000	12.000	375.073	661.628	859.601	177.101	859.601	177.101
8	40.000	50571	738	12.000	12.000	351.755	619.246	845.501	125.501	845.501	125.501
9	45.000	50571	738	12.000	12.000	309.212	542.639	797.176	54.676	797.176	54.676
10	50.000	50571	738	12.000	12.000	247.444	431.808	714.626	−35.373	714.626	−35.373
11	55.000	66004	768	13.250	13.250	166.321	286.621	597.721	−144.778	597.721	−144.778
12	60.000	83838	798	14.500	14.500	65.193	106.429	445.811	−274.188	445.811	−274.188
13	65.000	104166	828	15.750	15.750	−56.719	−109.547	258.116	−424.383	258.116	−424.383
14	70.000	127081	858	17.000	17.000	−200.197	−362.088	33.857	−596.142	33.857	−596.142
15	75.000	152677	888	18.250	18.250	−366.020	−651.975	−227.747	−790.247	−227.747	−790.247
16	80.000	181048	918	19.500	19.500	−554.968	−979.987	−527.477	−1007.477	−527.477	−1007.477
17	85.000	212288	948	20.750	20.750	−767.821	−1346.903	−866.112	−1248.612	−767.821	−1346.903
18	90.000	246490	978	22.000	22.000	−1005.359	−1753.505	−1244.432	−1514.432	−1005.359	−1753.505
19	95.000	283749	1008	23.250	23.250	−1268.362	−2200.572	−1663.217	−1805.717	−1268.362	−2200.572
20	100.000	324158	1038	24.500	24.500	−1557.611	−2688.884	−2123.247	−2123.247	−1557.611	−2688.884

(1) Dead load on spans *AB* and *BC*.
(2) Dead plus live load on spans *AB* and *BC*.
(3) Dead load on spans *AB* and *BC*; live load on span *AB* only.
(4) Dead load on spans *AB* and *BC*; live load on span *BC* only.
(5) Maxima moment.
(6) Minima moment.

TABLE 8-2. SECTION PROPERTIES, ECCENTRICITIES OF THE TENDON AND PRESSURE LINE AS WELL AS STRESSES DUE TO PRESTRESSING ALONE, SPAN *AB*.

		Section				Tendon		Prestress Only		
20th Pt.	Location (ft)	I (in.^{4})	Area (in.^{2})	Location (in.) CG Top	Location (in.) CG Bottom	Eccentricity (in.)	Prestress Force (K)	Eccentricity (in.) Pressure Line	Stress (ksi) Top	Stress (ksi) Bottom
0	.000	50571	738	12.000	12.000	.000	800.000	.000	1.084	1.084
1	5.000	50571	738	12.000	12.000	−1.320	800.000	−1.116	.872	1.295
2	10.000	50571	738	12.000	12.000	−2.520	800.000	−2.112	.683	1.484
3	15.000	50571	738	12.000	12.000	−3.780	800.000	−3.168	.482	1.685
4	20.000	50571	738	12.000	12.000	−4.680	800.000	−3.864	.350	1.817
5	25.000	50571	738	12.000	12.000	−5.700	800.000	−4.680	.195	1.972
6	30.000	50571	738	12.000	12.000	−6.480	800.000	−5.256	.086	2.081
7	35.000	50571	738	12.000	12.000	−7.140	800.000	−5.712	−.000	2.168
8	40.000	50571	738	12.000	12.000	−7.560	800.000	−5.929	−.041	2.209
9	45.000	50571	738	12.000	12.000	−7.920	800.000	−6.085	−.071	2.239
10	50.000	50571	738	12.000	12.000	−8.160	800.000	−6.121	−.078	2.246
11	55.000	66004	768	13.250	13.250	−6.600	800.000	−4.357	.341	1.741
12	60.000	83838	798	14.500	14.500	−4.560	800.000	−2.113	.710	1.294
13	65.000	104166	828	15.750	15.750	−2.400	800.000	.250	.996	.935
14	70.000	127081	858	17.000	17.000	.240	800.000	3.094	1.263	.601
15	75.000	152677	888	18.250	18.250	3.480	800.000	6.537	1.526	.275
16	80.000	181048	918	19.500	19.500	6.960	800.000	10.221	1.752	−.009
17	85.000	212288	948	20.750	20.750	10.680	800.000	14.145	1.950	−.262
18	90.000	246490	978	22.000	22.000	14.760	800.000	18.429	2.133	−.497
19	95.000	283749	1008	23.250	23.250	18.720	800.000	22.593	2.274	−.687
20	100.000	324158	1038	24.500	24.500	21.240	800.000	25.317	2.301	−.760

TABLE 8-3. PRESSURE LINE LOCATIONS TOGETHER WITH THE COMBINED TOP AND BOTTOM FIBER STRESSES FOR FOUR CONDITIONS OF LOADING, SPAN *AB*.

(1) Prestress and Load			(2) Prestress and Load			(3) Prestress and Load			(4) Prestress and Load		
Eccentricity (in.) Pressure Line	Stress (ksi) Top	Stress (ksi) Bottom	Eccentricity (in.) Pressure Line	Stress (ksi) Top	Stress (ksi) Bottom	Eccentricity (in.) Pressure Line	Stress (ksi) Top	Stress (ksi) Bottom	Eccentricity (in.) Pressure Line	Stress (ksi) Top	Stress (ksi) Bottom
.000	1.084	1.084	.000	1.084	1.084	.000	1.084	1.084	.000	1.084	1.084
.552	1.188	.979	1.841	1.433	.734	2.265	1.514	.653	.128	1.108	1.059
.937	1.261	.906	3.290	1.708	.459	4.138	1.869	.298	.088	1.100	1.067
.973	1.268	.899	4.165	1.874	.293	5.437	2.116	.051	−.299	1.027	1.140
1.080	1.289	.878	4.886	2.011	.156	6.583	2.333	−.165	−.616	.967	1.201
.779	1.232	.935	4.975	2.028	.139	7.096	2.431	−.263	−1.341	.829	1.338
.430	1.165	1.002	4.789	1.993	.174	7.335	2.476	−.308	−2.114	.682	1.485
−.086	1.067	1.100	4.211	1.883	.284	7.181	2.447	−.279	−3.056	.503	1.664
−.652	.960	1.207	3.359	1.721	.446	6.753	2.366	−.198	−4.046	.315	1.852
−1.447	.809	1.358	2.054	1.473	.694	5.872	2.198	−.030	−5.265	.084	2.083
−2.409	.626	1.541	.355	1.151	1.016	4.597	1.956	.211	−6.652	−.178	2.346
−1.862	.742	1.340	−.058	1.032	1.051	4.608	1.781	.301	−6.529	−.006	2.090
−1.135	.845	1.159	−.517	.930	1.074	4.573	1.635	.369	−6.226	.140	1.864
−.600	.893	1.038	−1.393	.797	1.134	4.121	1.464	.467	−6.115	.226	1.705
.091	.942	.922	−2.337	.682	1.182	3.601	1.317	.546	−5.848	.306	1.558
1.047	1.001	.800	−3.241	.590	1.210	3.121	1.199	.602	−5.315	.392	1.409
1.897	1.034	.707	−4.478	.485	1.257	2.309	1.070	.672	−4.890	.450	1.292
2.628	1.049	.638	−6.057	.370	1.317	1.153	.934	.753	−4.583	.485	1.202
3.349	1.057	.578	−7.873	.255	1.380	−.237	.801	.834	−4.287	.511	1.124
3.567	1.027	.559	−10.415	.110	1.476	−2.354	.639	.948	−4.492	.499	1.088
1.953	.888	.652	−15.016	−.137	1.678	−6.531	.375	1.165	−6.531	.375	1.165

(1) Dead load plus prestressing.
(2) Dead load plus prestressing plus live load on *AB* and *BC*.
(3) Dead load plus prestressing plus live load on Span *AB*.
(4) Dead load plus prestressing plus live load on Span *BC*.

(2) Dead and live load on both spans
(3) Dead load on both spans and live load on span AB only
(4) Dead load on both spans and live load on span BC only
(5) Maxima moment
(6) Minima moment

Based upon experience, a tendon trajectory, the ordinates of which are given in Table 8-2, is adopted for a trial, as is the effective prestressing force of 800 kips. The trajectory does not follow a mathematical curve but was selected by plotting. Using the trial trajectory of Table 8-2 and the force of 800 kips, the secondary moment due to prestressing is found to be 272 k-ft at B; there is none at A and C. With this value of the secondary moment it can be shown that the pressure line is located as shown in Table 8-2, as are the stresses due to prestressing alone. Combining the stresses due to prestressing with those due to the four loading conditions shown, results in the stresses given in Table 8-3.

A review of Table 8-3 will reveal that the stresses under dead load and prestressing are all compressive. The maximum value is 1541 psi. Hence, the beam cannot be critical with respect to initial stresses. Under the other three loading conditions it will be seen that the more critical stresses occur in span AB when the live load is on that span alone. The maximum compressive stress is 2476 psi and the greatest tensile stress is −308 psi. If the concrete strength at 28 days is specified to be 5500 psi, the maximum compressive stress would be $0.45\,f'_c$ and the tensile stress would amount to $4.2\sqrt{f'_c}$; both of these are well within the values permitted by ACI 318.

With the electronic computational devices which are currently available, it is believed that the most efficient method of designing continuous prestressed concrete members for service loads is by trial and error. If the assumed conditions of cross section, eccentricity, and prestressing force result in unacceptable stresses, the assumed dimensions can quickly and easily be altered and a new analysis made. One becomes proficient in this process with a small amount of experience.

8-9 Limitations of Elastic Action

As has been stated previously, prestressed-concrete continuous structures perform substantially as elastic structures under loads which do not result in stresses that exceed the normal working stresses permitted by recognized prestressed-concrete design criteria. Adequate experimental data are available to substantiate this.

Under normal conditions, when the design load is exceeded, the concrete stresses remain reasonably elastic up to the load which causes visible cracking in the structure. The first cracks, which are not visible to the unaided eye, do not materially affect the performance of the structure. As a matter of fact, when the load at which the first crack observed with the unaided eye during a test of a beam is plotted on a load-deflection curve, it is often below the point at which pronounced deviation of the tangent (or plasticity) to the elastic deflection curve take place.

In most continuous structures, the cracking load will not be reached simultaneously in all highly stressed sections, since the magnitude of the moments vary at different sections along the member and, when members are designed for moving live loads, the largest moments that may occur at each section under the assumed design loads do not occur under the same condition of loading. Furthermore, once the cracking load has been significantly exceeded at a particular section, there is a reduction in the effective moment of inertia, and hence, stiffness of the member in the cracked area. At loads which result in one or more areas of a beam being stressed substantially above cracking, the effective modulus of elasticity of the concrete in the cracked areas may be considerably lower than in areas that are still stressed in the elastic range–this further contributes to the reduction in the stiffness of the member in the cracked areas.

Because of the localized changes in stiffness of a continuous member subjected to significant overload, the distribution of moments are no longer proportional to the distribution of moments in the elastic range. This is explained by the fact that, after cracking has reached a significant degree at one or more areas in a beam, the moment that results from the application of additional loads is carried in greater proportion by the portion of the member that remains uncracked. The areas first to attain a highly cracked and highly stressed condition yield more upon the application of additional load than areas that remain uncracked, and hence, the cracked areas resist less of the additional loads than would be indicated by purely elastic analysis. The phenomenon is called "redistribution of the moments."

Redistribution of the moments is the phenomenon that results in continuous beams which are designed on a purely elastic basis, frequently, but not always, having very high factors of safety. Redistribution of the moments can be illustrated by considering the fixed beam of Fig. 8-20, which, when subjected to loads slightly above the cracking, has a distribution of moments that is virtually identical to the distribution that would result from an elastic analysis. When an additional increment of load is applied, the cracked sections are not as stiff in proportion to the uncracked sections as they were previously, and hence, the distribution of moments deviates from the elastic distribution. For redistribution to be complete, the sections which crack first must form completely plastic hinges. If plastic hinges formed at the ends of the beam of Fig. 8-20, the moments at the ends would not increase upon the application of additional load. The ends would simply rotate and the positive moment would increase as a result of the new load application.

The redistribution of moments takes place to various degrees. Redistribution is said to be complete when the various critical sections of a beam all attain a high degree of plasticity and attain the ultimate moments that would be indicated by

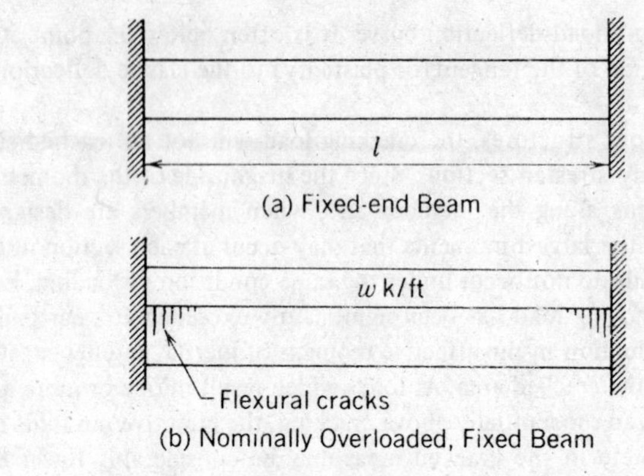

(a) Fixed-end Beam

(b) Nominally Overloaded, Fixed Beam

w k/ft

$M_n \cong \frac{wl^2}{12}$ $\quad$ $M_n \cong \frac{wl^2}{12}$

$M_p \cong \frac{wl^2}{24}$

(c) Moments in Nominally Overloaded, Fixed Beam

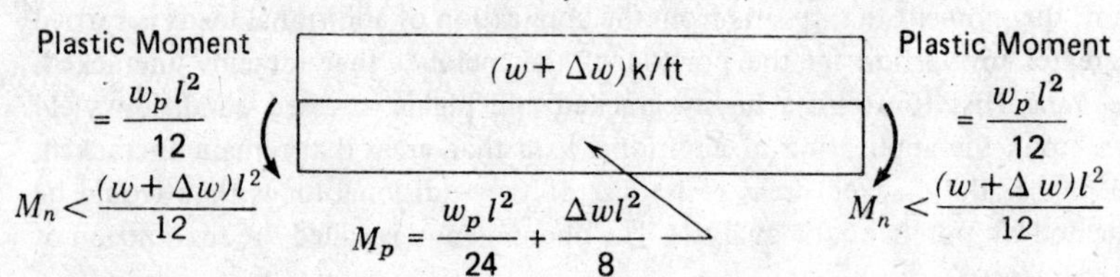

(d) Moments in Significantly Overloaded, Fixed Beam

Fig. 8-20 Variation in moments in a fixed beam subjected to various conditions of overload.

the flexural strength analysis developed in Sec. 5-2. Some beams, if loaded to destruction, would fail before redistribution is complete. It is not completely understood why this occurs.

Additional research is needed into the phenomena that control the redistribution of moments, but at the present time, it is believed the following characteristics improve the redistribution of moments that take place in a beam, loaded to destruction.

(1) Low values of the steel index (percentages of reinforcement).

(2) Good bond between the prestressing reinforcement and the concrete.

(3) The use of untensioned reinforcement, in areas of high moment in beams with very low values of steel index, and in beams that have unbonded tendons, in order to improve the cracking patterns at high loads.

(4) Prevention of large inclined shear cracks in areas of high moment, since such cracks reduce the ultimate moment and rotation capacities of the section.

The use of untensioned shear reinforcement in the areas of high shear and high moment is considered necessary.

(5) Redistribution appears to be more complete in members having large differences in flexural capacity at the various sections than it is in members in which the flexural capacity is nearly equal at all critical sections (Ref. 15).

The Building Code Requirement for Reinforced Concrete (ACI 318) (Ref. 6) permits limited redistribution of the moments in continuous beams. These provisions are as follows:

Continuous beams and other statically indeterminate structures shall be designed for adequate strength and satisfactory behavior under service loads.

Behavior shall be determined by elastic analysis, taking into account the reactions, moments, shears, and axial forces produced by prestressing, the effects of temperature, creep, shrinkage, axial deformation; restraint of attached structural elements, and foundation settlement.

If bonded reinforcement is provided at the supports in accordance with Section 18.9.2, negative moments due to design dead and live loads, calculated by elastic theory for any assumed loading arrangement, may be increased or decreased by not more than

$$20\left[1 - \frac{\omega_p + \dfrac{d}{d_p}(\omega_p - \omega')}{0.36\beta_1}\right] \text{percent} \tag{8-8}$$

These modified negative moments shall also be used for final calculations of the moments at other sections in the span corresponding to the same loading condition. Such an adjustment shall only be made when the section at which the moment is reduced is so designed that ω_p, $[\omega_p + d/d_p(\omega - \omega')]$, or $[\omega_{pw} + d/d_p(\omega_w - \omega'_w)]$, whichever is applicable, is not greater than $0.24\,\beta_1$.

The moments to be used in design shall be the sum of the moments due to reactions induced by prestressing (with a load factor of 1.0) and the moments due to design dead and live loads including redistribution as permitted in the above paragraph.

The definitions of the steel indices (ω_p, ω, and ω') and β_1 are as given in Sections 5-2 and 5-4.

From these provisions it will be seen the redistribution of moments would be a maximum value of 20% for a reinforcement index of 0 and a minimum value of 6.67% when the applicable steel index is 0.20.

8-10 Analysis at Design Loads

Flexural strength analysis for indeterminate prestressed concrete structures should be based upon an elastic analysis or upon an elastic analysis including the limited redistribution of moments according to Eq. 8-8 and the other requirements specified in Sec. 8-9 of this book (and ACI 318, Sec. 18.10).

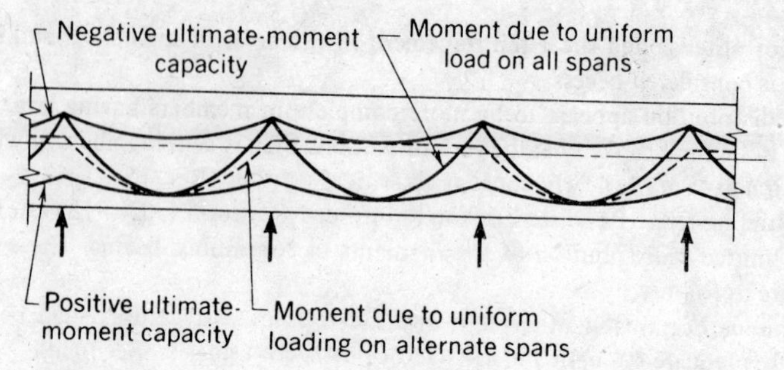

Fig. 8-21 Moment diagrams for uniform loads on all spans and on alternate spans of a multi-span beam superimposed on envelope of ultimate moment capacity.

In investigating the safety of a multispan continuous beam under design loading, all conditions of loading must be considered. The loading arrangement that would result in the collapse of the member due to negative moment may be quite different from the loading that would result in collapse due to positive moment or from positive and negative movement, simultaneously, in the same span. In addition, since reversal of moments occurs in continuous structures, it is possible to have critical positive moments develop under overload in areas of the beams which are normally stressed by negative moments. The opposite condition is also possible. Therefore, to facilitate the determination of the ultimate strength of a member, an envelope of the maximum and minimum moment capacities that can be developed at various sections in the member are computed and plotted as shown in the sketch of Fig. 8-21. The moment diagrams due to the various conditions of loading (with the appropriate load factors) are computed and plotted in the envelope. The moments can be distorted from those obtained by the elastic analysis as much as that given in the excerpt from the ACI Building Code given in Sec. 8-9, on projects which are governed by the Code. If the moment diagrams so computed fall within the limits of the ultimate moment envelopes, the design is satisfactory. If not, the design must be altered.

Previous to 1975, it was customary in the U.S. to ignore secondary moments due to prestressing when making a strength analysis. This is no longer the case. The secondary moments, which exist when the member is loaded in the elastic range, do not disappear as the load is increased and it approaches its capacity. To ignore the existence of the secondary moments in the strength analysis is the same as assuming a redistribution of moment in the amount of the secondary moment; it may very well be that this amount exceeds that permitted by the applicable design criteria.

ILLUSTRATIVE PROBLEM 8-5 Compute the ultimate moment for the beam of Problem 8-4, using the non-concordant tendon trajectory in Table 8-2. Assume

f'_c = 5500 psi, f_{pu} = 270,000 psi, A_{ps} = 5.00 in.^{2} The design load is to be taken as 1.4 D + 1.7 L in which D and L are the service dead and live loads, respectively.

SOLUTION: The design moments for various conditions of loading are shown in Table 8-4. The secondary moment, which is not factored, is included in the design (factored) dead load because both are of a "permanent" nature. These moments are used, together with the positive and negative moment capacities computed at various locations along the span, to construct the diagram shown in Fig. 8-22. An examination of Fig. 8-22 will reveal that for the condition of maximum design load on both spans (curve 1), the design moment diagram is well within the envelope of the positive and negative moment capacities. For the conditions of design dead load and secondary moment on both spans but with design live load on span AB only (Curve 2), the design moment curve slightly exceeds the positive moment capacity in span AB (about 2%), while in span BC the design moment curve is very close to being tangent to the negative moment capacity curve near midspan. One should not count on any redistribution of moment in a case such as this because the formation of a plastic hinge in either span will prevent use of the "excess" negative moment capacity at the interior

TABLE 8-4 Design Moments for I.P. 8-5.

	Moments (k-ft)					
Point	Sec. Moment and Design D. L. Moment	Design LL on Spans *AB* and *BC*	Design LL on Span *AB*	Design LL on Span *BC*	Design Maximum Moment	Design Minimum Moment
0	0	0	0	0	0	0
1	169	146	194	- 48	363	121
2	312	267	363	- 96	675	216
3	427	362	506	-144	933	283
4	516	431	624	-192	1140	324
5	578	476	716	-241	1294	338
6	613	494	782	-289	1395	324
7	621	487	824	-337	1445	284
8	602	455	839	-385	1441	217
9	556	397	829	-433	1385	123
10	483	314	794	-481	1277	2
11	383	205	733	-529	1116	- 146
12	255	70	647	-577	902	- 322
13	98	- 90	535	-625	633	- 527
14	- 89	- 275	398	-673	309	- 762
15	- 307	- 486	235	-722	- 73	-1028
16	- 558	- 722	46	-770	- 512	-1328
17	- 843	- 984	-167	-818	-1010	-1827
18	-1162	-1272	-407	-865	-1569	-2434
19	-1517	-1585	-672	-914	-2189	-3102
20	-1908	-1923	-962	-962	-2870	-3831

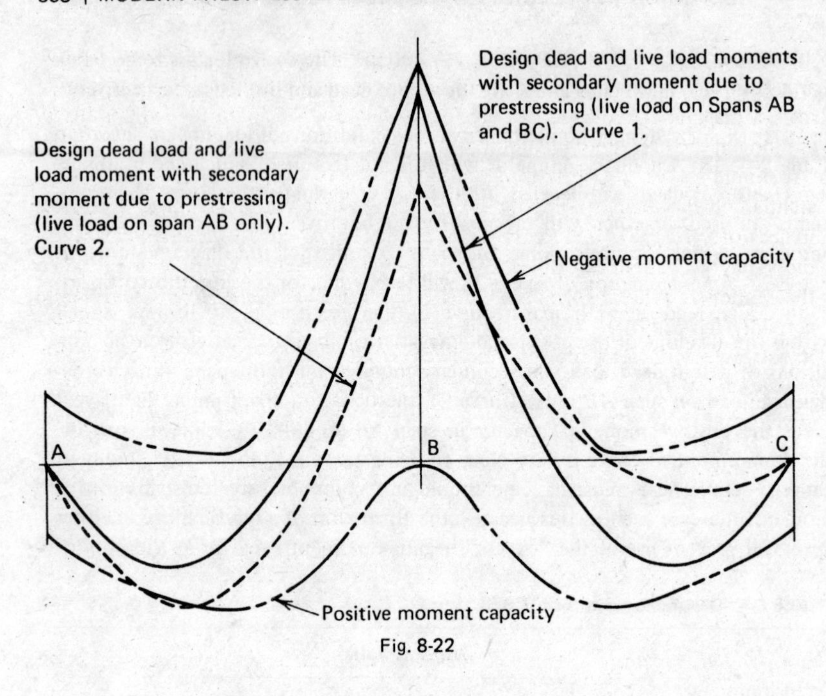

Fig. 8-22

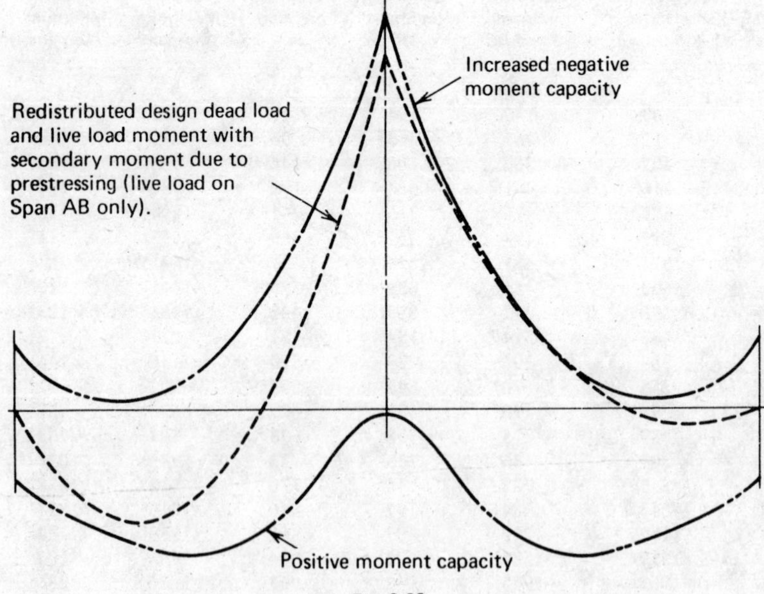

Fig. 8-23

support. If the negative moment capacity in the spans were increased by the provision of additional reinforcing in the upper portions of the beams near mid-span, a diagram as shown in Fig. 8-23 could be obtained and some redistribution of moment could be counted upon.

It should be recognized that in continuous beams, the steel indices can be substantially higher than is normally experienced in simple beams; in the latter, steel indices of 0.10 to 0.40 are commonly found. In continuous beams, the steel indices may be as high as 2.00, since the tendon may be very near the flange when the section is subject to moment reversal. Such highly reinforced areas would certainly fail suddenly, and for this reason, critical areas of this type should be investigated with great care and modified so that adequate safety of the structure is assured.

8-11 Additional Considerations

The effect of tendon friction during stressing has not been discussed in the preceding articles on the design of continuous beams. The fundamental principles that govern friction loss of post-tensioned tendons are the same for simple and continuous beams and, since it is more of a practical than theoretical consideration, the discussion of tendon friction is included in the discussion of post-tensioning methods (Chap. 15).

Tendon friction can result in substantial losses of stress, and for this reason, particularly in the design of continuous structures, the designer must select the tendon layout that will result in lower friction losses. The designer should estimate the friction loss as various layouts are studied and, when high losses are of significance and are unavoidable, the design should be made on the basis of a variable prestressing force.

The use of sharp bends or abrupt changes of slope in the tendon is generally avoided, since such sharp bends can result in significant secondary bending stresses in the tendons.

The angular change through which the tangent to the tendon passes has important influence on the friction loss which results during stressing, with larger losses resulting from the larger curvatures. From this standpoint, a tendon layout composed of two chords is preferable to a parabolic layout since the angular change with a parabola is twice that of the angular change obtained with chords.

In short-span, continuous prestressed structures, the dead load of the structure is small in comparison to the live load and, if the structure is subjected to a moving live load, the critical sum of the maximum and minimum moments is nearly the same as the simple beam moment for the same span. Therefore, the prestressing force required for such a structure is not significantly less than is required for the simple span. This greatly reduces the economy of materials that one would normally expect to result from the use of continuity. The advantages of less deflection, great resistance to lateral and longitudinal loads, among other things, are still attained through the introduction of continuity.

When the dead-load moment is a large portion of the total moment, the variation in moment in a continuous beam is less than is found in simple beams and, for this reason, the prestressing force is less for the continuous structure. This accounts for the fact that continuity is used on long-span structures to a much greater degree than in short-span structures.

The same general procedures of design and analysis that have been presented in this chapter are used in designing prestressed rigid frames. In such construction, special attention must be given to the moments that result from the axial shortening of the members due to prestressing, creep, shrinkage, and temperature.

8-12 Continuous Beams Utilizing Prestressed Beam Soffits

Simple flexural members composed of prestressed components and plain or reinforced-concrete components have been employed in many applications domestically and abroad. The most common application of this type of construction is in the use of prestressed bridge stringers used in combination with a cast-in-place slab to form a T beam in the completed structure (*see* Sec. 6-4). Other types of composite simple beams include those with cross sections, as illustrated in Fig. 8-24.

Composite beams composed of a precast, prestressed component and a cast-in-place component can be made continuous at nominal cost under some conditions, by including normal reinforcing steel as negative-moment reinforcing, as is illustrated in Fig. 8-25. This type of beam may be designed in such a manner that it is necessary to shore the precast soffit in place after erection, during placing, and curing of the cast-in-place concrete. The result of this procedure is that the entire dead load, as well as the live load, is carried by the composite section of the continuous beam. On the other hand, the precast component may be proportioned in such a manner that the entire dead load is resisted by the precast section alone and the composite beam is continuous for live loads alone.

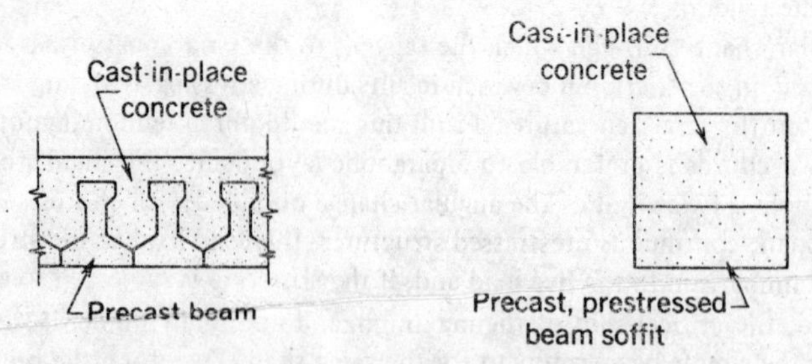

Fig. 8-24 Typical sections of two types of composite beams.

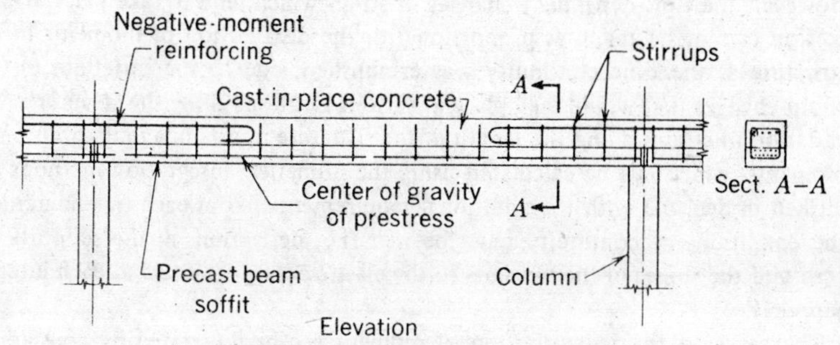

Fig. 8-25 Composite beams utilizing a precast soffit.

8-13 Continuous Beams Constructed in Cantilever

In Sec. 8-5, in which the assumptions generally made in the analysis of continuous prestressed beams are listed, it is assumed that effects of each cause of moments can be calculated independently and superimposed to attain the result of the combined effect of the several causes. This is the principle of superposition. This assumption is valid for a continuous member that is stressed after the complete structural scheme has been constructed. The assumption is not true for a structure constructed in increments with stresses imposed upon the structure during each construction phase.

In considering this effect, the variation in strain the concrete undergoes during the lifetime of the structure must be considered. As was pointed out in Secs. 6-2 and 6-3, the more sophisticated methods of analyzing the losses of prestress and the deflection of beams must take into account the initial strains to which the structure is subjected, as well as the time dependent changes in strain due to shrinkage, creep, and relaxation. For a continuous member that is cast-in-place in one operation and stressed in its final structural form, the effects of shrinkage and relaxation can reasonably be assumed to be the same at every point in the structure and, since the effect of creep is assumed to be directly proportional to the stress level in the concrete, the creep strain at every section in the structure at any point in time can be assumed to be in direct proportion to the distribution of moments in the structure at the time of stressing. This is not precisely true, since there will be some difference in the relaxation loss along the length of a tendon due to the variation in initial stress in the tendon. This variation results from friction during stressing and other causes.

In the case of members cast in segments, subjected to stresses while still in the form of determinate structural elements, and later rendered continuous and indeterminate by cast-in-place closure joints and additional prestressing, the instantaneous strains in the structure can still be considered to be in proportion to the distribution of moments in the structure at the time the structure was rendered continuous. (An example of this type of structure is shown in Fig. 8-3).

However, the time-dependent changes in strain which tend to take place at each section cannot be taken as proportional to the distribution of moments in the structure at the time continuity was established, due to the variations in free strain changes that would take place at various locations along the structure. The variation in strains along the length of the structure results in a redistribution of moments, which can be calculated using the numerical integration methods described in Sec. 6-3 with the additional requirement that at each time increment the conditions of continuity must be met (i.e. deflections at the supports are zero and the slopes of the tangents to the elastic curves are equal at each interior support).

The cause of the redistribution of moment can be illustrated by considering two cantilever beams as shown in Fig. 8-26. These beams are rendered continuous when the cast-in-place joint between them is constructed and the continuity tendons between the two cantilevers are installed. It should be apparent that if the two beams were not rendered continuous, the effect of creep would be to cause vertical deflection and a rotation of the ends of the beams with the passing of time. Because this rotation is prevented by the provision of continuity, a positive moment is created near midspan. The creep-caused positive moment is of significant magnitude and is accompanied by a reduction of the negative moment at the supports, which is normally of negligible magnitude.

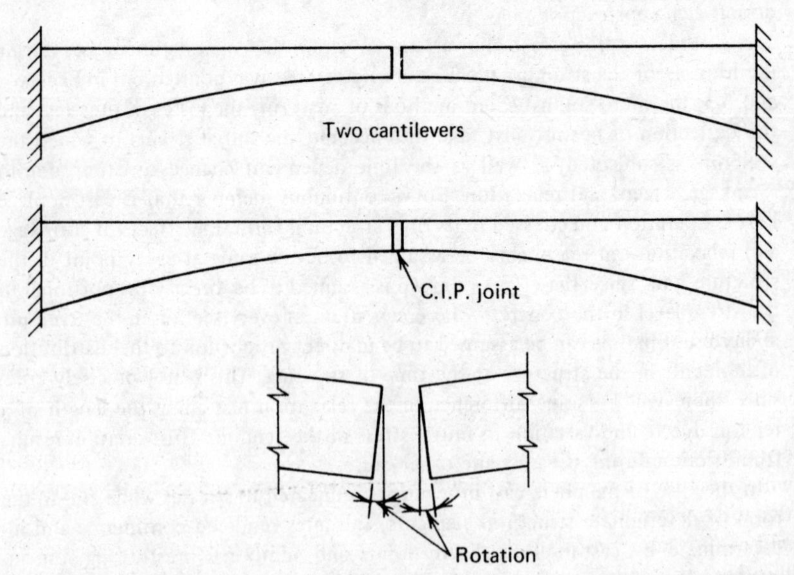

Fig. 8-26 Two cantilever beams used to illustrate the cause of moment redistribution due to creep.

REFERENCES

1. Muller, Jean, "Long-Span Precast Prestressed Concrete Bridges Built in Cantilever," pp. 705–740, Concrete Bridge Design, ACI Publication SP-23, Detroit, Michigan, 1969.

2. Muller, Jean, "Continuous Prestressed Concrete Structural Design," *Proc. Western Conf. on Prestressed Concrete*, 109–132 (Nov. 1952).

3. Saeed-Un-Din, K., "The Effect of Creep Upon Redundant Reactions in Continuous Prestressed Concrete Beams," *Mag. of Concrete Research*, **10**, 109 (Nov. 1958).

4. Muller, Jean, "Flexural Strength of Prestressed Concrete Continuous Structures," pp. 9–19. Paper presented at the Knoxville Convention of the A.S.C.E. (January 1956).

5. Guyon, Y., "The Strength of Statically Indeterminate Prestressed Concrete Structures," *Proc. of a Symposium on the Strength of Concrete Structures*, 305 (May 1956).

6. ACI Committee 318, Building Code Requirements for Reinforced Concrete" (ACI 318-83), Detroit, 1983.

7. Parme, A. L. and Paris, G. H. "Designing for Continuity in Prestressed Concrete Structures," *Journal of the American Concrete Institute*, **23**, 45–64 (Sept. 1951).

8. Muller, Jean, "Flexural Strength of Prestressed Concrete Continuous Structures," pp. 9–19. Paper presented at the Knoxville Convention of the A.S.C.E. (January 1956).

9. Guyon, Y., "The Strength of Statically Indeterminate Prestressed Concrete Structures," *Proc. of a Symposium on the Strength of Concrete Structures*, 305 (May 1956).

10. ACI Committee 318, "Proposed Revision of ACI 318–63: Building Code Requirements for Reinforced Concrete," *Journal of the American Concrete Institute*, **67**, No. 2, 77–186 (Feb. 1970).

11. ACI Committee 318," Building Code Requirements for Reinforced Concrete" (ACI 318–71) 1975 Supplement, p. 7, American Concrete Institute, Detroit, 1975.

CHAPTER 8–PROBLEMS

1. A slab 8.333 ft long has straight haunches at each end as shown in Fig. 8-10. The slab is rectangular in cross section and is solid. The slab depth is 14 in. at each end and is 7 in. deep for the central 3.33 ft. (Length of each haunch is 2.50 ft). The top surface of the slab is level. If the slab is to be stressed with a straight tendon that is parallel to the top surface of the slab, determine where the tendon must be placed if secondary moments due to prestressing are to be avoided.

SOLUTION: Assume a trial location, such as at 3.50 in. from the top surface, which results in no eccentricity in the center portion of the slab, and compute the rotations at the ends of the slab. Compute the moment required at each end of the slab required to cause an equal but opposite rotation at each end of the slab. This moment is the secondary moment for the tendon being on the trial location. The trial trajectory should be adjusted by an amount equal to

$M_{sec} \div P$. Adding this adjustment to the trial location will give location of the concordant tendon. The results can be checked by computing the rotation at the ends of the member with the adjusted trajectory; there should not be any rotations at the ends. In this case, the correct location is 4.55 in. from the top of the slab.

2. A prismatic beam has five equal spans of 50 ft each. It is prestressed by a straight tendon located at an eccentricity of e below the centroidal axis. Determine the location of the pressure line for prestressing alone.

SOLUTION: The pressure line locations are as follows:

End supports : e below the centroidal axis
First interior supports : 0.263 e above the centroidal axis
Second interior supports : 0.053 e below the centroidal axis

3. A continuous cast-in-place post-tensioned bridge, having the cross sectional dimensions shown in Example IV of Appendix B, is to have one span of 162 ft and another of 150 ft. The supports are hinged. The thickness of the bottom slab varies linearly from the thickness of 5.75 in. at 0.10L on each side of the bent to 12 in. at the center line of the bent. The tendon is to be placed on a path composed of compounded second-degree parabolas as shown in Example V of Appendix B. The initial force in the tendon after anchorage (*see* Sec. 15-6) and after elastic shortening of the concrete is as shown in the table below. Section properties and eccentricities of the tendons are also given in the table. Assume E_s = 27000 ksi and E_c = 3600 ksi. Determine the secondary moment due to prestressing as well as the ordinates of the pressure line for prestressing alone and for prestressing plus dead load.

SOLUTION: Based upon the data given in the table on p. 353, the moments due to dead load and due to prestressing at the bent are determined to be −26952 k-ft and +9567 k-ft, respectively. From these, the locations of the pressure line with and without dead load are found to be as given in Table 2 for Problem 8-3.

TABLE 1 Problem 8-3

			Section			Tendon	
20th Pts.	Location (ft)	(in.^{4})	Area (in.^{2})	Location (in.) CG Top	Location (in.) CG Bottom	Eccentricity (in.)	Prestress Force (K)
			SPAN NO. 1				
0	.000	7161480	8019	34.608	43.391	.000	5757.000
1	8.100	7161480	8019	34.608	43.391	−6.402	5778.000
2	16.200	7161480	8019	34.608	43.391	−12.067	5799.000
3	24.300	7161480	8019	34.608	43.391	−17.065	5820.000
4	32.400	7161480	8019	34.608	43.391	−21.396	5841.000
5	40.500	7161480	8019	34.608	43.391	−25.062	5863.000
6	48.600	7161480	8019	34.608	43.391	−28.060	5884.000
7	56.700	7161480	8019	34.608	43.391	−30.393	5905.000
8	64.800	7161480	8019	34.608	43.391	−32.059	5926.000
9	72.900	7161480	8019	34.608	43.391	−33.058	5947.000
10	81.000	7161480	8019	34.608	43.391	−33.392	5968.000
11	89.100	7161480	8019	34.608	43.391	−32.688	6001.000
12	97.200	7161480	8019	34.608	43.391	−30.579	6033.000
13	105.300	7161480	8019	34.608	43.391	−27.063	6065.000
14	113.400	7161480	8019	34.608	43.391	−22.142	6097.000
15	121.500	7161480	8019	34.608	43.391	−15.813	6088.000
16	129.600	7161480	8019	34.608	43.391	−8.079	6056.000
17	137.700	7714397	8450	36.484	41.515	2.938	6024.000
18	145.800	8169629	8882	38.100	39.899	15.100	5992.000
19	153.900	8543251	9313	39.496	38.503	24.926	5897.000
20	162.000	8847614	9744	40.698	37.301	28.938	5802.000
			SPAN NO. 2				
0	.000	8847614	9744	40.698	37.301	28.938	5802.000
1	7.500	8543251	9313	39.496	38.503	24.926	5882.000
2	15.000	8169629	8882	38.100	39.899	15.100	5983.000
3	22.500	7714397	8450	36.484	41.515	2.938	6016.000
4	30.000	7161480	8019	34.608	43.391	−8.079	6049.000
5	37.500	7161480	8019	34.608	43.391	−15.813	6082.000
6	45.000	7161480	8019	34.608	43.391	−22.142	6069.000
7	52.500	7161480	8019	34.608	43.391	−27.063	6036.000
8	60.000	7161480	8019	34.608	43.391	−30.579	6002.000
9	67.500	7161480	8019	34.608	43.391	−32.688	5969.000
10	75.000	7161480	8019	34.608	43.391	−33.392	5936.000
11	82.500	7161480	8019	34.608	43.391	−33.058	5915.000
12	90.000	7161480	8019	34.608	43.391	−32.059	5894.000
13	97.500	7161480	8019	34.608	43.391	−30.393	5873.000
14	105.000	7161480	8019	34.608	43.391	−28.060	5851.000
15	112.500	7161480	8019	34.608	43.391	−25.062	5830.000
16	120.000	7161480	8019	34.608	43.391	−21.396	5809.000
17	127.500	7161480	8019	34.608	43.391	−17.065	5787.000
18	135.000	7161480	8019	34.608	43.391	−12.067	5766.000
19	142.500	7161480	8019	34.608	43.391	−6.402	5745.000
20	150.000	7161480	8019	34.608	43.391	.000	5724.000

TABLE 2 Problem 8-3

Prestress Only			Prestress and Load		
Eccentricity (in.) Pressure Line	Stress (ksi)		Eccentricity (in.) Pressure Line	Stress (ksi)	
	Top	Bottom		Top	Bottom
.000	.717	.717	.000	.717	.717
-5.408	.569	.909	2.640	.794	.628
-10.087	.440	1.077	4.818	.858	.553
-14.106	.328	1.223	6.476	.907	.497
-17.465	.235	1.346	7.628	.943	.458
-20.166	.159	1.447	8.277	.965	.437
-22.206	.102	1.525	8.450	.973	.432
-23.588	.063	1.580	8.151	.968	.444
-24.309	.042	1.611	7.396	.950	.473
-24.370	.041	1.619	6.194	.919	.518
-23.773	.058	1.603	4.557	.875	.579
-22.166	.105	1.554	2.797	.829	.646
-19.161	.193	1.452	1.385	.792	.701
-14.759	.323	1.298	.332	.766	.744
-8.961	.496	1.091	-.346	.750	.773
-1.669	.710	.820	-.529	.743	.778
7.086	.962	.495	-.379	.744	.769
19.137	1.258	.092	1.870	.766	.652
32.343	1.578	-.271	4.014	.786	.557
43.420	1.816	-.520	2.253	.694	.573
48.724	1.895	-.596	-7.018	.408	.767
43.467	1.813	-.520	.631	.648	.614
32.369	1.576	-.272	1.104	.704	.641
19.158	1.256	.091	-2.183	.649	.782
7.104	.961	.493	-5.406	.596	.952
-1.655	.709	.819	-6.366	.571	.993
-8.900	.495	1.084	-6.802	.557	1.006
-14.700	.323	1.290	-6.667	.558	.996
-19.102	.194	1.443	-6.008	.574	.966
-22.109	.106	1.543	-4.843	.604	.919
-23.721	.059	1.593	-3.187	.648	.854
-24.323	.042	1.609	-1.487	.695	.790
-24.267	.043	1.601	-.068	.732	.737
-23.551	.063	1.570	1.059	.762	.694
-22.173	.102	1.515	1.892	.783	.662
-20.139	.159	1.438	2.406	.794	.641
-17.443	.234	1.338	2.599	.797	.632
-14.089	.327	1.215	2.462	.790	.635
-10.075	.438	1.071	1.976	.774	.649
-5.402	.566	.904	1.136	.747	.676
.000	.713	.713	.000	.713	.713

9 Direct Stress Members, Temperature and Fatigue

9-1 Introduction

The first portion of this chapter is devoted to a discussion of prestressed members subject to direct stress. With the possible exception of piles, prestressed concrete is not used extensively for members that are designed primarily to resist direct stress.

The second portion of the chapter is devoted to the consideration of fire resistance, the effect of nominal temperature variations, and fatigue of prestressed concrete.

9-2 Tension Members or Ties

In the application of rigid frames, trusses, certain types of continuous beam framing, and in some water-front structures, among other types of structures, it is necessary to include structural components subject to direct tensile stress alone. Such members are referred to as ties. It may be desirable to provide ties of prestressed concrete rather than of steel or reinforced-concrete for one or more reasons.

Prestressed ties generally deform less under load than ties of reinforced concrete or steel. In addition, the deformation of the ties can be controlled by the designer. For example, a steel or reinforced-concrete tie would normally be designed in such a manner that the stress in the steel would be of the order of 20,000 psi under full load and, if the modulus of elasticity of the steel is assumed to be 29×10^6 psi, the deformation of such a tie 1000 in. long would be

$$\text{Deformation} = \frac{20{,}000 \times 1000}{29 \times 10^6} = 0.69 \text{ in.}$$

Obviously the deformation of a steel or reinforced-concrete tie can be reduced by using a lower allowable stress.

A prestressed tie can be proportioned in such a manner that the concrete stress is confined within any desired limits, without affecting the amount of steel required and the total force the tie will develop, simply by varying the area of the concrete section. If the modulus of elasticity of the concrete is 4×10^6 psi and the concrete stress due to effective prestress alone is 2000 psi, the deformation of a tie 1000 in. long under full load (concrete stress = zero under full load) would be

$$\text{Deformation} = \frac{2000 \times 1000}{4 \times 10^6} = 0.50 \text{ in.}$$

By confining the concrete stress due to prestressing to a lower limit, the deformation can be reduced in direct proportion to the stress.

The use of prestressed ties in roof trusses may be preferred over the use of steel ties as a result of the greater fire resistance inherent in concrete members.

Prestressed ties, due to their lack of cracks, are less subject to deterioration by corrosion, and hence, may be preferred over steel and reinforced concrete ties in certain applications. Prestressed concrete ties offer significant advantages when used as soil or rock anchors or when used as ties in anchored retaining walls and bulkheads.

In using prestressed ties, the designer must consider the effects of creep and shrinkage on the deformation of the tie. If the tie is prestressed and immediately thereafter put into service, assuming the superimposed load and the prestressing force are nearly equal, the concrete stress in the tie would be near zero and there would be little deferred deformation due to creep; shrinkage would still continue, however.

If the tie were prestressed and stored for a period of a year or more before being put into service, a substantial amount of creep deformation would have taken place, and for service loads of short duration which were applied subsequently, the tie would elongate in proportion to the instantaneous modulus of elasticity of the concrete. For service loads of constant duration and of the same magnitude as the effective prestressing force, the concrete stress would be reduced to zero upon the application of the service load. The tie would instantly elongate by an amount determined by the instantaneous modulus of elasticity. The deformation (elongation) would continue until partial recovery of the original creep deformation had been obtained.

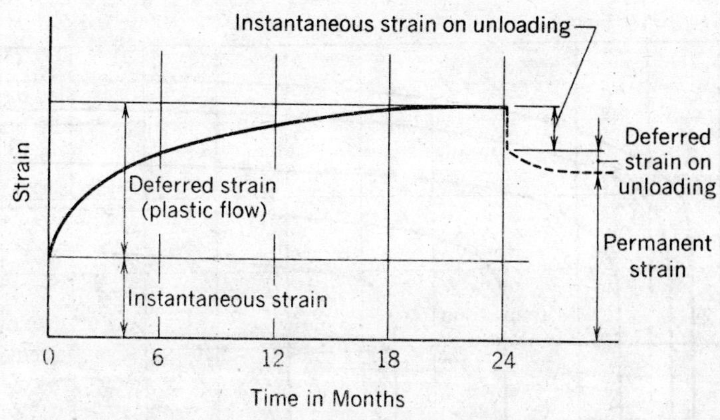

Fig. 9-1 Concrete strains under long-term loading and unloading.

Accurate prediction of the amount of deferred strain that would be recovered upon removal of the load can only be made if the properties of the concrete are known. The plastic strain is generally much lower for unloading than for loading and a residual strain remains in the concrete. This is illustrated in the strain-time diagram in Fig. 9-1 (*see* Ref. 1).

The total strains occurring in concrete subjected to constant sustained stresses applied at various ages are as illustrated in Fig. 9-2 (Ref. 2). Using the principles of superposition described in Sec. 6-2, one can employ curves such as shown in Fig. 9-2 to estimate the creep effects due to variation in loading. For example, assume a unit stress of 1000 psi is applied at the age of 28 days and held constant until the age of 91 days, at which time, it is completely removed. The strain vs time diagram would be as in Fig. 9-3. The curve of Fig. 9-3 is constructed using the curves from Fig. 9-2 with the assumption that the strain deformations upon loading or unloading at any particular age of the concrete take place at the same rate and are of the same magnitude, but of opposite sign.

It should be apparent that as the load is increased in a prestressed tie, the concrete stress reduces and the steel stress increases in direct proportion to the strain change in the concrete. Since the elastic modulus of concrete in tension is virtually the same as in compression, the action will continue until the tensile strength of the concrete is reached, at which time, the concrete will crack and the entire load must thereafter be carried by the steel alone. If the load is then reduced below the value of the effective prestress, the cracks will close and there will be a compressive force in the concrete which is equal to the difference between the effective prestressing force and the service load.

If the load resisted by the steel and concrete just prior to cracking is greater than the ultimate strength of the steel, the cracking load will be the ultimate load. This condition can exist in members prestressed with a very small percentage of reinforcing.

Consideration of the action of prestressed ties and the elastic-plastic nature of concrete leads one to the conclusion that the designer must consider the stress

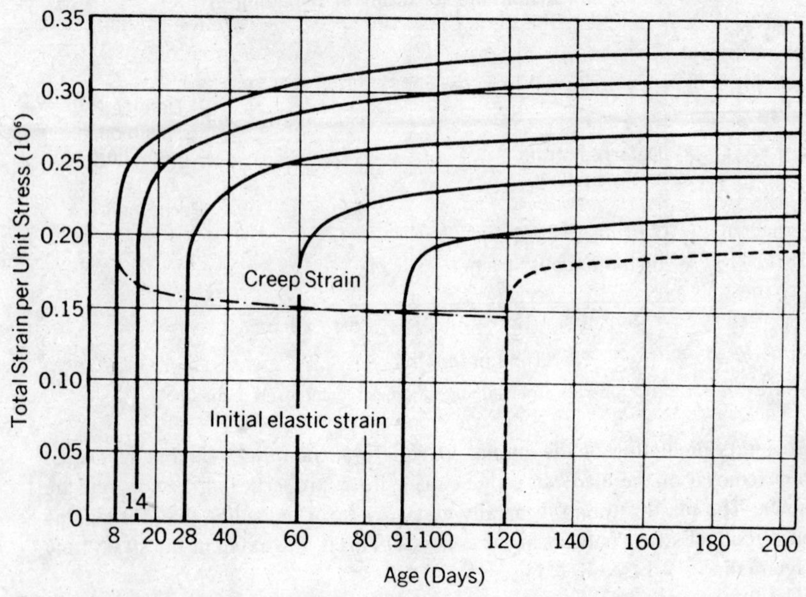

Fig. 9-2 Total strain due to constant sustained stress applied at various ages to a high-strength concrete stored at high humidity.

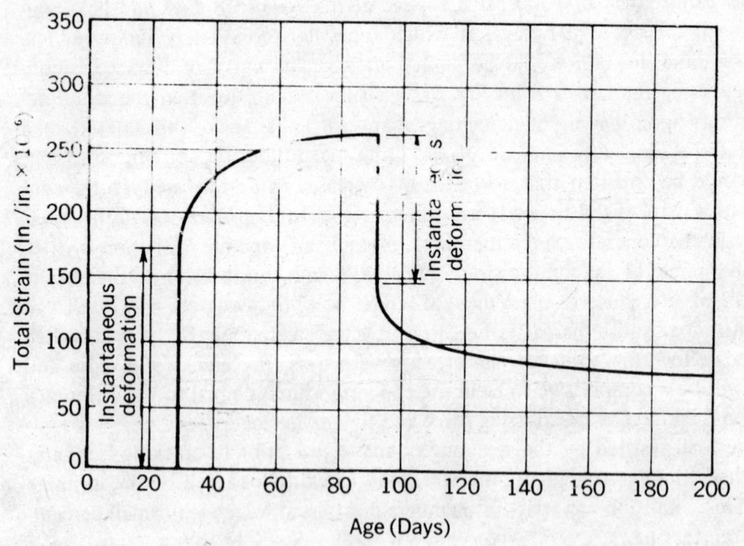

Fig. 9-3 Strain vs time diagram for concrete having creep characteristics of that in Fig. 9-2. For 1000 psi applied at 28 days and removed at 91 days.

level in the concrete and the duration of the applied loads in estimating the deformations that will result in the ties. Obviously, such deformation will result in deflection of a structure and may result in significant secondary stresses.

9-3 Columns and Piles

Prestressed concrete members that are axially loaded in compression are designed in the same manner as are reinforced concrete members. ACI 318, Sec. 18.11 (Ref. 3) provides that members with an effective prestress of less than 225 psi must have the minimum amount of non-prestressed reinforcement that is specified for reinforced concrete columns. There is no minimum amount of non-prestressed reinforcement required for columns with an effective prestress that is 225 psi or more.

The prestressing steel of axially loaded members is required by ACI 318 to be enclosed either by ties that are spaced at the least dimension of the column, but not more than 48 tie diameters, or by spirals that conform to the requirements for reinforced concrete columns.

Ultimate strength interaction diagrams for prestressed concrete members can be constructed using the principles of strain compatibility. The diagrams give the relationship between axial load capacity and moment capacity. Typically, they are of the shape shown in Fig. 9-4, from which it will be seen that they are similar to interaction diagrams for reinforced concrete columns.

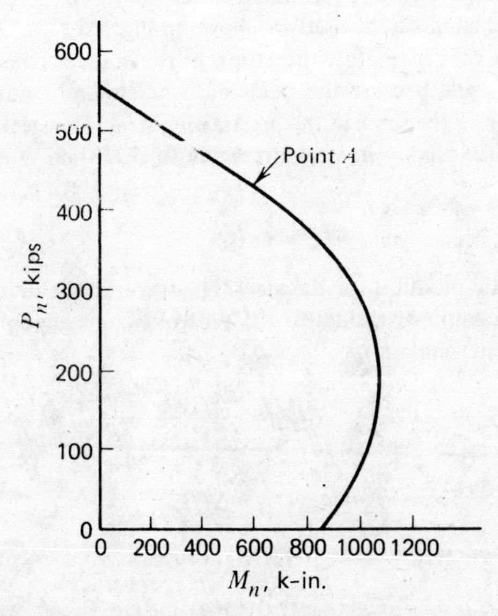

Fig. 9-4 Interaction diagram for the 14 in. square pile of I.P. 9-1.

An interaction diagram must be drawn for each column design; its construction is a function of the dimensions of the concrete and prestressing steel as well as the physical properties of the materials. To make an interaction diagram, one should have a stress-strain curve for the prestressing steel used in the column.

The intersect of the curve with the ordinate is the compressive strength capacity of the column under concentric load. It is usually taken to be the strength of the column when it is subjected to a uniform strain in the concrete of 0.002 in./in. The point of maximum moment capacity of the column is sometimes referred to as the balanced point. It is the combination of load and moment at which crushing of the concrete and yielding of the tension steel will occur simultaneously. The point where the axial load is equal to zero is the flexural capacity of the column section.

The capacity reduction factors specified in the ACI 318, Sec. 9.3, apply to axially loaded members. For axial loads greater than that at the balanced point, on members with normal ties (not spirals), ACI 318 provides that the capacity reduction factor (ϕ) be taken as 0.70. It also provides that in certain cases, ϕ can be linearly increased from 0.70 to 0.90 as P_n decreases from the balanced load to zero.

It will be noted that the interaction diagram is nearly linear between the intercept with the ordinate and point A in Fig. 9-4. For usual design computations, it is sufficiently accurate to assume this portion of the curve is linear; it actually has slight curvature.

The interaction curve can be constructed by computing the axial load and moment capacity of the section under different assumptions as to strain in the section. The assumed strains to be used are shown in Fig. 9-5.

For the condition of Fig. 9-5(a), the strain in the concrete due to the axial load is uniformly ϵ_0, and because the steel and concrete are bonded, the strain results in a reduction in the force in the prestressing steel. The steel is stressed in the elastic range so the change in steel stress due to the strain of ϵ_0 in the concrete is

$$\Delta f_s = -\epsilon_0 E_s$$

where E_s is the elastic modulus for the steel. The forces in the prestressing steel on the tensile and compressive sides of the centroidal axis are equal (for symmetrical reinforcement) and are

$$T_T = A_{ps}(f_{si} - \Delta f_s)$$

and

$$T_c = A'_{ps}(f_{si} - \Delta f_s)$$

The strength of the concrete is taken to be $k_3 f'_c$, and

$$C = (k_3)(f'_c)(b)(h)$$

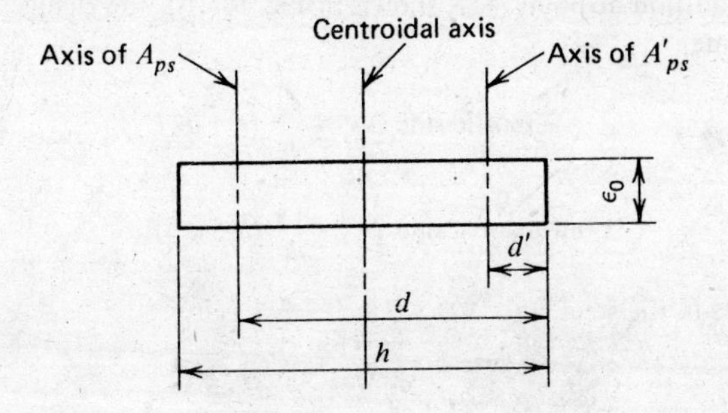

(a) Strain Distribution Assumed for Interaction Curve Point at $M_u = 0$

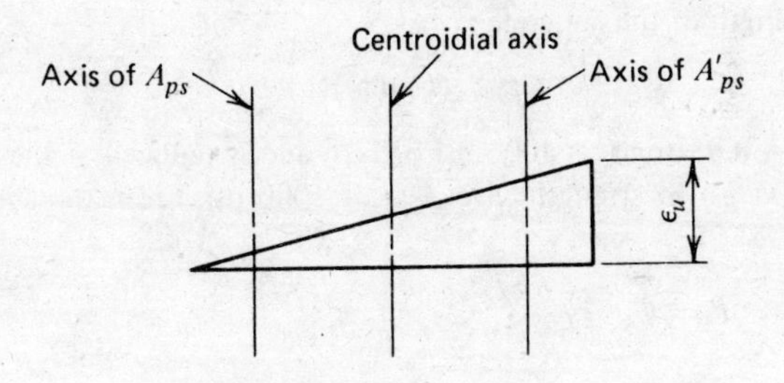

(b) Strain Distribution Assumed for Interaction Curve at Point A

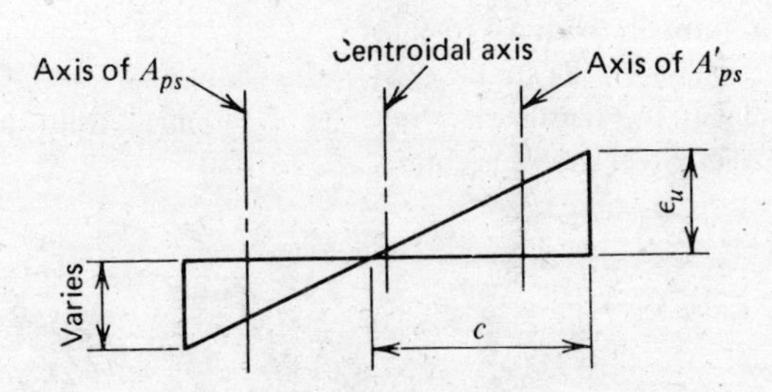

(c) Strain Distribution for Points Between Point A and Where $P_u = 0$

Fig. 9-5 Strain distribution for interactive curve.

where b and h are the width and depth of the concrete section, respectively. The strength factor, k_3, is normally taken to be equal to 0.85. The value of P_n for this condition of strain becomes

$$P_n = (k_3 f_c')(b)(h) - A_{ps}(f_{si} - \Delta f_s) - A'_{ps}(f_{si} - \Delta f_s)$$

$$P_n = C - T_T - T_c$$

For the condition at point A as shown in Fig. 9-5(b), the changes in the steel stresses become:

$$\text{tensile side } \Delta f_s = \frac{\epsilon_u}{h}(h - d)E_s$$

$$\text{Compressive side } \Delta f_s' = \frac{\epsilon_u}{h}(h - d')E_s$$

and the forces in the steel are

$$T_T = A_{ps}(f_{si} - \Delta f_s)$$

and

$$T_c = A_{ps}'(f_{si} - f_s')$$

while the strength of the concrete is

$$C = k_3(k_1)(f_c')(b)(h)$$

in which $k_1 = 0.85$ for $f_c' = 4000$ psi or less, and is reduced at the rate of 0.05 for each 1000 psi in strength in excess of 4000 psi. From the above, one can write

$$P_n = C - T_T - T_c$$

and

$$M_n = C(h/2 - k_2 h) + T_T(d - h/2) - T_c(h/2 - d')$$

The term k_2 is normally assumed to equal 0.42.

For the case where tensile strains exist, as shown in Fig. 9-5(c), the stress for a particular strain in the tensile steel must be determined from the stress-strain curve for the steel. Steel strains become

$$\epsilon_s = \epsilon_{se} + \epsilon_u \frac{d - c}{c}$$

$$\epsilon_s' = \epsilon_{se} - \epsilon_u \frac{c - d'}{c}$$

in which ϵ_{se} is the steel strain due to the prestressing of the tendons. The force in the tendons on the tensile side becomes

$$T_T = f_{su} A_{ps}$$

in which f_{su} is the steel stress corresponding to a strain of ϵ_s from the stress-strain curve for the prestressing steel. On the compression side, the steel continues to perform elastically and

$$T_c = A_{ps}'\left[f_{si} - E_s E_u\left(\frac{c - d'}{c}\right)\right]$$

and the force in the concrete section becomes

$$C = k_3(k_1)(f'_c)(b)(c)$$

and P_n is

$$P_u = C - T_T - T_c$$

The moment is

$$M_n = C(h/2 - k_2 c) + T_T(d - h/2) - T_c(h/2 - d')$$

For the case of $P_n = 0$, the intercept of the abscissa, a trial and error procedure must be used to find the value of c for which $P_n = 0$.

The procedure described is for the case of a rectangular column with the reinforcement confined to two rows. The same procedure is used for columns with more than two rows of steel; the relationships given above require modification to include the effects of the additional rows of steel.

Interaction diagrams for prestressed round columns can be made following the same general methods described above for rectangular columns. The relationship for the force in the concrete is as follows (Ref. 4).

$$C_c = \frac{0.85 f'_c D^2}{2}\left(\frac{\phi_1}{2} - \frac{1}{4}\sin 2\phi_1 - \frac{1}{(1.134k)^2}\left\{\cos^2\phi_2\left[\frac{\phi_1 - \phi_2}{2} - \frac{1}{4}(\sin 2\phi_1 - \sin 2\phi_2)\right] - \frac{2}{3}\cos\phi_2(\sin^3\phi_1 - \sin^3\phi_2) - \frac{1}{32}(\sin 4\phi_1 - \sin 4\phi_2 - 4\phi_1 + 4\phi_2)\right\}\right)$$

while that for the moment of the force in the concrete section is

$$M_c = \frac{0.85 f'_c D^3}{4}\left\{\frac{1}{3}\sin^3\phi_1 - \frac{1}{(1.134k)^2}\left[\frac{\cos^2\phi_2}{3}(\sin^3\phi_1 - \sin^3\phi_2) + \frac{\cos\phi_2}{16}(\sin 4\phi_1 - \sin 4\phi_2 - 4\phi_1 + 4\phi_2) + \frac{\cos^2\phi_1\sin^3\phi_1 - \cos^2\phi_2\sin^3\phi_2}{5} + \frac{\sin^3\phi_1 - \sin^3\phi_2}{7.5}\right]\right\}$$

As will be seen from Fig. 9-6 the angles ϕ_1 and ϕ_2 are functions of the strain diagram and the cross section of the column. The angle ϕ_1 varies with the distance between the neutral axis and the centroid of the column. Note that $\cos\phi_1 = (1 - 2k)$ when k is less than 1. The angle ϕ_1 is equal to π when k is greater than 1. The angle ϕ_2 is defined by the point on the strain diagram where the strain is equal to 0.0017 in./in. From Fig. 9-6 it will be seen that

$$y_2 = \left(\frac{1}{2} - k\right)D + kD\frac{\epsilon_1}{\epsilon_u} = D\left[\frac{1}{2} - k + k\frac{\epsilon_1}{\epsilon_u}\right]$$

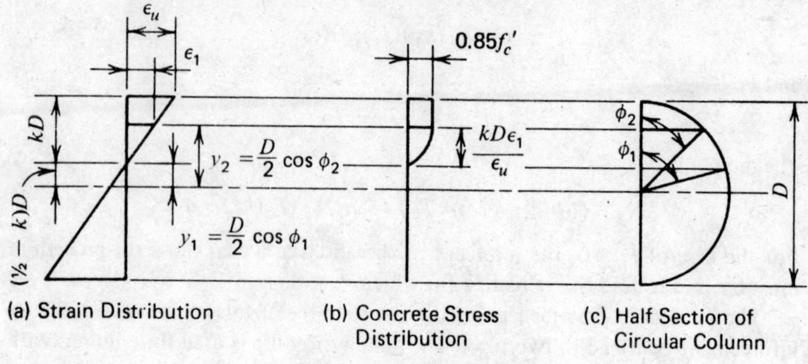

Fig. 9-6 Data for analysis of circular columns.

and

$$\cos \phi_2 = \frac{1}{2} - k + k\,\frac{\epsilon_1}{\epsilon_u}$$

The value of P_n for $M_n = 0$ should be computed using a concrete strain of 0.002 as was done for the rectangular column. Other points on the diagram can be found using the maximum flexural strain of 0.003. The effects of the prestressing steel, both for forces and moments, should be treated in the same manner as they are with rectangular sections.

It should be recognized that prestressed concrete columns have lower capacities than do reinforced concrete columns. The prestressing force causes compressive stresses in the concrete and hence reduces the capacity to carry external vertical load. In reinforced concrete columns, the steel is in compression and hence works with the concrete in resisting vertical load.

The effects of slenderness must be considered in the design of prestressed concrete columns. This can be done using the approximate methods contained in ACI 318, Sec. 10.11.

ILLUSTRATIVE PROBLEM 9-1 Construct the interaction curve for a square column having the following dimensions

$$b = h = 14.0 \text{ in.}$$
$$d' = 2.50 \text{ in.}$$
$$d = 11.50 \text{ in.}$$

The column is prestressed with four 0.5 in. strands, $f_{pu} = 270$ ksi, $f_{si} = 160$ ksi, and $E_s = 28{,}000$ ksi. The capacity reduction factor is to be taken as 0.70 for loads greater than the balanced load, and is to vary linearly from 0.70 to 0.90 between the balanced load and zero load capacity. Assume $\epsilon_0 = 0.002$ in./in., $\epsilon_u = 0.003$ in./in., $f_c' = 5000$ psi, and k_1, k_2 and k_3 are 0.80, 0.42 and 0.85, respectively.

SOLUTION: For the case of a concentric load,

$$\Delta f_s = \Delta f_s' = (0.002)(28000) = 56 \text{ ksi}$$
$$T_T = T_c = (2)(0.153)(160 - 56) = 31.8 \text{ kips}$$
$$C = (0.85)(5)(14)(14) = 833 \text{ kips}$$
$$P_n = 833 - (2)(31.8) = 769 \text{ kips}$$

For the case of zero strain on the tensile side and ϵ_u on the compressive side,

$$\Delta f_s = 0.003 \left[\frac{14.0 - 11.5}{14}\right] 28{,}000 = 15.0 \text{ ksi}$$

$$\Delta f_s' = 0.003 \left[\frac{14.0 - 2.5}{14}\right] 28000 = 69.0 \text{ ksi}$$

$$T_T = (2)(0.153)(160 - 15.0) = 44.37 \text{ kips}$$
$$T_c = (2)(0.153)(160 - 69.0) = 27.85 \text{ kips}$$
$$C = (0.80)(0.85)(5)(14)(14) = 666.4 \text{ kips}$$
$$P_n = 594.2 \text{ kips}$$
$$M_n = 666.4\,[7.0 - (0.42)(14)] + 44.4\,(11.5 - 7.0)$$
$$- (27.8)\,(7.0 - 2.5) = 821.0 \text{ k-in.} = 68.4 \text{ k-ft}$$

For the case of tensile strain on the tensile side of the section, and ϵ_u on the compressive side, with the strain in the prestressing steel due to the prestress

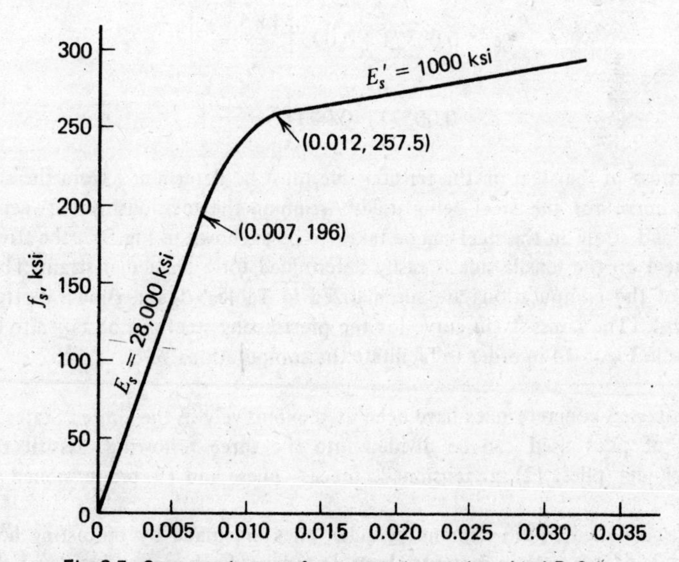

Fig. 9-7 Stress-strain curve for prestressing steel used in I.P. 9-1.

TABLE 9-1

	P_0 = 769.352 *k*		.70 X P_0 = 538.546 *k*		
c (in.)	P_n (k)	M_n (k-ft)	phi P_n	phi M_n	phi
14.000	594.184	68.393	415.928	47.875	.700
13.300	559.511	81.120	391.657	56.784	.700
12.600	524.688	92.250	367.281	64.575	.700
11.900	489.688	101.791	342.781	71.253	.700
11.200	454.478	109.749	318.134	76.824	.700
10.500	419.016	116.135	293.311	81.294	.700
9.800	383.248	120.962	268.273	84.673	.700
9.100	347.103	124.248	242.972	86.973	.700
8.400	310.488	126.014	217.341	88.210	.700
7.700	272.284	126.663	192.238	89.426	.706
7.000	235.960	124.870	172.851	91.473	.732
6.300	199.878	121.215	151.721	92.010	.759
5.600	163.608	115.806	128.529	90.976	.785
4.900	126.899	108.656	103.057	88.241	.812
4.200	89.453	99.739	75.019	83.646	.838
3.500	51.369	88.774	44.442	76.804	.865
2.800	12.705	75.246	11.329	67.096	.891
2.580	.000	59.126	.000	53.213	.900

$M_{u\max}$ = 126.752 k-ft at c = 7.858 in. 0.70 $M_{u\max}$ = 88.726 k-ft.

alone being equal to 160 ÷ 28,000 = 0.00571, strain in the steel become

$$\epsilon_s = 0.00571 + 0.003\left[\frac{11.5 - c}{c}\right]$$

$$\epsilon_s' = 0.00571 - 0.003\left[\frac{c - 2.5}{c}\right]$$

The stress in the steel on the tension side must be determined from the stress-strain curve for the steel being used. Assuming the relationship between the stress and strain in the steel can be taken to be as shown in Fig. 9-7, the stress in the steel on the tensile side is easily determined for a particular strain. The results of the computations are summarized in Table 9-1 and shown plotted in Fig. 9-4. (The stress-strain curve for the prestressing steel can be taken to be as shown in Fig. 5-14 in order to facilitate the computations.)

Prestressed-concrete piles have been used extensively in the United States. The types of piles used can be divided into the three following classifications (1) cylinder piles, (2) pre-tensioned, precast piles, and (3) pre-tensioned spun piles.

Post-tensioned, multi-element cylinder piles are made by precasting hollow cylinders of concrete in sections about 16 ft long. Each section has a wall thick ness of from 5 to 6 in. and holes are formed longitudinally through the walls at the time the sections are cast. After the precast sections have cured, they are aligned and the post-tensioning tendons are threaded through the holes in the walls and stressed and grouted in place. In this manner, piles up to 150 ft can be made.

Cylinder piles are also made pre-tensioned. In this process, the piles are either cast in conventional molds or may be cast in traveling molds which slipform the hollow sections.

The cylinder piles are normally made in diameters from 3 ft to 4 ft 6 in. and have been used with design loads up to 550 tons. The hollow shape is an efficient one for resisting axial loads, as well as for resisting bending moments that may be applied from any direction. Typical details and dimensions for cylinder piles are given in Fig. 9-8.

Pre-tensioned, precast piles have been made with square, triangular, octagonal, and round cross sections, both hollow and solid. Pre-tensioned piles have been used a great deal in the construction of waterfront structures and are fabricated in the normal manner used in pre-tensioned construction. This type of pile has been used more than cylinder or spun piles. Typical details and dimensions of square and octagonal prestressed piles are shown in Figs. 9-9(a) and 9-9(b).

It should be noted that the capacities of the piles given in Figs. 9-8 and 9-9 are based on stress and not on strain compatibility. Hence, these capacities are only applicable for concentric loads and where buckling is not a possibility.

Pre-tensioned spun piles are made in individual molds designed to resist the pre-tensioning force during the casting and curing of the pile. The manufacturing procedure consists of placing the tendons and reinforcing cage in steel molds, stressing the tendons, and placing the mold on revolving wheels that turn the mold as the concrete is placed. The centrifugal force compacts the concrete and forces the excess water from the plastic concrete. The pile is then cured and stripped from the mold.

Prestressed piles generally have better driving characteristics than reinforced-concrete piles. The prestressed piles seem to penetrate better and with less effort. In addition, prestressed piles can be made longer than is practical with reinforced-concrete piles, due to their lower dead weight and higher resistance to bending moments. Prestressed piles will also stand up well under adverse driving conditions, if they are properly designed and fabricated.

Pre-tensioned sheet piles have also been used on a number of projects in this country. Such piles are generally solid and rectangular in cross section, but have a tongue and groove to interlock them with adjacent piles in the completed structure.

The procedure used in the design of prestressed piles is no different than that employed in the design of columns that have axial load or combined axial load and bending. Experience has shown, however, that a minimum prestressing stress of from 700 to 900 psi is required in order to prevent the piles from cracking

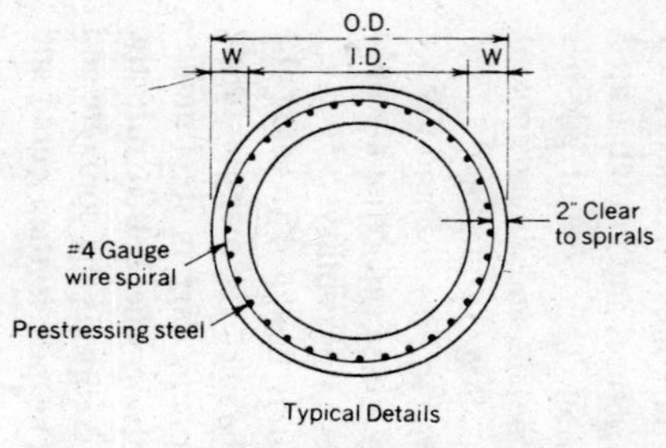

PILE PROPERTIES

Pile Size			Area A_c sq in.	Approx. Weight p/f (1)	Minimum Prestress Force (2) kips	Strands Per Pile Diameter (3)		Moment of/inertia in.^{4}	Section Modulus in.^{3}	Perimeter in.	Design Bearing Capacity (tons) Concrete Strength (4)	
OD in.	ID in.	W in.				7/16"	1/2"				5000 psi	6000 psi
36	26	5	487	508	414	24	18	60,000	3334	113	242	292
	24	6	565	590	481	28	21	66,100	3676	113	282	339
48	38	5	675	703	574	33	25	158,200	6593	151	337	405
	36	6	792	826	674	39	29	178,100	7422	151	396	475
54	44	5	770	802	655	38	28	233,400	8645	170	385	462
	42	6	904	940	769	44	33	264,600	9802	170	452	542

Fig. 9-8 Pre-tensioned cylinder pile cross section.

NOTES

(1). Weights are based on 150 lbs per cubic foot.

(2). Minimum prestressed force based on unit prestress of 850 psi after losses.

(3). Based on 7/16" and 1/2" strand with an ultimate strength of 31,000 lbs and 41,300 lbs respectively.

(4). Design bearing capacity based on 5000 psi and 6000 psi concrete and an allowable unit stress on the tip of the pile of $.2f'_cA_c$.

SQUARE PRESTRESSED PILES

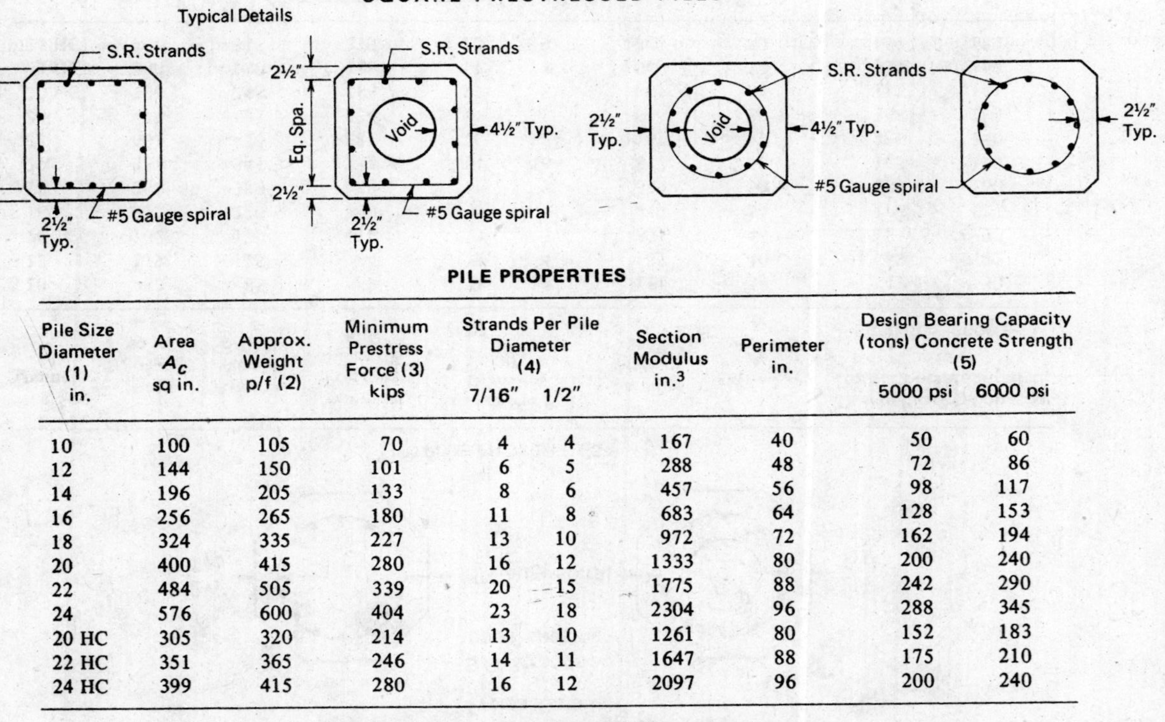

PILE PROPERTIES

Pile Size Diameter (1) in.	Area A_c sq in.	Approx. Weight p/f (2)	Minimum Prestress Force (3) kips	Strands Per Pile Diameter (4)		Section Modulus in.^{3}	Perimeter in.	Design Bearing Capacity (tons) Concrete Strength (5)	
				7/16"	1/2"			5000 psi	6000 psi
10	100	105	70	4	4	167	40	50	60
12	144	150	101	6	5	288	48	72	86
14	196	205	133	8	6	457	56	98	117
16	256	265	180	11	8	683	64	128	153
18	324	335	227	13	10	972	72	162	194
20	400	415	280	16	12	1333	80	200	240
22	484	505	339	20	15	1775	88	242	290
24	576	600	404	23	18	2304	96	288	345
20 HC	305	320	214	13	10	1261	80	152	183
22 HC	351	365	246	14	11	1647	88	175	210
24 HC	399	415	280	16	12	2097	96	200	240

Fig. 9-9(a) Typical pre-tensioned pile cross section. (See notes on p. 383).

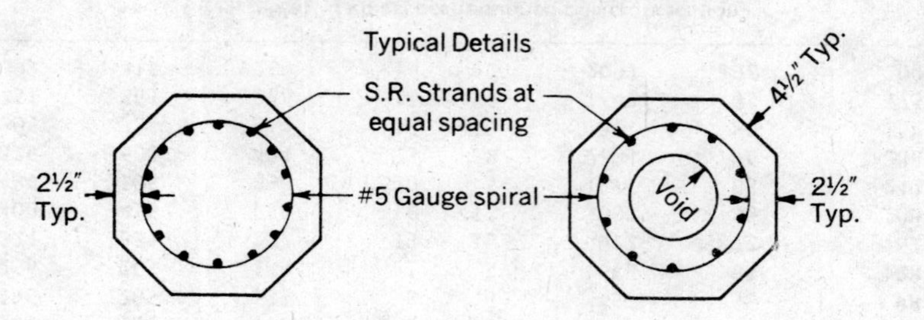

PILE PROPERTIES

Pile Size Diameter (1) in.	Area A_c sq in.	Approx. Weight p/f (2)	Minimum Prestress Force (3) kips	Strands Per Pile Diameter (4)		Section Modulus in.^{3}	Perimeter in.	Design Bearing Capacity (tons) Concrete Strength (5)	
				7/16"	1/2"			5000 psi	6000 psi
10	83	85	59	4	4	111	34	41	50
12	119	125	84	5	4	189	40	59	71
14	162	170	114	7	5	301	46	81	97
16	212	220	149	9	7	449	54	106	127
18	268	280	188	11	8	639	60	134	160
20	331	345	232	14	10	877	66	165	198
22	401	420	281	16	12	1167	72	200	240
24	477	495	334	19	15	1515	80	238	286
20 HC	236	245	166	10	8	805	66	118	141
22 HC	263	280	188	11	8	1040	72	134	160
24 HC	300	315	210	12	9	1308	80	150	180

Fig. 9-9(b) Typical pre-tensioned pile cross section.
(See notes on p. 383).

NOTES

(For both square and octagonal piles)

(1) Voids in 20″, 22″ and 24″ diameter hollow-core (HC) piles are 11″, 13″ and 15″ diameter, respectively, providing a minimum 4-1/2″ wall thickness. If a greater wall thickness is desired, properties should be increased accordingly.

(2) Weights based on 150 lb per cubic foot of regular concrete.

(3) Minimum prestress force based on unit prestress of 700 psi after losses.

(4) Bases on 7/16″ and 1/2″ high strength strand with an ultimate strength of 31,000 lbs and 41,300 lbs respectively. If regular strength strand is used, the number of strands per pile should be increased accordingly in conformance with strand manufacturer's tables.

(5) Design bearing capacity based on 5000 psi and 6000 psi concrete and an allowable unit stress on the tip of the pile of $.2f'_c A_c$. These bearing capacity values may be increased if higher strength concrete is used.

during driving. Cracking during driving has occurred on many projects and is believed to be the result of tensile stresses in the piles, due to the piles rebounding elastically from the driving hammer. This type of cracking is more apt to occur when driving is commenced (particularly in soft materials) and little tip resistance has been developed. The cracking can be controlled by using techniques compatible with the conditions at hand (Ref. 15).

9-4 Fire Resistance

Fire resistance, as determined by standard tests, is a measure of the ability of a structural element to prevent the spread of fire and to retain the necessary structural strength at elevated temperatures. The fire resistance of a member is generally expressed in hours and is indicative of the length of time the member can be subjected to a standard fire test without failing. Failure may result from the inability of the member to adequately perform any one of the following functions:

(1) Walls, floors, and roof elements must not allow a temperature rise, on the side opposite to the fire, which would be sufficient to allow inflammable material to ignite, either by conduction through the member or as a result of holes and cracks forming in the member, which would allow flames, hot air, or gases to pass through the member.

(2) All structural members must retain sufficient strength at elevated temperatures to assure safety to the occupants as well as the firemen who are fighting the fire. Severe structural deflection could result in cracking of the supported slabs or panels, which would cause them to fail according to the requirement described above.

(3) The elements must also retain their structural integrity under the action of a stream of water, during or immediately after exposure to the standard fire test.

The effects of elevated temperatures on the principal constituents of reinforced and prestressed concrete are interesting from an academic standpoint and can be generalized as follows:

Concrete: The coefficient of thermal conductivity is often assumed to be greater for high-strength concrete, such as is used in prestressed concrete, than for low-strength concrete, although the variation in this coefficient is large. The compressive strength of concrete can be assumed to be approximately 67 and 33% of the strength at room temperature when the temperature is raised to 750 and 1450°F, respectively. The modulus of elasticity, in terms of the modulus at room temperature, is reduced to approximately 50% at 750°F and 20% at 1450°F. Decomposition of the concrete is evident at about 1300°F (Ref. 6).

High-tensile steel wire: The strength of high-tensile steel generally varies as follows at elevated temperatures: (1) Slight increase in strength up to temperatures of 300°F; and (2) The ultimate tensile strength at 750°F is approximately 50% of the original ultimate tensile strength. The reduction in the elastic modulus of high-tensile steel is of the order of 6% at 400°F and 20% at 600°F. In addition, elevated temperatures result in a significant increase in the relaxation of the steel. (See Chapter 2.) The thermal coefficient of linear expansion is not a constant value for cold drawn wire at high temperatures. The expansion of wire heated to temperatures above 300°F is not entirely recovered upon cooling. Virtually all of the strength of wire is regained upon cooling from temperatures as high as 600°F, even though the strain is not. The strength upon cooling for wire heated as high as 750°F may be a considerable percentage of the original strength (Ref. 7).

Reinforcing steel: The strength of normal reinforcing steel is reduced to 50% of its original strength at temperatures of 950 to 1100°F.

In view of the above general properties, at elevated temperatures and after cooling from elevated temperatures, a prestressed structure that was exposed to a fire and did not fail could be expected to possess the characteristics indicated for the following maximum temperature conditions.

Above 400°F, but less than 600°F, the reduction in the effective prestress may reduce the resistance of the cooled structure to cracking, although the ultimate strength may not be materially affected, due to the recovery of the steel strength upon cooling.

Above 600°F, but less than 750°F, the structure would be expected to be badly cracked and have permanent deflection. The ultimate strength of the structure after cooling may still be adequate, since the regain of steel strength would be considerable and the maximum reduction in concrete strength would be of the order of one-third.

Above 750°F, the reduced steel strength and the loss of concrete compressive strength would very likely render the cooled structure unsafe.

In the past several years, there has been considerable research into the ability of prestressed concrete to resist the effects of fire. This research has resulted in the requirements summarized in Table 9-2. The requirements are in the 1982 Edition of the Uniform Building Code (Ref. 8). Grade A concrete is defined as concrete made with aggregates such as limestone, calcareous gravel, trap rock, slag, expanded clay, shale, slate silicons, or any other aggregates possessing equivalent fire resistive properties and containing 40% or less quartz, chert, or flint. Grade B concrete is all concrete other than Grade A concrete and includes concrete made with aggregates containing more than 40% quartz, chert, or flint. In addition, the Uniform Building Code lists the following requirements and interpretations:

Bonded prestressed concrete tendons: For members having a single tendon or more than one tendon installed with equal concrete cover measured from the nearest surface, the cover shall be not less than that set forth on Table No. 43-A.

For members having multiple tendons installed with variable concrete cover,

TABLE 9-2 Requirements of the 1982 Edition of the Uniform Building Code Stipulated in Table No. 43-A

Structural Parts to be Protected	Item Number	Insulating Material Used	Minimum Thickness of Insulating Material for Following Fire-Resistive Periods (in inches)			
			4 Hr.	3 Hr.	2 Hr.	1 Hr.
Bonded Pretensioned Reinforcement in Prestressed Concrete[5]	30	Grade A[6] concrete — Beams or girders	4[7]	3[7]	$2\frac{1}{2}$[7]	$1\frac{1}{2}$
		Grade A[6] concrete — Solid slabs[8]		2	$1\frac{1}{2}$	1
Bonded or Unbonded Posttensioned Tendons in Prestressed Concrete[5,10]	31	Grade A or B Concrete Unrestrained Members:				
		Solid Slabs[8]		2	$1\frac{1}{2}$	
		Beams and Girders[11]				
		8 in. wide		$4\frac{1}{2}$	$2\frac{1}{2}$	$1\frac{3}{4}$
		>12 in. wide	3	$2\frac{1}{2}$	2	$1\frac{1}{2}$
Bonded or Unbonded Posttensioned Tendons in Prestressed Concrete[5,10]	32	Grade A or B Concrete Restrained Members:[12]				
		Solid Slabs[8]	$1\frac{1}{2}$	1	$\frac{3}{4}$	
		Beams and Girders[11]				
		8 in. Wide	$2\frac{1}{2}$	2	$1\frac{3}{4}$	
		>12 in. wide	2	$1\frac{3}{4}$	$1\frac{1}{2}$	

[5] Where lightweight Grade A concrete aggregates producing concrete having an oven-dry weight of 110 pounds per cubic foot or less are used, the tabulated minimum cover may be reduced 25 percent, except that in no case shall the cover be less than $\frac{3}{4}$ inch in slabs nor $1\frac{1}{2}$ inches in beams or girders.

[6] For Grade B concrete increase tendon cover 20 percent.

[7] Adequate provisions against spalling shall be provided by U-shaped or hooped stirrups spaced not to exceed the depth of the member with a clear cover of 1 inch.

[8] Prestressed slabs shall have a thickness not less than that required in Table No. 43-C for the respective fire-resistive time period.

[9] For use with concrete slabs having a comparable fire endurance where members are framed into the structure in such a manner as to provide equivalent performance to that of monolithic concrete construction.

[10] Fire coverage and end anchorages shall be as follows: Cover to the prestressing steel at the anchor shall be $\frac{1}{2}$ inch greater than that required away from the anchor. Minimum cover to steel bearing plate shall be 1 inch in beams and $\frac{3}{4}$ inch in slabs.

[11] For beam widths between 8 and 12 inches, cover thickness can be determined by interpolation.

[12] Interior spans of continuous slabs, beams and girders may be considered restrained.

the average tendon cover shall be not less than that set forth in Table No. 43-A provided:

(1) The clearance from each tendon to the nearest exposed surface is used to determine the average cover.

(2) In no case can the clear cover for individual tendons be less than one-half of that set forth in Table No. 43-A. A minimum cover of $\frac{3}{4}$ in. for slabs and 1 in. for beams is required for any aggregate concrete.

(3) For the purpose of establishing a fire-resistive rating, tendons having a clear covering less than that set forth in Table No. 43-A shall not contribute more than 50 percent of the required ultimate moment capacity for members less than 350 square inches in cross-sectional area and 65 percent for larger members. For structural design purposes, however, tendons having a reduced cover are assumed to be fully effective.

Many types of prestressed concrete standard products, such as double T slabs, cored slabs, etc., have been tested for fire resistance and carry the approval of the Underwriters' Laboratories, Inc. (Ref. 19). In addition, a number of prestressed concrete structures have been subjected to actual fires. The performance of these structures has almost invariably been good.

Additional information on fire resistance of prestressed concrete can be found in Refs. 10, 11, and 12.

9-5 Normal Temperature Variation

The effect of nominal atmospheric temperature variations on the performance of prestressed structures and on the magnitude of the effective prestress is occasionally questioned by persons who are unfamiliar with prestressed concrete. Because the thermal coefficients of linear expansion for steel and concrete are of the same order (6.5×10^{-6} ft/ft/°F±), if the steel and concrete have the same temperatures at all times, there is no significant effect on the effective prestress for normal changes in temperature.

If the tendons are not bonded to the member, but are exposed to the atmosphere, it would be possible for the temperature of the tendons to be different from that of the concrete section, due to their difference in mass and exposure to heat sources. Under such conditions, the effect of temperature variations should be studied on the basis of estimated maximum temperature variations and the effect of these variations on the effective prestress.

Atmospheric temperature variations can result in significant stresses in structures prestressed by jacks rather than by tendons (*see* Sec. 1-3). The effect of temperature variations must be given careful consideration in this type of structure.

A prestressed member with either bonded or unbonded tendons will, of course, expand or contract with temperature variations. Provision should be made for thermal expansion and contraction in prestressed construction, just as it is in other types of construction, unless computations indicate the effect can be reasonably neglected. It should be kept in mind that changes in the length of a member can only cause stresses in a structure if the structure is restrained.

If a temperature gradient exists within a member and the temperature distribution is known, the thermal stresses in the section can be easily computed (Ref. 13). Priestley has shown if plane sections are assumed to remain plane, for a section as shown in Fig. 9-10 subject to the temperature distribution shown in Fig. 9-11, the thermally induced strains and stresses are as shown in Fig. 9-12. The equations for stress and for force and moment equilibrium are as follows:

$$f_y = E\left(\epsilon_1 + \epsilon_2 \frac{y}{d} - \alpha\, t_y\right) \tag{9-1}$$

$$\epsilon_1 + \epsilon_2 \frac{n}{d} = \frac{\alpha}{A} \int_0^d t_y b_y\, dy \tag{9-2}$$

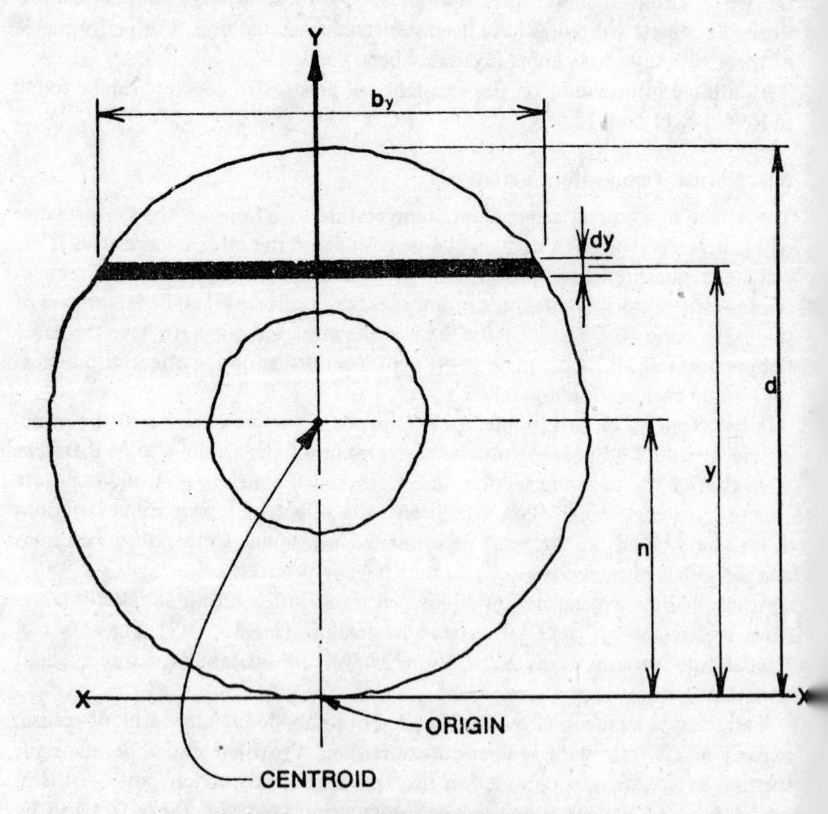

Fig. 9-10 Cross section of a tubular beam.

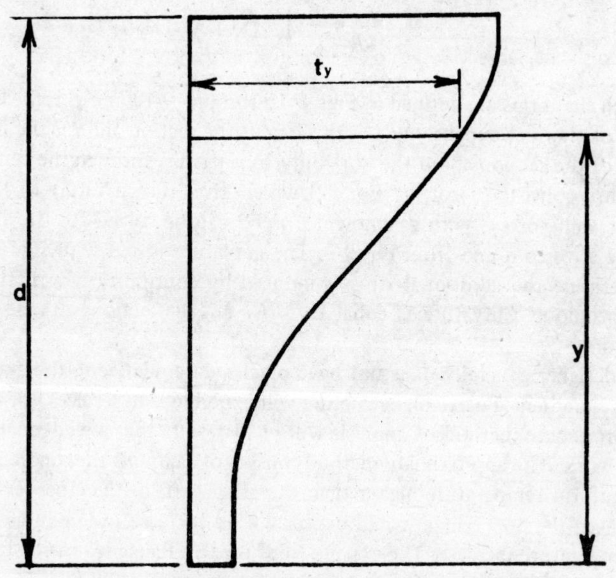

Fig. 9-11 Temperature gradient in beam of Fig. 9-10.

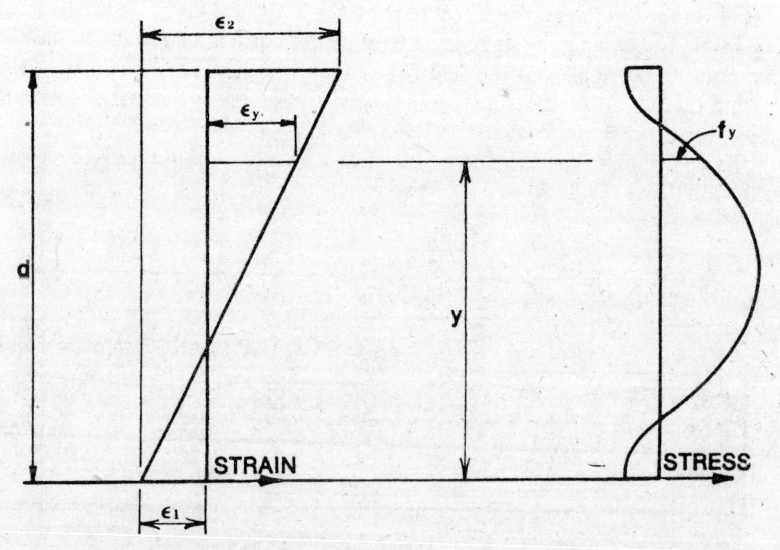

Fig. 9-12 Strain and stress distributions in beam of Fig. 9-10 due to temperature gradient of Fig. 9-11.

$$\frac{\epsilon_2}{d} = \frac{\alpha}{I} \int_0^d t_y(y - n)\, b_y dy \qquad (9\text{-}3)$$

in which the terms are defined in Figs. 9-10 through 9-12 and α is the linear coefficient of thermal expansion, A is the area of the section and I is the moment of inertia of the section about the horizontal axis passing through the centroid. The procedure consists of solving Eq. 9-3 which gives the curvature of the section and the value of ϵ_2. With ϵ_2 known, Eq. 9-2 can be solved for the value of ϵ_1 and the stresses found from Eq. 9-1. The curvature found with Eq. 9-3 is used to determine the additional stresses induced by continuity in structures which are continuous. Curvature is equal to M/EI and deflections can be computed therefrom.

The designer generally does not have precise data relative to the extreme temperature gradient a particular structure will experience in service. For this reason an approximate method of analysis will yield results that are adequate for most design work. The approximate method consists of assuming the top flange is raised to a uniform temperature higher than the other parts of the cross section. If unrestrained, the top flange would experience an increase in length as a result of the temperature increase. The expansion of the top flange is resisted by the webs and bottom flange.

The first step in computations with the approximate method consists of computing the forces that would exist in the top flange due to the increase in temperature if the flange were completely restrained. The force results in a compressive stress in the top flange. The effect of the force in the top flange on the section as a whole is computed by applying a tensile force equal in magnitude to the compressive force, applied at the centroid of the top flange. The sum of the stresses from the two forces gives the approximate stresses due to thermal effects in the section.

As an example, consider the bridge cross section shown in Fig. 9-13. Assume the top flange temperature is 30°F higher than that of the webs and bottom flange.

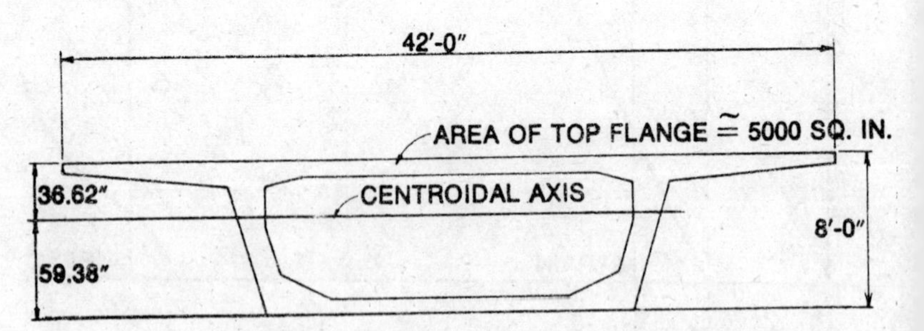

Fig. 9-13 Typical bridge cross section.

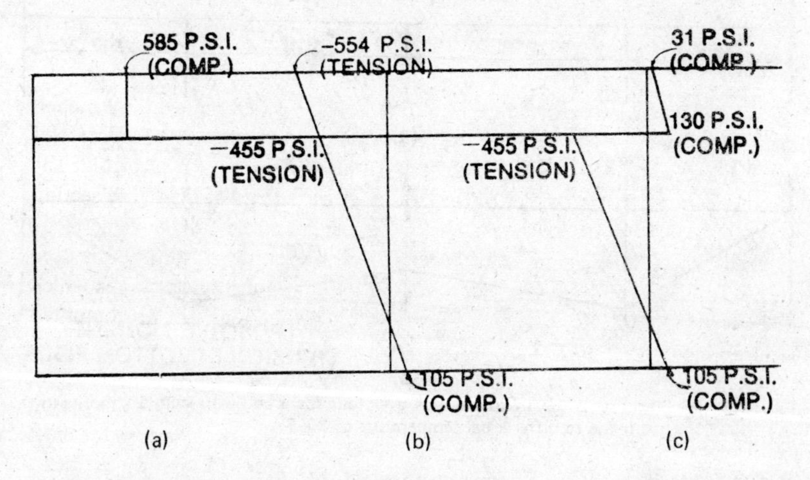

Fig. 9-14 Stresses in the bridge of Fig. 9-12, by approximate calculation, for a temperature differential of 30°F; (a) restrained top flange; (b) stresses due to tensile force; and (c) combined stresses.

The compressive stress in the top flange, if fully restrained, would be:

$$fc = (\Delta t)(\alpha)(Ec) \tag{9-4}$$

in which α is the linear coefficient of thermal expansion and Ec is the elastic modulus of the concrete. If $\alpha = 0.0000065$ and $Ec = 3{,}000{,}000$ psi, $fc = 585$ psi. The force in the top flange would be:

$$p = 585 \times 5000 = 2925 \text{ kips}$$

The tensile force applied 5.29 inches from the top of the section produces the stresses shown in Fig. 9-14(b) and the combined stresses are as shown in Fig. 9-14(c). These are the approximate stresses that would exist if the beam were a single simply supported span.

If made continuous over four spans as shown in Fig. 9-15(a), it can be shown that secondary moments and reactions as shown in Fig. 9-15(b) would exist. The secondary reactions are those required to prevent the beam from deflecting upward from its supports as a result of the thermal gradient. The secondary moment at the first interior support results in a stress of 314 psi (compression) in the top fiber and 510 psi (tension) in the bottom fiber. The net stress in the member due to the effect of the differential temperature would be those obtained by combining the stresses shown in Fig. 9-14(c) with those resulting from the secondary moment of Fig. 9-15(b).

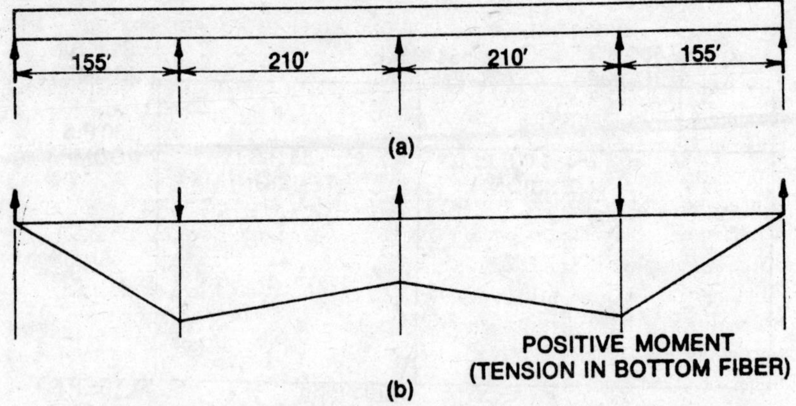

Fig. 9-15 (a) Elevation of beam continuous over four spans and (b) secondary moments in the beam due to differential temperature of 30°F.

9-6 Fatigue

Fatigue strength of structural elements is important if the elements are to be subjected to frequent reversals of stress or variations in stress. Resistance to fatigue is, therefore, an essential property for bridge members, but is not normally an important consideration in building construction.

Although the effect of fatigue on prestressed-concrete beams is not completely understood, there has been considerable research into the causes and types of fatigue failures which may result under the action of stress variations of various magnitudes.

Fatigue failures could be expected to occur in any of the following modes:

(1) Failure of the concrete due to flexural compression.

(2) Failure of the concrete due to diagonal tension or shear.

(3) Failure of the prestressing steel due to flexural, tensile-stress variations.

(4) Failure of pre-tensioned beams due to loss of bond stress.

(5) Failure of the end anchorages of post-tensioned beams.

Although the fatigue limit for concrete in compression alone is generally thought to be from 50 to 55% of the static compressive strength, no failures of prestressed flexural members due to compressive flexural stresses have been reported in the literature. Apparently the restriction of the concrete compressive stress in complete bridge structures to $0.40 f'_c$ results in adequate safety (Ref. 14). It should be pointed out that the fatigue limit mentioned above is for the load alternating from zero to the maximum value, which is a larger variation than is normally experienced in prestressed flexural members in actual service.

In composite bridge stringers ($f'_c = 5000$ psi) of moderately long span, the compressive stress in the top fibers of the precast section may vary from 1500 psi under the loading condition of dead load alone to 2000 psi under the loading condition of dead load plus live load and impact. In a similar manner the compressive stress in the composite flange (cast-in-place deck–$f'_c = 3000$ psi ±) may vary from zero to 900 psi. It is apparent that these ranges of stress variations and maximum values are considerably below the fatigue limit of the concrete in compression (0 to $0.50f_c'$) and for this reason, fatigue of the concrete should not be a problem under these conditions. The variation of stress in the compressive flange from zero to $0.40f_c'$, due to the application of live load plus impact alone, would only be expected to occur in short-span structures in which the dead load of the structure itself is relatively unimportant.

Fatigue failures due to diagonal tension or shear alone have apparently not been observed in prestressed concrete practice or research. Prestressed railroad ties, which are subjected to high-shear stresses, have shown distress when tested under repeated loads, as well as in service, but failure usually is caused by lack of adequate bond rather than insufficient shear or diagonal tensile strength. Railroad ties are very special elements, due to their high-shear loads and short-shear spans.

The majority of fatigue failures that have been found in testing prestressed beams have resulted from fatigue of the tendons. It appears that after the cracking load is reached, concentrations of stress or other phenomena related to the cracks develop in the tendons in the vicinity of the cracks and failure results. It must be emphasized, however, that most tests reveal that the fatigue resistance of prestressed-concrete beams is high and generally superior to conventionally reinforced concrete.

In several fatigue tests conducted abroad it was found that individual wires contained in the prestressing tendons failed by fatigue at points where they passed over spacers which were used to hold the wires in position (Ref. 15). This would lead one to suspect that the secondary stress that develops in the wire due to its passing over a spacer (usually a small diameter wire or bar) can be sufficiently severe to cause a point of fatigue weakness and possible fatigue failure. No failures in actual structures due to this effect are known, but it is believed that designers should avoid the use of spacers and should avoid the use of sharp bends in the tendons when possible, if fatigue is a consideration in the design.

Bond failures have been found in testing very short-span members, as was mentioned above. It appears that cracking of the beam sets up conditions which result in deterioration of the flexural bond between the tendon and the concrete as additional variations in the load are applied. When the flexural bond is destroyed from the point of cracking to the vicinity of the support, in the region where transfer bond is developed, failure ensues. (*See* Sec. 5-9.)

The available experimental data leads one to conclude that the types of tendons normally used domestically in pre-tensioned work provide adequate safety against bond failure for members of usual proportions.

There are indications that a light, hard coating of normal oxidation on the surface of the tendons improves the dynamic-bond properties, just as it improves the static bond properties (*see* Sec. 5-9 for other factors that affect bond stresses).

No reports of fatigue failures in the steel at the anchorages of post-tensioned members are to be found in the literature. This type of failure is extremely unlikely in bonded construction, since the grout is very effective in developing flexural bond stresses. This was demonstrated in one test in which the end anchorages were removed from the tendons of a grouted beam and the member was then subjected to a fatigue test. The results were satisfactory and failure resulted from fatigue of the tendons, and not from lack of bond.

In unbonded post-tensioned construction, the end anchorages could be subjected to some variation in stress under the action of variation in external load. This type of construction is not generally used in members to be subjected to frequent variations in stress; however, there are very little experimental data available on the performance of this type of construction under repeated loads.

Existing acceptance standards for unbonded tendons include fatigue test requirements (Ref. 10).

REFERENCES

1. Guyon, Y., *Prestressed Concrete*, p. 58, John Wiley and Sons, Inc., New York, 1953.
2. Ross, A. D., "Creep of Concrete Under Variable Stress," *ACI Journal*, **29**, No. 9, 739-758 (Mar. 1958).
3. ACI Committee 318, *Building Code Requirements for Reinforced Concrete* (ACI 318-83), American Concrete Institute, Detroit, 1983.
4. Portland Cement Association, "Ultimate Load Tables for Circular Columns," Portland Cement Association, Skokie, Ill., 1960.
5. Smith, E. A. L., "Tension in Concrete Piles During Driving," *Journal of the Prestressed Concrete Institute*, **5**, No. 1, 35-40 (Mar. 1960).
6. Guyon, op. cit. p. 101.
7. Hill, A. W. and Ashton, L. A., "The Fire Resistance of Prestressed Concrete," *Proc. World Conference on Prestressed Concrete*, A20-1-A20-8 (July 1957).
8. "Uniform Building Code," 1982 edition, International Conference of Building Officials, Whittier, Calif., 1982.
9. "Building Materials Directory, January 1982," Underwriters' Laboratories, Inc., Chicago, 1982.
10. *Post-Tensioning Manual*, Post-Tensioning Institute, 3rd edition, Phoenix, 1981.
11. *PCI Design Handbook*, Prestressed Concrete Institute, Chicago, 1971.
12. ACI Committee 423, "Recommendations for Concrete Members Prestressed with Unbonded Tendons" (proposed), American Concrete Institute, Detroit, 1983.
13. Priestly, M. J. N., "Design of Concrete Bridges for Temperature Gradients," *Journal of the American Concrete Institute*, Proceedings Vol. 75, No. 5, May 1978, pp. 209-217.
14. Hoffman, P. C. McClure, R. M. and West, H. H., "Temperature Study of an Experimental Segmented Concrete Bridge," *P. C. I. Journal*, Vol. 28, No. 2, March/April 1983, Prestressed Concrete Institute, Chicago.
15. Nordby, Gene M., "Fatigue of Concrete-A Review of Research," *Journal of the American Concrete Institute*, **30**, 210-215 (Aug. 1958).
16. Sawko, F. and Saha, G. P., "Fatigue of Concrete and Its Effect Upon Prestressed Concrete Beams," *Magazine of Concrete Research*, **20**, No. 62, 21-30 (Mar. 1968).

10 Cracking and Other Defects—Their Cause and Remedy

10-1 Introduction

Defects, generally in the form of cracks, honeycombing, and excessive deflection, often occur in what are otherwise well-designed and properly fabricated prestressed-concrete members. The purpose of this chapter is to describe the most common defects and suggest means of preventing their occurrence.

Flexural cracks that are fine and closely spaced are to be expected in prestressed-concrete flexural members in which the tensile stresses exceed the modulus of rupture, just as they are in reinforced concrete. Cracks of this type are considered to be normal and are not considered to be defects.

Prestressed concrete members that have minor defects can be repaired with the modern materials available with the expectation that the member will give many useful years of service. Recommended methods of repair are described in the following articles when applicable.

10-2 Cracking

Undesirable cracking occurs in prestressed members due to a variety of causes. Most cracks that occur in precast elements are not structurally important, but cracks should be avoided whenever possible. The following is a description of cracks which are commonly found in precast prestressed concrete members, as well as an explanation of their cause and methods of avoiding them.

Large Flexural Cracks

Most precast prestressed members are simple beams and are designed to be supported at their ends only. If members of this type are supported or have loads applied to them other than in the direction or at the locations intended by the designer, large flexural cracks may occur. Precast members with overhangs sometimes become supported at locations that are unintended, due to the deflection of the member at the time of stressing. This is illustrated in Fig. 10-1, in which a member with overhangs at each end is shown before and after prestressing. The upward deflection of the relatively long interior span may result in rotations at the intended point of support. These rotations cause the member to be supported at the extreme ends. The application of the reactions at the end of the member results in flexural cracking, as is shown in Fig. 10-1. This type of cracking can be avoided if the designer properly investigates all facets of the design of each individual member and properly specifies the permissible locations of supports during fabrication and erection, as well as the precautions to be taken to avoid difficulties of this type.

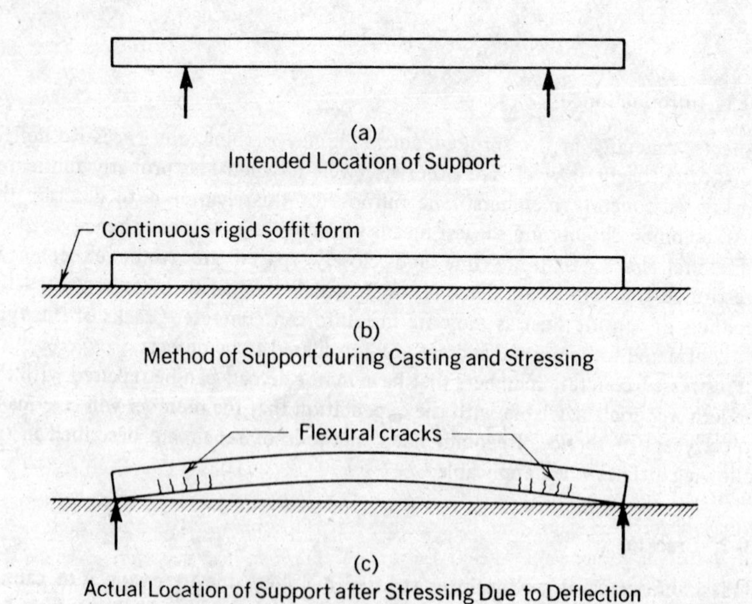

Fig. 10-1 Example of unintended mode of support of a beam with overhangs resulting from deflection due to prestressing.

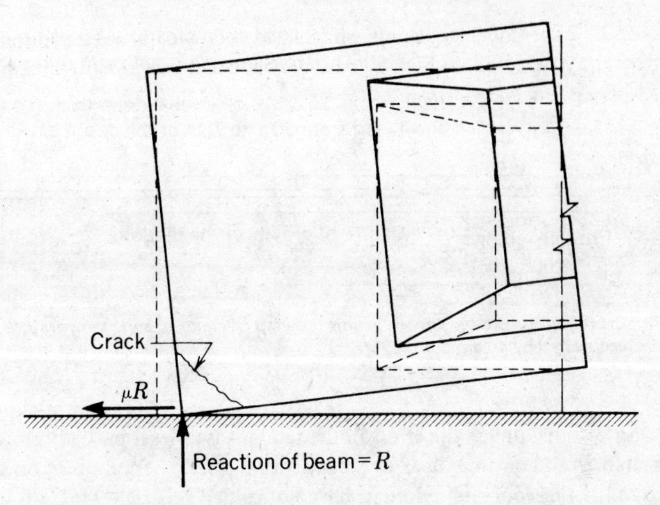

Fig. 10-2 Cracking of the end of a simple beam due to rotation and shortening. The position of the beam before stressing is shown by the broken lines. The coefficient of friction between the beam and the soffit form is denoted by μ.

Restraint at Time of Prestressing

At the time of prestressing, the ends of simple precast elements normally rotate and the member shortens as a result of the deformations caused by prestressing. If the soffit form at the time of stressing is rigid and the member becomes supported on the extreme end, the bottom corner of the member frequently cracks, as is shown in Fig. 10-2. This type of cracking is frequently aggravated by the additional localized tensile stresses resulting from the anchorage of the tendons. It can be avoided by removing a portion of the soffit form prior to stressing and by providing reinforcing in the ends of the member to strengthen the corners.

Longitudinal Temperature and Shrinkage Strains

In the manufacture of heat- or steam-cured pre-tensioned concrete products, if the steam curing is stopped without releasing the pre-tensioning force, the concrete members and the tendons exposed between the members cool and contract. The contraction results in an increase in the tension in the exposed tendons between the members, as a result of the ends of the tendons being anchored to the abutments of the pre-tensioning bench and being unable to yield. This tensile stress may be further aggravated by shrinkage of the concrete. The combined effect of these phenomena may result in the concrete members becoming cracked at intervals of 15 to 25 ft, as illustrated in Fig. 10-3, or may result in the tendons breaking at the anchorages where they are subjected to localized stress concentrations. For this reason, heat or steam curing should never be discontinued for

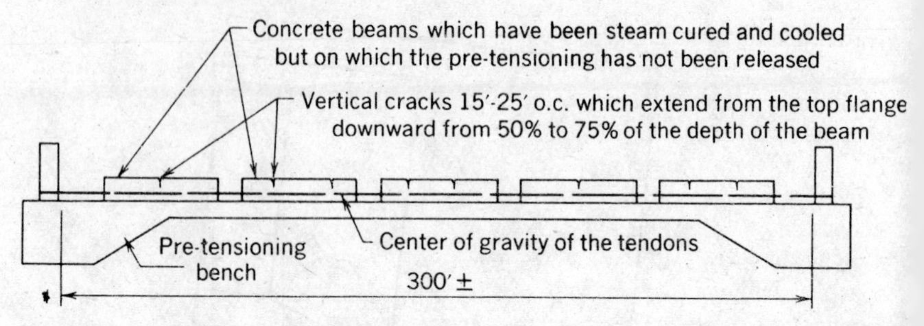

Fig. 10-3 Cracking in pre-tensioned beams due to shrinkage and temperature strains aggravated by steam curing.

any length of time, unless the stress in the tendons is at least partially released. If it is desired to discontinue heat or steam curing after a period of 24 or 30 hr of curing, and if the concrete cylinders have not gained sufficient strength to allow the full release of the prestress, the tendons should be partially released by an amount which is about 25% of the anticipated contraction of the tendons, which normally would take place if the entire prestress were released. It is best, although not always essential, that the forms be loosened before the partial release is made. In a similar manner, a breakdown in the curing facility may result in the cracking of the members, and in such an eventuality, under average conditions of design, a partial release should be made if the concrete strength exceeds 3000 psi.

It is also possible to have similar cracks form in post-tensioned members, due to similar causes (temperature change, shrinkage or both). This is particularly true in long, large members which are lightly reinforced. The cracks can be avoided by keeping the concrete continually wet until prestressing, since by so doing, shrinkage strains will not occur. Another method is to stress the member, either partially or completely, soon after curing is completed. In other words, long periods of drying or large temperature variations on unstressed members should be avoided.

Uneven End Bearing Cracking

This phenomenon can occur at the ends of precast members due to stress concentrations resulting from non-uniform bearing. The non-uniform bearing may be the result of the mating planes having been cast incorrectly, due to rotations at the ends of the members as a result of prestressing, temperature gradients in the member, or applied loads, due to lateral (torsional) rotations resulting from superimposed loads being applied eccentrically, or due to the surfaces not being smooth or being dirty (*see* Sec. 13-1).

The best means of avoiding cracking from these causes is by providing careful attention to construction and erection details. Surfaces intended to be in contact with each other should be observed during erection to be sure they are clean and will fit properly. Beds of mortar or flexible bearing pads are frequently used to insure uniform bearing.

The effects of superimposed loads should be considered and provisions made for them. L-shaped or inverted T-shaped beams that are to receive superimposed loads on one side only will tend to rotate towards the loaded side. Thus, the end of the beam or the supporting column may be subjected to high concentrated stresses and tend to crack. The torsional rotations may be avoided by using temporary shoring. Longitudinal rotations due to superimposed dead and live loads should also be considered at the ends of unrestrained beams. Bearing details should be such that edge loading is avoided.

In Section 9-5 the stresses and deformation of flexural members subject to temperature gradients within the members are discussed. If the ends of members subjected to temperature gradients are not fully restrained, they will experience rotations and translations as a result of the temperature gradients. Diurnal solar heating is frequently the cause of temperature gradients in members exposed to the sun. The exposure can be during construction and hence be temporary and of short duration or, in the case of bridge decks, building roof members, wall panels, etc., may be a permanent condition. It is imperative that the supports of concrete members, both temporary and permanent supports, be designed in such a way that the rotations and translations due to temperature gradients can take place without creating excessive stresses and cracking at the supports.

Form Settlement

If the soffit form of a member settles after the concrete has been placed and has attained its initial set, cracks similar to structural flexural cracks can occur. This type of cracking can be prevented by constructing the soffit forms on a suitable foundation

Flange-web Cracks

Cracks sometimes occur at the junction of the web and the flanges of precast members that are I or T shaped in cross section. The cracks are more common at the junction of the web and top flange than at the junction of the web and bottom flange. There are two probable causes for this. First, the settlement of the plastic concrete, which sometimes accounts for this type of cracking, is greater and tends to cause this type of cracking near the top flange where the top flange concrete tends to span across the settled web concrete. This effect does not exist near the bottom flange. Second, the slope of the fillet between the top flange and the web is generally flatter than the slope of the fillet at the bottom flange. This type of crack, which is illustrated in Fig. 10-4, is generally thought to not extend completely through the web of the member when it is the result of

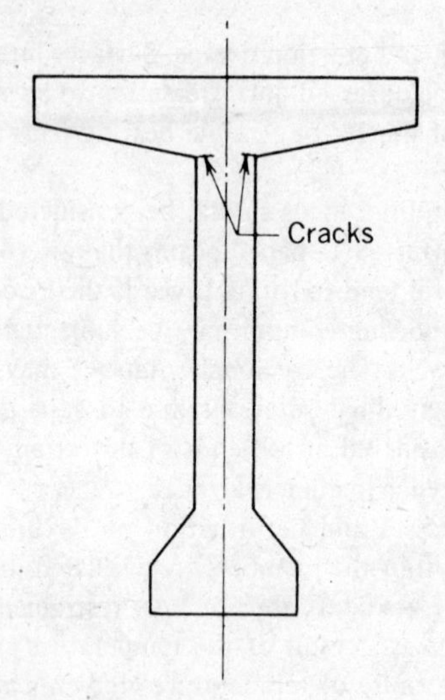

Fig. 10-4 Web-flange cracks due to concrete settlement.

concrete settling in the form, since the plastic top flange concrete will sag down and become reunited with the web concrete, except at the outside surfaces. If this is the cause, the possibility of these cracks occurring can be reduced by allowing the web concrete to settle before the top flange concrete is placed, using concrete of lower settlement and shrinkage characteristics and possible by re-vibrating the concrete in the vicinity of the potential cracks. These cracks might also be caused by form expansion, in which case, the cause of the form expansion should be eliminated or reduced as much as possible. Wood forms expand when they absorb water or are exposed to steam curing conditions. Steel forms will expand more rapidly than the concrete they contain if they are subjected to a rapid temperature increase.

End Block Cracks

Cracks in end blocks may occur in the loaded face. Cracks may also occur along the paths of the tendons due either to the tensile stresses resulting from the anchorage forces or to the edge distance not being sufficiently large. Control of end block cracking is best accomplished by providing reinforcing steel in accordance with the methods of Sec. 6-8 and by providing reasonable edge distance for the tendons. Concrete quality in the end blocks is sometimes poor, due to inadequate compaction in these areas. The poor compaction may be due to the end block being congested with reinforcing. It is generally preferred to obtain a high-quality, well-compacted concrete in the end block, even if it must be under-reinforced, rather than have an over-reinforced end block in which the concrete cannot be well compacted and, hence, will be of low quality.

Web Cracks Following Post-tensioning Tendons

Cracks of this type have been experienced from five different causes. If post-tensioning tendons are grouted during sub-freezing temperatures, the grout may freeze and the resulting increase in volume can cause tensile forces that crack the web. In European practice, some of the mixing water in the grout has been occasionally replaced with alcohol in order to lower the freezing point of the grout and eliminate this problem, although it is recognized that the quality of the grout suffers from this procedure. Some producers have adopted the practice of steam curing their beams for two or three days after grouting to avoid the problem. Cracks following the paths of the tendons can also result from settlement of the concrete below the ducts in thin webbed I- or T-shaped members, which is similar to one of the causes of web-flange cracks. This is best prevented by using members that do not have excessively thin webs and by using concrete mixes that have low settlement characteristics. Cracks following tendons of large dimension in rectangular ducts have been known to be the result of using high pressures during the injection of the grout. This cause of cracking can be eliminated by using lower pressures or by using smaller round ducts. Beams with over hanging ends, as are shown in Fig. 10-5, have been known to develop cracks along the tendons as a result of high tensile stresses acting along the reduced concrete section that follows the tendon. This situation is critical at the time of stressing when the large future reaction at the end of the overhang is not yet in position and the large shear force due to prestressing is acting without the reaction to counteract it. The shear force due to prestressing acts in the direction opposite to that which results from the applied reaction, and hence, the principal tensile stresses are more or less acting on the plane of the tendons anchored in the bottom flange. This condition is best controlled by stressing the tendons in stages with some of the tendons being stressed after the suspended span has been erected. Another cause of this type of cracking is illustrated in Fig. 10-6, in which it will be seen that the beams are prestressed with a combination of pre-tensioned and post-tensioned tendons, the latter being straight and near the bottom of the beam for a substantial portion of the length of the beam. These beams were reinforced with stirrups oriented in a plane parallel to the longitudinal axis of the beams. The stirrups did not extend transversely across the bottom flange of the beam. This type of crack can be controlled by having reasonably closely spaced reinforcing steel extend across the bottom flange near the soffit, as is shown in Fig. 10-7.

Restraint of Reinforcing Bars

Large diameter reinforcing bars placed in the concrete immediately under post-tensioning anchorages can result in cracking. This is the result of the large bar being rigid and unable to deform in a manner that is compatible with the highly stressed concrete. The concrete immediately under the anchorages is expected to

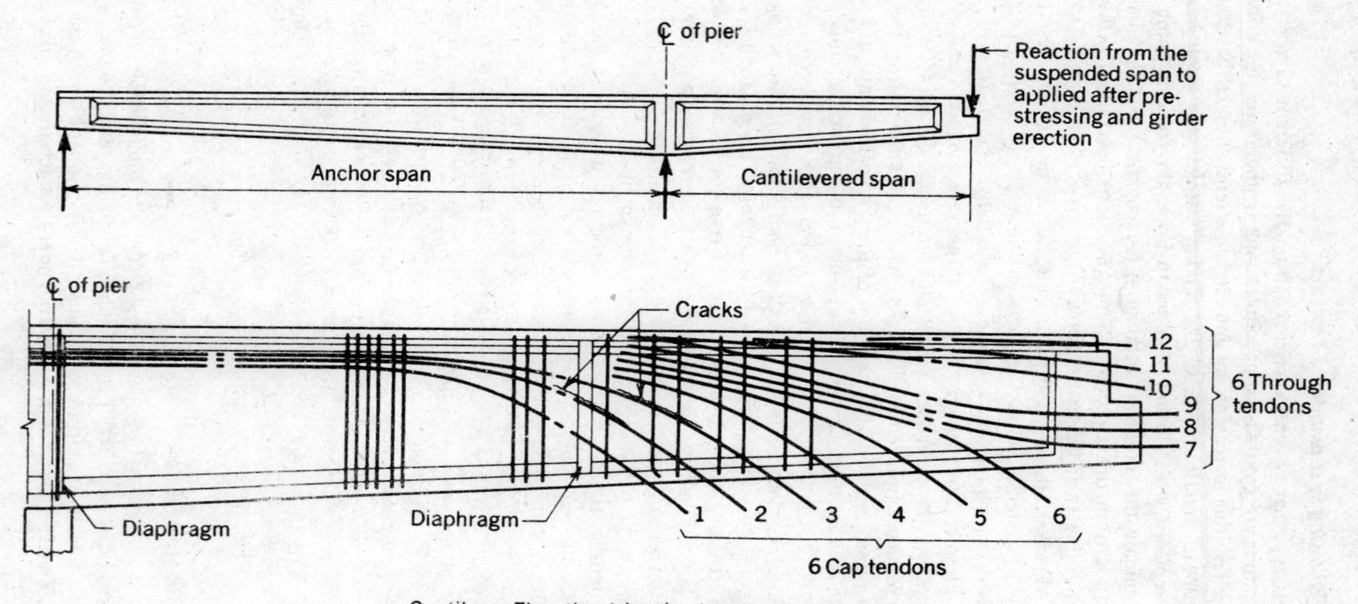

Fig. 10-5 Beam with overhanging end.

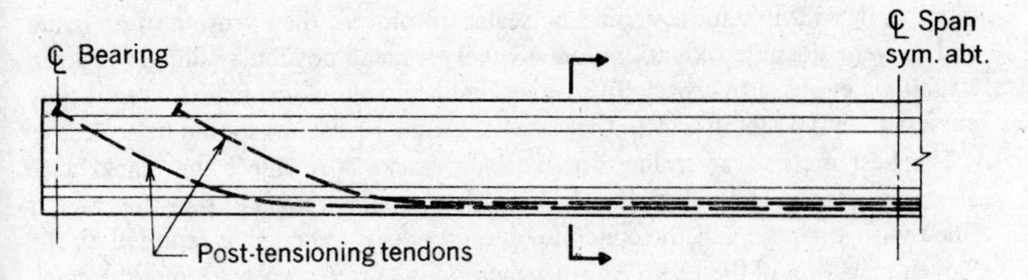

Half Elevation (stirrups and pretension tendon not shown)

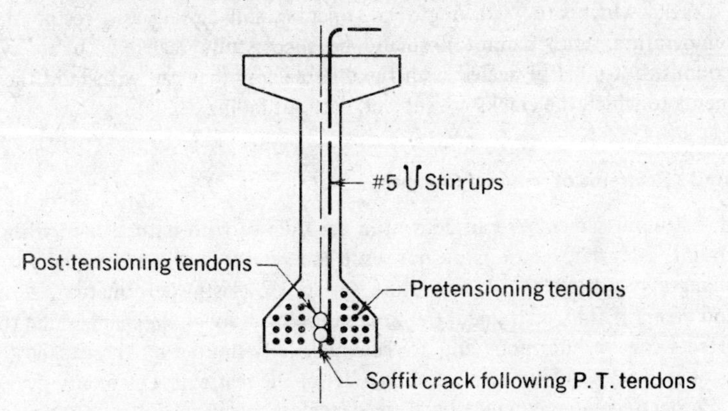

Fig. 10-6 Combined pre-tensioned and post-tensioned girder with soffit cracks.

undergo some localized plastic deformations. The use of large bars in the highly compressed anchorage areas should be avoided.

Most of the cracks described above, although not desirable, are not serious structurally. Difficulties due to deterioration are to be expected for normal conditions of service only if the cracks exceed 0.01 in. in width. Cracks which are

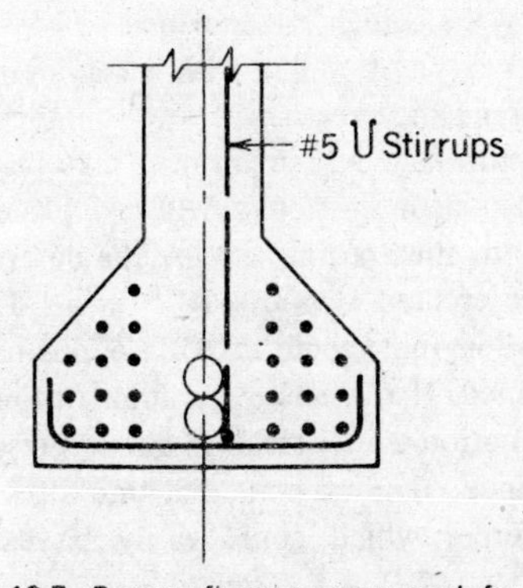

Fig. 10-7 Bottom flange transverse reinforcing.

0.01 in. in width or more should be sealed to prevent the intrusion of moisture and prevent possible oxidation, loss of steel area, and possible spalling. Cracks in structures exposed to especially adverse conditions should be sealed, even if they are less than 0.01 in. in width. Cracks which show rust stains should be sealed.

The best method of sealing non-working cracks is to inject the cracks with an epoxy resin of low viscosity. This is done in such a manner that the crack is filled with the resin and the concrete on each side of the crack is reunited by the "glueing" action of the resin. Another method is to rout a groove along the crack throughout its entire length and fill the groove with an epoxy compound.

Cracks which are "working" (i.e. opening and closing as a result of loads, temperature, etc.) cannot normally be successfully sealed with epoxy compounds, but must be sealed with flexible sealants that can withstand the movements to which the cracks are subject, without failing.

10-3 Restraint of Volume Changes

If a structural member can deform in accordance with natural laws without restraint, such deformations do not result in stresses. This applies equally to strain changes resulting from temperature variations, elastic deformation, shrinkage, and creep. If fully restrained, the forces developed by strain changes due to these effects can be enormous and are generally only limited by the capacity of the weakest portion of the structural member or the restraining elements.

Prestressed concrete members are considerably more critical with respect to restraint than is reinforced concrete, due to the fact the sections are generally crack free and creep deformations tend to shorten the length of the members, as does shrinkage and the elastic shortening due to prestressing. The creep deformation tends to change the deflection of normal reinforced concrete flexural members, but does not tend to shorten the member. For this reason, the designer must give particular attention to the problems of restraint when designing prestressed-concrete structures.

In Fig. 10-8, plan views of the structural framing of two buildings are shown. In one plan the shear walls, which are provided to give stability against lateral loads, are at the corners of the building with a maximum distance of 150 ft between the walls. Assuming the prestressed roof members are pre-cast, much of the shortening due to shrinkage and creep may take place before the members are erected and all of the elastic shortening will take place before erection.

For the purposes of this illustration, assume the deferred deformation for the members after they are erected is 600×10^{-6} in./in. Deferred deformation is defined as the deformation that would take place, due to volume changes, if the concrete were unrestrained. If the elastic modulus of the concrete were 4×10^6 psi, the deferred deformation would result in a unit stress of 2400 psi if the concrete were fully restrained. It is obvious that unit stresses of this order would develop tremendous forces which could easily be expected to exceed the

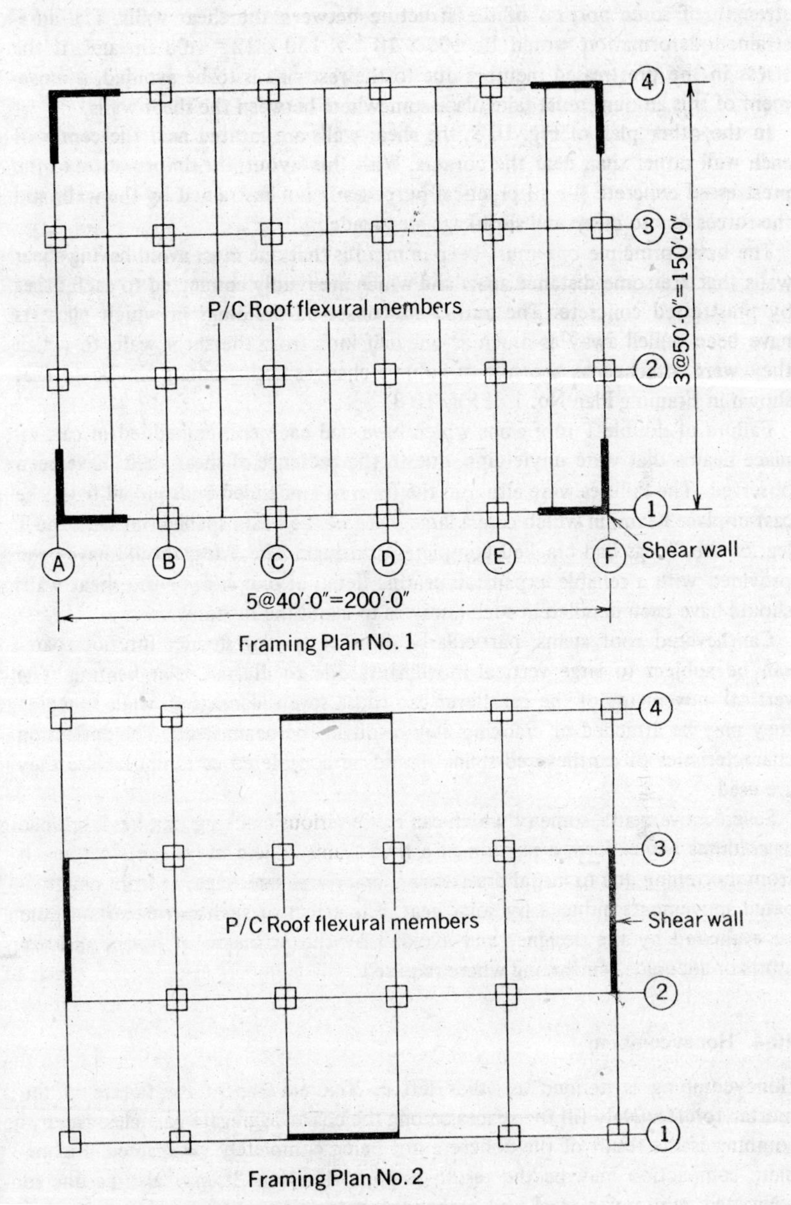

Fig. 10-8 Framing plans with different shear wall layouts.

strength of some portion of the structure between the shear walls. The unrestrained deformation would be $600 \times 10^{-6} \times 150 \times 12 = 1.08$ in. and, if the stress in the prestressed member due to the restraint is to be avoided, a movement of this amount must take place somewhere between the shear walls.

In the other plan of Fig. 10-8, the shear walls are located near the center of each wall rather than near the corners. With this layout, the deformation of the prestressed concrete for all practical purposes is not restrained by the walls and the forces due to creep and shrinkage are avoided.

The basic principle one must keep in mind is that one must avoid having shear walls that are some distance apart and which are rigidly connected to each other by prestressed concrete. The author has observed buildings in which pilasters have been pulled away as much as one-half inch from the shear walls to which they were attached as a result of volume changes with framing similar to that shown in Framing Plan No. 1 of Fig. 10-8.

Failure of double-T roof slabs which have had each end embedded in cast-in-place beams that were unyielding, due to the presence of shear walls, have been observed. The failures were either in the form of embedded ends pulled from the cast-in-place beam, in which case a large piece of the beam spalled out with the T leg, or the T legs had cracked completely through. The T legs should have been provided with a reliable expansion-bearing detail at one end, or the shear walls should have been detailed in such a way as to avoid the restraint.

Cantilevered roof spans, particularly those with long slender interior spans, can be subject to large vertical movements due to diurnal solar heating. The vertical movements of the cantilever can cause severe damage to walls to which they may be attached or cracking may result in the beam itself. The deflection characteristics of cantilevered spans should be considered carefully before they are used.

Severe movements, some of which can cause serious cracking, can be developed in columns which form a portion of a rigid frame. These movements can result from shortening due to initial prestressing, creep and shrinkage, or from unanticipated movements induced by solar heat. The effect of such movements should be evaluated by the designer and avoided by the provision of hinges and slip joints or adequate reinforcing where required.

10-4 Honeycombing

Honeycombing is defined as voids left in concrete due to the failure of the mortar to effectively fill the spaces among the coarse aggregate particles. Honeycombing is the result of the concrete not being completely compacted. Incomplete compaction may be the result of poor vibration. It may also be due to congested reinforcing steel and embedded items which restrict the placing of the concrete. Webs of I-shaped members that are too thin can also cause honeycombing, since thin webs make it difficult, if not impossible, to insert internal vibrators of sufficient size to properly compact the concrete.

Honeycombed areas should be repaired as soon as they are discovered using the methods given in the Concrete Manual published by the Bureau of Reclamation (Ref. 1). If the curing of the concrete is continued, by keeping it saturated, a good repair can be made that should not adversely affect the strength or durability of the member.

10-5 Buckling

Long narrow beams, which are often used in precast concrete construction in order to minimize their weight, have a tendency to buckle laterally during handling and when subjected to transverse load before they have been incorporated into the final structure. The beams normally become laterally braced by a slab or diaphragms in the completed structure. If a slender beam is loaded, there is a critical value of load that may be much lower than the load causing flexural or shear failure, at which, lateral deflection and rotation start to take place near the center of the span. The lateral deflection and rotation can cause the beam to collapse if it is allowed to develop without restraint. The dead load of the girder alone can be enough to cause this buckling.

The maximum spacing between lateral supports for the compression flange of a beam permitted in ACI 318 (Sec. 10.4.1) is 50 times the least width of the compression flange.

Crookedness or lateral deflection due to differential shrinkage or solar heating effects can be sufficiently large to result in significant lateral or torsional moments in narrow beams. Torsional moments due to these causes can be large enough to render a girder unstable. Experience has shown that narrow girders often have lateral deflections initially and normally require lateral bracing during hauling and erection.

10-6 Deflection

The deflection of prestressed members can result in construction difficulties in instances when it is significantly greater or less than that which was anticipated. In composite cast-in-place bridge construction, it is customary to use a detail as shown in Fig. 10-9. This detail anticipates the deflection of the girder not being exactly as computed and gives a means of compensating for the deviations from the computed value as well as for the variations in deflection between adjacent girders. Occasionally, a detail as shown in Fig. 10-10 is used, but this detail is not considered as good, since it frequently requires field adjustment of the finished grade.

Variation in deflection between adjacent elements in building construction can present difficulties in constructing structures of precast elements, unless provision is made for the variation. Roofing cannot be applied directly to precast concrete surfaces that have abrupt edges or joints between elements. If it is there is danger that the roofing will become damaged and will leak. Therefore

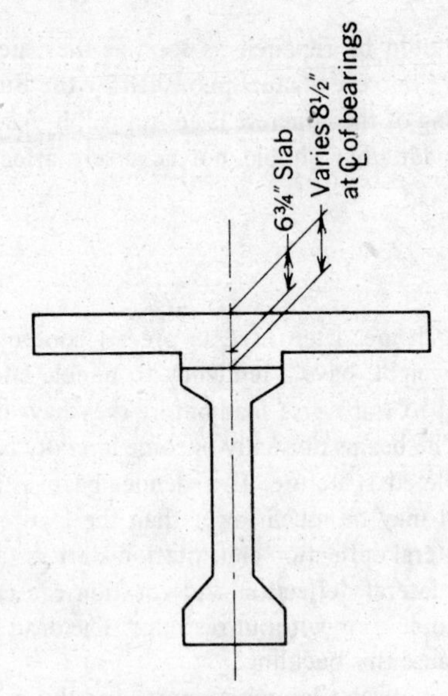

Fig. 10-9 Recommended bridge deck detail.

provision should be made to eliminate sharp edges or joints which could cause roof damage.

Floor construction often consists of precast elements over which a topping is placed in order to achieve a smooth, level wearing surface. Electrical conduits and other items are often embedded in the topping and the prudent designer should take this, as well as the estimated deflection and variation thereof, into account when specifying the minimum thickness of the topping.

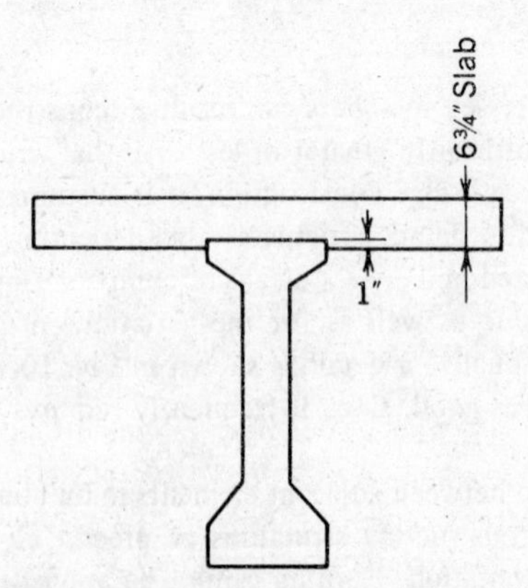

Fig. 10-10 Bridge deck detail not considered as good as that of Fig. 10-9.

The uniformity of deflection (or lack thereof) between members of common dimension and composition is undoubtedly the best measure of the quality control exercised in the production of prestressed concrete. Deflection is a measure of the quality of the concrete contained in a member. It is also a function of the prestressing force and its eccentricity. The existence of large variations in deflection between members of identical dimensions and materials must be interpreted as an indication of large variations in the properties of the materials used in the members, large dimensional variations between members, or large variation in the amount of prestressing force and moment.

A significant difference in age between members that are otherwise identical can also account for a difference in deflection between them.

Deflection and differential deflection tolerances, considered to be normal standards of the industry, have been published by the Prestressed Concrete Institute (Ref. 2). When stricter tolerances are required, the designer is expected to so specify; greater cost is expected to accompany reduced allowable tolerances.

10-7 Corrosion of Prestressing Steel

Prestressing steel is generally considered to be somewhat more sensitive to corrosion, than ordinary reinforcing steel, due to the fact that the individual wires or strands that are used in prestressing are frequently small in comparison to reinforcing bars. The best protection the steel can have is to be surrounded by cement-rich grout or concrete that is well compacted and impermeable. Grout and concrete are alkaline and steel cannot corrode when confined to an alkaline atmosphere.

When the tendons are not protected by sufficient cover of dense concrete, moisture can reach the steel under some conditions of service, in which case, corrosion may form on the tendons. The products of corrosion may cause cracks, staining of the concrete, and eventually spalling of the concrete.

The prestressing strands shown in Fig. 10-11 were first noticed by a spalled area on the top surface of the concrete. These particular strands were corroded completely through, due to the action of sea water which occasionally found its way to the upper surface of the concrete deck during heavy storms. The strands were embedded in a joint between precast slabs. The joint had been filled with a small-aggregate concrete, but it was not well compacted. Thus, the strands were not completely covered with cement mortar. It is interesting to note that on the same project, a post-tensioned pile had been damaged by a ship which collided with the structure, thus leaving the post-tensioned tendon, which was still encased in its sheath, exposed to the sea. The sheath was opened, the grout was removed, and the wires examined. The wires were found to be bright with no signs of oxidation, which clearly demonstrates the ability of grout to protect tendons.

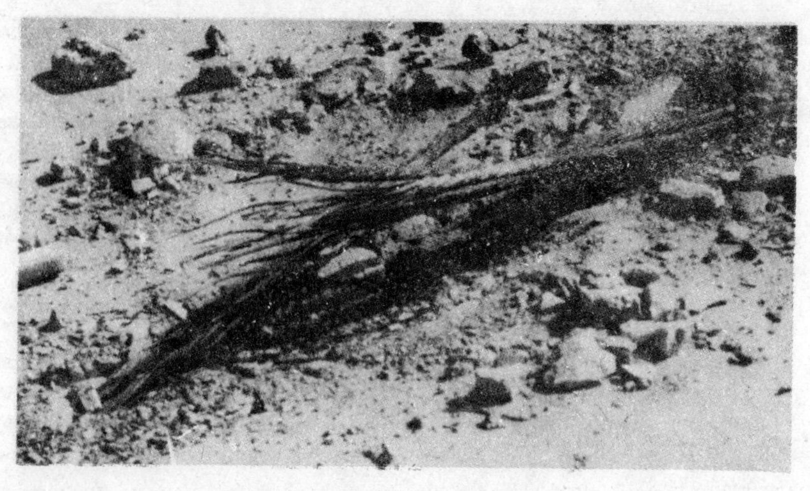

Fig. 10-11 Corroded post-tensioned tendon.

The action of the chloride ion present in some commercially available aggregates for making non-shrink concrete and mortar has been blamed for causing the corrosion at the end anchorages of some post-tensioning tendons. The non-shrink concrete was used to replace the concrete under a post-tensioning anchorage which crushed during stressing. It is believed the non-shrink concrete was porous and permitted rainwater to penetrate to the anchorage and tendon. This, in combination with the chloride ion and perhaps the metallic aggregate, resulted in the wires being severely corroded in the areas adjacent to the repairs. The use of non-shrink aggregates are not recommended in prestressed concrete construction.

Stress corrosion, which has been described in Sec. 2-8, has been blamed for the failure of prestressing wire on two projects in North America. One project consisted of prestressed concrete pipe in which some of the pipe concrete contained calcium chloride and some did not. Stress corrosion was found to varying degrees in all of the pieces of pipe made with concrete containing calcium chloride, while none was found in any of the pipe that did not have calcium chloride. From this experience, it is obvious that calcium chloride (and probably any material containing the chloride ion) should not be used in prestressed concrete (Ref. 3). On the second project, some post-tensioning wires were left ungrouted (after they had been stressed) for a period of several months due to certain difficulties that had been experienced in the construction. When the work was resumed, it was found, upon restressing of some of the previously stressed tendons, that many wires had corroded and broken. It is believed this would not have happened if the tendons had been grouted reasonably soon after stressing.

10-8 Concrete Crushing at End Anchorages

Occasionally the concrete in the vicinity of the end anchorages of post-tensioning tendons will not be of adequate strength or will not be properly compacted. Thus, the concrete fails by crushing at the time of stressing. This type of failure should not be confused with the splitting type of end block failures or cracking described in Sec. 6-8. The best means of avoiding this type of failure is to take whatever steps are necessary to ensure the concrete in the vicinity of the anchorages is properly compacted and of adequate quality.

Failures of this type can generally be satisfactorily repaired by removing all of the fractured concrete and placing new concrete of adequate strength, as required to restore the work to the original lines.

The extreme ends of post-tensioned members have been precast in order to facilitate placing concrete in extremely congested end blocks. Using this procedure, the end sections are cast in a horizontal rather than vertical position. This facilitates placing and vibrating of the concrete.

10-9 Deterioration

The concrete used in prestressed construction is subject to deterioration from chemical and mechanical action as is any other concrete. Higher-quality, dense concrete is normally used with prestressed concrete; hence its resistance to most forms of deterioration can be expected to be greater than that of lower-quality concrete. In addition, the fact that prestressed concrete can be made crack-free is believed to add to the general resistance to the effects of corrosion and erosion.

When prestressed construction is to be used in applications that will expose the structure to agents known to sometimes promote stress corrosion in high-strength steel, extra precautions should be taken to ensure that the concrete is dense, cement-rich, and that it remains crack free under all conditions of loading. The use of type II (modified) portland cement is common in prestressed members used in structures exposed to sea water.

10-10 Grouting of Post-tensioned Tendons

The equipment that is currently in use in mixing and injecting grout in post-tensioning units is much more efficient than that used when post-tensioning was first introduced in the United States. However, from time-to-time difficulty will still be experienced in grouting, due to such problems as the grout being improperly mixed or proportioned, cement lumps in the grout, improper flushing prior to injection, or unintended obstructions in the post-tensioning ducts. The frequent result is that one or more tendons may become partially grouted.

In order to salvage members with problems such as these, holes must be drilled into the post-tensioning duct with great care in order to not damage the tendons. The extent of the grouting and the location of obstructions can thus be determined. After this has been done, the tendons should be flushed with lime water and the tendons grouted, using the drilled holes as ports, with the procedures described in Sec. 15-8.

10-11 Damage Due to Couplers

Post-tensioning tendons are sometimes provided with couplers in order to accommodate a particular construction procedure or in order to make long tendons out of two or more short ones. The couplers are normally considerably larger in diameter than the tendons and they must be provided with a housing to prevent their being bonded to the concrete during concreting, as well as to provide space for the coupler to move during stressing. If, during stressing, the tendon is released due to the tendon breaking or some other mishap, or if the distance provided in the coupler housing is not sufficiently large, the coupler may come to bear on the concrete at the end of the housing. This may cause spalling of the concrete.

Care must be taken to be sure the housings for the couplers are sufficiently large and that the couplers are properly located in the housing.

If damage occurs as a result of the coupler bearing on the end of the housing, the fractured concrete should be removed and a patch applied to the member. Patches of this type must be applied with care. The use of epoxy mortars may be preferred because this type of damage would normally occur sometime after the curing of the concrete has ceased.

10-12 Wedge-Type Dead Ends

Wedge type post-tensioning anchorages at the non-jacking end of tendons must move when they are stressed, since the transverse forces that anchor the tendon can only develop as a result of movement of the wedge. Hence, when wedge type anchorages are used as dead end anchorages (non-jacking ends), provision must be made to ensure that the tendon and wedge can move during stressing. If the anchorage is embedded in concrete, provision must be made to prevent concrete from coming in contact with the wedge and tendon. The method employed in one post-tensioning system is illustrated in Fig. 10-12.

10-13 Looped or Pig Tail Dead Ends

Dead end anchorages have been made by embedding the ends of tendons in either a looped or in a "pig tail" shape, as shown in Fig. 10-13. The method has the advantage of saving the cost of an anchorage, but it must be done properly or the anchorage will fail as a result of excessive stresses on the concrete. When these methods are used, it is recommended the radius of the bend be computed

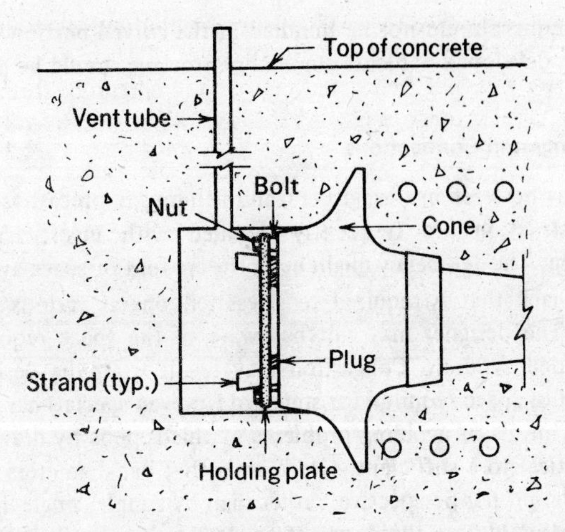

Fig. 10-12 Detail of embedded non-jacking end of a tendon with a wedge-type anchorage.

using the relationship for secondary stresses due to tendon curvature (Sec. 6-15). The bearing stress between the tendon and concrete should not exceed the cylinder strength of the concrete. In addition, it is recommended that the individual wires or strands be spaced apart as shown in Fig. 10-13, Section AA, in order to ensure that the concrete completely surrounds each wire or strand. The

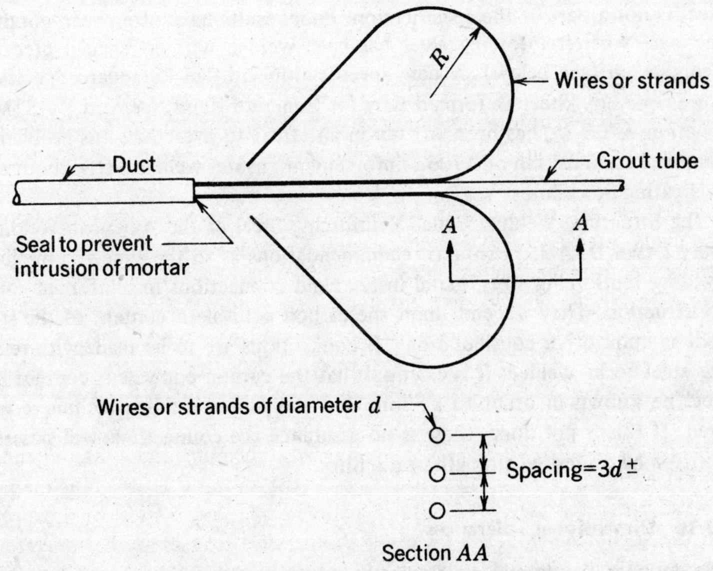

Fig. 10-13 Looped or "pigtailed" tendons.

wires or strands should not be bundled in the curved portion where the anchorage is being developed. At least nominal reinforcing should be provided.

10-14 Congested Connections

Connections between beams, girders, and columns in precast as well as in cast-in-place construction are frequently detailed with insufficient clearance for construction. The deficiency might be in the amount of space available for a post-tensioning jack that is required to stress tendons at various stages in the construction. The designer may not be aware of the space required for stressing post-tensioned tendons. Congestion may result from the designer overlooking the amount of space required for standard (or even special) bar bends.

The best means of avoiding problems of this type is by drawing details which may be critical to a sufficiently large scale that bar diameters and bends can be shown in the proper perspective, rather than by simple, single-line drawings. This should be done during the design stage. It is recommended that shop drawings always be made with details drawn to a scale that is sufficiently large to detect possible erection clearance problems before they actually occur. Actual bar diameters, rather than nominal diameters, should be used for reinforcing steel bars.

10-15 Inadequate Welding

The standard specification for deformed billet-steel bars for concrete reinforcement, ASTM designation A615, states in Section 1.4 that the weldability of the steel is not a part of the specification. Poor results have often been obtained in the past when reinforcing steel has been welded without special precautions being taken (see below). A new specification entitled "Standard Specification for Low-Alloy Steel Deformed Bars for Concrete Reinforcement," ASTM Designation A706-75, has been written in an effort to overcome this problem. The chemical composition of A706 reinforcement ensures weldability without special preheating procedures.

The Structural Welding Code–Reinforcing Steel of the American Welding Society (AWS D1.4-79) contains recommendations as to the type and methods of welding reinforcing steel, metal inserts, and connections in reinforced concrete construction. They depend upon the carbon equivalent content of the steel as well as upon other considerations. If connections are to be made with reinforcing steel to be welded, it is essential that the carbon equivalent content of the steel be known in order to be sure that the proper welding technique will be used. If this is not done, there is no assurance the connections will possess the anticipated or desired strength or ductility.

10-16 Dimensional Tolerances

The designer should give consideration to the dimensional tolerances that might be expected in the construction of prestressed concrete. Special allowances may be required to provide for the fact that concrete members cannot be made to exact dimension, as is the case of members made of other materials.

Cast-in-place prestressed concrete can be expected to be built within the same dimensional tolerances as one would expect for reinforced concrete construction. In the case of precast concrete, the designer should specify the maximum dimensional tolerances he is willing to accept. It should be recognized that exceptionally small permissible tolerances would be expected to increase the cost of precast members.

Dimensional tolerances, which may be specified by reference, are contained in Ref. 7 for cast-in-place construction and in Ref. 8 for pre-tensioned precast concrete members.

REFERENCES

1. U.S. Department of the Interior, Bureau of Reclamation, *Concrete Manual*, Government Printing Office, Washington, D.C., 1975.

2. Prestressed Concrete Institute. *PCI Design Handbook*, Prestressed Concrete Institute, Chicago, 1971.

3. Monfore, G. E. and Verbeck, G. J., "Corrosion of Prestressed Concrete Wire in Concrete," *Journal of the ACI*, **32**, No, 5, 491–515 (Nov. 1962).

4. *Specification for Deformed and Plain Billet-Steel Bars for Concrete Reinforcement* (ASTM A615), American Society for Testing and Materials, Philadelphia, 1980.

5. *Specification for Low-Alloy Steel Deformed Bars for Concrete Reinforcement* (ASTM A706), American Society for Testing and Materials, Philadelphia, 1980.

6. *Structural Welding Code-Reinforcing Steel* (AWS D1.4-79), American Welding Society, Miami, 1979.

7. ACI Committee 347. "Recommended Practice for Concrete Formwork." (ACI 347–78), *ACI Manual of Concrete Practice*, pp. 347-11-347-14, American Concrete Institute, Detroit, 1983.

8. Prestressed Concrete Institute. *PCI Design Handbook*, pp. 4-8-4-9, Prestressed Concrete Institute, Chicago, 1978.

11 Roof and Floor Framing Systems

11-1 Introduction

The subjects of prestressed and reinforced-concrete roof- and floor-framing systems are inseparable because identical concrete sections are used with each mode of reinforcing and because some framing schemes incorporate elements composed of each type of construction. For this reason, each will be considered in this discussion; however, the emphasis will be placed upon the prestressed members, and reinforced-concrete elements will only be discussed when they are used in lieu of or in combination with prestressed elements.

The desirable features of floor and roof systems frequently, but not always, include the following:

(1) Good performance at service loads with adequate strength for design loads.

(2) Economy in first cost and maintenance.

(3) Adaptability to long and short spans with minimum revision to the manufacturing facilities.

(4) Minimum total depth of construction.

(5) Ease in providing vertical openings of various sizes for elevators, stairwells, plumbing, skylights, etc.

(6) High stiffness of individual precast units (low deflection).

(7) Ease in developing diaphragm action of roof or floor structures for resisting horizontal loads (seismic, wind, etc.).

(8) Clean, attractive soffit that is smooth or nearly so and, therefore, can be left exposed.

(9) Stability of precast elements during manufacture, transportation, and erection at the job site.

(10) Low, uniform, stable deflection.

(11) High fire resistance.

(12) Good thermal and sound insulating qualities.

(13) Large and small daily production possible with minimum capital investment.

(14) Minimum erection time.

(15) Large equipment and skilled labor not required for erection.

(16) Minimum additional labor, forms, or welding required for joining the elements at the time of or after erection.

(17) Good acoustical properties if the soffit is to be left exposed.

(18) Precast elements should not be fragile.

It will be seen that no one framing scheme can possibly satisfy all of these requirements. The relative importance of the factors listed above will vary from job to job and in different areas of the country.

It is not possible to discuss each system and method of framing that have been used or produced as a standard product in this country. There have been many different methods and variations of these different methods. A directory of precast, prestressed-concrete producers and their products has been published by the Prestressed Concrete Institute (Ref. 1). The directory includes thirteen types of beams, girders, and joists; nine types of "stemmed" units (i.e. double-T beams, single-T beams, etc.); five types of members with continuous internally formed voids; eleven types of piles; and a variety of architectural and miscellaneous units. The PCI Design Handbook (Ref. 2) also contains valuable data regarding commonly used products. Therefore, this discussion must be limited to the general schemes which have received reasonably widespread acceptance. Furthermore, the discussion must be confined to general terms, since the many variations that have been made to the several basic schemes render specific limitations for dimensions and other factors nonexistent.

11-2 Double-T Slabs

Double-T slabs are used extensively in North America for both roof and floor construction. They are made in a variety of widths and depths, as illustrated in Fig. 11-1. When used as floor slabs, a concrete topping from 2 to 4 in. thick is usually placed over the top of the slabs. The topping concrete is generally designed to work compositely with the precast section. The topping also provides a means of obtaining a flat wearing surface. *It should be recognized that due to*

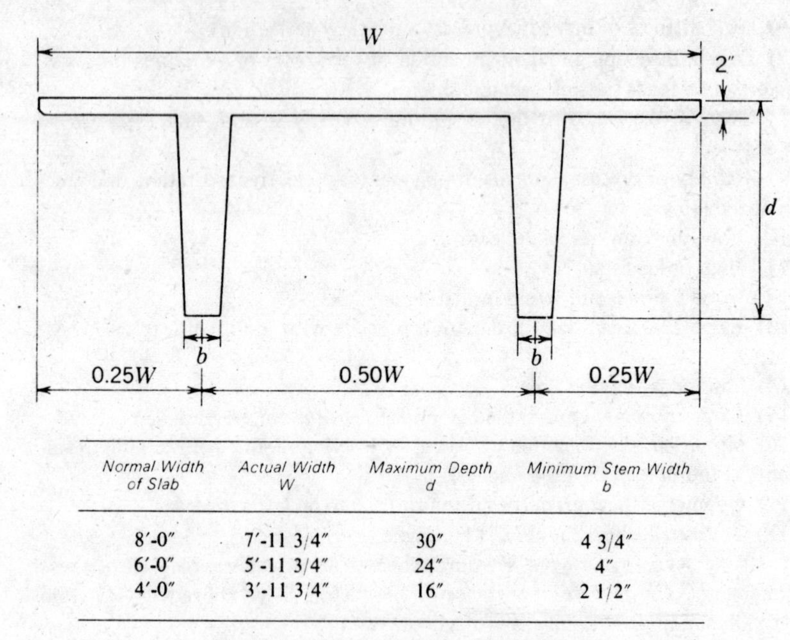

Normal Width of Slab	Actual Width W	Maximum Depth d	Minimum Stem Width b
8′-0″	7′-11 3/4″	30″	4 3/4″
6′-0″	5′-11 3/4″	24″	4″
4′-0″	3′-11 3/4″	16″	2 1/2″

Fig. 11-1 Typical double-T slab dimensions.

casting irregularities, variations in deflection, and other construction inaccuracies, the upper surface of erected double-T slabs cannot be expected to be flat and true to line. A concrete topping is generally not used in roof construction and insulation and roofing are applied directly to the precast slabs.

When a concrete topping is not used, it is normal practice to provide field-welded connections in the edges of the top flanges so that the slabs can develop diaphragm action. They also help to eliminate differential deflection between the flanges of adjacent double-T slabs. Differential deflections may result in the roofing being damaged. Fill material can be provided where necessary between adjacent members in order to eliminate abrupt variations in elevation and thereby avoid roof damage.

If the field-welded connections are to be used to resist earthquake-induced forces, they should be designed for four times the forces to which they would be subject under the code-specified lateral force unless tests have been made which demonstrate the connections have sufficient ductility (energy-absorbtion capability) to render this requirement unnecessary; in addition, the tests should demonstrate that the diaphragms formed through the use of such connections would not displace to such an extent that the overall stability of the structure under seismic forces might be questionable.

Double-T slabs are efficient structurally. The slab portion is quite thin and has well-balanced negative and positive moments. The legs of the joists or tees are

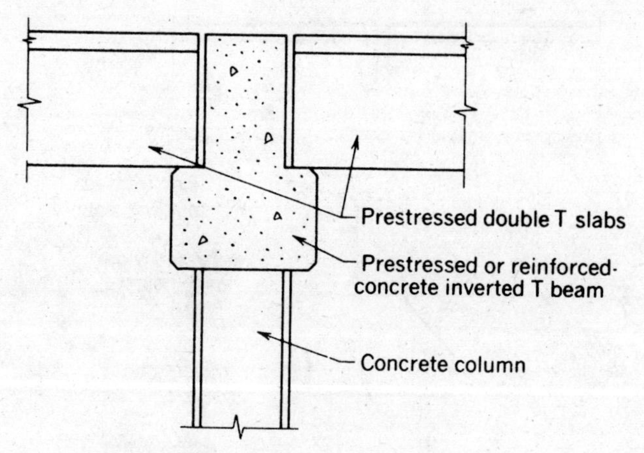

Fig. 11-2 Inverted T beam supporting double-T slabs.

thin and are normally highly stressed. The legs do not have significant torsional stiffness but this is not normally important in roof or floor construction. The slabs are relatively light when compared to other types of framing for the longer spans. There is little excess material in the slabs. A significant contribution to the structural efficiency of the slabs results from the fact that a very large portion of the total dead load is acting on the slab at the time of prestressing. Double-T slabs can be obtained at relatively low cost in virtually all sections of the country.

When used in applications that do not require beams to support the slabs, such as single-span commercial buildings with bearing walls, the depth of construction that is required for double-T slabs is quite low. When it is necessary to support the double-T slabs on beams and columns, the total depth of construction may become relatively great. This is particularly true if long spans are required in each direction, because, during erection, the slabs must be lowered vertically upon the supporting members and cannot be rotated in a horizontal plane into position, due to the width of the slabs. The inverted T beam, which is illustrated in Fig. 11-2, has been widely used with double-T slabs. The inverted T beam may or may not be prestressed and may or may not be made continuous in the completed structure. The inverted T beam does not have an efficient shape for long simple spans. For the longer spans, a large wide top flange is necessary to achieve the required flexural capacity. In applications where a cast-in-place topping will be provided, an adequate top flange may result. An efficient solution for long spans in each direction is the use of double-T slabs in combination with beams that have wide flanges, as is illustrated in Fig. 11-3.

The size of vertical openings that can conveniently be provided in double-T slabs is restricted to the clear width of flange between stems, unless special strengthening or intermediate support can be developed. The top flange of the

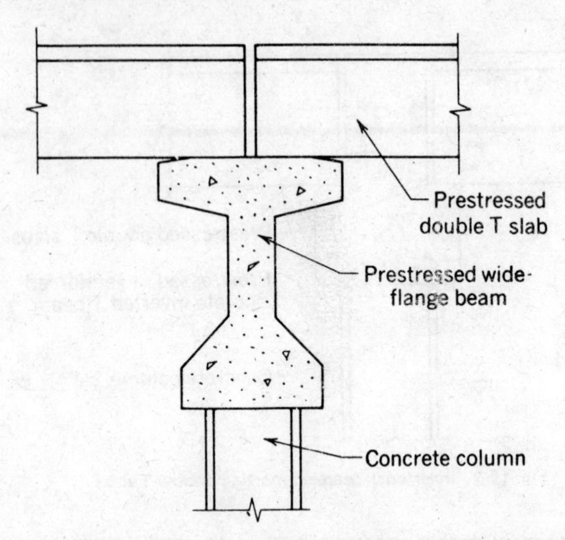

Fig. 11-3 Double-T slabs supported by prestressed wide flange beam.

slab is essential in developing ultimate moment capacity and, for this reason, the openings may have to be confined to the areas near the ends of the span where the moments are low. The soffits of double-T beams can be left exposed in many industrial and commercial applications.

Double-T slabs are stable during manufacture and erection, but they must be handled with reasonable care or the relatively fragile, outstanding flange can be cracked. Double-T slabs occasionally have large upward deflections due to prestressing. The deflection is sometimes not uniform from slab to slab and, in an attempt to minimize this undesirable feature, deflected tendons and partial prestressing are frequently used. When fabricated with a top flange of the order of 2 in. thick, as is frequently the case, the fire resistance of double-T slabs can be made to have a 2 hr rating if provided with the proper insulation and built-up roofing, if the stems are of adequate dimension. When provided with a cast-in-place concrete topping, a fire rating of 2 hr is more easily obtained.

Double-T slabs are normally made on pre-tensioning benches from 200 to 400 ft long, although some manufacturers have used individual molds which resist the pre-tensioning force during concrete placing and curing.

Because of the relatively large size of each double-T slab, the erection time normally required with these members is not excessive (on a unit-area basis) and equipment of moderate size is usually adequate. The amount of labor required to complete the structural roof or floor during and after erection varies from job to job and is dependent upon the amount of welding and other tasks required to complete the structure.

Some manufacturers have supplied this type of slab made of ordinary reinforced concrete rather than prestressed concrete. Virtually all double-T slabs

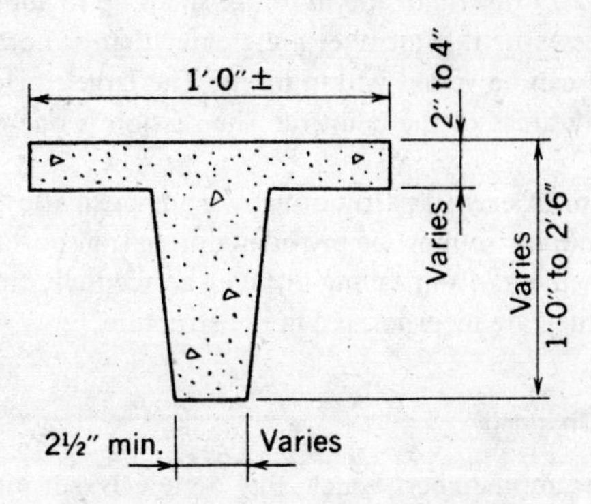

Fig. 11-4 T beam made in double-T mold.

are currently made pre-tensioned. Adequate results can be obtained with reinforced concrete on short and moderate spans, if camber is provided in the members, so that creep will not result in the members sagging.

Double-T slabs fabricated by reputable and skilled manufacturers are efficient in many applications. They not only perform well, but are also attractive in appearance.

11-3 Single-T Beams or Joists

Since-T-beams of two general types are used in roof and floor construction. The first of these is made from the same mold that is used in double-T beams and has the general dimensions shown in Fig. 11-4. The advantages and disadvantages of T members of this type are virtually identical to those of the double-T-slab.

The other type of T beam is made in a mold that allows the dimensions to be varied approximately, as is indicated Fig. 11-5. This member can be used on

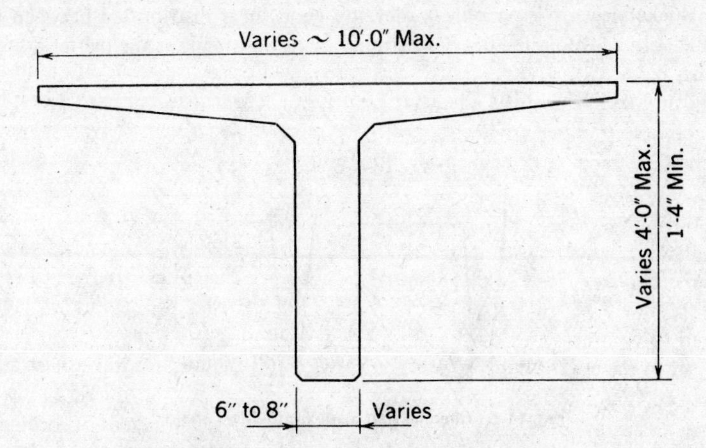

Fig. 11-5 Large T beam made in special mold.

roof spans up to 120 ft in length and in bridge spans up to about 60 ft. The web and flange thicknesses of this member are greater than is normally used in the double-T slabs and can be varied within limits. The large single T has been used extensively in many areas of the country. The section is one of high structural efficiency.

The single-T beam of each type is normally made prismatic. It is essential that some type of temporary support be provided during transportation and erection to prevent these members from falling or being accidentally tipped on their side until such time as they are incorporated in the structure.

11-4 Long-Span Channels

Long-span channels are members which, like double-T slabs, incorporate a thin slab with a leg in such a manner that the member can be used for relatively long spans without supplementary joists. Therefore, long-span channels are used to span from bearing wall to bearing wall or from beam to beam, just as the double-T slabs are. This single factor differentiates long-span channels from short-span channels (normally reinforced concrete) that require joists if the span exceeds 8 to 10 ft. Long-span channels can be used on somewhat greater spans than are possible with standard double-T slabs, as a rule, due to their being more narrow and having legs of comparable or larger size than a double T.

Like a single-T, some forms of channels are made from the same forms used to make double-T slabs, and hence, they have the same general attributes as double-T slabs. There is no need to consider this form of channel any further in this discussion, but the general dimensions that are used in some channels of this type are given in Fig. 11-6.

Channels of many other dimensions have been used. When the top flanges and legs are made thicker, the general strength is increased as is the fire resistance, stiffness, and cost. The deflection and the variation of deflection is less with channels of more ample proportions, such as are illustrated in Fig. 11-7, and with thicker legs, it is possible to develop good shear distribution between the members. In addition, the heavier sections are not as fragile as the lighter double-T slabs.

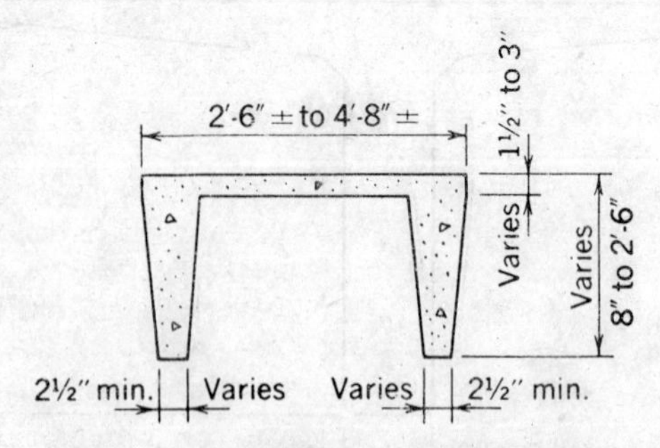

Fig. 11-6 Channel slab made in double-T mold.

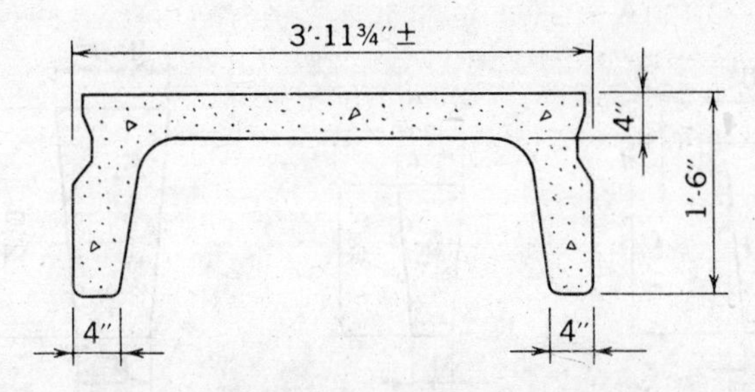

Fig. 11-7 Channel slab of moderate proportion.

End and intermediate diaphragms or flange stiffeners have been used in some types of prestressed and reinforced-concrete channels, but the provision of such secondary members greatly complicates the manufacturing facilities and techniques that must be employed. This is due to temperature changes and the shrinkage of the concrete, which have a tendency to make the precast member cling to the form or mold as does the elastic deformation of the concrete which results from the prestressing. These effects can be minimized by placing a contraction joint in the forms.

The channels have several theoretical or practical advantages over double-T slabs as can be seen from the above discussion. The advantages have not proved to be sufficiently important to justify the additional cost that results from the use of channels. Long-span channels are also made in reinforced concrete and such construction is very efficient and economical for moderate spans.

11-5 Prestressed Joists

Prestressed joists of various types and sizes are used with many types of deck materials, such as short-span channels, concrete plank, cast-in-place concrete, poured gypsum, and lightweight insulating roof materials composed of cement-coated wood fibers. Typical sizes and shapes of joists that have been used are shown in Fig. 11-8. This type of framing is without question the most versatile, since the cross-sectional shape of the joist does not restrict the maximum span on which it can be used to the same degree as in channel slab, double-T slab, and T beam construction. This versatility is the result of the joist spacing not being a fixed dimension.

The depth of construction required with this type of framing is somewhat greater than is required in comparable spans of channel slabs and double-T slabs when each is used in bearing-wall construction, but it may be somewhat less when interior girders are required. The relatively lower construction depth required when interior girders are used is a result of using girders that have a flat upper surface on the bottom flange on which the joists are supported. This type of girder, which is illustrated in Fig. 11-9, can be used without difficulty in joist construction; the joists can be rotated into position as a result of their relatively

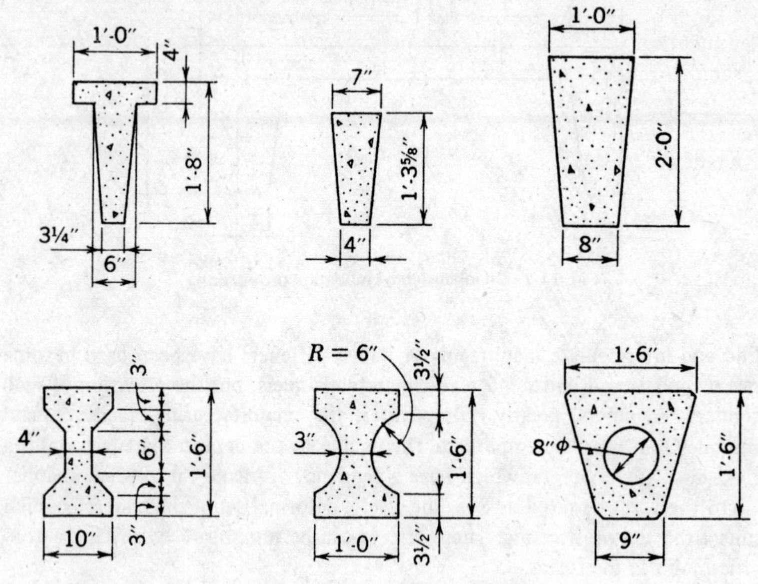

Fig. 11-8 Prestressed joists.

small width. As was explained previously, this cannot be done with double T and channel slabs, due to their width.

Another significant advantage to joist construction is the ease of allowing for vertical openings. Since the joist spacings are normally from 2.5 to 8.0 ft, the size of the vertical openings obtainable without special framing or strengthening is greater than that which is possible with most other types of precast framing.

Many of the other desirable characteristics of the previously listed structural elements are a function of the design of the joists and, therefore, escape generalization. Included in this category are the maximum span, the fire resistance, stability during handling, stiffness, appearance of the joist, and the

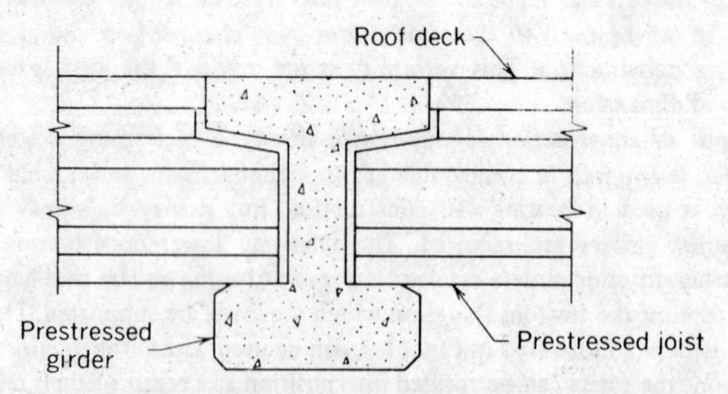

Fig. 11-9 Cross section through girder supporting prestressed joists.

magnitude and stability of the deflection. Well-designed joists, however, will be satisfactory in all of these respects.

The degree of diaphragm action that can be developed, the appearance of the soffit, the fire resistance, the insulating quality, as well as the erection time and labor required, are contingent upon the type of deck that is selected.

11-6 Solid Precast Slabs

Prestressed-concrete solid slabs of two types are possible: small, prestressed planks that are 2 to 4 in. thick, and larger pre-tensioned slabs that are 6 to 12 in. thick. The smaller plank has been used in this country on a few projects, but considerable difficulty was reported to have been experienced in controlling the deflection and straightness of the individual units. It is believed that the large variations in deflection are the combined results of the normal variation in the quality of the concrete, the shrinkage of the concrete, and the eccentricity of the tendons not being maintained as precisely as would be required in order to obtain a uniform product. Prestressed slabs of the order of 3 in. thick are used in some areas of the country as precast soffits. The precast soffits are erected and shored at the center, after which, a topping is placed. The resulting composite slab can easily be made continuous. The shoring gives a means of equalizing initial differential deflection and camber.

Large, solid prestressed slabs capable of carrying roof loads or nominal floor loads on spans to about 30 ft are considered quite practical if made partially prestressed. Although solid pre-tensioned slabs have not been used to a great extent in this country, hollow slabs and solid, post-tensioned slabs, which are discussed subsequently, have been used to a very significant degree. The same general types of structures could be made with each type of construction. The principal advantages of pre-tensioned solid slabs are the ease of manufacture, small depth of construction, and the smooth soffit. The latter results in the elimination of the need for a suspended ceiling in some structures. The principal disadvantages of solid slabs include the dead weight of the slabs, which can be partially offset by using lightweight concrete, and the difficulty of providing large vertical openings.

Solid slabs can be used with precast, prestressed beam soffits, as is illustrated in Fig. 11-10. The cast-in-place concrete, which is placed between the ends of

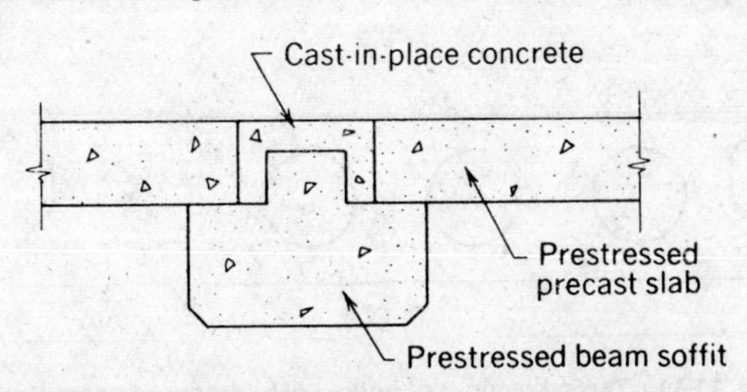

Fig. 11-10 Solid prestressed slabs with prestressed beam soffit.

the slabs, connects the slabs and the beam soffit with the result the beam has a T shape for all subsequently applied loads. In addition, in order to minimize deflections or to increase the strength, reinforcing can be extended from the slabs into the cast-in-place concrete to develop continuity.

11-7 Precast Hollow Slabs

Hollow slabs or "cored" slabs have been used to a very significant degree. The primary advantages and disadvantages of these types of elements are substantially the same as for the solid slabs discussed in Sec. 11-6. The primary difference is that the hollow slabs are lighter and structurally more efficient in the elastic range. Typical cross sections of the hollow slabs currently produced in the United States are illustrated in Fig. 11-11.

The slabs are normally made in continuous pieces and cut to the desired length with a saw, after the slabs have been cured. Sometimes, layers of slabs are cast one on top of the other (over a period of several days) until the stack is several slabs deep. The slabs cannot be cut or removed from the bench until the last slab that was cast has gained sufficient strength to allow the release of stress.

Other types of hollow slabs are made by casting the slabs in the normal manner and forming the voids in the slab by use of paper tubes, which are left in the member, or with metal or inflatable rubber tubes, which are removed from the slab after the slab has hardened.

11-8 Cast-in-Place Prestressed Slabs

Four general types of cast-in-place prestressed slabs are possible in concrete construction. There are one-way slabs, two-way slabs, flat plates and flat slabs. These different types of slabs are characterized by the following:

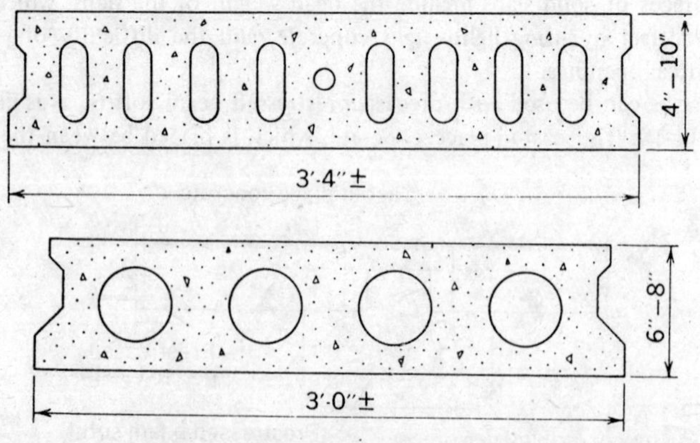

Fig. 11-11 Typical sections of hollow slabs produced domestically.

(1) *One-way slabs.* A one-way slab is supported by continuous supports extending across the entire width of the slabs. The supports may be beams, bearing walls, piers, or abutments. They may be positioned perpendicular to the longitudinal axis of the span or at an angle from the perpendicular to the longitudinal axis of the span (skew angle). One-way slabs may be simple spans or may be continuous over three or more supports.

(2) *Two-way slabs.* A two-way slab is generally square or rectangular in plan and is supported on all four sides by continuous supports. The supports are normally beams or bearing walls, and the slabs may or may not be continuous over the supports.

(3) *Flat slabs.* A flat slab is supported by a series of columns, which ar normally positioned on a square or rectangular pattern. There are no beam spanning from column to column, in either direction. Flat slabs have drop panel or capitals at the columns and may be statically determinate or continuous ove several column rows, and hence, statically indeterminate.

(4) Flat plates are the same as flat slabs except they have a flat soffit withou drop panels or capitals. This type of construction lends itself to the lift sla technique.

The design of one-way slabs, either simple or continuous spans is done b employing the principles set forth for simple and continuous flexural membe in Chapters 4 through 8. In order to simplify the design calculations, slabs ai normally analyzed for a 1-ft width rather than working with the actual widtl After the amount of prestressing steel and reinforcing steel required for a widt of 1 ft has been determined, it is a simple matter to extend these data to th actual slab width.

It should be apparent that transverse bending moments exist in one-way slab subjected to concentrated loads. The transverse reinforcing required for a specific condition may be dictated by the applicable design criteria or, in special cases, may be determined by the use of an elastic analysis. The elastic analysis of transverse bending moments in one-way slabs is beyond the scope of this book, and the interested reader is referred to Refs. 3 through 6.

Since 1976, the Uniform Building Code (Ref. 7) has contained a provision for the minimum permissible amount of non-prestressed reinforcement in one-way slabs prestressed with unbonded tendons. This provision, which is not contained in ACI 318, is as follows:

> One way, unbonded, posttensioned slabs and beams shall be designed to carry the dead load of the beam plus 25 percent of the unreduced superimposed live load by some method other than the primary unbonded post-tensioned reinforcement. Design shall be based on the strength method of design with a load factor and capacity reduction factor of one. All reinforcement other than the primary unbonded reinforcement provided to meet other requirements of this Section may be used in the design.

The design of two-way slabs should be done following the procedures given in the *Building Code Requirements for Reinforced Concrete*, ACI 318-83. The effects of the transverse loads and the prestressing should be analyzed separately and the effects superimposed. The equivalent loading due to the curvature of the prestressing should be computed, using the methods described in Chapter 8, and these loads used to compute the effects of prestressing. A strength analysis should be made.

The design of prestressed flat slabs is done by analyzing the slab in each of the two principal directions, as if it were a one-way slab. Because of the difference in the stiffness in the slab along the column lines (column strips) and the slab located between column lines (middle strips), it is customary to place a greater percentage of the prestressing tendons in the column strips than in the middle strips. The proportioning of the tendons between the column strips and the middle strips must be left to the judgment of the designer.

The spacing and distribution of the tendons in flat plates and flat slabs, specified in the ACI-ASCE Committee 423 report, "Tentative Recommendations for Concrete Members Prestressed with Unbonded Tendons" (Ref. 8), are as follows:

> 3.2.4.2.–For panels with length/width ratios not exceeding 1.33, the following approximate distribution may be used: simple spans; 55 to 60% of the tendons are placed in the column strip, with the remainder in the middle strip; continuous spans; 60 to 70% of the tendons are placed in the column strip. When length/width ratio exceeds 1.33, a moment analysis should be made to guide the distribution of tendons. For high values of this length/width ratio, only 50% of the tendons along the long direction shall be placed in the column strip, while 100% of the tendons along the short direction may be placed in column strip. Tests indicate that the ultimate strength is controlled primarily by the total amount of tendons in each direction, rather than by the tendon distribution. Some tendons should be passed through the columns or at least around their edges.
>
> 3.2.4.3–The maximum spacing of tendons in column strips should not exceed four times the slab thickness, nor 36 in. (91 cm), whichever is less. Maximum spacing of tendons in the middle strips should not exceed six times the thickness of the slab, nor 42 in. (107 cm), whichever is less.

At the time this book was written, the ACI-ASCE Joint Committee 423 had prepared and submitted for peer review a new report entitled "Recommendations for Concrete Members Prestressed with Unbonded Tendons." It is expected that this document will be approved by the members of ACI and ASCE and that it will replace the "Tentative Recommendations." The spacing and distribution of tendons recommended for two-way systems in these "Recommendations," illustrated in Fig. 11-12, consists of distributing the tendons in one direction and banding them together in the other. This procedure facilitates placing the

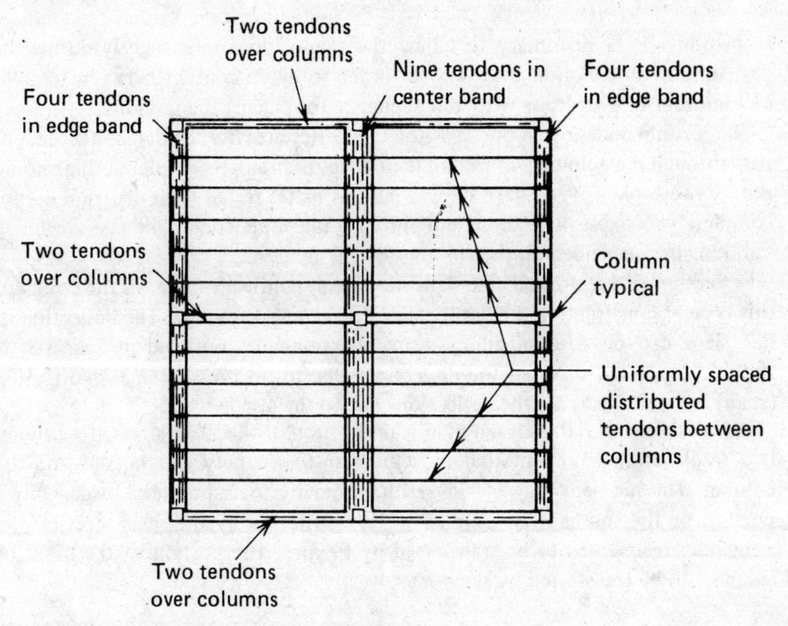

Fig. 11-12 Plan of flat plate slab showing spacing of tendons using bonded tendon distribution.

tendons and hence reduces construction cost. Specific recommendations for the detailing of banded tendons are as follows:

(1) The tendons required in the design strip of slab may be banded in one direction and distributed in the other. The design strips are defined as the portions of the slabs center-to-center of adjacent panels in each direction.

(2) In the distributed direction at least two tendons should be placed inside the design shear section at columns.

The "Recommendations" contain other details for non-prestressed bonded reinforcement, minimum quantities of non-prestressed reinforcement, tendon spacings and minimum amounts of average prestress. The interested reader should obtain a copy of the "Recommendations" before executing a design of this type.

It should be emphasized that no new theoretical considerations in addition to those previously presented are required in the analysis of prestressed flat slabs. The primary novel features in this type of design include the decisions as to what portion of the tendons will be placed in the middle strips and column strips, as well as the determination of tendon paths for the tendons in each direction.

The tendon paths generally conflict where the tendons cross near the columns and near the center of the slab panels. The designer normally specifies that all the tendons which run in the same direction are on the same path, as far as possible. It is customary to adjust the tendon locations slightly during the preparation of the shop drawings, in order to avoid conflict between tendons and in order to establish a workable sequence for placing the tendons.

The recommendation in the ACI-ASCE 423 Report (Ref. 8) that some tendons pass through the columns or around their edges is an important detail that should not be overlooked. Tests have shown that flat plates tested to destruction do not completely collapse if tendons are through the supporting columns; the failed slab remains suspended on the columns by the tendons.

Deflection of prestressed flat slabs may be a significant factor in the design of this type of construction, just as it is in other types of framing. The deflection of flat slabs can be determined by using the principles outlined in Sec. 6-3. It should be pointed out that the deflections due to prestressing in each direction (span) are additive as are the deflections due to the applied loads.

In flat slabs and flat plates, a significant portion of the shear stress at a column is a result of the moment that must be transferred between the slab and the column. The moment may be the result of gravity loads or lateral loads. Only a portion of the moment is assumed to be transferred by the shear stresses; the remainder is assumed to be transferred by flexure. The portion of the moment assumed to be transferred by shear is:

$$\gamma_v = \left(1 - \frac{1}{1 + \frac{2}{3}\sqrt{\frac{c_1 + d}{c_2 + d}}}\right)$$

in which

$c_1 + d$ = The effective width of the slab normal to the axis about which the moment acts. $c_1 + d$ is equal to the column width plus the effective depth of the slab, d. (See Fig. 11-13.)

$c_2 + d$ = The effective width of the slab parallel to the axis about which the moment acts. $c_2 + d$ is equal to the column thickness plus the effective depth of the slab, d.

d = Distance from the extreme compression fiber to the centroid of the tension reinforcement.

γ_v = Fraction of unbalanced moment transferred by eccentricity of shear at slab-column connections.

The dimensions $c_1 + d$ and $c_2 + d$ define the critical section.

The maximum shear stress is taken to be the sum of the shear stresses due to the design vertical force V_u and due to the design moment M_u. This can be written

$$v_u = \frac{V_u}{A_c} + \frac{\gamma_v M_u c}{J_c} \tag{11-1}$$

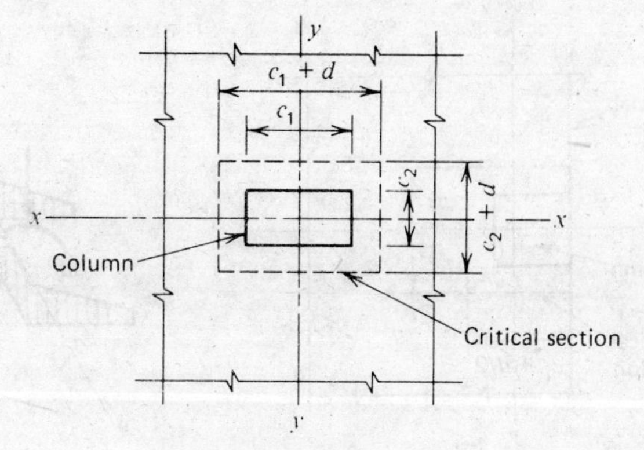

(a) Plan at Interior Column

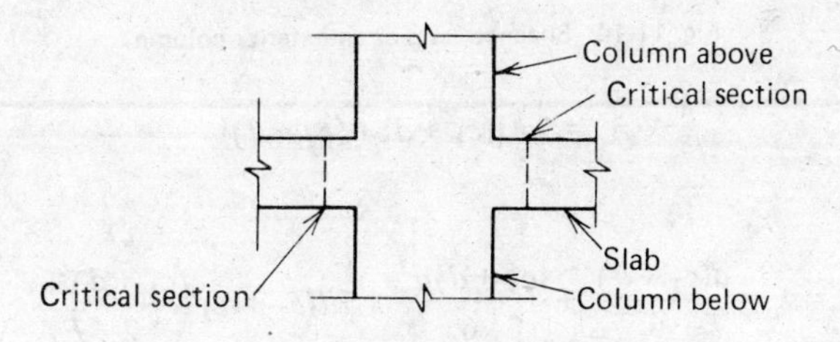

(b) Section at Interior Column

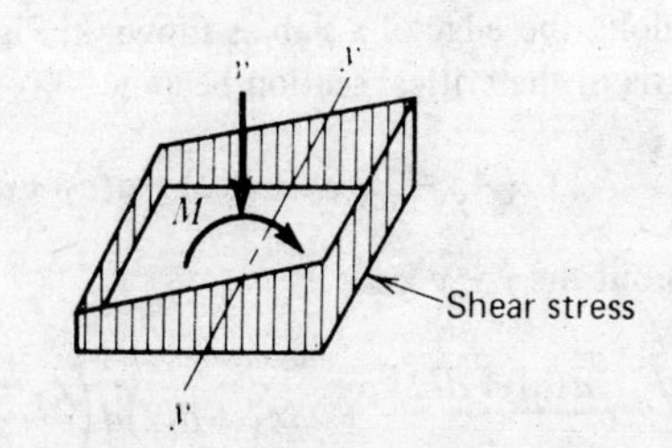

(c) Distribution of Shear Stresses

Fig. 11-13 Shear stresses at an interior column.

in which

A_c = Area of the critical section.

c = The distance to the fiber under consideration $\pm \frac{c_1 + d}{2}$ or $\pm c_2 + d$.

J_c = The polar moment of inertia for the critical section.

For an interior column with a moment rotating around the $y - y$ axis, A_c and J_c are

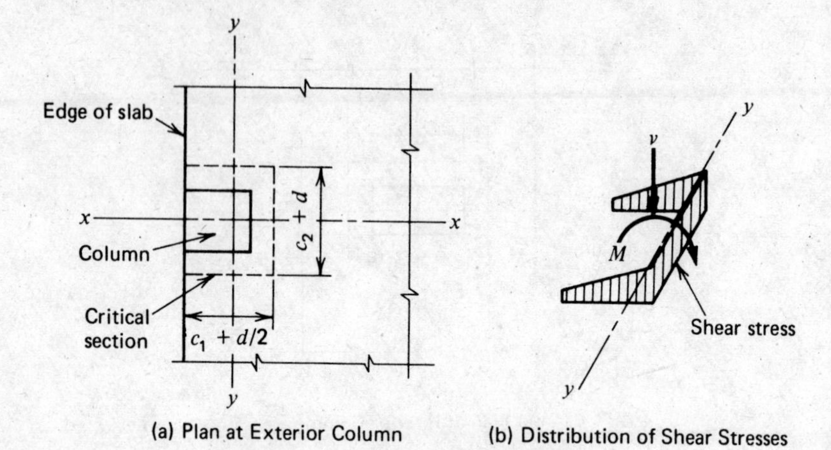

Fig. 11-14 Shear stresses at an exterior column.

$$A_c = 2d\ [(c_1 + d) + (c_2 + d)]$$

and

$$J_c = \frac{d(c_1 + d)^3}{6} + \frac{(c_1 + d)d^3}{6} + 2d(c_2 + d)\left(\frac{c_1 + d}{2}\right)^2$$

For a column along the edge of a slab as shown in Fig. 11-14, the area and polar moment of inertia of the critical section become:

$$A_c = 2d(c_1 + d/2) + d(c_2 + d)$$

For moments about the $y - y$ axis

$$J_c = \frac{(c_1 + d/2)d^3}{6} + \frac{d(c_1 + d/2)^3}{6} + 2(c_1 + d/2)d\left(\frac{c_1 + d/2}{2} - x\right)^2 + d(c_2 + d)x^2$$

where

$$x = \frac{d(c_1 + d/2)^2}{2d(c_1 + d/2) + (c_2 + d)d}$$

For moments about the $x - x$ axis

$$J_c = \frac{d(c_2 + d/2)^3}{12} + \frac{(c_2 + d/2)d^3}{12} + 2d(c_1 + d/2)\left(\frac{c_2 + d}{2}\right)^2$$

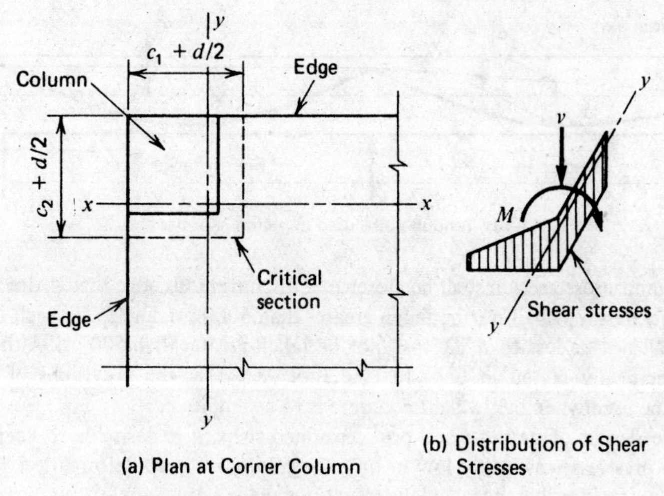

Fig. 11-15 Shear stresses at a corner column.

For a corner column (see Fig. 11-15)

$$A_c = [(c_1 + d/2) + (c_2 + d/2)]\, d$$

$$x = \frac{d(c_1 + d/2)^2}{2A_c}$$

$$y = \frac{d(c_2 + d/2)^2}{2A_c}$$

For moments about the $y - y$ axis

$$J_c = \frac{d(c_1 + d/2)^3}{12} + \frac{(c_1 + d/2)d^3}{12} + d(c_1 + d/2)\left(\frac{c_1 + d/2}{2} - x\right)^2 + d(c_2 + d/2)x^2$$

For moments about the $x - x$ axis

$$J_c = d(c_1 + d/2)\, y^2 + \frac{(c_2 + d/2)d^3}{12} + \frac{d(c_2 + d/2)^3}{12} + d(c_2 + d/2)\left(\frac{c_2 + d/2}{2} - y\right)^2$$

For prestressed slabs meeting the requirements of Section 18.9.3 of ACI 318, shear stresses computed with Eq. 11-1 must not exceed $v_c = \phi(3.5\sqrt{f'_c} + 0.3f_{pc} + V_p/b_o d)$, where b_o is the perimeter of the critical section, V_p is the vertical component of all effective prestress forces crossing the critical section, and f_{pc} is the average compressive stress in the concrete due to effective prestress forces only (after allowance for all prestress losses). When using this equation, no portion of

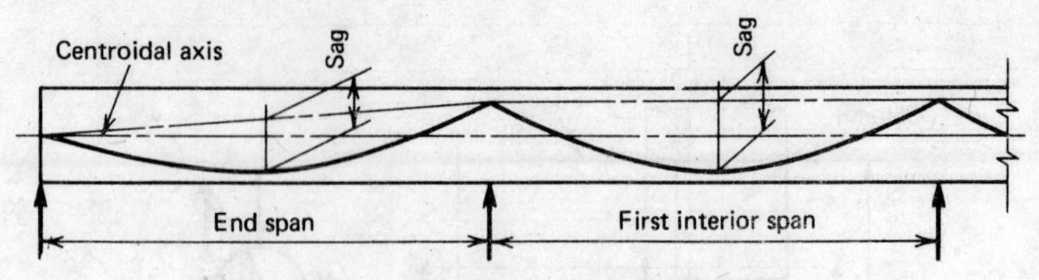

Fig. 11-16 Tendon paths used in preliminary design.

the column cross section shall be closer to a discontinuous edge than 4 times the slab thickness, f'_c shall not be taken greater than 5000 psi, and f_{pc} in each direction shall not be less than 125 psi, nor be taken greater than 500 psi. Although not specifically stated in Sec. 11.12.2.3 of ACI 318, the provisions of 11.2 should be used when lightweight concrete is to be employed.

In the design of cast-in-place post-tensioned slabs, it is desirable to keep the average prestress reasonably low in order to minimize creep deformation. (Considerable cracking has occurred in structures made with slabs of this type as a result of elastic shortening as well as from deferred strain (i.e., creep and shrinkage, see Sec. 10-3). Because of this, it is customary to permit tensile stresses in the concrete when under total service load. One design approach is to use the load balancing technique and proportion the prestress to balance the total dead load plus a small portion of the live load. Flexural capacity must be checked (Secs. 5-4 and 5-5) and non-prestressed reinforcement added if needed. In preliminary design (some engineers use this procedure in final design too), it is usual practice to assume the tendons are on paths as shown in Fig. 11-16; the need for counter-curvature in the tendons at the supports is ignored. With this assumption, the computations for load balancing are extremely simple (see Sec. 6-19).

It is normally necessary to terminate the tendons near mid-depth of the slab at the outermost end of the end span. The size of the end anchorages normally dictates this requirement. Hence, as will be seen from Fig. 11-16, it is possible to have a greater sag in the interior spans than in the end spans. This has significant bearing on the amount of prestressing required because e, in Eq. 6-10, is equal to the sag in a span. Because of this, economy frequently requires the

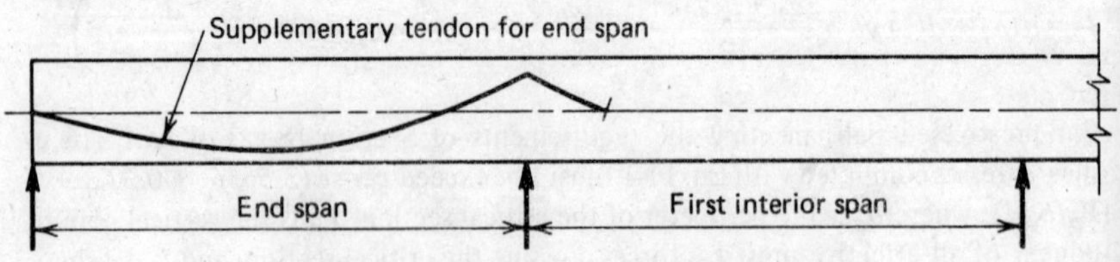

Fig. 11-17 Supplementary tendon for end spans.

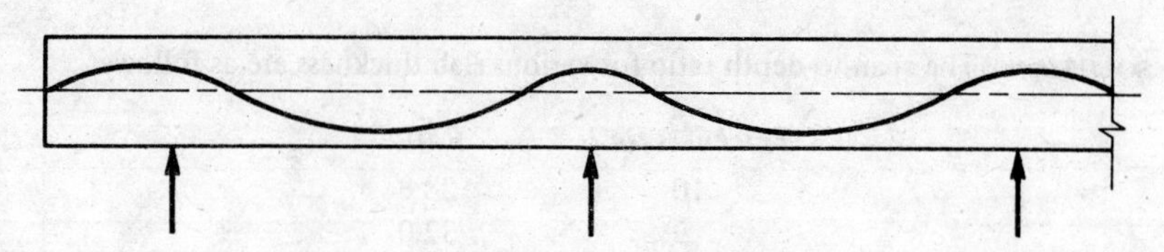

Fig. 11-18 Slab with overhanging end.

designer to proportion the tendons which extend the entire length of the member for the prestress required in the most critical of the interior spans. Additional short tendons, such as are shown in Fig. 11-17 are provided in the end span in order to achieve the higher prestress required therein (because of the smaller sag). On occasion, the end spans overhang the end supports and the tendons in the end spans can be placed on paths identical to those in the interior spans and the supplementary tendons are not required. See Fig. 11-18.

The ACI-ASCE Joint Committee 423, "Tentative Recommendations for Concrete Members Prestressed with Unbonded Tendons" (Ref. 8), contains the following relative to span-thickness ratios:

3.2.5 Deflection and Camber

3.2.5.1. Span-thickness ratios. For prestressed slabs continuous over two or more spans in each direction, the span-thickness ratio should generally not exceed 42 for floors and 48 for roofs. These limits may be increased to 48 and 52, respectively, if calculations verify that both short- and long-term deflection, camber and vibration frequency and amplitude are not objectionable.

3.2.5.2. Short- and long-term deflection and camber should be computed and checked for serviceability for all members.

The designer is frequently tempted to use a very thin slab with a view toward reducing dead load and thereby enhancing the economy of his design. It must be remembered that the amount of prestressing becomes greater as the slab thickness is reduced. This results in an increased cost for prestressing. In addition, the average prestress required is greater in the thinner slab and hence the deformations, both initial and deferred, will also be greater.

ILLUSTRATIVE PROBLEM 11-1 Prepare a preliminary design for a post-tensioned flat plate which has five spans of 24 ft in each direction. The floor slab is to be designed for a superimposed dead load of 25 psf and a live load of 50 psf. Assume the live load can be reduced at the rate of 0.08% per sq ft of tributary area but not more than 40%. Use $f'_c = 4000$ psi, $f_{pu} = 270$ psi, and assume $f_{se} = 165$ psi.

SOLUTION: The span-to-depth ratio for various slab thickness are as follows:

Thickness (in.)	*Ratio*
10	28.8
9	32.0
8	36.0
7	41.1
6	48.0

For reasons of better deflection, vibration, and deformation characteristics, adopt the 8 in. thickness. Thicknesses of 7 or 7.5 in. could also be used with reasonable performance under service loads.

Loads:

Slab	100 psf	
S.D.L.	25 psf	
T.D.L.	125 psf	
L.L.	30 psf	(reduced 40% because of the tributary area)
T.L.	155 psf	

Assume the slab will be post-tensioned with tendons having a diameter of 0.75 in. and that a clear cover of 1.5 in. is required. The theoretical sag of an interior span would be 8.00 - 2(1.50 + 0.38) = 4.25 in. For an end span, assuming the tendon terminates at mid-depth and is 1.88 in. from the bottom and top of the slab at midspan and interior support respectively, the midspan sag would be $(4.00 - 1.88) + \frac{1}{2}(4.00 - 1.88) = 3.18$ in. In order to balance the full dead load in an interior span, the required prestressing force from Eq. 6-10 would be

$$P = \frac{0.125 \times 24^2 \times 12}{8 \times 4.25} = 25.4 \text{ kpf}$$

and the average prestress is 354 psi. These levels of prestress are not unrealistic ing, the end span requires

$$P = \frac{25.4 \times 4.25}{3.18} = 33.9 \text{ kpf}$$

and the average prestress is 354 psi. These levels of prestress are not unrealistic and will be adopted for a first trial. The tendons will be detailed for paths shown in Fig. 11-19. The supplementary end-span tendons will terminate at the first quarter point of the first interior span, but will follow the same path as the full-length tendons. For the preliminary design, the prestressing force will be assumed to be constant and equal to 34.0 kpf for the end span and adjacent 2.40 ft of the first interior span. For the remaining length, the prestressing force will be assumed to equal 25.0 kpf.

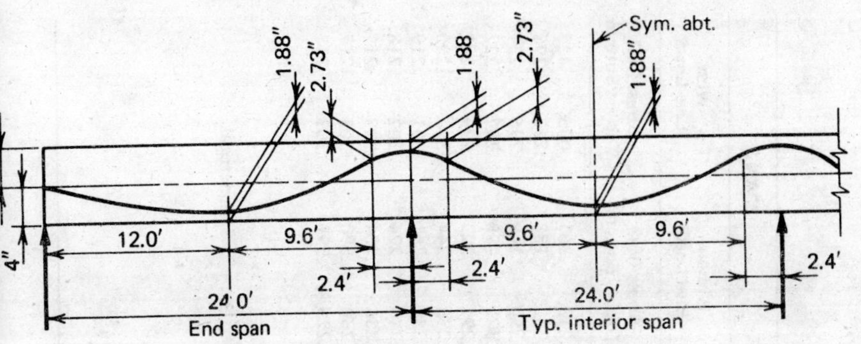

Fig. 11-19 Tendon layout for I.P. 11-1.

Using the methods described in Chapter 8, the secondary moments due to prestressing are found to be as shown below,

Support	*Sec. Moment k-ft*
End	0
First Interior	+1.43
Second Interior	+0.76

and the stresses in the end, first interior, and middle spans are found to be as summarized in Table 11-1 for the conditions of prestress plus dead load and prestress plus total load.

From the table it will be seen that, in spite of the fact the tendons have been analyzed with compounded second-degree parabollic paths rather than the simplistic paths of Fig. 11-16, the dead loads are reasonably well balanced (compare the top and bottom fiber stresses along the spans under dead load alone). Under total load, significant residual compression remains in the bottom fibers.

A preliminary strength analysis, based upon the use of unbonded tendons, shows that the moment capacity of the section where the tendon has effective depth of 6.12 in. is 12.7 k-ft/ft. The design load negative and positive moments for span three are -10.4 and 6.1 k-ft, respectively. Hence, it is apparent that the preliminary design is conservative for the following reasons:

(1) A high residual compression remains in the tensile fibers under service load.

(2) The service load deflections must be very small because the pressure line has little eccentricity under dead load plus effective prestress.

(3) The flexural strength of the slab exceeds the design strength at the points of maximum moment (supports).

For the above reasons, it should be apparent that a second trial design should be made with a reduced prestressing force. It would seem reasonable to make the second trial with prestressing forces reduced by the ratio of the critical

TABLE 11-1

		SPAN 1				SPAN 2				SPAN 3			
		Without Live Load		With Live Load		Without Live Load		With Live Load		Without Live Load		With Live Load	
		Stress (ksi)		Stress (ksi)		Stress (ksi)		Stress (ksi)		Stress (ksi)		Stress (ksi)	
20th PTs.	Location (FT)	Top	Bottom	Top	Bottom	Top	Bottom	Top	Bottom	Top	Bottom	Top	Bottom
0	.000	.354	.354	.354	.354	.341	.367	.170	.537	.213	.307	.085	.435
1	1.200	.378	.329	.408	.299	.450	.257	.320	.387	.332	.188	.242	.278
2	2.400	.397	.310	.453	.254	.430	.277	.337	.371	.350	.169	.295	.224
3	3.600	.410	.297	.488	.219	.224	.296	.163	.356	.322	.198	.297	.222
4	4.800	.418	.289	.514	.194	.205	.315	.173	.347	.297	.223	.299	.221
5	6.000	.420	.287	.529	.178	.190	.330	.182	.338	.276	.244	.300	.220
6	7.200	.417	.291	.536	.172	.179	.341	.191	.329	.259	.261	.301	.219
7	8.400	.407	.300	.532	.175	.171	.349	.200	.320	.246	.274	.302	.218
8	9.600	.393	.315	.519	.189	.167	.353	.208	.312	.236	.284	.303	.217
9	10.800	.372	.335	.496	.211	.167	.353	.216	.304	.230	.290	.303	.217
10	12.000	.346	.361	.463	.244	.171	.349	.224	.296	.228	.291	.303	.217
11	13.200	.323	.384	.430	.278	.179	.341	.232	.288				
12	14.400	.311	.396	.403	.304	.190	.330	.239	.280				
13	15.600	.311	.397	.384	.323	.205	.314	.247	.273				
14	16.800	.321	.386	.372	.335	.224	.295	.254	.266		Symmetrical		
15	18.000	.343	.364	.367	.340	.247	.272	.261	.259				
16	19.200	.376	.331	.369	.338	.274	.246	.267	.253				
17	20.400	.421	.287	.379	.328	.305	.215	.274	.246				
18	21.600	.476	.231	.396	.312	.339	.181	.280	.240				
19	22.800	.473	.234	.350	.357	.326	.194	.234	.285				
20	24.000	.341	.367	.170	.537	.213	.307	.085	.435				

design moments to the flexural strength at midspan and support. This ratio is:

$$\frac{10.4}{12.7} = 0.82$$

Based upon the above, effective prestresses of 20 kips/ft and 28 kips/ft are adopted in lieu of the previously used values of 25 and 34 kips/ft. The resulting stresses under total load are given in Table 11-2. A review of the stresses will reveal that nominal tensile stresses occur at the supports only under full service load. Hence, from an elastic design standpoint, the prestressing could be reduced even further.

The flexural strength of the members would be reduced to approximately 80% of those computed above for the original amount of prestressing. Hence, a small redistribution of moment will be required to conform to the strength requirements of ACI 318; this is perfectly acceptable in members with low percentage of reinforcement *provided sufficient bonded steel is present to assure control of cracking*. (See ACI 318, Sec. 18.10.4.)

At the option of the designer, the prestressing forces could be reduced even further. If this were to be done, strength requirements, as well as minimum

TABLE 11-2

		SPAN 1		SPAN 2		SPAN 3	
		Stress (ksi)		Stress (ksi)		Stress (ksi)	
20th PTS.	Location (ft)	Top	Bottom	Top	Bottom	Top	Bottom
0	.000	.291	.291	−.015	.598	−.063	.479
1	1.200	.363	.219	.145	.437	.102	.314
2	2.400	.424	.158	.192	.390	.180	.235
3	3.600	.473	.110	.071	.345	.213	.202
4	4.800	.510	.072	.107	.308	.242	.174
5	6.000	.535	.047	.140	.276	.266	.150
6	7.200	.550	.033	.168	.248	.285	.130
7	8.400	.552	.031	.191	.224	.301	.115
8	9.600	.542	.040	.211	.205	.311	.104
9	10.800	.521	.061	.226	.190	.318	.098
10	12.000	.488	.094	.236	.179	.320	.095
11	13.200	.451	.131	.243	.173		
12	14.400	.416	.166	.244	.171		
13	15.600	.383	.199	.242	.174		
14	16.800	.352	.230	.235	.181		
15	18.000	.324	.258	.224	.192		
16	19.200	.298	.285	.208	.208		
17	20.400	.274	.308	.188	.228		
18	21.600	.252	.330	.164	.252		
19	22.800	.175	.407	.094	.322		
20	24.000	−.015	.598	−.063	.479		

bonded reinforcement requirements of ACI 318, Sec. 18.9 (see Sec. 5-5), would require the provision of non-prestressed bonded reinforcement to supplement the prestressed reinforcement.

The tendons should be arranged in column- and middle-strips in accordance with the provisions of "Tentative Recommendations for Concrete Members Prestressed with Unbonded Tendons" (see above).

A deflection analysis should be made, as should a shear analysis for the slab-column connections.

11-9 Other Types of Framing

Precast and cast-in-place, prestressed, thin-shell or folded-plate construction has been used in the United States to a fairly significant degree. In Southern California, the cast-in-place type of framing has been used in the construction of many churches and in one large bakery. This type of framing is illustrated in Fig. 11-20.

Precast thin shells or folded plates have been used to a less significant degree. This type of element is considered to be very practical for long-span roofs and floors as well as for bridge members, which are discussed in the next chapter.

Fig. 11-20 Prestressed concrete folded-plate roof.

Thin-shell sections, which can be made from one mold with only minor adjustments, are illustrated in Fig. 11-21. Members of this cross section have been made with a vibrating screed that rides on the mold and screeds the top surface to the desired profile and thickness. The precast elements can be made pretensioned for lengths up to about 100 ft or even longer, if they can be handled and transported. They could also be precast in short sections and post-tensioned together at the job site when very long members, which cannot be conveniently transported, are required.

Cast-in-place, post-tensioned concrete has proved to be an economical mode of construction for parking structures in many parts of the country. The construction frequently consists of post-tensioned beams spanning 52 to 72 ft, spaced from 18 to 27 ft on centers. The slab between the beams is also frequently post-tensioned. The construction has often been done with light-weight concrete. Frequently, the beams are made continuous with the building columns, which results in a frame for vertical loads. Lateral loads are normally resisted by shear walls, although the frames can be proportioned to resist the lateral loads too.

Prestressed-concrete space frames have been used abroad in the construction of exceptionally long spans. These frames are very light and offer a good solu-

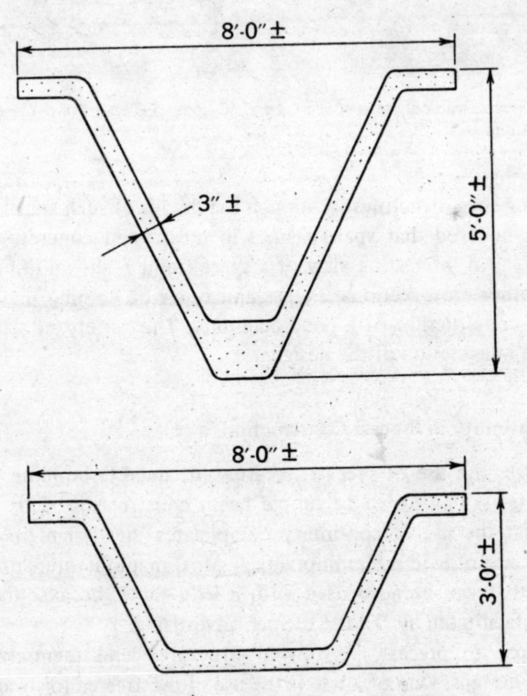

Fig. 11-21 Thin-shell folded-plate roof sections.

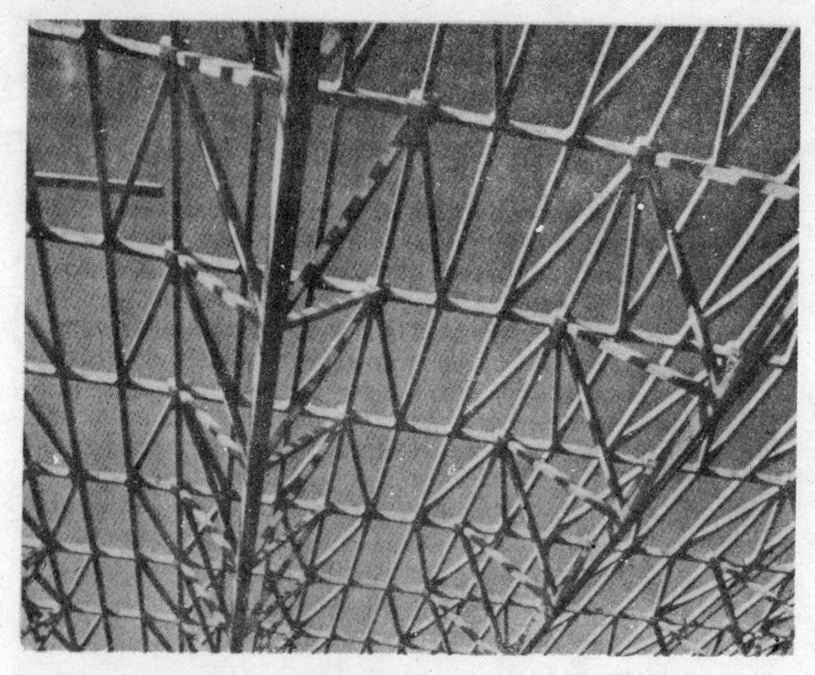

Fig. 11-22 Prestressed-concrete space frame. (*Courtesy Cement and Concrete Association, London.*)

tion for long-span structures. Although the details of such structures are complicated, it is believed that space frames in prestressed concrete are practical in this country. An interesting view of a space frame is shown in Fig. 11-22. There are many other cross-sectional shapes and types of framing that can be used in prestressed construction with good economy. The variety of schemes is limited only by the imagination of the designer.

11-10 Continuity in Precast Construction

By and large, the use of precast, prestressed-concrete building products in the United States is confirmed to simple beam construction. This may be due to the fact that the use of continuity complicates the design of a structure and renders the use of load tables impractical. Most manufacturers prefer to produce a product that can be advertised with a load table, because the selling of the products is facilitated by the use of such advertising.

Some types of precast, prestressed structures lend themselves to partially continuous designs. One of these is the use of prestressed joists and beam soffits made continuous by the placing of a composite reinforced-concrete deck.

Construction of this type is more complicated to design than simple beam construction, but it has good characteristics from the standpoints of fire resistance, deflection, and resistance to vertical and horizontal loads.

REFERENCES

1. "Precast, Prestressed Concrete Producers and Products," Prestressed Concrete Institute, Chicago, 1969.
2. *PCI Design Handbood*, Prestressed Concrete Institute, Chicago, 1971.
3. Kist, H. J. and Bouma, A. L., "An Experimental Investigation of Slabs, Subjected to Concentrated Loads," *Publications, International Association for Bridge and Structural Engineering*, **14**, 85–110 (1954).
4. Kawai, T., "Influence Surfaces for Moments in Slabs Continuous over Flexible Cross Beams," *Publications, International Association for Bridge and Structural Engineering*, **16**, 117–138 (1957).
5. Westergaard, "Computation of Stresses in Bridge Slabs Due to Wheel Loads," *Public Roads* (Mar. 1930).
6. Brotchic, J. F. and Wynn, A. J., *Elastic Deflections and Moments in an Internal Panel of a Flat Plate Structure–Design Information*, Commonwealth Scientific and Industrial Research Organization, Australia, 1975.
7. Uniform Building Code, International Conference of Building Officials, Whittier, California, 1976.
8. ACI-ASCE Committee 423, "Tentative Recommendations for Concrete Members Prestressed with Unbonded Tendons," *Journal of the American Concrete Institute*, **66**, No. 2, 81–86 (Feb. 1969).

12 Bridge Construction

12-1 Introduction

The discussion of the various factors that influence the design of bridges will be made with respect to simple bridge spans, unless otherwise stated. The same general principles apply in the design of continuous spans. In addition, this discussion is limited to highway bridges designed for normal truck loadings. The same principles apply to bridges designed for other purposes and types of loading. However, the span range in which each basic type of framing is most efficient may be altered if the ratio of the dead load to live load is appreciably different with the other types of live loading.

The basic configuration of the most efficient and economical structural elements in prestressed-concrete bridges for any specific structure is a function of the following:

(1) Span length.

(2) Live and impact loads to be carried by the bridge.

(3) Type of structure to be used (simple beams, continuous beams, etc.).

(4) Allowable stresses.

(5) Size of the structure (width, total length).

(6) Feasibility of various construction types and procedures, as controlled by the requirements of the bridge site.

The effects of each of these factors is discussed below, as a means of introducing the reader to the basic problems. In subsequent articles, the various basic types of highway-bridge constructions commonly used, as well as some less common yet economical modes of prestressed bridge construction, are discussed and the limitations of each are pointed out.

The length of a bridge span affects the design in three ways:

(1) The dead load of a bridge member, in proportion to the live and total loads, increases as the span is increased. Therefore, for very short spans, the live load for which the bridge must be designed is very nearly the total load and, for very long spans, the dead load is of much greater significance than the live load.

(2) The moment for which a flexural member must be designed is a function of the square of the span, while the shear is a direct function of the shear span and the load. Thus, for moving live loads, the shear force in short spans is very large in proportion to the bending moment and, in long spans, the bending moment is of much greater importance than the shear forces.

(3) The impact loads which must be included in the design, and which are usually considered as a function of the live load, are relatively smaller for long spans than for short spans.

These three factors obviously have great influence on the optimum cross section for bridge members to be used on different spans. Bearing in mind that relatively nominal tensile stresses are allowed in the precompressed tensile zone under the effect of the total load, it is apparent that the principles of elastic design are of somewhat greater significance in bridge design than in building design, where higher tensile stresses can normally be used. The strength of a bridge structure is very important and must be sufficiently large to ensure that the safety factor is greater than a specific minimum value.

For the purposes of this discussion, it will be assumed that no tensile stresses are to be permitted under total load.* Considering the maximum total moment relationship (See Section 4-4).

$$M_t = M_d + M_l = -P(e + r^2/y_b) \tag{12-1}$$

it becomes apparent that, since the dead load is very small for short span bridges, the relationship approaches

$$M_t = M_l = -P(e + r^2/y_b) \tag{12-2}$$

From this, it follows that when the short span structure is acted upon by dead load alone, the pressure line is located at a distance nearly equal to e below the center of gravity of the section. Therefore, the cross section of the member must provide a relatively large bottom flange to resist the prestressing force during the periods when the live load is not being applied. Solid slabs, hollow slabs, and beams with large top and bottom flanges, but with webs of good proportions, all

*Current bridge design criteria do permit significant tensile stresses under some circumstances (See Sec. 3-15).

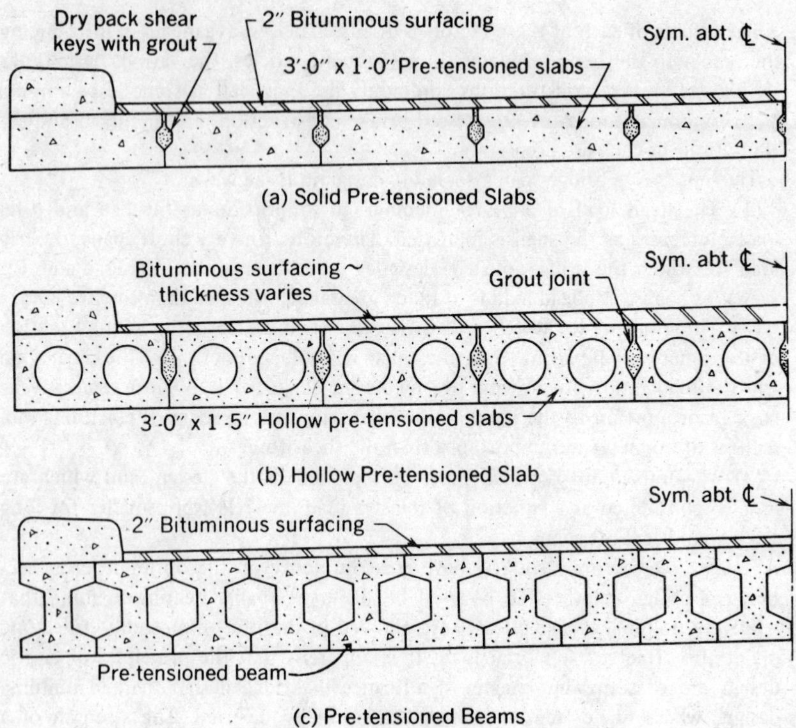

Fig. 12-1 Typical half-sections of short-span bridges.

satisfy these conditions. Bridge cross sections of this type are illustrated in Fig. 12-1.

When the span is very large, shear is less important and dead load is a large percentage of the total load. Consideration of the total moment relationship given in Eq. 12-1 and the location of the pressure line under the action of prestressing plus dead load will reveal that, under the condition of no live load, the pressure line will be some distance above the center of gravity of the tendons (the distance equals M_d/P). Because the dead load is acting at the time of stressing, a large, bottom flange is not required to resist the prestressing force during the periods in which the intermittent live load is not applied. In precast construction, unless the stressing is done in two or more stages, all of the dead load of the structure will not be acting at the time of prestressing, and a bottom flange of moderate size may be required to resist the prestressing force temporarily during construction, even for longer spans. In cast-in-place construction, on the other hand, T-shaped beams are often satisfactory for the long spans, since the entire dead load, with the possible exception of sidewalks and wearing surfaces, is acting at the time of stressing. Efficient cross sections for

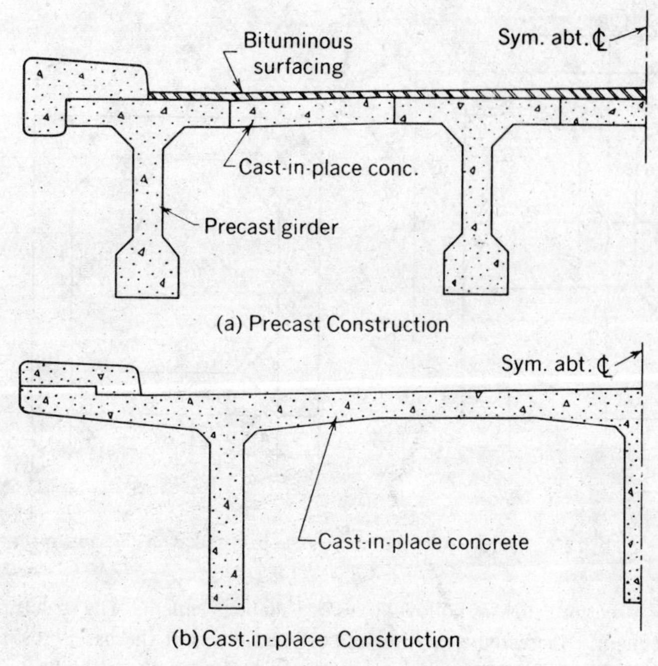

Fig. 12-2 Typical half-sections of long-span bridges.

simple prestressed bridges with relatively long spans approach those shown in Fig. 12-2 for cast-in place and precast construction.

It should be apparent that top flanges of large size are required or desirable in both long and short spans to develop adequate flexural strength and, in long spans, where bending moment is of greater importance than shear, a large top flange is necessary, due to elastic-design considerations.

The optimum cross sections for spans of moderate length are between the short-span solid or hollow slabs and the T beams which are optimum for very long spans. Therefore, either I beams or hollow boxes, such as are illustrated in Fig. 12-3, are most efficient for bridges of moderate spans.

It should be apparent that very heavy, live loads, such as are encountered in railroad bridges, would render the use of solid and hollow slabs practical for spans of moderate length, whereas very light, live loads, such as are encountered in the design of pedestrian bridges, would permit the use of T beams, even for relatively short spans.

Short span continuous structures often have significant reversals of moment due to live load. Also, there is normally a negative moment of great magnitute at the interior support sections and a significant positive moment at or near the center of the span between supports. Because of these factors, short-span continuous structures often require cross sections that are approximately symmetrical

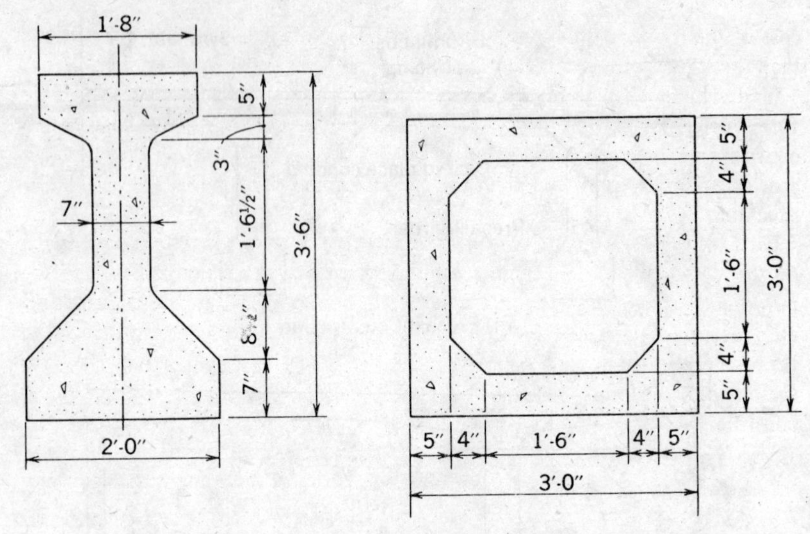

Fig. 12-3 Typical sections of prestressed beams used on moderate spans.

at each section, such as hollow boxes or I-shaped elements. The structural depth is frequently increased near the support sections, but the basic cross-sectional shape is usually maintained throughout the entire length of the beam. Continuous bridges of long spans are not subject to moment reversals and, because of the greater importance of dead loads, larger top rather than bottom flanges are more efficient in areas of positive moment, while the opposite is true at locations of negative moment.

The allowable stresses in the concrete section and in the prestressing steel will obviously have an influence upon the cross-sectional shape of prestressed elements that are feasible under specific conditions. The allowable concrete stress in the top fiber of precast sections has significant influence on the amount of prestressing steel required in pre-tensioned construction. The AASHTO bridge design criteria allows small temporary tensile stresses in the top fibers of precast elements (See Sec. 3-15).

It has been customary to limit the compressive stress to 0.40 f_c' in bridge design. This limitation of compressive stress, for beams of good cross-sectional proportions, is only a problem for girders of very long span. It is the author's opinion that in the future higher compressive stresses will be permitted in members of very long span, since there should be little risk of fatigue in such structures due to the fact that the variation in stress due to the live loads is so small.

On large bridges the designer can use precast or cast-in-place elements and techniques which are designed for the specific conditions and, hence, are the most economical in labor and materials. This is feasible on large jobs, since the cost of the forms and plant required to produce the special members may be less than the savings of materials and labor that results from the special design.

In designing small jobs, the designer must use elements that require simple forms or members that are standard products for the prestressing plants and contractors located in the vicinity of the job site. This is done so that the high cost of elaborate forms and plant facilities are not entirely amortized on the one small job.

Each bridge site is characterized by certain conditions that may dictate which of the design or construction procedures are feasible. Among these are the purpose of the bridge (i.e., grade separation, river crossing, railroad crossing, etc.), accessibility of the site from existing precasting plants, the required skew, the quality of labor and materials available near the site, etc. Any of these factors can obviously govern the feasibility of various types of framing or construction procedure.

The reader will readily understand that the factors briefly discussed above vary considerably from job to job and from one locality to another. The types of framing discussed in Secs. 12-2 and 12-3 are used almost without deviation for bridges of short and moderate spans in the United States, since the construction procedures that are feasible for structures of these types are less apt to be controlling factors in the design. When designing long-span bridges, the designer may be compelled to use cast-in-place or precast construction either to facilitate construction or to maintain minimum horizontal and vertical clearances during construction. The type of framing that may be economical for a bridge or grade separation structure in the United States may be far different than the type of framing that may be economical in some remote country where skilled labor and equipment may not be as available as in this country. In view of this, further attempts will not be made to generalize upon the influence of the various factors that control the selection of the bridge framing.

12-2 Short-Span Bridges

Short-span bridges, for the purposes of this discussion, will be assumed to have a maximum span of 45 ft. It should be understood that this is an arbitrary figure and that there is no definite line of demarcation between short, moderate, and long spans in highway bridges. As has been mentioned above, short-span bridges are most efficiently made of solid slabs, hollow slabs, or I beams, which are of generous proportions.

The solid slabs are most economical when used on short spans of the order of 20 ft or less. The slabs are usually precast 3 or 4 ft wide and have shear keys cast in the sides. After the members are erected and the joints between the slabs are grouted, the keys will transfer shear from one member to another.

Solid slabs may be of the type shown in Fig. 12-1, in which case, from $1\frac{1}{2}$ to 2 in. of asphaltic concrete, portland cement concrete, or bituminous paving material is applied to serve as a wearing surface and leveling course. Solid slabs of

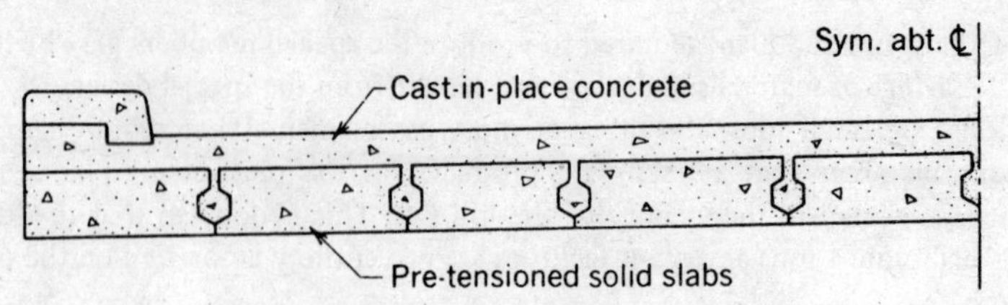

Fig. 12-4 Composite precast-slab construction.

the composite type, such as is illustrated in Fig. 12-4, can also be used efficiently. In the composite type of slab, the cast-in-place concrete serves both as the leveling course, the wearing surface, and the structural top flange. It can be used to develop continuity for live loads, if desired, by placing reinforcing steel in the cast-in-place concrete at the locations of negative moment. Because of the relatively high live-load moment and due to moment reversals, which are characteristic of short-span continuous construction, little is to be gained through the use of continuity in short-span bridges, with the exception of increased resistance to lateral loads, reduced deflections, and the elimination of deck joints.

The hollow slabs, which may have round or square voids and which are used on short spans, are generally made in units 3 ft wide and in depths from 18 to 27 in., although they can be made in any convenient width and depth. The hollow slabs are frequently used in bridges with spans from 20–50 ft. Longitudinal shear keys are used with the hollow slabs, just as with the solid slabs. Hollow slabs are not frequently used with composite, cast-in-place concrete toppings, but the use of some type of levelling course is normally required.

Some form of transverse tie is required in all types of bridges, in order to ensure that the members will not be spread apart by the application of the live load and that the proper distribution of the live load will be developed between the members. In slab construction, this tie most frequently consists of threaded steel bars placed through small holes which are formed transversely through the member during their fabrication. Nuts are then placed and tightened at each end of the bar. In some instances, the transverse tie consists of a post-tensioning tendon that is placed, stressed, and grouted after the slabs are erected. The traverse tie bar or post-tensioning tendon usually extends from one side of the bridge to the other and is placed along the skew. When large skew angles are encountered, the slabs are frequently tied together by connecting each unit to the next with short, tie bars that extend between two units only and are placed in holes perpendicular to the axis of the slabs. The short bars are offset as required by the skew. This detail is illustrated in Fig. 12-5.

Channel sections of various depths have also been used in several areas of the country. The channels are used in the same general manner that the solid and hollow slabs are used, with the single exception that a transverse tie of the el-

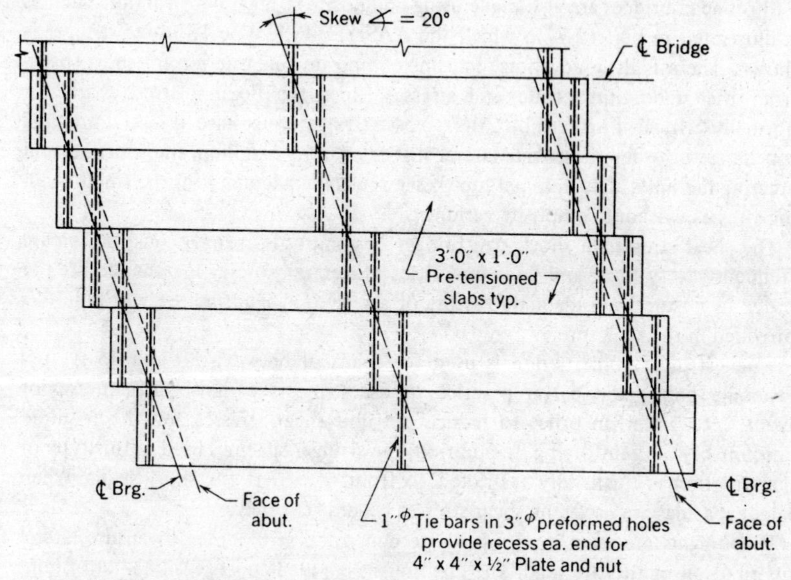

Fig. 12-5 Half-plan of skewed bridge showing staggered transverse tie-bar layout.

ements in channel-slab bridges is often developed by bolting the legs of the channels together, rather than by using a tie bar that extends across the entire width of the bridge. This is illustrated in Fig. 12-6.

The crown or superelevation of the roadway, when using any of the short-span bridges of the types discussed above, is developed in one of three ways. These are: (1) constructing the bearing seats on the abutments and piers on straight slopes, so that the slabs will be placed on a slope; (2) constructing the abutment and pier bearing seats level and developing the required crown or superelevation with the leveling course; and (3) constructing a level bearing seat for each slab unit, in a series of steps which result in the required superelevation or crown. Each of these methods has been illustrated in Figs. 12-1 through 12-6, and each has merits which require no further elaboration.

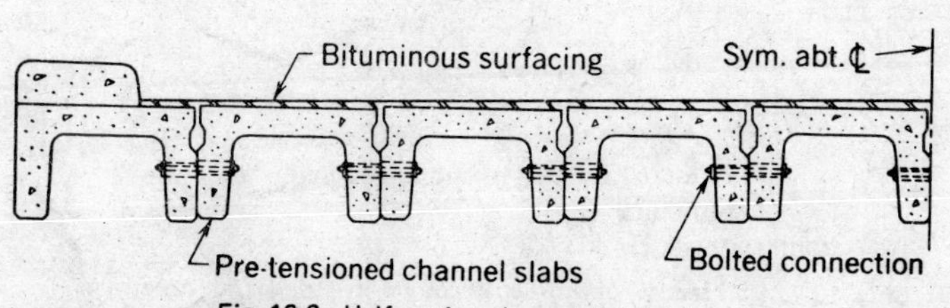

Fig. 12-6 Half-section channel-slab bridge.

Short-span bridges are also made using composite-stringer construction, such as is illustrated in Fig. 12-7, in which the AASHTO-PCI, type 1, bridge stringer is shown. There is little advantage in using composite construction for short spans, from the standpoint of flexural stresses, since the flexural stresses are not normally critical. Furthermore, the precast stringer must have at least a nominal top flange in order to provide lateral stiffness during handling, transporting, and erecting the units. The nominal top flange is usually adequate for the small bending stresses, without composite action.

The shear stresses in short-span bridges of composite-stringer construction are frequently very high, and large quantities of web reinforcing may be required in order to ensure that adequate factors of safety against ultimate shear failure is provided in the structure.

When stringer construction is used for spans of between 30 and 45 ft, it is generally considered better practice to use stringers with web thicknesses of from 7 to 10 in. in order to reduce the unit shear stresses and the required amount of web reinforcing. In addition, the stringer spacings used in this type of construction are generally restricted to from 4 to 6 ft for short spans. When larger spacings are used, the shear stresses become excessive.

Diaphragms are used in stringer-type construction in order to ensure lateral distribution of the live load. The diaphragms are made post-tensioned or of reinforced concrete. One diaphragm is usually placed at each end and at the center of the shorter spans. The diaphragms are normally placed along the skew, unless the skew angle is very great, in which case, the diaphragms may be staggered across the bridge in much the same manner as in steel bridges or in the hollow-slab bridge illustrated in Fig. 12-5.

No generalities will be given pertaining to the relative cost of each of the above types of construction, since cost is a function of many variables. However, it should be pointed out that the stringer type of construction requires a considerably greater depth of construction when compared with solid, hollow,

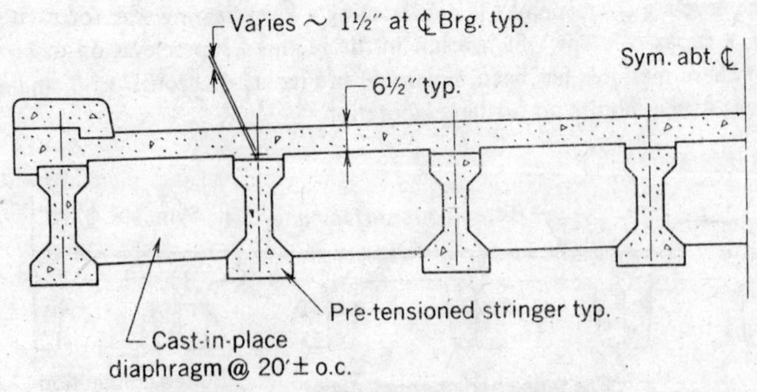

Fig. 12-7 Half-section of short-span composite-stringer construction.

or channel slabs. The stringer type of construction does not require a separate wearing surface, as do the slab types of construction, unless precast slabs are used to span between the stringers in lieu of a cast-in-place deck. The stringer construction generally results in the minimum total quantity of superstructure materials. The construction time required to complete a bridge after the precast members have been erected is greater with stringer framing than with the slab type of framing.

12-3 Bridges of Moderate Span

Again for the purposes of this discussion only, moderate spans for bridges of prestressed concrete shall be defined as being from 45 to 80 ft. Bridges in this span range can generally be divided into two types–stringer type bridges and slab type bridges. By far the greatest precentage of bridges encountered in American practice has been in the former category.

Stringer type bridges, which employ a composite, cast-in-place deck slab, have been used in virtually all areas of the country. For moderate spans, the AASHTO-PCI Stringers, types 2 and 3, are sometimes used. These stringers are normally used at spacings of the order of 5 to 6 ft. The cast-in-place deck is generally of the order of 6.5 in. thick. This type of framing is virtually the same as that used on composite-stringer construction for short-span bridges (illustrated in Fig. 12-7).

It should be pointed out that the AASHTO-PCI stringer types I through IV have relatively small flanges and the top flange is smaller than the bottom flange. The dimensions of the AASHTO-PCI stringers are given in Fig. 12-8.

Because of the small flanges, a relatively large depth of construction is required for any specific span when the AASHTO-PCI Stringers are used. This is not a disadvantage in some applications. A large stringer depth does result in a small prestressing force being required for a specific stringer spacing. In some instances, however, bridge construction depth is of prime importance and, in such instances, stringers with larger flanges, such as those illustrated in Fig. 12-9, can be used at spacings of from 6 to 8 ft with a significantly less depth of construction than that which would be required if stringers with small flanges were used.

Another important consideration is the size of the top flange. As has been pointed out, the dead weight of a structure, as well as the stringer alone, becomes greater as the span is increased. The significance of this can be best understood if a stringer with a smaller top flange than bottom flange is analyzed for various stringer spacings on a span of 70 to 80 ft., with composite construction. It will frequently be found that the bottom flange is adequate and that the capacity and spacing of the member is limited by the compressive stresses in the smaller top flange. If the span were only 50 ft and the same procedure were followed, it would be found that the bottom flange limits the design. The difference is due to the difference in the ratio of dead load to live

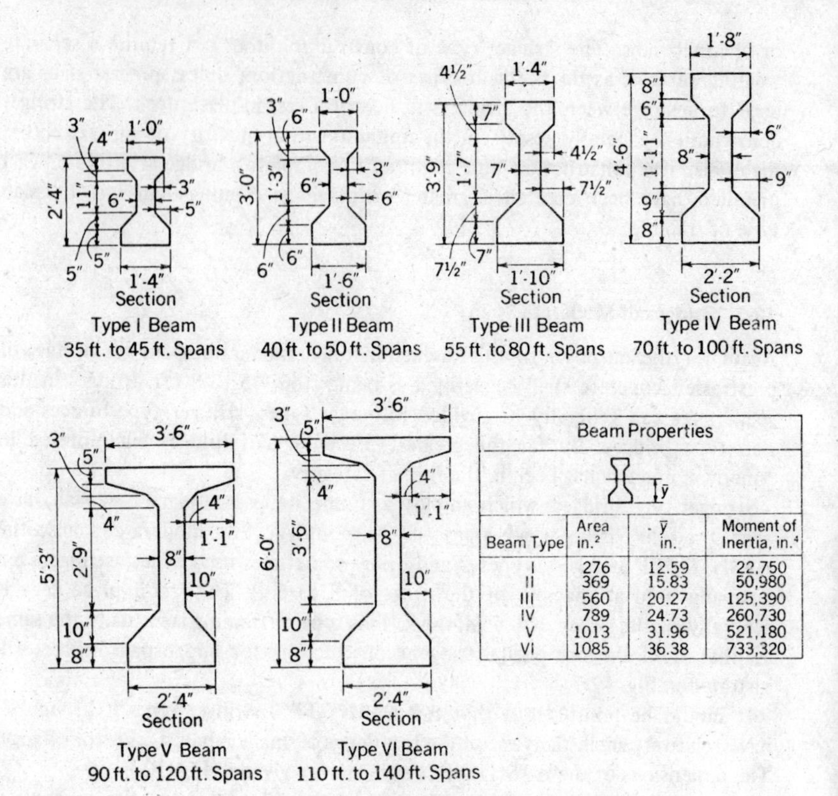

Beam Type	Area in.^{2}	$\bar{y}$ in.	Moment of Inertia, in.^{4}
I	276	12.59	22,750
II	369	15.83	50,980
III	560	20.27	125,390
IV	789	24.73	260,730
V	1013	31.96	521,180
VI	1085	36.38	733,320

Fig. 12-8 AASHTO-PCI standard bridge stringers.

load which occurs as the span is increased. This restriction can be avoided by selecting stringer shapes similar to that shown in Fig. 12-10, when the span is greater than 70 ft.

Composite stringer construction is also used with details similar to those shown in Fig. 12-11. In this type of framing, the top flange of the girder is reinforced or post-tensioned traversely and forms a portion of the deck of the completed superstructure. The slab between the flanges of the stringers can be either cast-in-place or precast. Another scheme is to cast the stringers with daps near the top—the daps are used to support precast slabs. This can be done with I-shaped stringers or with hollow stringers, such as shown in Fig. 12-12.

In order to eliminate the need for large bottom flanges on stringers in the span range of 50 to 70 ft, two-stage post-tensioning has been employed. This procedure consists of using a T-shaped beam in which 50 to 75% of the tendons can be stressed prior to removing the beams from the casting bed. After the deck has been cast, the remaining tendons are stressed. This type of construction is

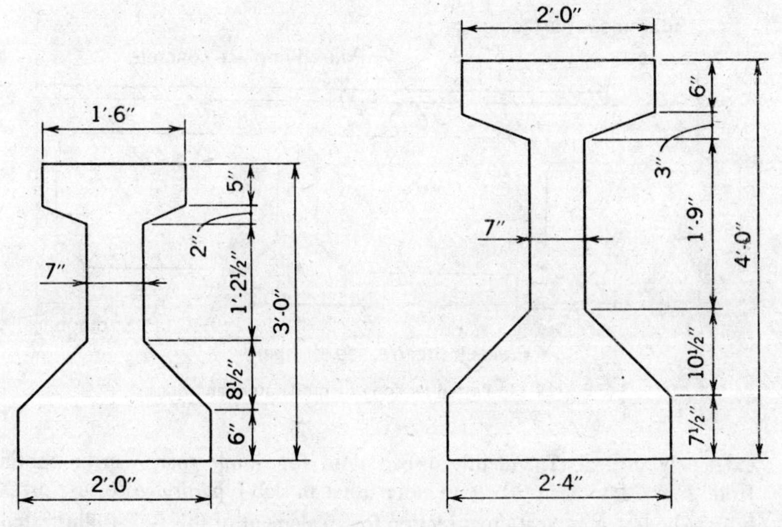

Fig. 12-9 Bridge stringers with large flanges.

restricted to post-tensioned (or combined pre-tensioned and post-tensioned) construction. The second post-tensioning, which cannot be done until the deck has attained reasonable strength, may prolong the required construction time.

Slab-type bridges in the moderate-span range have been made of solid, cast-in-place, post-tensioned concrete in applications where depth of construction is

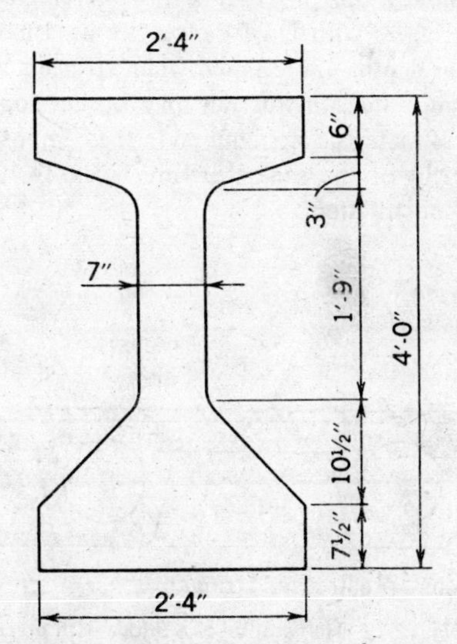

Fig. 12-10 Well-proportioned bridge stringer for span of 60–70 ft.

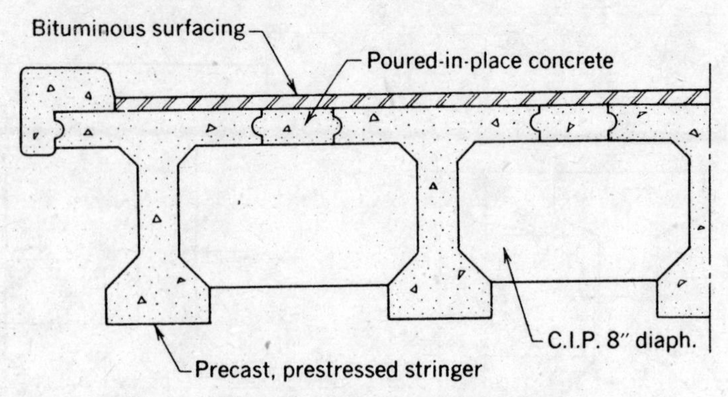

Fig. 12-11 Partial section of a moderate-span bridge.

extremely critical. The depth-to-span ratio for simple spans can be confined from 1 in 26 to 1 in 30. It is more usual in slab-type bridges to use precast, hollow boxes. A typical cross section for an element of this type is illustrated in Fig. 12-3.

Diaphragm details in the moderate-span bridges are generally similar to those of the short spans, with the exception that two or three interior diaphragms are sometimes used, rather than only one at midspan, as in the short-span bridge.

As in the case of short-span bridges, the minimum depth of construction in bridges of moderate span is obtained by using slab construction, which may be either solid- or hollow-box in cross section. Average construction depths are required when stringers with large flanges are used in composite construction. Large construction depths are required when stringers which have small flanges are used. The composite construction may be developed with cast-in-place or precast decks. Less materials are required in the composite type of construction, and the dead weight of the superstructure is less for the stringer construction than for the slab construction.

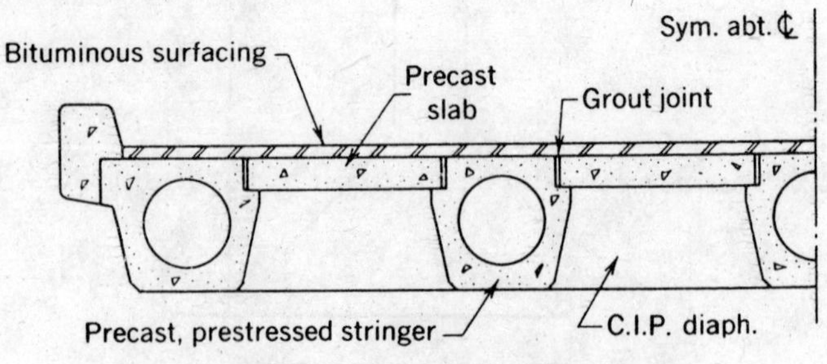

Fig. 12-12 Half-section of a bridge with precast slabs.

12-4 Long-Span Bridges

For bridge spans of the order of 100 ft, the same general types of construction used in bridges of moderate span are frequently used, with the singular exception of solid slabs which would be much too heavy for spans of this magnitude. The stringers are frequently spaced from 7 to 9 ft in bridges of this span range when composite-stringer construction is used. Due to the dead weight of such construction, precast hollow-box construction is generally employed for spans of this order only when the depth of construction must be minimized. Cast-in-place, post-tensioned, hollow-box bridges with simple and continuous spans are frequently used for spans of the order of 100 ft and longer.

Simple, precast, prestressed stringer construction would be economical in this country in spans up to 300 ft, under some conditions. However, only limited use has been made of this type of construction on spans in excess of 100 ft. For very long simple spans, the advantage of precasting is frequently nullified by the difficulties involved in handling, transporting, and erecting the girders, which may be of the order of 10 ft deep and weigh over 200 tons. The exceptions to this are on large projects on which all of the spans are over water of sufficient depth and character that the precast beams can be handled with floating equipment, when special girder launchers can be used (*see* Chapter 16), and when segmental construction is used (*see* Chapter 15).

Precast, long-span bridges may approach the general shape illustrated in Fig. 12-13. A very long, simple-span bridge of cast-in-place construction will normally approach the cross section shown in Fig. 12-14. These types of bridges may be more accurately described as girder bridges rather than stringer bridges. The general reduction (or elimination) of the size of the bottom flange, as well as the large top flange supplied in the cross-sections of Figs. 12-13 and 12-14, should be noted.

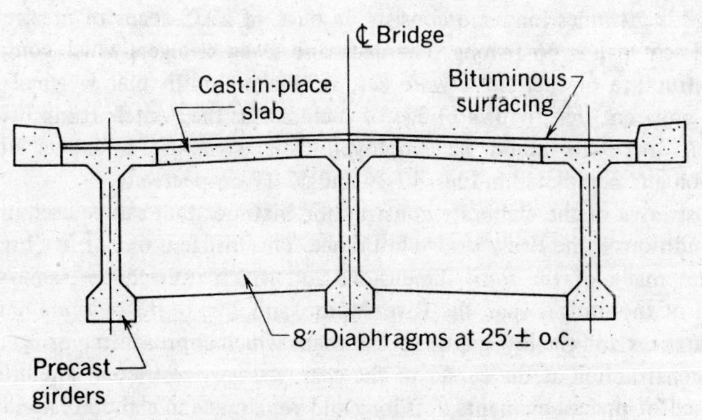

Fig. 12-13 Typical section of a long-span precast bridge.

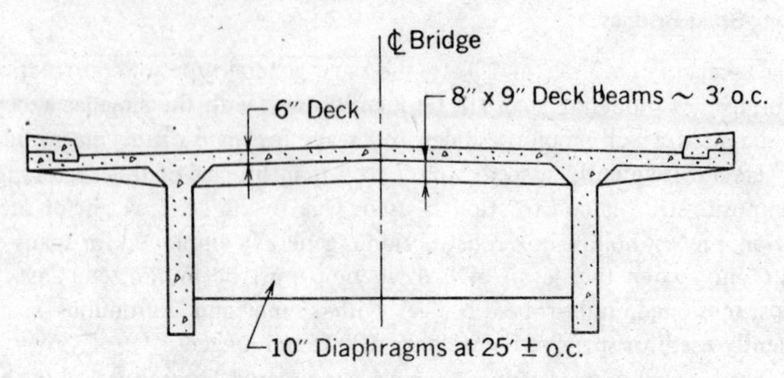

Fig. 12-14 Cross section of a long-span cast-in-place bridge.

The use of cast-in-place, post-tensioned, box-girder bridges has been significant. Typical examples of the cross sections used for structures of this type are those shown in Fig. 12-15. Although structures of these types are occasionally used for spans less than 100 ft, they are more frequently used for spans in excess of 100 ft and have been used in structures having spans in excess of 300 ft. Structurally efficient in flexure, especially for continuous bridges, the box-girder is torsionally stiff and hence is an excellent type of structure for use on bridges which have horizontal curvature. Some governmental agencies use this form of construction almost exclusively in urban areas where appearance from below, as well as from the side, is considered to be important.

12-5 Bridges of Special Types

Exceptional procedures and methods may be practical in the construction of bridges that are very large or that have other special and unique requirements.

An illustration of such a project is the Lake Pontchartrain Bridge in Louisiana. This bridge is 24 miles long and consists, in part, of 2235 spans of prestressed concrete. Each span is 56 ft long. The deck and seven stringers which compose the superstructure of each span were cast as one monolith that weighed 180 tons. It is apparent that, with a bridge of such size as the Pontchartrain Bridge, very special methods could be employed. This structure is shown under construction and completed in Figs. 12-16 and 12-17, respectively.

Also illustrative of the elaborate construction methods that can be used under special conditions is the Esbly Bridge in France. This bridge is one of five bridges which were made of the same dimensions and which utilized the same steel molds. All of the bridges span the River Marne, and, due to the required navigational clearances and the low grades on the roads which approach the bridge, the depth of construction at the center of the span was very restricted. The bridges were formed of precast elements, 6 ft long and were made in elaborate molds by first casting and steam-curing the top and bottom flanges in which the ends of

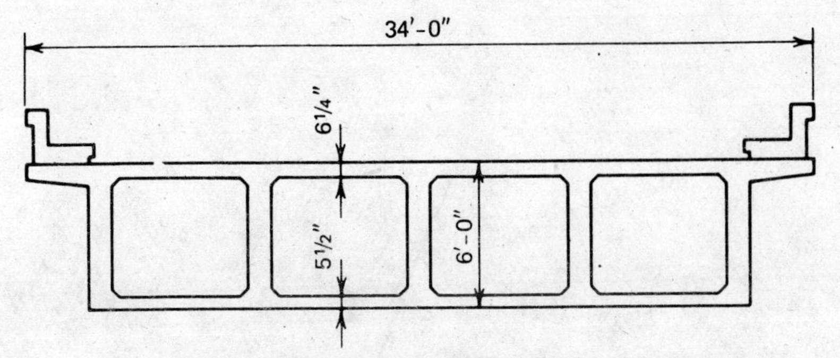

(a) Cross Section of a Box Girder Bridge with Vertical Exterior Webs

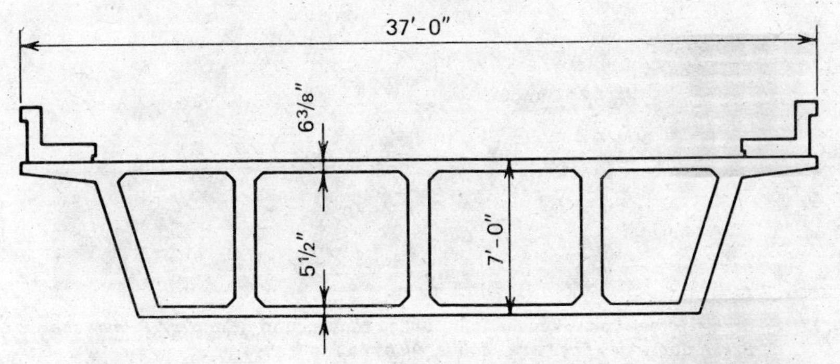

(b) Cross Section of a Box Girder Bridge with Inclined Exterior Webs

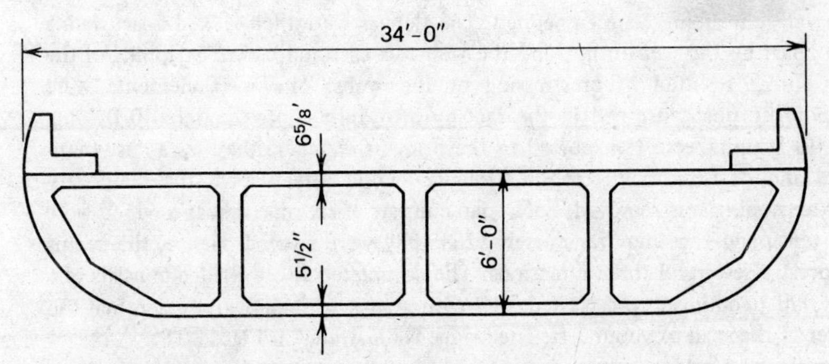

(c) Cross Section of a Box Girder Bridge with Curved Exterior Webs

Fig. 12-15 Typical cross sections of cast-in-place box-girder bridges.

Fig. 12-16 Bridge across Lake Pontchartrain, under construction. (*Courtesy Palmer and Baker, Consulting Engineers, Mobile, Alabama.*)

the web reinforcing were embedded. The flanges were then jacked apart, being held apart by the web forms, and the web was cast and cured. Stripping of the web forms resulted in prestressing of the webs. The 6 ft elements were temporarily post-tensioned in the factory into units approximately 40 ft long. The 40 ft units were transported to the bridge site, where they were raised into place and post-tensioned together longitudinally, after which, the temporary post-tensioning was removed. Each span consists of six ribs or beams which were post-tensioned together transversely after they were erected. Hence, the beams are prestressed in all three directions. The completed Esbly Bridge consists of a very flat, two-hinged, prestressed arch with a span of 243 ft and a depth at the center of the span of about 3 ft. The bridge is illustrated in Fig. 12-18.

Cantilevered bridge construction, in which the segments may be cast-in-place or precast, has been used extensively in Europe. Its use is on the increase in the U.S. An example of this type of construction is shown in Fig. 12-19. In cantilever bridge construction, the superstructure is constructed segmentally, starting from the piers and working each way until the construction meets at the center of the spans. Each segment is post-tensioned to the previously constructed portions of the superstructure as the construction proceeds.

Fig. 12-17 Completed bridge across Lake Pontchartrain. (*Courtesy Palmer and Baker, Consulting Engineer, Mobile, Alabama.*)

Fig. 12-18 The Esbly bridge across the River Marne in France. (*Courtesy Freyssinet Co., Inc., New York.*)

Fig. 12-19 Pierre Bénite Bridge in France. (*Courtesy Freyssinet Co., Inc. New York.*)

Many other elaborate framing schemes have been used in Europe. Most of these have not been considered practical in the American market, due to the relatively high cost of labor in the United States and the normal domestic practice of having separate organizations that are responsible for the engineering and construction. In Europe, where the firm that executes the construction of a project is often responsible for the design, there is less reluctance to use construction methods and procedures that are feasible with only one patented system, if such a system will result in the most economical solution for the particular structure. Time will tell if these construction methods will prove to be feasible in North America.

REFERENCES

1. Libby, J. R. and Perkins, N. D., *Modern Prestressed Concrete Highway Bridge Superstructures*, Grantville Publishing Co., San Diego, 1976.

13 Connections for Precast Members

13-1 General

The connections between precast members should be designed in such a manner that they are capable of withstanding the design (factored) vertical and horizontal loads for which the structure is proportioned, without failure, excessive deformation or rotation. It is normally preferable that the strength of the connections exceed those of the members connected. The details of the connection should be such that they are readily adjusted to accommodate construction tolerances. The tolerances to be considered are not only variations in length, width and elevation, but also possible deviation from the anticipated planes of bearing. Connections must also be detailed to provide for the necessary erection clearances, reinforcing bar bends, and clearances that may be required for special requirements such as post-tensioning tendons after erection.

In the interest of economy, connections should be as simple as possible and should be such that the precast members can be set and disconnected from the erection equipment quickly with the erected members being stable. The connections should be of such a nature that they are easily inspected after they are completed.

The designer should pay attention to the details of the connections in order to be sure the structure will act as has been assumed in the design of the individual elements. If the beams that connect to the opposite sides of a column are designed as simple beams, the connection details should not result in the members being continuous or partially continuous as a result of a cast-in-place reinforced slab or topping, or due to other mechanical fasteners. On the other hand, if the members have been designed as continuous under certain conditions of loading, the connections should be carefully detailed to achieve the intended continuity.

Flexural members undergo rotations and deflections due to the application of transverse loads. Rotations and deflection due to other effects such as temperature, creep, shrinkage, etc. may be encountered. Connection details, particularly for simple spans, should be made in such a manner the necessary rotations can take place without restraint and the risk of spalling from the member being supported unintentionally at an unreinforced edge, as is shown in Fig. 13-1.

Connections that incorporate welded reinforcing steel should be used with caution, since reinforcing steel frequently contains relatively high amounts of carbon, which necessitates special welding procedures. It is recommended that all welding be done in conformance with the recommendations of the American Welding Society (Ref. 1).

Booklets showing standard connection details are available from prestressed-concrete manufacturer's associations. [Refs. 2 and 3].

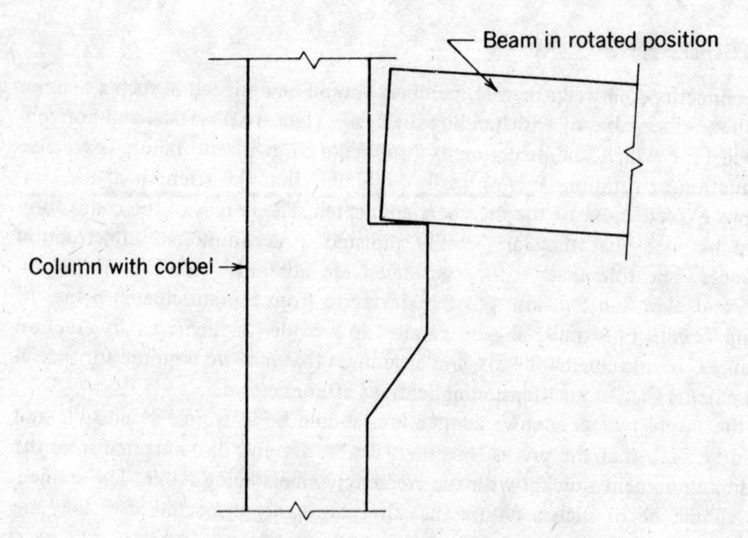

Fig. 13-1 Edge loading due to rotation of a simple beam.

13-2 Computation of Horizontal Forces

The computation of horizontal forces to be used in the design of connections for restrained members can be done by first determining the unrestrained change in length (Δ) of the member that would be expected to occur after the member has been erected and the connection effected. The force from the restraint must then be determined.

If a simple beam were fully restrained, such as being attached to two infinitely stiff shear walls as shown in Fig. 13-2, the restraining force R_o would be that which would cause an increase of strain equal to Δ in the bottom fibers and can be computed as follows:

$$R_O = \frac{E_c A \Delta}{L\left(1 + \frac{y_b^2}{r^2}\right)} \tag{13-1}$$

In the case of a single span supported vertically and restrained against translation, but not rotation, by two columns of equal stiffness (as shown in Fig. 13-3), the value for R_o can be computed as

$$R_O = \frac{3E_c I_c \Delta}{2H^3} \tag{13-2}$$

in which E_c and I_c are the elastic modulus and the moment of inertia of the column, respectively. For a multi-span frame containing a number of equal spans restrained by columns of equal stiffness, such as shown in Fig. 13-4, the force in the interior spans can be approximated by

$$R_i = R_O i(n + 1 - i) \tag{13-3}$$

in which n is the number of spans, i is the number of the span under consideration reckon from the end, and R_o is computed from Eq. 13-2. In Eqs. 13-1 and 13-2, the effective modulus can be used in computing R_o. (*See* Sec. 3-8.)

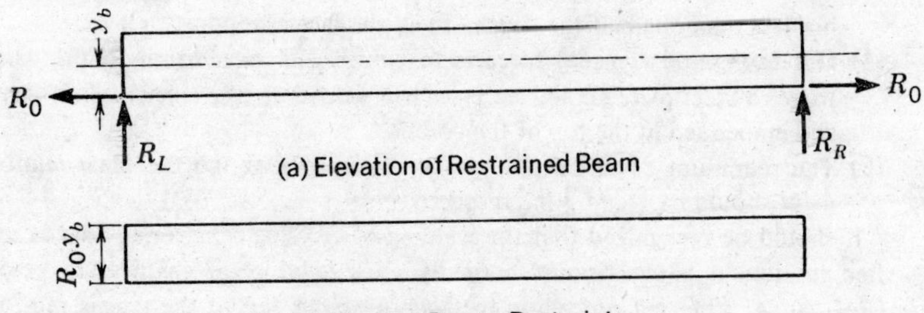

Fig. 13-2 Restrained simple beam.

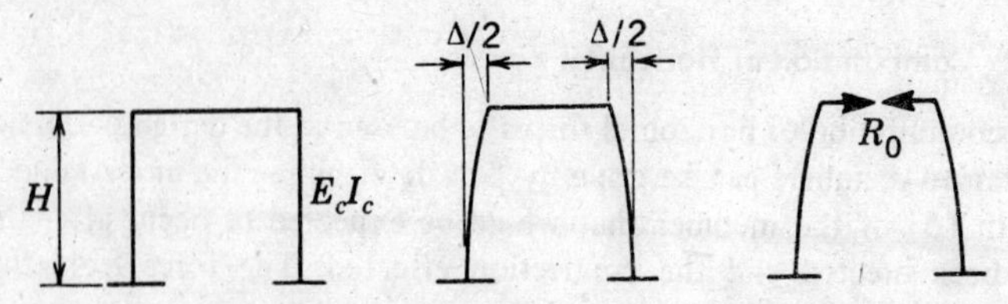

Fig. 13-3 Effect of beam shortening for a beam supported by two columns in which the beam ends are not restrained against rotation.

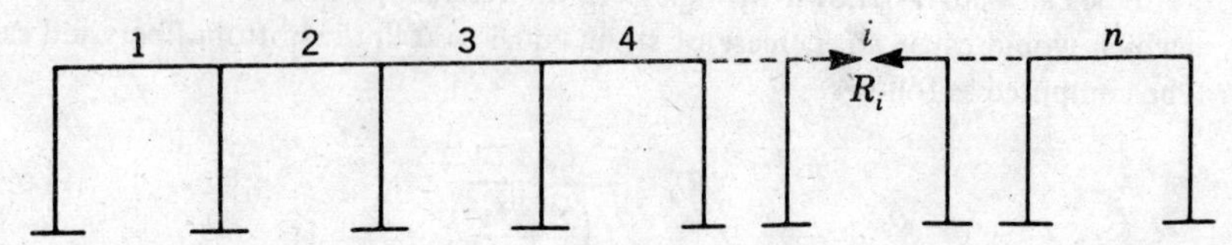

Fig. 13-4 Effect of restraint for a multi-span frame in which beam ends are not restrained against rotation.

In multi-span structures where difficulty is experienced in applying Eq. 13-3, the restraining force in interior spans can be roughly approximated as being equal to 50% of the applied prestressing force (Refs. 3 and 4).

13-3 Corbels

Corbels for the support of precast beams on columns or walls were studied by tests performed at the Portland Cement Association Laboratories (Ref. 5). These tests resulted, in part, in the recommendations shown in Fig. 13-5, which can be summarized as follows:

(1) The minimum distance from the bearing plate to the edge of the corbel should be 2 in.

(2) The tensile reinforcement should be welded to an anchorage bar of the same size as the tensile reinforcement.

(3) The depth of the corbel at the outside face of the bearing plate should be not less than one-half the depth (d) at the face of support.

(4) If a horizontal (tensile) force is to be resisted in addition to the vertical force, a steel plate should be provided welded to the tensile reinforcement and embedded in the top of the corbel.

(5) The minimum cover on the tensile reinforcement and the shear reinforcement should be $\frac{5}{8}$ and $\frac{3}{4}$ in., respectively.

It should be recognized that the welding of crossing reinforcing bars as specified in item 2 above, is *not* normally considered good engineering practice (Ref. 6). A preferred procedure for anchoring the end of the tensile reinforcement to the concrete would be to fillet weld the ends of the tensile reinforcement to flat steel plates or angles rather than to an anchorage bar.

The shear span, a, and the distance from the extreme fiber to the tensile reinforcement, d, to be used in the design of corbels are illustrated in Fig. 13-5. The provisions of ACI 318 relative to the design of corbels include the following:

(1) a/d shall not exceed unity.

(2) For $a/d = 0.50$ or less, the corbel may be designed as for a shear-friction connection (*see* Sec. 13-11), but the quantity and spacing of reinforcing specified herein shall be met.

(3) The percentage of tensile reinforcement, ρ, shall not exceed $0.13f'_c/f_y$.

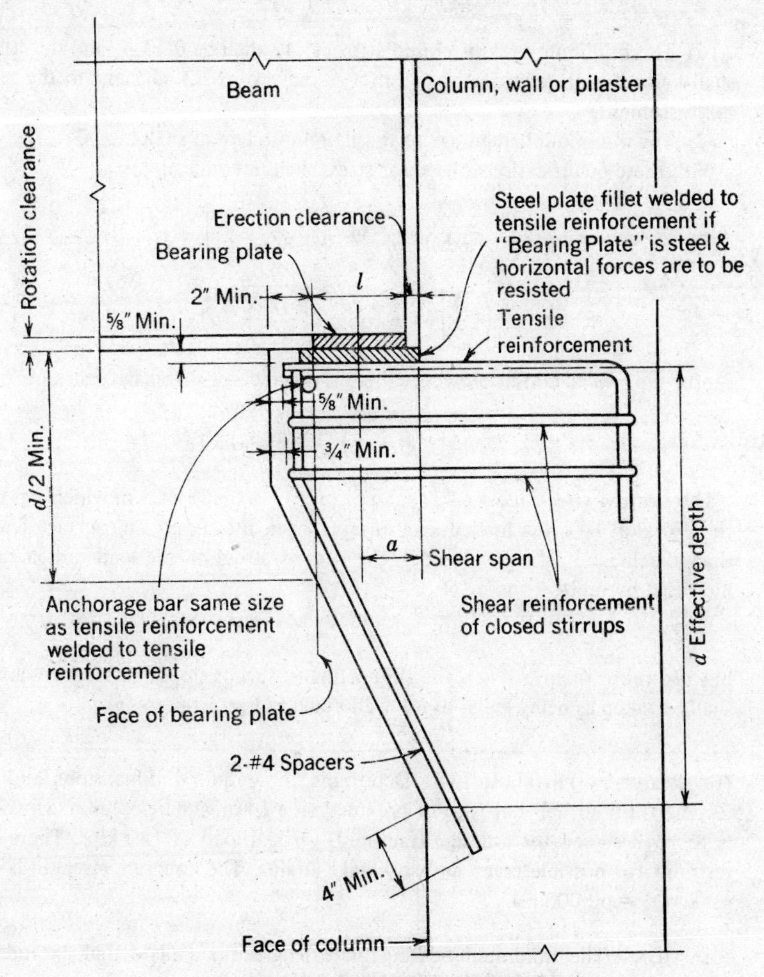

Fig. 13-5 Typical details for corbel design.

(4) The ratio of the design tensile force on the corbels (N_{uc}) to the design shear load (V_u) shall not be taken as less than 0.20.

(5) If provisions are made to prevent tension from restrained shrinkage and creep strains ($N_{uc} = 0$), the corbel will be subjected to shear and moment only and the combined percentage of reinforcement (ρ_v) shall not exceed $0.20f'_c/f_y$. The area of the shear reinforcement parallel to the tensile reinforcement A_h shall not exceed A_s. The combined percentage of reinforcement shall be taken as

$$\rho_v = \frac{A_s + A_h}{bd}$$

(6) The minimum area of closed stirrups, A_h shall be $0.50\,A_s$ and the stirrups shall be uniformly distributed within a distance of $2d/3$ adjacent to the tensile reinforcement.

(7) The minimum percentage of tensile reinforcement shall be $0.04f'_c/f_y$.

With these qualifications, the shear stress shall not exceed

$$V_n = \left[6.5 - 5.1\sqrt{\frac{N_{uc}}{V_u}}\right]\left[1 - 0.5\,\frac{a}{d}\right] \times \left(1 + \left[64 + 160\sqrt{\left(\frac{N_{uc}}{V_u}\right)^3}\right]\rho\right)\sqrt{f'_c}\,b_w d \qquad (13\text{-}4$$

or, for the special condition where there is no creep or shrinkage restraint

$$V_n = 6.5\left(1 - 0.5\,\frac{a}{d}\right)(1 + 64\rho_v)\sqrt{f'_c}\,b_w d \qquad (13\text{-}5$$

The bearing stress due to V_u shall not exceed $\phi(0.85f'_cA_1)$ in which ϕ is equa to 0.70 and A_1 is the loaded area except, when the supporting surface is wide than the loaded area on all sides, the bearing stress on the loaded area can b increased by multiplying by

$$\sqrt{A_2/A_1}$$

but not more than 2. When the supporting surface is sloped or stepped, area A shall be taken as being equal to $b'l'$ as defined in Fig. 13-6.

ILLUSTRATIVE PROBLEM 13-1 Determine the required dimensions and rei forcing for a corbel that is to be provided on a 14 in. square column. The corb is to be designed for a design (factored) vertical load of 140 kips. There is n restraint for possible creep and shrinkage strains. The concrete strength is 45(psi and f_y = 60,000 psi.

SOLUTION: The minimum bearing plate length for a plate that extends th width of the corbel can be computed as follows:

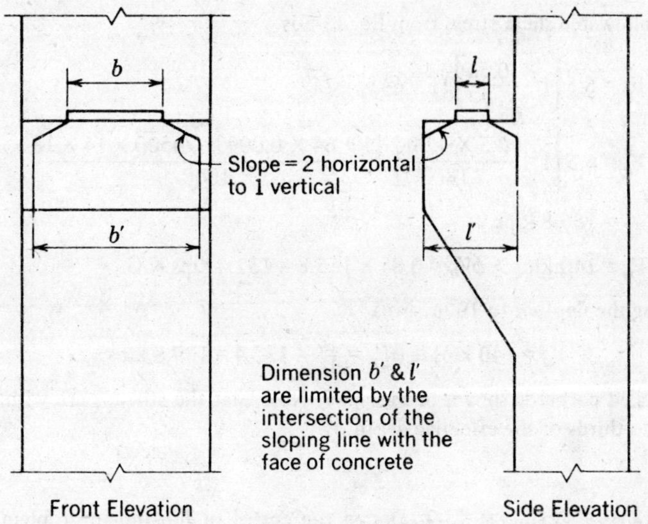

Fig. 13-6 Limits for computation of bearing stresses in accordance with ACI 318 for the condition of loaded area less than supporting area.

$$l = \frac{V_u}{0.85\phi bf_c'} = \frac{140{,}000}{0.85 \times 0.70 \times 14 \times 4500} = 3.73 \text{ in. Use } 4.5 \text{ in.}$$

Providing for an erection gap of 1 in. ±0.75, the maximum possible shear span a becomes

$$a = 1.75 + \frac{4.5}{2} = 4.00 \text{ in.}$$

The maximum combined reinforcement is

$$\rho_v = \frac{A_s + A_h}{bd} = 0.20\left(\frac{4{,}500}{60{,}000}\right) = 0.015$$

Adopting a trial dimension of 18 in. for d, $v_u = \dfrac{140{,}000}{14 \times 18} = 556$ psi and the maximum combined area of steel is:

$$A_s + A_h = 0.015 \times 14 \times 18 = 3.78 \text{ sq in.}$$

Try 5 No. 5 bars for A_s and 2 No. 4 closed stirrups for A_h

$$A_s = 1.55 \text{ in.}^2$$

$$A_h = 0.80 \text{ in.}^2 > 0.5\,A_s$$

$$\rho_v = \frac{2.35}{14 \times 18} = 0.0093 < 0.015$$

and the allowable shear stress from Eq. 13-5 is

$$v_u = 6.5\left[1 - \frac{0.5a}{d}\right][1 + 64\rho_v]\sqrt{f_c'}$$

$$V_n = 6.5\left[1 - \frac{0.5 \times 4.00}{18}\right]\frac{[1 + 64 \times 0.0093]\sqrt{4500} \times 14 \times 18}{1000}$$

$$= 155.8 \text{ kips}$$

$$V_u = 140 \text{ kips} > \phi V_n = 0.85 \times 155.8 = 132.4 \text{ kips N.G.}$$

Increasing the depth d to 19 in. gives

$$V_u = 140 \text{ kips} \cong \phi V_n = \tfrac{19}{18} \times 132.4 = 139.8 \text{ kips}$$

The detailed corbel is shown in Fig. 13-7. Note that the stirrups are within the upper two-thirds of the effective depth.

ILLUSTRATIVE PROBLEM 13-2 Design the corbel of Illustrative Problem 13-1 assuming V_u = 140 k and N_{uc} = 50 k, all other details remain the same.

$$a = 4.0 \text{ in.}$$

$$\frac{N_{uc}}{V_u} = 0.357$$

Assuming a value of d of 24 in., the shear stress due to V_u is $v_u = \dfrac{140{,}000}{14 \times 24} = 417$ psi and the maximum amount of tensile reinforcing is

$$A_s = \frac{0.13 f_c'}{f_y} bd = \frac{0.13 \times 4500 \times 14 \times 24}{60{,}000} = 3.28 \text{ in.}^2$$

Adopting 4 No. 8, A_s = 3.16 sq. in., ρ = 0.0094, a/d = 0.167, from Eq. 13-4, the allowable shear stress is

$$v_u = [6.5 - 5.1\sqrt{0.357}][1 - 0.5 \times 0.167][1 + (64 + 160\sqrt{0.357^3})0.0094][67.08]$$

$$= [3.45][0.917][1.92][67.08] = 408 \text{ psi}$$

$$V_u = 140 \text{ kips} > \phi V_c = 0.85 \times 408 \times 14 \times 24 = 116.5 \text{ kips}$$

$$\text{min. depth} = \frac{140}{116.5} \times 24 = 28.8 \text{ in.}$$

Therefore, use a depth of 30 in., 4 No. 8 bars for tensile reinforcing and 4 No. 4 closed stirrups, $A_h = 1.60 \text{ in.}^2 > 0.5A_s$, spaced at 4 in. o.c.

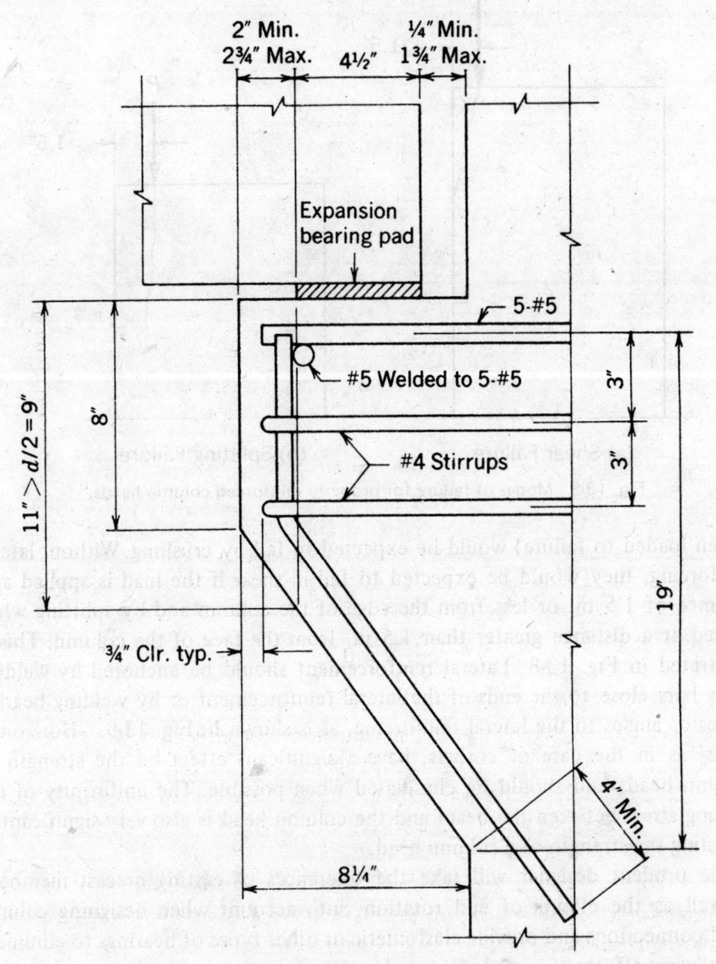

Fig. 13-7 Corbel for Prob. 13-1.

13-4 Column Heads

The provisions of the building codes relative to bearing stresses can be a critical limitation in the design of column-beam connections in precast construction. A study of the ultimate strength of column heads with various reinforcing details, loaded in various manners, was conducted by the Portland Cement Association (Ref. 7) in order to determine relationships for the accurate determination of the strength of these connections.

The study revealed that without restricting the bearing stress to conventional levels, column heads that are adequately reinforced with lateral reinforcing

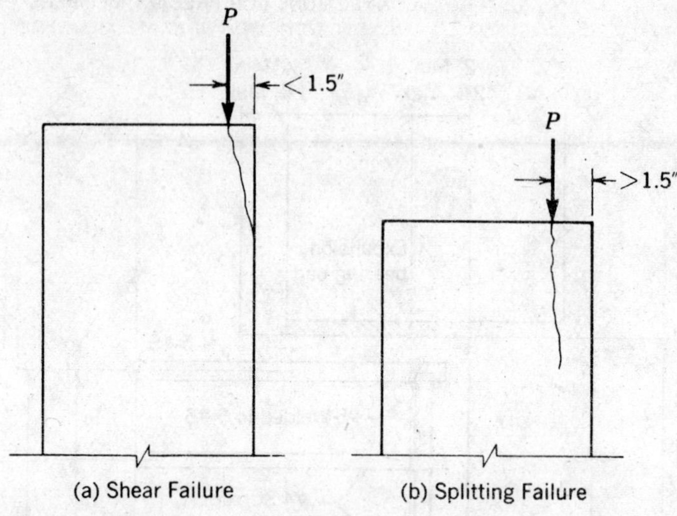

(a) Shear Failure (b) Splitting Failure

Fig. 13-8 Modes of failure for properly reinforced column heads.

(when loaded to failure) would be expected to fail by crushing. Without lateral reinforcing, they would be expected to fail in shear if the load is applied at a distance of 1.5 in. or less from the edge of the column and by splitting when loaded at a distance greater than 1.5 in. from the face of the column. This is illustrated in Fig. 13-8. Lateral reinforcement should be anchored by welding cross bars close to the ends of the lateral reinforcement or by welding bearing plates or angles to the lateral reinforcing, as is shown in Fig. 13-9.* Horizontal loads, as in the case of corbels, have a significant effect on the strength of column heads and should be eliminated when possible. The uniformity of the bearing stress between the beam and the column head is also very significant in affecting the strength of a column head.

The prudent designer will take the tolerances of casting precast members, as well as the effects of end rotation, into account when designing column head connections and provide elastomeric or other types of bearings to eliminate the adverse effects of irregularities and rotations.

The nominal bearing strength, of a laterally reinforced column head that is uniformly loaded in bearing, in which the horizontal forces are either eliminated accurately determined, and in which the product of sl is two or more can be computed by

$$v_c = (A)(B)(C) \tag{13-6}$$

*See Sec. 13-3 for a comment on welding of crossing reinforcing bars.

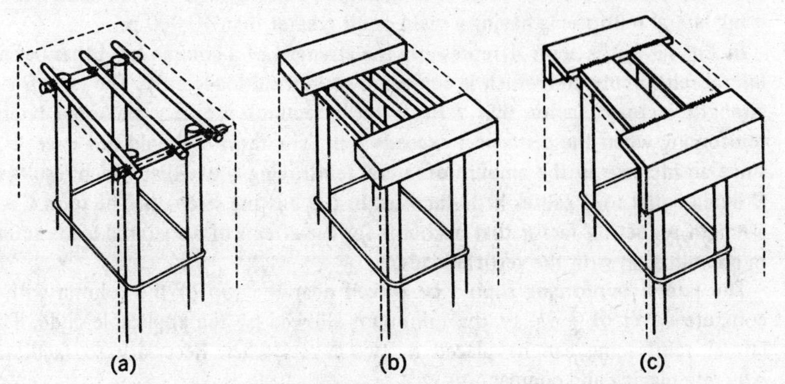

Fig. 13-9 Modes of welding lateral reinforcement in column heads.

in which

$$A = 69\sqrt{f'_c}\ \sqrt[3]{\frac{s}{l}} \tag{13-7}$$

$$B = 1 + C_1\sqrt{\frac{A_{sl}}{b}} \quad \text{(Max value = 2)} \tag{13-8}$$

$$C = \frac{1}{16}\,\frac{N_{uc}}{V_u}$$ when the lateral reinforcing is welded to transverse bars of equal size. (13-9)

$$C = \frac{1}{9}\,\frac{N_{uc}}{V_u}$$ when the lateral reinforcing is welded to embedded steel bearing plates or steel angles. (13-10)

and

A_{sl} = area of the lateral reinforcing (minimum yield strength = 40,000 psi) with a maximum value of 0.16 sq in. per inch of bearing plate width.

b = width of bearing plate.

C_1 = a constant which equals 0 when s is less than 2 in. and which is equal to 2.5 when s is 2 in. or more.

f'_b = concrete bearing strength in psi.

f'_c = specified cylinder strength in psi.

T_u/V_u = ratio of horizontal and vertical design loads (Factored).

s = distance from the edge of the column to the center line of the bearing plate in inches.

l = length of the bearing plate in inches.

It should be noted that very little increase in strength (if any) is achieved by using lateral reinforcing having a yield point greater than 40,000 psi.

In Eq. 13-6, the term A represents the strength of a column head that has no lateral reinforcing and which is subjected to vertical loads only. The term B is a strength increase factor that reflects the effects of properly anchored lateral reinforcing when the distance s exceeds 2 in. The term B should not exceed 2, since an increase in the amount of lateral reinforcing above that which results in B being equal to 2 causes little increase in the bearing strength. The term C is a strength reduction factor that accounts for the effects of horizontal loads acting in combination with the vertical loads.

The lateral reinforcing should be placed near the top of the column with a concrete cover of $\frac{5}{8}$ in. or the minimum allowed by the applicable code. The lateral reinforcing can be placed in two layers when necessary to facilitate concrete placing and compaction.

The nominal bearing strength computed with Eq. 13-6 should be used in comparing the design (factored) load to the strength as

$$V_u \leqslant \phi V_n$$

in which ϕ is taken to be 0.85 or less.

ILLUSTRATIVE PROBLEM 13-3 Design the column head for a 12 in. × 12 in. column that is to support two beams symmetrically placed about the center line of the column. Horizontal loads are eliminated. The design (factored) load for each beam is 200,000 lb, $f'_c = 4000$ psi, and the bearing pad width is 3 in.

SOLUTION: Provide a 1 in. gap between girders and use a distance s of 2.75 in. in order to provide equal edge distances for the beam and column without lateral reinforcing.

$$v_c = 69\sqrt{4000} \times \sqrt[3]{\frac{2.75}{3.00}} = 4239 \text{ psi}$$

$$\phi V_n = 0.85 \times 4239 \times 3 \times 12 = 129.700 \text{ lb}$$

Therefore, lateral reinforcing is required and the term B from Eq. 13-8 must equal

$$B = \frac{200{,}000}{129{,}700} = 1.54 = 1 + 2.5\sqrt{\frac{A_{sl}}{12}}$$

since $s > 2.0$ inches and

$$A_{st} = 12\left(\frac{1.54 - 1.0}{2.5}\right)^2 = 12\left(\frac{0.54}{2.5}\right)^2 = 0.560 \text{ in.}^2$$

Three No. 4 bars, which are provided with a No. 4 anchorage bar welded at each end, give a lateral reinforcing area of 0.60 sq in. and provide a good solution. Note that the actual ultimate bearing stress on the concrete with this solution would be 5560 psi.

13-5 Post-tensioned Connection

Continuity of beam-column connections can be obtained by utilizing the principle of segmental construction shown in Fig. 13-10. This connection has the advantage that continuity can be developed without the use of embedded metal bearing plates and field welded connections. The principal disadvantage of this detail is that it is necessary to use a mortar or concrete joint, unless the beams and columns can be cast one against the other, match-marked, and later erected in the same relative position. This type of joint may result in a slightly longer construction time than would be experienced with other types of connections and may induce slightly larger moments in the columns, due to beam shortening.

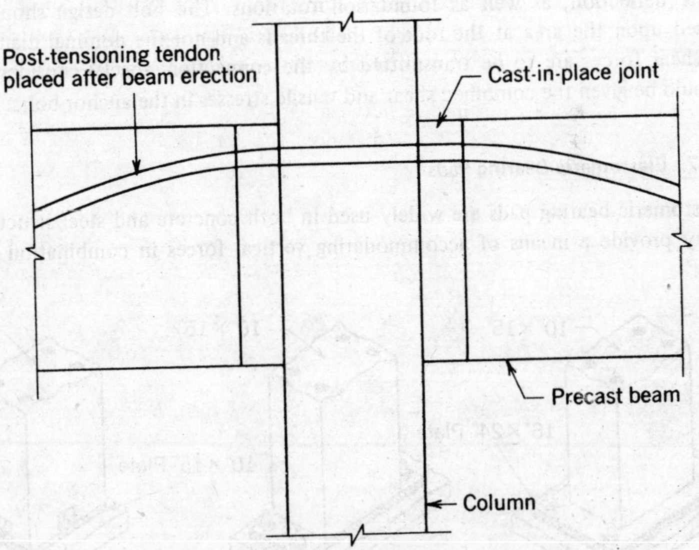

Fig. 13-10 Post-tensioned beam-column connection.

13-6 Column Base Connections

Column base connections shown in Fig. 13-11 are commonly used in precast concrete structures. Each detail has certain advantages relative to fabrication of the precast columns, but there is little difference in the structural performance of the types of details. The recessed column detail requires careful filling of the recesses with high quality concrete after erection if the full strength capacity of the connection is to be achieved. The welding of the reinforcing to the bearing plates and the grouting of the plates after erection must be done properly in order to obtain optimum results. It should be recognized that the capacity of this type of connection is generally considered to be limited by the allowable bearing stress in the weaker of the two concretes. If a compression force is considered to be transferred from the column reinforcing to the anchor bolts through the bearing plate, it may be rationalized that a force larger than that controlled by the concrete bearing stress may be allowed.

The provision of jam nuts gives an easy means of plumbing and adjusting the elevation of the columns at the time of erection.

Column connections of this type should be designed using the capacity reduction factor $\phi = 0.70$. The critical section in design of the extended base plate can be taken as the plane tangent to the column reinforcing steel nearest the anchor bolts that are loaded in tension.

The stiffness of the connection must include the effects of bolt elongation, plate deflection, as well as foundation rotations. The bolt design should be based upon the area at the root of the threads and not the nominal diameter. If shear forces are to be transmitted by the connection, special consideration should be given the combined shear and tensile stresses in the anchor bolts.

13-7 Elastomeric Bearing Pads

Elastomeric bearing pads are widely used in both concrete and steel structures. They provide a means of accommodating vertical forces in combination with

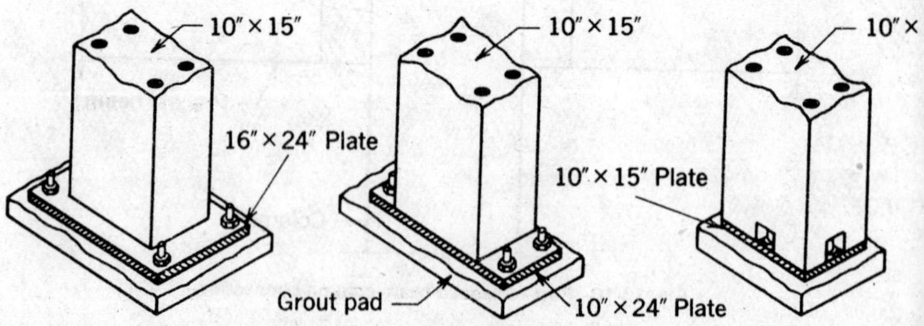

Fig. 13-11 Typical column base connections for precast columns.

horizontal displacements and rotations in all directions. This type of bearing is particularly interesting, since it functions in proportion to the shear modulus of the elastomer and to a shape factor and is not affected by friction, dirt or oxidation.

Some elastomeric bearing pads are solid elastomer, while others are formed of laminations of steel plates and elastomer. The provision of bonded steel plates between layers of elastomer reduces the bulging of the elastomer and, since shear stresses result from the bulges, higher loads can be carried on laminated pads. Plain elastomer pads are suitable for use where low stresses, due to vertical loads, can be accepted and where larger horizontal movements are not needed.

"Standard Specifications for Highway Bridges" (Ref. 8) includes a specification for the design of elastomeric bearing pads. This procedure is based upon a shape factor that is equal to the area of the loaded face divided by the side area that is free to bulge. The shape factor is used with curves available from the manufacturers of this type of bearing. The AASHTO specifications make the following additional limitations:

(1) T (the total effective elastomer thickness) = the summation of thickness of individual layers of laminated bearings.

(2) Maximum allowable compressive deflection under dead load due to non-parallel surfaces = $0.06\ T$.

		Plain Bearing	*Laminated Bearing*
(3) Minimum length of rectangular bearings (parallel to direction of translation)	=	$5T$	$3T$
(4) Minimum width of rectangular bearings (perpendicular to direction of translation)	=	$5T$	$2T$
(5) Minimum radius for circular bearings	=	$3T$	$2T$
(6) Total displacement due to temperature (sum of positive and negative displacements)	=	$0.5T$	
(7) Maximum allowable unit stress due to vertical load:			
Total load (without impact)		800 psi	
Dead load only		500 psi	

(8) Maximum allowable compressive deflection in a plain bearing or in any layer of a laminated bearing under dead plus live loads (without impact) is $0.07T$.

A typical plot of the relationship between the shape factor and compression strain for an elastomer having a hardness of 52 durometer is shown in Fig. 13-12.

The variation in shear modulus for elastomers of different hardnesses is given in Table 13-1 and the creep in compression is shown in Fig. 13-13.

TABLE 13-1 Modulus of Elasticity in Shear

50 Durometer Hardness	*60 Durometer Hardness*	*70 Durometer Hardness*
110 psi at 70°F	160 psi at 70°F	215 psi at 70°F
1.1 × 110 psi at 20°F	1.1 × 160 psi at 20°F	1.1 × 215 psi at 20°F
1.25 × 110 psi at 0°F	1.25 × 160 psi at 0°F	1.25 × 215 psi at 0°F
1.9 × 110 psi at −20°F	1.9 × 160 psi at −20°F	1.9 × 215 psi at −20°F

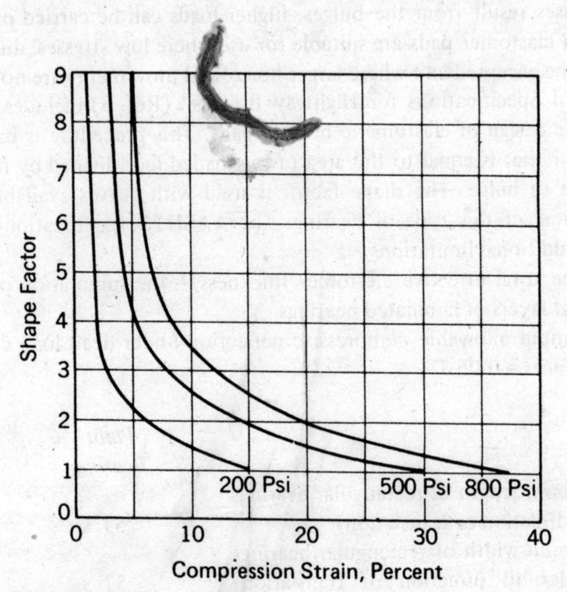

Fig. 13-12 Shape factor vs compression strain for elastomer of 52 durometer hardness loaded to various stress levels.

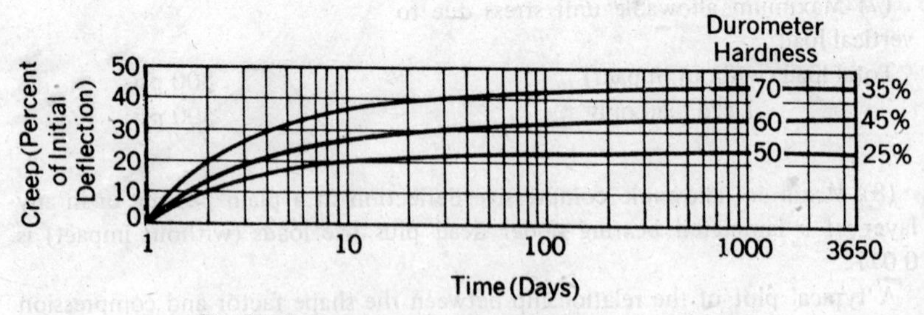

Fig. 13-13 Creep in compression of typical elastomers.

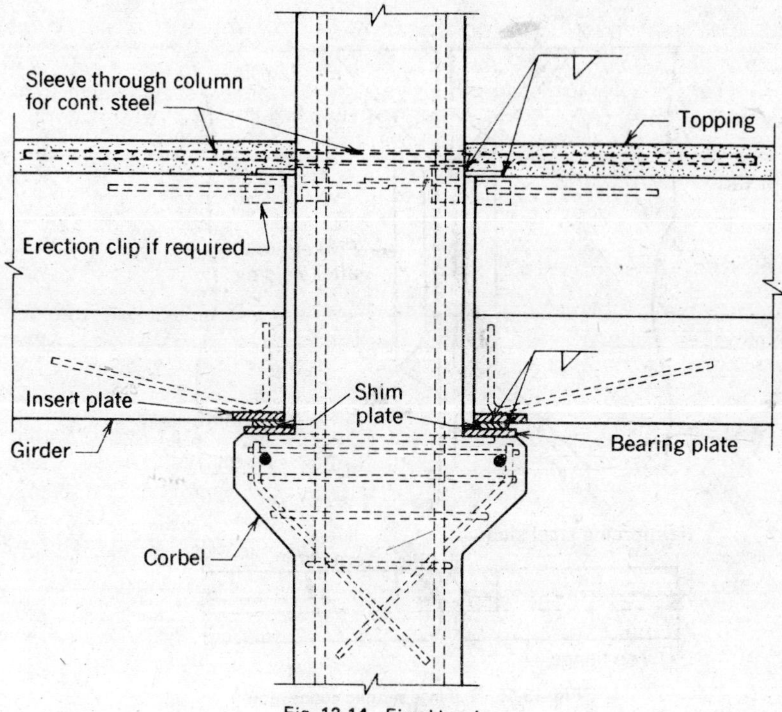

Fig. 13-14 Fixed bearing.

for this purpose is not common, since in order to be effective, the units must be restrained against movements that tend to separate the units. Although this is sometimes possible through the inclusion of continuous reinforcing or prestressing in the end joints (over the supporting beams or girders), the details can be involved and are not straight forward. The use of adhesives is not common for providing shear strength, since they are not resistant to heat, and hence, cannot be used structurally in fire-rated construction. A commonly used shear connector detail for double-T, single-T, and channel slabs is shown in Fig. 13-15. It should be noted that these details frequently include the welding of reinforcing steel, and hence, caution is required (*see* Sec. 10-15).

It should also be observed that the imposition of a force on the connection results in compression in some of the bars. Because the bars are positioned in thin webs, the bars cannot be tied and hence braced against buckeling, as is customary in reinforced concrete compression members. Connections of this type cannot be expected to exhibit great ductility, an important consideration in a seismic design.

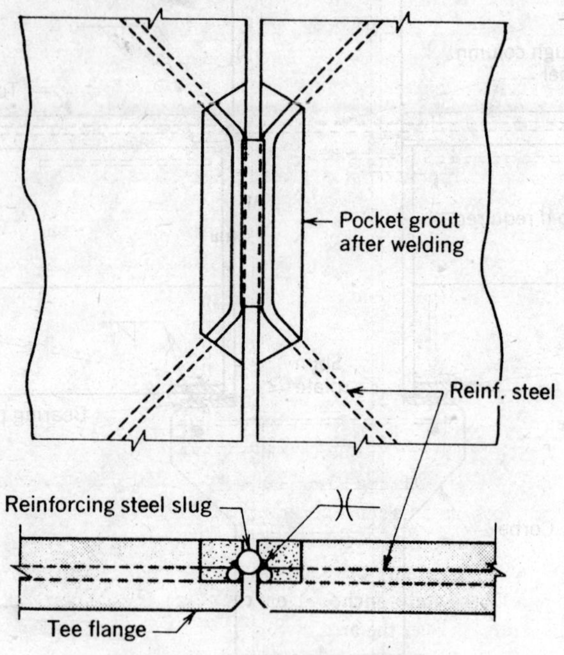

Fig. 13-15 T flange seismic connection.

Hollow core slabs (for building construction) made with some processes can be provided with metallic inserts that can be used to develop the required shear forces. Other manufacturing processes do not permit the inclusion of metal inserts and must normally be used with a structural topping in areas where lateral loads must be resisted.

Concrete shear keys can be used efficiently in bridges to provide the required lateral and longitudinal shear resistance for horizontal forces. The shear keys are provided between the end diaphragms and the bent or pier caps or at the abutments. The keys can be lined with thin (0.25 in.) expansion joint material (to prevent bond) where the keys are designed to be fixed against displacements, but free to rotate. Thicker expansion joint material or expanded polystyrene is frequently used to line the keys, which are designed to provide for displacements (expansion bearings) as well as rotations.

13-11 Shear-friction Connections

When it is inappropriate to consider shear as a measure of diagonal tension, ACI 318 provides the shear-friction concept as a means of designing. The method can be used in the design of corbels when the ratio of the shear span to the effective depth is 0.50 or less. There are many areas where the concept will find application in precast structural elements.

In the shear-friction concept, a crack is assumed to exist in the member in the location where one might expect the element to fail in shear. Reinforcing steel that crosses the crack is assumed to provide a force (normal to the crack) which develops a frictional resistance along the crack. If the reinforcing is present in sufficient quantity, the frictional resistance will be greater than the shear stress. The reinforcing should be approximately perpendicular to the crack.

The shear stress v_u is not permitted to exceed 800 psi or $0.2f_c'$ and the area of steel required is computed from

$$A_{vf} = \frac{V_u}{\phi f_y \mu} \tag{13-11}$$

in which the yield strength, f_y, cannot exceed 60,000 psi, $\phi = 0.85$ and, μ is equal to 1.4 for concrete cast monolithically, 1.0 for concrete placed against a clean, artificially roughened, hardened concrete (construction joints) surface, and 0.70 for concrete plated against clean, unpainted as-rolled steel.

If tensile stresses exist across the crack, additional reinforcement must be provided to resist them. In any case, the reinforcement must be placed in such a manner that it is adequately anchored on each side of the assumed crack and must be well distributed over the area.

REFERENCES

1. *Structural Welding Code–Reinforcing Steel*, American Welding Society, AWS 01.4-79, Miami, Florida, 1979.
2. "PCMAC Connections Manual," prepared by the Prestressed Concrete Manufacturers Association of California, Inc. (Oct. 1969).
3. PCI Committee on Connection Details. *PCI Manual on Design of Connections for Precast Prestressed Concrete.* Prestressed Concrete Institute, Chicago, 1973.
4. Burton, K. T., Corley, W. G. and Hognestad, E., "Connections in Precast Concrete Structures–Effect of Restrained Creep and Shrinkage," *Journal of the Prestressed Concrete Institute,* 12, No. 2, 18-37 (April 1967).
5. Kriz, L. B. and Raths, C. H., "Connections in Precast Concrete Structures–Strength of Corbels," *Journal of the Prestressed Concrete Institute*, 10, No. 1, 16-61 (Feb. 1965).
6. *Reinforcement Anchorages and Splices*, Concrete Reinforcing Steel Institute, Chicago, Ill., 1980.
7. Kriz, L. B. and Raths, C. H., "Connections in Precast Concrete Structures–Strength of Column Heads," *Journal of the Prestressed Concrete Institute*, 8, No. 6, 45-75 (Dec. 1963).
8. "Standard Specifications for Highway Bridges," 12th ed., American Association of State Highway Officials, Washington, D.C., 1977.

9. Libby, J. R. and Perkins, N. D., *Modern Prestressed Concrete Highway Bridge Superstructures*, Grantville Publishing Co., San Diego, 1976.

10. Rejcha, C., "Design of Elastomer Bearings," *Journal of the Prestressed Concrete Institute*, 9, No. 5, 62–78 (Oct. 1964).

11. "Sky Span Elastomeric Bridge Bearings Engineering and Standards Handbook," General Tire and Rubber Company, Wabash, Indiana.

12. "Design of Neoprene Bearing Pads," E. I. du Pont de Nemours and Company, Inc., Wilmington, Delaware.

13. Long, J. E., *Bearings in Structural Engineering*, John Wiley & Sons, New York, 1974.

14 Pre-tensioning Equipment and Procedures

14-1 Introduction

Pre-tensioning equipment and methods which are characteristic of American practice vary in many respects from the equipment and methods employed abroad. The basic principles in the design of prestressing facilities here and abroad are identical, but, because the normal type of products produced in domestic prestressing plants are larger than the average pre-tensioned products produced abroad, the prestressing facilities required domestically are also larger. No attempt will be made to describe the prestressing facilities which are typical of foreign practice, since a considerable amount of information on this subject is available in the literature. The scope of this chapter is confined to the discussion of equipment and procedures that are peculiar to pre-tensioning as practiced in the United States.

The stress in the pre-tensioning tendons must be maintained as nearly constant as possible during placing and curing of the concrete. This can be accomplished in two ways.

(1) The tendons are stressed and anchored to individual steel molds designed to withstand the prestressing force as well as the stresses which result from the plastic concrete.

(2) The tendons are restrained by a special device, called a pre-tensioning bed or bench, which restrains the pre-tensioning force and provides a level surface on which the forms are placed.

In addition to these devices, other equipment peculiar to pre-tensioned construction, including the mechanism used to stress and release the prestressing tendons, the forms, vibrators, and, tendon deflectors, are discussed in this chapter.

14-2 Pre-tensioning with Individual Molds

With the exception of a few firms which employ stress-resisting molds in the manufacture of double-T roof slabs and pre-tensioned spun piles, this technique is not considered practical in this country. Where pre-tensioned railroad ties and small joists for residential construction are produced in pre-tensioned concrete, this method has the advantage of allowing individual units to be mass produced with the products (and molds) moving through the plant in a production cycle, rather than requiring that the materials and plant be brought to the molds or forms, as is required with a pre-tensioning bench. In manufacturing spun pre-tensioned piles, this technique must be used, since there is no other practical means of spinning the mold and pre-tensioned tendons as a unit.

Another advantage of this method, when employed on small pre-tensioned products, is that the prestressing plant need not be as large and as elongated as is required in using a bench, since small pre-tensioned products in their individual molds can be stacked and need not be arranged in long rows. This advantage applies, but to a lesser degree, with large products that can only be handled with very large cranes and cannot be stacked very high, if at all.

14-3 Pre-tensioning Benches

Pre-tensioning benches are normally designed to withstand a specific maximum force applied at a specific maximum eccentricity. Therefore, it is customary, when stipulating the capacity of a pre-tensioning bench, to give the maximum permissible force (shear) and maximum permissible moment the bench can safely withstand. The maximum moment is normally expressed in terms of the bench proper (slab portion of the bench which extends between the uprights at the abutments) and not necessarily in terms of the top surface of the abutments, which may be recessed to accommodate the stressing mechanism.

Pre-tensioning benches are generally one of the following types:

(1) Column-type bench, which may serve as the mold or form, as well as the device which restrains the tendons.

(2) Independent abutment-type bench in which the independent abutments rely upon soil pressure, piling, or rock foundations for stability.

(3) Strut-and-tie-type bench.

(4) Abutment-and-strut-type bench.
(5) Tendon-deflecting-type bench.
(6) Portable benches.

Each of these types of benches has specific areas of application, and each will be discussed separately.

One additional definition that must be given before discussing the types of pre-tensioning benches pertains to the use to which the bench is to be subject. Some benches are designed to produce a specific product and may be termed a fixed bench. Other benches are designed to produce any type of product that is normally encountered in practice and are termed universal benches.

Column Benches

Column benches rely upon the column action of the bench alone to resist the prestressing force. The eccentricity of the prestressing force, with respect to the bench, must be confined to relatively low values in order to achieve economy with this type of bench. For this reason, the use of column benches is generally restricted to fixed benches designed to produce a single T, double T or pile. An example of a column-type bench, which is designed to produce double-T slabs, is shown in Fig. 14-1.

The column-type benches are generally designed using the Euler formula to compute the critical buckling stress. Adequate safety factors against both crushing of the concrete and buckling of the column must be allowed in the design. The dead weight of the bench alone is relied upon to prevent the column from buckling. The buckling of the bench could occur at the center of the bench by

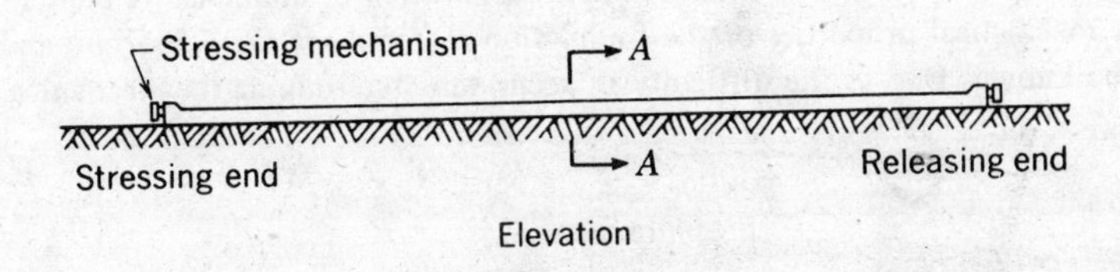

Elevation

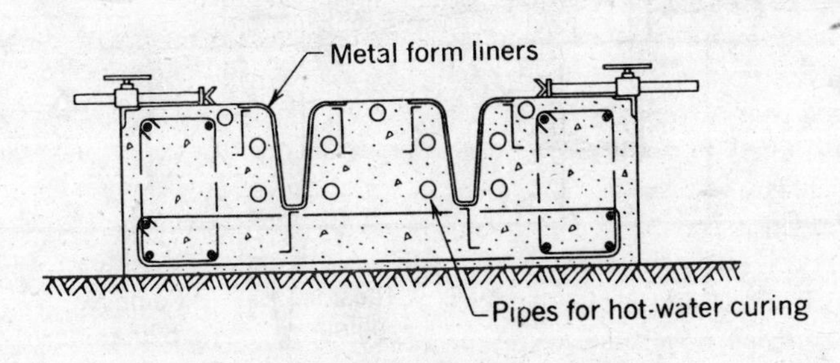

Section *A-A*

Fig. 14-1 Column-type double-T bench.

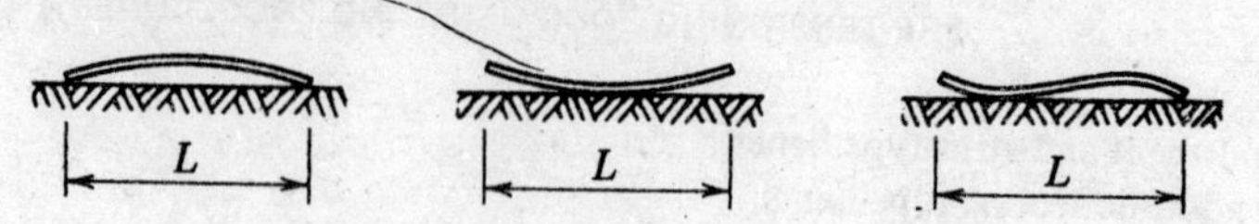

Fig. 14-2 Possible forms of column buckling.

buckling upward; at the ends of the bench which could buckle upward; or by a combination of both. These are illustrated in Fig. 14-2.

Column-type benches are not normally used for universal benches or in benches in which a relatively large eccentricity of the prestressing force must be accommodated.

Independent Abutment Benches

This type of pre-tensioning bench is composed of two large abutments that are structurally independent of each other as well as of the paving material used as a casting surface between the abutments. The abutments, when embedded in soil, may rely exclusively upon the weight of the abutment and passive soil pressure for stability, but it is more common to incorporate piling in the abutments, in order to increase the stability. Abutments of these two types are shown in Fig. 14-3.

When founded on sound rock, independent abutments can be formed by keying the abutments into the rock and, if necessary, increasing the resistance of the abutments to overturning by providing anchors that are embedded in the rock. This is illustrated in Fig. 14-4.

The design of independent abutments requires accurate knowledge of the character of the soil or rock on which the abutments are to be located. The effects of long-term loading, variation of loading, and variation of the moisture content on the mechanical properties of the foundation material are also important and must be known. Due to the difficulty in accurately determining the mechanical

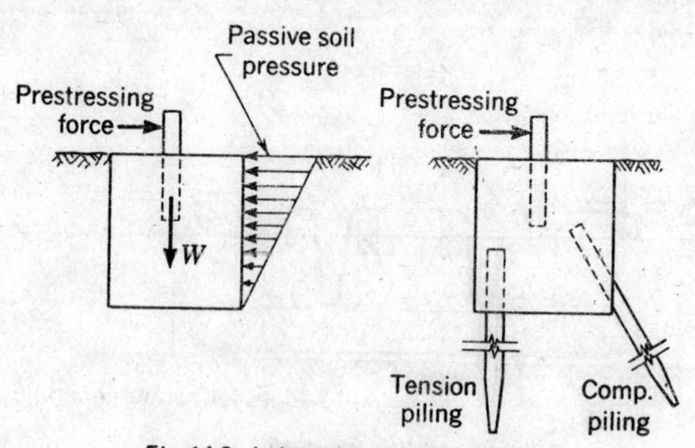

Fig. 14-3 Independent abutments in soil.

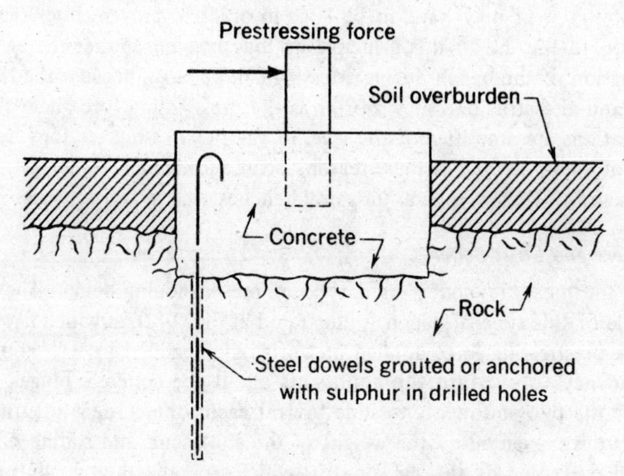

Fig. 14-4 Independent abutments on rock.

properties of the foundation material, this type of pre-tensioning facility is not frequently used. When the foundation material is adequate, this type of pre-tensioning bench is economical for long benches, since the casting surface is not a structural component and hence can be considerably lighter than is required for other types of benches.

Strut and Tie Bench

The principle of the strut and tie pre-tensioning bench is illustrated in Fig. 14-5. Examination of this illustration will reveal that the prestressing force (P) results in a tensile force in the tie (T) and a compressive force in the strut (C). The uprights are relied upon to distribute the three forces.

This type of bench can be used for large eccentricities and is adaptable to universal pre-tensioning benches for this reason. One serious objection to this type of bench, when large prestressing forces are used, is that the compressive force in the strut is larger than the prestressing force ($C = P + T$) and the dimen-

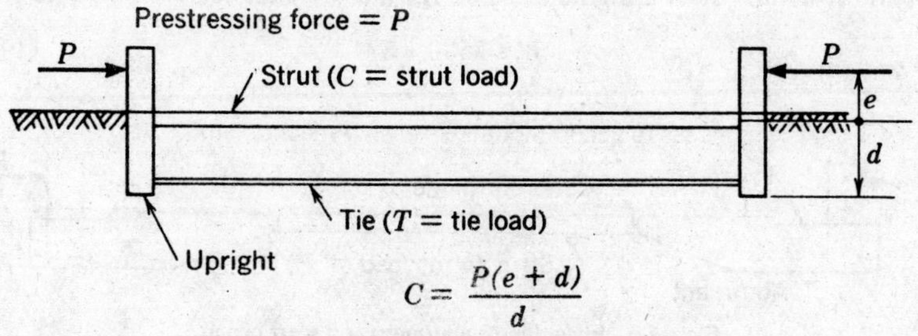

Fig. 14-5 Elevation of strut and tie bench.

sions of the strut may have to be large in order to prevent buckling. Another objection to the use of this principle on long benches is that the effect of the deformation of the bench during stressing can be large, because the tie becomes longer and the strut becomes shorter as the prestressing force is applied—these deformations are amplified at the level of the prestressing tendons by the lever action of the uprights. For these reasons, strut and tie benches are generally used only on short benches, such as universal benches used in laboratories.

Abutment and Strut Bench

This is the most frequently used type of pre-tensioning bench. The structural principle of this type of bench is illustrated in Fig. 14-6, in which it will be seen that the prestressing force applied on uprights embedded in each abutment has the tendency to overturn the abutments about the concrete hinges, as well as to force the two abutments to slide toward each other. The overturning of the abutment is prevented by the weight of the abutment, and sliding of the abutments is prevented by the slab or strut which separates the two abutments. The provision of the hinge between the abutments and the slab insures that the slab section is subjected to a direct axial force (alone) which is equal to the prestressing force in magnitude.

The design of this type of bench consists of determining the amount and shape of the concrete abutment that will provide an adequate safety factor against overturning, the reinforcing of the abutment, and the proportioning of the slab, which is generally designed as a plain concrete column. The abutment of this type of bench is usually made of heavily reinforced concrete or of post-tensioned concrete. The slab portion is generally reinforced with a welded wire fabric in order to help control cracking due to shrinkage.

The approximate quantity of concrete required for the two abutments of abutment- and strut-type benches can be computed from

$$Q = 35 + 0.06M \tag{14-1}$$

in which Q is the approximate quantity of concrete in cubic yards and M is the moment due to prestressing kip-feet for which the bench is designed. The reinforcing required can be taken at 75 lb per cubic yard. The approximate weight of the structural steel uprights required for the two ends can be computed from

$$W = 3500 + 6M \tag{14-2}$$

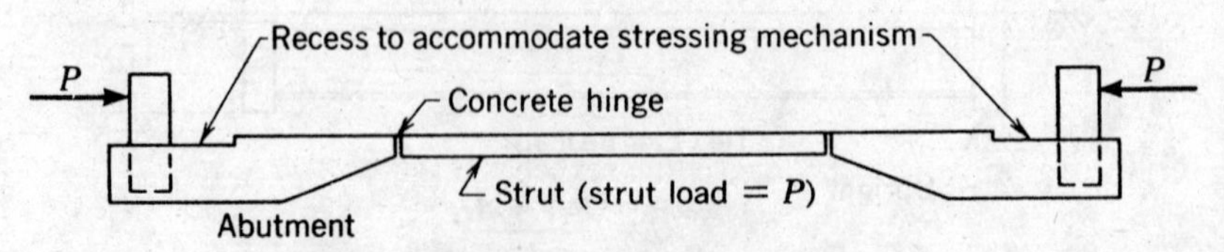

Fig. 14-6 Elevation of abutment and strut bench.

in which W is the approximate weight of the uprights in pounds. The quantity computed from Eq. 14-2 does not include the cross-beams, templates, pull rods or other components of the stressing mechanism that may be required.

Tendon-Deflecting Benches

Most of the methods used in deflecting the tendons in pre-tensioned beams and girders require that the tendons be held in the lower position by devices attached to the slab portion of the pre-tensioning bench. The tendons are held in the upper position by devices which bear on the slab portion of the bench. The slab is thus subjected to a series of vertical loads, as well as to the axial load associated with abutment and strut benches. This type of loading and the bench shape usually employed under such conditions are illustrated in Fig. 14-7.

Due to the large vertical forces, which may have to be applied at virtually any point along the bench in order to achieve the desired results in prestressed products produced with deflected tendons, the use of the concrete hinge, which is characteristic of the abutment and strut bench, is not feasible. Furthermore, since the slab portion of the bench is subject to combined bending and direct stress, the slab must be made of reinforced or prestressed concrete rather than the plain concrete used in abutment- and strut-type of benches. The cost of this type of bench is substantially greater than abutment and strut benches of equal capacity.

The quantity of materials required for the abutments of a tendon-deflecting bench can be approximated from Eqs. 14-1 and 14-2.

Portable Benches

Pre-tensioning benches that can be moved from job site to job site have been used to a limited degree. The portable benches may be of any of the above types and may be entirely or only partially portable. An example of a partially portable bench is a bench of the abutment- and strut-type in which the principal components of the abutment are portable and the strut and counterweight portions of the abutments are not moved.

A universal, portable pre-tensioning bench of the abutment- and strut type is a very practical piece of equipment for general contractors engaged in large highway and marine projects on which large quantities of pre-tensioned concrete

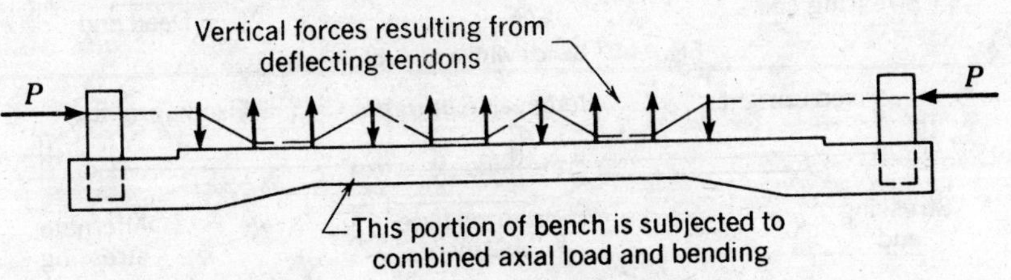

Fig. 14-7 Elevation of tendon-deflecting bench.

may be required in areas where there are no permanent pre-tensioning plants. In a similar manner, portable benches of the column type are occasionally feasible for job site fabrication of double-T and single-T roof members for large building projects.

In the design of universal pre-tensioning benches, it is often desired to provide a means for adjusting the length of the bench so that the waste of the pre-tensioning tendons can be minimized for different production problems and maximum of economy can be achieved in operating the bench. This has been accomplished by the use of pull-rod extensions in the stressing mechanisms, the mechanics of which will be apparent to the reader after considering the discussion of stressing mechanisms which follows this section. Another method used quite extensively is to provide a long dead-end abutment with several alternative positions for the dead-end uprights, as is illustrated in Fig. 14-8. The provision of an intermediate abutment with removable uprights has also been used successfully. Intermediate abutments designed to withstand the prestressing load from either direction, as illustrated in Fig. 14-9, have been used.

The maximum force used in the design of the pre-tensioning bench should be about 15% greater than the total initial force applied to the maximum number of tendons used with the bench. Experience has shown that the shrinkage of the concrete and the effect of temperature variations, which take place during curing and stripping of the concrete, result in the prestressing force being larger at the time of release than it is immediately after stressing. The increase in the stress is a function of: (1) the ratio of the length of the tendon, which is embedded in concrete, to the total length of the tendon between the uprights of the bench; (2) the type of cement; (3) the type of curing used; (4) the air temperature during stripping; and (5) other factors. The gain in stress in the tendons increases with the ratio of length of the embedded tendon to total tendon length. The gain is also affected by the curing time and is more severe when the air temperature during stripping is low.

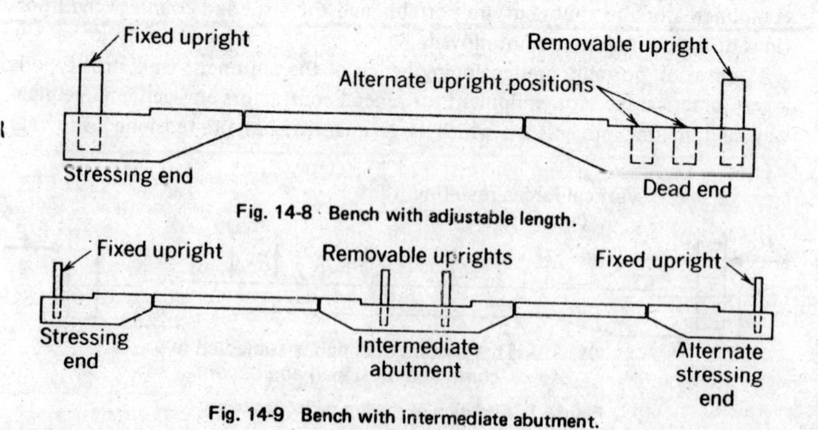

Fig. 14-8 Bench with adjustable length.

Fig. 14-9 Bench with intermediate abutment.

14.4 Stressing Mechanisms and Related Devices

The tensioning of the pre-tensioning tendons can be done by stressing each tendon individually or by stressing all of the tendons at one time. Each method has advantages, and the method used most frequently has significant influence on the design of the stressing mechanism. It is possible to design the stressing mechanism so that the tendons can be stressed with either procedure. In the majority of the plants the tendons are stressed individually. In addition, the stressing mechanism may be designed so that the pre-tensioning tendons can all be released simultaneously with hydraulic jacks, rather than by cutting the tendons.

If the stressing is to be done by stressing one strand at a time, the jack used for the stressing must normally have a stroke of 30 to 48 in. and a working capacity of from 10 to 20 tons, depending upon the length and size of the tendons to be stressed. If the tendons are to be released simultaneously, large jacks, the capacity of which depends upon the maximum prestressing force for which the bench is designed, must be provided for this purpose. The jacks which are used to release the bench do not normally require a stroke in excess of 6 in.

Releasing the tendons hydraulically all at one time has the advantage that the force can be transmitted to the concrete products slowly, and hence, shock or impact loading is avoided. The result is that the transmission length is minimized. In addition, with this type of equipment, the products can be partially released if this becomes necessary or desirable. The principal disadvantage of releasing the tendons with hydraulic jacks is that all of the strain is released at one end of the production line. The result is that the products at the releasing end tend to move away from the releasing end. The amount of movement is a direct function of the elastic deformation of the concrete products and unembedded tendons. When deflected tendons are used, releasing the tendons simultaneously may not be possible, since it may not be possible to release the tendon deflecting devices before the prestressing force is released, and hence, the necessary movement of the products can not take place freely. Another advantage of releasing all of the tendons at one time is that with this procedure the cutting of the tendons between precast units can be done without adhering to a strict schedule.

Releasing the tendons individually is generally done by cutting them one at a time with an acetylene torch (hydraulic cutters have been used). The tendons must be cut according to a strict sequence in order to avoid eccentric loading in the products and in order to prevent too many tendons being cut at one location, which will result in failure of the remaining tendons. In other words, the tendons at each end of each product in the line must be cut at approximately the same time.

When all of the tendons are to be stressed simultaneously, the jacks used to stress the tendons must be of large capacity. Unless the stressing is done in several increments, which is a procedure that is not recommended for normal operation, the stroke of the jacks must be of the order of 30 to 48 in. The same jacks used to stress the tendons can be used to release the pretensioning force, but, during the releasing operation, the stroke required is again generally less than 6 in.

It should also be mentioned here that, when the tendons are stressed individually, it is not necessary to apply an initial load to each tendon in order to equalize the length of the tendons, as is often required (by construction specifications) when all of the tendons are stressed simultaneously. Furthermore, if an anchor slips on one tendon during or after the stressing of the tendons, the tendon that has slipped can be simply restressed, if the tendons are being stressed individually. If, on the other hand, the tendons are all being stressed simultaneously, it is necessary to release all of the tendons and restress after reanchoring the slipped tendon.

It will be found that the cost of the hydraulic jacks is greater in installations designed to stress all of the tendons simultaneously than in installations designed for individual stressing of the strands. The total labor cost of stressing is about equal with both methods, if an initial force must be applied to the tendons when stressed simultaneously, but more time may be required for stressing the tendons individually in members in which there are a large number of tendons. The time factor may be important in some instances, since in the interest of safety, all work in the vicinity of a pre-tensioning bench must be stopped during the stressing operation.

The stroke of the jacks specified for any particular installation should be based upon the anticipated elongation of the steel during stressing plus an allowance for slack in the tendon and slack in the anchorages. The normal theoretical elongation for the pre-tensioning tendons is 7 to 8 in. per 100 ft of length. The slack and anchorage take-up, for which jacking stroke must be provided, is much larger in the case of single tendon stressing than in stressing all the tendons simultaneously, assuming an initial load is applied to the tendons individually, in the latter case, in order to equalize the length of the tendons. It is recommended that the jacks used for stressing tendons individually be 18 in. longer than is required for the theoretical, elastic elongation of the tendon. An extra 10 in. of stroke is recommended for jacks used for the simultaneous stressing of tendons which have been equalized in length by the application of a force equal to about 5% of the initial prestressing force. These recommendations apply for pre-tensioning application in which the maximum length of the tendons to be stressed is of the order of 300 ft.

Experience has shown that for the capacity, stroke, and use to which the hydraulic jacks are subjected in pre-tensioning installations, the jacks should be double acting rather than relying upon springs to return the pistons to the closed (or open) position. Furthermore, it is recommended that the jacks be designed to develop the maximum normal load that will be used with the jacks at a pressure of the order of 5000 to 6000 psi and that the jacks be guaranteed for intermittent service for pressures up to 10,000 psi. The piston rods of the jacks to be exposed during the stressing and releasing operations should be hard-chrome plated in order to protect the jack against corrosion and subsequent damage to the hydraulic seals in the jack, which often results from corrosion on the piston rods.

The hydraulic pumping unit used to operate the jacks should also be designed for a maximum intermittent operating pressure of 10,000 psi and a minimum continuous operating pressure of the order of 5000 to 6000 psi. The pumping unit should be designed as simply as possible, in order to avoid confusing the workmen, and at least one extra pressure gauge that is not used under normal operating conditions should be provided in the system for use in checking the calibration of the other gauges in the system. In addition, the unit should be made in such a manner that damaged pressure gauges can be easily removed and replaced.

The structural frames to which the tensioned tendons are attached and which project above the top surface of the abutments generally consist of uprights, pull rods, cross-beams, and templates. The purpose of these frames is to transfer the load that is in the pre-tensioning tendons to the abutments, as well as to provide a means of applying and releasing the prestress. Unless screw jacks or hydraulic jacks with threaded piston shafts are used to maintain the prestressing force during curing, the structural frames must have a positive means of connecting the tendons to the uprights, since simple hydraulic jacks cannot be relied upon to maintain a constant strain on the tendons over a long period of time.

There are several types of structural arrangements that can be used for stressing mechanisms, and the type best suited for a specific situation is dependent upon a number of factors including the capacity of the bench, the type products to be made on the bench (i.e., fixed or universal bench), and the method of stressing to be used. Rather than discuss each of these factors separately, three types of structural stressing and releasing systems are described. The designer must determine which type of arrangement is most suited to the needs of any particular installation, after carefully weighing the advantages and disadvantages of each system.

The first system to be considered is that which is shown in Fig. 14-10. It consists of cross-beams, templates, fixed uprights, and, at the stressing end, vertical beams and pull rods. The vertical beam at the stressing end is held away from the fixed uprights by a strut at the top and the concrete abutment at the bottom. The clear space which results between the vertical beams and the fixed uprights is used to accommodate nuts and plates that hold the force in the pull rods when the jacks are not in use. This system is shown set up with four, small-stroke, high-capacity jacks (100 ton) which have a hole through their center so that the pull rods can be placed through the jacks. This mechanism can be used to stress the tendons individually. The 15-ton jack is used in this procedure. Alternately, all of the tendons can be stressed simultaneously, either by using the four, small-stroke high-capacity jacks, as shown, and taking several increments of loading in

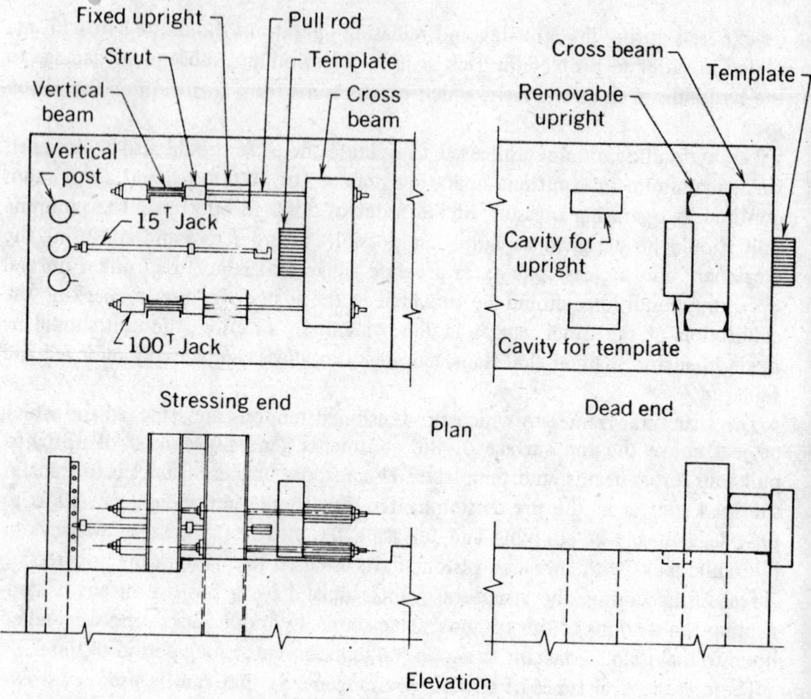

Fig. 14-10 Stressing mechanism, type 1.

order to obtain the required elongation of the tendons, or by using four, long-stroke high-capacity jacks and obtaining the entire elongation in one stressing increment. It must be pointed out that unless the center of gravity of the prestressing force is midway between the planes of the pull rods in each direction, the forces in the pull rods, and hence the pressures in the four jacks, will not be equal during stressing (if all tendons are stressed simultaneously) and releasing. This can result in the operation of the hydraulic system being somewhat complicated and dangerous if not controlled properly. From the plan view in Fig. 14-10, it will be seen that the area between the uprights is free of obstructions, which is essential when the tendons are to be stressed individually.

The system shown in Fig. 14-11 is designed for tensioning the tendons individually at one end and releasing them simultaneously at the other end. The stressing end consists of uprights, which support the template, and has provision for a long-stroke small-capacity stressing jack that may be used with an individual abutment or that may be of the type which bears directly on the template or anchorage device during stressing. The releasing end in this bench consists of a template, cross-beams, uprights, jacking yokes, and pull rods. The releasing is done with two jacks. If desired, the tendons can be stressed simultaneously with this bench by stressing at what is normally the releasing end. This sytem is less

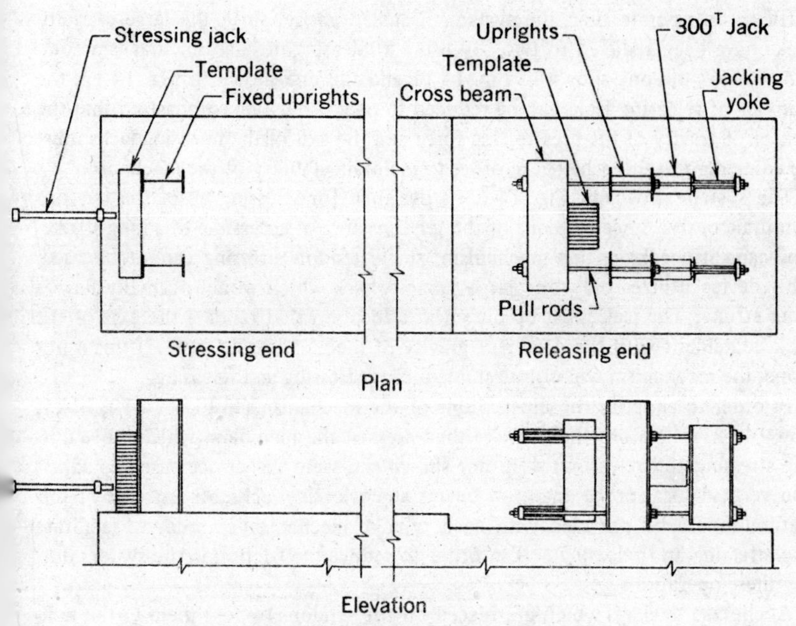

Fig. 14-11 Stressing mechanism, type 2.

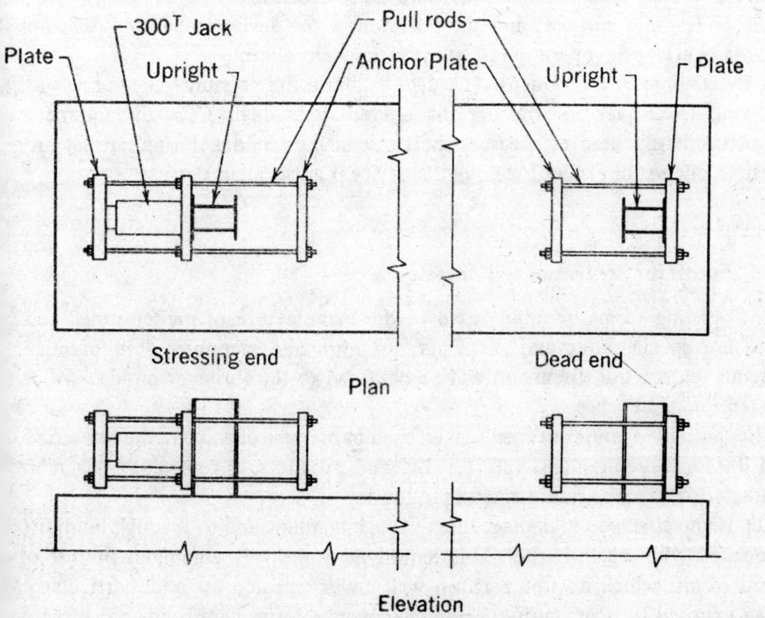

Fig. 14-12 Prestressing mechanism, type 3.

difficult to operate than the system discussed above, since the large-capacity jacks have been reduced to two. By using a releasing mechanism that is a combination of the one shown in Fig. 14-11 and the one shown in Fig. 14-12, the number of releasing jacks can be reduced to one. It must be emphasized that the center of gravity of the prestressing force and the axis of the releasing jacks must be coincident in this scheme, in order to maintain stability of the mechanism.

The system shown in Fig. 14-12 is designed for stressing all of the tendons simultaneously. Since the ends of the tendons are not accessible to a long-stroke, low-capacity jack in this mechanism, single tendon stressing cannot be used. This device utilizes only one large-capacity jack which would normally have a long stroke. The jack must be placed at a location that renders the axis of the jack coincident with the center of gravity of the prestressing force. If this is not done, the mechanism will not be stable during stressing and releasing.

In order to simplify the illustrations of the mechanisms in Figs. 14-10, 14-11, and 14-12, rollers or other devices that support the members (which move during stressing and releasing) were not shown nor were the devices used to adjust the vertical location of the cross-beams and releasing jacks. Accessories of this nature should be provided with each type of mechanism in order to facilitate the stressing and releasing and in order to reduce the friction in the system during these operations.

Anchorage devices, which are placed on the tendons to hold them to the templates during stressing of the tendons and during curing of the concrete, are also a part of the stressing mechanism. There are several satisfactory types of devices available for this purpose, and the selection of the device to be used does not materially affect the operation of the stressing mechanism.

A dynamometer is frequently used to calibrate the pressure gauge used with the long-stroke jack in stressing the tendons individually. The dynamometer is not ordinarily used to measure the stress in the tendons during normal production, due to the risk of breaking the device if an accident occurs.

14-5 Forms for Pre-tensioned Concrete

Stress-resisting forms or molds used in the manufacture of pre-tensioned concrete are special structural elements not normally encountered in practice. For this reason, this discussion will be confined to the forms or molds used on pre-tensioning benches.

The desirable features of forms to be used in pre-tensioned concrete are varied, and the form requirements vary for different products. In general, the desirable characteristics can be summarized as follows:

(1) High resistance to damage due to rough handling and to the high humidity associated with steam curing. This requirement normally eliminates the use of wood forms, which do not perform well under repeated use and, particularly, when exposed to steam curing. Although concrete forms have been used successfully, the lighter steel forms are generally preferred.

(2) Precision of the form units and dimensions. Since the forms are generally made in panels which can be connected together to form a large member, it is essential that the panels fit together precisely.

(3) Ease of handling. When erecting or stripping the form, it is essential that the individual pieces of the form that must be handled are not awkward to handle and remain in a generally upright position. This characteristic facilitates laying the forms on their backs so that they may be cleaned easily. In addition, it facilitates setting and adjusting the form in the precise position required during assembly.

(4) The form should be designed in such a manner that one side may be erected in the final position independently of the opposite side. This facilitates the layout of the member being made as well as the forming of special blockouts and transverse holes through the member, and the securing the web reinforcing and post-tensioning units, if any, in the proper location.

(5) Adjustability of the forms. The forms or components of the forms should be adjustable in such a manner that members of several shapes can be made from the form or form components.

(6) Form vibration. The forms should be sufficiently strong to withstand the effects of form vibration. Brackets or rails to facilitate placing form vibrators should be supplied with the forms.

(7) Rigid, structural soffit form. The soffit form must be rigid and must not deform during use, since such deformation results in the soffits of the products being curved and uneven. In addition, the soffit form should be a structural element to which the side form can be securely attached (anchored against lateral and uplift loads) in order to prevent the side forms from moving during placing of the concrete. This latter requirement is particularly significant in the manufacture of I shaped beams that have large, bottom flanges, since the uplift may be very large under such conditions.

(8) The forms should be made with a minimum of joints, and all joints should be as tight as possible in order to minimize leakage and bleeding.

Standard forms that satisfy the above requirements are produced by several firms. Custom-made forms that incorporate the above characteristics can be made by many fabricators.

An illustration of a type of custom-made form that incorporates the above characteristics is shown in Fig. 14-13. It should be noted from this illustration that the side form units are narrow and, when stripped, can be easily turned on their backs to facilitate cleaning of the forms. Each side form can be set up independently of the other by attaching the form to the concrete soffit and plumbing the side by use of the adjustable brace. The removable web forms can be replaced by web forms of various shapes, and the waste concrete soffit can be adjusted in height. In this manner, the side forms can be utilized in making a variety of beams. The continuous rail on each side form allows heavy form vibrators to be rolled to any position along the form where they can be firmly clamped to the form. This avoids the necessity of the workmen carrying the vibrator from location to location.

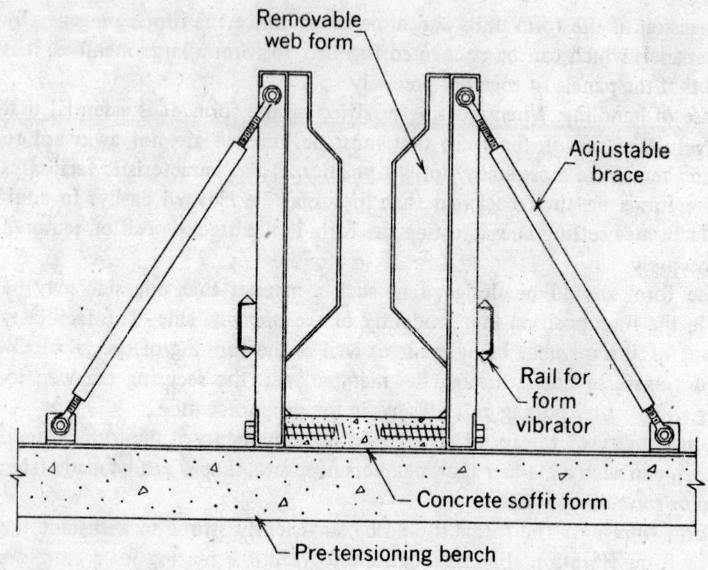

Fig. 14-13 Cross section of an adjustable steel form suitable for the manufacture of pre-tensioned concrete.

The form illustrated in Fig. 14-13 is only one of many possible solutions. The special conditions that exist in a particular prestressing plant or on a specific project may require other form details.

The end forms or bulkheads must have provision for the tendons to pass through them, and yet, the holes, slots, or grooves through which the tendons pass must not be so large (if left unplugged during placing of the concrete) to allow mortar to bleed out of the form. One obvious method of making the end bulkheads is to cut the bulkhead from a steel plate through which holes are drilled for each tendon. This type of bulkhead has two serious disadvantages: the tendons must be threaded through the holes when they are being placed in the bench; and, if the holes are small enough to prevent excessive leakage of the mortar during placing of the concrete, the threading and moving of the bulkheads into position along the bench is very difficult.

Bulkheads with large holes for the tendons have been used successfully by placing corks, which have holes through their centers for the tendons, in the holes to prevent mortar leakage and to form a recess in the end of the members to allow the patching of the ends of the members and prevent corrosion of the ends of the tendons. The placing of the corks in such bulkheads can be very difficult if there are many tendons in the members and if the tendons are closely spaced.

Bulkheads with slots in them for the tendons have also been used, but in such cases, it is usually necessary to tape the slots to prevent mortar leakage. The taping of the slots is often a slow and laborious operation.

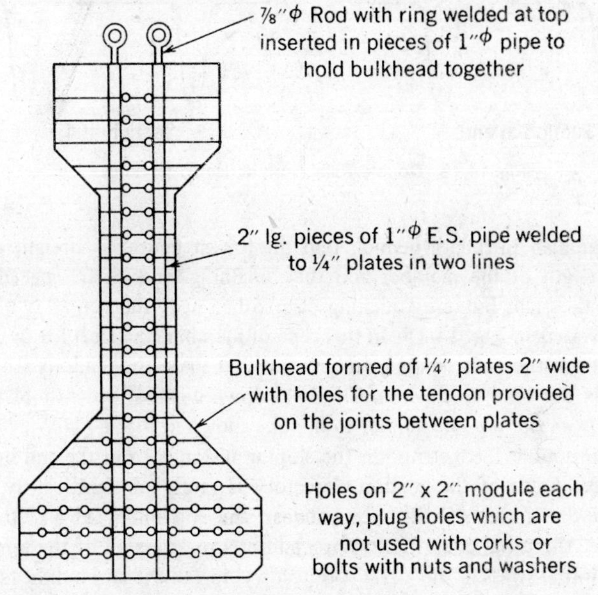

Fig. 14-14 Bulkhead for steel forms used in pre-tensioned construction.

Another type of bulkhead that has been used successfully is composed of a series of plates which can be connected with a removable bar assembly and which have holes provided at the joints between the plates. This type of bulkhead, which is illustrated in Fig. 14-14, does not allow significant mortar leakage and can be installed after the tendons have been placed and stressed.

14-6 Tendon-Deflecting Mechanisms

As has been stated previously, it is often desirable in long-span pre-tensioned members to have the tendons more eccentric near the center of the beam than at the ends of the beam. In manufacturing pre-tensioned products, the normal procedure is to stress the tendons from one abutment to the other, which results in the tendons being very straight. If the pre-tensioned tendons are to follow a path other than a straight one, additional forces must be made to act on the tendons to cause the trajectory of the tendons to be other than straight. The devices used to deflect the tendons from the straight line are called "tendon deflectors" or "tendon-deflecting mechanisms." Tendon-deflecting mechanisms frequently consist of devices which support the tendons in the high position and devices which hold the tendons in the lower position. These components are generally referred to as "hold-up devices" and "hold-down devices," respectively.

In applying the principle of deflected tendons to double-T roof and floor slabs, it is customary to deflect the tendons at one or two points near the

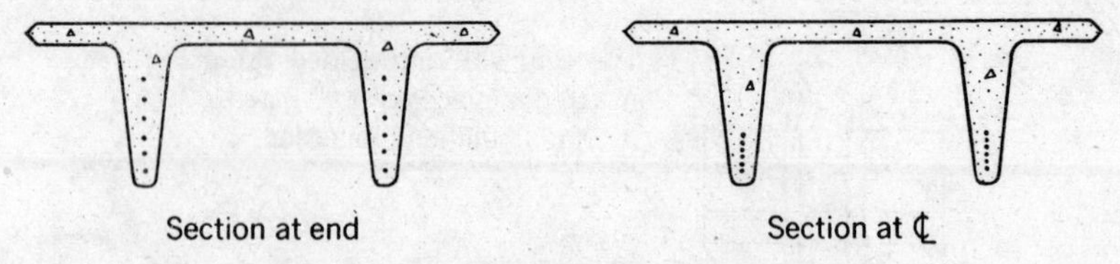

Fig. 14-15 Double T slab with deflected tendons.

center of the slab in such a manner that the lowest tendon is straight through the entire length of the member and the remaining tendons are spaced out at the end of the slab, but are touching each other near the center of the slab. This is illustrated in Fig. 14-15. In this type of member, the steel forms or bulkheads that form the ends of the members and that space the tendons at the ends also serve as the hold-up devices. The hold-down devices, which hold the tendons down from the top, are similar to the one shown in Fig. 14-16.

Because in double-T construction the sloping distance from the end bulkhead to the lowest point of the tendon trajectory is rarely of significantly greater length than the horizontal distance between the end bulkhead and the hold-down device, the usual construction procedure is to stress all of the tendons to the desired initial stress in the straight trajectory, install the end bulkheads, place the hold-down devices at the proper locations, and push the tendons down to the lower position with a hydraulic jack that is temporarily installed in the hold-down device. The tendons are held in the lower position by nuts or pins and the hydraulic jack is moved to the next hold-down device. The tendons are then jacked down at this location. The procedure is continued until the tendons are jacked down at all the hold-down devices.

After the concrete has gained sufficient strength to allow the prestress to be released, the hold-down devices are removed, after which, the pre-tensioning force is released. If released with jacks, the slabs which are nearest the releasing end move a few inches at the time of release. It is for this reason that it is necessary to remove the hold-down devices before releasing the pre-tensioning tendons. If this were not done, the hold-down devices would restrain the movement of the slabs, with the possibility of cracking the slab or damaging the hold-down devices.

After the slabs have been removed from the pre-tensioning bench, the metal rods or pipes, which extend into the slab to hold the tendons in the deflected position, are removed and the holes which are formed by these rods are filled with grout.

When the length of the tendon from the hold-up device to the hold-down device is significantly larger than the horizontal distance between these two devices, the initial stress applied to the tendons in the straight position should be lower than the stress that is desired initially in the deflected position. In this manner, the deflecting of the tendons raises the stresses in the tendons to the desired values.

In the case of bridge girders, because the vertical depth or amplitude of the

Fig. 14-16 Hold-down device shown spanning a double-T form. Workmen are in process of deflecting the tendons with a hydraulic jack. (*Courtesy of Food Machinery and Chemical Corp., Lakeland, Florida.*)

deflected tendon is generally large in proportion to the horizontal dimension of the sloping portion of the tendon, the additional stress which results in the deflected tendons (due to pushing the tendons into the deflected position) is generally quite important and must be taken into account when calculating the required initial stress that must be applied to the tendons.

At the present time, there is no standard method used to deflect the pre-tensioning tendons in bridge-girder construction. Several methods have been used or have been proposed for deflecting the tendons in bridge girders, and these methods are described in the following paragraphs.

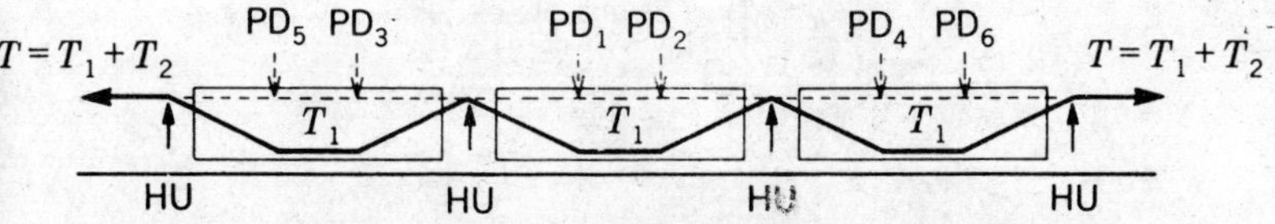

Initial tension T_1, in "up" position. Strand profile and final tension T by push down (PD), which increases strand tension by T_2.

Fig. 14-17 Jacking down at the hold-down points.

Jacking Down at the Hold-Down Points

The method is shown schematically in Fig. 14-17. The procedure followed in this sytem is substantially the same as that used in deflecting the tendons in double-T slabs, except the length of the deflected-tendon path may be quite

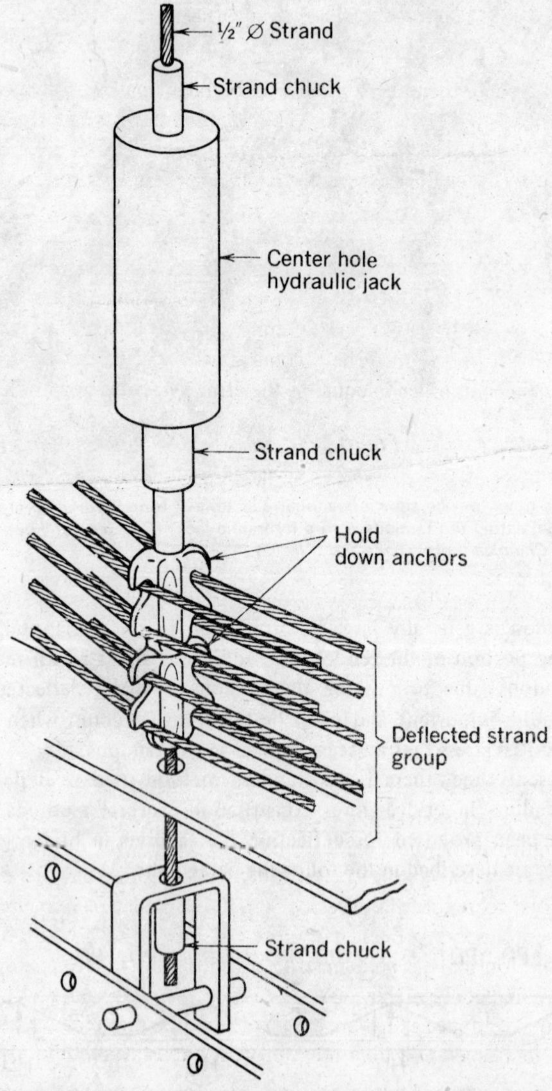

Fig. 14-18 Tendon hold-down device for use with jacking down at the hold-down points.

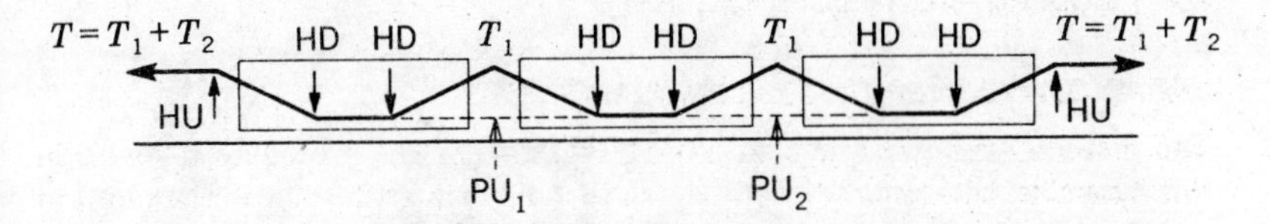

Initial tension T_1, in "down" position. Strand profile and final tension T by push up (PU), which increases strand tension by T_2.

Fig. 14-19 Jacking up at the hold-up points.

significantly greater than the length in the straight path, and hence, the initial prestress applied to the tendon must be adjusted accordingly. In addition, the end bulkheads are not normally sufficiently strong to act as the hold-up device and, as a result, special devices must be supplied for this purpose.

One means of deflecting the tendons in this manner is shown in Fig. 14-18. In this method, metal anchors provided in the upper surface of the prestressing bench are used to anchor the bottom end of the hold-down device and a center-hole jack is used to jack the tendons down. A strand chuck anchors the tendons in the deflected position. The strand chuck and the hold-down anchors are expended with this procedure. The jacking must be done according to a predetermined sequence in order to equalize the effects of friction along the bench.

Jacking Up at the Hold-Up Points

This procedure, which is shown schematically in Fig. 14-19, is virtually the same as that used in jacking down at the hold-down points. The principal difference is that the tendons are stressed in the lower position, attached securely to the pre-tensioning bench at the hold-down points by a device which extends through the soffit form, and the tendons are jacked up at the hold-up points (*see* Fig. 14-20).

After the members are removed from the bench, the cavity at the bottom of the beam at each hold-down point must be patched.

Stressing the Deflected Tendons Individually

Satisfactory results have been obtained in some instances by placing and stressing the tendons in the deflected trajectory. When this procedure is used, some means of reducing the effect of the friction which occurs between the tendons and the hold-down and hold-up devices must be employed. The methods that have been used to reduce the effect of the friction include stressing the tendons from each end, supplying rollers with needle bearing at each hold-down and each hold-up point during the stressing (the low friction rollers are removed after stressing is completed so they can be re-used many times), and applying a vibration to the tendons as they are stressed, which reduces the coefficient of friction at the hold-down and hold-up points. Again in this method, the hold-down

(a)

(b)

Fig. 14-20 (a) The expendable hold-down device attached to a removable portion of the soffit form, which is anchored to the pre-tensioning bench. (b) A general view of the deflecting mechanism showing the hold-down devices, deflected tendons, and the tower used to deflect the tendons.

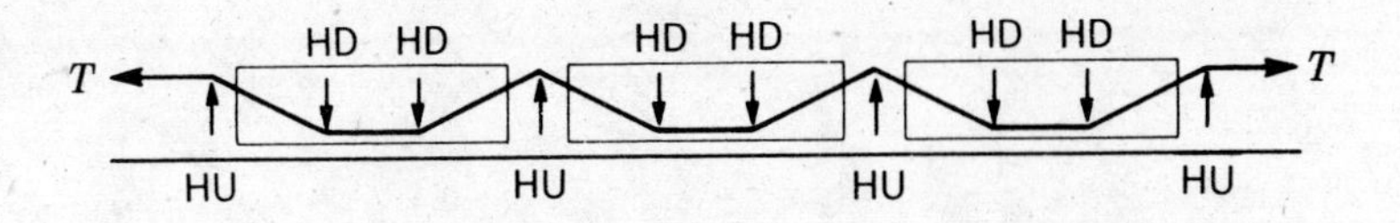

Single or multiple strand tensioning in "draped" pattern. Strand profile established by hold downs (*HD*) and hold ups (*HU*).

Fig. 14-21 Deflected or draped tendons stressed in the deflected position.

devices must extend through the soffit form, which results in a cavity that must be patched after the members are removed from the pre-tensioning bench. The scheme is illustrated in Fig. 14-21.

Each of the methods discussed above have advantages and disadvantages, and the method to be used on any particular project must be selected on the basis of the available equipment and the results that are desired. Significant disadvantages of the use of deflected tendons include the large capital investment in a stressing bed which is capable of deflecting tendons and the deflecting mechanisms which are required, the friction losses which occur along the deflected tendons, the secondary stresses in the tendons which result, from bending the tendons over small diameter pins and rollers (does not apply to all methods), and the extra labor required to deflect the strands.

15 Post-tensioning Systems and Procedures

15-1 Introduction

Post-tensioning materials equipment are available from a number of firms in the United States. These firms sell the necessary prestressing steel, ducts, anchorages and accessories, as well as either rent or sell the necessary stressing and grouting equipment. As a part of their service, the firms normally provide shop drawings for specific projects with the sale of their materials. In addition, when requested, they can normally supply field technicians who are qualified to instruct workmen in the proper methods of installing, stressing, and grouting their materials. Sometimes, the suppliers of post-tensioning materials sell the materials completely installed, stressed, and grouted.

The details of the systems available in the United States are continually changing. For this reason, no attempt will be made in this book to describe the precise details of the systems available at the time of this writing. The reader who is interested in learning the details of the individual systems is advised to contact the Post-Tensioning Institute.* A current list of suppliers of post-tensioning

*Post-Tensioning Institute, 301 West Osborn, Phoenix, Ariz. 85013.

materials who can be contacted for specific details, will be provided by the Institute.

The primary differences between post-tensioning and pre-tensioning from the construction viewpoint, can be enumerated as follows:

(1) Post-tensioning offers a means of making prestressed concrete, either in a plant or at the job site, without requiring a large capital investment in pre-tensioning facilities.

(2) Post-tensioned tendons allow the construction of cast-in-place structures which would not be feasible in pre-tensioned concrete.

(3) Post-tensioning tendons can be placed on curved paths easily and without large, special deflecting equipment.

(4) Friction losses during stressing of post-tensioned tendons must be considered in the design of such construction, since the friction losses may have significant influence on the performance.

(5) Each post-tensioned tendon must be stressed individually and this, in combination with the cost of the end anchorages, sheath, special equipment, and grouting of the tendons, results in the unit cost (cost per pound of effective prestressing force) of post-tensioned tendons being substantially greater than the cost of pre-tensioned tendons, in most instances.

(6) Post-tensioning anchorage devices and stressing equipment are often protected by patents and frequently require manufacturing techniques and facilities that are very precise. As a result, each system is generally available from only one supplier or representative of the manufacturer.

In addition to the brief description of the general types of post-tensioning systems available in this country, the general construction procedures recommended in post-tensioned construction are discussed in this chapter, as are the computation of friction losses, gauge pressures, and elongations to be attained during stressing. Special construction methods and devices unique to post-tensioned construction are also considered.

15-2 Description of Post-tensioning Systems

The post-tensioning systems commonly used in this country can be separated into four general categories, which are:

(1) Parallel wire systems.
(2) Multi-strand systems.
(3) Single strand systems.
(4) High-tensile strength bar systems.

The steel that comprises the tendons of these various systems has the general characteristics discussed previously in Chapter 2.

Parallel-wire systems currently available in the U.S. employ a number of wires, 0.250 in. in diameter, which are bundled or grouped into a unit referred to as a tendon or a cable. The term "parallel wire" may not be precisely accurate in describing the relative position of the individual wires in the tendons, since in some instances, the individual wires are bundled or grouped together without spacers and with no real attempt being made to ensure that the individual wires in a tendon are parallel. Hence, the term parallel wire serves as a means of differentiating systems that employ tendons composed of a number of wires bundled into a group from the other types of tendons.

The original Freyssinet system is one the first parallel-wire post-tensioning systems. It received its name from the eminent French engineer Eugène Freyssinet who invented the anchorage device used in this system. This system was introduced in this country when the use of linear prestressed concrete was just begining to gain impetus. The anchorage device used in this system consists of two parts, called the "female cone" and the "male plug." An illustration of early anchorage cones is given in Fig. 15-1. Anchorage cones anchor the wires by the friction which results from the wedge shape of the male plug and the hole in the female cone.

The original Freyssinet system, which utilized the concrete female cone and male plug together with tendons composed of wire, is no longer marketed in the United States. A more recent evolution in the Freyssinet system utilizes a cast steel anchorage that employs the same wedge principle to anchor tendons composed of a number of seven-wire or nineteen-wire strands. This anchorage is illustrated in Fig. 15-2.

Tendons composed of a number of strands are referred to as the multi-strand type; they are identical to the parallel wire tendons except that seven-wire and nineteen-wire strands are used in lieu of solid wires.

Special jacks are required when using the Freyssinet system. These jacks ac-

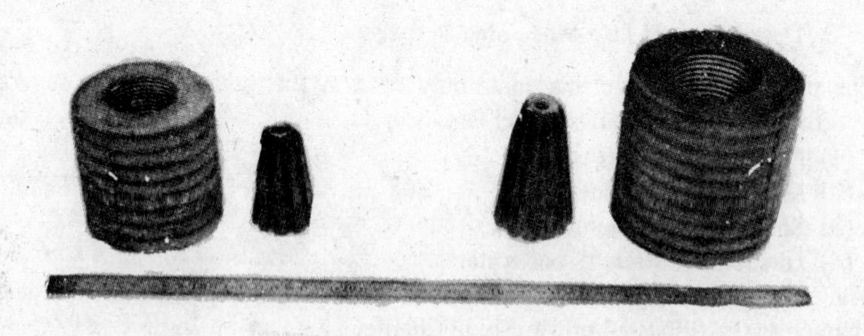

Fig. 15-1 Freyssinet anchorage cones. ***(Courtesy Freyssinet Co., Inc., New York.)***

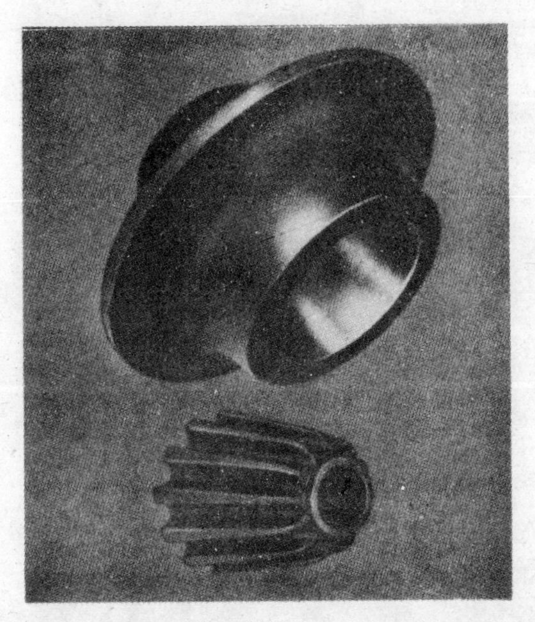

Fig. 15-2 Freyssinet anchorage of cast steel for use with tendons composed of 12 seven-wire strands of ½ in. diameter. (*Courtesy Freyssinet Co., Inc., New York.*)

tually consist of two rams connected in series. Steel wedges on the periphery of the main piston are used to attach the wires of the parallel wire tendons. Strand anchors are used in a similar manner with the strand tendons. After the wires or strands have been stressed by the introduction of hydraulic pressure to the main piston, the pressure is held in the main piston while pressure is applied to the secondary or plugging piston that forces the male plug into the cavity of the female cone. The stressing is completed by releasing the pressure in the plugging piston and then in the main piston. The jack is then disconnected from the tendon by removing the steel wedges. After stressing, the excess wire at the ends of the tendons are cut off and the tendon is grouted in place. A Freyssinet jack, connected to a wire tendon in stressing position as well as a tendon that has been stressed and one that has not been stressed, are shown in Fig. 15-3. The mechanics of the stressing sequence are shown in Fig. 15-4.

The Freyssinet anchorage cone is provided with a hole through the male plug for the introduction of grout after stressing. Before grouting, it is necessary to plug the openings between the wires, the male plug, and the female cone, since large quantities of grout escape through these openings and reduce the effectiveness of the grouting apparatus if this is not done.

Due to the necessity of a wedge moving longitudinally to develop lateral

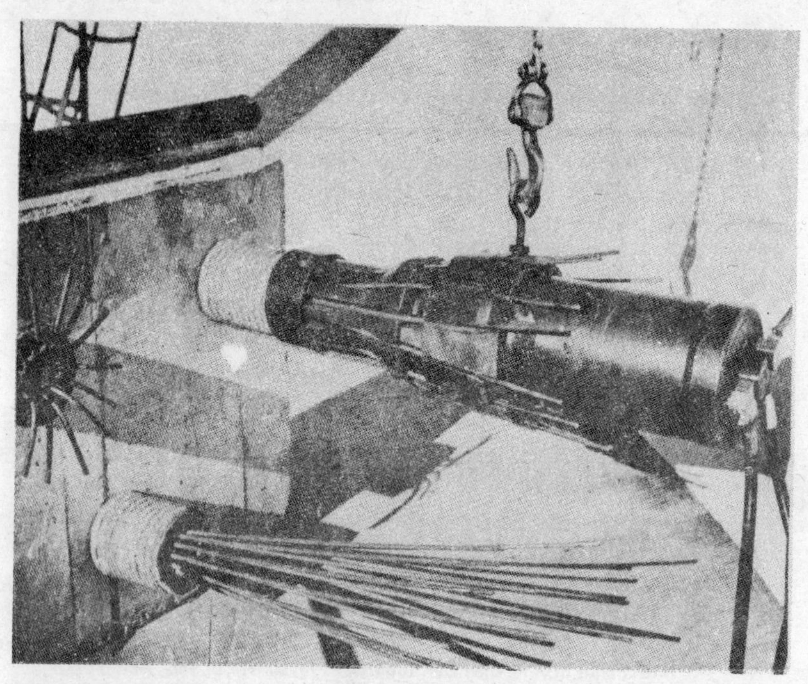

Fig. 15-3 Freyssinet jack in stressing position. (*Courtesy Freyssinet Co., Inc., New York.*)

forces, which is essential if a wedge or conical plug is to anchor a tendon by friction, a portion of the elongation of the tendon that results from stressing is lost when the prestressing force is transferred from the jack to the anchorage cone. This loss of elongation is increased in the case of a Freyssinet cone by the elastic expansion of the female cone which takes place upon anchoring the tendon. The entire loss of elongation which results from these actions is referred to as the *loss due to the elastic deformation* of the anchorage, or the *loss due to seating the anchorage.* The loss due to the elastic deformation of the anchorage can be of significance under certain conditions. The subject is discussed further in Sec. 15-6.

Several other multi-strand systems employing from one to as many as 48 seven-wire strands in each tendon are available in the United States. All of these systems rely upon wedges for the anchorage of the strands. The details of the anchorages, grouting procedure, and stressing equipment must be obtained from the individual supplier.

End anchorage of parallel-wire post-tensioning tendons is also achieved by cold-formed heads on the ends of 0.25 in. diameter wire. Several systems that utilize this principle are available in the United States. The system is used with a

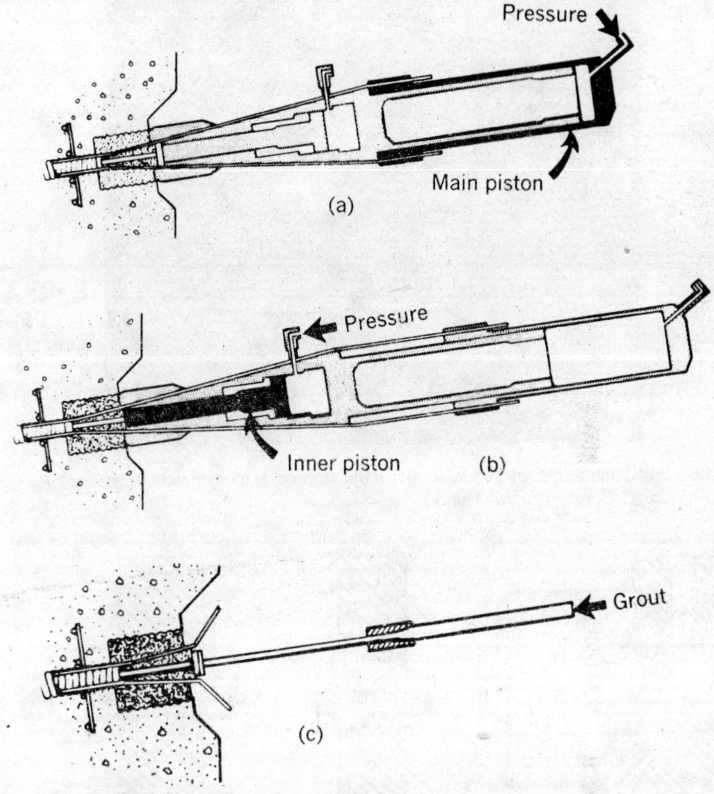

Fig. 15-4 Stressing sequence in the Freyssinet system. Showing (a) the stressing, (b) the plugging, and (c) the grouting of the tendon.

number of variations, but in all of them, the button heads bear on a stressing ring or stressing nut to which a hydraulic jack is attached for stressing. After stressing, the stressing ring is locked in the stressed position either with shims or with threaded nuts. This illustrated in Fig. 15-5.

Some years ago, tendons composed of a single large strand with either swaged or zinc-attached end anchorages were available in this country. These tendons were appealing from several technical viewpoints but they proved to be uneconomical in competition with the other types of tendons; for this reason they are no longer available.

Another type of post-tensioning tendon commonly used in this country is the high-tensile steel bar. Tendons of this type were first developed in England where they were called "Lee-McCall bars." Smooth high-tensile strength bars have been used with threaded nuts and couplers as well as with wedge-type

Fig. 15-5 End anchorages of tendons with cold-formed button heads. (*Courtesy Prescon Corp., Corpus Christi, Texas.*)

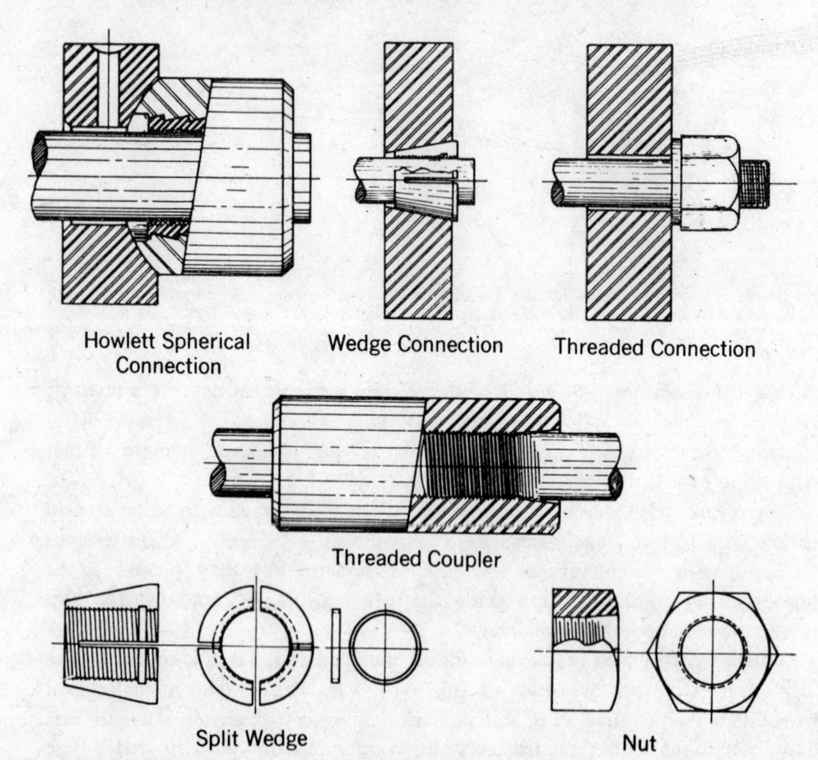

Fig. 15-6 Some of the types of anchorages and couplers used with the bar systems. (*Courtesy Rod, Inc., Berkeley, California.*)

anchorages as shown in Fig. 15-6. High-tensile strength bars are also available with deformations rolled onto their exterior surface, the deformations being rolled in the form of a helix to permit their use as threads in couplers and with nuts. The deformed bars are illustrated in Fig. 15-7.

It should be pointed out that couplers are required to join individual bars together at the job site to form the longer bar tendons. This affects the detailing and construction of longer members designed to be stressed with bar tendons.

The jack required to stress the button-head tendons and the high-tensile bar tendons is a simple center-hole ram or jack. This type of equipment is readily available in the United States in various sizes. Center-hole rams are much less expensive and lighter than a Freyssinet jack of equivalent capacity.

Each system of post-tensioning has certain advantages and disadvantages and no attempt is made here to discuss all of the systems or factors that may influence the designer or contractor in selecting a system for a particular project. The

Threadbar Plate Anchorage

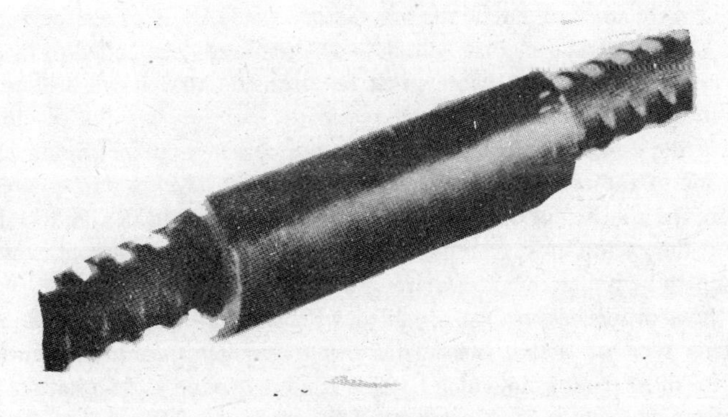

Threadbar Coupling

Fig. 15-7

designer should select and specify the systems which will best meet the requirements of each individual job. In most applications, however, all of the systems will work equally well and the controlling factor is the cost of the tendons and the labor required to install, stress, and grout the tendons.

15-3 Sheaths and Ducts for Post-tensioning Tendons

In prestressed construction, it is essential that the tendons do not become bonded to the concrete until they are stressed. In post-tensioned construction, since the tendon cannot be stressed until the concrete is hardened, it is necessary to ensure that the concrete is not in contact with the tendon during placing and curing of the concrete. This can be done by placing the tendon in a tube or covering it with a material that will prevent the concrete from coming in contact with the tendon. In addition, a hole can be formed through the concrete section with a form which may or may not be removable. The tendon can thus be inserted in the preformed hole after the concrete has hardened.

In virtually all cast-in-place post-tensioned building construction in the United States, the tendons are placed in sheaths and the complete unit is embedded in the concrete at the time the concrete is placed. A large portion of the post-tensioned concrete in bridges is constructed by preforming holes for the tendons through the members with rigid tubing. After the concrete has gained sufficient strength for stressing, the tendons are pulled into the preformed holes, stressed, and grouted. Preformed holes can also be made by placing rubber tubes, which are either inflated with water or air or which are stiffened with metal rods, in the member and withdrawing them after the concrete has set. When the tendons are to be factory-fabricated and are to be grouted after stressing, the sheath most commonly used is a flexible metal hose of interlocking construction. When the tendons are to be left unbonded after stressing, the usual practice has been to coat the tendons with a lubricating rust inhibitor and wrap the tendon with waterproof paper or plastic tubing.

When the wire or strand tendons are factory made, it is necessary that the sheath be of a construction that will allow the tendons to be coiled to facilitate handling and shipping. The sheath must be strong so that it will not become damaged in transit and so that it will retain its shape during the placing and vibrating of the concrete. If the sheath does not retain its shape during placing and vibrating of the concrete, a large friction loss will develop during stressing. In addition, the coefficient of friction between the sheath and the tendon should be low and the sheath must be impervious to cement, although it is not essential that the sheath be impervious to water.

All of these properties can be obtained with an interlocking, flexible, metal sheath. This type of sheath is available bright and galvanized. The former is adequate for most tendons in which friction is not expected to be a serious problem. A galvanized sheath is recommended for longer tendons, due to its lower coefficient of friction, and for tendons which may be subjected to a long exposure to moist air before stressing and grouting. The interlocking sheath is also available with an asbestos packing in the interlock which reduces the intrusion of cement and water during placing of the concrete.

Rigid metal tubing should be used for sheath on long tendons, since the friction characteristics of this type of sheath have proved to be very favorable.

The use of paper wrappings or plastic tubes for sheaths is limited to applications in which the tendons are to be left unbonded. When using paper and plastic for a sheath, the tendons can be coiled for ease of handling and shipping. Normal handling will not damage the sheath. Care must be exercised, however, to ensure that adequate coatings of rust inhibitors are applied to the tendons before the paper or plastic is applied. The rust inhibitor is essential to protect the tendon, as well as to ensure a low coefficient of friction between the tendon and sheath during stressing.

The use of preformed holes with the high-tensile bar tendons is not recommended. Due to the stiffness of the bars, preformed holes must be very straight, with no secondary curvature of significance, in order to allow the bar to be inserted. An exception to this is when the tendons can be used on straight trajectories and somewhat oversized ducts can be tolerated.

As stated above, preformed holes can be made with rubber tubes stiffened by inflating the tubes with air or water. Holes can also be preformed with rubber hoses that are stiffened with steel rods that are removed before the hoses are pulled out. The use of neoprene is preferred to natural rubber, due to the high resistance to deterioration when exposed to petroleum derivatives. When air or water is used in the tubes, extra care must be exercised in tying the tubes in place, since the inflated tubes are lighter than the plastic concrete and they tend to float when the concrete is placed and vibrated.

When using steel rods to stiffen neoprene hoses, the rods should be smooth (not deformed bars) and lubricated to facilitate inserting and removing them from the hoses. When this procedure is used, the steel rods are frequently made in two pieces–and one piece is removed from each end. Hoses that are in two pieces and that are taped together near the center have also been used, but there is a chance that the tape will loosen during placing of the concrete, in which case, the hole through the tube may not be continuous and the beam may have to be rejected. Furthermore, when the hose is in two pieces, each half must be pulled out simultaneously. If this is not done, the taped joint may hold just long enough for the pulling at one end to draw the half of hose at the other end into the beam.

Preformed holes made with removable rubber or neoprene hoses are usually only used on short tendons. This is because the friction losses with preformed holes in the concrete are relatively large. This is discussed in Sec. 15-5.

15-4 Forms for Post-tensioned Members

Post-tensioning is frequently used on projects in which only a few prestressed members are required and in which the cost of setting up an efficient pre-tensioning operation is greater than the savings that can be made by the use of pre-tensioning in lieu of post-tensioning. This situation frequently exists, even though the basic pre-tensioning facilities may be available. In post-tensioning, one set of side forms may be used with two or three soffit forms to obtain an efficient and economical production cycle. In pre-tensioning, if the same amount of forms are used, by moving them along the bench as the members are produced, the entire pre-tensioning bench (or line) will be tied up and the production will not be greater than is possible with post-tensioning. In addition, post-tensioning has been used frequently for the construction of "custom" beams for bridges in which only 10 or 12 beams are required. In such cases, the beams have often been made on the job site by the general contractor.

Wooden forms have been used in many post-tensioning applications, because the cost of steel forms is prohibitive unless they can be re-used many times. The wooden forms are not usually made adjustable, but are built to produce members of only one size. Well built wooden forms that are properly oiled can frequently be used ten or more times before they must be completely rebuilt or discarded. Lining the inside of wood forms with light-gauge metal reduces the amount of moisture which the wood absorbs from the plastic concrete and increases the life expectancy of the forms.

When using wood forms, the method of curing the concrete must be such that expansion of the wood due to the absorption of moisture will not crack the concrete. This is particularly dangerous with I-shaped beams and necessitates protecting the backs of the forms, as well as the front surfaces which are in contact with the concrete, when water curing is applied before stripping of the form. Steam curing is not recommended for members made in wood forms.

Steel forms are used in the construction of post-tensioned members when the number of re-uses which can be obtained from the forms justifies the higher form cost. In this case, the forms may be made adjustable or for only one shape of member, depending upon the circumstances.

The same basic characteristics that are desirable for forms used in the manufacture of pre-tensioning are desirable in the manufacture of post-tensioned concrete, with the possible exception of the adjustability. In manufacturing pre-tensioned concrete, it is frequently necessary to keep the forms as close to the top surface of the pre-tensioning bench as possible, in order to reduce the stress in the uprights and the tendency toward overturning the abutments. This condition does not exist in post-tensioned construction and the forms can be built elevated on small columns which have rubber vibration insulators that will allow maximum use and benefit from form vibration. This scheme is illustrated in Fig. 15-8. It will be seen that the form vibrator can be attached to the soffit form as well as the side forms. The vibrators may be moved along the length of the form as the member is cast or, if desired, vibrators may be positioned at various locations along the form rather than moving the vibrators during the placing of the

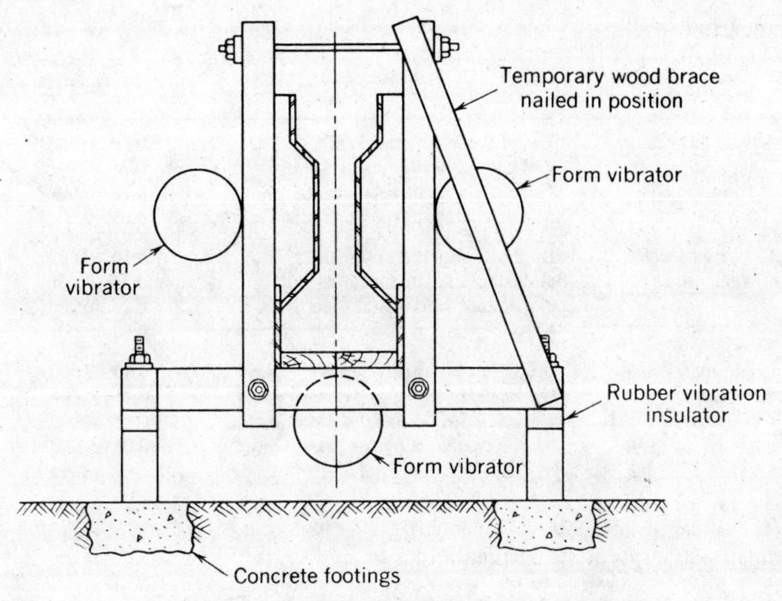

Fig. 15-8 Form designed for use with external vibrators.

concrete. When this type of vibration is used, it is normally necessary to supplement the form vibration with internal vibration. This method, which for all practical purposes makes the form into a combination form vibrating table, such as is used in the manufacture of small concrete products, is only feasible in precasting plants and on large casting operations conducted at the job site, due to the relatively high cost of the equipment.

15-5 Effect of Friction During Stressing

The variation in stress in post-tensioned tendons due to friction between the tendon and the duct during stressing is considered to be the function of two effects. These are the primary curvature or *draping* of the tendon, which is intentional, and the *secondary curvature* or wobble which is the unavoidable minor horizontal and vertical deviations of the tendon from the theoretical position. Coefficients for each of these effects, which are commonly specified by post-tensioning design criteria in this country, are listed in Table 15-1. Tests are frequently made with varying types of tendons and ducts or sheaths in order to determine the actual friction coefficients for materials under consideration for a specific application.

TABLE 15-1

Type of Tendon	Type of Duct	Design Values μ	K
Uncoated wire or large diameter strands	Bright flexible metal sheath	0.30	0.0020
	Galvanized flexible metal sheath	0.25	0.0015
	Galvanized rigid metal sheath	0.25	0.0002
	Mastic coated, paper or plastic wrapped	0.05	0.0015
Uncoated seven-wire strand	Bright flexible metal sheath	0.30	0.0020
	Galvanized flexible metal sheath	0.25	0.0015
	Galvanized rigid metal sheath	0.25	0.0002
	Mastic coated, paper or plastic wrapped	0.08	0.0014
Bright metal bars	Bright flexible metal sheath	0.20	0.0003
	Galvanized flexible metal sheath	0.15	0.0002
	Galvanized rigid metal sheath	0.15	0.0002
	Mastic coated, paper or plastic wrapped	0.05	0.0002

The stress at any point in a tendon can be determined by substituting the proper coefficients in the relationship.*

$$T_O = T_x e^{(\mu\alpha + KX)} \tag{15-1}$$

in which: T_O = the force at the jacking end, T_x = the force at point X feet from the jacking end, e = base of the Naperian logarithm, K = secondary curvature coefficient, μ = primary curvature coefficient, α = total angle change between the tangents to the tendon at the end and at point X (sum of the horizontal and vertical angles) in radians, and X = distance in feet from the jacking end to the point under consideration.

For low values of $\mu\alpha + KX$, Eq. 15-1 is approximately equal to:

$$T_O = T_x(1 + \mu\alpha + KX) \tag{15-2}$$

Eq. 15-2 is permitted for use with values of $\mu\alpha + KX$ as high as 0.3 by ACI318.

The variation in stress in a post-tensioned tendon stressed from one end, according to the above relationship, will be as is illustrated in Fig. 15-9. The maximum value of stress in the tendon is at the jacking end and the minimum value is at the dead end. If the tendon were stressed from each end simultaneously, the curve would be symmetrical about the center line and would have the shape of AB on each side.

The effects of friction during post-tensioning should be investigated by the designer at the time the trajectory of the post-tensioning tendons is selected. The computations will reveal the magnitude of the friction loss that can be expected

*The relationship can be written for unit stresses as follows:

$$f_o = f_x e^{(\mu\alpha + KX)}$$

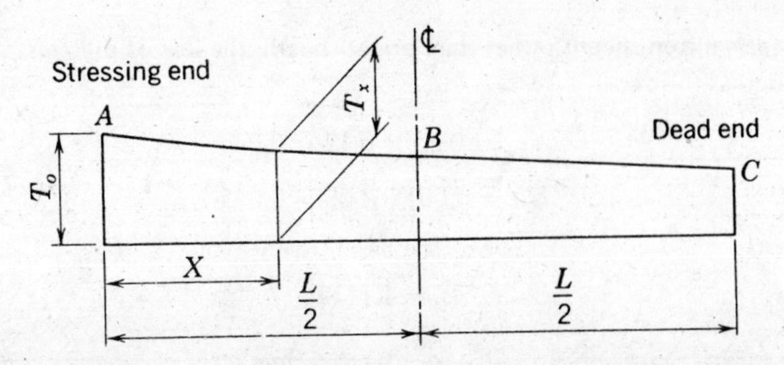

Fig. 15-9 Variation in stress due to friction in a tendon that is post-tensioned from one end.

and will allow the designer to determine if special precautions should be required in the specifications in order to reduce the friction loss. Special procedures to reduce the effects of friction include using galvanized sheath in lieu of bright sheath, using rigid sheath in lieu of flexible, reducing the curvature of the tendons, or using a water-soluble oil on the tendons. If used, the oil is removed by flushing the ducts with water before the tendon is grouted. The use of water-soluble oil to reduce the friction is generally employed only as a last resort when serious friction is encountered on the job site. It is not recommended that this procedure be relied upon during design.

ILLUSTRATIVE PROBLEM 15-1 Compute the force which would be expected to result at midspan of a tendon that is 100 ft long and which is on a parabolic curve having an ordinate of 3 ft at midspan. Assume the sheath is to be a bright flexible metal hose and that the tendons are to be composed of parallel wires.

SOLUTION: Since the tangent of the angle between the tangents to the tendon can be assumed to be numerically equal to the value of the angle expressed in radians, the value of α is found by

$$\alpha = \tan \alpha = \frac{4e}{L} = \frac{4 \times 3 \text{ ft}}{100 \text{ ft}} = 0.12$$

Using the coefficients from Table 15-1

$$= 0.30 \times 0.12 = 0.036$$

$$KX = 0.002 \times 50 \text{ ft} = \underline{0.100}$$

$$0.136$$

$$T_O = T_x e^{0.136} = 1.15 T_x$$

$$T_x = 0.873 T_O$$

With galvanized sheath rather than bright sheath, the loss of prestress would be computed as follows:

$$= 0.25 \times 0.12 = 0.030$$

$$KX = 0.0015 \times 50 = \underline{0.075}$$

$$0.105$$

$$T_o = 1.111T_x$$

15-6 Elastic Deformation of Post-tensioning Anchorages

As explained in Sec. 15-5, the variation in stress along a post-tensioned tendon at the time of stressing is assumed to follow a curve such as *ABC* in Fig. 15-10. The curve *EDBC* in this figure indicates the assumed variation in stress after the tendon has been stressed and anchored. It will be noted that a reduction in the stress at the end of the tendon resulted from the anchoring procedure. The reduction in stress occurs when the prestressing force is transferred from the jack to the anchorage device, at which time, a portion of the elongation of the tendon obtained during stressing is lost due to the deformation of the anchorage device. Although some *positive type* anchorage devices, such as the button-head systems, when properly applied, have no appreciable deformation of anchorage, the wedge or cone-type anchorages often deform significantly as the load is applied to them. Other types of anchorage which have components stressed in the plastic range may require several hours to reach the limiting value of anchorage deformation. The manufacturers of the anchorage devices that are to be allowed on any project should be consulted in order to determine the limits of the anchorage deformation that may be encountered in the field. If the computations indicate that the deformation may result in a significant reduction in stress of the tendons, the deformation should be measured in the field as a means of ensuring that the desired results are obtained.

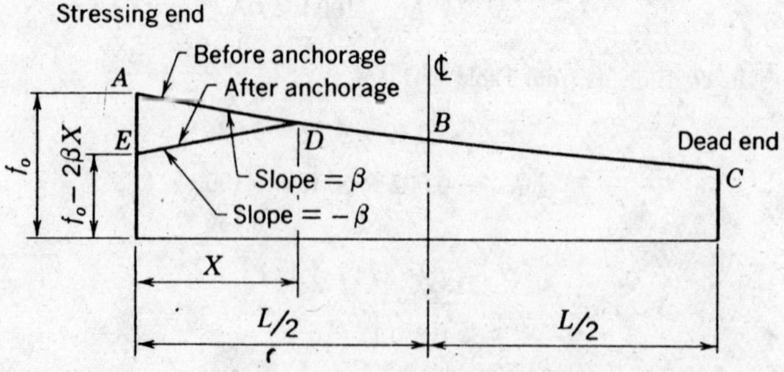

Fig. 15-10 Variation in stress in a post-tensioned tendon before and after anchorage. Condition I.

In order to simplify the computation of the effect of the anchorage deformation on the stress in the tendon, the curves *AB*, *AD*, *BC* and *DE* in Fig. 15-10 are assumed to be straight lines. The slopes of lines *AB* and *DE* are assumed to be of equal magnitude, but of opposite sign. Tests have shown that this assumption is approximately correct.

The state of stress indicated in Fig. 15-10 will be referred to as condition I. This condition is characterized by the fact that the stress at the center of the tendon is not affected by the deformation of the anchorage, which means that the length X is less than one-half of the length of the tendon. This condition generally occurs in long tendons or in members that have high friction, such as those with high primary curvatures. In some instances, it is desirable to stress tendons having stress distributions of this type from one end only, but, since this condition generally exists only in tendons of considerable length, they are more frequently stressed from both ends simultaneously.

The assumed variation in stress referred to as condition II is shown in Fig. 15-11. In this case, the stress at midspan is affected by the deformation of the anchorage, since the distance X is greater than one-half the length of the tendon. There is generally no advantage to be gained in stressing a tendon having this type of stress distribution from each end simultaneously.

The most severe effects from the deformation of the anchorage are found in short cables and those with very low friction. In this case, the stress at the dead end is reduced by the deformation of the end anchorage. This condition is characterized by the computed length of distance X being greater than the length of the tendon. The assumed distribution of stress for condition III is illustrated in Fig. 15-12.

Determination of the effect of anchorage deformation on the stress at midspan, which is generally the most critical section from the designer's viewpoint, will be developed. A similar procedure is used in structures having the critical sections at other locations. The procedure is as follows:

(1) Determine the ratio between the stress at the end and midspan resulting

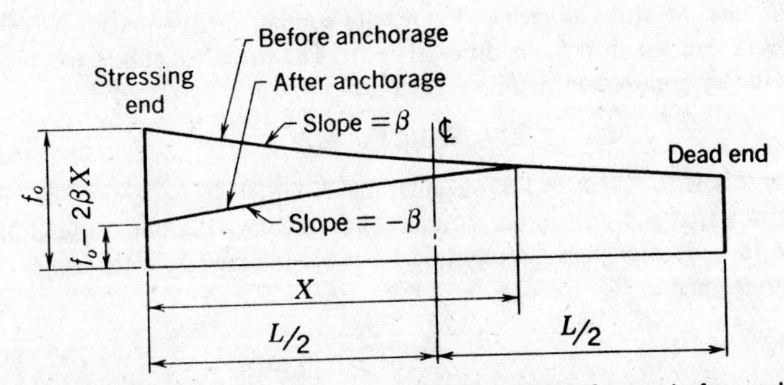

Fig. 15-11 Variation in stress in a post-tensioned tendon before and after anchorage. Condition II.

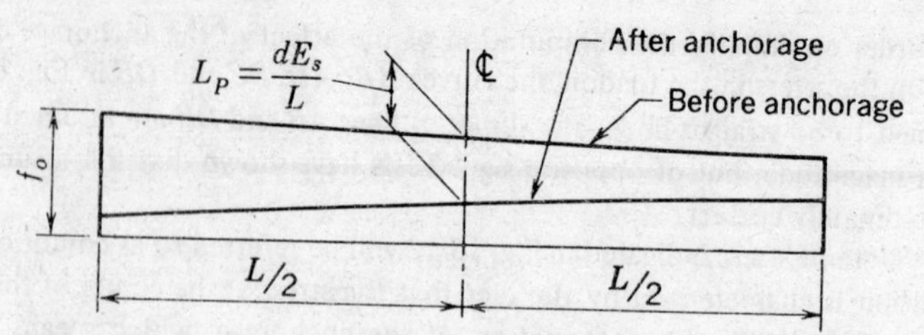

Fig. 15-12 Variation in stress in a post-tensioned tendon before and after anchorage. Condition III.

from the friction, as explained in Sec. 15-5. This can be expressed as

$$f_o = f_{℄} e^{\phi}$$

where

$$\phi = \mu\alpha + \frac{KL}{2}$$

(2) Assume the deformation of anchorage does not affect the stress at midspan of the beam (condition I) and compute the slope of the curve between midspan and stressing end as follows:

$$\text{Slope} = \beta = \frac{2f_{℄}(e^{\phi} - 1)}{L} \tag{15-3}$$

(3) Compute the length of the tendon on which the stress is reduced by the anchorage deformation

$$X = \sqrt{\frac{dE_s}{\beta}} \tag{15-4}$$

where X = length of tendon which is affected in in., β = slope in psi/in., E_s = modulus of elasticity of the steel in psi. and d = deformation of the anchorage in in.

(4) If X is less than $L/2$, condition I exists, the stress at midspan is not affected, and the stress at each end is readily computed if desired. If X is greater than $L/2$, but less than L, condition II exists. The stress loss at midspan resulting from the anchorage deformation is equal to

$$L_p = 2\beta(X - L/2) \tag{15-5}$$

If this loss is too great to be tolerated, higher initial stresses should be investigated in a trial and error procedure until a satisfactory solution is found. If X is found to be greater than L, condition III exists and the loss at the center of the tendon is equal to

$$L_p = \frac{dE_s}{L} \tag{15-6}$$

and the stress after anchorage at the center of the tendon can be computed directly.

It must be emphasized again that the anchorage deformation for some post-tension systems is very small and can be reasonably neglected. Anchorage deformations as high as 1 in. have been observed with wedge-type anchors under very unusual conditions. It is important to recognize that this phenomenon exists and that it must be taken into consideration during the design. This is particularly true when short tendons are to be used.

ILLUSTRATIVE PROBLEM 15-2 Determine the effect of the elastic deformation of the anchorage cone on the stress at midspan of a tendon 100 ft long with a maximum ordinate of 3 ft, if the desired stress at midspan is 165,000 psi, the deformation of the anchorage is 0.50 in., the modulus of elasticity of the steel is 28,000,000 psi, and a bright metal sheath is used. Also determine the effect if the sheath is galvanized rather than bright.

SOLUTION: The effect of friction for these conditions were determined in Prob. 15-1. Using the calculated value for e^{ϕ}, the computation of β and X becomes:

$$\beta = \frac{2 \times 165{,}000 \times 0.15}{1200 \text{ in.}} = 41.2 \text{ psi/in.}$$

$$X = \sqrt{\frac{0.50 \times 28 \times 10^6}{41.2}} = 584 \text{ in.} < 600 \text{ in.} = \frac{L}{2}$$

Therefore, condition I exists and the deformation of the anchorage of 0.50 in. does not reduce the stress in the tendon at midspan.

Using galvanized sheath:

$$\beta = 41.2 \times \frac{0.11}{0.15} = 30.2 \text{ psi/in.}$$

$$X = \sqrt{\frac{0.50 \times 28 \times 10^6}{30.2}} = 680 \text{ in.} > 600 \text{ in.} = \frac{L}{2}$$

The reduction in stress at midspan due to the deformation of the anchorage is:

$$L_p = 2\beta \left(X - \frac{L}{2}\right) = 2 \times 30.2 \times 80 = 4830 \text{ psi}$$

Therefore, the stress in the tendon at the center after the tendon is anchored is

$$f = 165{,}000 - 4800 = 160{,}200 \text{ psi}$$

If the stress at midspan is increased to 169,000 psi at the time of stressing, the values of β and X and the loss of the prestressing stress in the tendon at midspan become:

$$\beta = \frac{2 \times 169{,}000 \times 0.11}{1200} = 31.0 \text{ psi}$$

$$X = \sqrt{\frac{0.50 \times 28 \times 10^6}{31.0}} = 672 \text{ in.}$$

$$L_p = 2 \times 31.0 \times 72 = 4500 \text{ psi}$$

Therefore, if the stress at midspan is increased to 169,500 psi at the time of stressing, the deformation of the anchorage of 0.50 in. will reduce the stress to approximately 165,000 psi, which is the desired value.

It should be noted that the stress at the end of the tendon which is required to obtain 165,000 psi in the tendon at the center is equal to 165,000 × 1.15 = 190,000 psi when a bright sheath is used. When galvanized sheath is used, the stress at the end of the tendon which is required to obtain the desired stress at the center line after anchoring is equal to 169,500 × 1.11 = 188,000 psi. It will be seen that the required tendon stress at the end of the tendon is not materially reduced by the use of galvanized sheath in this case.

ILLUSTRATIVE PROBLEM 15-3 Compute the before anchorage (jacking) and initial (after anchorage) stresses for the tendon layout shown in Example V of Appendix B if the tendons are jacked to a force of 6460 kips at each end. The prestressing tendons consist of 208 strands each having an area of 0.153 sq in. The friction coefficients are $\mu = 0.25$ and $K = 0.0002$. The elastic modulus of the steel is 27,000 ksi and the anchorage set is 0.625 in. Determine the location of the maximum and minimum points of stress assuming the tendons to follow second-degree parabolas between the dimensioned points; assume linear variation of stress between these points.

SOLUTION:

Pt.	μ	α	K	Δx
A				
	0.25	0.0686	0.0002	81.00
B				
	0.25	0.1157	0.0002	64.80
C				
	0.25	0.1157	0.0002	16.20
D				
	0.25	0.1250	0.0002	15.00
E				
	0.25	0.1250	0.0002	60.00
F				
	0.25	0.074	0.0002	75.00
G				

The summary of the computations are shown in the table below and in the Fig. 15-13.

Pt.	H (ft)	L (ft)	Friction Factor	Jacking Stress (ksi)	Initial Stress (ksi)
A	3.610		1.0000	202.991	180.890
		81.000			
B	.833		1.0339	196.334	187.547
		35.343			
					191.941
		29.456			
C	4.583		1.0781	188.279	188.279
		16.200			
D	5.520		1.1133	182.320	182.320
		.788			
*			1.1153	181.992	
		14.211			
E	4.583		1.0797	187.992	187.992
		24.744			
					191.418
		35.255			
F	.833		1.0340	196.301	186.536
		75.000			
G	3.610		1.0000	202.991	179.846

*Point of minimum stress.

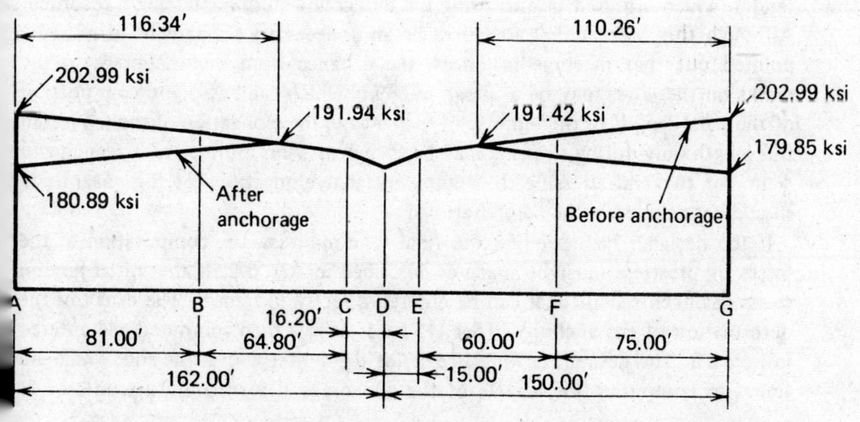

Fig. 15-13 Plot of stresses in the tendon of I.P. 15-3 before and after anchorage.

15-7 Computation of Gauge Pressures and Elongations

At the time the design of post-tensioned members is made, the computation of gauge pressures and elongations for post-tensioning should be made and summarized on the drawings, if the post-tensioning system is specified by the designer. In following this procedure, the designer is assured that the stressing can be achieved as intended. When the system of post-tensioning is not specified, the computation of gauge pressures and elongations must be supplied by the contractor for review and approval by the engineer. In this case, it is recommended that the summary of the stressing data be placed on the shop drawings.

The computation of gauge pressures and elongations are conveniently made and summarized in tabular form as illustrated in Table 15-2. Some systems have peculiarities of their own which are not included in the list of items in Table 15-2. These must be covered by special notes or instructions and will not be discussed here.

The length of the prestressing tendon used in the computations is normally taken as the horizontal projection of the tendon. For more precise computations or where there are large curvatures of the tendons, the length can be computed by:

$$L' = \left(1 + \frac{8}{3}\frac{a^2}{L^2}\right) L \qquad (15\text{-}7)$$

in which L is the horizontal projection of the tendon, a is the vertical deflection of the tendon and L' is the correct length. Provision is made in the table for the length to be recorded in both feet and inches, since the former length is required in the computation of the effect of friction and the latter is required in the computation of elongations.

Under item 5, the distance from the anchorage to the mark or marks on the tendon which are used in measuring the elongation during stressing is recorded. Although this may at first appear to be an unnecessary refinement, it must be pointed out that in some instances, the distance from the anchorage to the marks on the wires may be as great as 18 in., which can be a significant portion of the total length of the tendon. In such a case, the elongation of the tendon in this length only during stressing may be as high as $\frac{1}{8}$ in. for one-end stressing and $\frac{1}{4}$ in. for two-end stressing. It is apparent that elongations of this magnitude should be included in the computations.

If the designer has specified the final tendon stress, the computation of the losses of prestress must be made as described in Art. 6-2. If the initial tendon stress has been specified, it can be entered directly in item 9. The effect of the deformation of the anchorage (Sec. 15-6), if any, is then computed and entered in item 10. The desired, jacking stresses at the center and at the end, which are found in computing the effects of the anchorage deformation, are entered as

TABLE 15-2 Typical Calculation Sheet for Use in Computing the Gauge Pressures and Elongations in Post-tensioning

Item Number	*Tendon Numbers* 1	2	3
1. Type of tendon			
2. Area of steel (in.^{2})			
3. Length of tendon (ft)			
4. Length of tendon (in.)			
5. Distance from anchorage to marks (in.). Note: For one-end stressing, use only the length at stressing end. For two-end stressing use the length at both ends			
6. Total length of unit to be stressed (in.)			
7. Desired effective stress (psi)			
8. Losses of prestress.			
a. Elastic shortening*			
b. Relaxation, creep and shrinkage			
9. Desired stress before anchorage deformation (psi)			
10. Effect of anchorage deformation (psi)			
11. Desired jacking stress at center (psi)			
12. Anticipated jacking stress at jack (psi)			
13. Anticipated gauge pressure (psi)			
14. Maximum gauge pressure (psi)			
15. Elongation to be obtained (in.)			
a. Total			
b. Net			
16. Special instructions			

* The losses of prestress should include all the effects listed in Sec. 6-2. The loss due to elastic shortening is often left to be included in the stressing computations. This should be investigated for each job.

items 11 and 12, respectively. The value of the jacking stress at the end is not normally allowed to exceed 80% of the guaranteed, ultimate tensile strength of the tendon.

The gauge pressure to be anticipated during the stressing of a tendon is higher than the theoretical gauge pressure, due to internal friction of the jack and, in some systems, due to the friction of the tendons which must slide over the anchorage and the jack during stressing. This friction ranges from 2% to 15% and a reasonable estimate of the friction can only be obtained by calibrating the jack when all of the contributing factors are present. The data obtained in calibrating a jack can be plotted in a curve that shows the relationship between the gauge pressure and the steel stress. The gauge pressure that would be anticipated in obtaining the desired initial stress at the end of the tendon can be taken directly from such a calibration curve. This gauge pressure is termed anticipated, since it is computed on the basis of assumed friction coefficients and is subject to error. For this reason, the elongation of the tendon is generally used as the controlling measurement during stressing and the normal procedure is to apply the anticipated pressure and then check the elongation. If the elongation is not satisfactory, the pressure is increased until a satisfactory elongation is obtained. In order to ensure that the tendon will not be overstressed, a maximum permissible gauge pressure (corresponding to the maximum allowable jacking stress) is taken from the calibration curve. The maximum permissible gauge pressure is entered in item 14.

Before the reference marks used in measuring the elongation of the tendons are placed on the tendons, it is normal procedure to apply a small pressure of about 10% of the anticipated gauge pressure to the tendon in order to tighten up the jack and eliminate any slack that may be in the tendon. This value should be constant on the entire job in order to eliminate the possibility of the workmen using the incorrect value when changing from tendons of different size.

Since the initial pressure applied to the tendon results in a small stress in the tendon, the amount of elongation that is obtained during the application of the remaining force is the result of the deformation of the steel between the lower value of stress and the higher value. In addition, because there is a variation in stress in the tendon due to friction, an average value of stress must be used in determining the elongation that is to be obtained in the stressing. In the case of stressing from one end only, the average stress would be the stress at the center of the tendon in most instances, while, in the case of stressing from each end simultaneously, the average stress would occur approximately midway between the end and the center of the tendon. After the appropriate average stress is determined, the total elongation that would result from this average stress can be calculated from the stress-strain diagram for the steel that is to be used. The net elongation is determined by taking the increment of strain from the lower stress to the higher stress, for the steel at the average location in the tendon, and multiplying this value by the length of the tendon.

15-8 Construction Procedure in Post-tensioned Concrete

Although the construction details and the system of post-tensioning to be used will influence the construction procedure to be followed, general statements and precautions are considered of value. Careful planning and attention to details of construction in post-tensioned elements is important and can have material influence on the cost and performance of post-tensioned construction.

The normal procedure followed in the construction of a post-tensioned beam consists of erecting the soffit form, and one side form, placing the reinforcing steel, prestressing steel, end anchorages, erecting the remaining side form, placing and curing the concrete, stripping the forms, and finally stressing and grouting the post-tensioning tendons. In large productions of post-tensioned members, the reinforcing and post-tensioning steel may be tied together in a jig and set in the forms as a unit. On smaller jobs, the post-tensioning units are normally tied in place after the reinforcing steel cage is partially or completely assembled in the forms. The latter procedure is much the more common, and it must be emphasized that the erection of one side form, which can be used to facilitate layout and to secure the reinforcing cage, inserts, and blockout forms in place, expedites the assembly of post-tensioned members.

When post-tensioning tendons in sheaths are used, the tendons must be securely tied in place at close intervals and in such a manner that secondary curvatures of the tendons are minimized and the primary curvature of the tendon is close to the curve specified on the drawings. It must be emphasized that it is important for the location of the tendons at the center of simple flexural members to be close to that specified on the drawings. As was shown in Sec. 4-7, the tendon does not normally need to be placed within precise limits at points between the center of the beam and the ends in order to achieve a satisfactory post-tensioned flexural member. Hence, it is generally important that the tendon be placed on smooth curves that will minimize friction losses during stressing, but it is not of great importance if a tendon is not precisely on the specified trajectory.

No general statement can be made on the maximum spacing that should be permitted for the ties and supports that secure the post-tensioning tendons in place during placing of the concrete. This is a function of the type and size of tendons, as well as the arrangement of the reinforcing steel and required trajectory of the tendons.

Care must be exercised when tying the tendons, since it is possible to damage the sheath during the tying operation. This is particularly true when paper or plastic sheath is used. Damaged sheath may permit grout to enter the sheath and bond the tendon to the beam. This may render the stressing of the tendons impossible.

After the tendons and anchorages have been tied in place, they should be carefully inspected to be sure they are securely tied at all locations and that there is no possibility of mortar leaking into the sheath or anchorage device during placing and vibrating the concrete. Although mortar that leaks into a sheath or anchorage does not always result in the tendon being bonded and impossible to stress, the extra labor required to clean the anchorage before stressing can be started (or to overcome the high friction in the tendon which may be encountered during stressing) usually far exceeds the amount of labor required to properly seal the sheath and anchorage before concreting.

When ducts are to be preformed in the concrete, the procedure is very similar to that followed for tendons which are in sheath. Rubber ducts should not be tied so tightly to the reinforcing that they will be difficult to remove.

The manufacturers of some types of post-tensioning tendons recommend that the tendon be moved back and forth in the sheath during and after the placing of the concrete in order to be certain that the tendon is not bonded to the concrete by any mortar which may have leaked into the sheath. Although there is no technical objection to this procedure, it is not practical for all systems of post-tensioning and is considered unnecessary if sufficient attention is given to the prevention of leaks in the sheaths and anchorages before the concrete is placed in the forms. Rigid sheaths placed in the forms without the tendons in them, sometimes become damaged by the internal vibrators used in placing the concrete. This is most likely to occur in areas that are congested. The damage can be so great as to prevent the insertion of the tendons. As a means of guarding against this problem, it is recommended that the sheaths be inspected during the concreting operation to confirm that they remain open. This is easily done by pulling (with a small wire or cable) or pushing (with compressed air) a ball through the ducts during the time the concrete is being placed.

After post-tensioned members are cured, the tendons are usually stressed and grouted as soon as possible. The stressing is done according to the procedure recommended by the manufacturers of the post-tensioning anchorages and jacking equipment. In post-tensioning, the elongation of the tendons is generally used as the controlling measurement for ensuring proper stress at the center of a member. As has been explained, the elongation is a better measure of the stress at the center of the tendon than are the gauge pressure and the stress in the tendon at the end of the tendon.

The procedure used in grouting post-tensioning units also varies with the post-tensioning system being used. In general, the grouting ports and sheaths are very small and prohibit the use of large-grain sand in the grout. Very fine sand is used occasionally in order to reduce the quantity of cement required and in an attempt to reduce the shrinkage of the grout. Although the addition of an adequate quantity of very fine sand to mortar may result in the shrinkage of the mortar being only one-half of that which results from the use of neat cement, the addition of fine sand in the amounts which can be added to the grout for post-tensioning tendons will not materially reduce the shrinkage. Pozzolanic material (fly ash) is sometimes specified as an admixture for the grout, but little if any benefits are gained from this procedure, because the shrinkage of concrete is not reduced and may be increased by such admixtures. For these reasons, the grout used in domestic post-tensioning operations rarely contain sand.

The type of cement used in the grout for the post-tensioning ducts should be the same type used in the concrete, because a difference in the electrolytic properties of the cement in the grout and the basic concrete may result in the deterioration of the prestressing steel. Normally, a grout mixture of 4.0 to 4.5 gallons of water per sack of cement will result in a grout that can be easily injected. It is important that grouting tubes, hoses, ports, and valves do not leak during grouting. Leaks may result in plugging of the system and the trapping of water in the sheath. When small round sheaths are used, the grout can be introduced at pressures up to 100 psi. When large round or rectangular ducts are used, the grout pressure may have to be restricted to atmospheric pressure, in order to avoid cracking of the concrete section along the trajectory of the tendon.

During freezing weather, the grout that is injected into post-tensioning ducts may freeze and cause cracking of the concrete along the path of the post-tensioning tendons. This can be avoided by heating the member after grouting for 24 to 48 hr, in order to allow the grout to harden, or by replacing some of the mixing water in the grout mixture with alcohol in order to reduce the freezing point of the grout. The use of one part alcohol to eight parts of water has been recommended by some authorities as a satisfactory method of providing protection against freezing of the grout for temperatures as low as 20°F, without excessive sacrifice in the strength of the grout. The use of more alcohol will render the grout more resistant to freezing, but will also reduce the strength of the grout (Ref. 6).

If post-tensioned members are to be exposed to freezing weather before the tendons are grouted, the ducts or sheaths should be drained of all water that may collect in the sheath during placing and curing of the concrete. If this is not done and the water freezes, the members may be cracked, just as if grout had frozen in the ducts. Water should not be allowed to remain in the ducts of post-tensioning tendons for more than just a few days, even if there is not any danger of freezing, because corrosion of the tendons will generally proceed at a more rapid rate when the tendons are submerged or in a very moist atmosphere.

15-9 Construction of Multi-element Beams

Post-tensioned beams may be made of a number of precast elements that are post-tensioned together to form one structural unit. This procedure is used in order to facilitate handling, transportation, and forming large members. Types of multi-element beams are illustrated in Fig. 15-14. It will be seen that a member may be formed of many elements made in small forms on vibrating tables or of only a few large elements made with the same general procedure used in the construction of monolithic members. Preformed holes are generally used for the tendons in multi-element members. When preformed holes cannot be employed, because of the post-tensioning system to be used, the tendons must be coupled at each joint.

Members composed of many small elements normally must have mortar placed in the joints at the time the elements are aligned for stressing. The mortar is necessary to assure a uniform distribution of stress between the elements. The mortar joints present a difficulty in the assembly of the members, because it is necessary to prevent the mortar from entering the post-tensioning ducts and

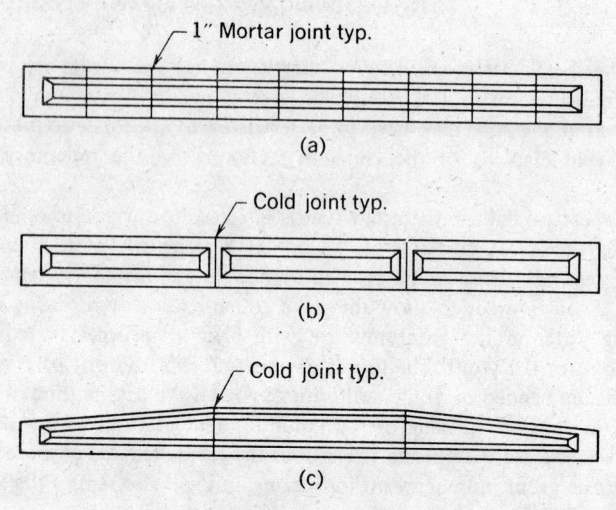

Fig. 15-14 Multi-element post-tension beams.

bonding to the tendons. Additionally, the placing of the mortar requires considerable labor. In many instances, the prestress can be applied to the multi-element beams immediately after the mortar is placed, since the total deformation of the plastic mortar is small, due to the relatively short length of the mortar joints. The thin mortar joints cannot fail due to prestressing, even if the mortar is partially plastic at the time of stressing.

The need of mortar joints in multi-element members can be eliminated by casting, first, one element, and, after the first element has hardened, casting the second element against the end of the first element. In this manner, the cold joint formed between the two elements must be a perfect fit, and yet the elements can be separated for handling and transportation purposes. A light coating of oil is often placed on the end of the elements to reduce the bond between the elements during manufacture. Shear keys, which provide vertical and horizontal projections of the end surfaces, are often provided at the ends of each element to facilitate aligning the elements in the field. Shear keys are not normally required due to structural considerations. The joints between the individual pieces of multi-element beams are frequently coated with epoxy resin at the time of assembly. This glues the pieces together. A bridge on which this procedure was used is shown in Fig. 15-15.

When multi-element beams composed of only a few pieces are assembled in the field, it is not necessary to set the elements close together and in precise alignment with the cranes or other handling equipment being used. It is generally easier and more expeditious to set the elements on wooden blocks, with a space

Fig. 15-15 Multi-element post-tensioned construction was used on the bridge between the mainland and the island of Oléron in France. (*Courtesy Freyssinet Co., Inc., New York.*)

of approximately 1 ft between the elements. This procedure facilitates threading the post-tensioning tendons in the ducts. The elements can be pulled together with the use of the post-tensioning jacks after the tendons have been inserted.

REFERENCES

1. Bumanis, Alfreds, "Friction Loss Study of 402-Ft. Tendons," *Journal of the Prestressed Concrete Institute*, **11**, No. 4, 57–63 (Aug. 1966).
2. Polivka, Milos, "Grouts for Post-tensioned Prestressed Concrete Members," *Journal of the Prestressed Concrete Institute*, **6**, No. 2, 28–38 (June 1961).
3. Committee on Grouting of Post-tensioned Tendons, "A Report of Field Experiences in Grouting Post-tensioned Prestressed Cables," *Journal of the Prestressed Concrete Institute*, **7**, No. 4, 13–17 (Aug. 1962).
4. Committee on Grouting of Post-tensioned Tendons, "Tentative Recommended Practices for Grouting Post-tensioned Prestressed Concrete," *Journal of the Prestressed Concrete Institute*, **5**, No. 2, 78–81 (June 1960).

5. Grouting Subcommittee of the Combined Western Concrete Reinforcing Steel Institute Post-tensioning Committee and the Prestressed Concrete Manufacturers Association of California Task Force on Post-tensioning, "Recommended Practice for Grouting Post-tensioning Tendons," (July 1967)–Tentative.

6. STUVO Committee on Grouting, "Some Facts About Grouting," *Proceedings of the Second Congress of the Federation Internationale de la Precontrainte*, Session 1, Paper No. 3, Amsterdam, Netherlands, 1955.

16 Erection of Precast Members

16-1 General

An important consideration in the use of precast concrete is the methods of erection that are feasible under the jobsite conditions. This obviously has significant effect on the cost of construction. Occasionally, high-rise buildings cannot be made at a reasonable cost with precast construction due to the inability of cranes to lift and place heavy precast members at the heights and reaches that are necessary. The large cranes required to erect the larger precast members are costly and are not readily available in all localities. In bridge construction, the use of cranes to erect precast members may not be feasible due to the reaches that are involved, due to the risk of floods, or due to other considerations. The designer must give careful consideration to all the factors that affect the methods that can be used in the construction of each particular structure and prepare his design to best suit job site conditions.

16-2 Truck Cranes

Truck cranes are often used to erect precast units in building and bridge construction. Although cranes with rated capacities of 100 tons or more are fre-

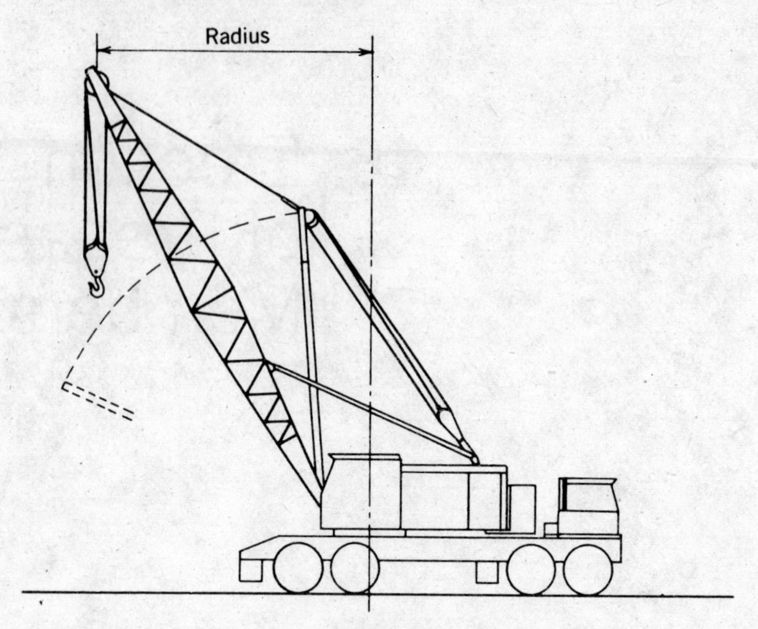

Fig. 16-1 Load radius for a truck crane.

quently used, the designer should be aware of the fact that the rated capacity is the maximum load the crane can lift with a relatively short boom and with the minimum possible load radius. The load radius is the horizontal distance measured from the vertical axis about which the crane cab rotates to the line of the load. This is illustrated in Fig. 16-1.

Capacities for cranes of 35 tons, 65 tons, and 82 tons are given in Tables 16-1, 2, and 3. It should be noted that for large loads on short radii, the capacity of a crane is limited by structural considerations, where as for loads at larger radii, the capacity is limited by tipping. From Table 16-3 it will be seen that a 82 ton crane with a 100 ft boom working on a load radius of 40 ft has an approximate capacity of from 21 to 23 tons on outriggers, depending upon whether the load is being lifted over the side or over the end. This clearly illustrates the fact that one must carefully investigate the equipment available in each locality and be certain the capacity will be adequate for the intended use.

It should also be noted that the capacity of a crane is less when it is on its tires and without the benefit of its outriggers. In addition, it is very important that a crane be exactly level when making lifts that are close to its capacity. If it is necessary to move a large precast member some distance with one or more cranes, the ground over which the cranes move must be very level and firm.

Precast building beams and girders can often be efficiently erected using truck cranes in combination with dollies, even when the building dimensions, site conditions or crane capacity preclude the possibility of the crane setting the

TABLE 16-1 Capacities Capacity of a 35 Ton Truck Crane with Angle Boom

***Capacities are based on machine equipped with Retractable High Gantry, 8 X 4 drive Carrier—9′ 0″ wide, 12:00 X 20 14-ply rating tires, Power Hydraulic Outriggers, 13,200#ctwt.**

BOOM				On Outriggers		On Tires	
Length	Radius	Angle	Point Ht. W	Rear	Side	Rear	Side
	10′	79°	40′ 9″	70,000*	70,000*	53,100	40,800*
	11′	77°	40′ 7″	70,000*	70,000*	46,300	37,800*
	12′	75°	40′ 3″	69,300*	69,300*	41,000	35,200*
	13′	74°	39′11″	64,500*	64,500*	36,800	32,300
35′	14′	72°	39′ 9″	60,300*	60,300*	33,300	29,200
	15′	70°	39′ 4″	56,600*	56,600*	30,400	26,500
	20′	61°	37′ 1″	42,100*	42,100*	21,000	18,100
	25′	51°	33′ 9″	32,600*	32,600*	15,900	13,600
	30′	40°	28′ 9″	26,500*	26,500*	12,600	10,700
	35′	24°	20′ 9″	22,200*	22,200*	10,400	8,800
	10′	80°	45′10″	70,000*	70,000*	53,000	40,500*
	11′	79°	45′ 8″	70,000*	70,000*	46,200	37,600*
	12′	77°	45′ 5″	69,000*	69,000*	40,900	35,000*
	13′	76°	45′ 2″	64,200*	64,200*	36,600	32,200
	14′	74°	44′11″	60,100*	60,100*	33,100	29,000
40′	15′	73°	44′ 7″	56,400*	56,400*	30,200	26,400
	20′	65°	42′ 8″	41,900*	41,900*	20,800	17,900
	25′	57°	39′10″	32,400*	32,400*	15,700	13,400
	30′	48°	36′ 1″	26,300*	26,300*	12,500	10,600
	35′	37°	30′ 7″	22,000*	22,000*	10,200	8,600
	40′	23°	21′10″	18,900*	18,900	8,600	7,200
	12′	80°	55′ 8″	68,600*	68,600*	40,600	34,500*
	13′	79°	55′ 5″	63,800*	63,800*	36,300	31,900
	14′	77°	55′ 3″	59,600*	59,600*	32,800	28,700
	15′	76°	54′11″	55,900*	55,900*	29,900	26,100
50′	20′	70°	53′ 5″	41,500*	41,500*	20,500	17,600
	25′	64°	51′ 4″	32,000*	32,000*	15,400	13,100
	30′	57°	48′ 7″	25,900*	25,900*	12,200	10,200
	35′	50°	44′11″	21,600*	21,600*	9,900	8,300
	40′	42°	40′ 2″	18,500*	18,500*	8,300	6,900
	50′	20°	23′ 9″	13,900	13,500	6,100	4,900
	13′	81°	64′ 9″	63,300*	63,300*	36,000	31,700
	14′	80°	64′ 6″	59,100*	59,100*	32,600	28,500
	15′	79°	65′ 3″	55,400*	55,400*	29,600	25,800
	20′	74°	63′11″	41,100*	41,100*	20,200	17,400
60′	25′	69°	62′ 3″	31,600*	31,600*	15,100	12,800
	30′	63°	60′ 1″	25,500*	25,500*	11,800	9,900
	35′	58°	57′ 3″	21,200*	21,200*	9,600	8,000
	40′	52°	53′ 9″	18,100*	18,100*	8,000	6,600
	50′	39°	43′10″	13,600*	13,300	5,800	4,600
	60′	18°	25′ 5″	10,500	10,200	4,300	3,400
	15′	80°	75′ 5″	53,100*	53,100*	29,400	25,500
	20′	76°	74′ 4″	40,700*	40,700*	19,900	17,100
	25′	72°	72′11″	31,200*	31,200*	14,800	12,500
	30′	67°	71′ 1″	25,100*	25,100*	11,500	9,600
70′	35′	63°	68′ 9″	20,900*	20,900*	9,300	7,700
	40′	58°	65′11″	17,700*	17,700*	7,700	6,300
	50′	48°	58′ 4″	13,200*	13,000	5,500	4,300
	60′	36°	47′ 2″	10,300	9,900	4,000	3,100
	70′	17°	26′11″	8,100	7,800	3,000	2,200
	20′	78°	84′ 8″	40,400*	40,400*	19,600	16,800
	25′	74°	83′ 4″	30,900*	30,900*	14,500	12,200
	30′	70°	81′ 9″	24,700*	24,700*	11,200	9,300
	35′	67°	79′ 9″	20,500*	20,500*	9,000	7,400
80′	40′	63°	77′ 5″	17,300*	17,300*	7,400	5,900
	50′	54°	71′ 3″	12,900*	12,700	5,100	4,000
	60′	45°	62′ 8″	9,900*	9,600	3,700	2,800
	70′	33°	50′ 3″	7,900	7,500	2,700	1,900
	80′	16°	28′ 5″	6,100*	6,000	1,900	1,200
	20′	79°	94′10″	36,300*	36,300*	19,300	16,500
	25′	76°	93′ 9″	30,500*	30,500*	14,200	11,900
	30′	73°	92′ 4″	24,400*	24,400*	10,900	9,000
	35′	70°	90′ 7″	20,100*	20,100*	8,700	7,100
90′	40′	66°	88′ 7″	16,900*	16,900*	7,000	5,600
	50′	59°	83′ 3″	12,500*	12,500	4,800	3,700
	60′	51°	76′ 2″	9,600*	9,400	3,400	2,400
	70′	42°	66′ 8″	7,500*	7,300	2,400	1,600
	80′	31°	53′ 2″	5,800*	5,700	1,600	900
	90′	15°	29′ 9″	3,900*	3,900*	1,000	400
	20′	80°	105′ 1″	32,400*	32,400*	19,000	16,200
	25′	77°	103′11″	27,000*	27,000*	13,900	11,600
	30′	74°	102′ 9″	22,800*	22,800*	10,600	8,700
	35′	71°	101′ 2″	19,300*	19,300*	8,400	6,800
	40′	68°	99′ 4″	16,300*	16,300*	6,700	5,300
100′	50′	62°	94′ 9″	12,100*	12,100*	4,500	3,400
	60′	55°	88′ 8″	8,800*	8,800*	3,100	2,100
	70′	48°	80′ 9″	6,600*	6,600*	2,000	1,200
	80′	40°	70′ 4″	5,000*	5,000*	1,300	600
	90′	30°	55′10″	3,600*	3,600*	700	—
	100′	14°	31′ 1″	2,300*	2,300*	200	—

Courtesy of Link-Belt Speeder.)

member into its final location. In this case, one or two cranes are used to hoist the precast member to the proper level and set it on dollies which then are used to move the member horizontally to the proper location. The dollies can be provided with hydraulic jacks to allow the members to be lowered into place after they are correctly positioned horizontally. This method does not work well on sloping floors such as those found in some types of parking garages.

TABLE 16-2 Capacities of a 65 Ton Truck Crane

***Capacities are based on machine equipped with Boom Gantry, 8 × 4 drive Carrier—11′ 0″ wide, 14:00 × 20 18-ply rating tires, Power Hydraulic Outriggers, 18,000 lbs. ctwt and 4,000 lbs. Bumper ctwt.**

BOOM				ON OUTRIGGERS		ON TIRES	
Length	Radius	Angle	Point Ht. W.	Rear	Side	Rear	Side
40′	12′	78°	45′ 11″	130,000*	130,000*	76,700*	62,310*
	13′	76°	45′ 7″	128,040*	127,140*	73,870*	58,450*
	14′	74°	45′ 4″	119,790*	119,080*	71,250*	55,010*
	15′	73°	45′ 0″	112,500*	111,960*	68,790*	51,930*
	20′	65°	43′ 1″	86,000*	85,930*	47,760	35,510
	25′	57°	40′ 5″	67,530*	66,680	35,540	26,140
	30′	48°	36′ 6″	54,380*	49,320	28,010	20,370
	35′	37°	31′ 1″	44,890*	38,810	22,880	16,460
	40′	23°	22′ 5″	37,240*	31,730	19,150	13,600
50′	13′	79°	55′ 11″	127,360*	127,110*	73,630*	58,370*
	14′	78°	55′ 8″	119,760*	119,050*	71,010*	54,930*
	15′	76°	55′ 5″	112,480*	111,930*	68,560*	51,850*
	20′	70°	53′ 11″	85,960*	85,900*	47,890	35,650
	25′	64°	51′ 10″	67,600*	66,990	35,650	26,250
	30′	58°	48′ 11″	54,490*	49,560	28,100	20,470
	35′	50°	45′ 5″	45,000*	39,030	22,980	16,560
	40′	43°	40′ 7″	38,150*	31,960	19,270	13,730
	50′	21°	24′ 4″	28,390	23,030	14,220	9,870
60′	15′	79°	65′ 10″	112,630*	112,070*	68,410*	51,920*
	20′	74°	64′ 5″	86,080*	85,990*	48,170	35,930
	25′	69°	62′ 8″	67,770*	67,420	35,850	26,450
	30′	63°	60′ 6″	54,670*	49,880	28,250	20,630
	35′	56°	57′ 8″	45,150*	39,270	23,100	16,680
	40′	52°	54′ 2″	38,270*	32,160	19,380	13,830
	50′	39°	44′ 4″	28,580	23,220	14,330	9,980
	60′	19°	26′ 1″	22,060	17,760	11,020	7,440
70′	20′	76°	74′ 10″	85,930*	85,840*	48,150	35,920
	25′	72°	73′ 4″	67,680*	67,530	35,810	26,410
	30′	68°	71′ 6″	54,600*	49,930	28,200	20,580
	35′	63°	69′ 2″	45,080*	39,300	23,050	16,630
	40′	58°	66′ 5″	38,190*	32,170	19,320	13,780
	50′	48°	58′ 11″	28,580	23,220	14,280	9,930
	60′	36°	47′ 10″	22,100	17,800	11,010	7,430
	70′	17°	27′ 7″	17,710	14,130	8,690	5,650
80′	20′	78°	85′ 0″	85,750*	85,660*	48,100	35,870
	25′	74°	83′ 10″	67,560*	67,560*	35,730	26,340
	30′	70°	82′ 2″	54,500*	49,940	28,120	20,490
	35′	67°	80′ 2″	44,980*	39,280	22,960	16,540
	40′	63°	77′-11″	38,090	32,140	19,220	13,690
	50′	54°	71′ 10″	28,530	23,170	14,180	9,840
	60′	45°	63′ 1″	22,050	17,760	10,930	7,350
	70′	33°	50′ 10″	17,700	14,120	8,640	5,600
	80′	16°	29′ 1″	14,540	11,470	6,920	4,270
90′	20′	79°	95′ 4″	81,510*	81,510*	48,030	35,800
	25′	76°	94′ 2″	67,420*	67,420*	35,640	26,250
	30′	73°	92′ 8″	54,380*	49,920	28,010	20,390
	35′	69°	91′ 0″	44,840*	39,240	22,840	16,430
	40′	66°	89′ 0″	37,950*	32,080	19,110	13,570
	50′	59°	83′ 2″	28,460	23,090	14,060	9,720
	60′	51°	76′ 7″	21,970	17,670	10,810	7,240
	70′	42°	67′ 1″	17,620	14,040	8,530	5,500
	80′	31°	53′ 8″	14,500	11,420	6,840	4,200
	90′	15°	30′ 6″	12,110	9,420	5,510	3,170
100′	25′	77°	104′ 5″	67,260*	67,260*	35,530	26,140
	30′	74°	103′ 2″	54,240*	49,890	27,890	20,270
	40′	68°	99′ 10″	37,790*	32,010	18,980	13,440
	50′	62°	95′ 2″	28,360	23,000	13,930	9,590
	60′	55°	89′ 1″	21,870	17,570	10,680	7,100
	70′	48°	81′ 2″	17,520	13,940	8,400	5,360
	80′	40°	70′ 10″	14,400	11,320	6,720	4,080
	90′	30°	56′ 5″	12,040	9,350	5,410	3,070
	100′	14°	31′ 10″	10,170	7,780	4,350	2,250
110′	25′	79°	114′ 8″	64,820*	64,820*	35,410	26,030
	30′	76°	113′ 6″	54,090*	49,840	27,760	20,150
	40′	70°	110′ 6″	37,630*	31,920	18,840	13,300
	50′	65°	106′ 5″	28,260	22,890	13,790	9,440
	60′	59°	101′ 0″	21,750	17,460	10,530	6,960
	70′	53°	94′ 2″	17,400	13,820	8,260	5,220
	80′	46°	85′ 7″	14,280	11,200	6,580	3,930
	90′	38°	74′ 5″	11,930	9,240	5,280	2,940
	100′	26°	59′ 0″	10,080	7,690	4,240	2,140
	110′	14°	33′ 0″	8,580	6,430	3,370	1,470

BOOM				ON OUTRIGGERS		ON TIRES	
Length	Radius	Angle	Point Ht. W.	Rear	Side	Rear	Side
120′	30′	77°	123′ 10″	53,930*	49,780	27,630	20,01[illegible]
	40′	72°	121′ 1″	37,460*	31,820	18,690	13,16[illegible]
	50′	67°	117′ 4″	28,140	22,780	13,630	9,29[illegible]
	60′	62°	112′ 6″	21,620	17,330	10,380	6,80[illegible]
	70′	56°	106′ 6″	17,270	13,680	8,100	5,06[illegible]
	80′	50°	99′ 0″	14,140	11,070	6,420	3,78[illegible]
	90′	44°	89′ 8″	11,790	9,100	5,130	2,79[illegible]
	100′	36°	77′ 8″	9,960	7,560	4,090	2,00[illegible]
	110′	27°	61′ 6″	8,470	6,320	3,250	1,35[illegible]
	120′	13°	34′ 2″	7,240	5,280	2,520	—
130′	30′	78°	134′ 0″	52,380*	49,720	27,490	19,8[illegible]
	40′	74°	131′ 6″	37,280*	31,720	18,540	13,0[illegible]
	50′	69°	128′ 1″	27,960*	22,660	13,470	9,1[illegible]
	60′	64°	123′ 10″	21,490	17,200	10,210	6,6[illegible]
	70′	59°	118′ 4″	17,130	13,550	7,940	4,9[illegible]
	80′	54°	111′ 8″	14,000	10,930	6,260	3,6[illegible]
	90′	48°	103′ 7″	11,650	8,960	4,960	2,6[illegible]
	100′	42°	93′ 7″	9,810	7,420	3,940	1,8[illegible]
	110′	35°	80′ 11″	8,340	6,190	3,090	1,2[illegible]
	120′	26°	63′ 10″	7,120	5,160	2,390	—
	130′	13°	35′ 4″	6,080	4,290	1,770	—
140′	30′	79°	144′ 2″	47,730*	47,730*	27,340	19,7[illegible]
	40′	75°	141′ 11″	37,090*	31,610	18,380	12,8[illegible]
	50′	71°	138′ 10″	27,760*	22,530	13,310	8,9[illegible]
	60′	66°	134′ 10″	21,360	17,060	10,050	6,4[illegible]
	70′	62°	129′ 11″	16,980	13,400	7,770	4,7[illegible]
	80′	57°	123′ 11″	31,850	10,780	6,090	3,4[illegible]
	90′	52°	116′ 8″	11,500	8,810	4,800	2,4[illegible]
	100′	46°	107′ 11″	9,660	7,270	3,770	1,6[illegible]
	110′	40°	97′ 4″	8,190	6,040	2,930	1,0[illegible]
	120′	33°	84′ 0″	6,980	5,020	2,230	—
	130′	25°	66′ 1″	5,960	4,160	1,640	—
	140′	12°	36′ 5″	5,070	3,420	1,110	—
150′	35′	78°	153′ 5″	39,180*	38,750	21,980	16,[illegible]
	40′	76°	152′ 4″	34,890*	31,500*	18,220	12,[illegible]
	50′	72°	149′ 4″	27,320*	22,400	13,150	8,[illegible]
	60′	68°	145′ 8″	21,210*	16,920	9,880	6,[illegible]
	70′	64°	141′ 1″	16,840*	13,250	7,600	4,[illegible]
	80′	59°	135′ 8″	13,700*	10,630	5,920	3,[illegible]
	90′	55°	129′ 2″	11,350*	8,660	4,620	2,[illegible]
	100′	50°	121′ 5″	9,510*	7,120	3,600	1,[illegible]
	110′	45°	112′ 1″	8,030*	5,880	2,760	—
	120′	39°	100′ 11″	6,820*	4,870	2,070	—
	130′	32°	86′ 11″	5,770*	4,010	1,470	—
	140′	24°	68′ 4″	4,810*	3,290	—	—
	150′	12°	37′ 6″	4,060*	2,640	—	—
160′	35′	79°	163′ 7″	36,070*	36,070*	21,830	15,[illegible]
	40′	77°	162′ 6″	31,450*	31,390	18,060	12,[illegible]
	50′	73°	159′ 10″	25,070*	22,270	12,980	8,[illegible]
	60′	69°	156′ 5″	20,110*	16,770	9,710	6,[illegible]
	70′	65°	152′ 2″	16,680*	13,100	7,430	4,[illegible]
	80′	61°	147′ 2″	12,760*	10,470	5,740	3,[illegible]
	90′	57°	141′ 2″	10,520*	8,500	4,450	2,[illegible]
	100′	53°	134′ 5″	8,660*	6,960	3,420	1,[illegible]
	110′	48°	125′ 11″	7,110*	5,720	2,590	—
	120′	43°	116′ 2″	5,830*	4,710	1,890	—
	130′	38°	104′ 5″	4,770*	3,860	1,300	—
	150′	23°	70′ 5″	3,130*	2,500	—	—
	160′	11°	38′ 6″	2,540*	1,940	—	—

(Courtesy of Link-Belt Speeder.)

TABLE 16-3 Capacities of a 82 Ton Truck Crane With Tubular "Hi-Lite" Boom

A	W Boom Point Height	On Outriggers Side	On Outriggers Rear	On Tires Side	On Tires Rear
8°	45' 7"	164,000*	164,000*	77,590*	119,850*
6°	45' 4"	153,080*	153,380*	72,250*	115,550*
5°	45' 2"	143,280*	143,560*	67,540*	104,770
3°	44'11"	134,610*	134,890*	63,400*	95,160
6°	43' 0"	102,000*	103,260*	45,470	64,770
7°	40' 4"	80,280*	81,300*	33,570	48,620
8°	36' 7"	65,000*	65,850*	26,250	38,580
8°	31' 4"	51,800	55,060*	21,290	31,720
4°	23' 0"	42,390	45,540*	17,670	26,720
9°	55' 8"	143,850*	143,850*	72,200*	115,270*
8°	55' 6"	140,420*	140,420*	67,510*	105,000
7°	55' 4"	134,590*	137,210*	63,360*	95,370
°	53'10"	101,890*	103,260*	45,640	64,920
4°	51' 8"	80,370*	81,400*	33,700	48,740
3°	49' 0"	65,090*	65,940*	26,370	38,690
°	45' 5"	52,060	55,160*	21,410	31,840
3°	40'10"	42,660	47,210*	17,820	26,860
4°	34' 6"	35,930	41,100*	15,090	23,070
2°	25' 0"	30,850	34,680*	12,930	20,070
9°	65' 6"	130,000*	130,000*	63,530*	95,860
4°	64' 4"	102,090*	103,380*	45,980	65,230
9°	62' 7"	80,580*	81,610*	33,940	48,960
4°	60' 5"	65,240*	66,100*	26,560	38,860
3°	57' 8"	52,360	55,290*	21,560	31,970
2°	54' 2"	42,910	47,330*	17,940	26,980
9°	44' 6"	31,060	36,340*	13,050	20,190
0°	26'10"	23,890	29,090*	9,850	15,750
6°	74' 7"	100,500*	100,500*	45,960	65,200
2°	73' 2"	80,490*	81,520*	33,900	48,910
8°	71' 5"	65,150*	65,990*	26,500	38,800
°	69' 1"	52,400	55,180*	21,500	31,910
°	66' 4"	42,930	47,220*	17,890	26,910
8°	58'11"	31,070	36,270*	13,000	20,130
5°	47'11"	23,930	29,060*	9,840	15,740
8°	28' 6"	19,110	23,920*	7,590	12,610
3°	84'11"	95,000*	95,000*	45,910	65,130
4°	83' 8"	80,350*	81,380*	33,820	48,820
°	82' 1"	65,000*	65,860*	26,410	38,700
°	80' 1"	52,380	55,040*	21,400	31,800
°	77'10"	42,890	47,080*	17,780	26,800
°	71' 8"	31,020	36,120*	12,900	20,030
°	63' 2"	23,880	28,940*	9,750	14,640
4°	51' 1"	19,100	23,850*	7,540	12,550
°	30' 0"	15,630	20,000*	5,860	10,230
°	95' 1"	90,000*	90,000*	45,820	65,040
°	94' 0"	77,030*	77,030*	33,720	48,700
°	92' 7"	64,840*	65,690*	26,290	38,570
°	90'11"	52,340	54,870*	21,270	31,670
°	88'11"	42,830	46,910*	17,650	26,670
°	83' 8"	30,940	35,950*	12,770	19,890
°	76' 8"	23,790	28,760*	9,620	15,510
°	67' 2"	19,010	23,690*	7,420	12,430
°	54' 0"	15,580	19,890*	5,780	10,150
°	31' 6"	12,970	16,920*	4,480	8,350
°	104' 4"	72,720*	72,720*	33,590	48,570
°	103' 0"	64,640*	65,490*	26,160	38,430
°	101' 6"	52,270	54,670*	21,130	31,520
°	99' 8"	42,750	46,720*	17,500	26,520
°	95' 1"	30,830	35,760*	12,620	19,730
°	89' 1"	23,680	28,580*	9,470	15,350
°	81' 4"	18,700	23,500*	7,270	12,280
°	71' 0"	15,470	19,710*	5,640	10,010
°	56'10"	12,880	16,770*	4,380	8,240
°	32'10"	10,840	14,400*	3,340	6,810

Boom L	Boom R	Boom A	W Boom Point Height	On Outriggers Side	On Outriggers Rear	On Tires Side	On Tires Rear
110'	25'	79°	114' 6"	68,000*	68,000*	33,460	48,430
	30'	76°	113' 5"	62,600*	62,600*	26,010	38,280
	35'	73°	112' 0"	52,190	54,470*	20,980	31,370
	40'	71°	110' 5"	42,650	46,510*	17,340	26,350
	50'	65°	106' 4"	30,710	35,560*	12,450	19,570
	60'	59°	101' 0"	23,550	28,370*	9,300	15,180
	70'	53°	94' 2"	18,760	23,300*	7,100	12,120
	80'	46°	85' 7"	15,340	19,510*	5,480	9,840
	90'	38°	74' 7"	12,760	16,580*	4,220	8,090
	100'	29°	59' 5"	10,740	14,250*	3,220	6,690
	110'	14°	34' 1"	9,090	12,290*	2,370	5,520
120'	30'	77°	123' 7"	59,140*	59,140*	25,860	38,120
	35'	75°	122' 5"	52,100	54,270*	20,810	31,200
	40'	72°	120'11"	42,540	46,290*	17,180	26,180
	50'	67°	117' 2"	30,590	35,320*	12,280	19,390
	60'	62°	112' 6"	23,410	28,150*	9,130	15,010
	70'	56°	106' 6"	18,610	23,080*	6,930	11,940
	80'	50°	99' 1"	15,190	19,300*	5,310	9,670
	90'	44°	89'10"	12,610	16,370*	4,050	7,920
	100'	36°	77'11"	10,590	14,040*	3,050	6,520
	110'	27°	61'11"	8,970	12,130*	2,230	5,380
	120'	14°	35' 5"	7,610	10,510*	1,530	4,410
130'	30'	78°	133'11"	55,590*	55,590*	25,690	37,950
	35'	76°	132' 8"	51,160*	51,160*	20,640	31,020
	40'	74°	131' 5"	42,420	45,400*	17,000	26,000
	50'	69°	128' 0"	30,450	35,100*	12,100	19,210
	60'	64°	123' 8"	23,260	27,930*	8,940	14,820
	70'	59°	118' 4"	18,460	22,850*	6,750	11,750
	80'	54°	111' 8"	15,030	19,070*	5,120	9,480
	90'	48°	103' 7"	12,450	16,160*	3,870	7,730
	100'	42°	93' 8"	10,430	13,830*	2,880	6,340
	110'	35°	81' 2"	8,820	11,920*	2,060	5,210
	120'	26°	64' 4"	7,480	10,320*	----	4,260
	130'	13°	36' 7"	6,350	8,900*	----	3,430
140'	30'	79°	144' 1"	52,170*	52,170*	25,530	37,780
	35'	77°	143' 0"	47,450*	47,450*	20,470	30,850
	40'	75°	141'10"	42,300	42,790*	16,820	25,820
	50'	71°	138' 8"	30,300	34,780*	11,910	19,020
	60'	66°	134' 8"	23,100	27,680*	8,760	14,630
	70'	62°	129'10"	18,290	22,610*	6,560	11,560
	80'	57°	123'11"	14,860	18,840*	4,930	9,290
	90'	52°	116' 8"	12,280	15,930*	3,680	7,540
	100'	46°	108' 0"	10,260	13,600*	2,690	6,150
	110'	41°	97' 6"	8,650	11,700*	1,880	5,020
	120'	34°	84' 2"	7,320	10,100*	----	4,080
	130'	25°	66' 7"	6,200	8,750*	----	3,270
	140'	13°	37' 8"	5,240	7,480*	----	2,570
150'	35'	78°	153' 4"	44,330*	44,330*	20,290	30,660
	40'	76°	152' 1"	39,730*	39,730*	16,640	25,630
	50'	72°	149' 2"	30,160	31,690*	11,720	18,830
	60'	68°	145' 7"	22,940	25,900*	8,560	14,430
	70'	64°	141' 1"	18,130	21,410*	6,360	11,360
	80'	59°	135' 7"	14,680	16,860*	4,740	9,090
	90'	55°	129' 2"	12,100	14,410*	3,490	7,340
	100'	50°	121' 5"	10,090	12,230*	2,490	5,960
	110'	45°	112' 2"	8,470	10,360*	1,680	4,830
	120'	39°	101' 1"	7,150	8,880*	----	3,890
	130'	32°	87' 2"	6,040	7,610*	----	3,090
	140'	24°	68'10"	5,090	6,510*	----	2,400
	150'	12°	38'10"	4,250	5,610*	----	1,780

Boom L	Boom R	Boom A	W Boom Point Height	On Outriggers Side	On Outriggers Rear	On Tires Side	On Tires Rear
160' ②	35'	79°	163' 6"	41,120*	41,120*	20,110	30,480
	40'	77°	162' 5"	36,900*	36,900*	16,450	25,440
	50'	73°	159' 8"	29,270*	29,270*	11,530	18,630
	60'	69°	156' 4"	22,780	24,000*	8,370	14,240
	70'	65°	152' 1"	17,960	19,740*	6,160	11,160
	80'	61°	147' 1"	14,510	15,660*	4,540	8,890
	90'	57°	141' 2"	11,920	13,140*	3,290	7,140
	100'	53°	134' 2"	9,900	11,020*	2,290	5,760
	110'	48°	126' 0"	8,290	9,280*	1,490	4,630
	120'	43°	116' 4"	6,960	7,770*	----	3,690
	130'	38°	104' 7"	5,860	6,530*	----	2,890
	140'	31°	90' 1"	4,910	5,500*	----	2,210
	150'	24°	70'11"	4,100	4,600*	----	1,610
	160'	12°	39'11"	3,370	3,920*	----	----
170' ②	35'	79°	173' 8"	38,110*	38,110*	19,930	30,290
	40'	78°	172' 7"	34,270*	34,270*	16,270	25,250
	50'	74°	170' 1"	27,170*	27,170*	11,340	18,430
	60'	71°	166'11"	22,100*	22,100*	8,170	14,030
	70'	67°	163' 1"	17,780	19,920*	5,960	10,960
	80'	63°	158' 5"	14,330	15,770*	4,330	8,690
	90'	59°	153' 0"	11,740	13,100*	3,080	6,940
	100'	55°	146' 7"	9,720	10,840*	2,090	5,550
	110'	51°	139' 1"	8,100	8,980*	----	4,420
	120'	47°	130' 5"	6,780	7,460*	----	3,480
	130'	42°	120' 2"	5,670	6,240*	----	2,690
	140'	37°	107'11"	4,730	5,200*	----	2,010
	150'	30°	92'10"	3,920	4,310*	----	----
	160'	23°	73' 0"	3,200	3,520*	----	----
	170'	12°	40'11"	2,560	2,820*	----	----
180' ②	40'	78°	182'11"	31,110*	31,110*	16,070	25,060
	50'	75°	180' 6"	25,100*	25,100*	11,140	18,230
	60'	72°	177' 6"	20,350*	20,350*	7,970	13,830
	70'	68°	173'11"	16,600	19,360*	5,760	10,750
	80'	65°	169' 6"	14,140	15,550*	4,130	8,480
	90'	61°	164' 6"	11,550	12,710*	2,880	6,730
	100'	58°	158' 7"	9,530	10,480*	1,890	5,340
	110'	54°	151' 8"	7,910	8,700*	----	4,210
	120'	50°	143'10"	6,580	7,240*	----	3,280
	130'	45°	134' 8"	5,480	6,030*	----	2,490
	140'	41°	123'11"	4,540	4,990*	----	1,810
	150'	35°	111' 2"	3,730	4,100*	----	----
	160'	30°	95' 6"	3,020	3,320*	----	----
	170'	22°	75' 0"	2,400	2,640*	----	----
	180'	11°	41'11"	1,830	2,010*	----	----

of Link-Belt Speeder.)

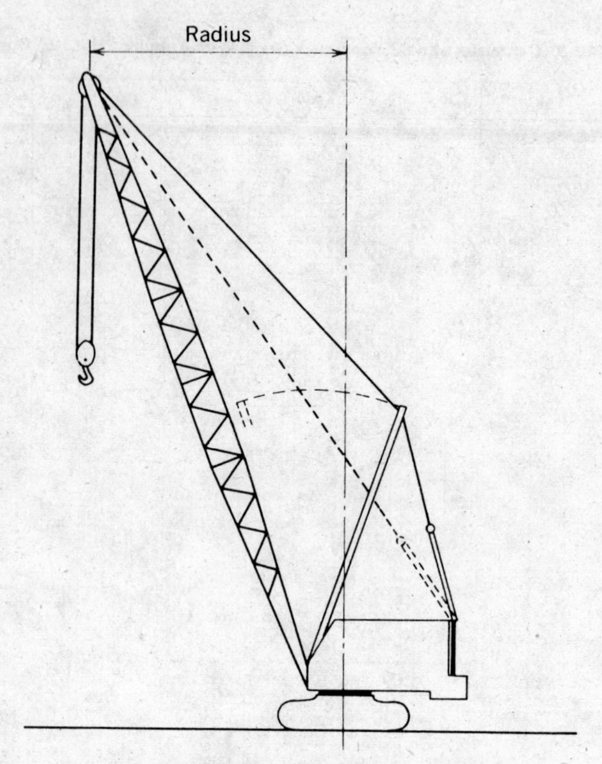

Fig. 16-2 Load radius for a crawler crane.

16-3 Crawler Cranes

Crawler cranes having high capacities are occasionally used to erect precast members. Crawler cranes must be moved from site to site on trailers, and for this reason, they are generally more costly and more difficult to use in developed areas. The load radius for a crawler crane is measured in a manner similar to that of a truck crane (*see* Fig. 16-2)

The capacities of a 103.5 ton crawler crane are given in Table 16-4. Crawler cranes working under maximum lifts, like truck cranes, must be level and be on firm ground.

16-4 Floating Cranes

Floating cranes with very large capacities are available in some localities. The girder shown being erected in Fig. 16-3 weighs 230 tons and is approximately 200 ft long. Some floating cranes have capacities as high as 600 tons. The

TABLE 16-4 Capacities of a 103.5 Ton Crawler Crane With Tubular "Hi-Lite" Boom.

OM		W Boom Point Height	Ctwt. "A"	Ctwt. "AB"
adius	Angle			
13'	80°	56' 0"	207,000*	207,000*
14'	79°	55'10"	185,090*	207,000*
15'	78°	55' 8"	160,470*	207,000*
16'	77°	55' 6"	144,660	200,000*
17'	76°	55' 3"	129,310	187,910
18'	74°	54'10"	116,830	169,980
19'	73°	54' 7"	106,490	155,050
20'	72°	54' 4"	97,780	144,720
25'	66°	52' 5"	69,010	102,660
30'	59°	49'10"	52,910	79,140
35'	53°	46' 5"	42,600	64,090
40'	45°	42' 1"	35,420	53,620
50'	25°	27' 8"	26,000	39,940
14'	81°	66' 0"	185,090*	201,160*
15'	80°	65'10"	160,470*	197,940*
16'	79°	65' 8"	144,660	197,500*
17'	78°	65' 5"	129,310	187,910
18'	77°	65' 2"	116,830	169,980
19'	76°	64'11"	106,490	155,050
20'	75°	64' 9"	97,780	144,720
25'	70°	63' 2"	69,010	102,660
30'	65°	61' 1"	52,910	79,140
35'	60°	58' 6"	42,600	64,090
40'	54°	55' 2"	35,420	53,620
50'	41°	45'11"	26,000	39,940
60'	23°	29' 9"	20,410	31,690
16'	81°	75'11"	144,660	179,500*
7'	80°	75' 8"	129,310	179,000*
18'	79°	75' 5"	116,830	169,980
9'	78°	75' 3"	106,490	155,050
20'	77°	75' 0"	97,780	144,720
25'	73°	73' 8"	69,010	102,660
30'	68°	71'11"	52,910	79,140
35'	64°	69' 9"	42,600	64,090
40'	60°	67' 1"	35,420	53,620
50'	50°	60' 0"	26,000	39,940
60'	38°	49' 6"	20,410	31,690
70'	21°	31' 8"	16,510	26,000
7'	81°	85' 9"	129,310	164,000*
8'	80°	85' 6"	116,830	162,000*
9'	79.5°	85' 5"	106,490	155,050
0'	79°	85' 3"	97,780	144,720
5'	75°	84' 1"	69,010	102,660
0'	72°	82' 7"	52,910	79,140
5'	68°	80' 9"	42,600	64,090
0'	64°	78' 6"	35,420	53,620
0'	55°	72' 7"	26,000	39,940
0'	46°	64' 5"	20,410	31,690
0'	35°	52' 9"	16,510	26,000
0'	20°	33' 5"	13,620	21,800
0'	81°	95' 8"	106,490	147,000*
0'	80°	95' 5"	97,780	139,290*
5'	77°	94' 5"	69,010	102,660
0'	74°	93' 1"	52,910	79,140
5'	70°	91' 5"	42.600	64,090
	67°	89' 6"	35.420	53,620
	60°	84' 5"	26,000	39,940
	52°	77' 8"	20,410	31,690
	43°	68' 7"	16,510	26,000
	33°	55'10"	13,620	21,800
	18°	35' 1"	11,370	18,560
	81°	105' 7"	97,780	127,550*
	78°	104' 8"	69,010	102,660
	75°	103' 6"	52,910	79,140
	72°	102' 0"	42,600	64,090
	69°	100' 3"	35,420	53,620
	63°	95'10"	26,000	39,940
	56°	90' 0"	20,410	31,690
	49°	82' 4"	16,500	25,980
	41°	72' 5"	13,580	21,770
	31°	58' 9"	11,350	18,540
	17°	36' 8"	9,550	15,970

BOOM			W Boom Point Height	Ctwt. "A"	Ctwt. "AB"
Length	Radius	Angle			
110'	25'	79°	114'10"	69,010	102,660
	30'	77°	113' 9"	52,910	79,140
	35'	74°	112' 5"	42,600	64,090
	40'	71°	110'11"	35,420	53,620
	50'	65°	106'11"	26,000	39,940
	60'	60°	101' 9"	20,410	31,670
	70'	54°	95' 2"	16,390	25,870
	80'	47°	86'10"	13,480	21,660
	90'	39°	76' 1"	11,260	18,460
	100'	30°	61' 6"	9,500	15,920
	110'	17°	38' 2"	8,040	13,830
120'	25'	80°	125' 0"	69,010	101,600*
	30'	78°	124' 0"	52,910	79,140
	35'	75°	122'10"	42,600	64,090
	40'	73°	121' 5"	35,420	53,620
	50'	68°	117'10"	26,000	39,940
	60'	63°	113' 2"	20,250	31,540
	70'	57°	107' 4"	16,260	25,740
	80'	51°	100' 1"	13,350	21,540
	90'	45°	91' 0"	11,140	18,330
	100'	37°	79' 6"	9,400	15,810
	110'	29°	64' 1"	7,970	13,760
	120'	16°	39' 7"	6,760	12,030
130	25'	81°	135' 2"	69,010	93,280*
	30'	79°	134' 3"	52,910	79,140
	35'	77°	133' 2"	42,600	64,090
	40'	74°	131'10"	35,390	53,620
	50'	70°	128' 7"	25,960	39,900
	60'	65°	124' 4"	20,110	31,400
	70'	60°	119' 1"	16,110	25,600
	80'	55°	112' 8"	13,210	21,390
	90'	49°	104' 9"	11,000	18,190
	100'	43°	95' 0"	9,260	15,670
	110'	36°	82'10"	7,840	13,640
	120'	27°	66' 7"	6,670	11,940
	130'	15°	41' 0"	5,640	10,490
140'	30'	80°	144' 7"	52,910	79,140
	35'	77°	143' 5"	42,540	64,030
	40'	75°	142' 3"	35,260	53,460
	50'	71°	137' 4"	25,820	39,750
	60'	67°	135' 4"	19,950	31,240
	70'	62°	130' 6"	15,960	25,440
	80'	57°	124' 8"	13,050	21,230
	90'	52°	117' 8"	10,840	18,030
	100'	47°	109' 2"	9,100	15,520
	110'	41°	98'11"	7,700	13,490
	120'	35°	86' 0"	6,530	11,800
	130'	26°	68'11"	5,540	10,380
	140'	15°	42' 3"	4,600	9,140
150'	30'	80°	154' 7"	52,900	75,890*
	35'	78°	153' 8"	42,400	63,900
	40'	76°	152' 6"	35,120	53,320
	50'	72°	149' 9"	25,600	39,600
	60'	68°	146' 2"	19,790	31,080
	70'	64°	141' 9"	15,790	25,280
	80'	60°	136' 5"	12,880	21,060
	90'	55°	130' 1"	10,670	17,870
	100'	51°	122' 6"	8,940	15,350
	110'	45°	113' 6"	7,530	13,320
	120'	40°	102' 7"	6,370	11,650
	130'	33°	89' 0"	5,390	10,230
	140'	26°	71' 3"	4,540	9,020
	150'	14°	43' 6"	3,780	7,950
160'	30'	81°	164' 9"	52,810	70,120*
	35'	79°	163'10"	42,290	63,780
	40'	77°	162' 9"	34,990	53,190
	50'	74°	160' 2"	25,520	39,460
	60'	70°	156'10"	19,640	30,930
	70'	66°	152' 9"	15,640	25,120
	80'	62°	147'10"	12,730	20,910
	90'	58°	142' 0"	10,520	17,710
	100'	53°	135' 2"	8,780	15,190
	110'	49°	127' 1"	7,370	13,170
	120'	44°	117' 7"	6,210	11,490
	130'	38°	106' 1"	5,240	10,080
	140'	32°	92' 0"	4,400	8,880
	150'	25°	73' 6"	3,670	7,830
	160'	14°	44' 9"	3,000	6,900

BOOM			W Boom Point Height	Ctwt. "A"	Ctwt. "AB"
Length	Radius	Angle			
170'	35'	80°	174' 0"	42,170	61,340*
	40'	78°	173' 0"	34,860	53,060
	50'	75°	170' 7"	25,380	39,310
	60'	71°	169' 6"	19,490	30,780
	70'	67°	163' 8"	15,480	24,970
	80'	64°	159' 1"	12,570	20,750
	90'	60°	153' 9"	10,360	17,550
	100'	56°	147' 5"	8,620	15,040
	110'	51°	140' 1"	7,220	13,010
	120'	47°	131' 7"	6,060	11,340
	130'	43°	121' 6"	5,090	9,930
	140'	37°	109' 6"	4,250	8,730
	150'	31°	94'10"	3,530	7,700
	160'	24°	75' 8"	2,890	6,790
	170'	13°	46' 0"	2,300	5,960
180'	35'	80°	184' 2"	42,040	55,650*
	40'	79°	183' 3"	34,720	51,970*
	50'	75°	180'11"	25,230	39,160
	60'	72°	178' 0"	19,340	30,630
	70'	69°	174' 5"	15,320	24,810
	80'	65°	170' 2"	12,410	20,590
	90'	62°	165' 2"	10,200	17,390
	100'	58°	159' 5"	8,460	14,880
	110'	54°	152' 8"	7,060	12,850
	120'	50°	144'10"	5,900	11,180
	130'	46°	135'10"	4,930	9,770
	140'	41°	125' 4"	4,100	8,580
	150'	36°	112'10"	3,380	7,550
	160'	30°	97' 7"	2,750	6,640
	170'	23°	77' 9"	2,180	5,840
	180'	13°	47' 1"	1,660	5,110
190'	35'	80°	194' 4"	41,900	49,950*
	40'	79°	193' 5"	34,580	46,080*
	50'	76°	191' 3"	25,080	39,010
	60'	73°	188' 7"	19,180	30,470
	70'	70°	185' 1"	15,160	24,650
	80'	67°	181' 2"	12,240	20,430
	90'	63°	176' 6"	10,030	17,220
	100'	60°	171' 0"	8,290	14,710
	110'	56°	164'10"	6,890	12,680
	120'	53°	157' 8"	5,730	11,010
	130'	49°	149' 5"	4,760	9,610
	140'	45°	140' 0"	3,940	8,420
	150'	40°	129' 0"	3,220	7,390
	160'	35°	116' 0"	2,590	6,490
	170'	30°	100' 2"	2,040	5,690
	180'	23°	79' 9"	1,540	4,980
	190'	13°	48' 3"	1,070	4,320
200'	40'	80°	203' 7"	34,430	45,200*
	50'	77°	201' 6"	24,920	35,830*
	60'	74°	198'11"	19,020	30,310
	70'	71°	195' 9"	14,990	24,480
	80'	68°	192' 0"	12,080	20,260
	90'	65°	187' 7"	9,860	17,050
	100'	61°	182' 6"	8,120	14,540
	110'	58°	176' 8"	6,720	12,510
	120'	55°	170' 1"	5,560	10,840
	130'	51°	162' 6"	4,590	9,440
	140'	47°	153'11"	3,770	8,240
	150'	43°	144' 1"	3,050	7,220
	160'	39°	132' 8"	2,430	6,320
	170'	34°	119' 2"	1,000	5,530
	180'	29°	102'10"	1,380	4,830
	190'	22°	81' 9"	—	4,190
	200'	12°	49' 4"	—	3,600

of Link-Belt Speeder.)

Fig. 16-3 Floating crane erecting a 230 ton girder on Crossbay Parkway Bridge in New York City.

Fig. 16-4 Girder launcher erecting a 70 ton, 130-ft long girder.

principal limitation on the use of floating cranes is the availability of sufficient depth of relatively calm water on which the crane can work.

16-5 Girder Launchers

Large precast bridge girders are sometimes erected with girder launchers such as the one shown in Fig. 16-4. The launchers are generally steel trusses, although aluminum has been used. For shorter spans with lighter girders, steel beams may be used rather than trusses.

There have been many different variations in the construction of launchers, but they all work on the same basic principle of moving over the already erected girders and cantilevering out to the next pier in order to provide an overhead structure that is sufficiently strong to support the necessary erection loads. The sequences in erecting a bridge with a girder launcher is illustrated in Fig. 16-5.

It should be noted that the launcher shown in Fig. 16-5 is approximately 1.7

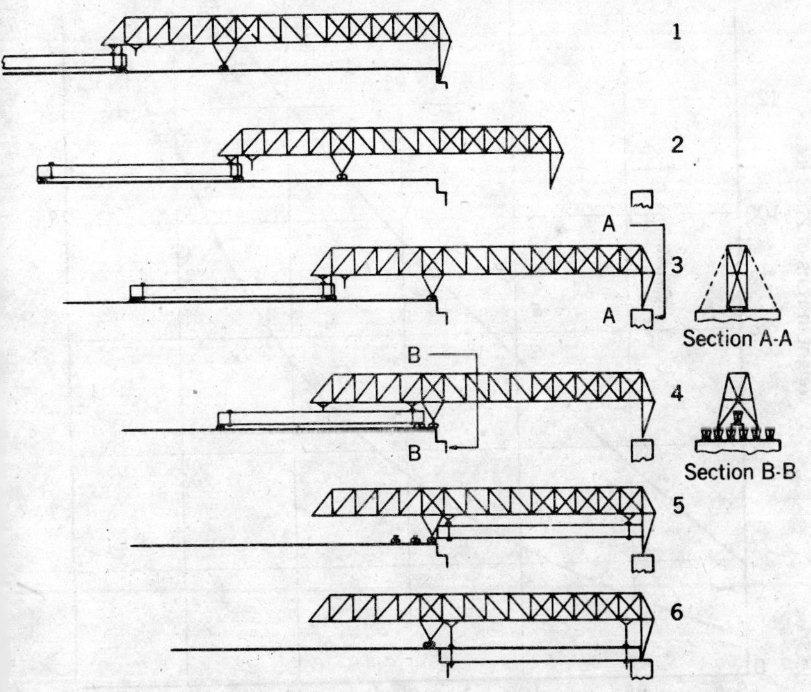

Fig. 16-5 Sequence in using a girder launcher. (1) Launcher assembled with girder connected for a counterweight. (2) Launcher and girder being moved into place for erecting first span. (3) Launcher in place for erecting first span. (4) Girder connected to launcher at outboard end. (5) Girder supported by launcher. (6) Girder erected.

times the length of the girder that is to be erected. A girder is used for a counterweight when the launcher is being positioned.

The approximate weight of concrete bridge girders suitable for normal launcher erection (wide top flanges) and that of the launching truss alone is shown in Fig. 16-6. The weight of the dollies, hoists, and other necessary equipment is approximately equal to that of the truss.

Girder launchers capable of erecting girders 220 ft long weighing 270 tons have been designed and proved to be economical. For the longer and heavier girders, guyed launchers having a central tower, as shown in Fig. 16-4 are considered to be most economical. For erecting shorter spans, cantilevered

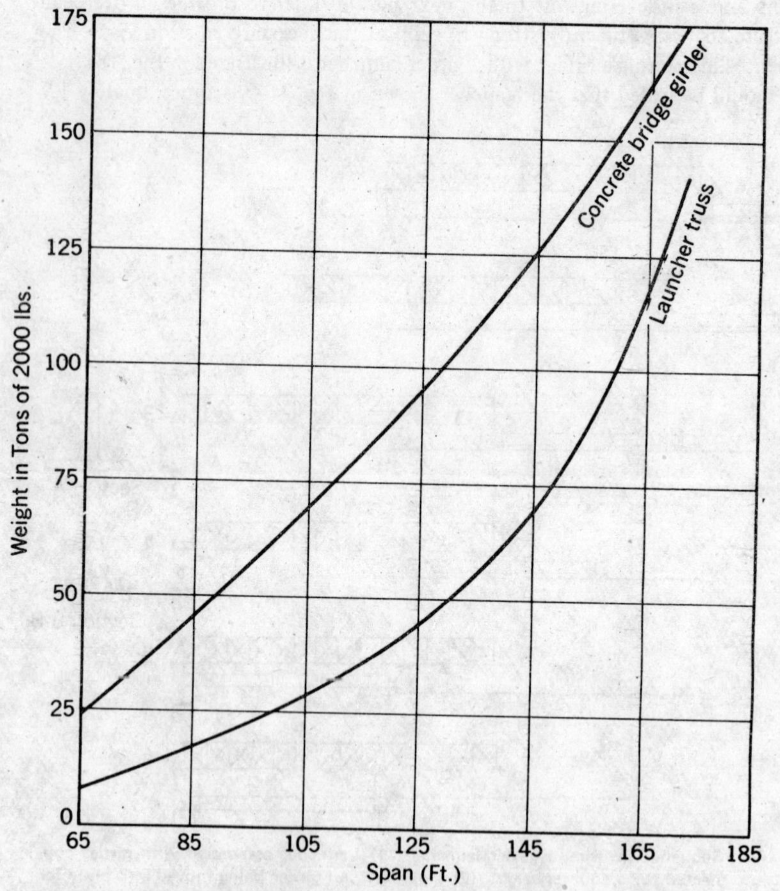

Fig. 16-6 Approximate bridge girder and launch-truss weights for various spans.

Fig. 16-7 Cantilevered truss launcher.

trusses, without the central tower as shown in Fig. 16-7, have proved to be economical and efficient.

16-6 Falsework

Falsework can be used to erect precast bridge girders economically under certain conditions. The falsework may consist of wood or steel towers which support steel or wood beams. The precast girders are normally rolled on to the beams and then slid transversely into place. Alternately, the girders may be precast in segments, assembled, stressed, and grouted on falsework. This later procedure eliminates the need of moving the assembled girder.

Many methods of sliding girders transversely have been used. The details of the scheme used on the Bamako Bridge are shown in Fig. 16-8 (Ref. 3).

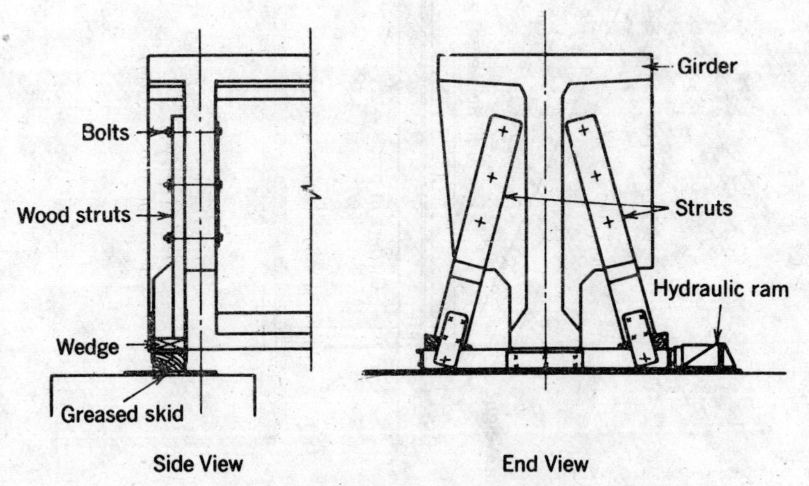

Fig. 16-8 Bamako bridge apparatus for transverse sliding of girder.

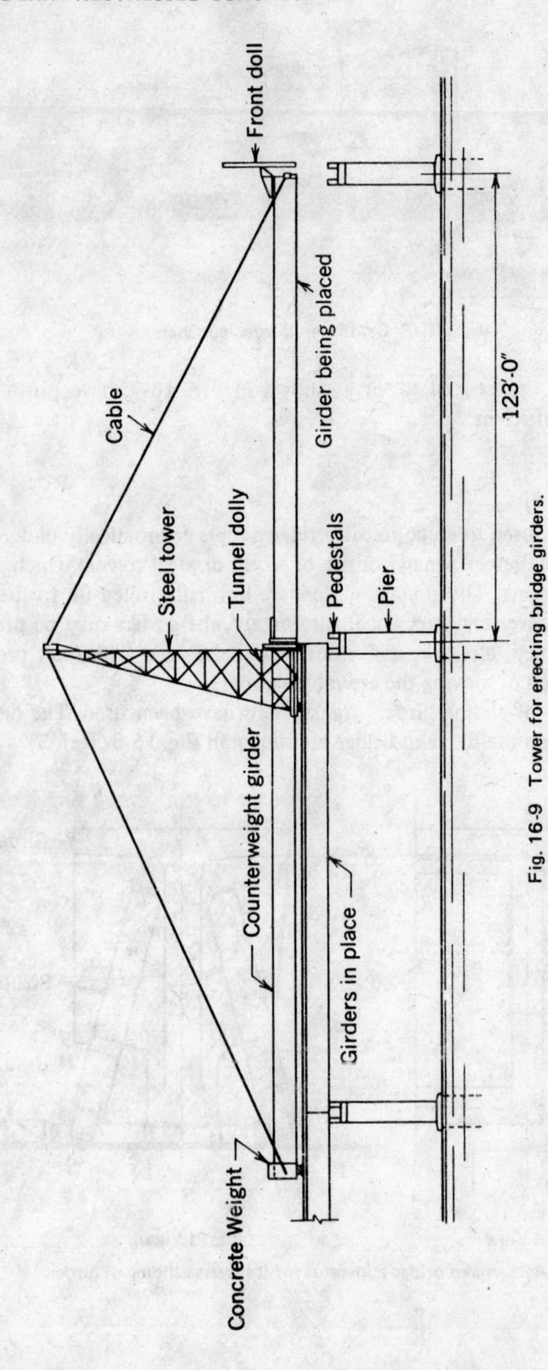

Fig. 16-9 Tower for erecting bridge girders.

16-7 Cable Ways and Highlines

Cable ways and highlines are occasionally considered for use in erecting precast concrete bridge girders. The cost of cableways is very high and they are not a type of device that is moved at low cost.

Cableways find their use on dam construction projects where they can be installed in a single location and used for a period of several months or years in handling loads of nominal magnitudes (say 25 tons maximum). Sites which particularly lend themselves to cableways are normally deep narrow gorges. Rather large sags with relatively small (at least not very large) spans are used.

Cableways could conceivably be used to good advantage to erect precast segments in cantilever construction, but it is doubtful they are practical for use in erecting complete girders of anything but short spans.

16-8 Towers

Rolling towers, as illustrated in Fig. 16-9, offer a means of erecting bridge girders of short to moderate length. The scheme utilizes a girder as counterweight in combination with an additional nominal counterweight. It has the disadvantage that the girders over which the tower moves during erection must bear the weight of two girders, the counterweight, and the tower complete with rigging. The result of this disadvantage is that the method is not feasible for the erection of long span girders where dead load is so important.

REFERENCES

1. Société Technique Pour L'Utilisation de la Precontrainte, Private Communication (July 7, 1966).

2. Stergiou, Paul, "Launcher Erects Post-Tensioned Girders," *Civil Engineering*, pp. 66–67 (Aug. 1967).

3. Muller, Jean, "Engineering Feats With Post-Tensioning," *Journal of the Prestressed Concrete Institute*, 5, No. 4, 12–40 (Dec. 1960).

Appendices

A. Recommendations for Estimating Prestress Losses

B. Excerpts from Standard Specifications for Highway Bridges

C. Condensed Method for Computation of the Loss of Prestress

Appendix A

Recommendations for Estimating Prestress Losses

Prepared by

PCI Committee on Prestress Losses

H. KENT PRESTON
Chairman

JAMES M. BARKER
HENRY C. BOECKER, JR.*
R. G. DULL
HARRY H. EDWARDS
TI HUANG
JAIME IRAGORRI
R. O. KASTEN†
HEINZ P. KORETZKY
PAUL E. KRAEMER‡
DONALD D. MAGURA‡
F. R. PREECE
MARIO G. SUAREZ
PAUL ZIA

* Replaced by Mario G. Suarez.
† Replaced by R. G. Dull.
‡ Previous Chairmen.

This PCI Committee report summarizes data on creep and shrinkage of concrete and steel relaxation, and presents both a general and a simplified design procedure for using these data in estimating loss of prestress after any given time period. A Commentary explains the design provisions. Detailed design examples for pretensioned and post-tensioned concrete structures explain the procedures.

CONTENTS

COMMITTEE STATEMENT

This recommended practice is intended to give the design engineer a comprehensive summary of research data applicable to estimating loss of prestress. It presents a general method whereby losses are calculated as a function of time.

This report contains information and procedures for estimating prestress losses in building applications. The general method is applicable to bridges, although there are some differences between it and the AASHTO *Standard Specifications for Highway Bridges* with respect to individual loss components.

A precise determination of stress losses in prestressed concrete members is a complicated problem because the rate of loss due to one factor, such as relaxation of tendons, is continually being altered by changes in stress due to other factors, such as creep of concrete. Rate of creep in its turn is altered by change in tendon stress. It is extremely difficult to separate the net amount of loss due to each factor under different conditions of stress, environment, loading, and other uncertain factors.

In addition to the foregoing uncertainties due to interaction of shrinkage, creep, and relaxation, physical conditions, such as variations in actual properties of concrete made to the same specified strength, can vary the total loss. As a result, the computed values for prestress loss are not necessarily exact, but the procedures here presented will provide more accurate results than by previous methods which gave no consideration to the actual stress levels in concrete and tendons.

An error in computing losses can affect service conditions such as camber, deflection, and cracking. It has no effect on the ultimate strength of a flexural member unless the tendons are unbonded or the final stress after losses is less than $0.5f_{pu}$.

It is not suggested that the information and procedures in this report provide the only satisfactory solution to this complicated problem. They do represent an up-to-date compromise by the committee of diverse opinions, experience and research results into relatively easy to follow design formulas, parameters, and computations.

CHAPTER 1—GENERAL ASPECTS RELATED TO PRESTRESS LOSSES

1.1—Tensioning of Prestressing Steel

1.1.1—Pretensioned construction

For deflected prestressing steel, loss *DEF*, occurring at the deflecting devices, should be taken into account.

1.1.2—Friction in post-tensioned construction

Loss due to friction in post-tensioned construction should be based upon wobble and curvature coefficients given below, and verified during stressing operations. The losses due to friction between the prestressing steel and the duct enclosure may be estimated by the following equation:

$$FR = T_o[1 - e^{-(Kl_{tx} + \mu\alpha)}] \qquad (1)$$

When $(Kl_{tx} + \mu\alpha)$ is not greater than 0.3, the following equation may be used:

$$FR = T_o(Kl_{tx} + \mu\alpha) \qquad (2)$$

Table 1 gives a summary of friction coefficients for various post-tensioning tendons.

1.2—Anchorage

Loss *ANC*, due to movement of prestressing steel in the end anchorage, should be taken into account. Slip at the anchorage will depend upon the particular prestressing system utilized and will not be a function of time. Realistic allowance should be made for slip or take-up as recommended for a given system of anchorage.

1.3—Transfer of Prestress

Loss due to elastic shortening may be calculated according to the provisions in this recommended practice. The concrete shortening should also include that resulting from subsequent stressing of prestressing steel.

1.4—Effect of Members in Structures

Loss of prestress of a member may be affected by connection to other structural elements or composite action with cast-in-place concrete. Change in prestress force due to these factors should be taken into account based on a rational procedure that considers equilibrium of forces and strain compatibility.

Table 1. Friction coefficients for post-tensioning tendons.

Type of tendon	Wobble coefficient, K, per foot	Curvature coefficient, μ
Tendons in flexible metal sheathing		
Wire tendons	0.0010 - 0.0015	0.15 - 0.25
7-wire strand	0.0005 - 0.0020	0.15 - 0.25
High strength bars	0.0001 - 0.0006	0.08 - 0.30
Tendons in rigid metal duct		
7-wire strand	0.0002	0.15 - 0.25
Pre-greased tendons		
Wire tendons and 7-wire strand	0.0003 - 0.0020	0.05 - 0.15
Mastic-coated tendons		
Wire tendons and 7-wire strand	0.0010 - 0.0020	0.05 - 0.15

CHAPTER 2—GENERAL METHOD FOR COMPUTING PRESTRESS LOSSES

2.1—Scope

2.1.1–Materials

2.1.1.1–*Lightweight concrete*–Lightweight aggregate concrete with a unit weight between 90 and 125 lb per cu ft where the unit weight varies because of replacement of lightweight fines with normal weight sand.

2.1.1.2–*Normal weight concrete*–Concrete with an approximate unit weight of 145 lb per cu ft where all aggregates are normal weight concrete aggregates.

2.1.1.3–*Prestressing Steel*–High strength prestressing steel that has been subjected to the stress-relieving process, or to processes resulting in low relaxation characteristics.

2.1.2– Prestressed units

Linearly prestressed members only. Excluded are closed sections prestressed circumferentially.

2.1.3–Curing

2.1.3.1–*Moist cure*–Impermeable membrane curing or other methods to prevent the loss of moisture from the concrete.

2.1.3.2–*Accelerated cure*—Curing in which the temperature of the concrete is elevated to not more than 160F for a period of approximately 18 hours, and steps are taken to retain moisture.

2.1.4–Environment

Prestressed concrete subjected to seasonal fluctuations of temperature and humidity in the open air or to nominal room conditions is covered.

The values for *UCR* and *USH* are based on an average ambient relative humidity of 70 percent.

2.2—Total Loss

2.2.1–Pretensioned construction

$$TL = ANC + DEF + ES + \sum_{t_1} (CR + SH + RET)$$

2.2.2–Post-tensioned construction

$$TL = FR + ANC + ES + \sum_{t_1}^{t} (CR + SH + RET)$$

2.3—Loss Due to Elastic Shortening (*ES*)

Loss of prestress due to elastic shortening of the concrete should be calculated based on the modulus of elasticity of the concrete at the time the prestress force is applied.

$$ES = f_{cr}(E_s/E_{ci}) \quad (5)$$

2.3.1–Pretensioned construction

In calculating shortening, the loss of prestress shall be based upon the concrete stress at the centroid of the prestressing force at the section of the member under consideration.

This stress, f_{cr}, is the compressive stress due to the prestressing force that is acting immediately after the prestress force is applied minus the stress due to all dead load acting at that time.

2.3.2—Post-tensioned construction

The average concrete stress between anchorages along each element shall be used in calculating shortening.

2.4—Time-Dependent Losses (General)

Prestress losses due to steel relaxation and creep and shrinkage of concrete are inter-dependent and are time-dependent. To account for changes of these effects with time, a step-by-step procedure can be used with the time interval increasing with age of the concrete. Shrinkage from the time when curing is stopped until the time when the concrete is prestressed should be deducted from the total calculated shrinkage for post-tensioned construction. It is recommended that a minimum of four time intervals be used as shown in Table 2.

When significant changes in loading are expected, time intervals other than those recommended should be used. Also, it is neither necessary, nor always desirable, to assume that the design live load is continually present. The four time intervals above are recommended for minimum non-computerized calculations.

2.5—Loss Due to Creep of Concrete *(CR)*

2.5.1—Loss over each step

Loss over each time interval is given by

$$CR = (UCR)(SCF)(MCF) \times (PCR)(f_c) \quad (6)$$

where f_c is the net concrete compressive stress at the center of gravity of the prestressing force at time t_1,

Table 2. Minimum time intervals.

Step	Beginning time, t_1	End time, t
1	Pretensioned anchorage of prestressing steel Post-tensioned: end of curing of concrete	Age at prestressin of concrete
2	End of Step 1	Age = 30 days, c time when a men ber is subjected t load in addition t its own weight
3	End of Step 2	Age = 1 year
4	End of Step 3	End of service li

taking into account the loss of pr stress force occurring over the pr ceding time interval.

The concrete stress f_c at the time shall also include change in applie load during the preceding time i terval. Do not include the fact *MCF* for accelerated cured concret

2.5.2—Ultimate creep loss

2.5.2.1—*Normal weight concre (UCR)*

Moist cure not exceeding 7 days:

$$UCR = 95 - 20E_c/10^6 \geqq 11$$

Accelerated cure:

$$UCR = 63 - 20E_c/10^6 \geqq 11$$

2.5.2.2—*Lightweight concrete (UC*

Moist cure not exceeding 7 days:

$$UCR = 76 - 20E_c/10^6 \geqq 11$$

Accelerated cure:

$$UCR = 63 - 20E_c/10^6 \geqq 11 \quad ($$

Table 3. Creep factors for various volume to surface ratios.

Volume to surface ratio, in.	Creep factor *SCF*
1	1.05
2	0.96
3	0.87
4	0.77
5	0.68
>5	0.68

Table 4. Creep factors for various ages of prestress and periods of cure.

Age of prestress transfer, days	Period of cure, days	Creep factor, *MCF*
3	3	1.14
5	5	1.07
7	7	1.00
10	7	0.96
20	7	0.84
30	7	0.72
40	7	0.60

Table 5. Variation of creep with time after prestress transfer.

Time after prestress transfer, days	Portion of ultimate creep, *AUC*
1	0.08
2	0.15
5	0.18
7	0.23
10	0.24
20	0.30
30	0.35
60	0.45
90	0.51
180	0.61
365	0.74
End of service life	1.00

Table 6. Shrinkage factors for various volume to surface ratios.

Volume to surface ratio, in.	Shrinkage factor *SSF*
1	1.04
2	0.96
3	0.86
4	0.77
5	0.69
6	0.60

2.5.3—Effect of size and shape of member (SCF)

To account for the effect of size and shape of the prestressed members, the value of SCF in Eq. (6) is given in Table 3.

2.5.4—Effect of age at prestress and length of cure (MCF)

To account for effects due to the age at prestress of moist cured concrete and the length of the moist cure, the value of MCF in Eq. (6) is given in Table 4. The factors in this table do *not* apply to accelerated cured concretes nor are they applicable as shrinkage factors.

2.5.5—Variation of creep with time (AUC)

The variation of creep with time shall be estimated by the values given in Table 5. Linear interpolation shall be used between the values listed.

2.5.6—Amount of creep over each step (PCR)

The portion of ultimate creep over the time interval t_1 to t, PCR in Eq. (6), is given by the following equation:

$$PCR = (AUC)_t - (AUC)_{t_1} \quad (11)$$

2.6—Loss Due to Shrinkage of Concrete (*SH*)

2.6.1—Loss over each step

Loss over each time interval is given by

$$SH = (USH)(SSF)(PSH) \quad (12)$$

2.6.2—Ultimate loss due to shrinkage of concrete

The following equations apply to

both moist cured and accelerated cured concretes.

2.6.2.1—*Normal weight concrete (USH)*

$$USH = 27{,}000 - 3000E_c/10^6 \quad (13)$$

but not less than 12,000 psi.

2.6.2.2—*Lightweight concrete (USH)*

$$USH = 41{,}000 - 10{,}000E_c/10^6 \quad (14)$$

but not less than 12,000 psi.

2.6.3—Effect of size and shape of member (SSF)

To account for effects due to the size and shape of the prestressed member, the value of *SSF* in Eq. (12) is given in Table 6.

2.6.4—Variation of shrinkage with time (AUS)

The variation of shrinkage with time shall be estimated by the values given in Table 7. Linear interpolation shall be used between the values listed.

2.6.5—Amount of shrinkage over each step (PSH)

The portion of ultimate shrinkage over the time interval t_1 to t, *PSH* in Eq. (12), is given by the following equation:

$$PSH = (AUS)_t - (AUS)_{t_1} \quad (15)$$

2.7—Loss Due to Steel Relaxation *(RET)*

Loss of prestress due to steel relaxation over the time interval t_1 to t may be estimated using the following equations. (For mathematical correctness, the value for t_1 at the time of anchorage of the prestressing steel shall be taken as 1/24 of a day so that log t_1 at this time equals zero.)

Table 7. Shrinkage coefficients for various curing times.

Time after end of curing, days	Portion of ultimate shrinkage, *AUS*
1	0.08
3	0.15
5	0.20
7	0.22
10	0.27
20	0.36
30	0.42
60	0.55
90	0.62
180	0.68
365	0.86
End of service life	1.00

2.7.1—Stress-relieved steel

$$RET = f_{st}\{[\log 24t - \log 24t_1]/10\} \times [f_{st}/f_{py} - 0.55] \quad (16)$$

where

$$f_{st}/f_{py} - 0.55 \geqq 0.05$$
$$f_{py} = 0.85\, f_{pu}$$

2.7.2—Low-relaxation steel

The following equation applies to prestressing steel given its low relaxation properties by simultaneous heating and stretching operations.

$$RET = f_{st}\{[\log 24t - \log 24t_1]/45\} \times [f_{st}/f_{py} - 0.55] \quad (17)$$

where

$$f_{st}/f_{py} - 0.55 \geqq 0.05$$
$$f_{py} = 0.90\, f_{pu}$$

2.7.3—Other prestressing steel

Relaxation of other types of prestressing steel shall be based upon manufacturer's recommendations supported by test data.

CHAPTER 3—SIMPLIFIED METHOD FOR COMPUTING PRESTRESS LOSSES

3.1—Scope

Computations of stress losses in accordance with the General Method can be laborious for a designer who does not have the procedure set up on a computer program. The Simplified Method is based on a large number of design examples in which the parameters were varied to show the effect of different levels of concrete stress, dead load stress, and other factors. These examples followed the General Method and the procedures given in the Design Examples.

3.2—Principles of the Simplified Method

3.2.1–Concrete stress at the critical location

Compute f_{cr} and f_{cds} at the critical location on the span. The critical location is the point along the span where the concrete stress under full live load is either in maximum tension or in minimum compression. If f_{cds} exceeds f_{cr} the simplified method is not applicable.

f_{cr} and f_{cds} are the stresses in the concrete at the level of the center of gravity of the tendons at the critical location. f_{cr} is the net stress due to the prestressing force plus the weight of the prestressed member and any other permanent loads on the member at the time the prestressing force is applied. The prestressing force used in computing f_{cr} is the force existing immediately after the prestress has been applied to the concrete. f_{cds} is the stress due to all permanent ((dead) loads not used in computing f_{cr}.

Table 8. Simplified method equations for computing total prestress loss *(TL)*.

Equation number	Concrete weight		Type of tendon			Tensioning		Equations
	NW	LW	SR	LR	BAR	PRE	POST	
N-SR-PRE-70	X		X			X		$TL = 33.0 + 13.8f_{cr} - 4.5f_{cds}$
L-SR-PRE-70		X	X			X		$TL = 31.2 + 16.8f_{cr} - 3.8f_{cds}$
N-LR-PRE-75	X			X		X		$TL = 19.8 + 16.3f_{cr} - 5.4f_{cds}$
L-LR-PRE-75		X		X		X		$TL = 17.5 + 20.4f_{cr} - 4.8f_{cds}$
N-SR-POST-68.5	X		X				X	$TL = 29.3 + 5.1f_{cr} - 3.0f_{cds}$
L-SR-POST-68.5		X	X				X	$TL = 27.1 + 10.1f_{cr} - 4.9f_{cds}$
N-LR-POST-68.5	X			X			X	$TL = 12.5 + 7.0f_{cr} - 4.1f_{cds}$
L-LR-POST-68.5		X		X			X	$TL = 11.9 + 11.1f_{cr} - 6.2f_{cds}$
N-BAR-POST-70	X				X		X	$TL = 12.8 + 6.9f_{cr} - 4.0f_{cds}$
L-BAR-POST-70		X			X		X	$TL = 12.5 + 10.9f_{cr} - 6.0f_{cds}$

Note: Values of *TL*, f_{cr}, and f_{cds} are expressed in ksi.

3.2.2—Simplified loss equations

Select the applicable equation from Table 3 or 9, substitute the values for f_{cr} and f_{cds} and compute TL or f_{se}, whichever is desired.

3.2.3—Basic parameters

The equations are based on members having the following properties:

1. Volume-to-surface ratio = 2.0.
2. Tendon tension as indicated in each equation.
3. Concrete strength at time prestressing force is applied:
   3500 psi for pretensioned members
   5000 psi for post-tensioned members
4. 28-day concrete compressive strength = 5000 psi.
5. Age at time of prestressing:
   18 hours for pretensioned members
   30 days for post-tensioned members
6. Additional dead load applied 30 days after prestressing.

Compare the properties of the beam being checked with Items 1 and 2. If there is an appreciable difference, make adjustments as indicated under Section 3.4.

It was found that an increase in concrete strength at the time of prestressing or at 28 days made only a nominal difference in final loss and could be disregarded. For strength at prestressing less than 3500 psi or for 28-day strengths less than 4500 psi, an analysis should be made following Design Example 1.

Wide variations in Items 5 and 6 made only nominal changes in net loss so that further detailed analysis is needed only in extreme cases.

3.2.4—Computing f_{cr}

$$f_{cr} = A_s f_{si}/A_c + A_s f_{si}\, e^2/I_c - M'e/I_c \qquad (18)$$

Table 9. Simplified method equations for computing effective prestress (f_{se}).

Equation Number	Concrete weight		Type of tendon			Tensioning		Equations
	NW	LW	SR	LR	BAR	PRE	POST	
N-SR-PRE-70	X		X			X		$f_{se} = f_t - (33.0 + 13.8f_{cr} - 11f_{cds})$
L-SR-PRE-70		X	X			X		$f_{se} = f_t - (31.2 + 16.8f_{cr} - 13.5f_{cds})$
N-LR-PRE-75	X			X		X		$f_{se} = f_t - (19.8 + 16.3f_{cr} - 11.9f_{cds})$
L-LR-PRE-75		X		X		X		$f_{se} = f_t -$ ([illegible].5 $+ 20.4f_{cr} - 14.5f_{cds})$
N-SR-POST-68.5	X		X				X	$f_{se} = f_{st} - (29.3 + 5.1f_{cr} - 9.6f_{cds})$
L-SR-POST-68.5		X	X				X	$f_{se} = f_{st} - (27.1 + 10.1f_{cr} - 14.6f_{cds})$
N-LR-POST-68.5	X			X			X	$f_{se} = f_{st} - (12.5 + 7.0f_{cr} - 10.6f_{cds})$
L-LR-POST-68.5		X		X			X	$f_{se} = f_{st} - (11.9 + 11.1f_{cr} - 15.9f_{cds})$
N-BAR-POST-70	X				X		X	$f_{se} = f_{st} - (12.8 + 6.9f_{cr} - 10.5f_{cds})$
L-BAR-POST-70		X			X		X	$f_{se} = f_{st} - (12.5 + 10.9f_{cr} - 15.7f_{cds})$

Note: Values of f_t, f_{st}, f_{cr}, and f_{cds} are expressed in ksi.

3.2.5—Tendon stress for pretensioned members

Except for members that are very heavily or very lightly* prestressed, f_{si} can be taken as follows:

For stress-relieved steel

$$f_{si} = 0.90\, f_t \quad (19)$$

For low-relaxation steel

$$f_{si} = 0.925\, f_t \quad (20)$$

3.2.6 Tendon stress for post-tensioned members

Except for members that are very heavily or very lightly* prestressed, f_{si} can be taken as

$$f_{si} = 0.95\,(T_o - FR) \quad (21)$$

3.3—Equations for Simplified Method

3.3.1—Total prestress loss

The equations in Table 8 give total prestress loss TL in ksi. This value corresponds to TL shown in the summaries of Design Examples 1 and 3.

3.3.2—Effective stress

The equations in Table 9 give effective stress in prestressing steel under dead load after losses. This value corresponds to f_{se} shown in the summary of Design Example 1.

As shown in the summary of Design Example 1, the stress existing in the tendons under dead load after all losses have taken place is the initial tension reduced by the amount of the total losses and increased by the stress created in the tendon by the addition of dead load after the member was prestressed. The increase in tendon stress due to the additional dead load is equal to $f_{cds}\,(E_s/E_c)$.

$$f_{se} = f_t - TL + f_{cds}(E_s/E_c) \quad (22)$$

3.3.3—Explanation of equation number

The equation number in Tables 8 and 9 defines the conditions for which each equation applies:

1. The first term identifies the type of concrete.
   N = normal weight = approximately 145 lb per cu ft
   L = lightweight = approximately 115 lb per cu ft

2. The second term identifies the steel in the tendon:
   SR = stress-relieved
   LR = low-relaxation
   BAR = high strength bar

3. The third term identifies the type of tensioning:
   PRE = pretensioned and is based on accelerated curing
   POST = post-tensioned and is based on moist curing

4. The fourth term indicates the initial tension in percent of f_{pu}:
   For *pretensioned* tendons it is the tension at which the tendons are anchored in the casting bed before concrete is placed.

   For *post-tensioned* tendons it is the initial tension in the tendon at the critical location in the concrete member after losses due to friction and anchor set have been deducted.

---

*When f_{cr} computed by Eq. (18) using the approximations for f_{si} is greater than 1600 psi or less than 800 psi the value of f_{si} should be checked as illustrated in Design Example 2.

3.4—Adjustment for Variations from Basic Parameters

3.4.1—Volume-to-surface ratio

Equations are based on $V/S = 2.0$

V/S ratio	1.0	2.0	3.0	4.0
Adjustment, percent	+3.2	0	−3.8	−7.6

Example: For $V/S = 3.0$, decrease TL by 3.8 percent.

3.4.2—Tendon stress

3.4.2.1—*Pretensioned tendons*

Pretensioned tendons are so seldom used at stresses below those shown for the equations in members where the final stress is important, that examples covering this condition were not worked out. Design Example 1 can be followed if necessary.

3.4.2.2—*Post-tensioned tendons*

Equations are based on $f_{si} = 185{,}000$ psi. If f_{si} is less than 185,000 psi reduce the total stress loss:

For stress-relieved strands

$$\Delta TL = 0.41\,(185{,}000 - f_{si}) \quad (23)$$

For low-relaxation strands

$$\Delta TL = 0.09\,(185{,}000 - f_{si}) \quad (24)$$

For high strength bars which are based on $f_{si} = 0.70\,f_{pu}$

$$\Delta TL = 0.09\,(0.70\,f_{pu} - f_{si}) \quad (25)$$

If f_{si} is greater than the value used in preparing the equations, ΔTL will be a negative number and will therefore increase the value of TL. Note that f_{si} is limited to a maximum of $0.70\,f_{pu}$ by ACI 318-71.

COMMENTARY

In this report a wide range of data has been assimilated to develop a general method for predicting loss of prestress, but including specific numerical values. In addition, creep and shrinkage of concrete and steel relaxation are presented as functions of time. By calculating losses over recommended time intervals, it is possible to take into account the interdependence of concrete and steel information.

It must be emphasized that losses, per se, are not the final aim of calculations. What is determined is the stress remaining in the prestressing steel. The stress remaining, however, must be evaluated using rational procedures.

The notation commonly used in the ACI Building Code is adopted wherever possible. For new terms descriptive letters are used.

References are listed chronologically. Not all the references listed were used in developing the numerical recommendations. They are included because of their value in understanding time-dependent behavior.

Chapter 1—General Aspects Related to Prestress Losses

1.1—Friction losses during post-tensioning are estimated using familiar equations, but with up-dated coefficients (see Reference 30). No known systematic study has been made c losses that occur at deflecting devices in pretensioned construction. The provision is to warn that friction at these points may produce conditions where the desired steel stresses are not reached.

1.2–Seating losses are particularly important where the length of prestressing steel is short. For this condition, tolerances in seating deformation should not be overlooked.

1.3–The effect of post-tensioning of each individual tendon on previously anchored tendons should be considered. This applies, of course, when pretensioned and post-tensioned systems are combined.

1.4—Many problems have occurred because of unaccounted restraint and the effects of volume changes of concrete cast at different times. There are several references that give techniques for calculating these effects (References 6, 9, 10, 12, 20, 29, 30). This reminder is included here, even though losses are but one factor influenced by structural integration.

Chapter 2—General Method for Computing Prestress Losses

This section presents the range of data studied and, consequently, the range of applicability. Extrapolation has been avoided beyond documented data. In effect, this also shows where additional research is needed. Some practices and conditions common to certain areas of the country cannot be incorporated because no information is available. It is in this situation that experience, engineering judgment, and local sources of information are depended upon.

2.2–Total loss of prestress is the sum of losses due to individual factors. Eqs. (3) and (4) list the factors to be taken into account for each type of construction. The terms *ANC*, *DEF*, and *FR* are defined in Section 1. The remaining terms are defined in Section 2. Losses due to creep, shrinkage, and steel relaxation are the sum of losses during each time interval described in Section 2.4.

2.3–It is not desirable to be "conservative" and assume a low value for the modulus of concrete. The estimated modulus should reflect what is specified for minimum concrete strength and what is specified for permissible variation of concrete strength.

2.4–The step procedure is recommended to realistically approach the actual behavior of prestressed concrete. By this technique, it is possible to evaluate loss of prestress with change in time and change in stress. What is done here is to take into account the interdependence of one deformation on the other. Specific steps are outlined in succeeding sections.

2.5–Ultimate creep is the total amount of shortening measured on standard 6 x 12-in. cylinders. Fig. 1 illustrates that values differ according to the type of cure and the type of concrete (References 2, 14, 15, 21, 24). Ultimate creep is affected by relative humidity as shown in Fig. 2 (References 3, 7, 8, 11, 17). In standard tests, specimens are stored at 50 percent relative humidity. Average relative humidity over the majority of the United States is 70 percent (Reference 18). Eqs. (7), (8), (9), and (10), therefore, are based on data shown in Figs. 1 and 2.

Fig. 3 presents the relationship between creep and the size and shape of the test specimen (see References 7, 11, 17, 19). To apply standard creep data to actual members, a creep factor *SCF* is introduced. Values of *SCF* versus volume to surface ratios are listed in Section 2.5.3.

Similarly, age at loading and extent of moist-cure affect the amount of ultimate creep. Data in Fig. 4 (see Referenecs 3, 4, 11, 14) illustrate the trend. For moist-cured concretes, *UCR* is modified by the factor *MCF* in Section 2.5.4.

A generalization of creep-time data (References 2, 15, 24) was developed to take into account the rate of shortening due to creep. A typical creep curve is shown in Fig. 5. The portion of ultimate creep *AUC* at a given time is listed in Table 5. The amount of creep *PCR* in a given time interval is simply the difference of the amount of creep at the beginning and end of the time interval.

The terms, *SCF*, *MCF*, and *PCR* are non-dimensional. *UCR* was developed from creep data expressed as strain in millionths per psi concrete stress. By multiplying these data by the steel modulus of elasticity, unit steel stress per unit concrete stress was obtained.

Therefore, in Eq. (6), steel stress is obtained by multiplying concrete stress f_c by the product of these four non-dimensional terms. As stated in Section 2.5, f_c is the net concrete stress that results from at least full dead load and possibly some portion of the live load. The amount of live load, if any, present for extended periods is left to the engineer's judgment.

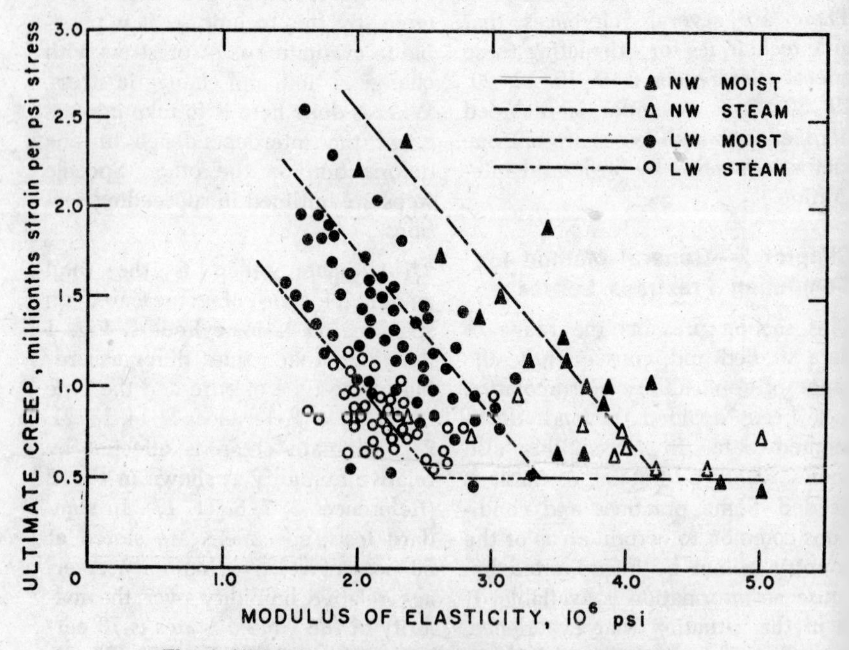

Fig. 1. Ultimate creep versus modulus of elasticity.

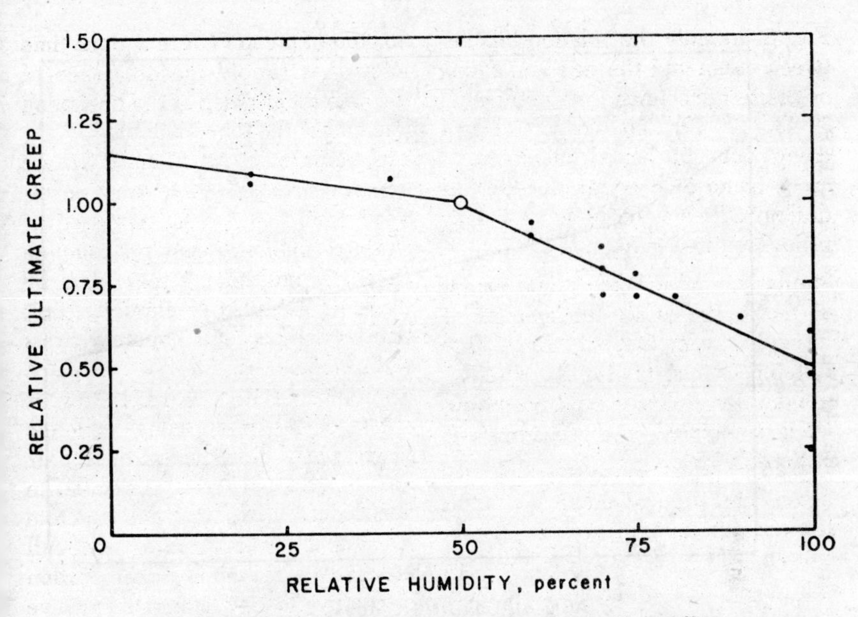

Fig. 2. Relative ultimate creep versus relative humidity.

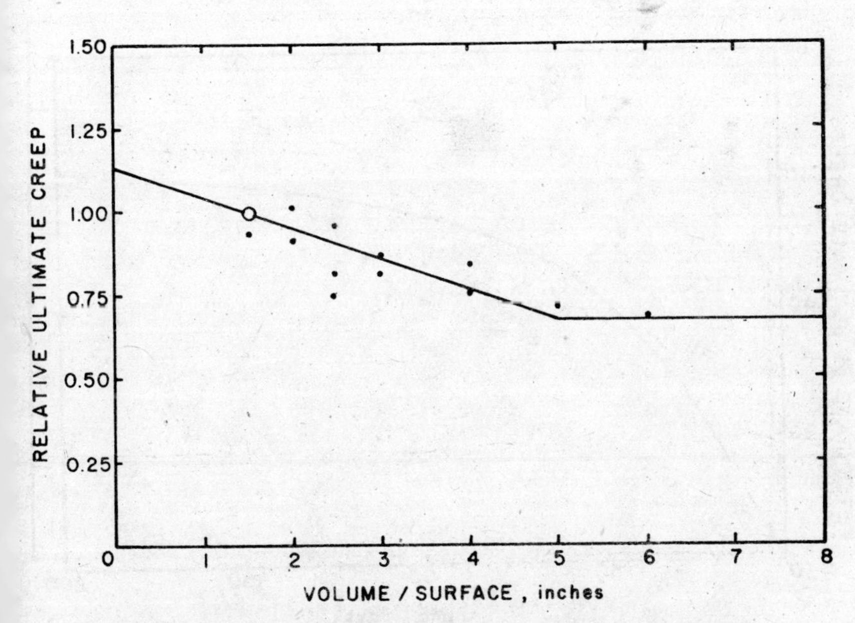

Fig. 3. Relative ultimate creep versus volume to surface ratio.

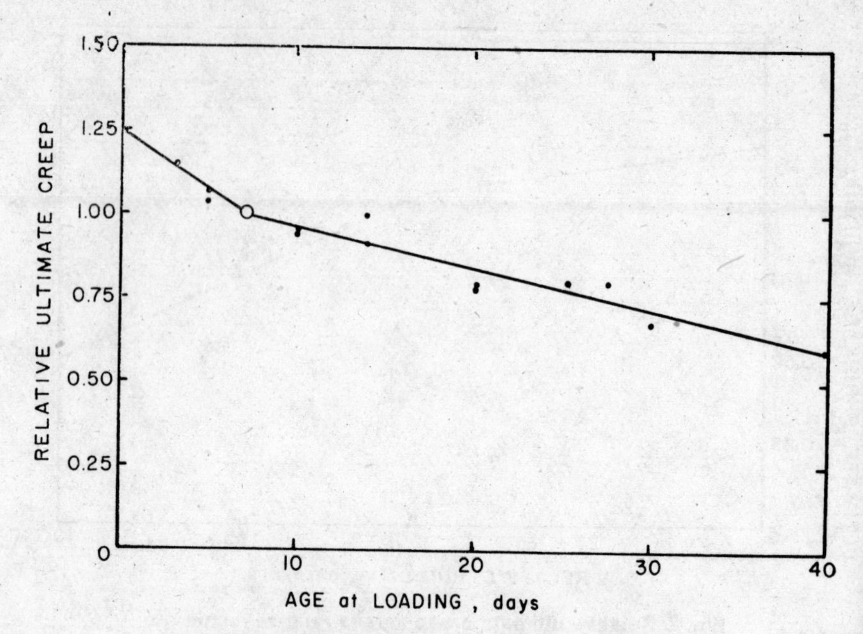

Fig. 4. Relative ultimate creep versus loading age.

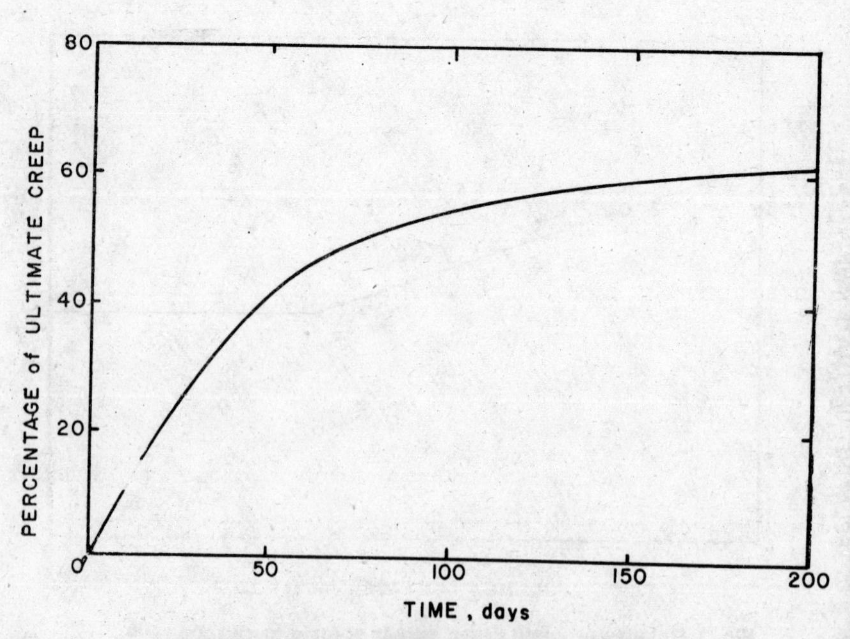

Fig. 5. Percentage of ultimate creep versus time.

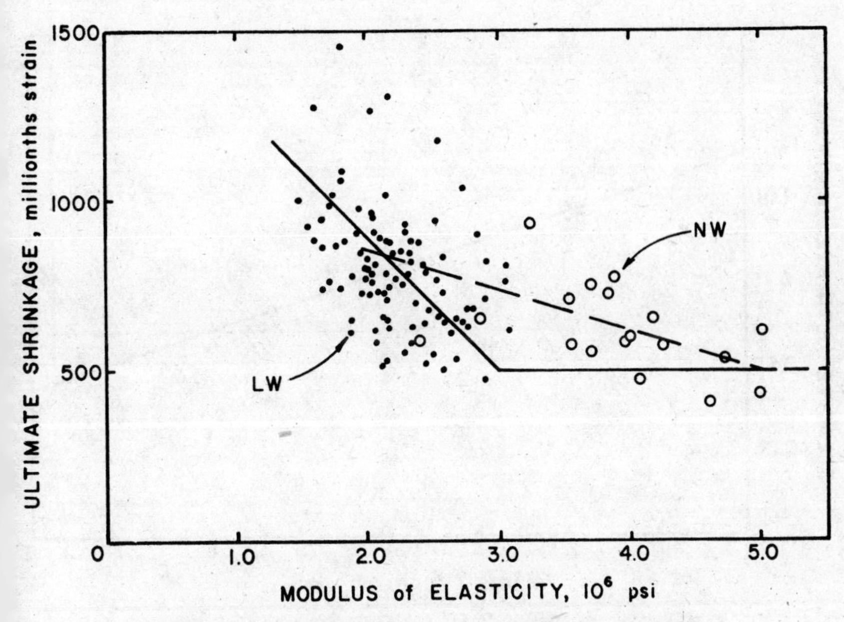

Fig. 6. Ultimate shrinkage versus modulus of elasticity.

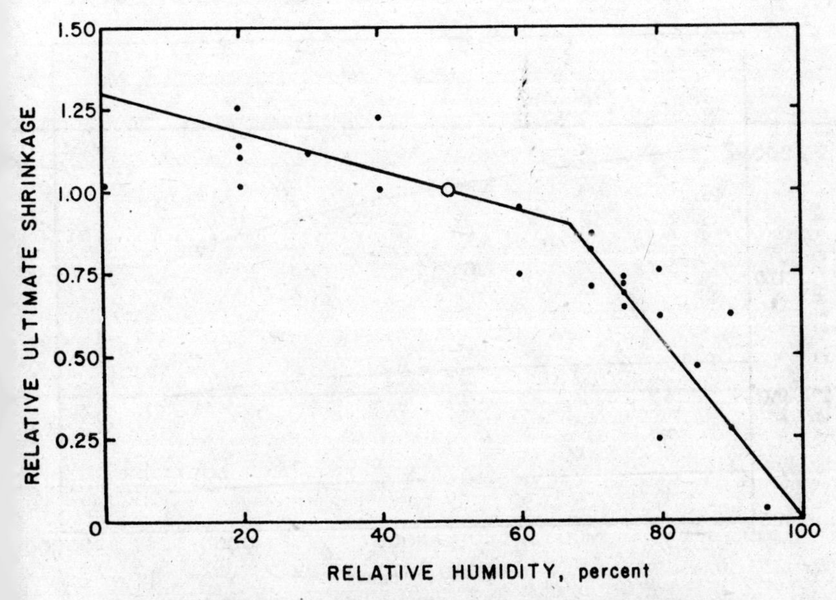

Fig. 7. Relative ultimate shrinkage versus relative humidity.

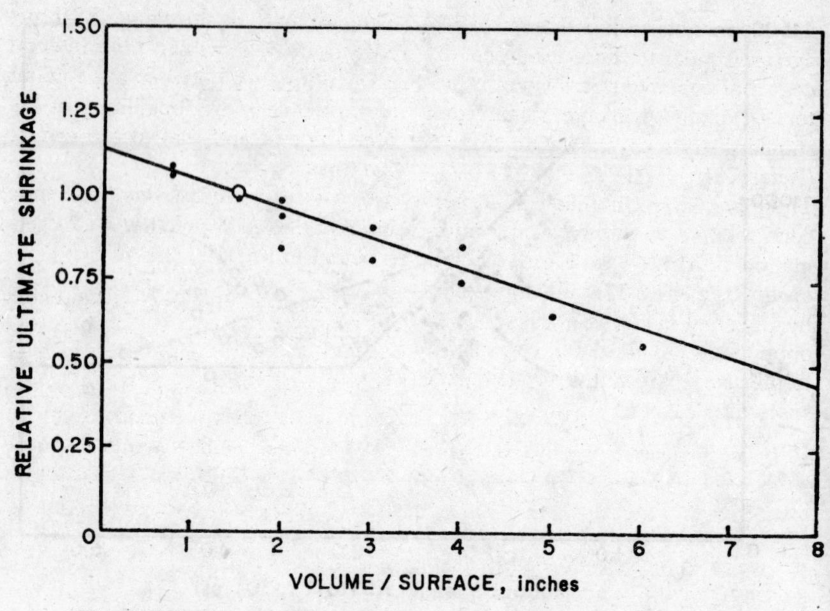

Fig. 8. Relative ultimate shrinkage versus volume to surface ratio.

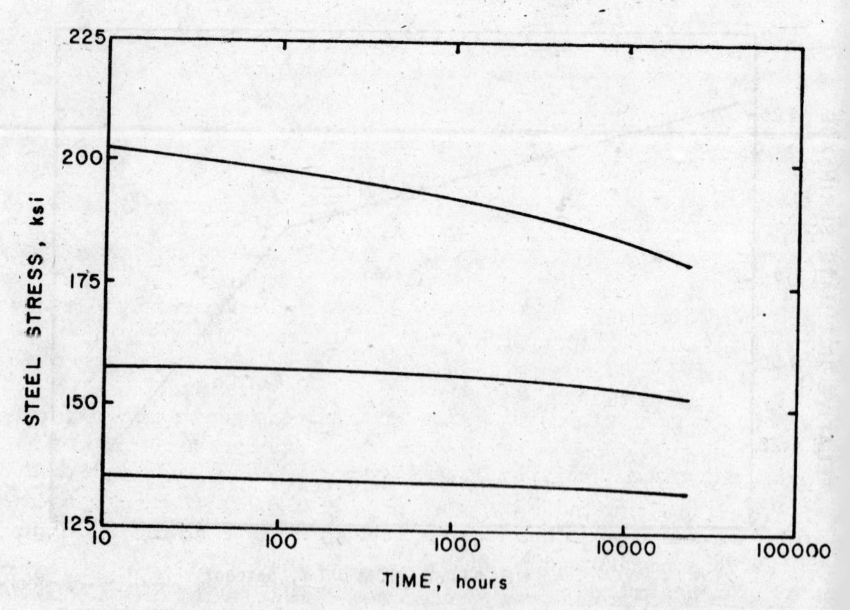

Fig. 9. Prestressing steel stress versus time for stress-relieved steel.

2.6—Comments on loss due to shrinkage are similar to those for creep, except that concrete stress is not a factor. Ultimate shrinkage strain from standard tests is shown in Fig. 6 (References 1, 2, 14, 15, 21, 24). These data were modified to 70 percent relative humidity using information illustrated in Fig. 7 (References 3, 7, 8, 11, 17), and multiplied by the steel modulus of elasticity to obtain Eqs. (13) and (14). *USH* is the loss in steel stress due to shrinkage shortening.

USH is influenced by the size and shape of the prestressed member. Shrinkage volume/surface data are presented in Fig. 8 (References 7, 11 17, 19). The size factor *SSF* is given in Table 6. By multiplying *USH* by *SSF*, standard test data can be applied to actual prestressed members. The variation of shrinkage with time was generalized using data in References 2, 15. and 24. This information is given in Table 7 as the portion of ultimate shrinkage *AUS* for a specific time after the end of curing. The amount of shrinkage *PSH* occurring over a specific time interval is the difference between the amount of shrinkage at the beginning of the time interval and that at the end of the time interval.
The loss of prestress over one time interval due to shrinkage of concrete is stated in Eq. (12).

2.7—Eqs. (16) and (17) (References 13, 32) give the loss of prestress due steel relaxation. The time t_1 in Step 1, listed in Section 2.4, is at anchorage of the prestressing steel. In pretensioned construction, where elevated temperatures are used in curing, losses during curing can be studied more closely using References 5, 22, and 26.
Fig. 9 shows typical steel relaxation with time under constant strain. It is seen that losses are less for lower initial stresses. This illustrates that by taking into account concrete shortening and steel relaxation over a previous time period, subsequent losses are less than that under assumed constant strain.

NOTATION

A_c = gross cross-sectional area of concrete member, sq in.
A_s = cross-sectional area of prestressing tendons, sq in.
ANC = loss of prestress due to anchorage of prestressing steel, psi
AUC = portion of ultimate creep at time after prestress transfer
AUS = portion of ultimate shrinkage at time after end of curing
CR = loss of prestress due to creep of concrete over time interval t_1 to t, psi
DEF = loss of prestress due to deflecting device in pretensioned construction, psi
e = tendon eccentricity measured from center of gravity of concrete section to center of gravity of tendons, in.
E_c = modulus of elasticity of concrete at 28 days taken as $33w^{3/2}\sqrt{f_c'}$, psi
E_{ci} = modulus of elasticity of concrete at time of initial prestress, psi
E_s = modulus of elasticity of steel, psi

ES = loss of prestress due to elastic shortening, psi

f_{cds} = concrete compressive stress at center of gravity of prestressing force due to all permanent (dead) loads not used in computing f_{cr}, psi

f_c = concrete compressive stress at center of gravity of prestressing steel, psi

f_c' = compressive strength of concrete at 28 days, psi

f_{ci}' = initial concrete compressive strength at transfer, psi

f_{cr} = concrete stress at center of gravity of prestressing force immediately after transfer, psi

f_{pu} = guaranteed ultimate tensile strength of prestressing steel, psi

f_{py} = stress at 1 percent elongation of prestressing steel, psi

f_{se} = effective stress in prestressing steel under dead load after losses

f_{si} = stress in tendon at critical location immediately after prestressing force has been applied to concrete

f_{st} = stress in prestressing steel at time t_1, psi

f_t = stress at which tendons are anchored in pretensioning bed, psi

FR = friction loss at section under consideration, psi

I_c = moment of inertia of gross cross section of concrete member, in.^{4}

K = friction wobble coefficient per foot of prestressing steel

l_{tx} = length of prestressing steel from jacking end to point x, ft

MCF = factor that accounts for the effect of age at prestress and length of moist cure on creep of concrete

M_{ds} = moment due to dead weight added after member is prestressed

M' = moment due to loads, including weight of member, at time prestress is applied to concrete

P = final prestress force in member after losses

P_o = initial prestress force in member

PCR = amount of creep over time interval t_1 to t

PSH = amount of shrinkage over time interval t_1 to t

RE = total loss of prestress due to relaxation of prestressing steel in pretensioned construction, psi

REP = total loss of prestress due to relaxation of prestressing steel in post-tensioned construction, psi

RET = loss of prestress due to steel relaxation over time interval t_1 to t, psi

SCF = factor that accounts for the effect of size and shape of a member on creep of concrete

SH = loss of prestress due to shrinkage of concrete over time interval t_1 to t, psi

SSF = factor that accounts for the effect of size and shape of a member on concrete shrinkage

t = time at end of time interval, days

t_1 = time at beginning of time interval, days

T_o = steel stress at jacking end of post-tensioning tendon, psi

T_x = steel stress at any point x, psi

TL = total prestress loss, psi

UCR = ultimate loss of prestress due to creep of concrete, psi per psi of compressive stress in the concrete

USH = ultimate loss of prestress due to shrinkage of concrete, psi

w = weight of concrete, lb per cu ft

α = total angular change of post-tensioning tendon profile from jacking end to point x, radians

μ = friction curvature coefficient

DESIGN EXAMPLES

The following three design examples were prepared solely to illustrate the application of the preceding recommended methods. They do not necessarily represent the real condition of any real structure.

Design aids to assist in calculating prestress losses are included in the *PCI Design Handbook* (see Reference 35). The aids will reduce the calculations required. However, detailed study of losses and time-dependent behavior will follow the steps outlined in the design examples.

The first example applies the general method to a pretensioned double-tee and the second example uses the simplified method for the same member. The third example problem illustrates the general method for a post-tensioned structure.

In these examples it is assumed that the member geometry, load conditions, and other parameters have been defined. Consequently, the detailed moment and stress calculations are omitted.

DESIGN EXAMPLE 1

Pretensioned Double Tee

Reference: *PCI Design Handbook*, p. 3-33.

Data: Double-tee section 10LDT32 + 2. Strand pattern 128-D1.
Steam cured, ligthweight double-tee (115 lb per cu ft) with 2-in. topping of normal weight concrete (150 lb per cu ft).
The beam is designed to carry a live load of 40 psf over a 70-ft span.

Required: Calculate the losses at the critical section, taken as 0.4 span in the *PCI Design Handbook.* $f'_{ci} = 3500$ psi, $f_c' = 5000$ psi

Section properties:

Non-composite

$A = 615$ in.^{2}
$I = 59{,}720$ in.^{4}
$y_b = 21.98$ in.
$y_t = 10.02$ in.
$Z_b = 2717$ in.^{3}
$Z_t = 5960$ in.^{3}
Weight: 491 lb per ft

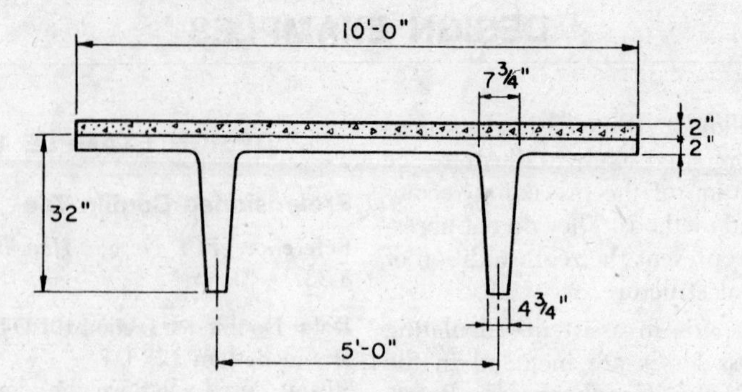

Design Example 1. Cross section of double-tee beam.

Composite

$I = 83{,}001$ in.^{4}
$y_{bc} = 25.40$ in.
$y_{tc} = 8.60$ in.
$Z_b = 3268$ in.^{3}
$Z_t = 9651$ in.^{3}
Weight: 741 lb per ft

The beam is prestressed by twelve ½-in. diameter 270-grade strands, initially tensioned to 0.70 f_{pu}.

Eccentricity of strands:

At ends = 12.98 in.
At center = 18.73 in.
$f_{py} = 230$ ksi

Transfer at 18 hours after tensioning strand, topping cast at age 30 days.

Losses—Basic data

$f'_{ci} = 3500$ psi
$E_{ci} = 115^{1.5}(33\sqrt{3500})$
$= 2.41 \times 10^6$ psi
$f'_c = 5000$ psi
$E_c = 115^{1.5}(33\sqrt{5000})$
$= 2.88 \times 10^6$ psi

Volume to surface ratio = 615/364
= 1.69

$SSF = 0.985$
$SCF = 0.988$

$UCR = 63 - 20(E_c/10^6)$
(but not less than 11)
$= 63 - 20(2.88) = 11$

$USH = 41{,}000 - 10{,}000(E_c/10^6)$
(but not less than 12,000)
$= 41{,}000 - 10{,}000(2.88)$
$= 12{,}200$ psi

At critical section
$e = 12.98 + 0.8(18.73 - 12.92)$
$= 17.58$ in.

$(UCR)(SCF) = 10.87$

$(USH)(SSF) = 12{,}017$ psi

Stage 1: Tensioning of steel to transfer

$t_1 = 1/24$ day
$t = 18/24$ day
$f_{st} = 189{,}000$ psi

$RET = f_{st}[(\log 24t - \log 24 t_1)/10] \times [f_{st}/f_{py} - 0.55]$
$= 189{,}000[(\log 18)/10][189/230 - 0.55]$
$= 6450$ psi

Dead load moment at 0.4 span
$M_{DL} = w(x/2)(L - x)$
$= (491/1000)(28/2)(70 - 28)$
$= 289$ ft-kips

Stress at center of gravity of steel due to M_{DL}

$f_c = [289,000(12)/59,720]\,17.58$
$= 1020$ psi (tension)

Assume $ES \approx 13$ ksi, then
$f_{si} = 189.0 - 6.45 - 13.0$
$= 169.55$ ksi
$P_o = 169.55(12)(0.153)$
$= 311.3$ kips

Stress at center of gravity of steel due to P_o:
$f_e = 311,300/615 +$
$311,300\,(17.58^2/59,720$
$= 2117$ psi (compression)
$f_{cr}^{\prime} = 2117 - 1020 = 1097$ psi
$ES = f_{cr}(E_s/E_c)$
$= 1097(28.0/2.41)$
$= 12,750 \text{ psi} \approx 13 \text{ ksi (ok)}$
$SH = CR = 0$

Total losses in Stage 1 =
$6450 + 12,750 = 19,200$ psi

Stage 2: Transfer to placement of topping after 30 days

$t_1 \quad = 18/24$ day
$t \quad = 30$ days
$PCR = 0.35$
$PSH = 0.42$
$f_{st} \quad = 189,000 - 19,200$
$= 169,800$ psi
$RET = 169,800\ [(\log 720 - \log 18)/10] \times [169.8/230 - 0.55]$
$= 5119$ psi

$CR = 10.87(0.35)(1097) = 4173$ psi

$SH = 12,017(0.42) = 5047$ psi

Total losses in Stage 2
$5119 + 4173 + 5047 = 14,339$ psi

Moment due to weight of topping
$250(28/2)(70 - 28) = 147,000$ ft-lb

Stress at center of gravity of steel due to weight of topping
$147,000(12)(17.58)/59,720 = 519$ psi

Increase in strand stress due to topping
$519(28.0/2.88) = 5048$ psi

Strand stress at end of Stage 2
$169,800 - 14,339 + 5048 = 160,509$ psi

Stage 3: Topping placement to end of one year

$t_1 \quad = 30$ days
$t \quad = 1 \text{ year} = 365$ days
$PCR = 0.74 - 0.35 = 0.39$
$PSH = 0.86 - 0.42 = 0.44$

$f_{st} \quad = 160,509$ psi
$RET = 160,509\,[(\log 8760 - \log 720/\log 720/10] \times [160.5/230 - 0.55]$
$= 2577$ psi

$f_c \quad = 2117(160,509/169,550) - 1020 - 519$
$= 465$ psi
$CR = 10.87(0.39)(465) = 1971$ psi
$SH = 12,017(0.44) = 5287$ psi

Total losses in Stage 3
$2577 + 1971 + 5287 = 9835$ psi

Summary of steel stresses at various stages (Design Example 1)

Stress level at various stages	Steel stress, ksi	Percent
Strand stress after tensioning and deflection ($0.70f_{pu}$) ..	189.0	100.0
Losses:		
Elastic shortening = 12.75		6.7
Relaxation: 6.45+ 5.12 + 2.58 + 2.54 = 16.69		8.8
Creep: 4.17 + 1.97 + 0.97 = 7.11		3.8
Shrinkage: 5.05 + 5.29 + 1.68 = 12.02		6.4
Total losses, TL	48.57	25.7
Increase of stress due to topping	5.05	2.7
Final strand stress under total dead load (f_{se})	145.48	77.0

Stage 4: One year to end of service life

$t_1 \quad = 1$ year
$t \quad =$ end of service life (say 40 years)
$PCR = 1 - 0.74 = 0.26$
$PSH = 1 - 0.86 = 0.14$
$f_{st} \quad = 160{,}509 - 9835 = 150{,}674$ psi
$RET = 150{,}674\,[(\log 350{,}400 - \log 8760)/10] \times [(150.7/230) - 0.55]$
$= 2537$ psi
$f_c \; = 2117(150{,}674/169{,}550) - 1020 - 519$
$= 343$ psi
$CR = 10.87(0.26)(343) = 969$ psi
$SH = 12{,}017(0.14) = 1682$ psi

Total losses in Stage 4 =
$2537 + 969 + 1682 = 5188$ psi

DESIGN EXAMPLE 2

Application of Simplified Procedure to Design Example 1

Compute f_{cds}

$f_{cds} = eM_{ds}/I$
$= 17.58(147)(12)/59{,}720$
$= 0.519$ ksi

Compute f_{cr}

$f_{cr} = A_s f_{si}/A_c + A_s f_{si} e^2/I_c + M' e/I_c$
$f_{si} = 0.90 f_t = 0.90(189) = 170.1$ ksi
$f_{cr} = 1.84(170.1)/615 + 1.84(170.1)(17.58)^2/59{,}720 - 289(12)(17.58)/59{,}720$
$= 0.509 + 1.620 - 1.021$
$= 1.108$ ksi

Equation L-SR-PRE-70 from Table 8 is
$TL = 31.2 + 16.8 f_{cr} - 3.8 f_{cds}$
$= 31.2 + 16.8(1.108) - 3.8(0.519)$
$= 31.2 + 18.61 - 1.97$
$= 47.84$ ksi

Adjustment for volume to surface ratio = 1.69

Use a straight-line interpolation between adjustment values for $V/S = 2.0$ and $V/S = 1.0$

Adjustment = (0.31)(3.2) = + 0.99%

Net $TL = 1.0099(47.84) = 48.31$ ksi
In Design Example 1, $TL = 48.57$ ksi
Difference = 0.26 ksi

Compute f_{se}

To find f_{se} in accordance with discussion under Section 3.32, and stress in tendons due to dead load applied after member was prestressed.

This stress is equal to
$f_{cds}(E_s/E_c) = 0.519(28/2.88) = 5.05$ ksi
$f_{se} = 189 - 48.31 + 5.05 = 145.74$ ksi

Note that f_{se} can also be computed from the equations shown in Table 9.

Equation L-SR-PRE-70 from Table 9 is
$f_{se} = f_t - (31.2 + 16.8 f_{cr} - 13.5 f_{cds})$
$= 189 - (31.2 + 16.8 \times 1.108 - 13.5 \times 0.519)$
$= 189 - (31.2 + 18.61 - 7.01)$
$= 189 - (42.8)$

An adjustment for variations in the basic parameters should be applied to the quantity in parentheses. In this case, adjust for a V/S of 1.69. The adjustment is +0.99 percent. The adjusted quantity becomes
$1.0099(42.8) = 43.22$
$f_{se} = 189 - 43.22 = 145.78$ ksi

Checking the assumed value of f_{si}*:*

In the application of the simplified method to Design Example 1, the value of f_{si} was assumed to be 170.1 ksi.

The following procedure can be used to check the accuracy of this assumed value.

For this example the exact value of f_{si} is the initial stress of 189 ksi reduced by strand relaxation from tensioning to release and by loss due to elastic shortening of the concrete as the prestressing force is applied.

From Section 2.7.1, the relaxation loss in a stress-relieved strand is

$RET = f_{st}\,[(\log 24t - \log 24t_1)/10] \times [f_{st}/f_{py} - 0.55]$

For stress-relieved strand
$f_{py} = 0.85(270) = 229.5$ ksi

By definition in Section 2.7.1, when time is measured from zero, $\log 24t_1 = 0$

$$RET = 189[(1.255 - 0)/10][189/229.5 - 0.55] = 6.49 \text{ ksi}$$

Stress loss due to elastic shortening of concrete

$$ES = (E_s/E_c)f_{cr} = (28/0.24)1.108 = 12.93 \text{ ksi}$$

Then, $f_{si} = 189 - 6.49 - 12.93 =$ 169.58 ksi
and $0.90\, f_t = 170.10$ ksi

Therefore, there is a $170.10 - 169.58 =$ 0.52 ksi stress error in f_{si}.

Consequently, in this particular case there is no need for a second trial.

As an example, assume a large error in the estimated f_{si}, say 10 ksi, and check its effect. The strand relaxation will not change.

Therefore, the change in ES will be

$$\Delta ES = (10/170.1)12.93 = 0.76 \text{ ksi}$$

If desired, the original estimate of f_{si} can be adjusted by 10 ksi and f_{cr} can be recalculated. One such cycle should always give an adequate accuracy.

DESIGN EXAMPLE 3

Post-Tensioned Unbonded Slabs

The following is a procedure for calculating the prestress losses in the longitudinal tendons which extend from end to end of the slab (see sketch showing floor plan and tendon profiles).

Data

$w = 150$ lb per cu ft
f_c' (28 days) $= 4000$ psi

Prestressed at age 4 days
$f_c' = 3000$ psi

Moist cured 7 days.

Loads

7½-in. slab = 94 psf
Superimposed load = 60 psf
The tendon profile shown is designed to balance 85 psf.

Friction Loss (FR)

The slab is prestressed by 270-grade, ½-in. diameter strand, pregreased and paper wrapped.

Coefficient of friction, $\mu = 0.08$
Wobble coefficient, $K = 0.0015$
$f_{py} = 230$ ksi.

Angular changes along tendon will be:

$$\theta_{AB} = 2(2.5)/[12(12)] = 0.0347 \text{ radians}$$

$$\theta_{BC} = \theta_{EF} = \theta_{FG} = \theta_{KL} = 2(4.0)/[12(9.6)] = 0.0694 \text{ radians}$$

$$\theta_{CD} = \theta_{DE} = \theta_{GH} = \theta_{HK} = 2(1.0)/[12(2.4)] = 0.0694 \text{ radians}$$

Angular change between A and L

$$\alpha = 0.0347 + 4(0.0694) + 4(0.0694) = 0.59 \text{ radians}$$

FR at L (middle of length of slab)

$$= T_o[1\text{-}e^{-(KL+\mu\alpha)}]$$
$$= T_o[1\text{-}e^{-\{(0.0015)(60)+(0.08)(0.59)\}}]$$
$$= T_o[1\text{-}e^{-(0.090+0.047)}]$$
$$= 0.128\, T_o$$

The distribution of frictional loss is not uniform, but nearly proportional to $(K + \mu\alpha/L)$. However, the variation of strand stress before anchoring is approximately as shown on p. 72.

Anchorage loss (ANC)

Anchorage set in a single strand anchor
= 1/8 in.
= Shaded area in diagram × $(1/E_s)$
Area = (1/8)29,000 = 3625 ksi-ft
= 302 ksi-in.

The maximum strand stress after seating of anchorage occurs x ft from end,

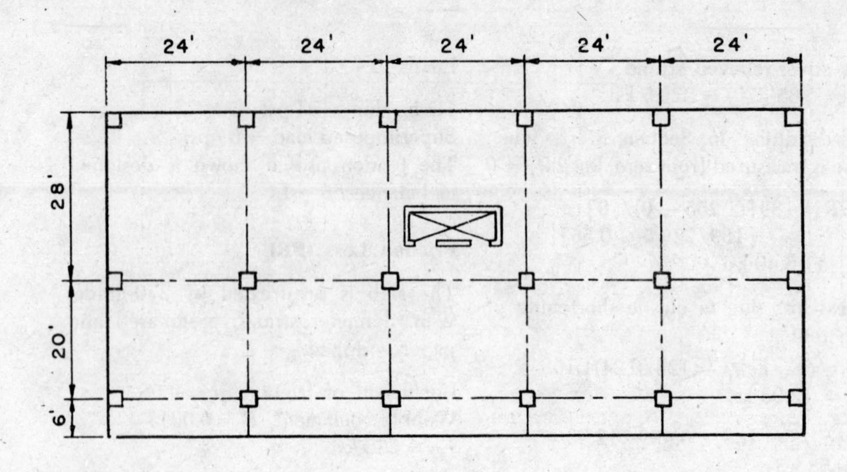

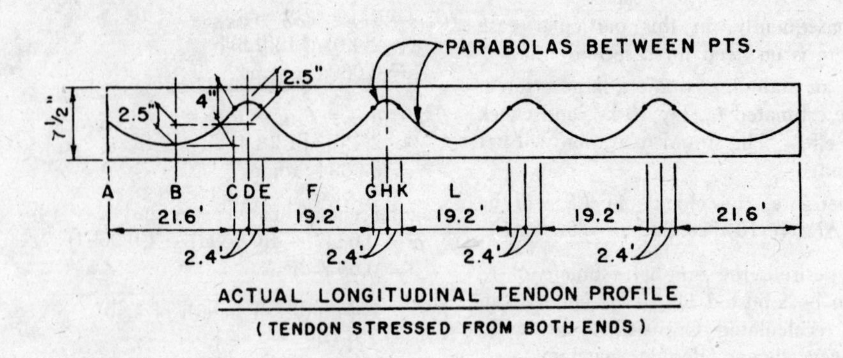

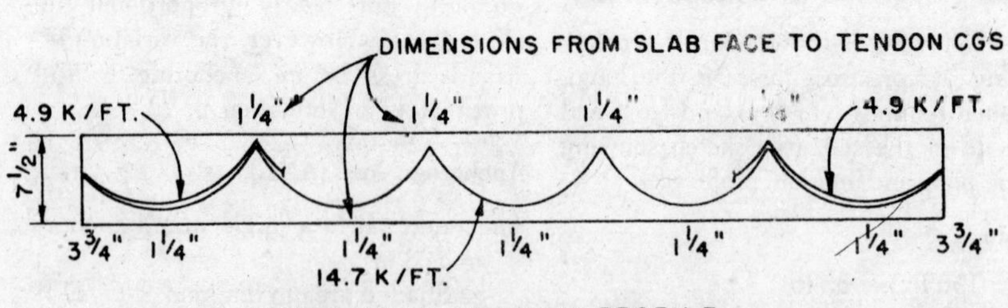

Design Example 3. Plan and tendon profiles of post-tensioned unbonded slab.

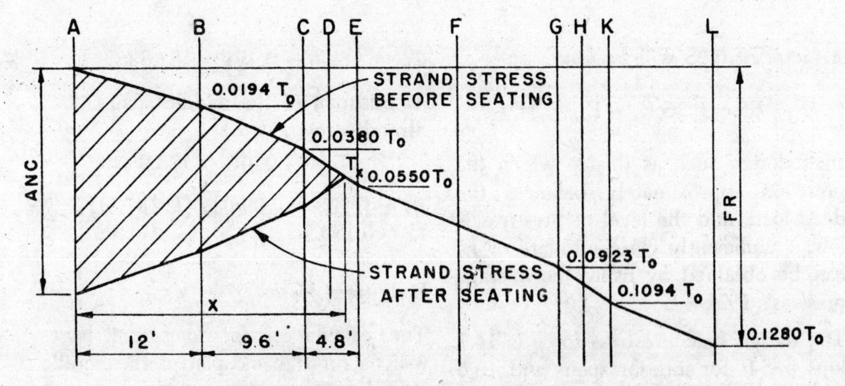

Approximate variation of strand stress.

and this stress T_x must not exceed $0.70 f_{pu} = 189$ ksi.

$$T_o - T_x = 0.0380\, T_o + (0.0550 - 0.0380)T_o(x - 21.6)/4.8$$

$$\text{Area} = (T_o - T_x)12 + (0.9806\, T_o - T_x)21.6 + (0.9620\, T_o - T_x)(x - 12)$$

Therefore

$$T_x = 0.9620 T_o = 0.0170 T_o (x - 21.6)/4.8 = 189 \text{ ksi}$$

$$(T_o - T_x)12 + (0.9806 T_o - T_x)21.6 + (0.9620 T_o - T_x)(x - 12) = 302 \text{ ksi-ft}$$

These equations can be solved by trial and error.

Approximate solution:

$x = 25.5$ ft
$T_o = 200$ ksi
$T_x = 192.4 - 2.8 = 189.6 \approx 189$ ksi

$$\text{Area} = 124.8 + 140.8 + 37.8 = 303.4 \approx 302 \text{ ksi-ft (ok)}$$

For initial end tension before anchorage
$T_o = 200$ ksi ($\approx 0.74 f_{pu}$)

Maximum stress after anchorage
$T_x = 189.6$ ksi

$ANC = 2(T_o - T_x) = 20.8$ ksi
$T_A = 200 - 20.8 = 179.2$ ksi
$FR = 0.128\, T_o = 25.6$ ksi
$T_L = 200 - 25.6 = 174.4$ ksi

Average stress after anchorage:
$T_B = 183.2$ ksi, $T_c = 186.9$ ksi

$$T_{av} = (T_o/60)(0.5)[(0.896)(12) + (0.916)(21.6) + (0.935)(13.5) + (0.948)(4.8) + (0.945)(20.1) + (0.908)(24.0) + (0.891)(14.4) + (0.872)(9.6)] = 182.8 \text{ ksi}$$

Elastic shortening loss (ES)

In post-tensioned structural members, the loss caused by elastic shortening of concrete is only a fraction of the corresponding value in pretensioned members.

The fraction varies from zero if all tendons are tensioned simultaneously to 0.5 if infinitely many sequential steps are used.

In a slab, strands are spaced far apart and it is unlikely that the stretching of one strand will affect stresses in strand other than those immediately neighbo ing.

A factor of 0.25 will be used.

$$ES = 0.25(E_s - E_{ci}) f_{cr}$$

In a design such as this in which the prestress approximately balances the dead load, and the level of prestress is low, a sufficiently close estimate of f_{cr} can be obtained by using the average prestress P/A.

The design final prestress force is 14.7 kips per ft for interior spans and 19.6 kips per ft for end spans. Assuming a long-term prestress loss of 15 percent, the initial prestress force will be 17.3 kips per ft and 23.1 kips per ft, respectively. The average strand stress after anchorage is 182.8 ksi.

Therefore, the required area of steel for the end spans is

$$A_s = 23.1/182.8 = 0.126 \text{ sq in. per ft}$$

This required area is supplied by ½-in. diameter strands spaced at 14 in.

$$A_s = 0.131 \text{ sq in. per ft}$$

Every fourth strand will be terminated in the first interior span, leaving an A_s of 0.098 sq in. per ft.

The actual initial prestressing forces are:

End span
$0.131(182.8) = 23.9$ kips per ft

Interior span
$0.098(182.8) = 17.9$ kips per ft

The average concrete stresses are 266 and 199 psi, respectively.

$$f_{cr} \approx (1/120)[(266)(48) + (199)(72)] = 226 \text{ psi}$$

$$E_{ci} = 33w^{1.5}\sqrt{f_{ci}'} = 33(150)^{1.5}\sqrt{3000} = 3.32 \times 10^6 \text{ psi}$$

$$ES = 0.25(226)(29/3.32) = 494 \text{ psi}$$

After all the strands have been tensioned and anchored:

$$T_{av} = 182.8 - 0.494 = 182.3 \text{ ksi}$$

At midspan of the middle span
Strand stress (at L)

$$174.4 - 0.494 \approx 173.9 \text{ ksi}$$

$$f_{cr} \approx [(173.9)(0.098)]/[(12)(7.5)] \approx 0.189 \text{ ksi}$$

Long-term losses

The calculation for long-term losses will be for the midspan of the middle span (at Section L).

Stage 1: To 30 days after prestressing

Relaxation:

$t_1 = 1/24$ day
$t = 30$ days
$f_{st} = 173.9$ ksi
$f_{st}/f_{py} = 0.756$

$$RET = f_{st}[(\log 24t - \log 24t_1)/10] \times [(f_{st}/f_{py}) - 0.55] = 173{,}900(0.2857)(0.206) = 10{,}230 \text{ psi}$$

Creep:

$CR = (UCR)(SCF)(MCF)(PCR)(f_c)$
$UCR = 95 - (20E_c/10^6)$
but not less than 11 psi
$E_c = 33(150)^{1.5}\sqrt{4000} = 3.83 \times 10^6$ psi
$UCR = 95 - 76.6 = 18.4$ psi

V/S ratio $= 0.5$(slab thickness) $= 0.5(7.5) = 3.75$ in.

$SCF = 0.80$
$MCF = 1.07$ (estimated)
$(UCR)(SCF)(MCF) = 15.75$
$f_c = f_{cr} = 189$ psi
$PCR = 0.35$
$CR = 15.75(0.35)(189) = 1042$ psi

Shrinkage:

$SH = (USH)(SSF)(PSH)$
$USH = 27{,}000 - (3000\,E_c/10^6)$
but not less than 12,000

$USH = 27{,}000 - 11{,}490 = 15{,}510$ psi
$V/S = 3.75$ in.
$SSF = 0.79$
$(USH)(SSF) = 12{,}270$ psi
Time after end of curing = 27 days
$PSH = 0.402$
$SH = 12{,}270(0.402) = 4933$ psi

Total losses in Stage 1:

$RET + CR + SH = 10{,}230 + 1042 + 4933 = 16{,}205$ psi

Tendon stress at end of Stage 1

$173{,}900 - 16{,}205 = 157{,}695$ psi

Concrete fiber stress

$189(157{,}695/173{,}900) = 171.4$ psi

Stage 2: To 1 year after prestressing

Relaxation:

$t_1 = 30$ days
$t = 1$ year $= 365$ days
$f_{st} = 157{,}695$ psi
$f_{st}/f_{py} = 157.7/230 = 0.671$
$RET = 157{,}695[(\log 8760 - \log 30)/10] \times [(0.671 - 0.55)] = 2070$ psi

Creep:

$f_c = 171.4$ psi
$PCR = 0.74 - 0.35 = 0.39$
$CR = 15.75(0.39)(171.4) = 1053$ psi

Shrinkage:

$PSH = 0.86 - 0.402 = 0.458$
$SH = 12{,}270(0.458) = 5620$ psi

Total losses in Stage 2:

$2070 + 1053 + 5620 = 8743$ psi

At end of Stage 2, tendon stress at Section L

$157{,}695 - 8743 = 148{,}952$ psi

Concrete fiber stress

$189(148{,}952/173{,}900) = 161.9$ psi

Stage 3: To end of service life (taken as 50 years)

Relaxation:

$t_1 = 1$ year $= 365$ days
$t = 50$ years $= 18{,}250$ days
$\log 24t - \log 24t_1 = 1.699$
$f_{st} = 148{,}952$ psi
$f_{st}/f_{py} = 0.634$
$RET = 148{,}952(0.1699)(0.084) = 2126$ psi

Creep:

$f_c = 161.9$ psi
$PCR = 1 - 0.74 = 0.26$
$CR = 15.75(0.26)(161.9) = 663$ psi

Shrinkage:

$PSH = 1 - 0.86 = 0.14$
$SH = 12{,}270(0.14) = 1718$ psi

Summary of steel stresses at various stages (Design Example 3)

Stress level at various stages		Steel stress, ksi	Percent
Tensioning stress at end		200	
Average stress after seating		182.8	
Middle section stress after seating		174.4	100.0
Losses:			
Elastic shortening	0.5		0.3
Relaxation	14.4		8.2
Creep	2.7		1.6
Shrinkage	12.3		7.0
Total losses after seating		29.9	17.1
Final strand stress at middle section without superimposed load		144.5	82.9

Total long-term losses:

$RET = 10,230 + 2070 + 2126 = 14,426$ psi

$CR = 1042 + 1053 + 663 = 2,758$ psi

$SH = 4933 + 5620 + 1718 = 12,271$ psi

Total losses $= 29,355$ psi

This example shows the detailed steps in arriving at total losses. It is not implied that this effort or precision is required in all design situations.

The *PCI Post-Tensioning Manual* provides a table for approximate prestress loss values which is satisfactory for most design solutions. The value recommended for slabs with stress-relieved 270-kip strand is 30,000 psi which compares with the calculated value of 29,355 psi.

Final tendon stress at Section L
$173.9 = 29.4 = 144.5$ ksi

Percentage loss after anchorage
$(174.4 - 144.5)/174.4 = 17.1$ percent but greater than 15 percent (assumed initially)

Assuming the same percentage loss prevails over the entire tendon length, the average prestressing forces after losses are 19.8 and 14.8 kips per ft, respectively, which are adequate when compared with the design requirements. Therefore, revision is not needed.

Appendix B

Excerpts from Standard Specifications for Highway Bridges of the American Association of State Highway and Transportation Officials, Twelfth Edition, 1977 (and Interim Specifications, 1978–1981).

1.6.7—LOSS OF PRESTRESS

(A) Friction Losses

Friction losses in post-tensioned steel shall be based on experimentally determined wobble and curvature coefficients, and shall be verified during stressing operations. The values of coefficients assumed for design, and the acceptable ranges of jacking forces and steel elongations shall be shown on the plans. These friction losses shall be calculated as follows:

$$T_0 = T_x \times e^{(KL+\mu\alpha)*}$$

When $(KL + \mu\alpha)$ is not greater than 0.3, the following equation may be used:

$$T_0 = T_x \times (1 + KL + \mu\alpha)$$

The following values for K and μ may be used when experimental data for the materials used are not available:

Type of Steel	Type of Duct	K/ft.	μ	(K/m)
Wire or ungalvanized strand	Bright Metal Sheathing	0.0020	0.30	0.0066
	Galvanized Metal Sheathing	0.0015	0.25	0.0049
	Greased or asphalt-coated and wrapped	0.0020	0.30	0.0066
	Galvanized rigid	0.0002	0.25	0.0007
High-strength bars	Bright Metal Sheathing	0.0003	0.20	0.0010
	Galvanized Metal Sheathing	0.0002	0.15	0.0007

Friction losses occur prior to anchoring but should be estimated for design and checked during stressing operations. Rigid ducts shall have sufficient strength to maintain their correct alignment without visible wobble during placement of concrete. Rigid ducts may be fabricated with either welded or interlocked seams. Galvanizing of the welded seam will not be required.

*The terms in the equation are defined as follows:

e = base of Naperian logarithms.
K = friction wobble coefficient per foot of prestressing steel.
T_0 = steel stress at jacking end.
T_x = steel stress at any point x.
μ = friction curvature coefficient.
α = total angular change of prestressing steel profile in radians from jacking end to point x.

(B) Prestress Losses

(1) Loss of prestress due to all causes, excluding friction, may be determined by the following method.* The method is based on normal weight concrete and one of the following types of prestressing steel: 250 or 270 ksi, (1724 or 1862 MPa), seven-wire, stress-relieved strand; 240 ksi (1655 MPa) stress-relieved wires; or 145 to 160 ksi (1000 to 1103 MPa) smooth or deformed bars. For data regarding the properties and effects of lightweight aggregate concrete and low-relaxation tendons, refer to documented tests or see authorized suppliers.

TOTAL LOSS

$$\Delta f_s = SH + ES + CR_c + CR_s$$

where Δf_s = total loss excluding friction in psi. (MPa)
SH = loss due to concrete shrinkage in psi. (MPa)
ES = loss due to elastic shortening in psi. (MPa)
CR_c = loss due to creep of concrete in psi. (MPa)
CR_s = loss due to relaxation of prestressing steel in psi. (MPa)

(a) SHRINKAGE

Pretensioned Members

$$SH = 17{,}000 - 150\,RH$$
$$= (117.21 - 1.034\,RH)$$

Post-tensioned Members

$$SH = 0.80\,(17{,}000 - 150\,RH)$$
$$= 0.80\,(117.21 - 1.034\,RH)$$

where RH = average annual ambient relative humidity in percent (See Figure 1.6.7)

(b) ELASTIC SHORTENING

Pretensioned Members

$$ES = \frac{E_s}{E_{ci}} f_{cir}$$

Post-tensioned Members**

$$ES = 0.5\,\frac{E_s}{E_{ci}} f_{cir}$$

*Should more exact prestress losses be desired, data representing the materials to be used the methods of curing, the ambient service condition and any pertinent structural detail should be determined for use in accordance with a method of calculating prestress losses tha is supported by appropriate research data.

**Certain tensioning procedures may alter the elastic shortening losses.

where

E_s = modulus of elasticity of prestressing steel strand which can be assumed to be 28×10^5 psi ($.193 \times 10^6$ MPa).

E_{ci} = modulus of elasticity of concrete in psi (MPa) at transfer of stress which can be calculated from:

$$E_{ci} = 33w^{3/2}\sqrt{f'_{ci}} \text{ or } (.0428w^{3/2}\sqrt{f'_{ci}})$$

where w is in lb/ft^3 (kg/m^3) and f'_{ci} is in psi (MPa)

f_{cir} = concrete stress at the center of gravity of the prestressing steel due to prestressing force and dead load of beam immediately after transfer, f_{cir} shall be computed at the section or sections of maximum moment. (At this stage, the initial stress in the tendon has been reduced by elastic shortening of the concrete and tendon relaxation during placing and curing the concrete for pre-tensioned members, or by elastic shortening of the concrete and tendon friction for post-tensioned members. The reductions to initial tendon stress due to these factors can be estimated, or the reduced tendon stress can be taken as $0.63 f'_s$ for typical pretensioned members.)

(c) CREEP ON CONCRETE
Pretensioned and Post-tensioned Members.

$$CR_c = 12f_{cir} - 7f_{cds}$$

where

f_{cds} = concrete stress at the center of gravity of the prestressing steel due to all dead loads except the dead load present at the time the prestressing force is applied.

(d) RELAXATION OF PRESTRESSING STEEL*
Prestensioned Members
250 to 270 ksi Strand (1724 to 1862 MPa)

$$CR_s = 20{,}000 - 0.4\,ES - 0.2\,(SH + CR_c)$$

$$\text{or } (137.9 - 0.4\,ES - 0.2\,(SH + CR_c))$$

Post-tensioned Members
250 to 270 ksi Strand (1724 to 1862 MPa)

$$CR_s = 20{,}000 - 0.3\,FR - 0.4\,ES - 0.2\,(SH + CR_c)$$

$$\text{or } (137.9 - 0.3\,FR - 0.4\,ES - 0.2\,(SH + CR_c))$$

*The relaxation losses are based on an initial stress of $0.70 f'_s$.

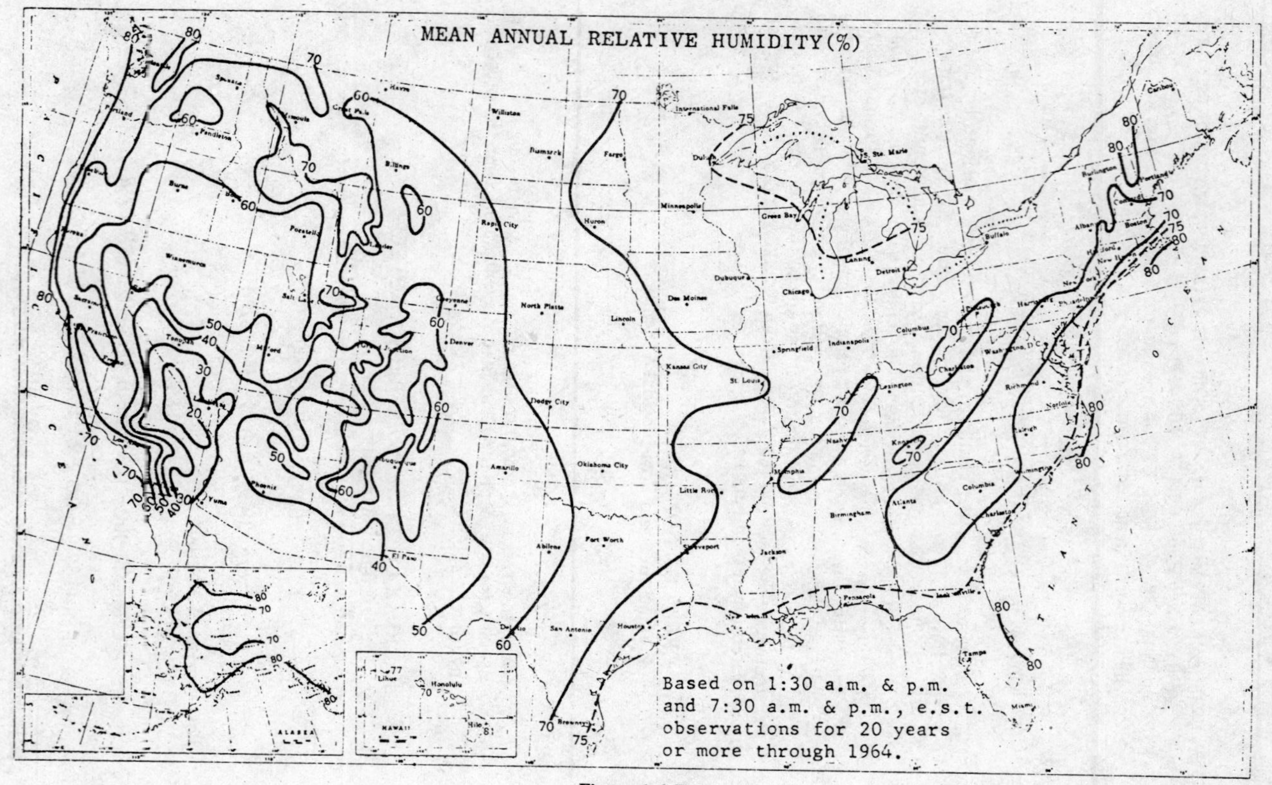
MEAN ANNUAL RELATIVE HUMIDITY(%)
Based on 1:30 a.m. & p.m.
and 7:30 a.m. & p.m., e.s.t.
observations for 20 years
or more through 1964.
International Falls
Minneapolis
Green Bay
Chicago
Dubuque
Des Moines
North Platte
Lincoln
Kansas City
St. Louis
Springfield
Indianapolis
Columbus
Lexington
Nashville
Memphis
Little Rock
Oklahoma City
Dodge City
Amarillo
Fort Worth
Abilene
San Antonio
Houston
Jackson
Birmingham
Atlanta
Columbia
Raleigh
Norfolk
Richmond
Buffalo
Boston
Albany
Detroit
Lansing
Williston
Bismarck
Fargo
Huron
Rapid City
Denver
Cheyenne
Billings
Spokane
Pendleton
Burns
Pocatello
Winnemucca
Phoenix
Yuma
El Paso
Portland
Honolulu
Hilo
Lihue
HAWAII
ALASKA

Figure 1.6.7

240 ksi Wire (1655 MPa)

$$CR_s = 18{,}000 - 0.3\ FR - 0.4\ ES - 0.2\ (SH + CR_c)$$

$$\text{or } (124.10 - 0.3\ FR - 0.4\ ES - 0.2\ (SH + CR_c)$$

145 to 160 ksi Bars (1000 to 1103 MPa)

$$CR_s = 3000\ (20.68\ \text{MPa})$$

where

FR = friction loss stress reduction in psi (MPa) below the level of 0.70 f_s' at the point under consideration, computed according to Article 1.6.7(A).

ES, SH, and CR_c = appropriate values as determined for either pre-tensioned or post-tensioned members.

(2) In lieu of the preceding method, the following estimates of total losses may be used for prestressed members or structures of usual design. These loss values are based on use of normal weight concrete, normal prestress levels, and average exposure conditions. For exceptionally long spans, or for unusual designs, the method in Article 1.6.7(B)(1) or a more exact method shall be used.

TYPE OF PRESTRESSING STEEL	TOTAL LOSS	
	$f_c' = 4{,}000$ psi (27.58 MPa)	$f_c' = 5{,}000$ psi (34.47 MPa)
PRETENSIONING		
Strand		45,000 psi (310.26 MPa)
POST-TENSIONING*		
Wire or Strand	32,000 psi (220.63 MPa)	33,000 psi (227.53 MPa)
Bars	22,000 psi (151.68 MPa)	23,000 psi (158.58 MPa)

*Losses due to friction are excluded. Friction losses should be computed according to Article 1.6.7(A).

COMMENTARY ART. 1.6.7(B)†

Introduction

The subject revision to Section 1.6.7(B)–Prestress Losses in the 1973 AASHTO Specifications for Highway Bridges is based largely on consideration of research results (1,2)* made available subsequent to the adoption of the current specification, and on a proposed revision of the specifications with respect to losses in post-tensioned bridges presented at the 1972 AASHTO regional Bridge Committee meetings (3).

Precise determination of prestress losses in a given situation is very complex and requires detailed information on the materials to be used, the methods of

*Numbers in parenthesis refer to references listed at the end of the commentary.

†Based upon the 1975 Interim Specifications Bridges.

curing, the ambient exposure conditions, and other detailed construction information that usually is not available to designers. For large and/or special structures such as precast or cast-in-place segmental cantilever bridges it may be appropriate or necessary to obtain this information, in order to maintain control of the geometry of the bridge during construction. For most structures, precise calculation of losses is not necessary, and it is possible to calculate reasonably accurate and satisfactory approximations of prestress losses for general design use.

The necessary level of precision in calculation of losses is subject to both technical and subjective evaluation. It has long been recognized that the value used for loss of prestress does not affect the ultimate strength of a bridge, but is important with respect to serviceability characteristics such as deflection, camber, stresses, and cracking (4). The value used for prestress loss also has an influence on economy in that higher values of losses require somewhat more prestressing steel. However, it has been shown that both the serviceability and economy of prestressed bridges are relatively insensitive to sizeable variations in the value assumed for prestress losses in typical designs (5). None-the-less, there is disagreement between agencies and designers as to what degree of precision is necessary in the calculation of losses, and there is some ambiguity in the research results on this subject, even when attempts are made to relate the research to a common base.

As a result of the above considerations, the subject revision of the AASHTO prestress loss specifications provides for three optional methods of loss calculation as follows:

1) The lump sum loss values in Section 1.6.7(B)(2).
2) The more precise and detailed provisions of Section 1.6.7(B)(1).
3) The use of still more exact procedures for loss calculation under the provisions of the footnote on page one of the proposed revision.

A detailed commentary of considerations related to the first two approaches is presented below. Due to the broad nature of calculations and procedures that might be involved in calculating losses under the provisions of the footnote as noted under the third option above, the discussion of this approach is not included in the commentary.

Section 1.6.7(B)(1)

As compared to the existing specifications, the revision adds consideration of three types of tendons used for post-tensioned structures: 250 or 270 ksi (1724 or 1862 MPa), seven-wire, stress-relieved strand; 240 ksi (1655 MPa), stress-relieved wires; or 145 to 160 ksi (1000 to 1103 MPa) smooth or deformed bars.

The total loss is specified as the direct addition of losses due to shrinkage, elastic shortening, concrete creep, and steel relaxation. Addition of these losses in this manner is an approximation because the losses are to some extent interrelated. As will be discussed below under Section 1.6.7(B)(1)(d), Relaxation of Prestressing Steel, the inter-relationship of losses is considered in the formulas given for relaxation of prestressing steel.

The loss format of addition of four loss components is similar to the procedure outlined in the "Criteria for Prestressed Concrete Bridges" published by the U.S. Department of Commerce, Bureau of Public Roads in 1954 (6). However, the values related to each of the four components in the loss equation have been modified to more accurately reflect the research that has occurred in each area since 1954.

Section 1.6.7 (B)(1)(a)—Shrinkage

The two equations for loss due to concrete shrinkage give similar shrinkage losses to those included in the table in the current specifications. However, the equations have the advantage of eliminating the abrupt jump in shrinkage loss between the three humidity ranges. A map of the United States showing the average annual ambient relative humidity is presented in Fig. 1 (7). This map is to be used as the basis for determining average annual ambient relative humidity for use in calculating shrinkage loss.

The ultimate shrinkage strain of small specimens (8, 9) was assumed to be 550×10^{-6} at 50 percent relative humidity. This is an average value for shrinkage strain, some concretes may inherently shrink somewhat more and others somewhat less. Research indicates that shrinkage strains are reduced 10 to 40 percent by steam curing (10).

Research (11) has shown that shrinkage is related to the size or thickness of the member. This factor is usually expressed as the Volume/Surface ratio of the member. In development of the subject equations an average Volume/Surface ratio of 4 inches (.102 m) was assumed, and the ultimate shrinkage strain was reduced by a factor of 0.77. The volume/surface ratio is effectively one-half the average member thickness and it therefore represents the distance from the drying surface to the center of the member.

Shrinkage loss is also a function of the average ambient relative humidity of the environment. The amount of shrinkage reduces as the humidity increases. In some cases concrete stored under water has been observed to expand slightly. In developing the existing AASHTO specifications, humidity shrinkage factors were assumed as follows (12):

Average Ambient Relative Humidity Percent	Humidity Shrinkage Coefficient
100-75	0.3
75-25	1.0
25-0	1.3

Based on the above assumptions and on assumed modulus of elasticity for steel of 28×10^6 psi ($.193 \times 10^6$ MPa), shrinkage losses were calculated for the table in the existing specifications as the product of the four interior columns in the table below:

Humidity	Shrinkage Strain	Steel Modulus-psi (MPa)	Volume/Surface Factor	Humidity Factor	Shrinkage Loss-psi (MPa)
100-75	550×10^{-6}	28×10^6 ($.193 \times 10^6$)	0.77	.3	3,540 (24.41)
75-25	550×10^{-6}	28×10^6 ($.193 \times 10^6$)	0.77	1.0	11,800 (81.36)
25-0	550×10^{-6}	28×10^6 ($.193 \times 10^6$)	0.77	1.3	15,250 (105.15)

These shrinkage loss values were rounded to 5,000, 10,000 and 15,000 for use in the existing (1973 edition) AASHTO specifications.

Multiplication of the shrinkage loss equation for post-tensioned design by the factor 0.8 is in consideration of the fact that post-tensioning is not accomplished until the concrete has reached an age of at least seven days at which time 20 percent of the shrinkage is assumed to have occurred. This is a very conservative assumption on time of stressing relative to concrete age for most bridges. A general curve relating shrinkage (or creep) to time is presented in Fig. C1(13). Use of some other reduction factor in the equation for shrinkage loss for post-tensioned applications might be considered if the designer has knowledge that the elapsed time from concrete placement to stressing will be significantly greater than seven days.

In the discussion relative to the subject revision of the loss specification, some expressed the view that the basic ultimate shrinkage strain of 550×10^{-6} was too low, while others suggested that it was too high, and still others felt the value to be satisfactory. This value was retained, and the tabular values in the existing specification were replaced by the two linear equations for reasons described above.

1.6.7(B)(1)(b) ELASTIC SHORTENING.

The table of losses due to elastic shortening in the current AASHTO Specification has been deleted in the revision, and the formula for elastic shortening loss has been generalized by including the ratio of the modulus of elasticity of steel to the modulus of concrete at release. The formula in the current specifications uses a value of 7 for the ratio of the moduli of elasticity.

A significant revision is in the definition of the concrete stress to be used in calculating elastic shortening loss. In the current specifications, this value is given the nomenclature f_{cr} where f_{cr} = average concrete stress at the center of gravity of the prestressing steel at time of release. The reason for using the average value of concrete stress (along the length of the beam) was to avoid the necessity of precise calculation of stresses at each point in a beam *before* a value of the elastic shortening loss could be obtained. It was intended that the average stress value could be estimated or obtained from the table values so that it would not be necessary for designers to use a trial and error process of design iterations in order

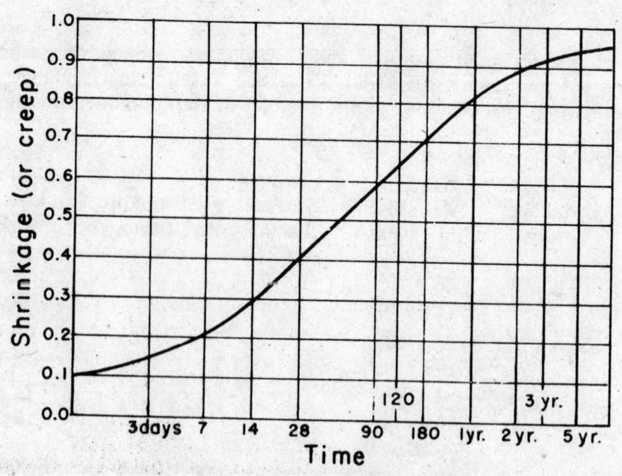

FIG. C1—Shrinkage or creep vs. time

to obtain a value for the elastic shortening loss. Use of average stress along the center of gravity of the steel is a conservative approximation since the average stresses and related elastic shortening is greater than the stress and elastic shortening at the points of maximum moment.

The current specifications were criticized for use of average concrete stresses in calculating both elastic shortening and concrete creep losses. It was pointed out that this process was theoretically incorrect and that the actual elastic shortening loss (and creep loss) is dependent on the concrete stress *at* the section under consideration. In development of the proposed revision, it was agreed that the definition of stress for use in elastic shortening loss calculations would be more precise if related to the section under consideration, and it was acknowledged that this increase in precision might require a trial and error design process. To minimize the amount of additional design calculation, the final specification wording related the concrete stress, f_{cir}, for use in the elastic shortening formula to "...the section or sections of maximum moment," rather than the section under consideration. For a simple span precast design, this change results in a requirement of loss calculation at one point only, whereas relating the definition of f_{cir} to the section under consideration would have required a separate loss calculation at each point where stresses were investigated. For simple span pretensioned designs, it should be understood that the intent of the proposed revision as written is that elastic shortening loss (and concrete creep loss) is to be calculated only at the point of maximum moment, and that this value is to be used as the loss value throughout the length of the beam. For continuous post-tensioned bridges, the intent of this portion of the specification is that the loss value be calculated at each controlling point of maximum positive or negative moment along the length of the structure.

It should be noted that the stress in the tendon for pretensioned products for use in calculating f_{cir} should be reduced from the initial stress value by an amount to allow for elastic shortening of the concrete and for steel relaxation prior to release of the tendon force to the member.* Recognizing that this further complicates the iterative process for the designer, an approximate value of the reduced tendon stress of 0.63 f_s' is provided. This approximation relates to tendons stressed initially to 0.70 f_s'. Note that initial overstress is permitted so that the initial tendon stress after transfer to concrete is not greater than 0.70 f_s'; this requires more accurate estimates for initial losses to ensure the correct tendon stress at transfer. It is more complicated to obtain the reduced tendon force for calculation of f_{cir} for post-tensioned applications because the reduction depends on both elastic shortening and tendon friction. In addition, the use of initial tendon stresses in excess of 0.70 f_s' is quite common in post-tensioned applications to compensate for anchor seating loss. If the designer is willing to sacrifice some accuracy for convenience in calculation; a value of 0.65 f_s' might be used in calculating f_{cir} for post-tensioned designs. For structures utilizing long tendons where the friction losses become sizeable, this approximation would become quite conservative.

For post-tensioned members with only one tendon, there would be no elastic shortening loss because the member shortening takes place prior to tendon anchorage. However, most post-tensioned applications involve multiple tendons. When multiple tendons are used, the first tendon stressed and anchored sustains elastic shortening loss due to stressing of all subsequent tendons. The last tendon stressed sustains no elastic shortening loss. As a result of this, the formula for elastic

---

*See Figures C2 and C3 for steel relaxation losses for constant strain specimens of stress relieved and stabilized 270k(1.2 MN) strand.

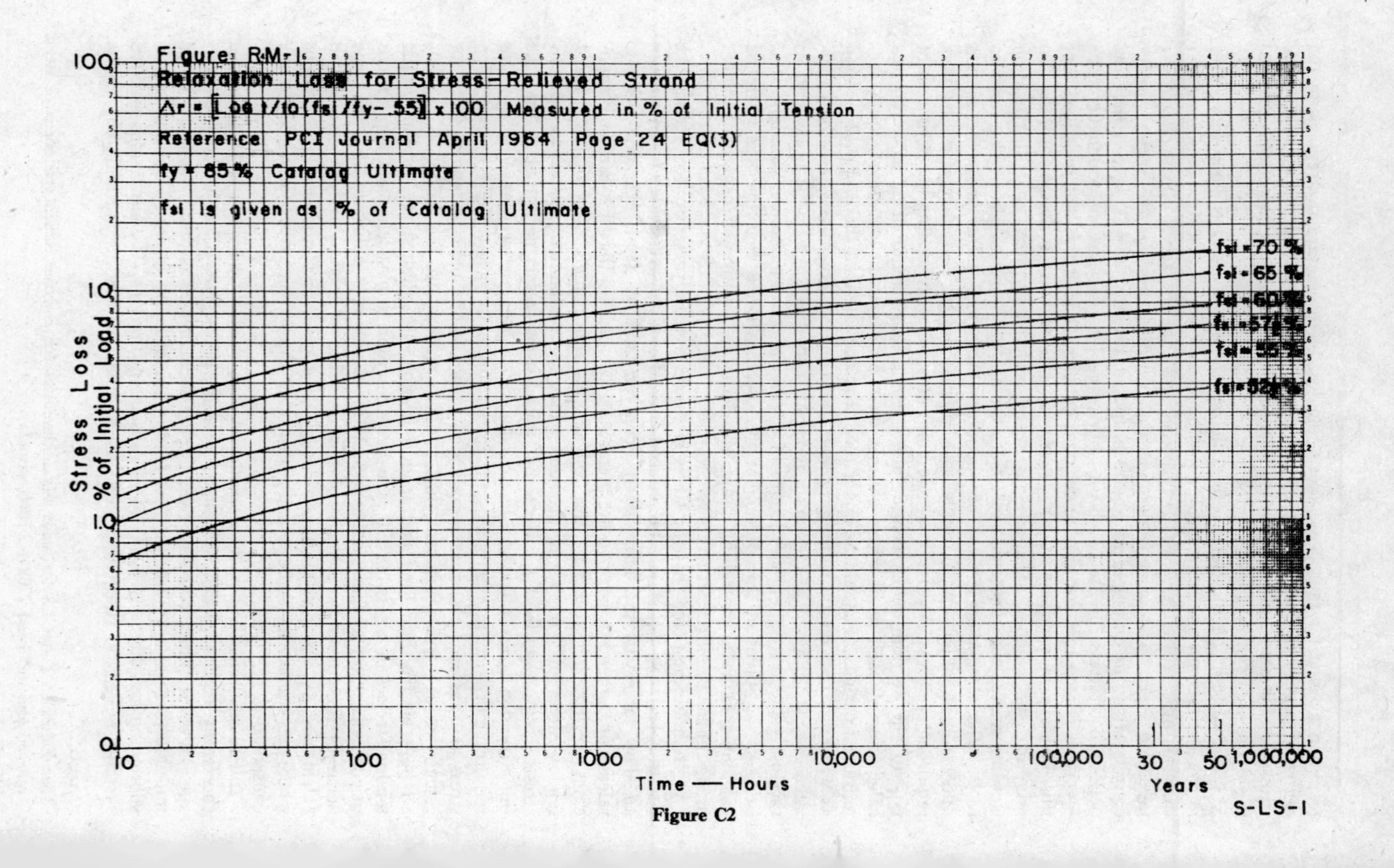
Figure: RM-1
Relaxation Loss for Stress-Relieved Strand
Δr = [Log t/10 (fsi/fy - .55)] x 100 Measured in % of Initial Tension
Reference PCI Journal April 1964 Page 24 EQ(3)
fy = 85% Catalog Ultimate
fsi is given as % of Catalog Ultimate
fsi = 70%
fsi = 65%
fsi = 60%
fsi = 57½%
fsi = 55%
fsi = 52½%
Stress Loss
% of Initial Load
100
10
1.0
0.1
10
100
1000
10,000
100,000
1,000,000
30
50
Years
Time — Hours
S-LS-I

Figure C2

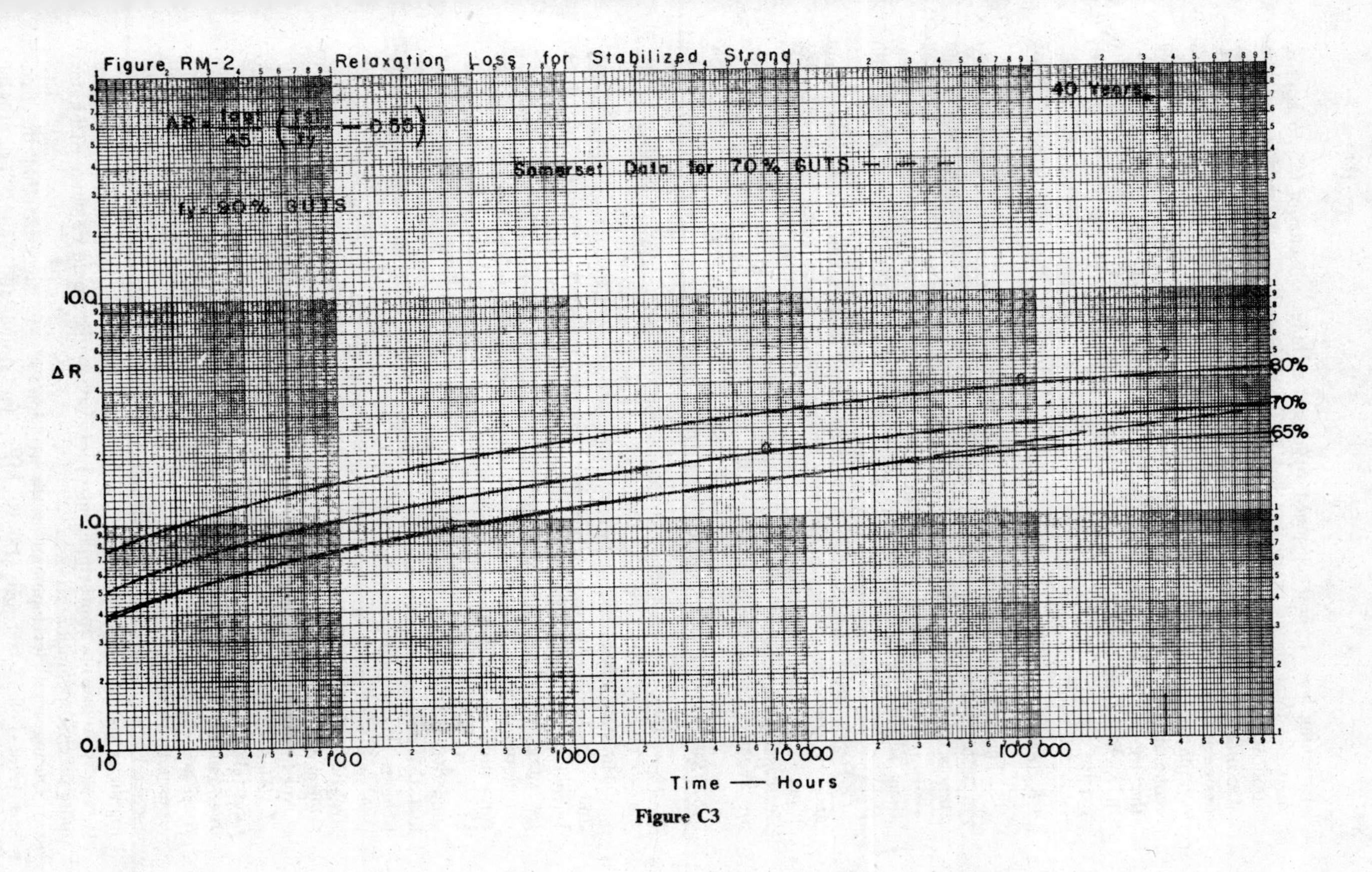

Figure RM-2 Relaxation Loss for Stabilized Strand
$\Delta R = \frac{\log t}{45}\left(\frac{f_{si}}{f_y} - 0.55\right)$
f_y - 90% GUTS
Somerset Data for 70% GUTS — — —
40 Years
ΔR
10.0
1.0
0.1
80%
70%
65%
10
100
1000
10 000
100 000
Time —— Hours

Figure C3

shortening loss for post-tensioned structures includes a factor of 0.5 to approximate the total elastic shortening loss as the average loss in all tendons. In some applications, the total elastic shortening loss is eliminated by temporarily overstressing each tendon by the amount of elastic shortening anticipated. While the designer should be aware of this option for special structures, this procedure does complicate the stressing operation and it should only be used where the potential economies in tendon material justify the additional work in the field, and where the stressing crew is sufficiently trained that a different stress level can be specified in each tendon with reasonable assurance that the stressing will be properly carried out.

1.6.7(B)(1)(c) CREEP OF CONCRETE

The formula in the proposed revision for loss due to concrete creep is based on both field measurements and analytical studies conducted at the University of Illinois (1). This formula utilizes the same value of f_{cir} as defined and discussed above in the section on elastic shortening loss. This results in the same complications relative to calculating concrete creep loss as were discussed above with respect to elastic shortening.

The quantity f_{cds} related to the permanent loads applied after prestressing and represents the change of concrete fiber stress at c.g.s. caused by these loads. Usually, the application of load causes the concrete fiber stress at c.g.s. to decrease in magnitude, and thereby reduces the loss due to concrete creep; hence the negative sign in the formula for CR_c.

1.6.7(B)(1)(d) RELAXATION OF PRESTRESSING STEEL

New formulas are added in the section on steel relaxation relative to the various types of steel used for tendons of post-tensioned structures.

The steel relaxation loss formulas for post-tensioned structures utilizing strand and wire tendons include terms reflecting the influence of friction loss. It is intended that steel relaxation losses for post-tensioned structures be computed at the same points of maximum positive and negative moment as were used for calculation of elastic shortening and concrete creep loss. For post-tensioned structures utilizing bar tendons, a flat value of steel relaxation loss of 3000 psi (20.68 MPa) is specified. Representatives of companies supplying bar tendons have suggested that the steel relaxation loss for bars should also consider the friction loss reduction and other loss modification factors used for strand and wire tendons. While this was acknowledged as theoretically correct, it was considered that such a formula was an unnecessary refinement in view of the very nominal steel relaxation loss specified for bar tendons. This loss value for bar tendons is an arbitrary figure developed on consideration of constant strain steel relaxation loss curves for both smooth and deformed bars (3).

All formulas for steel relaxation loss consider, in an approximate way, the inter-relationship between steel relaxation and other losses. The higher the losses due to elastic shortening, concrete creep, and shrinkage, the lower will be the steel stress and the related steel relaxation loss. This inter-relationship can be evaluated accurately by use of time-related curves for losses due to steel relaxation, shrinkage, and concrete creep, and by use of a step-by-step integration process considering the inter-relationship of the losses. The formulas in the proposed revision were developed to provide approximations of loss values that would be obtained by this more precise integration process.

During final development of the revision, it was suggested that the base loss values of 20,000 psi (137.90 MPa) for strand and 18,000 psi (124.11

MPa) for wire should be modified to reflect the actual constant strain steel relaxation loss for wire and strand which is about 15 percent for tendons stressed to an initial stress of 0.70 f_s'. These values would be about 28,300 psi (195.12 MPa) for strand, and about 25,200 psi (173.75 MPa) for wire. It was felt that this would put the equations in a more direct relationship to research results from constant strain strand tests. While there was agreement that such a procedure would be preferable from a technical standpoint, it was noted that all the modification factors in the formulas would have to be increased so that the net losses would remain the same as those resulting from the proposed equations. Further, it was noted that studies would have to be made to develop appropriate values for the reduction coefficients in the formulas. In view of these considerations, it was decided to retain the equations as previously developed with base loss values of 20,000 psi (137.90 MPa) for strand and 18,000 psi (124.11 MPa) for wire.

In application of the friction loss reduction in the equations for steel relaxation loss for post-tensioned wire and strand tendons, note that the value of *FR* to be used is limited to the reduction in stress level below stress level of 0.70 f_s'. Post-tensioning tendons are commonly stressed in excess of 0.70 f_s' (usually to about 0.74 f_s') to allow for anchor seating loss. While this process often results in some small portion of the tendon length retaining a stress slightly in excess of 0.70 f_s' immediately after anchoring, the equations for steel relaxation in wire and strand tendons are considered applicable. However, due to this over stressing, it is necessary to limit the value of *FR* in the equation to the reduction below a stress of 0.70 f_s' (rather than the total friction loss) in order to retain the basic presumption of the equations that the initial stress is 0.70 f_s'.

Steel relaxation increases rapidly with increases in initial stress above 0.70 f_s', and it decreases rapidly as stresses go below 0.70 f_s'. The decreases in stress are considered in the equations. It is emphasized, however, that the equations are not accurate where initial stresses at points of maximum moment exceed 0.70 f_s'. Where such over stressing is specified, an adjustment (increase) should be made in the steel relaxation loss values.

1.6.7(B)(2) LUMP SUM VALUES FOR PRESTRESS LOSS

Nearly all prestressed concrete bridges now in service were designed using the lump sum loss values of 35,000 psi (241.32 MPa) for pretensioned applications and 25,000 psi (172.37 MPa) plus friction for post-tensioned applications. In general the performance and serviceability characteristics of these bridges have been excellent. The revision of prestress losses in the 1971 AASHTO Interim Specifications was motivated by an awareness that research had shown some of the initial assumptions made in developing the initial lump sum loss values might be unconservative. This was particularly true with respect to steel relaxation loss, but applied to some extent to shrinkage and concrete creep losses. However, the basic intent of the provisions of the 1971 AASHTO Interim Specifications on prestress losses was to permit designers to develop lump sum losses that considered the prestress level and exposure conditions of the structure as well as the results of research then available. Some designers inferred an accuracy, or a requirement for precision, in the 1971 Interim Specifications that was not intended. As has been noted above, precise calculation of prestress losses is quite involved and requires detailed information on materials, exposure, and construction procedures not generally available to designers. Further, such precise calculations are usually not warranted from the standpoints of structural economy, serviceability, or strength.

There are numerous factors in the design of prestressed concrete bridges, as with bridges of other materials, that are included in the specifications for the convenience of the designer with the understanding that these factors are only approximations of actual behavior. Some of the provisions of Section 3—Distribution of Loads in the 1973 AASHTO Specifications exemplify such approximations. There are other similar approximations in the specifications, and in regard to prestressed concrete bridge design, a significant conservative approximation exists in the general use of the gross concrete section to compute section properties. Use of transformed section properties for all loads applied after prestressing is perfectly valid technically, and can result in an increase in the bottom section modulus in the order of 10 to 15 percent.

In view of the generally satisfactory performance of bridges designed using lump sum loss values, the conservative approximations in other portions of the specifications, and the convenience to the designer, it was considered desirable that the revision should include lump sum loss values for "prestressed members or structures of usual design." The introductory paragraph to the section indicates that structures of usual design include those of normal weight concrete, normal prestress levels, and average exposure conditions. In addition, it is suggested that the more precise procedures of Section 1.6.7(B)(1) should be used for structures of exceptionally long span or unusual design. The table of lump sum values includes concrete strengths of 4000 and 5000 psi (27.58 and 34.47 MPa). However, the lump sum loss values may be used for bridges with concrete strengths 500 psi (3.45 MPa) above or below the values of 4000 psi and 5000 psi (27.58 and 34.47 MPa) listed in the table headings. Thus, the range of concrete strengths covered by the lump sum loss values may be considered to extend from 3500 psi to 5500 psi (24.13 to 37.92 MPa). These lump sum loss values were developed on the assumption that the full allowable compressive stress of concrete is required by stress conditions during the life of the member. For structures or elements where the concrete strength is determined on the basis of the nominal minimums in the specifications, and where full utilization of the allowable compressive stresses does not occur during the life of the structure, the lump sum loss values will be somewhat conservative. Such conditions often occur when an over-size section is used for a short span due to the esthetic consideration of matching the depth of section used on a longer adjacent span.

EXAMPLE APPLICATIONS

EXAMPLE I (See Fig. E1)

Compute the prestress losses for the bridge in Fig. E1 which utilizes 54 in. I-beams with 16-½ in. diameter 270 k stress relieved strands at an eccentricity of 22.2 in. (from Illinois Department of Transportation Beam Charts). The structure to be built in Illinois has two equal spans of 70 ft. which are made continuous for composite dead load and live load by means of continuity reinforcement in the deck slab and diaphragm at the center pier.

Properties

Concrete: $f_c' = 5000$ psi (28 day strength)
$f_{ci}' = 4000$ psi (Release strength)

Initial steel stress: 0.7×270 ksi $= 189$ ksi

Section Properties

Beam: Area $= A = 599$ in²
Moment of inertia $= I = 213{,}078$ in⁴
Center of gravity $= \bar{y} = 24.97$ in. from bottom
Weight $= W_B = 0.625$ k/ft.

Composite (Transformed Section)

Moment of inertia $= I' = 530{,}866$ in⁴
Center of gravity $= \bar{y}' = 40.9$ in. from bottom

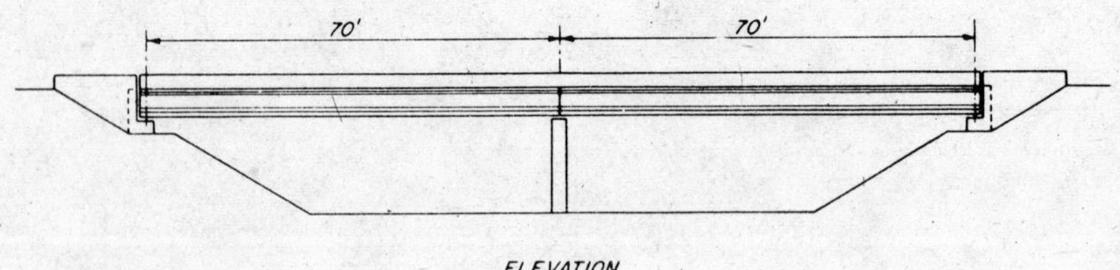

ELEVATION

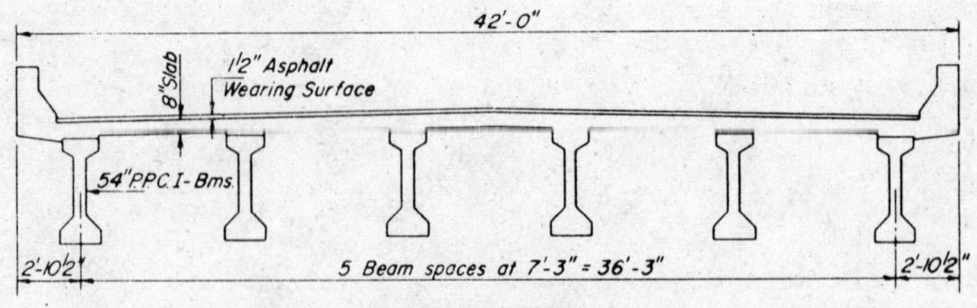

CROSS SECTION

Figure E1

Loss Calculation

Total Loss $= \Delta fs = SH + ES + CR_c + CR_s$

Shrinkage (SH)

$$SH = 17{,}000 - 150\ RH$$

In accordance with the U.S. Weather Bureau Map, the mean annual relative humidity for Illinois is about 70 percent.

$$SH = 17{,}000 - 150 \times 70 = \underline{6{,}500 \text{ psi}} \text{ (44.82 MPa)}$$

Elastic Shortening (ES)

$$ES = (E_s/E_{ci})\, f_{cir} \quad E_s = 28 \times 10^6 \text{ psi}$$
$$E_{ci} = 33\ W^{3/2} \sqrt{f_{ci}} = 33 \times 150^{3/2} \sqrt{4000} = 3.8 \times 10^6 \text{ psi}$$

$$f_{cir} = \frac{F}{A} + \frac{Fe^2}{I} - \frac{M_n db \times e}{I}$$

where:

F = Initial prestress force adjusted for relaxation between time of stressing and release of strands (assume 18 hours), and elastic shortening.
$M_n db$ = Moment due to the dead load of the beam at the point of maximum moment. In this structure the maximum moment occurs at midspan.
$F = A_s \times f_s$

Assume $F = 16 \times .153 \times .63 \times 270 = 416.4$ kips

$$M_n db = 0.125\ W_b L^2 = 0.125 \times .625 \times 70^2 = 382.8 \text{ ft-kips}$$

$$f_{cir} = \frac{416.4}{599} + \frac{416.4 \times 22.2^2}{213{,}078} - \frac{382.8 \times 12 \times 22.2}{213{,}078}$$

$$f_{cir} = 0.695 + 0.963 - 0.479 = 1{,}179 \text{ psi}$$

$$ES = \frac{28 \times 10^6}{3.8 \times 10^6} \times 1179 = 8687 \text{ psi}$$

Assume strand release 18 hours after stressing. From Fig. C2 the relaxation loss equals 3.5 percent of the initial stress of 189 ksi = $.035 \times 189{,}000 = 6{,}615$ psi.

Check initial force assumption of $0.63\ f_s'$

$$189{,}000 - 8687 - 6615 = 173{,}698 \text{ psi}$$
$$173{,}698/270{,}000 = 0.643\ f_s'$$

This is probably close enough to $0.63\ f_s'$ assumption but recalculate one time

Assume $F = 16 \times .153 \times .643 \times 270 = 425$ kips

$$f_{cir} = \frac{425}{599} + \frac{425 \times 22.2^2}{213{,}078} - \frac{382.8 \times 12 \times 22.2}{213{,}078}$$

$$f_{cir} = 0.709 + 0.983 - 0.479 = 1{,}213 \text{ psi}$$

$$ES = \frac{28 \times 10^6}{3.8 \times 10^6} \times 1{,}213 \text{ psi} = \underline{8938 \text{ psi}} \quad (61.62 \text{ MPa})$$

Check assumed force

$$189{,}000 - 8938 - 6615 = 173{,}447 \text{ vs } 173{,}447 \text{ vs } 173{,}698$$

assumed

$$173{,}447/270{,}000 = 0.642 f_s' \text{ OK.}$$

Creep of Concrete (CR_c)

$$CR_c = 12 f_{cir} - 7 f_{cds}$$

$$f_{cds} = \frac{M_n{}^s e}{I} + \frac{M_c{}^d [e + (\bar{y}' - \bar{y})]}{I'}$$

$M_n{}^s$ = Moment due to slab weight at point of maximum moment

$M_n{}^s = 0.125 \times W_s \times L^2 = 0.125 \times 150 \times .67 \times 7.25 \times 70^2 =$ 446 ft-kips

$M_c{}^d$ = Moment due to composite dead loads at the point of maximum moment

$M_c{}^d = C\, w_e\, L^2$ where C = Influence Coefficient

Asphalt $= .019 \times 7.25 = 0.138$ k/ft.

Curb, Parapet, Rail = 0.133 k/ft.

We = 0.271 k/ft.

$$M_c{}^d = .0625 \times 0.271 \times 70^2 = 83 \text{ ft-kips}$$

$$f_{cds} = \frac{446 \times 12 \times 22.2}{213{,}078} + \frac{83 \times 12\,[22.2 + (40.9 - 24.97)]}{530{,}866}$$

$$f_{cds} = 0.558 + 0.072 = 0.630 \text{ ksi}$$

$$CR_c = 12 \times 1213 - 7 \times 630 = \underline{10{,}146 \text{ psi}} \ (69.95 \text{ MPa})$$

Relaxation of Prestressing Steel (CR_s)

$$CR_s = 20{,}000 - 0.4\, ES - 0.2\, (SH + CR_c)$$

$$CR_s = 20{,}000 - 0.4 \times 8938 - 0.2\, (6500 + 10{,}146)$$

$$CR_s = 20{,}000 - 3575 - 3329 = \underline{13{,}096} \ (90.29 \text{ MPa})$$

Total Loss

$$\Delta f_s = SH + ES + CR_c + CR_s$$

$$\Delta f_s = 6500 + 8938 + 10{,}146 + 13{,}096 = \underline{\underline{38{,}680 \text{ psi}}} \ (266.68 \text{ MPa})$$

EXAMPLE II (See Fig. E2)

Compute the prestress losses for a 21″ deep by 3′-0″ wide precast voided deck slab bridge in Illinois spanning 45′-0″. The section requires 18-7/16″ diameter 270k strands (from Illinois Department of Transportation beam charts). The deck will be waterproofed with a membrane and asphalt weighing 19 pounds per square foot.

Beam Properties

Area	$= A$	$= 502$ in.^{2}
Weight	$= W_B$	$= 550$ lb/linear ft.
Moment of Inertia	$= I$	$= 25{,}007$ in.^{4}
Eccentricity	$= E$	$= 7.5$ in.
Design Span	$= L$	$= 45$ ft.

Concrete Properties

$$f_c' = 5{,}000 \text{ psi at 28 days}$$
$$f_{ci}' = 4{,}000 \text{ psi at release}$$
$$E_{ci} = 3.8 \times 10^6 \text{ psi}$$

Total Losses

$$\Delta f_s = SH + ES + CR_c + CR_s$$

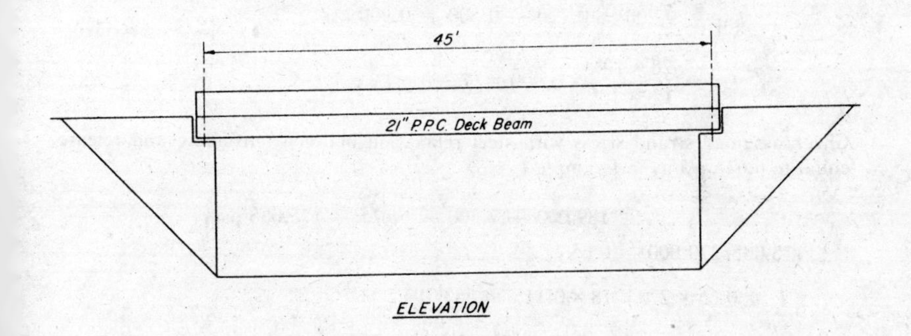

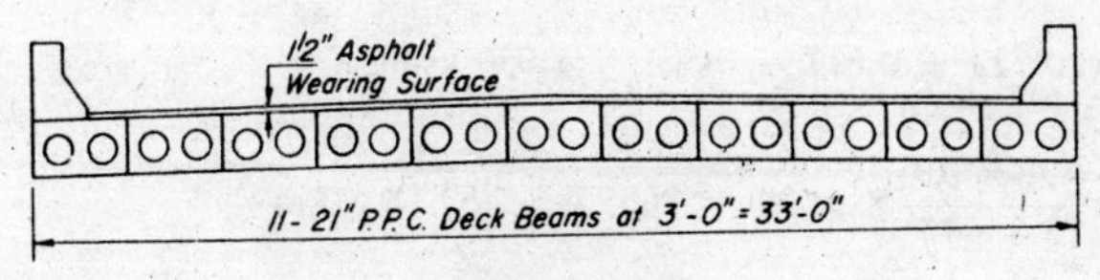

Figure E2

Shrinkage (SH)

$$SH = 17{,}000 - 150\, RH$$

From U.S. Weather Bureau Map for Mean Annual Relative Humidity, RH for Illinois is about 70 percent.

$$SH = 17{,}000 - 150 \times 70 = \underline{6500 \text{ psi}} \quad (44.82 \text{ MPa})$$

Elastic Shortening (ES)

$$ES = \frac{E_s}{E_{ci}} \times f_{cir} = \frac{28 \times 10^6}{3.8 \times 10^6} \times f_{cir}$$

$$f_{cir} = \frac{F}{A} + \frac{Fe^2}{I} - \frac{M_n db \times e}{I}$$

Assume strand stress $= 0.63\, f_s'$

$$F = 0.63 \times 270 \times 18 \times 0.115 = 352 \text{kips}$$

$M_n db$ = Moment due to beam dead load

$$M_n db = 0.125 \times 550 \times 45^2 = 139.2 \text{ft-kips}$$

$$f_{cir} = \frac{352}{502} + \frac{352 \times 7.5^2}{25{,}007} - \frac{139.2 \times 12 \times 7.5}{25{,}007}$$

$$f_{cir} = 0.700 + 0.790 - 0.500 = 0.990 \text{ ksi}$$

$$ES = \frac{28 \times 10^6}{3.8 \times 10^6} \times 0{,}990 = 7{,}300 \text{ psi}$$

Check assumed strand stress with steel relaxation between tensioning and release equal to 6,615 psi as in Example I

$$f_s = 189{,}000 - 7{,}300 - 6{,}615 = 175{,}085 \text{ psi}$$

$$175{,}085/270{,}000 = 0.65\, f_s'$$

$$F = 0.65 \times 270 \times 18 \times 0.115 = 363 \text{kips}$$

$$f_{cir} = \frac{363}{502} + \frac{363 \times 7.5^2}{25{,}007} - \frac{139.2 \times 12 \times 7.5}{25{,}007}$$

$$f_{cir} = 0.723 + 0.817 - 0.501 = 1.039 \text{ ksi}$$

$$ES = \frac{28 \times 10^6}{3.8 \times 10^6} \times 1.039 = \underline{7655 \text{ psi}} \quad (52.78 \text{ MPa})$$

Check assumed strand stress

$$f_s = 189{,}000 - 7655 - 6615 = 174{,}730 \text{ psi OK}$$

Creep of Concrete (CR_c)

$$CR_c - 12\, f_{cir} - 7\, f_{cds}$$

Moment due to additional dead load — $M_A D$

$$W = .019 \times 3 = .057 \text{ k/linear ft.}$$

$$M_A D = .125 \times .057 \times 45^2 = 14.43 \text{ ft-kips}$$

$$f_{cds} = \frac{M_A D \text{ [illegible]}}{\text{[illegible]}} \text{[illegible]} \frac{14.43 \times 12 \times 7.5}{25{,}007} = 0.052 \text{ ksi} = 52 \text{ psi}$$

$$CR_c = 12 \times 1{,}039 - 7 \times 52 = \underline{12{,}104 \text{ psi}}\ (83.45 \text{ MPa})$$

Relaxation of Prestressing Steel (CR_s)

$$CR_s = 20{,}000 - 0.4\, ES - 0.2\, (SH + CR_c)$$

$$CR_s = 20{,}000 - 0.4 \times 7655 - 0.2\, (6500 + 12{,}104)$$

$$CR_s = 20{,}000 - 3062 - 3721 = \underline{13{,}217 \text{ psi}} \quad (91.13 \text{ MPa})$$

Total Loss

$$\Delta f_s = 6500 + 7655 + 12{,}104 + 13{,}217 = \underline{39{,}476 \text{ psi}} \quad (272.18 \text{ MPa})$$

EXAMPLE III (See Fig. E3)

Check the prestress losses for a 17 in. deep by 3 ft. wide voided deck beam used on a span of 30 ft. on a country road near Yuma, Arizona. No wearing surface will be applied. From Illinois Department of Transportation beam charts, 7-½ in. diameter 270 k strands are required.

Beam Properties

Weight	$= W_B$	$=$	470 lbs per
Span	$= L$	$=$	30 ft.
Eccentricity of prestressing steel	$= e$	$=$	6.46 in.
Area	$= A$	$=$	431 in^2
Moment of Inertia	$= I$	$=$	13,274 in.^{4}

Concrete Properties

$$f_c' = 5000 \text{ psi at 28 days}$$
$$f_{ci} = 4000 \text{ psi at release}$$
$$E_{ci} = 3.8 \times 10^6 \text{ psi}$$

Total Losses

$$\Delta f_s = SH + ES + CR_c + CR_s$$

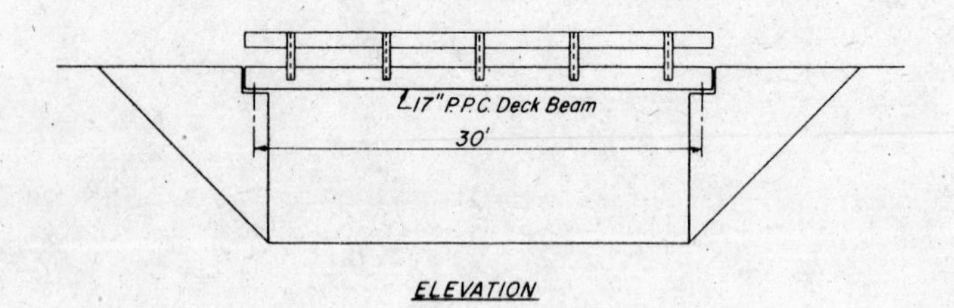

11-17" PPC Deck Beams at 3'-0" = 33'-0"

CROSS SECTION

Figure E3

Shrinkage (SH)

$$SH = 17{,}000 - 150\,RH$$

From U.S. Weather Bureau Map for Mean Annual Relative Humidity, *RH* for Yuma, Arizona is about 30 percent.

$$SH = 17{,}000 - 150 \times 30 = \underline{12{,}500 \text{ psi}} \quad (86.18 \text{ MPa})$$

Elastic Shortening (ES)

$$ES = \frac{28 \times 10^6}{3.8 \times 10^6} \times f_{cir}$$

$$f_{cir} = \frac{F}{A} + \frac{Fe^2}{I} - \frac{M_{n}db \times e}{I}$$

Assume $F = A_s \times 0.63\, f_s'$

$$F = 7 \times 0.153 \times 0.63 \times 270 = 182.2^{\text{kips}}$$

$$M_{n}db = 0.125 \times 0.470 \times 30^2 = 52.8^{\text{ft-kips}}$$

$$f_{cir} = \frac{182.2}{431} + \frac{182.2 \times 6.46^2}{13{,}274} - \frac{52.8 \times 12 \times 6.46}{13{,}274}$$

$$f_{cir} = 0.422 + 0.573 - 0.308 = 0.687$$

$$ES = \frac{28 \times 10^6}{3.8 \times 10^6} \times 0.687 = 5062 \text{ psi}$$

Check assumption of prestress force. Assuming strand release 30 hours after tensioning, the initial stress is reduced about 4 percent (see Figure C2).

$$0.04 \times 189{,}000 = 7{,}560 \text{ psi}$$

$$f_s = 189{,}000 - 5062 - 7560 = 176{,}378$$

$$176{,}378/270{,}000 = 0.653\, f_s'$$

Revise force:

$$F = 0.653 \times 270 \times 7 \times 0.153 = 188.8^{\text{kips}}$$

$$f_{cir} = \frac{188.8}{431} + \frac{188.8 \times 6.46^2}{13{,}274} - \frac{52.8 \times 12 \times 6.46}{13{,}274}$$

$$f_{cir} = 0.438 + 0.594 - 0.308 = 0.724 \text{ ksi}$$

$$ES = \frac{28 \times 10^6}{3.8 \times 10^6} \times 0.724 = \underline{5335 \text{ psi}} \quad (36.78 \text{ MPa})$$

Check assumption of prestress force:

$$f_s = 189{,}000 - 5335 - 7560 = 176{,}105 \text{ psi OK}$$

Concrete Creep (CR_c)

$$CR_c = 12\, f\text{cir} - 7\, fcds$$
$$fcds = 0$$
$$CR_c = 12 \times 724 = \underline{8688 \text{ psi}} \quad (59.90 \text{ MPa})$$

Relaxation of Prestressing Steel (CR_s)

$$CR_s = 20{,}000 - 0.4ES - 0.2\,(SH + CR_c)$$
$$CR_s = 20{,}000 - 0.4 \times 5335 - 0.2\,(12{,}500 + 8{,}688)$$
$$CR_s = 20{,}000 - 2134 - 4237 = \underline{13{,}629 \text{ psi}} \quad (93.98 \text{ MPa})$$

Total Loss

$$\Delta f_s = 12{,}500 + 5335 + 8688 + 13{,}629 = \underline{\underline{40{,}152 \text{ psi}}} \quad (276.84 \text{ MPa})$$

EXAMPLE IV

Calculate prestress losses for a post-tensioned box girder bridge built in Illinois with a simple span of 162 ft. Typical section as shown below.

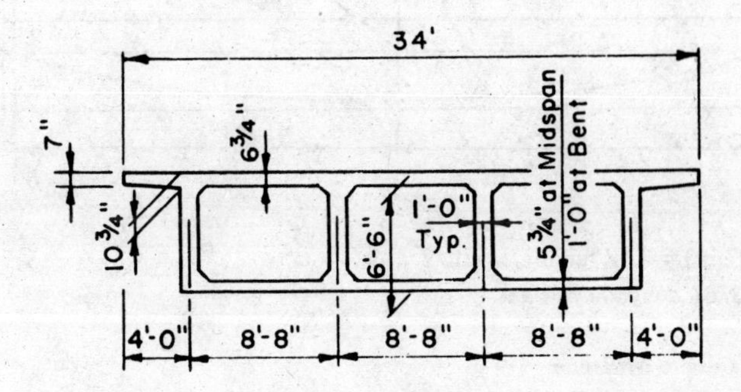

TYPICAL SECTION FOR EXAMPLES IV & V

Section Properties

Area	$= A$	$= 55.7$ ft.^{2}
Moment of Inertia	$= I$	$= 346$ ft.^{4}
Center of Gravity from top	$= Y_t$	$= 2.89$ ft.
Center of Gravity from bottom	$= Y_b$	$= 3.61$ ft.

Concrete Properties

$f_c' = 4500$ psi at 28 days

$f_{ci}' = 3500$ psi at time of stressing

$E_{ci} = 3.6 \times 10^6$ psi

Design Parameters

Prestressing steel = 270k ½" diameter strand

Permissable concrete tensile stress $= 6\sqrt{f_c'} = 6\sqrt{4500} = 402$ psi

Anchor seating loss = ¼ in.

Effective Prestressing force required $= 6930^k$

Friction coefficient $= \mu = 0.25$

Wobble coefficient $= K = 0.0002$

Stress tendon from one end only

Tendon Profile

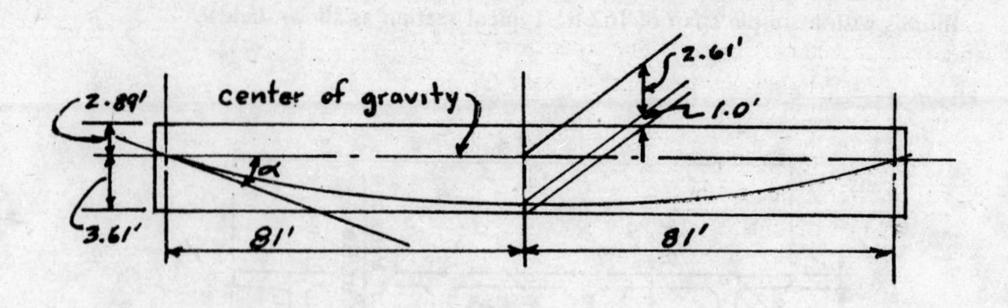

α (in radians) = 2 × tendon sag/L/2
α = 5.22/81 = .0644 radians

Friction Loss to Midspan

AASHTO 1.6.7(A)

$$T_O = T_X \times e^{(kl + \mu\alpha)}$$

when $(kl + \mu\alpha) \leqslant 0.3$

$$T_O = T_X \times (1 + kl + \mu\alpha)$$

$(kl + \mu\alpha) = 0.0002 \times 81 + 0.25 \times .0644$

$(kl + \mu\alpha) = 0.0162 + 0.0161 = 0.0323$

$$T_O \text{ at midspan} = \frac{T_O}{1 + 0.0323} = .97\ T_O$$

Assume tendons initially stressed at 0.70 f_s' at midspan after anchor seating loss and friction loss 0.70 × 270 = 189 ksi.
Friction loss to midspan = 0.03 × 0.72 × 270 = 5.8 ksi

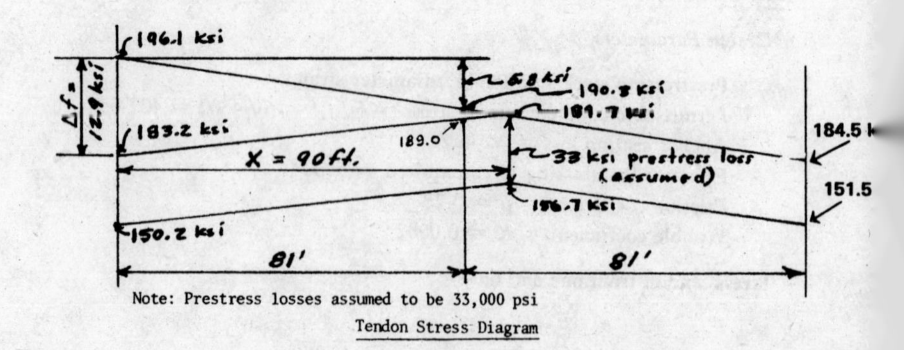

Note: Prestress losses assumed to be 33,000 psi

Tendon Stress Diagram

$$\Delta f^* = \sqrt{E\Delta L d/3L}$$
$$X^* = L\Delta f/2d$$

where Δf = Change in stress due to anchor set, ksi
d = Friction loss in length L, ksi
X = Length influenced by anchor set, ft.
L = Length to point where loss is known, ft.
ΔL = Anchor set in inches = ¼ in.
E = Modulus of elasticity, ksi

$$\Delta f = \sqrt{\frac{28 \times 10^3 \times .25 \times 5.8}{3 \times 81}} = 12.9 \text{ ksi}$$

$$X = \frac{12.9 \times 81}{2 \times 5.8} = 90 \text{ ft.}$$

With tendon stress at midspan equal 189 ksi (0.7 × 270) at time of anchorage, the stress after prestress losses becomes 156 ksi, and the 6930 kip effective force requires the following number of ½ inch diameter 270 k strands.

$$6930/156 \times .153 = 290 \text{ — ½ inch, 270k strands}$$

Force at midspan at time of anchorage:

$$189 \times .153 \times 290 = 8386 \text{kips}$$

Calculate Prestress Losses

$$\Delta f_s = SH + ES + CR_c + CR_s$$

Shrinkage (SH)

$$SH = 0.80\,(17{,}000 - 150\,RH)$$
$$= 0.80\,(17{,}000 - 150 \times 70) = \underline{5200 \text{ psi}} \quad (35.85 \text{ MPa})$$

Elastic Shortening (ES)

$$ES = 0.5 \times E_s/E_{ci} \times f_{cir}$$

$$M^D = 55.7 \times .150 \times 162^2/8 = 27{,}408 \text{ft-kips}$$

Dead load stress at 0.5 span at c.g. of tendon:

$$f_{DL} = \frac{27{,}408 \times 1000 \times 2.61}{144 \times 346} = 1435 \text{ psi tension}$$

---

*See "Post-Tensioned Box Girder Bridges—Design and Construction" published jointly by PCI and CRSI, pp 39-41 incl. for derivation of similar equations.

Stress due to post-tensioning at midspan at c.g. of tendon:

$$f_{ps} = \frac{8380 \times 1000 \times 2.61^2}{144 \times 346} + \frac{8380{,}000}{55.7 \times 144}$$

$$f_{ps} = 1145 + 1045 = 2190 \text{ psi}$$
$$f_{cir} = 2190 - 1435 = 755 \text{ psi}$$

$$ES = 0.5 \times \frac{28 \times 10^6}{3.6 \times 10^6} \times 755 = \underline{2936 \text{ psi}} \quad (20.24 \text{ MPa})$$

Concrete Creep (CR_c)

$$CR_c = 12 f_{cir} - 7 f_{cds}$$

Assume $f_{cds} = 0$ This is conservative, the dead load stresses due to curb and parapet weight could be used for f_{cds}, and this would slightly reduce the prestress loss due to concrete creep.

$$CR_c = 12 \times f_{cir} = 12 \times 755 = \underline{9060 \text{ psi}} \quad (62.47 \text{ MPa})$$

Relaxation of Prestressing Steel (CR_s)

$$CR_s = 20{,}000 - 0.3FR - 0.4ES - 0.2(SH + CR_c)$$

$FR = 0$ since prestressing steel is stressed to $0.7 f_s$ at midspan at time of anchorage

$$CR_s = 20{,}000 - 0.4 \times (2936) - 0.2(5200 + 9060)$$
$$CR_s = 20{,}000 - 1174 - 2852 = \underline{15{,}974 \text{ psi}} \quad (110.14 \text{ MPa})$$

Total Prestress Loss

$$\Delta f_s = 5200 + 2936 + 9060 + 15{,}974 = \underline{\underline{33{,}170 \text{ psi}}} \quad (228.70 \text{ MPa})$$

Very close to 33,000 psi lump sum value assumed.

EXAMPLE V

Calculate prestress losses for a post-tensioned box girder bridge built in Illinois with two continuous spans of 162 ft. — 150 ft. The typical section is shown in Example IV as are the section properties at midspan, the concrete properties and design parameters. The section properties at the pier (due to thickening bottom slab to 12 inches) are as follows:

Section Properties at Pier

Area	$= A$	$= 67.7$ ft²
Moment of Inertia	$= I$	$= 426$ ft⁴
Center of Gravity from Top	$= Y_t$	$= 3.40$ ft.
Center of Gravity from Bottom	$= Y_b$	$= 3.10$ ft.

Assume Prestress Losses = 32,000 psi

Angle change in 162 ft. span tendon profile (see Tendon Profile Diagram)

$$\left.\begin{aligned} AB &= \frac{2 \times 2.777}{81} = 0.069 \\ BC &= \frac{2 \times 3.75}{64.8} = 0.116 \\ CD &= \frac{2 \times .937}{16.2} = 0.116 \end{aligned}\right\} \alpha = 0.301$$

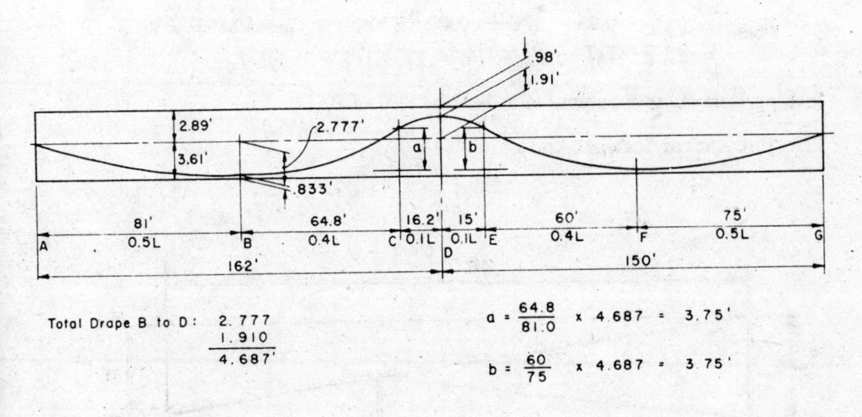

TENDON PROFILE

Stress calculations indicate that top fiber stress at pier controls the tendon force requirements and a P_j of 6460 kips is required.

$$.75 \times 270 \times .153 = 31 \text{ k/strand}$$
$$6460/31 = 209 \text{ strands required}$$
$$209/4 \text{ webs} = 52 \text{ strands/web}$$

Angle change in span 2 (see Tendon Profile Diagram)

$$\left.\begin{aligned} DE &= \frac{2 \times .937}{15} = 0.125 \\ EF &= \frac{2 \times 3.75}{60} = 0.125 \\ FG &= \frac{2 \times 2.77}{75} = 0.074 \end{aligned}\right\} \alpha = 0.324$$

Span 1

$$T_{bent} = T_o e^{-(\mu\alpha + kx)} \qquad \mu = 0.25, \quad k = .0002$$
$$T_{bent} = T_o e^{-(0.25 \times .301 + .0002 \times 162)} = T_o e^{-(.075 + .032)}$$
$$= T_o e^{-0.107} = T_o/e^{0.107} = T_o/1.113 = 0.897\, T_o$$

Span 2

$$T_{bent} = T_o e^{-(0.25 \times .324 + .0002 \times 150)} = T_o e^{-(.081 + .030)}$$
$$= T_o e^{-.111} = T_o/e^{.111} = T_o/1.117 = 0.895\, T_o$$

Jacking from A, assume tension at bent $= .897$ of P_j

Effect of Anchor Seating (Approximate)

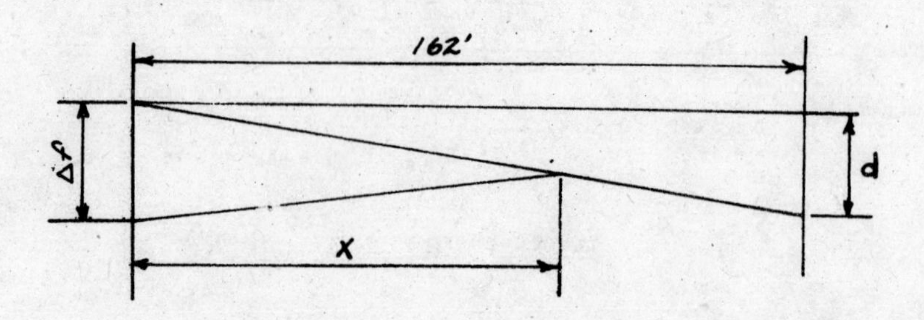

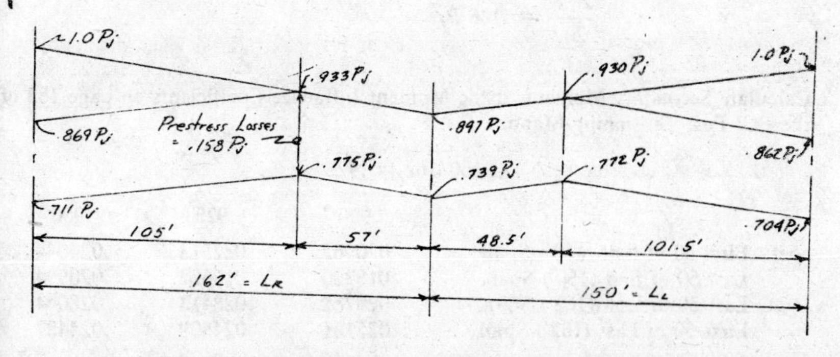

Prestress Losses = 32 ksi = 32/202.5 = .158 P_j

TENDON FORCE DIAGRAM

Assume anchor set = 5/8″ (max)

$$f_s' = 270 \text{ ksi}$$

$$f\text{jack} = 270 \times .75 = 202 \text{ ksi}$$

$$f\text{bent} = 202 \times .897 = 181.5 \text{ ksi}$$

$$d = 20.5 \text{ ksi}$$

$$x = \sqrt{\frac{E\,(\Delta L)\,L}{12\,d}}$$

$$x = \sqrt{\frac{27 \times 10^3 \times 162 \times 0.625}{12 \times 20.5}} = 105'$$

$$\Delta f = 2\text{d}x/L$$

$$\Delta f = \frac{2 \times 20.5 \times 105}{162} = 26.5 \text{ ksi}$$

$$\frac{26.5}{270 \times .75} = 0.131\ P_j$$

Span 2

$$x = \sqrt{\frac{27 \times 10^3 \times 150 \times .625}{12 \times 20.5}} = 101.5$$

$$f = \frac{2 \times 20.5 \times 101.5}{150} = 27.8 \text{ ksi}$$

$$\frac{27.8}{202} = .138\ P_j$$

Calculate Secondary Moments using Moment Influence Coefficients on page 183 of the PCI Post-Tensioning Manual.

$$150/162 = .925$$

		.900	.925	.950
coef	First 50 of first (150′) Span	.020982	.022513	.024044
	Last 50 of first (150′) Span	.018320	.019657	.020994
	Last 50 of Last (162′) Span	.028782	.028413	.028044
	First 50 of Last (162′) Span	.025131	.024809	.024487

$$W_e = \frac{2\ Pd}{b^2 L_e^2}$$

First Span (150′)

$$W_e = \frac{2 \times .728\ P_j \times 2.777}{.25 \times 150^2} = .000719\ P_j$$

$$W = \frac{2\ Pc}{(1 - b)(1 - b - a)\ L_e^2}$$

$$W = \frac{2 \times .76\ P_j \times 4.687}{(.5)(.4)\ 150^2} = .001583\ P_j$$

Second Span (162′)

$$W_e = \frac{2 \times .734\ P_j \times 2.777}{.25 \times 162^2} = .000621\ P_j$$

$$W = \frac{2 \times .76\ P_j \times 4.687}{(.5)(.4)\ 162^2} = .001357\ P_j$$

$\Sigma[\text{w} \times \text{coef}] \times 162^2$

$$.000719\ P_j \times .022513 = .0000162\ P_j$$
$$.001583\ P_j \times .019657 = .0000311\ P_j$$
$$.000621\ P_j \times .028413 = .0000177\ P_j$$
$$.001357\ P_j \times .024809 = \underline{.0000337\ P_j}$$
$$M = .0000987\ P_j \times 162^2 = 2.59\ P_j$$

Secondary Moment (M_S) at pier:

$$M_s = M - P_j \times e = 2.59\ P_j - .739\ P_j \times 1.91 = 2.59\ P_j - 1.41\ P_j = 1.18\ P_j$$

Check value of prestress losses at 0.4 span and bent

$$\Delta fs = SH + ES + CR_c + CR_s$$

Shrinkage Loss:

$$SH = 0.8\,(17{,}000 - 150 \times 70) = \underline{5{,}200 \text{ psi}} \quad (35.85 \text{ MPa})$$

Elastic Shortening Loss

$$ES = 0.5\left(\frac{E_s}{E_{ci}} f\text{cir}\right)$$

$$E_{ci} = 33 \times 150^{3/2}\sqrt{3500} = 3.6 \times 10^6 \text{ psi}$$
$$E_s = 28 \times 10^6$$

at 0.4 Span Friction Loss = $(.4 \times .103 = .0412\, P_j) = 8.3$ ksi
at Pier, Friction Loss = $.103\, P_j = .103 \times .75 \times 270 = 20.9$ ksi

$$P_j = 0.75 \times 270 \times 209 \times .153 = 6475^{\text{kips}}$$

Dead Load Moment at 0.4 span = 16,600ft-kips
Use tendon stress = $0.65\, f_s'$ to calculate f_{cir}:

$$F = 0.65 \times 270 \times 209 \times .153 = 5612^{\text{ft-kips}}$$
$$y \text{ to c.g. of strand at 0.4 span} = 0.96 \times 2.77 = 2.66 \text{ ft.}$$

Prestressing moment at 0.4 span when force equals 5612kips is:

$$M_{p/s} = 5612 \times 2.66 - M_s$$
$$M_{p/s} = 5612 \times 2.66 - 0.4 \times 1.18\,(6475) = 11{,}872 \text{ft-kips}$$

$$f_{ps} = \frac{-5612 \times 1000}{55.7 \times 144} - \frac{11{,}872 \times 2.66 \times 1000}{346 \times 144} = -1334 \text{ psi}$$

$$f_{dl} = \frac{16{,}600 \times 2.66 \times 1000}{346 \times 144} = +886 \text{ psi}$$

$$f_{\text{cir}} \text{ at 0.4 span} = -448 \text{ psi}$$

$$ES \text{ at } 0.4 = .5\left(\frac{28 \times 10^6}{3.6 \times 10^6}\right) \times 448 = \underline{1742 \text{ psi}} \quad (12.01 \text{ MPa})$$

$$y \text{ to c.g. strand at pier} = 3.40 - 0.98 = 2.42 \text{ ft.}$$

Dead load stress at c.g. strand at pier:

$$f_{dl} = \frac{26{,}200 \times 1000 \times 2.42}{144 \times 426} = +1034 \text{ psi}$$

$$M_{ps} = 5612 \times 2.42 + 1.18 \times 6475 = 21{,}222 \text{ft-kips}$$

$$f_{ps} = \frac{5612 \times 1000}{144 \times 67.7} + \frac{21{,}222 \times 2.42 \times 1000}{144 \times 426}$$

$$f_{ps} = 575 + 837 = 1412 \text{ psi}$$

$$f_{cir} \text{ at pier} = -1412 + 1034 = -378 \text{ psi}$$

$$ES \text{ pier} = .5 \left(\frac{28 \times 10^6}{3.6 \times 10^6}\right) \times 378 = \underline{1470 \text{ psi}} \quad (10.14 \text{ MPa})$$

Concrete Creep (CR_c) assume $f_{cds} = 0$ (conservative)

$$CR_c = 12\, f_{cir} - 7\, f_{cds}$$

at pier=

$$CR_c = 12 \times 378 = \underline{4536 \text{ psi}} \quad (31.27 \text{ MPa})$$

at 0.4 span =

$$CR_c = 12 \times 448 = \underline{5376 \text{ psi}} \quad (37.01 \text{ MPa})$$

Relaxation of Prestressing Steel (CR_s)

$$CR_s = 20{,}000 - 0.3FR - 0.4ES - 0.2\,(SH + CR_c)$$

Note: FR applicable *only* where it reduces tendon stress below $0.70 f_s'$.

$FR = 0$ at 0.4 span

CR_s at 0.4 span $= 20{,}000 - 0.4\,(1742) - 0.2\,(5200 + 5376)$

CR_s at 0.4 span $= 20{,}000 - 697 - 2115 = \underline{17{,}188 \text{ psi}}$ (118.51 MPa)

FR at pier $= (0.70 - 0.897 \times .75)\, 270 = 7.3$ ksi

CR_s at pier $= 20{,}000 - 0.3\,(7{,}300) - 0.4\,(1470) - 0.2\,(5200 + 4536)$

CR_s at pier $= 20{,}000 - 2190 - 588 - 1947 = \underline{15{,}275 \text{ psi}}$ (105.32 MPa)

Summation of Prestress Losses	at 0.4 span		at pier	
Shrinkage (SH)	(35.85)	5200	5200	(35.85)
Elastic Shortening (ES)	(12.01)	1742	1470	(10.14)
Concrete Creep (CR_c)	(37.07)	5376	4536	(31.27)
Relaxation of Prestressing Steel (CR_s)	(118.51)	17,188	15275	(105.32)
Total Prestress loss excluding friction	(203.44 MPa)	29,506 psi	26,481 psi	(182.58 MPa)

Note losses are less than the 32,000 psi assumed in the design, particularly at the pier which is the point that controlled the required tendon force. Design would be re-cycled with prestress losses of 26,500 psi assumed at pier with approximately 3½ percent reduction in tendon requirements (5500/157000 × 100 = 3.50 percent). If design had not utilized the allowable tension of $6\sqrt{f_c'}$ the calculated prestress losses would have been somewhat higher.

EXAMPLE VI (See Fig. E4)

Compute the prestress losses for the bridge in Fig. E4 which utilizes 54 in. I-beams with 38-½ in. diameter 270 k stress relieved strands at an eccentricity of 23.2 in. (from Colorado Department of Highways/G54 Beam Charts). The structure to be built near Denver has two equal spans of 100 ft. which are made continuous for composite dead load and live load by means of continuity reinforcement in the deck slab and diaphragm at the center pier.

Properties

Concrete: $f_c' = 6000$ psi (28 day strength)
$f_{ci}' = 5000$ psi (Release strength)
Initial steel stress: 0.7×270 ksi $= 189$ ksi

Section Properties

Beam: Area $= A = 631$ in^2
Moment of inertia $= I = 242{,}585$ in^4
Center of gravity $= \bar{y} = 26.67$ in. from bottom
Weight $= WB = 0.657$ k/ft.

Composite (Transformed Section)
Moment of inertia $= I' = 533{,}211$ in^4
Center of gravity $= \bar{y}' = 38.52$ in. from bottom

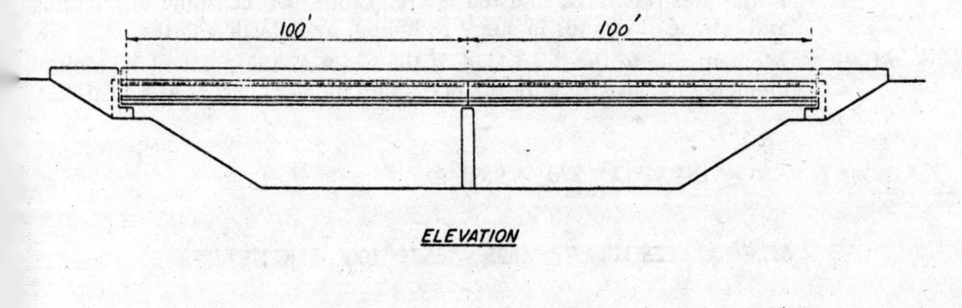

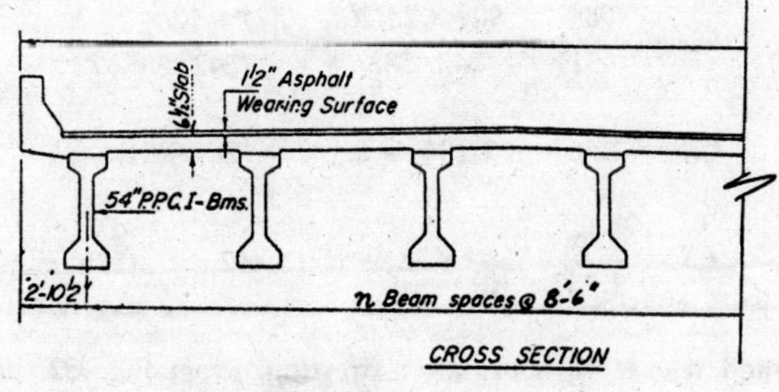

Figure E4

Loss Calculation

$$\text{Total Loss} = \Delta f_s = SH + ES + CR_c + CR_s$$

Shrinkage (SH)

$$SH = 17{,}000 - 150\ RH$$

In accordance with the U.S. Weather Bureau Map, the mean annual relative humidity for the Denver, Colorado, Metropolitan area is about 60 percent.

$$SH = 17{,}000 - 150 \times 60 = \underline{8{,}000\ \text{psi}} \quad (55.16\ \text{MPa})$$

Elastic Shortening (ES)

$$ES = \frac{E_s}{E_{ci}} \times f_{cir} \qquad E_s = 28 \times 10^6\ \text{psi}$$

$$E_{ci} = 33\ (W)^{3/2}\sqrt{f_{ci}'} = 33 \times (150)^{3/2}\sqrt{5000} = 4.287 \times 10^6\ \text{psi}$$

$$f_{cir} = \frac{F}{A} + \frac{Fe^2}{I} - \frac{M_n{}^{db} \times e}{I} \quad \text{where:}$$

F = Initial prestress force adjusted for relaxation between time of stressing and release of strands (assume 18 hours), and elastic shortening.

$M_n{}^{db}$ = Moment due to the dead load of the beam at the point of maximum moment. In this structure the maximum moment occurs at midspan.

$F = A_s \times f_s$

Assume $F = 38 \times .153 \times .63 \times 270 = 989\text{kips}$

$$M_n{}^{db} = .125\ W_b L^2 = .125 \times .657 \times 100^2 = 821\text{ft-kips}$$

$$f_{cir} = \frac{989}{631} + \frac{989 \times 23.2^2}{242,\ 585} - \frac{821 \times 12 \times 23.2}{242,\ 585}$$

$$f_{cir} = 1.567 + 2.194 - 0.942 = 2.819\ \text{ksi}$$

$$ES = \frac{28 \times 10^6}{4.287 \times 10^6} \times 2819 = 18{,}412\ \text{psi} \quad (126.95\ \text{MPa})$$

Assume strand release 18 hours after stressing. From Fig. C2 the relaxation loss equals 3.5 percent of the initial stress of 189 ksi = .035 × 189,000 = 6,615 psi.

Check initial force assumption of $0.63\ f_s'$

$$189{,}000 - 18{,}412 - 6615 = 163{,}973\ \text{psi}$$

$163{,}973/270{,}000 = 0.607\ f_s'$ This is probably not close enough to $0.63\ f_s'$ assumption, therefore recalculate one time.

Assume $F = 38 \times .153 \times .607 \times 270 = 953$ kips

$$f_{cir} = \frac{953}{631} + \frac{953 \times 23.2^2}{242{,}585} - \frac{821 \times 12 \times 23.2}{242{,}585}$$

$$f_{cir} = 1.510 + 2.114 - 0.942 = 2.682 \text{ ksi}$$

$$ES = \frac{28 \times 10^6}{4.287 \times 10^6} \times 2{,}682 \text{ psi} = \underline{17{,}517 \text{ psi}} \quad (120.78 \text{ MPa})$$

Check assumed force

$$189{,}000 - 17517 - 6615 = 164{,}868 \text{ vs } 163{,}890 \text{ assumed: OK}$$

Creep of Concrete (CR_c)

$$CR_c = 12\, f_{cir} - 7\, f_{cds}$$

$$f_{cds} = \frac{M_n{}^s e}{I} + \frac{M_c{}^d e + (\bar{y}' - \bar{y})}{I'}$$

$M_n{}^s$ = Moment due to slab weight at point of maximum moment
$M_n{}^s = 0.125 \times W_s \times L^2 = 0.125 \times .150 \times .54 \times 8.5 \times 100^2 = 860$ ft-kips
$M_c{}^d$ = Moment due to composite dead loads at the point of maximum moment
$M_c{}^d = Cw_eL^2$ where C = Influence Coefficient

$$\text{Asphalt} = .019 \times 8.5 = 0.162 \text{ k/ft.}$$

$$\text{Curb, Parapet, Rail} = \underline{0.200 \text{ k/ft.}}$$
$$W_e = 0.362 \text{ k/ft.}$$

$$M_c{}^d = .0625 \times 0.362 \times 100^2 = 226 \text{ ft-kips}$$

$$f_{cds} = \frac{860 \times 12 \times 23.2}{242{,}585} + \frac{226 \times 12\,[23.2 + (38.25 - 26.67)]}{533{,}211}$$

$$f_{cds} = 0.987 + 0.178 = 1.165 \text{ ksi}$$

$$CR_c = 12 \times 2{,}682 - 7 \times 1{,}165 = \underline{24{,}029 \text{ psi}} \quad (165.67 \text{ MPa})$$

Relaxation of Prestressing Steel (CR_s)

$$CR_s = 20{,}000 - 0.4\, ES - 0.2\,(SH + CR_c)$$
$$CR_s = 20{,}000 - 0.4 \times 17517 - 0.2\,(8000 + 24{,}029)$$

$$CR_s = 20{,}000 - 7007 - 6406 = \underline{6587 \text{ psi}} \quad (45.42 \text{ MPa})$$

Total Loss

$$f_s = SH + ES + CR_c + CR_s$$

$$f_s = 8000 + 17517 + 24{,}029 + 6587 = \underline{\underline{56{,}403}} \quad (388.89 \text{ MPa})$$

Note: This example uses 6000 psi concrete which is beyond the range covered by the table of lump sum values for prestress losses. Use of higher strength concretes to increase span range of girders normally results in higher values of prestress losses, and, for this reason, calculation of losses is required in these cases.

Appendix C

Condensed Method for Computation of the Loss of Prestress Based Upon the Recommendations of ACI Committee 209 and D. E. Branson

1. For prestressed concrete flexural members not having a composite topping or slab, the ultimate loss of prestress can be estimated from:

$$\Delta f_{si} = nf_{ci} + k_r \lambda n f_{ci} C_u + \frac{k_r \epsilon_{sh} E_s}{\zeta} + \Delta f_{sr} + mf_{cs} + \frac{mf_{cs}\beta_s k_r C_u}{\zeta} \quad \text{(C-1)}$$

in which

$$f_{ci} = \frac{P_0 k_s}{A} + \frac{M_d e_{ps}}{I} \quad \text{(C-2)}$$

and

A_g = Gross area of the concrete section
A_t = Gross area of the transformed section, such as an uncracked prestressed concrete section
A_s = Area of non-prestressed steel
A_{ps} = Area of prestressing steel
$b_{11} = n\rho_{ps}(k_s)(1 + \eta C_u)$

$$b_{12} = (n\rho)\,\frac{1 + (e_{ps})(e_s)}{r^2}\,(1 + \eta C_u)$$

C_u = Ultimate creep ratio
E_s = Elastic modulus of the steel
e_s = Eccentricity of non-prestressed reinforcement
e_{pL} = Eccentricity of the pressure line due to prestressing alone
e_{ps} = Eccentricity of the prestressing reinforcement
f_{cs} = Concrete stress at center of the steel due to a cast-in-place slab or a superimposed dead load
Δf_{sr} = Loss of stress in prestressing steel due to relaxation
f_{si} = Initial stress in the prestressing steel
f_y = Yield strength of the prestressing steel measured as the stress at 0.1% offset
I = Moment of inertia of the transformed section, such as an uncracked prestressed concrete section

$k_r = \dfrac{1}{1 + b_{12}}$, or when $\dfrac{A_s}{A_{ps}} \leqslant 2$, K_r can be approximated from $\dfrac{1}{1 + (A_s/A_{ps})}$

$k_s = 1 + \dfrac{(e_{ps})(e_{pL})}{r^2}$; for statically determinate members this becomes $1 + \dfrac{e_{ps}^2}{r^2}$

M_d = Moment due to dead load
m = Modular ratio for steel and beam concrete at time the slab is cast
n = Modular ratio at time of prestressing
P_i = Initial prestressing force

P_0 = Prestressing force after transfer (after elastic loss)

$r^2 = \dfrac{I}{A}$

ΔP_s = Loss of prestress at time slab is cast (excluding the initial elastic loss)
ΔP_u = Total loss of prestress (excluding the initial elastic loss)
ϵ_{sh} = Ultimate shrinkage strain (see Sect. 3-10)
$\zeta = 1 + (n\rho_{ps}k_s)(1 + \eta C_u) = 1 + b_{11}$
η = Relaxation coefficient. This coefficient is intended to take into account the effect of creep under decreasing load. It ranges from 0.75 to 0.90. An average value is 0.88
λ = Factor that is a function of the ratio of ultimate loss of prestress (excluding the initial elastic loss) ΔP_u, and the initial prestressing force (after elastic loss) P_0

$$\lambda = 1 - \Delta P_u/2P_0$$

λ can normally be taken to be from 0.91 to 0.88 for normal and lightweight concretes respectively. (It can be corrected if the first computation of the loss of prestress shows that a different value would be more appropriate). (See Table C-4)

$\rho = \dfrac{A_s}{bd}$, ratio of non-prestressed tensile reinforcing

$\rho_p = \dfrac{A_{ps}}{bd}$, ratio of prestressing reinforcing

$\rho_{ps} = \dfrac{A_{ps}}{A_g}$, ratio of prestressing steel area to gross concrete area

In computing f_{ci} with Eq. C-2, the initial prestressing force P_i can be used rather than the force after transfer, P_0. If this is done, the computed loss will be greater than if the value of P_0 is used. P_0 can be computed from

$$f_{ci} = \frac{P_i k_s + AM_d e/I}{A + nA_{ps}k_s} \tag{C-3}$$

The value of P_0 can be checked from

$$P_0 = P_i - nf_{ci}A_{ps} \tag{C-4}$$

Note that eccentricities are negative when below the centroidal axis, and dead load moments causing tension in the bottom fibers are positive.

2. For composite construction (shored or not shored) the ultimate loss of prestress is estimated from

$$\Delta f_{si} = nf_{ci} + nf_{ci}\alpha_s k_r C_u \lambda' + nf_{ci} k_r C_u [\lambda - \alpha_s \lambda'] \frac{I}{I'}$$

$$+ \frac{k_r \epsilon_{sh} E_s}{\zeta} + \Delta f_{sr} + mf_{cs}$$

$$+ \frac{[mf_{cs}\beta_s k_r C_u]}{\zeta} \frac{I}{I'} + \Delta f_{sds} \qquad \text{(C-5)}$$

Terms in the above, not previously defined in this Appendix, are as follows:

I' = Moment of inertia of the composite transformed section with due regard to modular ratio of beam and slab concrete (see Section 6-4)

f_{sds} = Concrete stress at the level of the center of gravity of the steel due to differential shrinkage and creep

$$f_{sds} = \frac{mQy_{cs}e_c}{I'}$$

e_c = Eccentricity of the prestressing steel in the composite section

Q = The force generated by the differential shrinkage and creep

y_{cs} = Distance from the centroid of the composite section to centroid of the slab

α_s = Ratio of the creep ratio for the concrete of the beam at the time the slab is cast to the ultimate creep ratio ($\alpha_s = t^c/d + t^c$ from Eq. 3-18)

β_s = Creep correction factor for beam concrete age when loaded (see Fig. 3-12)

λ' = Factor that is a function of the loss of prestress at time slab is cast (excluding the initial elastic loss) ΔP_s, and the initial prestressing force (after elastic loss) P_0

$$\lambda' = 1 - \frac{\Delta P_s}{2P_0}$$

λ' can be taken to be between 0.95 and 0.93 for normal and lightweight concretes respectively, when the time interval between prestressing and slab casting is between 3 weeks and 1 month. For a time interval of between 2 and 3 months, the λ' can be taken to be between 0.93 and 0.92 for normal and lightweight concretes, respectively. (See Table C-4.)

Losses of prestress occurring between the time of prestressing and when the ultimate values are attained can be estimated from Eqs. C-1 and C-5 by using intermediate values for C_u, ϵ_{sh}, and Δf_{sr}. These values can be determined from Eqs. 2-1, 3-17, and 3-18.

For critical members, a more detailed study can be made based upon the complete recommendations of ACI Committee 209 (Ref. 1) or those of Branson (Ref. 2). One must bear in mind, however, that loss of prestress computations based upon relationships such as these must be considered to be estimates and not precise computations.

The following average modular ratios are based on f'_{ci} = 4000 to 4500 psi for both moist-cured (M.C.) and steam-cured (S.C.) concrete and type 1 cement, and up to f'_c = 6360–7150 psi and 6050–6800 psi for moist-cured and steam-cured concrete at the age of 3 months. The values are equally applicable to grades 250 and 270 prestressing strands (ASTM A 416). (Ref. 2.)

TABLE C-1.

	Modular Ratio	Nor. Wt. (w = 145)		Sand-Lt. Wt. (w = 120)		All-Lt. Wt. (w = 100)	
		M.C.	S.C.	M.C.	S.C.	M.S.	S.C.
At release of prestress	$n =$	7.3	7.3	9.8	9.8	12.9	12.9
For the time between prestressing and slab casting							
= 3 weeks,	$m =$	6.1	6.2	8.1	8.3	10.7	10.9
1 month,		6.0	6.2	8.0	8.2	10.5	10.7
2 months,		5.9	7.9	8.2	8.2	10.3	10.6
3 months,		5.8	6.0	7.7	8.0	10.2	10.5

The term α_s is the term $\dfrac{t^c}{d + t^c}$ from Eq. 3-18, and can be taken for all types of concrete (normal or lightweight with types I or III cement, which is either moist- or steam-cured) as follows:

TABLE C-2.

For the time between prestressing and slab casting		α_s	β_s
=	3 weeks,	$\alpha_s = 0.38$	$\beta_s = 0.85$
	1 month,	0.44,	0.83
	2 months,	0.54,	0.78
	3 months,	0.60,	0.75

In the above listing, β_s is the creep correction factor defined above. (Ref. 2.)

The value of f_{sr}, the loss of prestress due to relaxation of the prestressing steel, can be estimated from Table C-3. Note that f_y is the stress at an offset of 0.1%, *not* the stress at a 1% extension.

TABLE C-3. Ultimate Loss of Prestress Due to Steel Relaxation by Eq. 2-1.

f_{si}/f_y	Stress-Relieved Steel %	Low-Relaxation Steel %
0.60	2.5	0.6
0.65	5.0	1.1
0.70	7.5	1.7
0.75	10.0	2.2
0.80	12.5	2.8
0.85	15.0	3.3
0.90	17.5	3.9

The values in the above table are based upon $f_{si} = 190$ ksi and $t = 10^5$ hours. The relaxation loss for low-relaxation steel is taken as 22% of that of stress-relieved steel. (Ref. 2)

Under usual conditions, the values of b_{11} and b_{12} can be taken as 0.25 (Ref. 2). Hence, under usual conditions ζ can be taken as 1.25. In addition, for low amounts of non-prestressed reinforcing, $(A_s/A_{ps} \leqslant 2)$, the value of k_r can be taken as unity. (For larger amounts of non-prestressed reinforcement, k_r can be taken as 0.80.) With these values, together with usual values of P_u/P_0, Eq. C-1 and C-5 can be written (for normal weight concrete) as follows

$$\Delta f_{si} = nf_{ci}(1 + 0.91 C_u) + 0.8 \epsilon_{sh} E_s + \Delta f_{sr} + mf_{cs}(1 + 0.8 \beta_s C_u) \quad \text{(C-6)}$$

and

$$\Delta f_{si} = nf_{ci}\left[1 + 0.91 \alpha_s C_u + C_u(0.91 - 0.95 \alpha_s)\frac{I}{I'}\right]$$

$$+ 0.8 \epsilon_{sh} E_s + \Delta f_{sr} + mf_{cs}\left[1 + 0.8 \beta_s C_u \frac{I}{I'}\right] \quad \text{(C-7)}$$

Similar simplified expressions can be written for lightweight concrete.

TABLE C-4. Typical Loss of Prestress Ratios for Concretes of Different Weights.

Ratio	Normal Wt.	Sand-Lt. Wt.	All-Lt. Wt.
$\Delta P_s/P_0$–3 wk to 1 mth between prestressing and sustained load application, including composite slab	0.10	0.12	0.14
$\Delta P_s/P_0$–for 2 to 3 months between prestressing and sustained load application, including composite slab	0.14	0.16	0.18
$\Delta P_u/P_0$	0.18	0.21	0.23

REFERENCES

1. ACI Committee 209, "Prediction of Creep, Shrinkage, and Temperature Effects in Concrete Structure," *Designing for Effects of Creep, Shrinkage and Temperature in Concrete Structures*, American Concrete Institute, Detroit, 1971.

2. Branson, D. E., "The Deformation of Non Composite and Composite Prestressed Concrete Members," *Deflections of Concrete Structures*, American Concrete Institute, Detroit, 1974, 83–127.

Example C-1. For the steam cured normal weight concrete double-tee beam shown in Fig. C-1, determine the loss of prestress. The material strengths are:

$$f_{pu} = 270 \text{ ksi}$$
$$f'_{ci} = 4000 \text{ psi}$$
$$f'_c = 5000 \text{ psi}$$

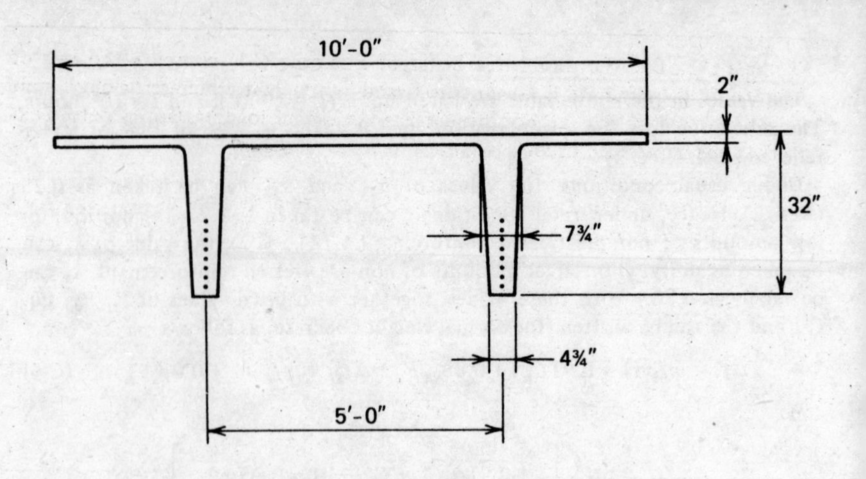

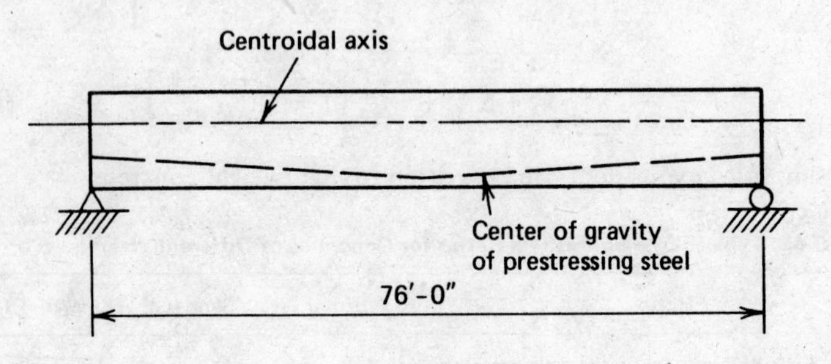

Fig. C-1.

Section properties, areas, and loadings are:

$A_g = 615$ sq. in. $\quad y_b = -21.98$ in. $\quad I_g = 59{,}720$ in.^{4}
$w_d = 641$ plf $\quad w_1 = 300$ plf $\quad A_{ps} = 2.14$ sq. in.
$f_{si} = 189$ ksi $\quad$ Span = 76.0 ft.

At midspan, $e = -18.48$ in. At end, $e = -11.12$ in.

Other parameters are:

Ambient relative humidity = 70%
$n = 7.3$, $C_u = 1.88$, $\epsilon_{sh} = 546$ micro-inches per in.
$E_s = 28{,}000$ ksi $\quad \Delta P_u/P_0 = 0.18$
$1 + b_{11} = 1.25 \quad f_{sr} = 0.075$*

SOLUTION:

Elastic loss =	8.19 ksi
Creep loss =	14.01 ksi
Shrinkage loss =	12.23 ksi
Relaxation loss =	14.18 ksi
Total loss	48.61 ksi

*Note that the initial steel stress equals 0.70 of f_{pu} and hence $f_{si}/f_y > 0.70$. The loss due to relaxation would normally be expected to be greater than 7.5% under these conditions.

Example C-2. For the double-tee beam of Example C-1, compute the loss of prestress if the concrete is steam cured sand-lightweight concrete having a unit weight of 120 pcf. The section properties, areas, and loadings are as before except w_d = 530 plf. The other parameters become:

$$n = 9.8, \quad C_u = 1.88, \quad \epsilon_{sh} = 546 \text{ micro-inches per in.}$$
$$\Delta P_u/P_0 = 0.21, \quad 1 + b_{11} = 1.25, \quad f_{sr} = 0.075^*$$

SOLUTION:

Elastic loss	13.15 ksi
Creep loss	22.13 ksi
Shrinkage loss	12.23 ksi
Relaxation loss	14.18 ksi
Total loss	61.69 ksi

Example C-3. For the single-tee beam of Fig. C-2, made of lightweight concrete (120 pcf), determine the loss of prestress under the following conditions:

f_{pu} = 270 ksi, f'_{ci} = 4000 psi, f'_c = 5000 psi
A_g = 782 sq. in., y_b = −35.19 in., I_g = 168,970 in.^{4}
$e_{ps} = e_s$ = −30.44 in., w_d = 652 plf
w_{sdc} = 100 plf (applied at age of 1 month)
w_1 = 300 plf (one-third assumed sustained and applied at age of 1 month)
A_{ps} = 2.45 sq. in., A_s = 1.22 sq. in.
f_{si} = 189 ksi Ambient relative humidity = 70%

Other parameters are:

$$n = 9.8, \quad m = 8.2, \quad C_u = 1.80, \quad \epsilon_{sh} = 513 \text{ micro-inches per in.}$$
$$\Delta P_u/P_0 = 0.21, \quad \beta_s = 0.83, \quad k_r = 0.74$$
$$1 + b_{11} = 1.25$$

SOLUTION:

	Ignoring A_s	Including A_s
Elastic loss	11.63 ksi	11.63 ksi
Creep loss	18.74 ksi	13.87 ksi
Shrinkage loss	11.49 ksi	8.50 ksi
Relaxation loss	14.18 ksi	14.18 ksi
Elastic gain, sustained load	−4.43 ksi	−4.43 ksi
Creep gain, sustained load	−5.29 ksi	−3.91 ksi
Total	46.32 ksi	39.84 ksi

---

*(Same as for Ex. C-1).

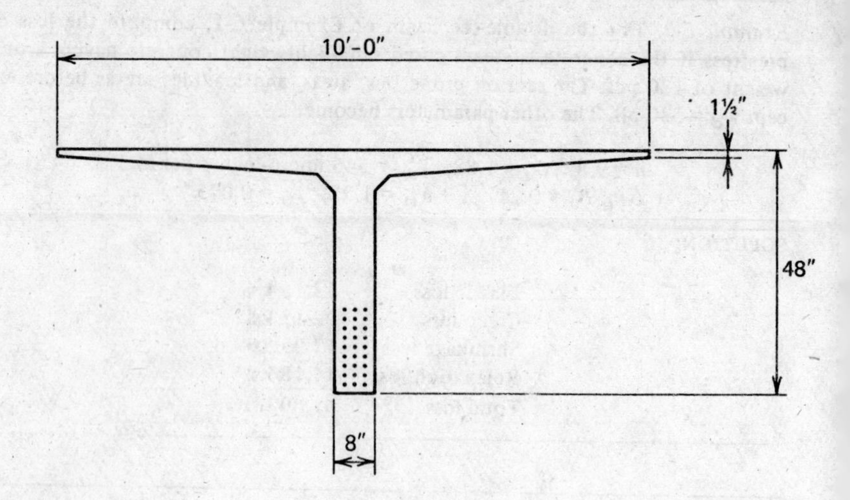

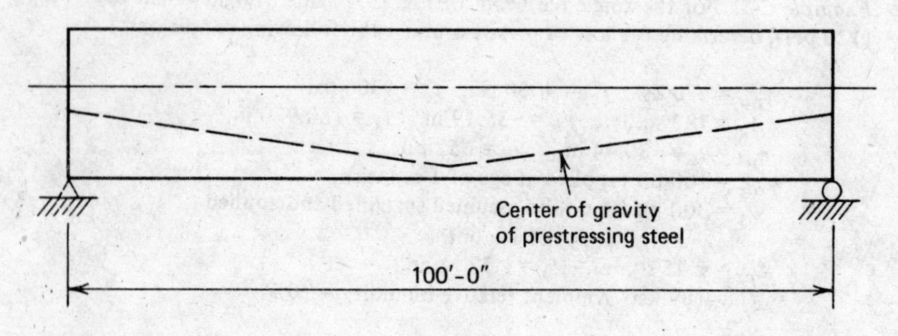

Fig. C-2.

Example C-4. Determine the loss of prestress for the unshored normal weight concrete AASHTO Type IV girder with a composite slab as shown in Fig. C-3.

Assume: $f_{pu} = 270$ ksi

For the precast girder, $f'_{ci} = 4000$ psi, $f'_c = 5000$ psi

For the cast-in-place slab, $f'_c = 4000$ psi

$$\frac{\text{E girder}}{\text{E slab}} = 1.07$$

Section properties and loads:

Precast section:

$A_g = 789$ sq. in., $y_b = -24.73$ in., $I_g = 260{,}740$ in.^{4}

$e_{ps} = -21.26$ in., $w_d = 822$ plf

Composite section:

$I_g = 629{,}890$ in.^{4}, $y_b = -38.91$ in.

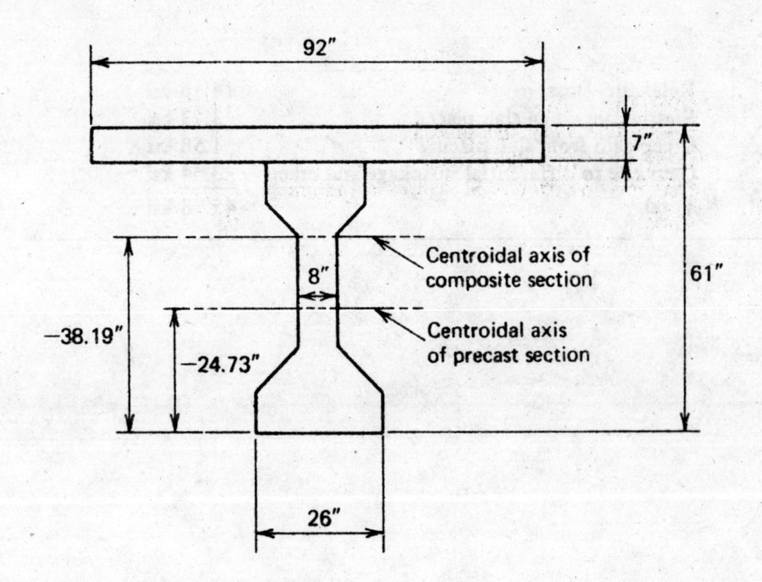

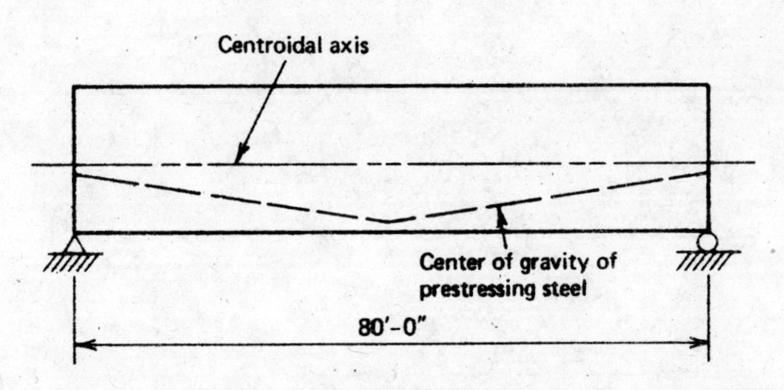

Fig. C-3.

Other parameters:

$A_{ps} = 4.90$ sq. in., $f_{si} = 189$ ksi
Ambient relative humidity = 60%
$n = 7.3$, $m = 6.1$, $C_u = 1.64$, $\epsilon_{sh} = 487$ micro-inches per in.
$\Delta P_s/P_0 = 0.14$
$\Delta P_u/P_0 = 0.18$
$\alpha_s = 0.54$, $\beta_s = 0.78$, $1 + b_{11} = 1.25$

SOLUTION:

Elastic loss	**14.10 ksi**
Creep loss before slab placed	**11.61 ksi**
Creep loss after slab placed	**3.93 ksi**
Shrinkage loss	**10.91 ksi**

Relaxation loss	14.18 ksi
Elastic gain when slab placed	−3.73 ksi
Creep gain from slab placing	−1.58 ksi
Gain due to differential shrinkage and creep	−3.44 ksi
Total	45.98 ksi

Index